T0310602

Computer Organisation and Architecture

Computer Organisation and Architecture

Evolutionary Concepts, Principles, and Designs

Pranabananda Chakraborty
An Eminent IT Professional and Senior Visiting Professor

CRC Press
Taylor & Francis Group
Boca Raton London New York

CRC Press is an imprint of the
Taylor & Francis Group, an **informa** business
A CHAPMAN & HALL BOOK

First edition published 2021
by CRC Press
6000 Broken Sound Parkway NW, Suite 300, Boca Raton, FL 33487-2742

and by CRC Press
2 Park Square, Milton Park, Abingdon, Oxon, OX14 4RN

ISBN: 978-0-367-25573-2 (hbk)
ISBN: 978-0-429-28845-6 (ebk)

Typeset in Palatino
by codeMantra

Visit the eResources Website: http://routledge.com/9780367255732

Contents

Part II High-End Processor Organisation

Preface

Computer technology, especially its design, organisation and architecture, is relentlessly expanding both in width and in depth, mainly due to continuous innovations in the area of electronic engineering over the last few decades that ultimately have been able to introduce various types of sophisticated computer components differing in size, cost, and performance. Despite such a rapid pace of change in technology, certain fundamental concepts are still found to be consistent and applied throughout. The primary goal of this book is to provide a comprehensive view of the evolution in the overall design principles and internal details of computers from zero generation to the stage of supercomputers describing the fundamental concepts of their organisation and architecture. Major emphasis is placed on explaining elementary principles when viewed from the engineering angle and subsequent modern implementations of computers. The material of this book has been mostly developed as a self-contained text with a self-study approach of computer design and architecture at an introductory level, primarily from a hardware standpoint with topic-wise modern implementations provided as a case study. Many examples with their solutions have been included in this book to exhibit the specific characteristics and the performances of each individual resource. This edition is intended for use at the undergraduate level for computer engineering, information technology, electrical engineering, electronics engineering, computer science, BCA, MCA, and similar other courses. As prerequisites, this book expects from the readers almost nothing beyond some familiarity with the basic knowledge of computers, computer programming, and binary numbers, especially the fundamentals of digital logic, which are advantageous. The details of electronics encompassed with this subject have been deliberately avoided and intentionally dropped from the coverage in the main body of this book. However, modern electronic components that are fabricated using recent innovative technologies employed in modern machines have been included in this book to explain the topics which lie in the domain of advanced architecture. No topics in this book are original, but originality is maintained in its representation. Many – but by no means all – topics, while discussed additionally, cover the issues related to their performance with special emphasis on their implementations by way of presenting examples drawn from commercially available respective representative machines, both contemporary and historical.

This book is composed of ten chapters providing the details from fundamental concepts to most modern design principles that are consistently applied to both computer organisation and computer architecture. The objective is perhaps more focused here in describing the basic principles involved in a comprehensive computer organisation rather than the design of a full processor. However, because computer organisation must be designed to implement a particular architectural specification, a thorough treatment of the organisation requires a detailed examination of the architecture as well. The first seven chapters cover the basic principles of computer organisation, operation, and performance, and the emergence of various types of computers considering all their generations as evolved till today. The remaining three chapters deal with innovative RISC (reduced instruction set computer) architecture; high-end processors' architecture and organisation; massive parallel processing systems; and, finally, supercomputers.

Organisation of this Book

This book demystifies a reasonable balance between the traditional methods used in conventional computer organisation and contemporary approaches used in modern computer architecture and its organisation. The traditional approaches are used in organizing independent computer systems, each comprising a single CPU (central processing unit), a hierarchical memory system consisting of different types of memory realized with diverse technology located at different levels in the computer system, and numerous types of I/O (input/output) systems. All are interconnected by buses of various types (or interconnection networks) providing necessary ports. Modern approaches are, however, dealt with the internal architecture of modern CPUs in the form of scalar/pipelined, superpipelined, superscalar, multithreaded, and multicore. The modern approaches also include the organisation and architecture of computer systems consisting of multiple CPUs (multiple processing elements) and shared/distributed memory systems that are formed by interconnecting multiple CPUs on a single board (multiprocessor) as well as by networking multiple independent computers (multicomputers) located apart. The traditional computer organisation and its related aspects are organised in Chapters 1–7, devoted to describing the fundamental organisational concepts of all resources present in a generic computer system. Modern approaches in computer architecture and its organisation are, however, described in Chapters 8–10, offering the state-of-the-art of this book. However, the structure of the parts, and independency and interdependency between the chapters of this book are shown in Figure I.1.

Salient Features of this Edition

The areas of computer organisation and architecture have experienced continuous innovations and notable improvements, particularly on a scale far greater than could have been anticipated. A lot of new developments in the architectural design of computers as well as its organisational approaches have evolved, mainly due to the rapid advancement in electronic engineering and technology that eventually introduces tiny-sized faster intelligent components while continuously slashing down the cost of these components. As a result, numerous affordable design concepts have emerged using more hardware packages, and many design issues have been thought of – some of them having already been implemented in practice as observed in the recent product releases. This edition attempts to address and include all these advancements as far as possible within the confinement of the limited available space, while maintaining a broad and comprehensive coverage of the entire field with possibly no loss in rhythm, incorporating the input and reviews of the Author's previous work by a number of experts in this area engaged in academic institutions as well as in IT industry. Responding to the suggestions they offered as well as those received from users, a lot of tightened modifications have taken place with narrative clarifications accompanied by suitable illustrations. In addition, many other issues in the meantime penetrate into the world of computers that made an impact in the user market, which in turn forces the manufacturers to redefine their trade-off. The net effect is the ultimate emergence of many new current topics of commercial importance, and an attempt is thus here made to include most of them appropriately in almost all relevant chapters.

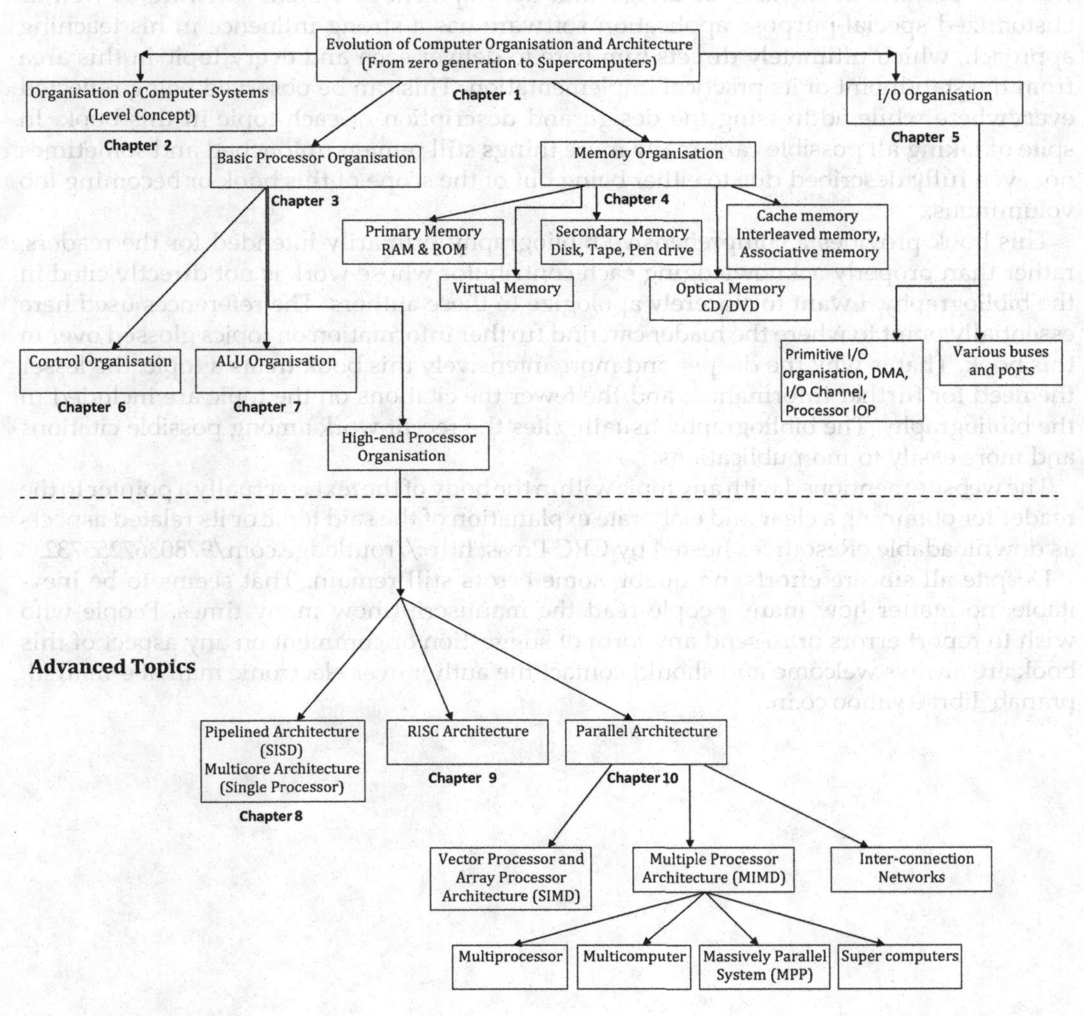

FIGURE I.1
The structure and independency/interdependency between chapters of this edition.

As usual, with each new edition, there is a constant struggle with the publisher to maintain a realistic balance between a reasonable page count and the number of new materials to be included. To attain this objective and to accommodate the minimum amount of contemporary important new materials, the needed room is created as usual by simply eliminating the relatively outdated items and tightening the narratives. Yet, many currently important materials and vital issues that deserve to be addressed have still remained outside the present delivery. The author deeply expressed his inability to counter such an unavoidable situation.

This book has been organised based on the author's long experience in teaching with this area in many engineering colleges and universities as well as in imparting industrial training to the senior software specialists in corporate sectors, including large multinationals, as part of his regular industry assignment. Moreover, his experience in industry with the projects in the area of design and development of system software as well as customized special-purpose application software has a strong influence in his teaching approach, which ultimately directs him here to deliver each and every topic in this area from the standpoint of its practical implementation. This can be observed being reflected everywhere while addressing the design and description of each topic in this book. In spite of taking all possible care, many more things still remain untouched and sometimes not even fully described due to either being out of the scope of this book or becoming too voluminous.

This book provides a comprehensive bibliography primarily intended for the readers, rather than properly acknowledging each contributor whose work is not directly cited in the bibliography. I want to sincerely apologize to these authors. The references used here essentially point to where the reader can find further information on topics glossed over in this book. That is why the deeper and more intensively this book treats a topic, the lesser the need for further information, and the fewer the citations on the topic are included in the bibliography. The bibliography usually cites the recent work among possible citations and more easily found publications.

The website mentioned with any topic within the body of the text is actually a pointer to the reader for obtaining a clear and elaborate explanation of the said topic or its related aspects as downloadable eResources hosted by CRC Press: http://routledge.com/9780367255732.

Despite all sincere efforts, no doubt, some errors still remain. That seems to be inevitable, no matter how many people read the manuscript how many times. People who wish to report errors or to send any form of suggestion or comment on any aspect of this book are always welcome and should contact the author over electronic mail at e-mail id: pranab_libra@yahoo.co.in.

Acknowledgements

Before tendering my thanks to individuals, let me first congratulate the community engaged in research and innovative developments in the area of computer organisation and its advanced architectures for their outstanding accomplishments in recent years. All my intentions and endeavours always attempt to document some of these superb accomplishments to include in the current presentations. I wholeheartedly adore the algorithms, protocols, techniques, and mechanisms developed by the community, not only for their practical merits but also for their good underlying science and implementation engineering. I find myself privileged for getting the opportunity to learn and write about them. It is at the same time hereby regretted that both this book and the author himself eventually ran out of space and time to include many of these excellent works.

Much of the useful works carried out by different authors are not directly included in the bibliography, which has been primarily organised for the readers rather than providing a comprehensive one to properly acknowledge each contributor. It intends to refer to where the readers can find further information on topics glossed over in this book. That is why, the deeper and more intensively this book treats a topic, the lesser the need for further information, and the fewer the citations on the topic are included in the bibliography. As more recent publications in turn provide references to earlier works, but not vice-versa, the bibliography attempts to cite more recent works among possible citations and also more easily found publications.

I also owe to the contemporary authors of this discipline who have made an impact on my delivery by way of lending their own ideas in many different areas of this subject that have been ventilated through their respective books. The author has no hesitation and is feeling free in expressing his gratitude and paying homage to them in this regard. As always, I wish to express my thanks to many people, including my reputed students who now engage in the teaching profession; my colleagues from industrial IT sectors; and my seniors in different dignified professions, including academics. Their constant chase and continuous pressing attitude insisted me to write this book. Once again, I find myself indebted to all of them.

Last but not least, I am obliged to my family members for their active assistance as being my best friends, including my wife, son, and daughter-in-law, who urged me to keep going on with the preparation of the manuscript by way of making some useful internal arrangements when I was running out of time and getting far behind the schedule.

Author

Pranabananda Chakraborty received his BSc degree in Physics from the University of Calcutta, India, and MSc degree in Applied Mathematics from Jadavpur University, India. He carried out his research work in Computer Science at the University Colleges of Science and Technology, Calcutta University, India. From 1978 to 1979, he was engaged in the design and development of a Fortran compiler for mainframe IBM Systems while he worked in the Software Export Division, ORG Systems, Baroda, India. From 1979 to 1983, he worked in a 100% Software Export Organisation with principals NCR Corporation, Dayton, Ohio, United States, in the area of system software development for different mainframe environments. From 1984 to 1995, he served as the regional manager at Computronics India, one of the largest organisations in this country dealing with large mainframe systems and software export activities. From 1996 to 2001, he worked in an organisation as a chief consultant dealing with a number of projects sponsored by the World Bank and Government of India. From 2005 to 2007, he was with Techno Campus, IT Sector, Calcutta, dealing with numerous system assignments, including instructional research and development, and a chief operation officer (COO) of the operations of corporate training in the area of computer science for large multinationals. From 2007 to 2017, he was in charge of the entire IT operations of different organisations in India, handling different projects running on various platforms ranging from mainframe systems to LAN-based system (cluster architecture) models. Apart from currently holding a senior position in IT industry, his teaching experiences span over nearly four decades (from 1983 onwards) in parallel as a visiting professor till date in a number of science and engineering colleges (both government and reputed private sectors); universities, including Birla Institute of Technology and Science (BITS), Pilani, India; and other famous technical institutes. In addition, he has continuously conducted and imparted high-end corporate training on academic subjects of core computer science sponsored by different bodies, including large multinationals and leading-edge R&D institutions. His research interests include computer architecture and also real-time systems, aiming mainly on their typical characteristics and behaviour.

Part I

Fundamental Computer Organisation

1

Computer and Its Environment

LEARNING OBJECTIVES

- To study the evolution of computer organisations and operating systems (OSs), and their numerous developments through different generations.
- To gain a clear understanding of the basic system structures and different types of relevant system software.
- To describe the salient features of different generations of computers, using representative computer systems of each generation up to supercomputers.
- To provide an overview of various hardware technologies and pioneering concepts.
- To describe the vibrant technological innovations in the area of microprocessors using family concept.
- To give an idea about multiple-processor architectures along with their numerous forms and different topologies.
- To explain the concepts of embedded systems and to describe in brief the generic real-time systems.
- To focus on the various forms and different issues of recent multicore processors.
- To introduce different types of operating systems – network operating system (NOS), distributed operating system (DOS), and real-time operating system (RTOS) – to handle different computing environments.

1.1 History in Short

The history of the evolution of the computer is itself very scintillating. A journey through it gives an overview of the evolutionary track with respect to computer structure, its design, and its function. It also helps to understand how its initial versions have been gone through many vibrant inventions and sparkling innovations to arrive at its present form (Augarten, S.).

The rapid pace in change that took place in the advancement of computer technology over this period to bring about the formation of today's computer architecture and organisation has been identified and narrated. These changes entail almost all aspects of computer science, starting from the underlying simple semiconductor technology through the most advanced integrated circuit (IC) technology. These technologies are presently used to fabricate numerous advanced computer components. Assembling of these superior components has eventually led to the extent of increasing use of ultimate parallel architecture concepts. It is interesting to note that despite radical changes in the concept of computer architecture and design, as well as the rapid pace in technological evolution by this time, certain fundamental concepts that emerged long back, still by and large, are consistently applied throughout, even at the present age of supercomputers. The implementation of the ongoing concepts existing at any point of time over this period is, however, observed to be decided mostly by the current state of technology and cost/performance, in order to fulfil certain primary prevailing objectives. Sometimes the implementation of an innovative design remains awaiting for the introduction of a befitting technology. This chapter, although not a complete one, highlights only the selective salient features of the ongoing evolution in the technology to realize numerous advanced computer components that summarily set the stage of a revolution in the design, organisation, and architecture of the computer system along with its inseparable part, the OS, as a whole.

1.2 Computer Organisation and Architecture

Computers are often described, categorized, and distinguished in terms of their *design*, *organisation*, and *architecture*. Computer design is mostly concerned in the formulation of the specifications of the proposed computer system considering primarily the different hardware components and the types of hardware facilities to be provided, the type of applications to be normally run, the target user community and their financial capability, the maximum permissible cost of the system and its allied performance, and similar other related aspects. All these when considered together lead to build up a computer system, which is called ***computer design*** (Blaauw).

After the selection of all the appropriate operational hardware units, the issues are now the relative placement of these units, and the way how the units are to be interconnected with one another for smooth operation as intended – all these when considered together lead to conceive a model of an abstract machine, which is known as ***computer organisation***.

Computer architecture involves both hardware organisation and the behaviour of the computer as experienced by the user. This includes the number of registers, instruction set, instruction format, addressing modes, the techniques for addressing memory, and similar other attributes that have a direct impact on the logical execution of a program. The architectural design also includes the specification of various functional models (like CPU (central processing unit), caches, buses, microcodes, and physical memory, etc.) and structuring of these models and their interoperations that eventually lead to essentially realize a complete computer system to achieve the desired *system performance*, which is the ultimate goal in the study of computer architecture.

Nevertheless, the architecture of the computer entails many other different issues, including the *processor design* that itself constitutes a central and very important element of computer architecture. That is why, various functional elements of a processor must be designed, interconnected without much affecting the existing organisation, and interoperated in such a way that desired *processor performance* can be achieved.

In summary, it can be concluded that whenever a specific hardware system is built up, a lot of strategies must be framed in advance, mainly with certain aspects in regard to each individual resource to be used, their relative placements, as well as the interconnections to be made between them so as to lead to a totality, commonly referred to as *computer organisation* (Langholz, G., et al.). The organisation of the computer must be designed in such a way that a particular architectural specification can be judiciously implemented. In fact, a thorough treatment of organisation requires a detailed examination of the architecture as well.

1.3 Hardware and Software: An Introductory Concept

A computing system consists of hardware, software, and programming elements, which are considered as its basic components. The electronic circuit, which is the lowest level of a computer system, consists of tangible objects such as processor, memory, input–output (I/O) devices, cables, integrated circuits ICs, printed circuit board, and power supplies, which all form the computer ***hardware***. The basic hardware elements in any computer system, be it a tiny microcomputer, minicomputer, supermini, mainframe, or gigantic supercomputer, are *processor, memory, I/O devices*, and *interconnecting buses*. Depending on certain basic parameters (such as size, capacity, capability, power, and speed) of these individual resources and considering the characteristics of the interconnections made between resources, the computers are accordingly classified. These resources, however, are normally driven individually by the respective software.

Software essentially consists of the systematic steps (algorithms) to be executed to perform a predefined task, and their computer representation in the form of a collective meaningful sequence of suitable instructions is called a ***program***. Software of any kind, however, is essentially the programs of any types and never includes the physical devices on which it resides. Software associated with computer systems truly defines and determines the ways that actually drive the hardware resources of the system. It is a key factor that unfolds the actual strength of the hardware by driving it properly to extract the best potential out of it. In fact, the joint effort of the software and the hardware manifests the real performance of the system. Software is mainly differentiated according to its purpose and broadly classified into two distinct categories, namely ***application software*** and ***system software***. Application software is developed using mostly high-level programming languages that mainly cater to the specific needs of the concerned user.

Some common programs that are developed to drive, control, and monitor the operations of the computing system resources as and when they are required to make computers better adapt to the needs of their users are historically called the ***system software***. Numerous I/O devices also require device-dependent programs that control and monitor the smooth operation of the devices during an I/O operation. These programs are essentially the *system software* known as ***device driver*** or sometimes called ***IOCS*** (input–output control system). All these programs are mostly written using a *low-level language*, such as *assembly language* and *binary language*, which are very near to the machine's (hardware's) own language or have the pattern so that the machine resources can be directly accessed from the user level. Nowadays, they are often developed also using a high-level language (HLL) like "C". Common system software, in particular, is very general and covers a broad spectrum of functionalities. It mainly comprises three major subsystems: (i) *language translators and runtime supporting systems for a programming language (compiler, assembler, loader)*, (ii) *utility systems*, and (iii) operating sustems. Some system software programs, such as *graphic library*, an *artificial intelligence, image processing*, and *expert system*, are specific to a particular application area and are rather not very common in others. The *OS, compiler, assembler, loader,* and to some extent *the utilities* are mainly required to commit physical hardware (machine) resources to bind with the application program for its execution. The optimal design of these software programs based on the *architecture and organisation of the underlying hardware,* their offered facilities, and lastly their effectiveness ultimately determine the efficiency of the hardware utilization and the programmability of the computer system as a whole. Figure 1.1, however, illustrates a conceptual representation of an overall computing environment when viewed from a user end in respect of the relative placement of hardware and the different types of software as already mentioned, including the OS.

With continuous steady advancement in computer technology, a different form other than that of traditional hardware and software gradually emerges, historically known as the ***firmware***, which is assumed to be an intermediate form present in between hardware and software, and indeed, a form of low-level software embedded in electronic hardware devices (normally in IC chips). This software is usually (not necessarily) infused at the time of hardware fabrication (manufacturing), or sometimes implanted later into the specified hardware. The programs as written are expected to be never changed, or at most are

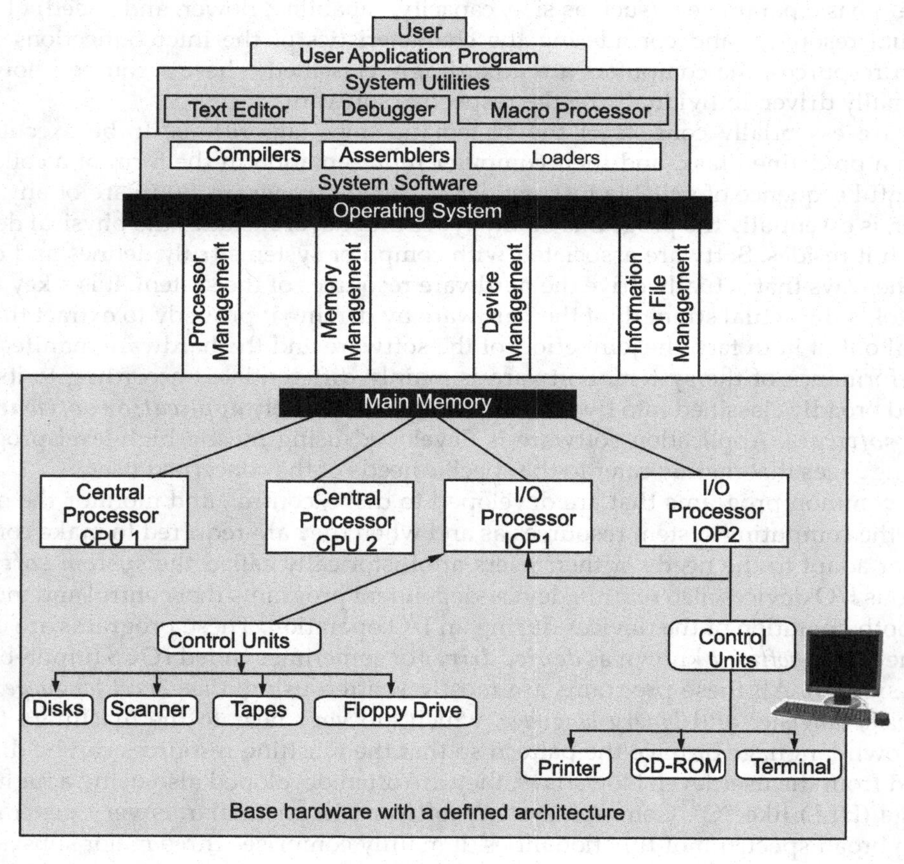

FIGURE 1.1
Level-wise position of OS, system software, and hardware organisation.

rarely modified. Nowadays, many appliances and instruments, process control systems, toys, etc. extensively use firmware for their control while in action. The programs in the firmware remain unchanged even when the power is off. *Microprogram* in many computers, in essence, is also a firmware.

1.4 Hardware and Software: Their Roles and Characteristics

Computer operation is a joint venture of both software and hardware. Any operation being performed by the software can also be directly built into the hardware. On the other hand, any instruction executed by the hardware can also be simulated in the software. In fact, hardware and software in a computer system are logically equivalent. To what extent the hardware will be involved and the software will participate in any operation is a major criterion at the time of designing and organizing a computer system. The decision in this regard that dictates to put certain functions in hardware and others in software is mostly determined by a number of factors such as speed, reliability, cost, target user group, frequency of execution of such functions, its expected rate of modifications and enhancements, and similar many other factors. There exist as such no prescribed boundary lines between the domain of hardware and the region of software, and there exist also no hard and fast rules about the effect that this must go into the hardware and that this must be explicitly programmed to place into the software. Designers and architects having different objectives and targets may often exercise different decisions to fulfil their requirements in order to reach their desired goals.

Computers at the early stage had a clear distinction between hardware and software. Frequently carried-out operations were all executed at the low level (hardware). The emergence of microprogramming concept (described in Chapter 6) and later its successful implementation indicated the dominance of ***reverse trend***. What was previously executed at the hardware level, say, an ADD operation, is now implemented at a relatively higher level, i.e. the microprogramming level in which a microprogram interprets an ADD instruction which is carried out in a series of small steps. This once again emphasizes the fact that there are not any such hard and fast rules which can govern that what must be in the hardware and what should be kept in the software. No such partition line, however, physically exists in between.

Further technological advancement once again influenced the design and architecture of computer system that insisted it to again ***swing back*** to the previous concept: the operation is to be executed at the lowest level, i.e. at the hardware level. In fact, the boundary as perceived between hardware and software is constantly changing and apparently looks like an arbitrary one. Today's software may be tomorrow's hardware. This is observed in today's ***embedded system***, where software is assumed to be embedded in hardware. The architecture of the computer system when viewed from the angle of *level concept* supports this fact once again where a low-level software is found very close to the hardware level. The ultimate outcome of this continuous back-and-forth swing in philosophies led to the eventual emergence of *two different schools* with *two distinct and different approaches*: the ***CISC*** **(complex instruction set computer)** ***architecture*** using microprogramming approach and the ***RISC*** **(reduced instruction set computer)** ***architecture*** using hardware approach.

1.5 Evolution of Computers: Salient Milestones

The birth of today's computer is not an isolated or sporadic event. It has been developed from many sources all over this globe and has been gradually moderated and fine-tuned with the invention, discovery, and development of the other related areas in electronics and electrical and mechanical engineering. History of computations leads us to the origin date back to a period as early as 2000 BC. Greeks and Romans at that time used a kind of calculating device known as ***abacus*** (a Latin word meaning *flat surface*) that consisted of stones manipulated on the flat surface to perform relatively sophisticated decimal arithmetic (base 10) calculations with digit-by-digit carry propagation since its storage capacity is limited to one digit. The abacus was invented in China in pre-Christian times, sometimes around 500 BC, and is thought of as the earliest mechanical computer.

The regime of mechanical computers, **the zero generation**, is assumed to have its origin date back to the beginning of seventeenth century with numerous remarkable innovative machines that came from many scientists and technocrats throughout the world over a considerable duration of time. All these machines were built up mainly with mechanical components, such as rotating cogwheels, and gears etc., and later electrical parts were incorporated into these machines.

1.5.1 The Generation of Computers: Electronic Era

The sensational invention of vacuum tube by Lee Dee Forest in 1906 has radically changed the computer world when vacuum tube came in use in the realization of computer system replacing the traditional concept implemented by electromechanical components. Electronic computers have then gone through many inventions and innovations during the past seven decades depending mostly on the outcomes of electronics advancements and its related areas that were successfully implemented in the design and realization of computers. These progressions, however, have been divided and categorized in terms of ***generation***. A new generation is declared only when a sharp breakthrough in the existing technology of both hardware and software is observed. New generations, however, have introduced new hardware and software technologies, and also inherited all the important features of its previous generations.

The **first-generation computers (1945–1954)** are demarcated with the use of vacuum tubes and relay memories interconnected with insulated wires, requiring much air conditioning due to awfully heat generation while in operation, and are very slow. Representative system of this generation was ENIAC (Electronic Numerical Integrator And Calculator) developed in 1946 by John Mauchley and Presper Eckert at Moore school, University of Pennsylvania, with Dr. Von Neumann as a consultant of this project. Later, EDVAC (Electronic Discrete Variable Automatic Computer), the first stored-program computer, was built by the Moore School between 1947 and 1950 based on Neumann's idea, and it became operational in 1951.

1.5.1.1 Von Neumann Architecture: Stored-Program Concept

Following the success of ENIAC and later of EDVAC with stored-program concept, Dr. Von Neumann and his colleagues A.W. Burks and H. H. Goldstine published a series of papers from 1946 to 1948 that for the first time clearly described and set up the basic architecture and organisation of a general-purpose computer system, including its logic design and programming aspects. The design principle as proposed was so fundamental in concept that it is still considered to be quite modern in its formation. The concept as ventilated, however, had a long-bearing and a far-reaching influence, mainly due to its ideal organisation of the basic resources (CPU, main memory, and I/O devices) along with the implementation of stored-program concept that eventually made the computer fully automatic (Burks, A. W., H. H. Goldstine, and J. Von Neumann.). All the computers in this globe were subsequently designed and developed later, and by and large, even modern computers of today mostly follow the basics of this principle. Figure 1.2 shows the typical architectural design of computer, which is derived from Neumann's idea and consists of the following components:

1. A CPU comprises arithmetic–logic unit (ALU) and a centralized program control unit for sequencing and executing the instructions given by the user;
2. A main memory unit stores information in the form of instructions and data;
3. An I/O unit consists of secondary memory units, information feeding unit (input), and result-receiving unit (output).

One of the notable features of this design was the placement of I/O systems, which were kept outside the core domain of the proposed form.

Neumann's concept as just described envisaged the computer design to be with a centrally located single memory for storing both program and data. The proposed Von Neumann architecture is sometimes also known as ***Princeton architecture***.

In contrast, another well-known concept in the computer design is what is popularly known as ***Harvard architecture*** that visualizes the computer design consisting of separate memories for storing program and data individually. Program memory and data memory can even be of different widths, type, etc. Program and data can be simultaneously fetched from these two different memories in one cycle, by separate control signals, namely "program memory read" and "data memory read". Implementation of such an architecture, e.g., includes Harvard Mark I computer.

Dr. Von Neumann along with his colleagues then developed a new computer system based on their proposed design (1946–1948) using fundamental resources (CPU, memory, and I/O devices), popularly known as ***IAS computer*** at the *Institute of Advanced Study* (IAS), Princeton University.

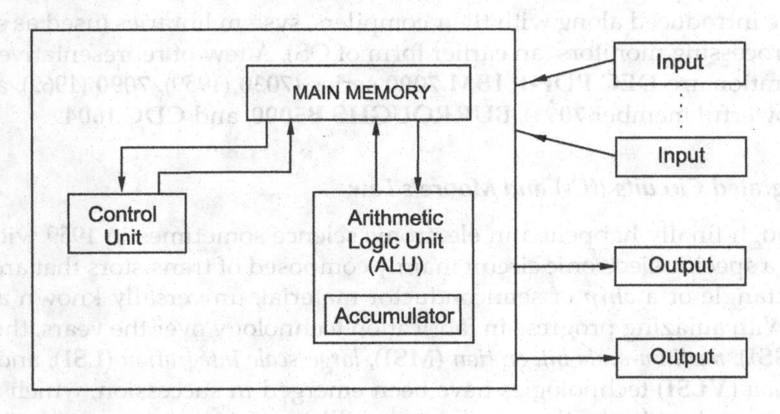

FIGURE 1.2
Basic design of Von Neumann machine.

1.5.1.1.1 Limitations: Von Neumann Bottleneck

In Neumann's proposed design, the relatively slower single memory centrally located for storing both program and data to be sequentially accessed by faster CPU consumes a huge amount of time to move instructions and data between CPU and main memory, and also, to a lesser extent, between main memory and I/O devices. The reason is that the existing speed disparity between the faster CPU and the relatively slower main memory causes the CPU to be remained idle over a longer duration for the data to arrive from the slower memory. Moreover, as technology gradually advances, the speed of the CPU has spurted enormously, while the speed of the memory also increases but quite moderately. As a result, the speed disparity between CPU and memory also constantly increases with passing days, which now seriously causes a critically notable degradation in the overall system performance. It appears that the technological advancement here turns out apparently as a curse and not a bliss, at least in this regard.

In addition, the program and data were rather mixed in a single memory (device) in Neumann's machines, requiring strictly sequential accesses, and were not allowed to store them in separate memory areas (or devices) that could otherwise made then possible to simultaneously access program and data in parallel. This sequential-access approach is used to consume more time to complete an execution as instruction and data are normally accessed one after another. All these together gave rise to such a crucial situation at the time of designing standard computers (Neumann concept-based) that it is sometimes referred to as ***Neumann bottleneck.***

For more details, see the website: http://routledge.com/9780367255732.

1.5.1.2 Second-Generation Systems (1955–1965)

The **second-generation computers (1955–1965)** commenced with fundamental changes introducing semiconductors (transistors) in place of vacuum tubes. As a result, the hardware of the computer became more powerful, immensely simplified with lowering in size and cost, less requirement of power, and consequently a sharp decrease in the generation of heat and thereby less requirement of air conditioning. Above all, the functioning of computer became more reliable with lowering of its total running overhead and maintenance. The architectural design also experienced radical changes with the introduction of index registers, floating-point arithmetic, multiplexed memory access, and I/O processors. High-level languages such as FORTRAN (introduced by IBM in 1956), COBOL, and ALGOL were introduced along with their compilers, system libraries (used as subroutine), and batch-processing monitors (an earlier form of OS). A few of representative computers of this generation are **DEC PDP-1, IBM 7000** series (**7030** (1959), **7090** (1962), and the last and most powerful member **7094)**, **BURROUGHS B5000**, and **CDC 1604**.

1.5.1.3 Integrated Circuits (ICs) and Moore's Law

A breakthrough finally happened in electronic science sometimes in 1959 with a notable invention of a specific electronic circuit mainly composed of transistors that are fabricated in a tiny rectangle or a ***chip*** of semiconductor material, universally known as *integrated circuit* (**IC**). With amazing progress in fabrication technology over the years, the *small-scale integration* (**SSI**), *medium-scale integration* (**MSI**), *large-scale integration* (**LSI**), and *very-large-scale integration* (**VLSI**) technologies have been emerged in succession, which are applied to ICs containing hundreds, thousands, and millions of transistors, respectively. ICs of these different categories (SSI, MSI, LSI, and VLSI) are roughly classified based on their *density,* which is again defined either as the number of transistors included in the IC chip, or as the number of logic gates per IC chip where a typical logic gate is composed of about five/six transistors. The impact of VLSI technology on computer design and application has been observed to be profound. At present, even its higher version *very-high-speed integrated circuit* (**VHSIC**) or *ultra-large-scale integration* (**ULSI**) is extensively used in our usual high-speed computers and also in supercomputers. The dominant technology being used in realizing transistors and other circuits within an IC, including *dynamic random-access memory* (**DRAM**), microprocessors, and other VLSI-based ICs, is what is known as ***complementary MOS*** or *CMOS* technology. At present, more advanced versatile technologies such as ***ECL, FPGA*** (*field-programmable gate array*), and ***GaAs*** (gallium arsenide) have been introduced in this regard and are now extensively used.

Moore's law: The growth in IC density constantly improved as time went on, and the number of transistors that could be accommodated on a single chip was observed doubling every year. This has been pointed out by Gordon Moore, *cofounder of Intel,* in 1965. He predicted that this pace would continue even in the forthcoming days, and promulgated a formulation in this regard, which is later known as *Moore's law* (Borkar, S).

It is interesting to note that Moore's law is borne out by and large throughout all the generations that evolved by this time till today with relentless increase in the speed of processor chip. A new generation of such chips has started to appear almost with a regular interval of every *three years* with *four times* as many transistors available on the new chip, and the performance as observed has improved four or five times every three years or so, since the introduction of Motorola MC 68000 and Intel X-86 family of microprocessors launched sometimes in 1978. The introduction of new *memory chips* (of course, with DRAM and its modern variants) also happens to continue at regular intervals, and they used to appear almost with the same pace as of processor chips with quadrupled capacity every three years. All these together, however, significantly confirm the remarkable predictions as proposed by Moore long back in regard of the technological trends of the forthcoming days.

For more details about Moore's law, see the website: http://routledge.com/9780367255732.

1.5.1.4 Third-Generation Systems (1965–1971): The MSI Era

The emergence of ICs and subsequent use of SSI/MSI chips in computer systems is considered a renaissance that eventually gave birth of a new generation with a revolutionary change in the existing design, organisation, and architecture of computers. Pipelining in the design of CPU apart from constantly increasing speed of the CPU itself, and introduction of speedy **cache memory** for quick access of information by CPU, reduced the speed gap between the main memory and CPU, thereby improving the overall execution speed, which were the remarkable developments however. In addition, memory size and speed also notably increased. **Virtual memory** technique has been introduced to minimize the memory–space scarcity problem. **Microprogramming** approach has been implemented in the CPU design to make it flexible and more versatile. Significant improvements in the performance of numerous I/O devices such as large disk with high-speed drive, high-speed printers, and terminals have also been observed. Implementation of multiprogramming with timesharing to interleave (overlap) CPU and I/O activities across multiple-user programs was one of the most distinctive features in OS design. In addition, many other novel ideas have been injected in the design and development of contemporary OSs that ultimately were able to almost fully extract this more-potential hardware capability on behalf of the user.

Family concept: The *family concept* was planned and initiated first by **IBM** with **System 360** family ("3" stands for third generation, and "60" is the 1960 decade) and later with **System 370** family, forming a set of similar and compatible machines (the same architecture but different models) released in succession over the years with a wide range of price and related performance (different implementations of the same architecture), eventually put an immense impact on the entire industry that immediately caught on instantly (Gifford, D). Within a few years, most successful other computer manufacturers of those days came out with their own different families of versatile computer systems, each of them equipped with all its own salient features of this generation, and became enormously successful in the area of educational institutions and in commercial and R&D organisations. Some notable of them are **CDC 6600/7600, UNIVAC, NCR 8100, 8200, 8500 series, DEC PDP, Burroughs 6700 and 6800 series, and many others.**

Many of the family-concept features proposed by IBM through the introduction of its System 360, however, have been subsequently taken into granted as a standard in the industry for both small and large computers. Presently, this family concept has been found to be implanted by **Motorola** and **Intel** in the 1980s in their **MC68000** and **X-86** line of microprocessor products, respectively.

For more details about the third generation, see the website: http://routledge.com/9780367255732.

1.5.1.4.1 Virtual Machine (VM)

To fulfil varieties of needs and continual increasing desires of the then users, the computer architecture was relentlessly upgraded with constantly emerging more advanced hardware. Hence, the new OS with changed design was often required to drive this advanced hardware, either to be developed afresh or by repeated modification over the existing older one. This, in turn however, invited another critical problem. That is why, a user's program while was runnable under an old OS, but became unusable under the new OS (enhanced version) without modification, and this modification might be again sometimes quite extensive and expensive as well. This situation was faced by IBM in particular, since many different versions of OS that came from IBM were introduced in quick succession for their 360 and 370 series of machines, and those versions were very similar, but not fully compatible. An IBM installation then had different versions of OS in the machine and was facing a lot of difficulties due to often switching of OS to meet all its upcoming user's needs and demands. To avoid many such operational difficulties resulting from frequent changes in OS caused by very often updates in hardware architecture, IBM extended the existing hardware architecture of its S/360 and later of S/370 series of computers to a form popularly known as the *virtual machines* (*VMs*), and accordingly developed a special form of OS, such as **DOS/VM** and **OS/VM.**

The ultimate objective of VM was to multiplex the entire system resources between the users in such a way that each user was under the illusion of having an undivided access to all the machine's resources. In other words, the users appeared to have a separate copy of the entire machine on their own, and each such copy was termed a ***virtual machine.*** Each VM was logically separated from all others; consequently, it could be controlled and run by its own separate OS. This led to the system organisation as depicted in Figure 1.3 where several different OSs were found to be concurrently used over a single piece of hardware. The heart of this system known as *virtual machine monitor* (*VMM*) runs on the bare hardware (physical hardware) and creates the required VM interface.

In spite of having several merits, *two major drawbacks* have been observed in the design of VM. *Firstly,* the *costs* of the hardware and the hardware interface being used were very

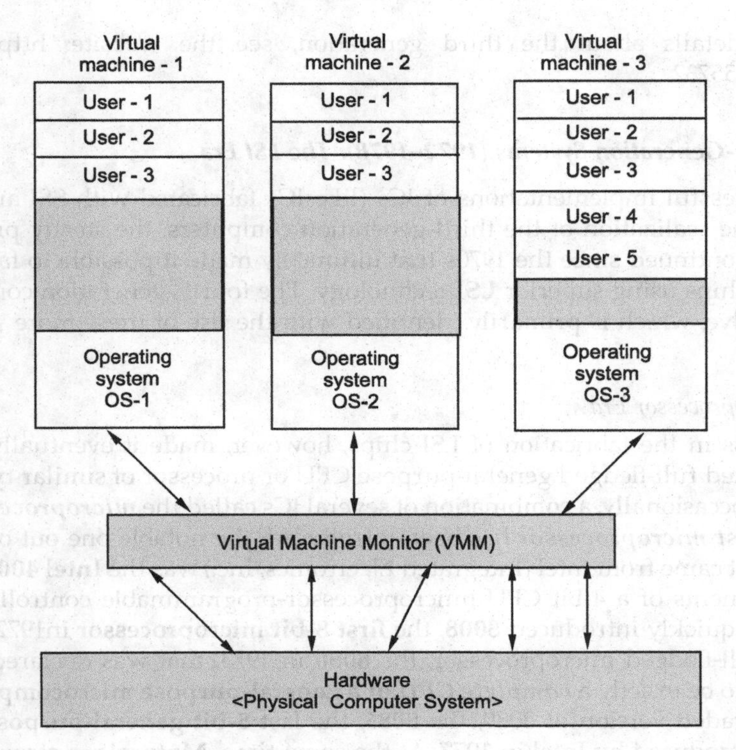

FIGURE 1.3
Schematic diagram of a VM and role of VMM.

high in those days. *Secondly*, any *fault* or *failure* in the common hardware interface (VMM), or in the single piece of basic hardware, on which the different VMs with their own OSs were concurrently running, would cause a severe breakdown in the operation of every machine, thus leading to a total collapse of the entire system. In addition, VMM interface is still a complex program and was not that simple to realize to obtain at least a reasonable performance. As a result, VM gradually fizzled out, but this concept and its successful implementation eventually left behind an immense impact that opened a new horizon in the architectural design of computer systems and its organisations (especially the innovation of computer networks) in the days to come.

1.5.1.4.2 Minicomputers: DEC PDP-8

DEC started minicomputer culture in this generation, and subsequently, a whole new industry came up with DEC introducing its first minicomputer **PDP-1** in 1961 and then a series of other PDPs, notably **DEC PDP-8** launched in April 1965. The salient features and major innovations out of many other nice features found in PDP-8 were a single bus (Unibus) structure, the **omnibus**, a **plug-and-play** support for different modules to connect, and a data transfer rate from memory roughly of 1.26 words/μs. The system was cheaper enough within an affordable limit with reduced overhead and could be placed on the top of a lab table requiring less air-conditioning facilities. The net effect was that DEC by that time was able to form a multibillion dollar industry, achieved a marketplace that was formerly reserved for IBM computers, and as such became a strong challenger of IBM, probably, was in the second position, just behind IBM.

For more details about the third generation, see the website: http://routledge.com/9780367255732.

1.5.1.5 Fourth-Generation Systems (1972–1978): The LSI Era

After the successful implementations of ICs (like ICs fabricated with SSI and MSI technologies) in the realization of the third-generation computers, the steady progress of IC technologies continued since the 1970s that ultimately made it possible to fabricate more advanced IC chips using superior **LSI** technology. The fourth-generation computers thus started to evolve, which is primarily identified with the use of these more sophisticated LSI chips.

1.5.1.5.1 Microprocessor Era

Rapid progress in the fabrication of LSI chips, however, made it eventually possible to have a tiny-sized full-fledged general-purpose CPU or processor of similar types on such a single IC or occasionally, a combination of several ICs called the ***microprocessor***. In 1971, the world's ***first microprocessor*** has been introduced; the notable one out of many such processors that came from Intel (Integrated Electronics, Inc.) was the **Intel 4004** containing all the components of a 4-bit CPU (microprocessor-programmable controller) on a single chip. Intel quickly introduced **8008**, the **first 8-bit microprocessor** in1972, and finally launched a full-fledged microprocessor, the **8080** in 1973, that was declared as the first one designed to be exactly a ***complete CPU* of a general-purpose microcomputer system**. Soon, an upgraded version of 8080, the **8085**, the last **8-bit general-purpose microprocessor**, was introduced by Intel in 1977. At the same time, **Motorola**, a strong challenger of Intel in semiconductor industry in those days, introduced the product **6800, an 8-bit machine** comparable to 8080 and widely used as an embedded controller in the industrial equipment. Many other reputed companies also launched their 8-bit microprocessor product with their own distinctive features, and the notable ones are **Zilog** (Z-80, a machine language compatible with Intel 8085), **Fairchild** (F-8), **National Semiconductor** (IMP-8), and **Rockwell International** (PPS-8).

Incidentally, the **first personal computer (PC), MITS Altair 8800,** was introduced sometimes in 1974. The BASIC language interpreter, written for this Altair 8800 computer, was developed in 1975 by **Bill Gates** and **Paul Allen**, the founders of Microsoft Corporation.

1.5.1.5.1.1 The Advanced Microprocessor Due to relentless progress of the IC technology with improved chip density, the internal architecture and the organisational approach in the design of one-chip microprocessors (Carter, J.) have been radically changed with increased complexity and performance as was reflected in the emergence of a powerful general-purpose **16-bit** microprocessor in 1978, the **Intel 8086** (Brey). Within a year or so (1979), it released another 16-bit microprocessor with a different internal architecture and organisation, the **Intel 8088**, a single-user PC product, fall into **PC/XT** class. In 1982, Intel launched **Intel 80286**, a new 16-bit microprocessor belonging to **PC/AT** (advanced technology) that eventually led to an ultimate ***eightfold increase*** in the execution speed for many instructions when compared to that of 8086/8088 (Brey. Barry B.). **Motorola 68000** introduced in 1979 was a radically different new 16-bit CPU chip with no compatibility with its earlier obsolete product 6800 or 6809. Although the chip provides 16-bit-wide data bus, all the user-accessible registers are 32bits wide. Thus, 68000 was essentially a midway (a hybrid) in between 16-bit and forthcoming 32-bit architectures. Due to the presence of many important and excellent features, this product (the first member of MC 68000 series) was considered to be no less than, not at all inferior to, and in fact, might be a little ahead than, any other contemporary products launched by different manufacturers, including Intel, of those days.

1.5.1.5.2 Semiconductor Memory (Chip-Based): Different Generations

Apart from the main objective of IC circuit technology to fabricate the processor either on a single chip or by a combination of a set of chips, it was also found that the same technology could be exploited to fabricate advanced form of new memories. **Fairchild, Inc.** produced the first relatively capacious IC-based semiconductor *non-destructive* memory of 256 bits in 1970. Improvements in memory technology since then continued and have passed afterwards through a series of **13 such generations,** namely 1K, 4K, 16K, 64K, 256K, 1M, 4M, 16M, 64M, 256M, 1G, 4G, and as of today 16 Gbits on a *single chip* ($1K = 2^{10}$, $1M = 2^{20}$, $1G = 2^{30}$). It is interesting to note that each subsequent generation has offered four times the storage density of the previous generation, but maintaining the size of the chip almost remained unchanged. The cost per bit also steadily declines with a notable decrease in access times.

With the introduction of more advanced (LSI) technology, it then became possible to have smaller, faster machines with a memory of larger capacity, and that too is against a relatively reduced price that consequently changes the nature of computers and that of computing world in less than a decade. Although large expensive systems and costly mainframe computers remain a part of the landscape, the computer has also been brought out to the reach of the "end user" with the introduction of office machines (advanced minicomputers) and single-user PCs.

1.5.1.5.3 Advanced Minicomputers: DEC PDP-11

Receiving an immense success of PDP-8 in the user domain, sooner DEC launched its enhanced version, a 16-bit successor of PDP-8, the **PDP-11** in the mid-1970s, thus maintaining a total downward compatibility with all its existing other members of this family while providing a comparatively powerful machine instruction set and a relatively large address field suitable for storage and handling information. The important parameters of PDP-11, such as the cost, performance, size, and overhead expenditure, were so attractive that it became enormously successful, both in commercial and in educational and R&D sectors. Continuous advancements in technologies, however, led to the constantly growing power of 16-bit minicomputer's functional operations, supported by increasing size of low-cost memory with many different types of available I/O devices that eventually made the system configuration immensely powerful offering multiuser/multi terminal support to handle the continuously emerging large application areas. The regime of DEC with its flagship PDP family thus steadily continued that eventually converted into a significant and convincing lead over the other minicomputer manufacturers.

1.5.1.5.4 Large Systems

The classification by generation, especially in the area of *large systems,* henceforth becomes less clear and not as much meaningful, although numerous new products, including important software as well as improved communications with related hardware, have been constantly introduced and included as technology rapidly progressed. Many computer manufacturers, including IBM, NCR, DEC, and BURROUGHS, used to offer a series of upgraded models belonging to one of its particular family, all with the same architecture but with continually improved different organisations to make use of the ongoing advanced technologies, resulting in the difference in their cost and performance. A representative large system in this generation is **IBM 3033** introduced in 1977, a fully compatible one with the existing S/370 processors, but with superior performance. Out of many of its salient features, one is the use of an improved ***instruction pipelining scheme***, which is at present followed universally by almost all modern CPUs as a basic design methodology. Another remarkable feature of 3033 is its capability of accommodating **multiple processors** additionally by way of configuring it with an add-on CPU making it two tightly coupled CPUs connected by a specialized hardware module known as the *multiprocessor communication unit* (MCU).

For more details about the fourth generation, see the website: http://routledge.com/9780367255732.

1.5.1.6 Fifth-Generation Systems (1978–1991): The VLSI Era

The relentless efforts in continuous technological improvements used in the fabrication of ICs ultimately arrived at VLSI sometimes in the first part of the 1980s. The fifth-generation computers are mainly demarcated by the use of VLSI-based high-density and high-speed processors, and capacious memory chips, and are also enriched with many other salient features. A one-chip 32-bit microprocessors were introduced by reputed vendors. Memory chip using VLSI technology came in 1986 with a capacity of 1Mbit (2^{20} bit). By this time, ***microcontrollers have*** also been radically enhanced containing all the elements of a complete computer (CPU, memory, and I/O connections) in a single chip (or a few chips together) that eventually came in use as an **embedded system** in many modern appliances, such as photocopiers, automatic teller machine (ATM), medical instruments, automated manufacturing equipment, and entertainment equipment and similar many others.

1.5.1.6.1 The Modern Microprocessors

The steadily increasing user density along with continuous evolution of numerous applications areas, including recently introduced GUI (graphical user interface)-based software, however, demanded new microprocessors with more speed and richer instruction set, more addressable memory, and wider data paths for faster internal communication. Both Intel and Motorola adequately addressed this state of affairs that eventually gave birth of 80386 series and 80486 series (Intel X-86 series) from Intel in 1985 onwards, and side by side, MC68020 and MC68030 (Motorola 68000 series in 1979) from Motorola onwards. AMD (advanced micro device) has also produced a *triple-clocked version* processor that ran even faster with a clock speed of *120 MHz* and a bus speed of *40 MHz*.

1.5.1.6.2 Personal Computers (PC)

By the 1980s, the costs of hardware had dropped drastically providing low-cost and high-computing powered microprocessor in tiny size, and the microcomputer culture began. The PC was initially organised based on Intel 8088/8086, which was subsequently replaced with passing of days by forthcoming releases of 80286/80386/80486/80586 and Pentium series of 16/32/64-bit microprocessor's families. Other manufacturers started producing PC using almost the same organisation as in IBM machines (IBM-PC compatible), but the components used by different manufacturers were obviously different. Irrespective of the brand (manufacturer), all the PCs are mostly handled by PC DOS (IBM), later by the MS DOS, (Microsoft) or MAC OS developed by Apple Corporation, and nowadays with different advanced intelligent members of Windows family introduced in succession developed by Microsoft Corporation, apart from popular Linux and Unix.

PCs with their tiny size and enormous single user-interactive processing capability when being attached with other large increasingly powerful computers (server), and to be used as communication terminals, they are called the ***workstation*** (client). The resulting gigantic client–server model thus formed can then simultaneously handle multiple interactive users, voluminous transactions, and large database processing that eventually have replaced the large mainframe computer centre of yesteryear. Usually, when PCs are interconnected with one another, they formed a ***network of computers,*** and also ***computer networks***, depending on how they are connected and in which manner they are being operated.

1.5.1.6.3 Superminis: High-End Minicomputers

By this time, minicomputers with improved performance at reduced cost consolidated their own position, particularly in R&D and educational institutions as well as in small-/medium-scale commercial organisations, especially in *real-time applications* like air traffic control or factory automation. In the late 1970s, the more powerful 32-bit supermini computer, maintaining the same line of its predecessor 16-bit minicomputer, was introduced with larger size of memories and more intelligent peripheral devices to support more multiusers to work simultaneously. DEC launched its most powerful supermini family popularly known as *VAX series.* IBM also introduced its supermini series known as *AS400* (later called *I-series*, and now known as *P-series*) and then started to continuously release its more advanced and more powerful models using state of the art, which is still dominating and at present is widely used in many application areas.

The superminis were employed in moderate-sized industry, and also in many other prominent areas, apart from being used as network **file servers**. The real differences between a mainframe and a supermini are the I/O capacity and their applications for which they are intended. Superminis were normally used for ***multiuser interactive applications,*** whereas most mainframes were usually targeted for large batch jobs or ***on-line transaction processing*** (OLTP) such as banking or airline reservation systems in which bulk volume of data in the form of large databases were involved and constantly in use. Presently, these superminis are found in supporting larger computers as well when hooked up with them. Nowadays, most of the superminis, including AS 400 (now called *P-series*), are commonly used as a powerful stand-alone system for both scientific and commercial applications as well as a server in a network of computers.

1.5.1.6.4 Large Systems: Mainframe and Supercomputers

Computer hardware entered a new generation using faster LSI/VLSI chips with diverse architectural approach for *multiprocessing* as well as *parallel computing* using **multiple processors** built into the system with shared and distributed memory. **Vector processor** and **array processor** with *multiple processing elements* are introduced. Multiprocessor architectures have been started to gradually evolve, and employed in the design of computers that consequently gave rise to a different architectural form of emerging computers of the forthcoming days known as the ***supercomputers.***

Several notable features that have been enhanced and worthy to mention for these large mainframe systems are *higher execution speeds, pipelined ALU organisation* for *vector processing, the* ***use of microprocessor*** *instead of hard-wired logic in I/O control units,* and most remarkably *the use of* ***multiple CPUs.*** These machines were specially designed to maximize the number of **FLOPS** (<u>*Fl*</u>oating-point <u>*Op*</u>erations per <u>*S*</u>econds). However, CPU speed less than 1 *gigaflop/sec* was not considered to be a member of supercomputer category.

Fifth generation is summarily highlighted with having advanced processors, capacious memory systems, high-density packaging, communication channels, and numerous I/O systems being built up with advanced emerging technology. Large mainframes and supercomputers in this generation can eventually consolidate their own positions, since many application areas gradually evolved by this time that could not be possible by any minicomputers to handle. Highly parallel architectures have been then used in organizing supercomputers, and were only useful and effective at that time on a small range of sophisticated problems requiring parallel processing. Many popular comparable representative mainframe large systems of this generation introduced by different vendors have also been observed. Only a few of them are listed as follows:

- Fujitsu Facon VP-200 having 500 million flops;
- CRAY 1 and then CRAY X-MP and Y-MP that came during this time;
- CDC CYBER 205, which is also a contemporary large computer system;
- IBM 3090 that came at later stage;
- DEC VAX 9000;
- BBN TC 2000;
- CONVEX C 3800 series.

Many revolutionary innovations in the area of ***system software,*** including *distributed OS, multiprocessor OS, special languages and compilers,* and various *software tools,* were developed to extract the potential power of the underlying hardware and to effectively realize parallelism that is required to handle large commercial applications as well as the ***real-time applications.***

1.5.1.6.5 Embedded Systems

By the mid-1980s, it became feasible to replace most of the commonly used expensive external system components by a modern form of low-cost *microcontroller* or by a *microprocessor,* which eventually became the norm rather than the exception for many electrical and electronic devices. This emerging new form of components is commonly known as the *embedded system.* It could be normally defined as a special-purpose computing system (not a general-purpose computer) designed to perform one or a few dedicated functions, often with real-time computing constraints. It is usually a combination of computer hardware and software, and perhaps additionally, it contains mechanical or other parts that are embedded as part within a larger device or system or product which serves a more general purpose. For example, an embedded system in an automobile provides a specific function as a subsystem of an entire car. Some embedded systems are found to be made tightly coupled with environment and thus impose many critical real-time performance constraints that must be met, for reasons such as safety and usability; others may have low or almost no performance requirements as such, only to simplify the system hardware, mainly for the reduction in costs. If the embedded system is designed for managing multiple activities simultaneously, it is often a comparatively complex one with many different attachments, depending on the services that it intends to provide.

The embedded system usually comprises limited computer hardware resources: CPU of ordinary 8- or 16-bit *microprocessors* or *microcontrollers,* little memory, small or non-existent keyboard and/or screen, and may have many more on-chip (or off-chip) cheaper peripherals with reduced size that can mainly interact with the outside world. Software architectures of the embedded systems may be of different types ranging over a wide spectrum of varieties, including simple control loop, interrupt controlled system, cooperative multitasking, pre-emptive multitasking or multithreading, monolithic kernels, microkernels, exokernels, and exotic custom OS. They keep adequate provisions so that additional software components can be included, if required. Embedded system often uses *compilers, assemblers, debuggers,* and also some other *software tools* to develop its own software. The program instructions written for the embedded systems are referred to as ***firmware*** and are usually stored in *read-only memory* (ROM) or *flash memory* chips.

Since the embedded system targets to attain specific objectives and is mostly dedicated to specific tasks, they naturally have widely varying *requirements and constraints,* mainly size, responsiveness to different environmental conditions (temperature, pressure, etc.), different models of computation, human interface (complex GUIs), reliability and performance, and above all, the lifetimes.

Embedded systems encompass a broad range of applications covering many important aspects of modern life. Some of the notable ones are given in the following.

Consumer electronics (mobile phone, digital camera, DVD players, GPS receivers, etc.), **office automation** (printers, monitors, scanners, fax machine, photocopiers, etc.), **household appliances** (microwave oven, washing machine, dishwasher, and advanced HVAC (heating, ventilation, and air conditioning) system that uses networked thermostats to more accurately and efficiently control temperature that can change by time of day and season), **home automation** (wired and wireless networking that can be used to sense and control lights, climate, security, audio/visual, etc.), **transportation systems** (automobiles, electric vehicles, hybrid vehicles to maximize efficiency and reduce pollution, automatic four-wheel drive, flight systems, and many others), **advanced avionics** (inertial guidance systems and GPS receivers), **telecommunications systems** (telephone switches, routers, and network bridges to route data), **wireless sensor networking** (WSN) (with the use of a new class of self-contained and battery-controlled miniature wireless devices called *motes* that can couple full wireless subsystems to sophisticated sensor which can measure many different things beyond count in the physical world and act on this information through IT monitoring and control systems), and **medical equipment** (electronic stethoscopes for amplifying sounds, and various medical imaging techniques such as PET, SPECT, CT scan, and MRI for non-invasive internal inspections).

Embedded systems for very-high-volume production are often implemented by **SoC (system-on-a-chip)**, which contains a complete system consisting of (multiple) processors, multipliers, caches, and interfaces on a single chip. The SoCs can be implemented as an **ASIC** (*application-specific integrated circuit*) or using a **FPGA** [see Chapter 2]. Nowadays, the embedded systems are extensively used, mainly to control many commonly used modern devices. Also, most of them are often mass-produced, benefiting from economies of scale.

For more details about the fifth generation, see the website: http://routledge.com/9780367255732.

1.5.1.7 Sixth-Generation Systems (1991–Present): The ULSI Era

Sooner by 1990, it became possible using VLSI technology to fabricate a CPU and main memory, or even all the electronic circuits of a computer on a single low-cost IC that can be mass-produced. By the mid-1990, the manufacturers could fabricate the entire CPU of a system/360-class large computer along with part of its main memory on a single IC. *Today's CPUs are mostly microprocessors, implying that their physical implementation is a single VLSI/ULSI chip,* which can be used in machines of a new class ranging from portable PCs to supercomputers that contain thousands of such CPUs. The triumph of the sixth generation began from 1991 and is continuing till date.

1.5.1.7.1 Computer Networks: Client–Server Model

The immense success of Virtual Machine 370 introduced by IBM (VM370) planted a seminal design concept that manifests in interconnecting a collection of autonomous computer systems capable of communication and cooperation between one another by way of using their hardware and software tools (communication links and protocols). This arrangement is popularly known as the *computer networks,* a typical variant of a class called the *multicomputers.* Each such autonomous machine, usually called the *host,* running under its own OS is capable to handle its own users executing local applications, and constitutes the computational resources to the networks as well.

When these small intelligent machines/workstations (client) are normally interconnected among themselves, and also with a comparatively larger and powerful time-sharing machine (server) using relatively high-speed connections (LAN or WAN) in a network environment, they form a low-cost cluster, called the *client–server model.* This arrangement, however, enables each member system (workstation/client) to perform their own jobs individually at their own computer, to share and enjoy both the hardware and software resources available in the entire arrangement, and leaves the main computer (server) to be simultaneously shared by a larger number of other users. The outcome was found to be a remarkable one in the design of computer hardware configurations that eventually led the computing practices to follow a completely different path. Computer network sometimes is loosely called a *distributed computer system,* since it carries a flavour of a true distributed computing system comprising hardware composed of loosely bound multiple processors (not multiprocessors).

For more details about the computer networks, see the website: http://routledge.com/9780367255732.

1.5.1.7.2 Distributed Systems: Multicomputers

Advancements in computer technology continued, notably in two main areas in parallel, namely the technology that promoted the low-cost small computers with a computing power of a decent-sized mainframe (large) and also the technology that upgraded the networking [introduction of ATM (asynchronous transfer mode)] in both LAN and WAN environments. The potential outcome of these two emerging technologies when combined made it possible to easily connect many powerful computing systems together, comprising a large number of CPUs in this arrangement by high-speed network interconnections. This framework is commonly called the ***distributed computing systems,*** another typical variant of a class sometimes called ***multicomputers.*** Nowadays, even large systems with their own networks are used as *sites* replacing the relatively smaller nodes, and then using the same communication methodology forms the *massively distributed computing system.* It is totally different from traditional ***centralized systems*** that consisted of a single CPU, its memory, peripherals, and some terminals. Distributed computing systems that use different forms of distributed OSs, a radically dissimilar software, are commonly referred by the term ***true distributed system*** or simply ***distributed system.*** Distributed system, however, has been explained in detail in Chapter 10.

For more details about the multicomputers, see the website: http://routledge.com/9780367255732.

1.5.1.7.3 *Large Systems: Multiprocessors*

Due to the advent of more powerful VLSI technology, one-chip powerful microprocessor and larger-capacity RAM at reasonable cost have eventually emerged in the mid-1980s that radically changed the traditional definition and the architecture of the **multiprocessor** (a system having tightly connected multiple CPUs equipped with usually a single shared virtual address space) system. Large-scale multiprocessor architectures then started to emerge with multiple memories that are now distributed with the processors. Here, each CPU can access its own *local memory* quickly, but access to other memories connected with other CPUs is also possible, but is relatively slower. That is why, these physically separated memories can be addressed as one logically shared address space, meaning that any memory location can be addressed by any processor, assuming that it has the correct access rights. This, however, does not discard the fundamental shared memory concept of multiprocessor rather than broadening the existing idea. These machines are historically called the *distributed shared memory systems* (DSM) or *scalable shared memory architecture* model using NUMA (non-uniform memory access). DSM architecture can be considered as ***loosely coupled multiprocessor***, or sometimes also commonly referred to as the ***distributed computing systems***, in contrast to its counterpart shared memory multiprocessor UMA (uniform memory access), considered as ***tightly coupled multiprocessor*** or often called the ***parallel processing system***.

Usually, the *parallel processing system* consists of a small number of processors that can be effectively and efficiently employed, but constrained by the bandwidth of the shared memory resulting in limited scalability. It tends to be used to work on a single program (or problem) to achieve the maximum speed-up. *Distributed computing system* (loosely coupled multicomputers) can contain any number of interconnected processors with as such no limits (theoretically) and hence is more scalable. It is primarily designed to allow many users to work together on many unrelated problems, but occasionally in a cooperative manner that involves sharing of resources. Unlike multiprocessor, the processors of multicomputer can be located far from each other in order to cover a wider geographical area. While multiprocessor, for being more tightly coupled than multicomputer, can exchange data nearly at memory speeds, but some fibre optic-based multicomputer of today is also found to work very closely at memory speeds. Therefore, the terms *tightly coupled* and *loosely coupled* although bear some useful concepts, but any distinct demarcation between them is difficult to maintain because the design spectrum is really a continuum.

In summary, the large systems of this generation with ULSI/VHSIC processors were able to carry out massively parallel processing (MPP) at its innermost level using currently available VLSI silicon, GaAs technologies along with optical technologies. The minimum target of this generation is to attain a speed of teraflops (10^{12} FLOPs). Scalable architectures have been introduced to perform heterogeneous processing to solve large-scale problems involving voluminous databases of diverse nature exploiting the services of a network of heterogeneous computers with shared virtual memories. Diverse spectrums of elegant architectural evolutions with these ULSI/VHSIC chips have the noteworthy characteristics, which are as follows:

MPP;

Scalable and latency-tolerant architectures;

Multiprocessors with shared access memory, such as

UMA,

NUMA,

Cache-only memory architectures (COMA)

Multicomputers using distributed memories with multiple address spaces.

Computers that emphasize massively data parallelism, etc.

are only a few, apart from many others that have been achieved by this time. All these machines, however, have already reached the targeted teraflops (10^{12} FLOP) performance.

Machines of this generation are also used to solve *real-life applications*, which include computer-aided design of VLSI circuits, large-scale database management systems, artificial intelligence, weather forecast modelling, ballistic missile control, oceanography, pattern recognition, and crime control just to name a few. Both the scalable computer architectures and parallel processing computer architectures are, however, expanding steadily to negotiate the forthcoming grand challenges of computing in a better way.

MPP systems of this generation that have been represented, such as Fujitsu (VPP 500), Cray Research (MPP), NEC SX series, HITACHI S-810/20, SR 8000, Thinking Machines Corporation (TMC/CMs), CONVEX C3800, IBM 390 VF, and last but not least, the gigantic IBM z-series,

For more details about the large systems, see the website: http://routledge.com/9780367255732.

1.5.1.7.4 Supercomputers

A supercomputer, at the time of its introduction in the 1960s, has been simply defined in terms of speed of calculation involving processing capabilities. Today's supercomputers offer extensive vector processing and data parallelism that differ from the usual parallel processors, and can be broadly classified as **pipelined vector machines** having a few powerful processors equipped with adequate vector hardware and **SIMD machines (see Chapter 10)** having a large number of simple processing elements (array processor) that put more thrust on **massive data parallelism**. Supercomputers are mainly used for highly computation-intensive tasks such as *weather forecasting*, *climate research* (including research into *global warming*), *oceanography* (to predict any unforeseen unnatural event), *molecular modelling* (computing the structures and properties of chemical compounds, biological macromolecules, polymers, and crystals, etc.), *physical simulations* (such as simulation of airplanes in *wind tunnels*, simulation of the detonation of *nuclear weapons*, and research into *nuclear fusion*), *cryptanalysis*, and many other similar problems. Military and R& D scientific agencies are mainly among the heavy users. Although there were many companies that introduced their own supercomputers or nearly supercomputers, a few of the notable ones of today are as follows:

- The Intel ASCI Red/9632 launched sometimes in 1999;
- The NEC Earth Simulator introduced in 2002, the fastest supercomputer till the beginning of 2004;
- The Intel supercomputer systems (the Paragon);
- The IBM Blue Gene/L, the largest supercomputer of today.

For more details about the supercomputers, see the website: http://routledge.com/9780367255732.

1.5.1.7.5 Real-Time Systems

Availability of numerous low-cost sophisticated modern components using more advanced hardware technology eventually facilitates emerging numerous application areas, including a specific type of applications, which are now extensively used in some popular specialized domain known as the *real-time application systems*. Examples of such applications include embedded applications (household appliance controllers, mobile telephones, programmable thermostats), real-time databases, multimedia systems, signal processing, robotics, air traffic control, process control systems, telecommunications, industrial control (e.g. SCADA), radar systems, missile guidance and many similar ones. These applications require different types of actions following mostly different approaches, commonly known as the *real-time computing*, which is becoming an increasingly important discipline nowadays.

A ***real-time application*** *can thus be defined as a program that should respond to activities in an external system within a maximum duration of time specified by the external system.* If the execution of the application takes too much time to respond to, or to complete, the needed activity, a failure can occur in the external system. Such applications thus require a *timely response* with response time *smaller* than the response requirement of the system. Hence, this application system is usually executed with a different approach known as the ***real-time computing***, which may be defined as the type of computing in which the correctness of the system depends not only on the logical result of the computation, but also on the time at which the results are generated (response requirement). A *real-time system* is thus entirely different from the traditional multitasking/multiuser system. It is often tightly coupled to its environment with the real-time constraints imposed by the environment. However, a real-time system can be roughly classified into three broad categories:

- Hard real-time systems (e.g. aircraft control);
- Firm real-time systems (e.g. banking);
- Soft real-time systems (e.g. video on demand).

Since a real-time system, hardware-wise and software-wise, is a totally different one from the traditional general-purpose computing system, its design is encompassed with several different issues, mainly *architectural issues, resource management issues,* and *software-related issues.* In addition, it must ensure *predictability* in instruction execution time, *speedy memory access, fast context switching,* and *prompt interrupt handling.* This architecture usually avoids caches and superscalar features, but provides enough support for fast and reliable communication. Thus, the basic principles encompassed with real-time systems are inevitably of numerous dimensions based on specific objectives to achieve. With few exceptions, several techniques are available to implement each of these basic principles. One should always make use of the specific techniques and related mechanisms in the design of a particular RTOS to ensure that only the respective policies which entail the underlying real-time requirements will be ultimately implemented.

For more details about the real-time systems, see the website: http://routledge.com/9780367255732.

1.5.1.7.6 Yesterday's and Today's Microprocessors

Following the introduction of 80486DX4 in the 1990s, Intel launched a downward compatible **80586** in 1993 labelled as **P5.** Intel then decided not to use the number anymore with its product to avoid the problems related to copyright of a number. This product, however, then is historically known as *Pentium*, a generic name of a family of microprocessors. Side by side, Motorola also introduced its downward compatible 68040 processor of 68000 series.

With two introductory versions, **Intel Pentium** family of microprocessors came out continuously in succession with many different upgraded and newer versions using constantly improved more innovative designs supported by many salient features as days passed on. The newer versions of Pentium, such as **Pentium Pro, Pentium II**, **Xeon**, and finally **Pentium III** released in succession from 1995 up to 2000 with increasing processor speed attaining up to 1 GHz, providing *multimedia extensions*, or *MMX* instructions. All these processors were especially optimized for 32-bit code with a specific target to equip it for the server market and thus were often bundled with **Windows NT** rather than with normal version, like Windows 98. **The Motorola 68040** was introduced maintaining a downward compatibility with all its existing predecessors, and it was enriched with many additional significant features that ultimately upheld this processor to a much higher level than all its compatible contenders. **The Motorola 68060**, the last member of MC 68000 family, was launched in the mid-1990s with many organisational upgrades, mainly a basic four-stage with additional two-stage *pipelined superscalar processor of degree 3*, and new fabrication features to mostly address and accommodate the constantly growing ***embedded system*** market. The **AMD** processors (like the **Athlon**) also used IA-32 architecture (introduced by Intel) over quite a considerable period in its products for use in advanced PCs as well as in workstations. These processors eventually attained a performance level that was nearly comparable to the earlier version of Intel Pentium 4.

Motorola, a strong competitor of Intel in this line holding a share of nearly 50% of the microprocessor market over many years with their marvellous products (MC 68000 series), finally decided to shift its activities from the then existing business line and started to put more thrust on a newer upcoming most promising area what is known as the *mobile communication*. As a result, Motorola has sold out its microprocessor division to a company, now called *Freescale Semiconductors, Inc.* As a result, Intel, the only giant of earlier days, remains, still continues exerting more efforts, and today almost monopolizes and captures a major share of the entire desktop and notebook market.

For more details about the Pentium series and Motorola processors, see the website: http://routledge.com/9780367255732.

1.5.1.7.7 Multicore Concept: Performance Improvement

The performance gains can be mainly achieved with increasing *processor speed* and enhancing *clock rate* by way of increasing *logic density* of the processor chip. But the basic physical limits under the currently available technology that would be viable for the *clock speed* and *IC density* to attain at best (to keep the generated heat manageable) for achieving even more performance gains have, by far, already reached very close to these limits, leaving very little scope at present to further go ahead. Side by side, the existing computer design is constantly reviewed and redefined to organise the basic resources (processor, memory, peripherals) of numerously different speeds after making a reasonable balance between them to keep the overall performance improvement ongoing. But still, the basic design with the fundamental resources of even today's versatile computers remains virtually the same as those of the early IAS computer of odd 60 years back. Thus, there can hardly be any revolutionary change that can happen at present in relation to computer design and its organisation from its existing form.

It has been observed that the processor power has relentlessly raced ahead at breakneck speed continuing for about 15 years since the late 1980s, and this tremendous raw speed of the processor could never be properly exploited to yield the expected performance. Two noteworthy design approaches, namely ***pipelining*** and ***superscalar architecture*** (see Chapter 8), in the area of instruction execution logic have evolved to make use of the potentials of the processor and were then made in-built within a processor design. These two approaches have been then continuously enhanced to obtain more performance gain, and they have already reached nearly their limits and now approaching almost a point of diminishing returns. It seems that any further appreciable enhancement in this direction is hardly possible to achieve; at best, simply a relatively modest progress can only be made. Another important implemented strategy that increases the performance beyond what can be achieved simply by increasing clock speed is the *increase in cache capacity and usage*, which is essentially an organisational enhancement. As chip density is constantly increased by this time, more room is now available within the chip that is now dedicated to incorporate larger size of more speedy caches inside the chip, and that again with multiple on-chip levels (now three levels). Further significant improvement in this area seems to be likely not possible.

It is now evident that the highest benefits that can be accrued from all these approaches as mentioned, which contribute their shares in yielding performance improvement, have already reached almost their respective limits, and the clock speed cannot be further increased to a much higher rate. This ultimately puts the designers to have now turned to a completely different fundamentally new approach so that the overall performance can be further enhanced within the limit of the present constraints. The principal idea behind this new approach is to build multiple processors on the same chip, also referred to as *multiple cores*, or *multicore*, that provide higher potential to further enhance performance without increasing the clock rate as such (Gibbs, W). Thus, the concept of building multiple cores within a processor chip with larger on-chip caches (to resolve the critical power issue in the chip) is eventually recognized and identified as one of the best possible acceptable solutions within the premises of current technical aspects (scenario) to obtain even faster microprocessors.

The **IA-32** architecture continued with the first version of **Pentium 4** processor belonging to P6 architecture released in late 2000 with higher speed of 3.2GHz and even more. Its later versions called the *Pentium D* (*Dual Core*) and subsequently *Core 2* were available at speeds of up to 3.2 GHz using 0.045-micron or 45-nm fabrication technology. Subsequently, **Pentium 4e** and later **Pentium 4 Extreme Edition** were then introduced. Sometimes in 2001, Intel and Hewlett–Packard (HP) then jointly implemented **IA-64** (64-bit) microprocessor architecture, called the ***Itanium***, with many salient features, appropriate to negotiate different prevailing situations (Krishnaiyer, R., et. al.). Both Pentium 4 and Core 2 have been then modified to include a 64-bit core and multiple cores. The notable advancement with this technology is the introduction of a new concept, called ***multithreading***. In 2002, Intel and HP jointly released a new microprocessor architecture in the line of Itanium, called *EPIC* (explicitly parallel instruction computing), which is essentially designed for the server market, and may or may not trickle down to the personalized home/business market in the near future (Evans, J, et al.). Subsequently, Intel launched **dual** and **quad core and even higher core** versions, but in the near future, the number of cores will likely to be increased to eight or even sixteen using even finer fabrication technology, with stepwise increment in cores, which is supposed to be an acceptable alternative solution in the current scenario to provide even faster microprocessors.

It is interesting to note that microprocessors have evolved much faster and gradually became more complex. This is observed in the rapid growth of Intel X-86 family that provides an excellent illustration in relation to the continuous advancement of microprocessor technology that happened over the past odd 30 years. The 8086, introduced sometimes in 1978, was the only microprocessor with a clock speed of 5 MHz and had

29,000 transistors. A quad-core Intel Core 2 introduced in 2008 operates at 3.2 GHz, has a speed-up of a factor of roughly 600, and has 820 million transistors, which is about 28,000 times as many as the 8086. Yet, the Core 2 has a slightly larger package than the 8086 and has a comparable cost.

All these discussions as presented in regard to the constant developments of microprocessor technology, and subsequently, the continuous evolution of various flagship microprocessor products based on existing technology introduced by different manufacturers from time to time, including two giants Motorola and Intel, can be summarily organised in a tabular form with a comparative study. This may be a handy tool that can be used as a good indicator to describe how computer technology has gradually progressed over this period, in general.

For more details about the multicore, see the website: http://routledge.com/9780367255732.

1.5.1.8 Grand Challenges: Tomorrow's Microprocessors

Although it is seemingly difficult to make any accurate prediction in regard to the forthcoming evolution in microprocessor technology, the current trends, however, in this area can at best envisage a realistic path that can pronounce without doubt the success of Intel family which should continue for quite a next few years. What may occur is perhaps a change in RISC technology, but more likely would step forward in the line of this new upcoming *multithreading* technology, accommodating even more processors in parallel within the framework of the ongoing architecture (at present, seven or more such processors). As the clock speed seemed to have already picked to its limit, and the surge to multiple cores has begun, about the only major change in the Pentium line in the near future will probably be the inclusion of a wider memory path (128 bits) and increasing memory speed. Side by side, a new technology is also required in the area of mass storage (secondary device) to cope the constantly increasing high-speed components constituting today's faster computer system. Flash memory could be a solution, because its write speed is comparable to hard disk.

We have arrived, and are now passing through, the sixth generation. A journey through all these generations reveals that each of the first two generations lasted more or less over a period of ten years using the yields of contemporary development and advancement in electronics. From the beginning of the third generation, the renaissance actually started giving needed impetus resulting in radical improvement both in hardware architecture and organisation and in sophisticated software design and development. Resurgence continued with constantly increasing high pace culminating to the emergence of many innovative concepts, and sophisticated technological developments throughout the entire fourth and fifth generations that can be titled as golden period of computer science and technology. Sixth generation, however, equally maintains the pace of improvements following the last generation's contributions with more refined technological development and innovative architectural improvement.

1.6 Evolution of Operating System and System Software: Their Roles

The basic concept and the definition of the OS together with the definition of software and its broad classification into two distinct categories, namely *application software* and *system software*, including *utilities*, have been previously discussed. The role of the software in, and its contribution to, creating a versatile computing environment with the available hardware has already been narrated in brief. The relationship that exists in between this software and the underlying hardware through the OS as well as their relative locations with respect to the hardware is shown in Figure 1.1. In fact, the forms and designs of the OS mostly depend on, and are determined by, the underlying architecture of the computer system that will be ultimately driven by OS. Consequently, there exists a one-to-one inseparable relationship between the OS and the underlying computer architecture. That is why, with the continuously evolving and constantly advanced innovative hardware architecture, the designs and issues of OS have also been constantly reviewed, redefined, and redesigned, and thus progressed by this time through gradual evolution, generation-after-generation, for the sake of staying to be exactly matched with the underlying continuously evolving advanced hardware technology. Language processors, assemblers, loaders, database systems, library routines, debugging aids, application-oriented modules, etc. although are not included within the domain of the OS, but also have equally progressed by this time, and they are on a par with the continuous advancement of OS , because they are always using the facilities provided by the underlying system (see Fig. 1.1). In fact, the performance of the OS sets the stage for the concert of the hardware and associated software constituting the entire computing environment.

Initially, the **first-generation computers** had ***no OS*** (**zero-generation OS**), but by the early 1950s, **punched card** has been introduced and the **first-generation OS** has been implemented. There was hardly any system software support compared with today's modern machine. By the mid-1950s, with more powerful **second-generation computer system**, the *system programs* in the form of **utilities** have been developed for most frequently used functions to manage files and to control I/O devices. The **second-generation OS**, known as the *batch system* or also called the *monitor,* was then introduced along with **assemblers** and **compilers** (FORTRAN). *Job control language* (JCL) has been developed, which was required to initiate the OS at the time of job submission. The **third-generation computers** could be considered as *a major breakthrough* due to their relatively advanced architecture and design with more powerful emerging hardware technologies and facilities equipped with more intelligent befitting **third-generation OS** that offered initially **multiprogramming**, which further upgraded to **interactive multiprogramming** to enable the user to *interact* with the system through the *terminals* during runtime. This OS has been once again upgraded to the **multiuser system** so that multiple users could now work simultaneously with their own different jobs in this centrally located single-processor hardware using some sort of *computer terminals* attached with this system. Another variant of this type of OS is the **multiaccess system** that allowed simultaneous access to a *single program* by multiple users (not multiuser). This approach, however, opened an important line of development in the area of **OLTP**, such as railway reservation system and banking system, where users could efficiently execute queries or updates against a database through hundreds of supporting active terminals under the control of a *single program.* Inclusion of ***virtual memory*** and subsequently addition of ***cache memory*** in the hardware system, however, have once again modified the architecture and the design of the existing machines, and consequently, the forms and designs of the existing OS too were accordingly upgraded to handle the changed situation. Such modified OSs were ***DOS/VS*** and ***OS/VS*** ("VS" stands for virtual storage), which came from IBM.

1.6.1 Modern Operating Systems

Over the years, a gradual evolution of more advanced hardware organisation and architecture emerged due to rapid pace of technology with the key hardware outcomes such as multicomputer systems, multiprocessor machines, sophisticated embedded systems, real-time systems, high-speed communication channels used in network attachments, and increasing the size of more speedy and variety of memory storage devices, which all greatly increased the machine speed as a whole. In the application domain, the introduction of more related intelligent software, multiuser client–server computing, multimedia applications, Internet and Web-based applications, applications using distributed systems, cloud computing, real-time applications, and many others have made an immense impact in the structure and design of evolving OSs. To manage these advanced machines and upcoming sophisticated environment, OSs also have been thus constantly progressed, thereby conceiving new thoughts and approaches, and formulating and incorporating a number of new design elements to organised the OS afresh that ultimately culminated in a major change in the existing concept, forms, design issues, structure, as well as in its nature. These modern OSs (either new OS or new releases of existing OS), however, properly fit and can adequately manage new developments in hardware to extract its highest potential, are also conducive to new applications, and befit to negotiate increasingly numerous potential security threats.

That is why, beyond the third generation of OS along with continuous releases of its enhanced versions, making any sharp demarcation between generations of OSs is seemingly difficult to realize, and in fact, there is a less general agreement on defining generations of OSs as such. Consequently, the classification of OSs by generation to drive this constantly changing environment becomes less clear and less meaningful. It could be summarily said that the scientific, commercial, and special-purpose applications of new developments ultimately resulted in a major change in OS in the early 1980s and that the outcomes of these changes are still in the process of being worked out. However, some of the notable breakthroughs in approaches that facilitated redefining the concept, structure, design, and development of OS to realize numerous modern OS to drive the constantly emerging advanced architecture in the arena of both scientific and commercial uses are being mentioned here.

Interactive single-user PC was initially driven by **PC-DOS** and then by **MSDOS**, which later included **Windows** as an *added platform* to incorporate GUIs. Finally, **Windows NT-based OS**, a full-fledged stand-alone and more advanced OS, was ultimately launched. Introduction of **computer networks** deserved a different type of OS to manage the network of individual computers known as the ***network operating system*** (**NOS**) or sometimes called the ***network file system*** (**NFS**), apart from the OSs (similar or dissimilar) that are separately installed in the individual member computer systems to manage them. *Client–server* (or *workstation–server*) model uses another widely used variant of NOS, a *more sophisticated and complex OS*, to manage all the resources as a single entity extracting their highest potential, known as the ***multiuser interactive OS*** that offered *coarse-grained distribution*. This approach summarily exhibited a *tremendous strength* with an excellent *cost/performance* ratios that led the computing practices to follow a completely different path.

The practice of using networks of computers following the concept of an ***open system*** (a collection of independent computer systems ranging from expensive supercomputers to cheaper PCs can now be interconnected using LAN or WAN, and standard interfaces to build up a robust distributed computing system) can now be driven by an *OS*, which will cast this arrangement to its user simply a single coherent system, yet will control the operations of multiple machines present in the framework in a well-integrated manner. This type of OS is known as the *distributed operating system*, which is a common OS shared by a network of computers (not computer networks) or is used to drive a computer system having multiple processors (NUMA model multiprocessor). Generic distributed OS is, however, committed to address a spectrum of common functionalities, which are distribute computations, real distributions and sharing of resources and components, computation speed-up, smooth communications, scalability (incremental growth), fault tolerance, and reliability. Distributed OS, however, differs in forms as well as in issues when used in networks of computers (multicomputers) from those when used in the multiprocessor system. The representative distributed OSs that came from IBM around 1995 onwards are MVS, MVS/XA, MVS/ESA, OS/390, and z/OS. **Real-time system** is another type of computing system used to handle *real-time applications* that follow mostly a different type of computing known as the *real-time computing* that processes a huge number of events in terms of bursts of thousands of interrupts per second, in particular possibly without missing a single event. Such systems require a different type of OS for its management known as the ***real-time operating system*** (*RTOS*).

For more details about the modern operating systems, see the website: http://routledge.com/9780367255732.

1.7 Genesis of Computer Organisation and Architecture

The development in the design of computer organisation and its architecture has been evolved over a considerable long period of time. From the regime of mechanical calculators sometimes in the seventeenth century, the next odd 150 years elapsed with a lot of development, enhancement, and ultimately the first concept of a general-purpose program-controlled computer, which was conceived by Charles Babbage in the nineteenth century, but was not finally implemented until the 1940s. The first major step was the inclusion of electronic technology discarding the then mechanical components that have eventually passed through generations after generations, and at present, the sixth generation is prevailing. Each generation is identified as a breakthrough in technology from its recent past generation and has been distinguished by its major characteristics. It is interesting to note that despite rapid technological advances, both in hardware and in software, the design in laying the logical structure of the computers as proposed by Von Neumann, and others in 1940s, has improved rather slowly. Sometimes an innovation in design awaits the arrival of a suitable technology for its implementation. A change in design and architecture of computer often demands high cost in modifications of existing application programs, or sometimes even requires developing them afresh. Once the system software along with its underlying particular computer hardware becomes popular and widely acceptable, it is observed that the users are very reluctant to switch to other computers requiring radically different software. Market forces also play an important role for a particular design feature to evolve. Large manufacturers by their dominance in the market also promote certain features to appear.

A brief detail of the major distinguished characteristics of different generations is summarized in a tabular form in Table 1.2 given in the website: http://routledge.com/9780367255732.

1.8 Summary

From the very early days, the computer organisation and its architecture have been continuously developed aiming always towards building up a computer system with even more processing power and enough capacity using the contemporary available technology that continuously strives to fulfil the constantly growing requirements of the user community as a whole with highest possible performance at reasonable cost.

Starting from the basic organisation of the modern computer as proposed by John Von Neumann and others which is still being followed, by and large, even in the design of today's large supercomputer, the computer organisation and architecture have then gone through many innovations and improvements giving rise to architecturally different generations starting from the zero generation to the current sixth generation. Each such generation is demarcated by radical change in the use of more advanced emerging hardware technology, which is caused mainly due to the rapid progression in the arena of electronic engineering and technology. Emergence of semiconductor technology followed by IC technology and subsequently the introduction of VLSI technology has had a profound impact, which is considered as the main driving force for the formation of computer architecture and design, from the single-chip microprocessor with pipelined and superscalar architecture and the high-capacity RAM chip to the proliferation of large-scale network of computers, such as computer networks, distributed computing systems, cluster architecture, and special-purpose real-time systems. The relentless improvement in chip fabrication technology, however, supplemented the scenario with eventually giving rise to a spectacular development in CPU architecture: the *multicore architecture*, which consists of multiple CPU (core) within a single-processor chip (*chip multiprocessor*). Continuous development and rapid advancement in all the areas of hardware and software technology are still in progress. This suggests that major advances in computer design will still continue in days to come.

This chapter has attempted to summarily consider many aspects of computer organisations, its architectures, and operations, including the different types of OSs that are required to drive various kinds of the underlying architectures. Much of the terminology needed to deal with the subject is introduced, and an overview of some of the important concepts has been presented. The subsequent chapters will provide a rather complete explanation of these terms and concepts, and will put the various parts of this chapter into proper perspective.

Exercises

1.1 Define computer hardware and computer software. Show with examples in the light of these terms, the differences between general-purpose and special-purpose computer. Explain the concept of the dual nature of hardware and software.

1.2 What are the main features observed in Babbage's analytical engine that have along bearing in the design of modern computers?

1.3 Discuss the salient features of Von Neumann concept in computer design. Show with diagram the implementation of this concept in a real computer system which has been produced in those days. What are the limitations of Von Neumann concept (also known as *Von Neumann bottleneck*)?

1.4 Discuss briefly Princeton architecture and Harvard architecture.

1.5 State and explain the principles that led to the development of IAS computer. Show with diagrams the different components of IAS computer and explain the steps of its operation.

1.6 What is meant by the term *stored-program concept*? Harvard-class machines use separate memories for program and data, while Princeton-class machines exploit single memory for both program and data. Discuss the advantages and disadvantages of these two classes. Which class do you consider the most widely used and why?

1.7 Let A = X(1), X(2), …., X(100) and Y = Y(1), Y(2), …., Y(100) be two vectors (one-dimensional arrays) consisting of 100 numbers and each that are to be added to form an array Z such that Z(I) = X(I)+Y(I) for I = 1, 2, …., 100. Using the IAS instruction set, write a program for this problem.

1.8 What is meant by "generation" of a computer? What are the inventions and developments that laid the foundation of respective generations? What are the distinctive main features that categorized different generations of computer?

1.9 What are the primary resources of a computer system? Describe the functions performed by each of the following components of a computer system: CPU, main memory, I/O processor, OS, compiler, and utilities.

1.10 The terms *software compatibility* and *hardware compatibility* are commonly used in computer architecture. What do they mean? Discuss their role in the evolution of computers.

1.11 Explain Moore's law. Discuss its significance and implications in making predictions on the technological developments of ICs of the forthcoming days.

1.12 What are the key characteristics that must be present while injecting the family concept in a series of computers or processors?

1.13 "The OS of a third-generation mainframe computer like IBM 360/370 series was considered versatile in comparison with its predecessor, the second generation": state the generic notable features they have included in their design.

1.14 What is meant by LSI technology and VLSI technology? Describe their influence on the design and application of both general-purpose and special-purpose computers.

1.15 Define the concepts of VM and virtual memory, and describe briefly the differences between them.

1.16 Define microprocessor. Why is it so called? "Today's microprocessors are immensely powerful". Discuss the main features that made them so.

1.17 Briefly describe the main architectural features with examples that distinguish between microcomputer, minicomputer, and mainframe computer.

1.18 "The demarcation line being drawn in the past between mainframe (large), mini-, and microcomputer systems is much less valid today": justify the statement.

1.19 Define the embedded systems. Discuss the importance of this system both from the angle of computer designers and from the users' point of view.

1.20 What are the key distinguishing features of a supercomputer and those of a large mainframe system?

1.21 What is meant by the multicore concept used in the present day's microprocessor technology? Enunciate the distinguishing features present in the products developed on the basis of this concept.

1.22 Define the real-time application. How does it differ from a non-real-time one?

1.23 Define the real-time computing. State the features that make it different from the conventional computing.

1.24 "A real-time system is said to be entirely different from the conventional multitasking/multiuser system": justify the statement in the light of defining a real-time system.

1.25 State the major design issues that are encompassed in the real-time system development.

1.26 "OS is meant to be often closely coupled with the underlying architecture of the computer system": justify the statement briefly describing the role played by the OS in this regard.

1.27 Discuss the salient and distinctive features that are possessed by a generic third-generation OS.

1.28 Multiprogramming is found in various forms. What are those mainly? How do they differ from one another?

1.29 Discuss, in the light of computer architectures and related OSs, the differences that exist between network of computers and computer networks.

1.30 Define the distributed OS. How does it differ from the NOS?

1.31 What are the main objectives that must be fulfilled by a distributed OS? Name some application areas that are conducive to the distributed OS.

1.32 RTOS is something different than its counterpart conventional multitasking OS: in which ways it differs.

1.33 What is the basic design philosophy being followed while the policy and mechanism of a generic RTOS is framed?

1.34 Explain why you consider that the RTOS is getting more and more importance with passing of days.

Suggested References and Websites

Alpert, D. and Avnon, D. "Architecture of the Pentium microprocessor." *IEEE Micro*, vol. 13, no. 3, pp. 11–21, June 1993.

Augarten, S. *Bit by Bit: An Illustrated History of Computers*. New York: Tickno and Fields, 1984.

Blaauw, G. and Brooks, F. *Computer Architecture: Concepts and Evolution*. Reading, MA: Addison-Wesley, 1997.

Brey. Barry B., *The Intel Microprocessors*, 8th ed. Pearson Education Inc., 2009.

Farrell, J. J. "The advancing technology of Motorola's microprocessors and microcomputers." *IEEE Micro*, vol. 4, no. October 1984, pp. 55–63, October 1984.

Henning, J. "SPEC CPU2006 Benchmark Descriptions." *Computer Architecture News*, September 2006.

Prasad, N. S. *IBM Mainframes: Architecture and Designs*. New York: McGraw-Hill, 1989.

Schaller, R. "Moore's law: Past, present, and future." *IEEE Spectrum*, vol. 34, no. 6, pp. 52–59, June 1997.

Charles Babbage Institute: Provides the history of computers and links to different sources of information.

Top 500 Supercomputer site: Provides architecture and organisation of current supercomputer products.

Intel website: intel.com

2

Computer System Organisation

LEARNING OBJECTIVES

- To explain level concept and to introduce the hierarchical levels in computer system Organisation.
- To discuss the relative locations of different levels with their constituents and the specific responsibilities.
- To introduce processor level and its design approaches involving major components and their functions.
- To describe the modular design of computer systems.
- To explain the standard method used for the evaluation of computer system performance.
- To explore register level, its combinational components and sequential components (e.g. multiplexers, decoders, registers etc.), and their roles in information processing.
- To introduce gate level, the lowest and the most fundamental level comprising most elementary components (e.g. different types of gates and flip–flop) that operate on Boolean algebra.

The operation of a digital computer is based on the storage and processing of data submitted by a user in the form of a set of instructions to be executed. This chapter gives an overview of a computer system when looked at from the end of a user and follows a level hierarchy: i.e. from the top up to certain levels, the user participation continues, and beyond that, the system's involvement starts and then follows a level hierarchy down to the digital logic level – the computer's own real hardware. From this point, the design process for digital systems begins with mainly *three* basic levels of abstraction in the descending order: the *processor level*, the *register level*, and the *gate level*. The main thrust of this chapter is to explain the characteristics of the design process, examine the design at the processor and register levels, and provide many aspects of digital logic as a building block that may facilitate the study of higher levels to implement the various functions of the computer. The basic elements (different gates) from which all digital computers are constructed are amazingly simple. This chapter describes how these different elements are used to construct various combinational and sequential circuits in the digital logic.

2.1 Modular Design Levels

Digital computer system design invariably follows a modular approach consisting of mainly *three* major levels which are in the descending order: the ***processor level,*** the ***register-transfer level,*** and the ***gate level*** or ***digital logic level.*** There is also another level under the gate level, which is called the ***device level*** consisting of individual components, like the working principle of a transistor, resistor, capacitor, etc. The boundaries between the levels are not truly well defined, and hence, it is quite natural that while describing a particular level, it may include components from more than one level. These concepts and techniques of constructing machines as a series of levels, and the details of some important levels, will ultimately guide the designer in formulating and constructing new computers, or designing further new levels.

With the level concept, the type of information processed becomes more complex as one moves gradually from a lower level to a higher level. In fact, the higher the level, the more will be the complexity. At the lowest level, i.e. at the gate level, the fundamental unit is *bit,* and the processing is carried out on individual or random ***bits.*** Here, the fundamental operation is only switching the gate, which is typically measured in *nanoseconds.* At the register level, the information is organised in the unit of word (consisting of a number of bits) into ***words,*** representing instructions, numbers, character strings (literals), etc. Here, for an elementary operation, the time taken is of the order of *microseconds.* At the next higher level, i.e. at the processor level, the unit of information is ***block of words,*** representing programs, sets of data, etc. Here, the time taken is of the order of *milliseconds.*

We can now elaborate this level concept to get a clear understanding how a computer works. Right from the time of submission of a program from a user's end up to the completion of its execution, it passes through these different levels S_1, S_2, …, S_k, step by step, where each level is performing its specific responsibility. The lowest level, of course, of any computer is the electronic circuit that can execute a limited set of simple, easy basic operations. To each such operation, there must have one or more related instructions. These primitive instructions form a language by which it is possible to communicate with the computer at this level. Such a language is called the *machine language.* At the time of designing a computer, it should be decided what instructions (operations) need to be included in its machine language, and as this level is the bottom-most level called the *machine level,* in our discussion, this level can also be termed as the *digital logic level.* Each of the levels above this digital logic level is on the top of another and is entrusted with certain respective operations to be executed. Hence, each such level S_k has a new set of instructions of the corresponding level based on its predecessor, which is more convenient to use than the set of instructions of a level lying beneath it. The set of instructions used by a particular level S_k form a language L_k of that level. A hypothetical design of a computer using this approach as a series of levels (layers) is depicted in Figure 2.1. The top-most level is the highest language level, which is most sophisticated and complex. This level-wise arrangement in designing a computer system can now be accomplished either by a *top-down* or by a *bottom-up* approach. Here, we will use a *top-down approach* for the sake of convenience and for better understanding.

In Figure 2.1, the highest level (**level 6**) at this moment is the user who writes programs using ***high-level languages*** such as C, COBOL, C++, FORTRAN, etc. which need to be *translated* (or *interpreted*) to the languages of level 4 or level 5, by suitable translators, popularly known as ***compilers*** (*interpreters*). Next lower level, **level 5**, is known as the ***assembly language level,*** a symbolic form for one of the underlying languages found in lower

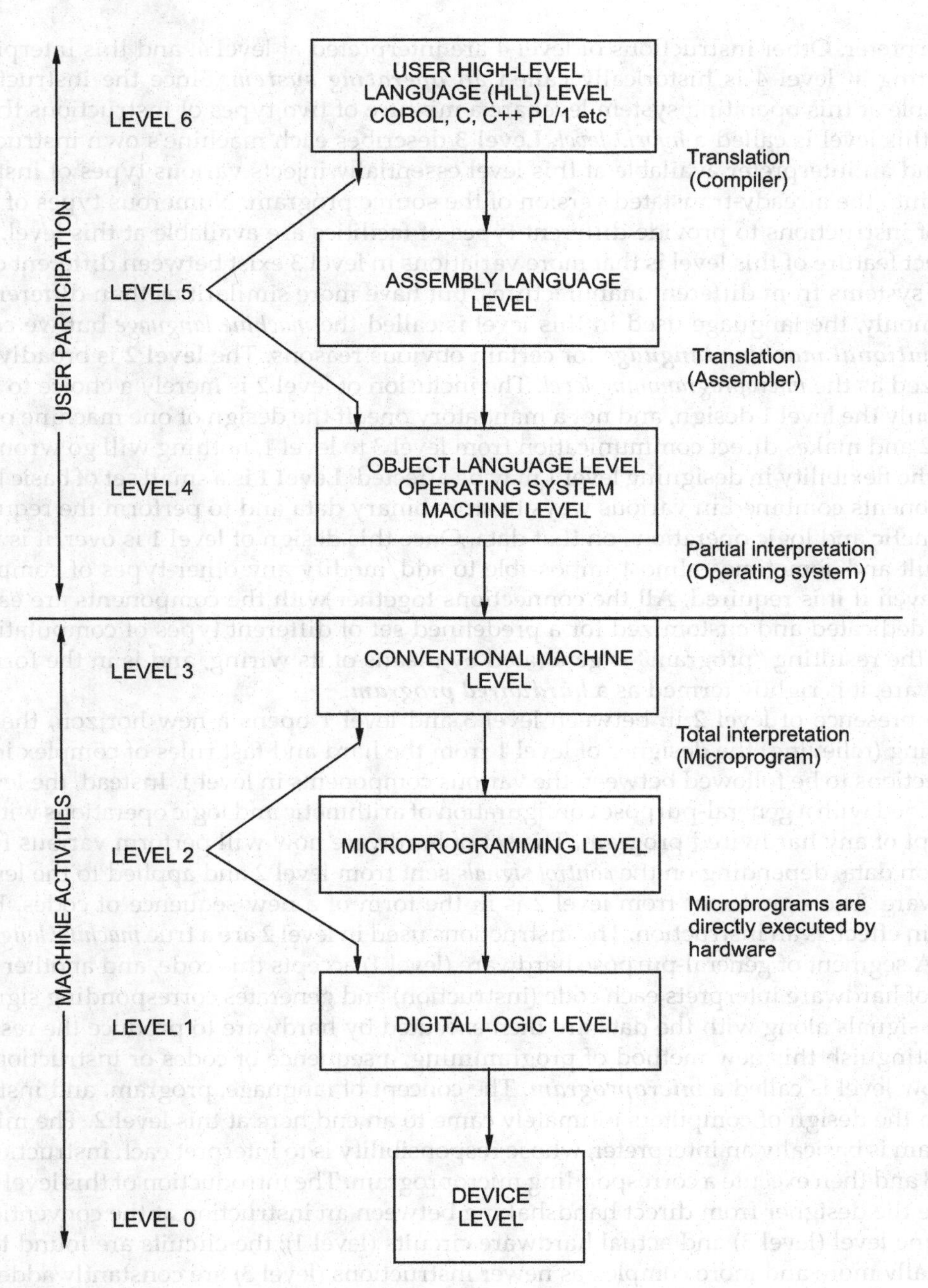

FIGURE 2.1
Level-wise structure of computer design. Most of the modern computers follow this concept.

levels that needs a translator used for the translation of the assembly language program, which is called an ***assembler***. The next lower level, **level 4**, is basically a ***hybrid* level** and can be considered as a *bridge* between the user of the outside world and the machine's own world. Most of the instructions in the language of level 4 are also found in the language of the level just below this, i.e. the level 3, and are directly interpreted by the level

3 interpreter. Other instructions of level 4 are interpreted at level 4, and this interpreter executing at level 4 is historically called an ***operating system***. Since the instructions available at this operating system level are a mixture of two types of instructions that is why, this level is called a *hybrid level*. **Level 3** describes each machine's own instruction set, and an interpreter available at this level essentially injects various types of instructions into the already-translated version of the source program. Numerous types of new sets of instructions to provide different types of facilities are available at this level. The distinct feature of this level is that more variations in level 3 exist between different computer systems from different manufactures, but have more similarities than differences. Commonly, the language used in this level is called the *machine language* but we call it ***conventional machine language*** for certain obvious reasons. The **level 2** is broadly categorized as the *microprogramming level*. The inclusion of level 2 is merely a choice to support only the level 1 design, and not a mandatory one. If the design of one machine omits level 2 and makes direct communication from level 3 to level 1, nothing will go wrong; at best, the flexibility in designing level 1 may be affected. **Level 1** is a small set of basic logic components combined in various ways to store binary data and to perform the required arithmetic and logic operations on that data. Once this design of level 1 is over, it is very difficult and sometimes almost impossible to add/modify any other types of computation, even if it is required. All the connections together with the components are essentially dedicated and customized for a predefined set of different types of computations. Since the resulting "program" is generated by means of its wiring, and is in the form of hardware, it is rightly termed as a ***hardwired program***.

The presence of level 2 in between level 3 and level 1 opens a new horizon, thereby releasing (relieving) the designer of level 1 from the hard and fast rules of complex interconnections to be followed between the various components in level 1. Instead, the level 1 is realized with a general-purpose configuration of arithmetic and logic operations with no concept of any hardwired program. The set of hardware now will perform various functions on data, depending on the *control signals* sent from level 2 and applied to the level 1 hardware. The signal sent from level 2 is in the form of a new sequence of codes. Each code, in effect, is an instruction. The instructions used in level 2 are a true *machine language level*. A segment of general-purpose hardware (level 1) accepts this code, and another segment of hardware interprets each code (instruction) and generates corresponding signals. These signals along with the data are then executed by hardware to produce the results. To distinguish this new method of programming, a sequence of codes or instructions at this low level is called a ***microprogram***. The concept of language, program, and instruction in the design of computers ultimately came to an end here at this level 2. The microprogram is basically an interpreter, whose responsibility is to interpret each instruction of level 3 and then execute a corresponding microprogram. The introduction of this level is to relieve the designer from direct handshaking between an instruction at the conventional machine level (level 3) and actual hardware circuits (level 1); the circuits are found to be gradually more and more complex as newer instructions (level 3) are constantly added to handle numerous upcoming situations. Moreover, there is no flexibility between level 3 and level 1; i.e., any change in an instruction in the conventional level when the hardware circuit design is over, even if it is truly required, is absolutely impossible to incorporate. Computers from different manufacturers have identical microprogramming levels; several similarities exist in the design of this level, but they differ in implementation, depending mostly on the design of the underlying hardware circuit. The format of these instructions and the way they are being expressed also differ substantially from one machine to another machine.

All these levels together facilitate the design to be more modular, and present the computer Organisation and architecture in a lucid structured framework. From the computer designer's point of view, the lowest level, level 1, is the ***digital logic level*** or ***gate level,*** which consists of fundamental hardware objects like gates. Building of gates with fundamental analog components such as transistors ICs (integrated circuits) etc. is not within the scope of our discussion of design concepts. We are only concerned about whether each gate has one or more digital inputs (either 0 or 1) and provides some output after executing some simple function over this input. For the sake of convenience, we just mention our level concept up to level 0, which one can consider as the *fundamental component level,* like the working principle of a transistor, resistor, capacitor, etc. while forming a gate, but any more details on this are beyond the scope of our present discussion.

2.2 Methods of Design

As shown in Figure 2.1, from the highest level up to level 3, different types of users will interact with the computer in suitable ways, and the way they will interact with the computer depends on the design of level 3 and down the lower levels. From *level 3* and downwards, there is no user involvement, and the computer design starts from this level and can be then carried out hierarchically downwards at several levels of abstraction. Normally, three such levels are recognized:

- The *processorlevel,* also known as the *architecture level;*
- The *register level;*
- The *gate level,* also known as the *logic level.*

It would be wise and perhaps most natural to proceed from higher to lower levels of design because this approach corresponds to a progression of successively greater levels of detail. Thus, we prefer to proceed through the following steps in the hierarchical order:

i. Describing the *processor-level* structure of the system and its specifications;
ii. Describing the *register-level* structure of each component type being used;
iii. Describing the *gate-level* structure of each component type.

2.2.1 The Processor Level

Level 3 considered as the highest level in the computer design hierarchy is commonly known as the ***processor level*** or ***system level,*** which consists of a set of components or modules of three basic types, namely *CPU* (*central processing unit* (*processor*)), *memory,* and *I/O* (input–output) *devices* that communicate with one another. In effect, a computer can be looked at as a network of these basic modules connected by a collection of certain paths called the *interconnection structure* or *buses.* Since the performance of this interconnection structure, like the other three basic modules, often determines the performance of the entire computer system as a whole, that is why, the designers often treat this interconnection structure as the *fourth* basic resource of the computer system. A schematic diagram of these major components along with their interconnections is shown in Figure 2.2.

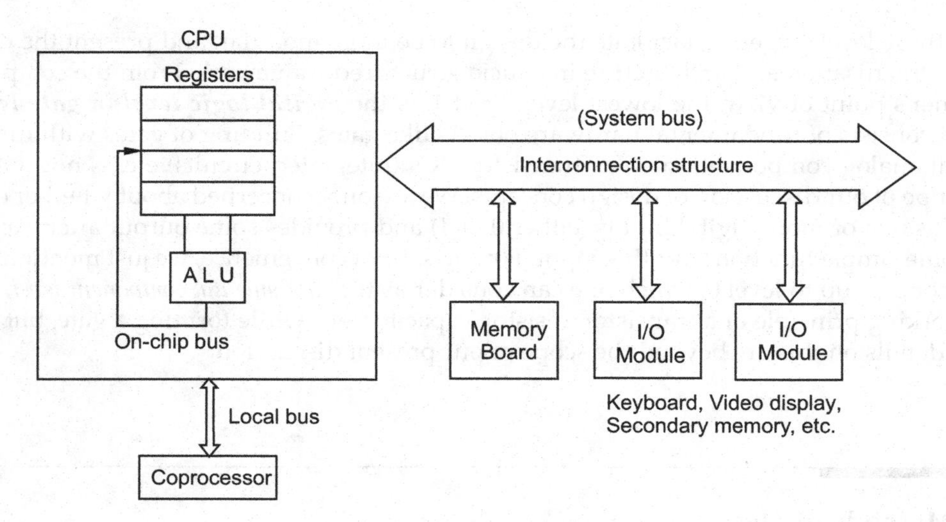

FIGURE 2.2
Primary resources of a computer system and their interconnections.

CPU is a rich resource and is the *brain* of the computer. Every CPU has its own instruction set, which determines the set of operations that the proposed CPU can perform. It is composed of several distinct parts, namely the *control unit, the arithmetic and logic unit* (ALU), and a number of *registers,* which are connected internally in various ways. Most computer systems have one such CPU called the ***uniprocessor systems***. Computer systems having two or more CPUs called the ***multiprocessors*** are also common today. The details of CPU Organisation and its working, however, are described in Chapter 3.

Memory is a module that temporarily holds or stores information (both instruction and data) which is ready for execution. This module is usually more precisely called the *main memory,* to distinguish it from external storage, peripheral devices, I/O devices, or secondary memory. Main memory is internal to the system and involves in CPU operation. It is relatively fast, directly connected to the CPU, and even controlled by the CPU. It is comparatively costlier and relatively faster than any other secondary memories. The details of different types of memories, their Organisations, and their working, however, are described in Chapter 4.

I/O appears to be functionally similar to memory when viewed from the end of a computer system with only two operations, *read* and *write*. Its main task is only to migrate data from outside world into the computer, and from the computer to the outside world. Many of the physical I/O devices are electromechanical devices, and hence, their speed of operation is slow compared to the speed of the other three resources. The designs and Organisations of I/O and interconnection buses have been discussed in detail in Chapter 5.

For a brief description about each of these processor-level components, see the website: http://routledge.com/9780367255732.

2.2.1.1 Design Approach

Processor-level design is essentially the starting point of a full-fledged computer design involving all the fundamental resources, but it is ultimately realized in the next lower level, the *register level,* and more precisely at an even lower level, the *gate level*. The formal design concept as depicted in Figure 2.2 shows a basic structure that is followed for a general-purpose computer design, which was subsequently implemented by different manufacturers with their own objectives. Due to continuous advancement of electronic technology, the computer architecture also constantly progresses in parallel from one generation to the next generation, but *this fundamental design concept remains by and large the same* and differs mainly in the way the interconnection is made between the resources, their numbers, speed, capacity, and finally their increased capability. Modern generations, however, further enhanced this design concept by including more than one CPU (multiprocessor) in the computer, but keeping the fundamental design concept almost remaining the same. With passing days, the design structure is constantly maturing more and more, and consequently becoming more complex, as enhanced architecture gradually evolves giving rise to the design concept of *networks of computers* (multicomputers). In this design, different computers (compatible or dissimilar) can be hooked up together by a communication network so that they can share between themselves the potential resources (both hardware and software) and also the activities present on the network using *message passing* mechanisms. All the basic aspects of multiprocessors and multicomputers, including their modern variants, namely, the distributed systems, and the cluster architecture, have been, however, discussed in Chapter 10.

A brief detail of the constantly evolving computer system design from the very early days up to the most modern form is given in the website: http://routledge.com/9780367255732.

2.2.1.2 Performance and Related Factors

The performance of a computer simply means how quickly and efficiently it can execute a user program, and does not necessarily depend only on the raw speed of the processor; rather, it greatly depends on the performance of each level (Figure 2.1) and on their coordinated actions during the execution of the program. Operating system that drives the resources, and interfaces between the users' outside world and the computer's inner world, also plays a vital role in the performance of the computer system. However, to yield ***better performance,*** it primarily requires

- A small useful machine instruction set (level 3);
- As much faster hardware as possible (level 1 and level 2);
- And above all, the smooth coordination between all these levels;
- Optimization in compiler design.

After the submission of a program, till its execution is completed, the *total time* consumed is considered as a ***measure of performance*** of the entire computer system. This time essentially depends on several architectural factors apart from many other related aspects, and in general, it is *affected* by the following factors:

- The speed of the processor;
- The memory response time;
- The speed of the attached I/O devices;
- The time taken by the interconnection structures (system bus and other buses);
- The intelligence and capability of the operating system that monitors the entire computer system activities in handling the entire process.

When we would discuss the ***performance of the processor***, we should consider only the duration of time when the processor is active. The processor time usage mainly depends on the design of the hardware involved consisting of mainly the CPU and memory being connected by a bus. The time required to execute program instructions (user instructions) involving instructions and data to move mostly between the CPU and relatively slower main memory, and to a lesser extent, between memory and I/O devices, greatly affects the performance of the processor. Hence, in almost all machines of today, a faster cache memory is used in between the processor and the main memory to mainly minimize most of the processor's visits to relatively slower main memory, which is comparatively time-consuming. Since the cache memory always contains the most currently referenced instructions and data issued and used by the processor, the processor-cache handshaking is mostly successful and is much faster than processor–main memory transfer, thereby reducing the processor time consumption to a great extent. In addition, the inclusion of the I/O processor (described in Chapter 5) ultimately releases the CPU completely from all rudimentary I/O-related processing activities, thereby allowing the processor to make use of its time to do its own useful work in parallel while the I/O activities are in process. The design, operation, and performance of different types of memories using different technologies present in the memory systems have been separately discussed in Chapter 4.

2.2.1.3 Processor Clock

A rough estimation of CPU speed is the number of *basic* (machine) operations that it can perform per unit of time. Such operations are controlled by a regular stream of signals issued by a central timing mechanism, called the ***system clock***, and such signals are called ***ticks*** or ***beats***. The speed of the clock is its ***frequency (rate)*** f, measured in number of ticks (generally in millions) per second. The unit of this frequency is usually megahertz (MHz), and nowadays, it is also measured in gigahertz (GHz). For example, the clock frequency of 200 MHz means that the clock can provide 200×10^6 (million) cycles per second. Similarly, 10^9 (billion) cycles is denoted by gigahertz. Between two consecutive ticks, the time duration is known as the ***clock cycle*** or ***clock period*** (t). For example, the clock frequency (f) of 200 MHz means that the clock can provide 200 million cycles (200×10^6) in 1 second. Therefore, one clock cycle will have the following time duration (clock period, t):

$$t = \frac{1}{200 \times 10^6}\,\text{s} = \frac{10^3}{200 \times 10^9}\,\text{s} = \frac{5}{10^9}\,\text{s} = 5\,\text{ns}$$

It is now clear that $t = 1/f$; i.e., *the clock period is the inverse of the clock frequency or clock rate, and vice versa.*

To execute a machine instruction, the processor divides the actions to be performed into a sequence of basic steps (operations) such that each step can be completed in one clock cycle. For example, a computer clocked at 500 MHz can perform one basic operation in the clock period, $t = 1/(500 \times 10^6) = 0.002 \times 10^{-6}$s. $= 0.002\,\mu\text{s} = 2$ ns. Complicated operations like operations on floating-point numbers can, however, require more than one clock cycle to complete the execution. Clock speeds of 1 gigahertz (1 GHz or 1000 MHz) and beyond are now feasible using faster versions of current CMOS technology and the other emerging more advanced technologies such as VLSI (very-large-scale integration) and ULSI (ultra-large-scale integration) in IC fabrication.

2.2.1.4 Performance Assessment: A Rough Estimation

Performance of instruction execution is measured experimentally by computing the average runtimes of representative or ***benchmark programs*** written in a high-level language that can reflect the real applications more closely to provide the performance figure more accurately. Let a representative source program A written in some high-level language generate a corresponding machine language object program in which actual N *machine language instructions* are to be executed. Let S be the average number of basic steps required to execute one machine instruction, where each basic step is assumed to be completed in one clock cycle. Therefore, each instruction requires S cycles on an average to complete its execution. These S cycles can be otherwise treated as *cycles per instruction* (*CPI*). So, for N instructions, the total number of steps required is $N \times S$, and this requires a total of $N \times S$ clock cycles. If the clock frequency (rate) is f cycles per second, then the program execution time, T, is given by:

$$T = \frac{(N \times S)}{f}\,\text{s} \tag{2.1}$$

But T will not be as simple as is found in Equation (2.1) if the speed-up factors such as pipeline concept, superscalar architecture (simultaneous execution of many instructions in an overlapped manner), and other factors are taken into account. T can be determined accurately only by the measurement of A's actual runtime or by a simulated execution.

T can also be made related to some basic parameters of the computer's architecture and its implementation. One such important parameter is the ***number of instructions (average) executed per second***, which is denoted by IPS. Then,

$$\text{IPS} = \frac{N}{T}, \text{or } T = \frac{N}{\text{IPS}}\,\text{s} \tag{2.2}$$

Although CPU performance is sometimes graded by a common measure known as ***CPI*** (*cycles per instruction*), it is a fact that different machine instructions, for many different reasons, may require different number of clock cycles to complete their execution.

However, if *"f"* is the frequency *of the clock in millions*, then $f \times 10^6$ cycles is obtained in 1 s. If IPS is the number of instructions executed per second, then IPS requires $f \times 10^6$ cycles. Thus, each instruction requires the number of cycles which is called *CPI*, and that is expressed as:

$$\text{CPI} = \frac{(f \times 10^6)}{\text{IPS}} \tag{2.3}$$

Substituting (2.3) in (2.2), we get

$$T = \frac{(N \times \text{CPI})}{f \times 10^6}\,\text{s} \tag{2.4}$$

Equation (2.4) comprising *three* apparently distinct factors can be used as a basis in roughly estimating the performance of a computer system. It is apparent that in order to reduce T, both N and S (CPI) should be reduced and f should be increased.

CPU performance can be sometimes scaled in terms of its raw power. The performance index in this regard is expressed in terms of millions of instructions executed per second, which is denoted by *MIPS,* where MIPS = IPS × 10^6. Clearly, by Equation (2.3), MIPS = f/CPI. It should be made clear that although MIPS and MIPS rating roughly bear the same meaning, they are not the same. MIPS rating to measure the performance of a processor is defined as the rate at which instructions are executed, but expressed as MIPS. This means that in deriving a MIPS rating, the unit is MIPS (i.e. to be divided by 10^6). Thus,

$$\text{MIPS rating} = \frac{f}{\text{CPI} \times 10^6} \text{MIPS} \tag{2.5}$$

To be very precise, what we commonly use is MIPS rating in which MIPS is used as a unit. MIPS rating, however, may not always reflect the actual performance scenario since simple instructions do better for all time. Similar to MIPS, there is another performance index called *MFLOPS* (Million FLoating-point Operations Per Second). Floating-point operation is, however, considered as one of the most complex tasks a processor often performs. MFLOPS is defined as

$$\text{MFLOPS} = \text{Number of floating-point operations per second} \times 10^6$$

$$\text{MFLOPS rate} = \frac{\text{Number of floating-point operations}}{\text{Execution time} \times 10^6} \tag{2.6}$$

Both MIPS and MFLOPS performance index are often mentioned by vendors of computer systems to give a notion about the speed of their processor while promoting their products. But a straight comparison of either MIPS rating or MFLOPS rating on different processors does not tell the whole actual story about their performance. It is often found, particularly when a large system is taken into consideration, that a processor with a lower MIPS rating can easily outperform a processor having higher MIPS (as is found in IBM mainframe systems when compared to those of HP, Burroughs, etc.). In fact, MIPS and MFLOPS rates vary with respect to a number of components, including clock rate (frequency), the instruction count (N), and the CPI of a given machine. Each of these components is, in turn, influenced by some other factors, as is discussed next. That is why, instruction execution rate (such as MIPS and MFLOPS) has proven to be not enough and is not considered a convincing approach while comparing the performance of different architectures. Moreover, the value as presented by the vendor usually indicates the highest performance that the processor can yield under ideal situations. In practice, these values are seldom achievable.

2.2.1.4.1 *Influencing Factors*

From Equation (2.4), it is observed that the performance mostly depends on the three basic elements, namely CPI, f, and N. These three elements individually, in turn, are mostly influenced by the three separate factors, namely architecture and Organisation, hardware technology being employed, and software being used, respectively. It is interesting to note that all the parameters we have discussed so far that collaboratively determine the performance are not necessarily independent. This means that changing one parameter may have an impact on another. Increasing or decreasing the value of any parameter will not impact the performance unless the overall value of T is reduced. That is why, a processor having a 700 MHz clock rate does not necessarily imply a better performance than a 500 MHz processor because it may have a different value (higher value) of CPI.

2.2.1.5 Design Principles: CISC and RISC

The basic principle to be followed at the time of designing a CPU, however, has been successfully explained and outlined by Equations (2.1) and (2.4). Continuous improvement in the advancement of VLSI technology tends to constantly increase *f*, and influences the performance of all types of computers and their designs. Hence, involvement of the parameter *f* in this discussion at present can be kept set aside. The remaining two fundamental parameters that are involved in determining the CPU performance, namely *N* and CPI, came under the focus of the CPU designers. As a result, two different philosophies ultimately evolved in the CPU design approach.

One group of CPU designers started to follow a specific philosophy that was inclined to somehow reduce *N*, even at the cost of increasing CPI. The corresponding products based on this design principle are the microprogram-based processors known as ***CISC*** (*complex instruction set computers*) processors. Another school of thought believed in a different philosophy that put more thrust on CPI and attempted to rigorously reduce it, sacrificing *N*, even if *N* were to increase by such an act. The corresponding products based on this approach are the ***RISC*** (*reduced instruction set computers*) processors.

To realize more improved performance of their processors, CISC designers have further attempted to decrease the increasing CPI as far as possible, which is being achieved with a technique known as ***pipelining***. The RISC designers, on the other hand, have always attempted to decrease the existing *N*, which is being achieved using *optimizing compilers*. It is emphasized at this juncture that CISC and RISC should not be treated as different classes of processor; rather, they can be referred to as the outcome of two different design principles and techniques.

2.2.1.6 Speed-Up Approach

The usual norms followed in the designs of modern computers, including microcomputers, incorporate a number of speed-enhancing features, although most of these features were already implemented mainly in mainframe computers of the late 1960s. Some other features are often included even in the design of the processors. The inclusion of a fast small **cache memory** in between CPU and main memory (M) in the architecture has been significantly minimized most of the processor's frequent visits to relatively slower main memory, thereby substantially enhancing the performance of the processor and in turn resulting in an overall improved system performance. This cache is often placed partly or wholly on the same chip as the CPU (on-chip), and/or on the mother board in between CPU and main memory (off-chip). Another notable enhancing feature implemented in the design of CPU is a technique known as the *pipelining* that increases the overall performance by overlapping the execution of successive instructions. The pipelining approach provides *instruction-level parallelism*. The ultimate target of the pipelining approach is to *virtually reduce the value of S* (CPI) and to attain the ideal value of *S* (= 1) as close as possible to minimize the overall value of T. The processor performance (effective MIPS rate) has been made further increased by way of replicating these instruction pipelines to form multiple parallel pipelines in the design of a processor (instruction-processing circuits) to overlap the simultaneous execution of more instructions. The CPU with this capability is called the ***superscalar***, and this superscalar design offers ***machine-level parallelism***. Almost all high-performance processors of today are designed to operate following this line of approach. Pipeline and superscalar designs of CPU have been separately discussed in detail in Chapter 8. Another speed-enhancing feature is the use of an ***optimizing***

compiler at the time of object code generation that can smoothly convert the compiler-generated code directly to the underlying instruction set of the target processor so that the product $N \times S$ can be kept at a minimum. This will ultimately reduce the total number of clock cycles required to execute a program. The compiler may sometimes rearrange (reorder) program instructions to yield better performance, of course, without affecting the semantics (logic) of the program. This approach, however, has been religiously applied in the RISC culture and is being considered as one of the secrets in the success of fast processing of RISC processors.

A brief detail of these approaches is given in the website: http://routledge.com/9780367255732.

2.2.1.7 Performance Measurements

Performance evaluation of computers can be primarily described in terms of the characteristics of only CPU, although many real programs are often I/O-bound, and many powerful machines also typically have only rudimentary I/O. The computer designers often estimate the performance by evaluating the effectiveness of salient features. From the user's end, one of the parameters that ultimately tells about the performance of a computer is the execution time T, which again depends on the clock frequency f measured in gigahertz (megahertz). The other performance indicators may be MIPS, which is the average execution speed of instructions measured in MIPS. Another indicator that may be considered is the average number of CPU clock cycles required to execute each instruction. Unfortunately, all these parameters when individually coupled with the other architectural features cannot always truly reflect the desired level of evaluation.

To evaluate the performance comparison rationally, the accepted norms of today are to select and run a set of agreed-upon standardized unbiased real application programs that will hardly be impacted by the compiler, operating system, and other factors while benchmarking the performance. Such selections and subsequent publication of representative application programs have been carried out by a non-profit Organisation called ***System Performance Evaluation Corporation (SPEC).*** SPEC also identified the different application domains with their respective test programs and provided the test results using many popular commercially established renowned computers. The domain covers the range from compilers, database applications, numerical matrix processing used in nuclear physics, crystal structure in quantum chemistry, and also to the programs used in astrophysics. These benchmark programs were developed sometimes in 1989 for the evaluation of general-purpose computers, and after that, those were constantly modified and upgraded to be remained on a par with the latest technological developments. Finally in 2000, a suite of benchmark programs has been published.

The computer under test compiles the *benchmark program* and also runs it. The total time required is measured. The same program is compiled and run on a selective reference computer. In the 1995 publication named as *SPEC CPU 95,* SPEC used the SUN SPARC station 10/40 as the reference computer for this purpose. Later, in the 2000 publication known as *SPEC CPU 2000,* SPEC used Ultra SPARC 10 workstation as the reference computer. The rating of the computer under test is given by:

$$\text{SPEC rating} = \frac{\text{Total time required on the reference computer}}{\text{Total time required on the computer under test}}$$

A SPEC rating of 100 indicates that the computer under test is 100 times as fast as the reference computer used in this benchmarking. This test is usually repeated for all types of different programs covering the selective domain in the SPEC suite, and finally, the geometric mean of all results is calculated. This approach, to a large extent, can be acceptable, because this rating measurement covers all the factors influencing the performance, including the effect of the compiler, the role of the operating system, the processor, the memory, and its management techniques (paging algorithm, page size, etc.) of the computer under test. The combined effect of all these factors, at least, provides something which is a close estimate of computer performance in a real-time environment.

SPEC benchmarking method has been getting continuously upgraded, keeping up with the constant evolution of high-speed processors and improved computer design methodology. By this time, after 2000, the different methodologies used in computer design have reached a reasonably standard form that most of the manufacturers follow by and large, but the processor designs continue to improve and their speeds continuously increase with no indication of any let-up. SPEC introduced an advanced benchmark suite in 2006 known as *SPEC CPU 2006*, which is essentially an industry-standard suite for processor-intensive applications. It is found to be adequate to estimate performance for applications that devote most of their time in doing computations (CPU operations) rather than in executing I/O activities. This suite has been developed based on existing applications that have already been ported to a diverse spectrum of platforms by the SPEC industry members. The benchmark suite consists of 17 floating-point programs written in C, C++, and Fortran, and 12 integer programs written in only C and C++. The suite consists of programs containing about 3 million lines of code. Incidentally, this is the fifth generation of processor-intensive suites introduced by SPEC replacing all its earlier versions, including SPEC CPU 95 and SPEC CPU 2000.

A few other analytical methods for performance evaluation also exist. Notable among them is the ***queuing theory*** – a statistical approach used in applied probability theory. But their usefulness is limited. Apart from these approaches, experimental methods using computer-based simulation are also extensively used on an actual system.

2.2.2 The Register Level

The register level or *register-transfer level* design is based approximately on the level of details as seen by a system's programmer. It is concerned with processing of information grouped into words (not individual primitive bits forming the words). The fundamental components used in this level are mostly word-oriented devices. These devices are basically small *combinational circuits* or *sequential circuits* used to process or store the information in the units of words. The standard combinational components and sequential components used in this level range from general-purpose devices to more specialized circuits dedicated for particular purposes (such as adders decoders etc.). The devices that are most commonly used in this level are broadly classified into the components which are described below.

2.2.2.1 Combinational Components

a. **Multiplexers**: used for computing general combinational functions as well as for routing of data.
b. **Decoders**: used for analysing and checking code and code conversion.

c. **Encoders**: used for creating and generating codes.
d. **Adders**: used for addition and subtraction operations.
e. **ALU**: used for numerical and logical computations.

2.2.2.2 Sequential Components

a. **Parallel registers**: used for temporary storage of information.
b. **Shift registers**: used for serial–parallel conversion as well as for temporary storage of information.
c. **Counters**: used for generating timing signal as well as to implement control, and for other different purposes.
d. **Programmable logic devices (PLDs)**: used to implement both general combinational functions and general sequential functions.

Various types of components at the *register level* are connected together to form circuits. This connection path is constructed by a set of lines referred to as *buses*, and the number of lines to be associated is usually (but not necessarily) the word size of the respective components being connected. This level of abstraction is, however, further decomposed into the next lower level, i.e. the *gate level* or *logic design level* where the design components are the basic logic gates along with some fundamental logic functions.

2.2.2.3 General Representation

The register-level components are commonly represented in the circuit by blocks usually with an abbreviated description of their behaviour. As shown in Figure 2.3, the block is associated with a number of signal lines; each line represents something that is involved in the working of the block. A single signal line in the figure can sometimes represent a set

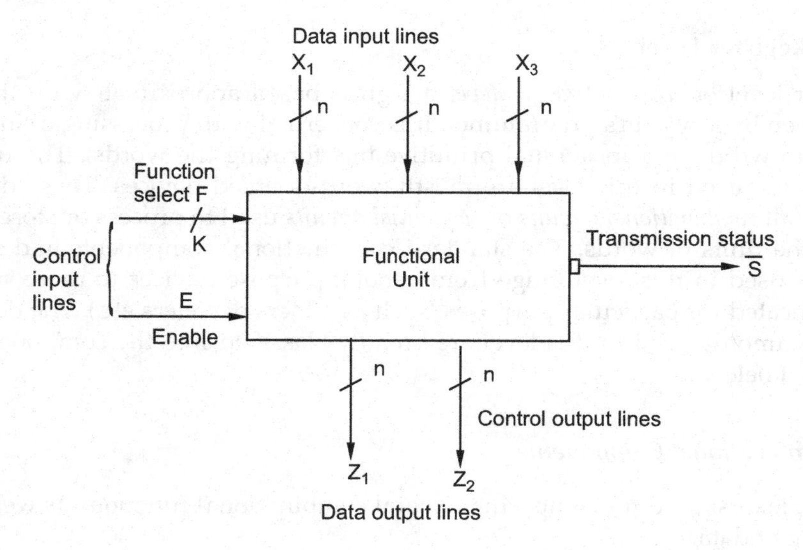

FIGURE 2.3
General block representation of a register-level component.

of $n > 1$ bits of information in parallel. This is represented by placing a "/" in the line and writing "*n*" to indicate that *n* number of lines is present. Different types of lines, such as **control lines** and **data lines**, are attached with the block according to its requirements. The control line indicates the type of operation to be performed by the line when the line is in its *active* or *asserted* state. This line, however, again fall into *two broad categories*: the ***select line*** that specifies one of several possible operations that the unit is to perform and the ***enable line*** that specifies the condition or time for a selected operation to be performed. The **data lines** are essentially I/O lines attached with the unit in the form of *data input lines* to enter data for the selected operation, and *data output lines* provide the result of the selected operation coming out of the unit.

The operation performed at this level is in the unit of words and not on the individual bits where each word is formed with *m* bits, and the input word length, however, may or may not be the same as the output word length in some situations. Most of the operations performed at this level are *numerical* rather than *logical*. However, many of the logical operations associated with this level appear to be somewhat complex and do not reflect the properties of the gate-level operations. We will only define here both the generic combinational and sequential components, and also the ways they are described.

For more details, see the website: http://routledge.com/9780367255732.

2.2.2.4 Combinational Circuits

Combinational circuits implement the essential functions that a digital computer offers. A combinational circuit is an interconnected set of gates (even lower-level components) whose output at any time is a function of only the input present at that time. As with a single gate, here also the appearance of the input is followed almost immediately by the generation of the output, with only gate delays. This circuit provides no memory ("memory–less") or state information except for the special case of ROM. In general terms, a combinational circuit consists of *n* binary inputs and *m* binary outputs. Similar to a gate, a combinational circuit can be defined in essentially three ways:

i. **Truth table**: For each of the 2^n possible combinations of the input signals, the binary value of each of the *m* output signals is listed.
ii. **Graphical symbols**: The interconnected layout of gates is exhibited.
iii. **Boolean equations**: Each output signal is expressed as a Boolean function of its input signals.

The different types of commonly used combinational components employed in the implementations of computer circuits are ***multiplexers, decoders, encoders, adders, ALU,*** etc. Adders and ALU, however, have been described in Chapter 7.

Each of these components with their various implementations, excluding adders and ALU, is illustrated in detail in the website: http://routledge.com/9780367255732.

2.2.2.5 Sequential Circuits

A new class of digital circuits have been developed in which some memory elements are introduced. In these type of circuits, an output at t_{n+1} is a function not only of the available inputs at t_{n+1}, but also of the output of the circuit at t_n. In these circuits, a portion or all the outputs of time t_n are stored in the memory elements and are fed back to the inputs of the circuit along with the external inputs at t_{n+1} to produce a new set of inputs for the circuit at t_{n+1}. These feedback connections mark the difference between a "memoryless" circuit and circuit with memory. With the introduction of memory elements as components in digital circuits, an additional parameter *time* is introduced and must be considered when dealing with these types of circuits. In effect, the outputs can now be obtained one after another depending upon information released from the memory with time. Hence, these types of circuits are called ***sequential***. A sequential circuit is essentially one in which the outputs follow a predetermined sequence of states, with a new state occurring each time a clock pulse occurs. The sequential circuit, however, makes use of combinational circuits as well. The most commonly used sequential components using sequential circuits employed in the realization of computer circuits are ***parallel registers, shift registers, counters,*** and ***PLDs***.

A brief description of the various implementations of each of these components is, however, found in the website: http://routledge.com/9780367255732.

2.2.2.5.1 Programmable Logic Device (PLDs)

One of the major digital system categories that implement numerous specific desired applications is called the *application-specific integrated circuits* (*ASICs*). Four subcategories of ASIC devices are available to realize advanced digital systems: PLDs, *gate arrays, standard cell*, and *full custom*. A brief description of each of these subcategories is as follows.

As multiplexer can be used to realize switching functions, ***PLDs*** are another class of components consisting of circuits containing many gates or other general-purpose cells as switching elements. Many digital circuits today are implemented using PLDs. These devices are not like microcomputers or microcontrollers that "run" the program of instructions. Instead, they are configured electronically, and their circuits are "wired" together electronically to form a logic circuit. This programmable wiring can be thought of as thousands of connections that are either connected (i.e. 1) or not connected (i.e. 0). However, configuring of these devices by any manual means, placing 1s and 0s in the grid, will not only be most tedious and time-consuming, but also likely to be prone to errors. PLDs actually allow most of these tedious steps to be automated by a computer and PLD *development software*. The only job now left to the circuit designer is to identify inputs and outputs, specify the logical relationship in the most convenient manner, and select a programmable device that is capable of implementing the desired circuit at the lowest cost. The concept behind fabricating PLDs is summarily simple: put as many logic gates as possible in a single IC, and control the interconnections of these gates electronically. The use of programmable logic, however, improves the efficiency of the design and development process, and that is why, most modern digital systems of today are implemented in this way.

The elements in the PLDs are interconnected to form a circuit, and this circuit can be then configured or *programmed* to implement any desired combinational or sequential functions to be used for several other purposes. Figure 2.4 shows a schematic block diagram of a PLD, which has n input variables (x_1, x_2, ..., x_n) and m output functions (f_1, f_2, ..., f_m). We assume that the function f_k is realized as a *sum-of-product* (*SOP*) *terms* involving the input variables. The input variables (x_1, x_2, ..., x_n) are presented in true and complemented form to the AND array, where up to j product terms are formed. These are later gated into the OR array, where the desired output functions are formed.

The general architecture of PLD is shown in Figure 2.5 using two input variables A and B in true and complemented forms.

A common method of connecting one of many signals entering a circuit network to one of many signal lines exiting the network is a switching matrix. This concept is illustrated in Figure 2.6. A matrix is simply a grid of conductors (wires) arranged in rows and

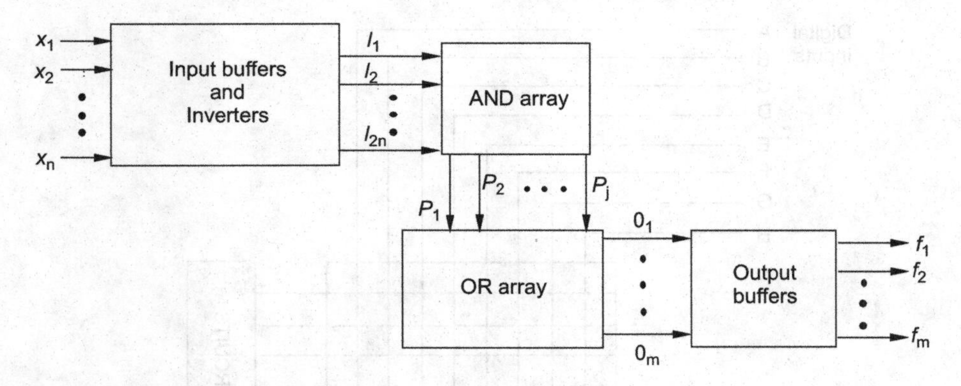

FIGURE 2.4
A schematic block diagram for a PLD.

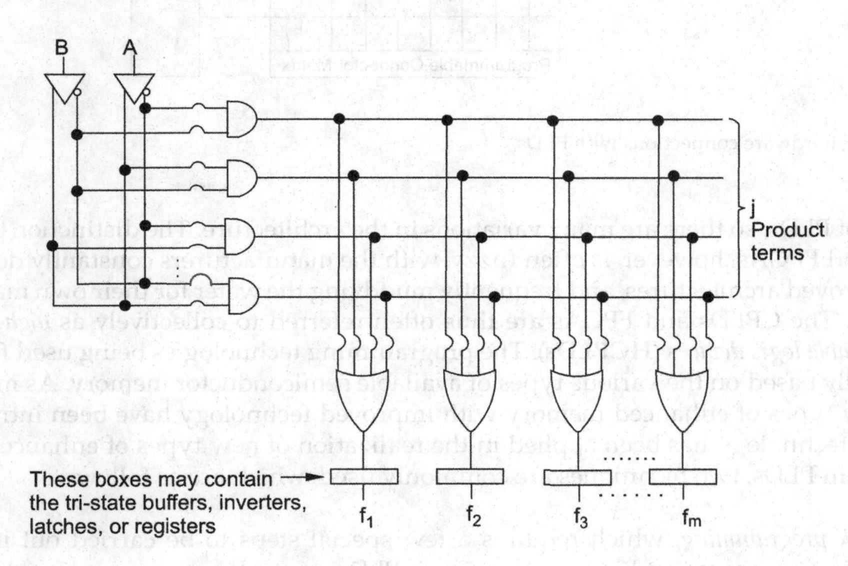

FIGURE 2.5
Schematic diagram of a general architecture of PLD using two input variables

columns. Input signals are connected to the columns of the matrix, and the output signals are connected to the rows of the matrix. At each intersection of a row and a column is a switch that can electrically connect that row to that column, thereby connecting a particular input to its corresponding output. The switches that connect rows to columns can be mechanical switches, fusible links, electromagnetic switches (relays), or transistors. This is the general structure used in many applications, like memory devices.

Since PLDs use a switch matrix, this is often referred to as the *programmable array*. By deciding which intersections are to be connected and which ones are not (Figure 2.6), we can "program" the way the inputs are to be connected to the outputs of the array. Various important PLD architectures will now be discussed briefly.

Generally, PLDs can be described as being one of three different types: simple programmable logic devices (**SPLDs**), complex programmable logic devices (**CPLDs**), or field-programmable gate arrays (**FPGAs**). There are several manufacturers with many different

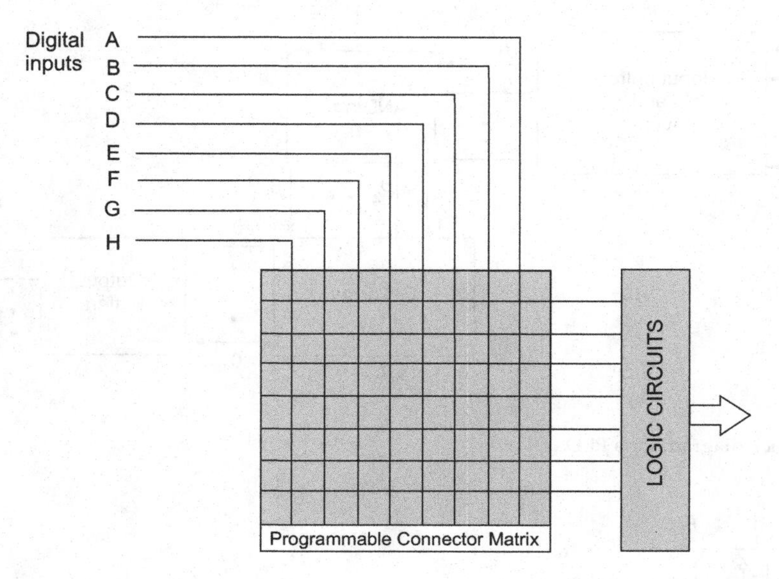

FIGURE 2.6
Configuring hardware connections with PLDs.

families of PLDs, so there are many variations in the architecture. The distinction between CPLDs and FPGAs, however, is often fuzzy, with the manufacturers constantly designing new, improved architectures, and frequently muddying the water for their own marketing purposes. The CPLDs and FPGAs are thus often referred to collectively as *high-capacity programmable logic devices* (**HCPLDs**). The programming technologies being used for PLDs are actually based on the various types of available semiconductor memory. As more and more new types of enhanced memory with improved technology have been introduced, the same technology has been applied in the realization of new types of enhanced PLDs. To program PLDs, two techniques are commonly used, which are as follows:

- *Mask programming,* which requires a few special steps to be carried out in the fabricating process of IC chips containing PLDs;
- *Field programming,* which is carried out by the designers or end users *in the field* with the help of small, low-cost programming units. Some of these PLDs are erasable to allow the same IC to be reprogrammed many times. This feature is effectively exploited by the designers at the time of developing and debugging a prototype of a new product.

2.2.2.5.1.1 Types of PLDs PLDs are classified depending on whether we program

i. The array of AND gates;
ii. The array of OR gates;
iii. Both the arrays;
iv. Only the outputs of the OR gates.

Based on the nature of above type of programming, there are ***four*** different types of architectural designs of PLDs:

i. **Programmable logic array (PLA):** In this type of PLD, the inputs of both arrays of AND and OR gates need programming.
ii. **Programmable array logic (PAL):** In this type of PLD, the inputs of the array of the AND gates are only programmed to generate the simple product terms, but the inputs of the OR gates are permanently hardwired; i.e., the OR gates are not programmable. PAL is a registered trademark of Advanced Micro Devices (AMD).
iii. **Programmable read-only memory (PROM):** In this type of PLD, the inputs of AND gates are permanently hardwired, but the inputs of the OR gates are programmed by blowing the fuse links of those OR gates that are required, and the remaining OR gates with unblown fuses are then used to connect the others.
iv. **Generic array logic (GAL):** This type of PLD is configured much like a PAL, but in this type, the outputs can be programmed in a variety of different ways (e.g. registered, combinational, or tri-stated). Unlike standard, low-density PAL which is one-time programmable, the GAL chip, on other hand, uses an EEPROM array (located at row and column intersections) to control the programmable connections to the AND matrix, allowing them to be erased and reprogrammed reasonably many times. In addition to the AND and OR gates used to produce the SOP functions, some GAL chips (an advanced version of conventional PAL chip) contain optional flip–flops for register and counter applications, tri-state buffers for the outputs, and control multiplexers used to select the various modes of operation. Consequently, they can be used as generic, pin-compatible replacements for most PAL devices.

A brief description of components of only **FPGA**s is illustrated here.

The various implementations of each of simple programmable logic devices (SPLDs), such as **PLA** and **PAL**, and also complex programmable logic devices (CPLDs) are, however, described in detail in the website: http://routledge.com/9780367255732.

2.2.2.5.1.2 Field-Programmable Gate Arrays (FPGAs) A more powerful class of PLDs has been developed to overcome the limitations of PAL chips where each output pin is provided for each SOP circuit. They are known as the *field-programmable gate arrays* (*FPGAs*) and were introduced sometimes in the mid-1980s. Gate arrays are ULSI circuits that offer hundreds of thousands of gates. The desired logic functions are realized by making interconnections of these prefabricated gates. A custom-designed mask for the specific application determines the gate interconnections, much like the stored data in a mask-programmed ROM. For this reason, they are often referred to as the *mask-programmed gate arrays* (MPGAs). Individually, these devices are less expensive than PLDs of compatible gate count, but the custom programming process executed by the chip manufacturer is very expensive and requires a great deal of lead time.

Figure 2.7 shows a conceptual block diagram of an FPGA. FPGA consists of a two-dimensional array of general-purpose logic circuits called the *cells* or *logic blocks* whose functions are programmable (white boxes in Figure 2.7). These cells are linked to one another by some *interconnects* (programmable buses), which consist of segments of wire and programmable switches (shaded boxes in Figure 2.7). This allows a high degree of routing flexibility on the chip. The cell types are not restricted to gates only; they are actually small multifunction circuits capable of *realizing all Boolean functions* of a few variables. A cell may even contain one or two flip–flops. Input and output buffers are provided for access to the

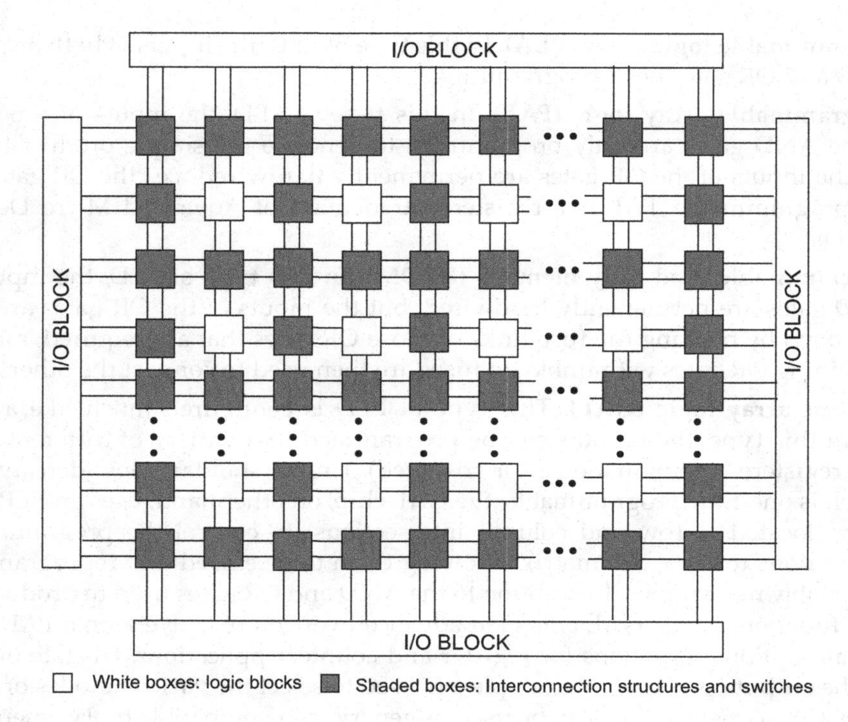

FIGURE 2.7
A conceptual schematic block diagram of an FPGA.

pins of the chip. FPGAs like all other PLDs are equally suitable for implementing small-scale manufacture and prototype designs.

Numerous types of designs exist for the logic blocks and the interconnect structures. *Two types* of logic cells are found in FPGAs. One type may be just a simple ***multiplexer-based circuit*** capable of implementing the logic functions as being realized by a multiplexer. Figure 2.8 shows one such cell of a four-input, 1-bit multiplexer with an AND and OR gate added. Another popular design is based on ***PROM table-lookup memories.*** This design uses a simple lookup table (LUT) as a logic block. For example, an LUT of four-input can be implemented in the form of a 16-bit ($2^4 = 16$) memory circuit in which the truth table of a logic function is stored. Each memory bit corresponds to one particular combination of these input variables. Such a LUT can be programmed to implement any functions of four variables. The versatility of a logic block may be further enhanced by including flip–flops in the blocks.

FPGAs also have a few fundamental characteristics that are shared. They typically consist of many relatively small and independent programmable logic modules that can be interconnected to create larger functions. Each module can usually handle only up to four or five input variables. Most FPGA logic modules utilize an LUT approach to create the desired logic functions using the available programmable logic modules. The logic modules are not associated with any I/O pin. Instead, each I/O pin is connected to a programmable I/O block that in turn is connected to the logic modules with selected routing lines. The I/O blocks can be configured to provide input, output, or bidirectional capability, and built-in registers can be used to latch incoming or outgoing data. A general architecture of FPGAs in this regard is shown in Figure 2.7.

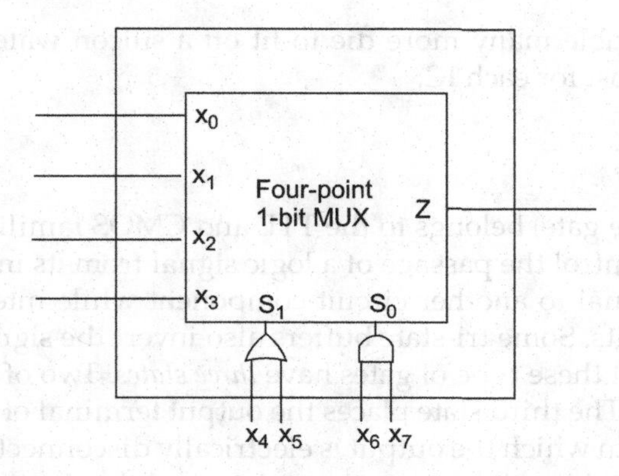

FIGURE 2.8
A representative basic cell of an FPGA.

Apart from having the logic blocks, many FPGA chips contain a number of memory cells that may be used to implement RAM and ROM components in the applications of **system-on-a-chip** (embedded systems).

Functionally, an FPGA chip is much advanced and more flexible than its counterpart devices such as a PAL or a CPLD, and can implement much larger logic networks consisting of up to a range of million logic gates. Another notable feature of FPGA is its relatively faster speed.

The functional capability of FPGAs has increased dramatically. They are now playing a dominant role in the area of very complex logic networks design. In addition, the FPGA implementations are always attractive in terms of cost. Moreover, the lead time needed to implement a specific digital design using an FPGA is very small compared to the cost of designing a custom chip. There still remain many other useful features that FPGAs possess. However, we restrict ourselves at this point from entering into any further details as this area lies entirely in the jurisdiction of basic electronics engineering. Interested readers can learn more about this subject from any modern textbook of logic design.

2.2.2.5.1.3 Standard-Cell ASIC Standard-cell ASICs use predefined logic function building blocks called the *cells* to realize the desired digital system. The IC layout of each cell is designed beforehand, and a library of available cells is stored in a computer database. The requisite cells are laid out for the desired application, and the interconnections between the cells are determined. Design costs for standard-cell ASICs are relatively higher because all IC fabrication masks that define the components and interconnections must be custom-designed. Greater lead time is also needed for the creation of additional masks. Standard cells do have a significant advantage over gate arrays. The cell-based functions that are designed happen to be much smaller than equivalent functions when realized in gate arrays, which consequently facilitates higher-speed operations and cheaper fabrication costs in general.

2.2.2.5.1.4 Full-Custom ASIC Full-custom ASICs are considered the ultimate ASIC choice. As the name implies, all components (transistors, resistors, and capacitors) and the interconnections between them are designed fully custom-specific by the IC designer. These design efforts require considerable time and costs, but can result in yielding ICs that can operate at the highest possible speed and require the smallest die (individual IC chip) area.

Smaller IC die sizes enable many more die to fit on a silicon wafer, which significantly lowers the fabrication cost for each IC.

2.2.2.6 Tri-State Buffers

A *tri-state buffer* (tri-state gate) belongs to the TTL and CMOS families and is a circuit that is used to essentially control the passage of a logic signal from its input to output in order to provide an input signal to another circuit component while interconnecting different register-level components. Some tri-state buffers also invert the signal as it goes through.

The name implies that these type of gates have *three states.* Two of the states produce the normal 0 and 1 signals. The third state places the output terminal of the buffer into a high-impedance (Hi-Z) state in which the output is electrically disconnected from the input that is supposed to drive it; it is as though someone detached the data output from the rest of the circuit with a wire-cutter. In reality, the output terminal is not an exact open circuit in the Hi-Z state, but has output impedance of several mega-ohms relative to the ground. The connection can be subsequently restored in a few nanoseconds by creating appropriate situations. Tri-state logic gates have active pull-up, lower output impedance, and faster switching time, and also the outputs can be tied together.

The traditional logic symbology has no special notation for tri-state outputs. Figure 2.9 shows a generalized tri-state buffer with the notation used in the IEEE/ANSI symbology to indicate a tri-state output. The *control input e* that controls the operation of the buffer is also called the *enable input.* When $e = 1$, the output y has the same logic value as the input x. When $e = 0$, the output is placed in the high-impedance Z. For this circuit in particular, the enable input e is called the *active high enable input* because when e is high, the circuit behaves normally but when e is low, the normal function of this circuit is disabled as the output y cannot respond to the input x. Similar tri-state gates with *active low enable input* can also be designed, in which when $e = 0$, the output y has the same logic value as the

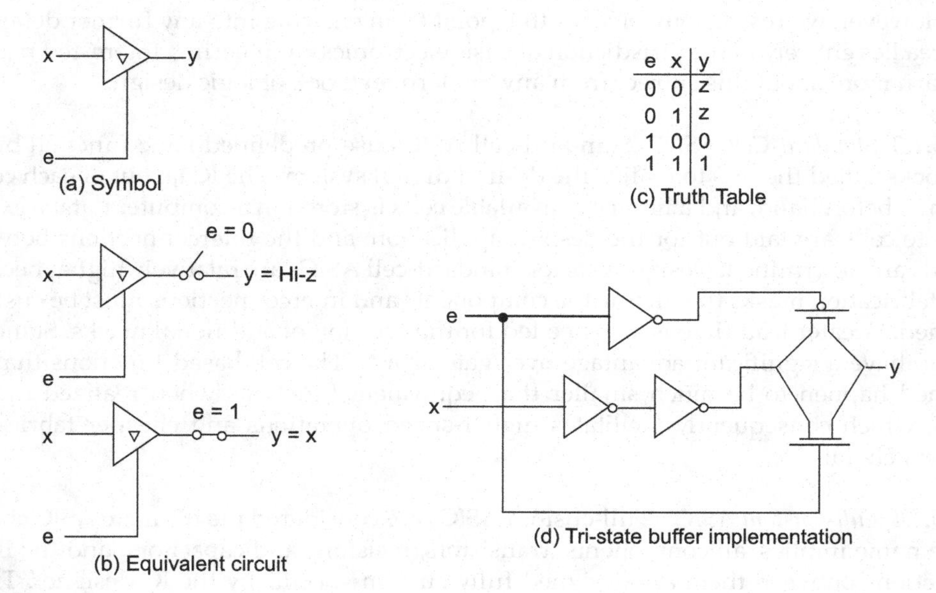

FIGURE 2.9
Tri-state buffer with active-high enable input.

input *x*. When $e = 1$, the output is placed in the high-impedance Z. The examples of such a tri-state buffer are IC chip 74125 and 74LS125.

Tri-state ICs: In addition to tri-state buffers, many ICs are designed with tri-state outputs. For example, the 74ALS374 is an octal D-type FF (flip–flops) register IC with tri-state outputs. This means that it is an 8-bit register of D-type FFs whose outputs are connected to tri-state buffers. This type of register can be connected to common bus lines along with the outputs from other similar devices to allow efficient transfer of data over the bus. Other types of logic devices that are available with tri-state outputs include decoders, multiplexers, A-to-D converters, memory chips, and microprocessors.

2.2.3 The Gate Level

In the area of computer design, both processor-level and register-level designs play a major role and are certainly of primary interest. Unfortunately, design at these levels while being considered an art mostly depends on the end use that the computer should achieve, and hence demands the skills and experiences of the concerned designers. In fact, there are no hard and fast rules in these designs. These levels are, in turn, built on the next lower level, the *gate level*, the most primitive level that has a rather substantial foundation, like Boolean algebra dealing with binary digits. The gate-level design is encompassed with various types of gates along with all their possible combinations to process *binary variables*. The gates are simple, memoryless processing elements, while the flip–flops are the bit-storage devices. A gate is essentially an electronic circuit of fundamental elements such as transistors and capacitors (Figure 2.1, device level) that produces an output signal, which is a simple Boolean operation on its input signals in terms of *bits* (binary digits), 0 and 1.

All these fundamental gates with their *functions, truth tables, and characteristics are individually* given in the website: http://routledge.com/9780367255732.

2.2.3.1 *Basic Memory Components: Latches and Flip–Flops*

An essential component of every computer and the majority of applications of digital logic require the storage of information. All the register-level components as we discussed earlier used for building a full-fledged computer are either combinational logic circuits or sequential logic circuits. By way of adding memory elements to a combinational circuit in the form of 1-bit storage elements, we obtain a sequential logic circuit. These memory devices may be mainly of two types: **latches** and **flip–flops**. Either of these two memory devices can store a single bit, i.e. 1 or 0 in it. Several such basic units can be cascaded to form a memory of large capacity. Commercially such basic memory units are available in the form of IC chips. We deliberately restrict here from the detailed discussions of many different forms of these two types of memory devices, namely latches and flip–flops. The interested readers may consult any standard textbook on digital electronics for more details in this area.

A brief description of different types of these basic components of a memory system to see how they work, and how they are combined to produce large memories, is illustrated in the website: http://routledge.com/9780367255732.

2.2.4 Genesis of Digital Systems

The entire chapter has enunciated the various hardware choices that are now available to the digital system designers with a better perception of numerous digital hardware alternatives. Any desired circuit functionality can now be realized by using several different types of this digital hardware. Both standard logic devices and PLDs can be used to realize the same functional blocks. Microcomputers and DSP (digital signal processing) can also often be employed with the necessary sequence of instructions (i.e. the application program) to produce the desired circuit functions. The trade-off in design engineering takes into account many different factors, which include the operation speed of the circuit, the cost of production, system power consumption, system size, amount of lead time to design the product, the complexity of the system thus designed, and similar other related aspects that can affect the design process to a large extent. In fact, most complex digital designs of today are realized with a mix of different categories of several hardware components. At the time of designing a digital system, many deserving critical trade-offs between the various types of available hardware have to be weighed in order to arrive at a final decision.

Figure 2.10 illustrates a schematic representation of a digital system family tree. This identifies most of the currently available hardware choices that can be useful in sorting out the many categories of digital devices. The graphical representation in the figure has not included all the exhaustive details – some of the more complex device types with many additional subcategories and older, obsolete device types have been deliberately omitted here for the sake of clarity. However, the major digital system categories that are commonly in use today include standard logic, ASICs, and microprocessor/DSP devices.

Standard logic devices consist of *three* major families, namely TTL, CMOS, and ECL, which are considered the basic functional digital components (gates, multiplexers,

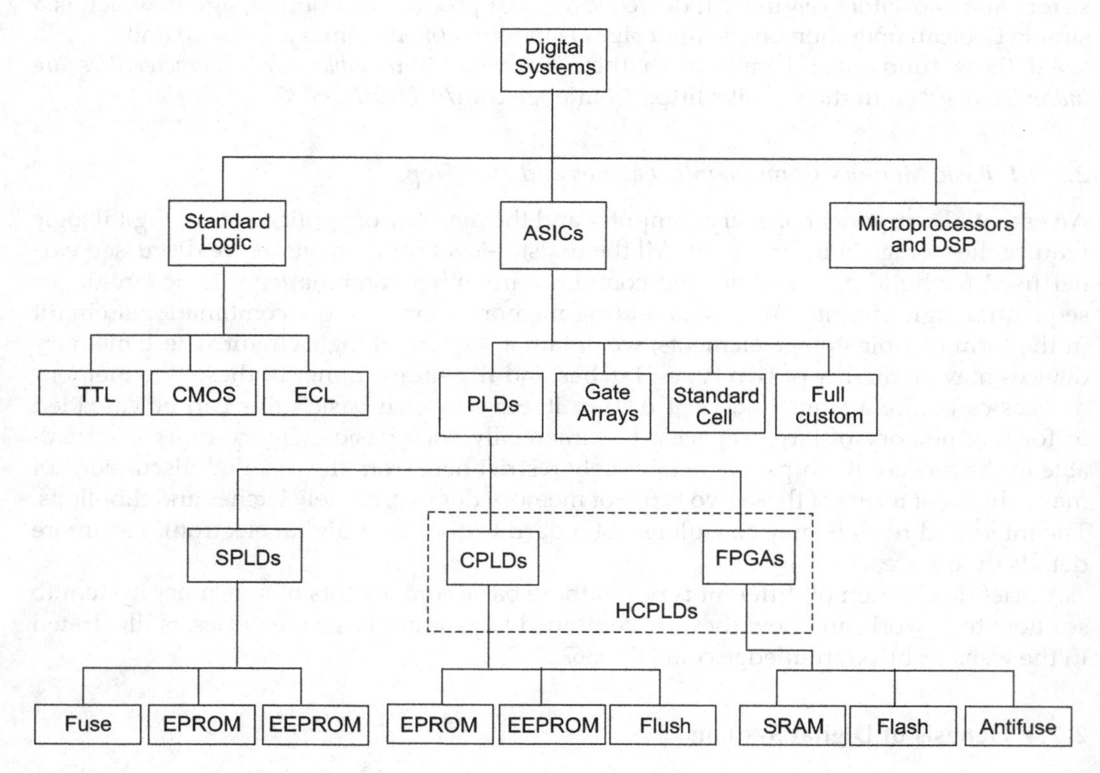

FIGURE 2.10
A schematic tree diagram representing digital system family.

decoders, registers, counters, flip–flops, etc.) that are available mostly as SSI (small-scale integration) and MSI (medium-scale integration) chips. TTL is a matured dominant technology over many years of use, but new designs of today although seldom apply TTL logic, yet many digital systems still employ TTL devices. CMOS, however, is probably today's most popular standard logic device family, primarily due to one of its distinct characteristics of low power consumption. ECL technology is a natural choice for higher-speed designs. Although these inexpensive devices are still available and equally useful today if the intended design is considered not much complex, but they are not found to be very lucrative for fulfilling the design requirements of complex applications. Incidentally, modern technologies capable of replacing the standard logic devices continue to evolve, especially at lower costs requiring lesser chip area at the time of fabrication. All these and other similar supporting factors consequently had an impact that eventually led the standard logic devices to be dropped from favour at the time of designing applications, the majority of which are mostly complex by nature.

The second digital system category called *microprocessor/digital signal processing* (*DSP*) consists of numerous types of functional blocks that are mostly program-controlled to realize different types of specific applications. As these devices are truly software-controlled, they normally offer a great deal of flexibility, in a way that by only changing the operating program, the same device can be used at once for other applications. One of the major drawbacks of such type of devices lies in its relatively slow operational speed for being controlled by software. This speed is always less than any digital system design that is based on hardware realizations.

The third one is probably the major digital system category referred to as *application-specific integrated circuits* (*ASICs*). This broad category today represents almost all the modern hardware design realization for a majority of digital systems. As the acronym implies, an IC is designed here to implement a specific desired application. All *four* subcategories of ASIC devices that are available to realize numerous digital systems have been already explained previously in the respective sections.

A brief description of this digital system family is given in the website: http://routledge.com/9780367255732.

2.3 Summary

The Organisation of a computer system can be conceived in terms of a series of levels or abstraction maintaining a level hierarchy in which each level in the hierarchy with more or less fixed tasks and responsibilities performs a specified range of operations at minimum cost, and relies on its just upper level in the hierarchy. All machine activities are usually obtained by means of the operations carried out by three basic levels, namely *the processor level, the register level,* and – the lowest one – *the gate level*. The type of components to be used in each such level is mostly standardized, but each of these three levels can be realized in many different ways that entirely depend on the specific Organisation being followed in building up a computer system. Processor-level system is absolutely well defined, and comprises CPUs and other processors, memories, I/O devices, and interconnection networks. Its performance is measured and evaluated by a non-profit Organisation named *System Performance Evaluation Corporation*, often abbreviated as SPEC. Register level consists of combinational devices (such as multiplexers, decoders, adders, and word gates), sequential devices (such as registers, counters etc.), and various general-purpose programmable elements (including PALs, PLAs, FPGAs, and ROM). The gate level is the lowest level in the architectural design of computing systems. This level comprises logic gates as basic functional components that operate on a well-defined theory based on Boolean algebra. Unlike combinational circuits, sequential circuits have memory, which is usually built from gates, and 1-bit storage elements (flip–flops) that store the circuit's state and are synchronized by means of clock signals.

Exercises

2.1 Discuss the contribution of level concept in the architectural design of the computer. Is it possible for a multilevel computer to have the device level and the digital logic level as not the lowest levels? Justify.

2.2 What is the difference between translation and interpretation? In what sense are hardware and software equivalent? Not equivalent? Do you think that level concept in the design of computers is still pertinent even in the design of modern computers of today? Explain.

2.3 "Introduction of microprogramming level in computer design opened a new horizon": discuss.

2.4 Discuss the methodology being commonly followed in the design of computers.

2.5 What are the standard combinational components and sequential components used at the register level of computer design?

2.6 What is the use of a multiplexer? Construct a 16-to-1-line multiplexer with two 8-to-1-line multiplexers and one 2-to-1-line multiplexer. Use block diagrams for the three multiplexers.

2.7 A digital computer has 16 registers, each with 32 bits. The registers are connected by a common bus which is constructed with multiplexers.

   a. How many selection inputs are there in each multiplexer?
   b. How many multiplexers are there in the bus?
   c. What size of multiplexers is needed?

   [Hint: b) Each multiplexer transfers one bit of the selected register. The number of multiplexers needed to construct the bus is equal to n, the number of bits in each register. Here, it is 32.

   c. The size of each multiplexer must be $k \times 1$, where k is the number of the registers, since it multiplexes k data lines, each from a register.

2.8 Design a 5×32 decoder using four 3×8 decoders (with enable inputs) and one 2×4 decoder.

2.9 State and explain the role being played by an encoder in the design of a computer. Draw the logic diagram of a 2-bit encoder, a circuit with four input lines, exactly one of which is high at any instant, and two output lines whose 2-bit binary value tells which input is high.

2.10 What is the difference between serial and parallel transfer? Using a shift register with parallel load, explain how to convert serial input data to parallel output and parallel input data to serial output.

2.11 The content of a 4-bit register is initially 1001. The register is shifted five times to the right with the serial input being 11011. What is the content of the register after each shift?

2.12 Why counter is considered an important component at the time of building a computer? Explain with a diagram the operation of a counter.

2.13 What does PLD stand for? How are the circuits reconfigured electronically in a PLD? What is actually being "programmed" in a PLD?

2.14 What is the contribution of PLDs in the circuit design of modern computers? What are the different types of PLDs commonly used in computer circuits?

2.15 State with diagram the different states of a tri-state buffer and explain its operation. What are the applications of tri-state buffer? Show how computer bus can be organised using tri-state buffer.

Suggested References and Websites

Armstrong, J. R. and Gray, F. G. *Structured Logic Design with VHDL*. Englewood Cliffs, NJ: Prentice-Hall, 1993.

Hamacher, C., Vranesic, Z. G., and Zanky, S. G. *Computer Organisation*, 5th ed. New York: McGraw-Hill, 2002.

Langholz, G., Francioni, J., and Kandel, A. *Elements of Computer Organisation*. Englewood Cliffs, NJ: Prentice-Hall, 1989.

Mano, M. *Logic and Computer Design Fundamentals*. Upper Saddle River, NJ: Prentice-Hall, 2004.

Siewiorek, D. P., Bell, C. G., and Newell, A. *Computer Structures: Reading and Examples*. New York: McGraw-Hill, 1982.

Tanenbaum, A. S. *Structured Computer Organisation*, 4th ed. Upper Saddle River, NJ: Prentice-Hall, 1999.

System Performance Evaluation Corporation: web page: www.spec.org.

Standard benchmarks are used to measure and compare the performance of different computer systems.

3

Processor Basics – Structure and Function

LEARNING OBJECTIVES

- To explain the fundamental concepts and basic structure of a generic register-organised CPU (central processing unit) and stack-organised CPU.
- To study the main tasks of a CPU and the way it operates.
- To learn about a generic machine instruction set (instruction set of a generic CPU), including multimedia instruction set, and the constituent elements of a machine instruction, instruction format, and design criteria.
- To introduce the types of register-organised CPUs and their organisation.
- To study numerous addressing schemes and different types of addressing modes, their significance and implications in CPU design.
- To present a representative IA-32 instruction set processor (ISP) architecture with an overview of IA-64 architecture, including multimedia extension (MMX).
- To describe the different types of machine instructions, including MMXs and the related operations.
- To demonstrate a representative stack-organised processor architecture.

3.1 Introduction

This chapter is aimed at explaining the fundamental aspects of the *processing unit*, commonly referred to as *CPU* (*central processing unit*), and sometimes is even called an *instruction set processor* (*ISP*), which performs data-processing operations and coordinates the activities of the other resources. However, the CPU is essentially built with **three** major parts. The *Control unit* supervises and monitors various elements of the system, including those of the CPU, to manage numerous functions and the operations. The *arithmetic–logic unit* (ALU) performs the simple logic and arithmetic functions required to execute the instructions. The *register set* usually stores intermediate data and thereby assists the execution of an instruction. The internal organisation of a CPU and how it performs a variety of functions while executing instructions of a program will now be explained in detail.

The basic design of the CPU primarily entails the various functions that the proposed CPU will offer. To realize each such function, some instructions (or simply an instruction) need to be executed. The set of all such instructions thus required to obtain each and every predefined function of the CPU is called the *machine instruction set*. A description of a computer's machine instruction set thus provides a strong basis for explaining its CPU structure, architecture, and organisation as well as its functions. We will now examine the usual formats of generic machine instructions, the various types of operands associated with different types of operations that may be specified by machine instructions, as well as how to specify and include these operands and operations in machine instructions. With a rapid development in the field of electronics, the architecture and organisation of *processors* have, however, continuously evolved over the years aiming towards a gradual increase in performance that eventually led to the introduction of *pipelined organisation* and subsequently *superscalar architecture* of the CPU for even more increased performance. Both *pipeline* organisation and *superscalar architecture* of CPUs will be discussed separately in Chapter 8. This chapter is dedicated mainly towards explaining the structure, function, and the basic principles involved in designing a conventional CPU, which are mostly common to all types of processors.

3.2 Processor (CPU) Organisation

A computer is used to solve users' problems submitted in the form of a program consisting of high-level instructions which are first converted into the *machine language* of computer's primitive instructions, and then these machine language instructions are directly executed by the electronic circuits built in the form of numerous digital systems. The design of these machine instructions (to be as simple as possible and totally fulfil the performance requirements of the computer) and the respective electronic arrangements to execute these instructions are of primary importance and constitute the decisive factors while framing the architecture and organisation of the computer system and of CPU to be realized. To understand the organisation of the CPU, let us first consider the requirements that are associated with the CPU.

3.2.1 Fundamental Concepts

The primary and ultimate function of the CPU and other ISPs is to do the actual work by executing the set of instructions (programs) which are either available in an external main memory or stored in an external secondary memory. These instructions are brought into the main memory at the time of execution. The responsibilities that the CPU must perform to execute the stored set of instructions (programs) one after another are as follows:

- **Fetch instructions**: The CPU must read instructions from memory one at a time.
- **Decode instructions**: The instruction thus fetched must be interpreted to determine what action is required and that action is then to be taken.

- **Fetch data**: The execution of an instruction may require operands that are read as data from memory or from an I/O module.
- **Execute instructions**: The instruction is then executed; it may require performing some arithmetic or logical operations on data. This execution may often involve several operations, which mostly depends on the nature of the instruction being executed.
- **Write data**: The results of an instruction execution may require data to be written to a targeted memory location or to an I/O module.

Figure 3.1 shows a simplified view of the CPU, indicating its connection to the rest of the system via the system bus. The major components of the CPU are as follows.

The *control unit* is responsible for fetching instructions and data from main memory and decoding the instructions (analyse to determine their types). The control unit also supervises, monitors, and controls the movements of data and instructions into and out of the CPU and also controls the operation of CPU. The ALU performs numerous operations (such as additions, Boolean, etc.) needed to execute the instructions. The CPU also contains a small, high-speed memory used to temporarily store data, intermediate results, and certain control information. This memory comprises a number of *registers*, each of which has certain predefined functions.

The number of registers present in a CPU, together with its internal organisation, plays a critical role in the organisation of the CPU and also influences the corresponding machine instruction set design. This instruction set thus available in a CPU determines to a large extent the performance and the versatility of a CPU. The usual trade-off is, of course, cost

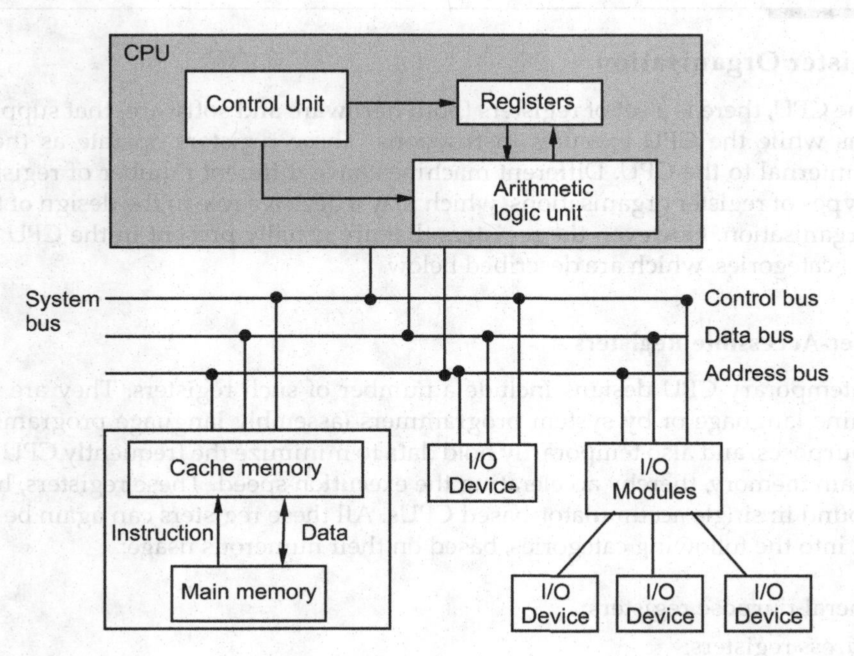

FIGURE 3.1
Schematic diagram of CPU with system bus.

versus speed. Among many other important factors that can influence the speed of the CPU, one is, of course, how many registers are present within the CPU. Based on the number of hardware registers available in a CPU, most computers fall into one of three types of CPU organisations:

1. *Accumulator-based CPU* (*single accumulator organisation*);
2. *General-register organisation CPU* (*multiple registers*);
3. *The stack-organised CPU.*

Each of these three organisations (described separately in the subsequent sections) gives rise to a totally different approach in CPU design with its own merits and drawbacks, mostly depending on the environment and the applications where it would be used. It is, after all, the prerogative of the designer to select a particular design approach taking into account the trade-off criteria and the ultimate target that the machine should meet. While the design approaches as mentioned in (1) and (2) follow the **Von Neumann design** concept, the design approach as mentioned in (3) is a total departure from this fundamental design concept and is called a *non-Von Neumann machine.* The most interesting feature is that in spite of having a strong demarcation line in these various types of CPU organisations, some computers combine or can combine features from more than one organisational structure mentioned above, to achieve their specific design target. The earlier Intel 8085 processor is an example of such an organisation.

3.3 Register Organisation

Within the CPU, there is a set of registers (both hardware and software) that support CPU operations while the CPU executes instructions. These registers operate as the fastest memory internal to the CPU. Different machines have different number of registers and various types of register organisations, which play a decisive role in the design of the CPU and its organisation. However, the registers that are usually present in the CPU fall into two broad categories, which are described below.

3.3.1 User-Accessible Registers

Most contemporary CPU designs include a number of such registers. They are used by the machine language or by system programmers (assembly language programmer) for various purposes, and also temporarily hold data to minimize the frequently CPU visits to slower main memory, thereby accelerating the execution speed. These registers, however, are not found in single accumulator-based CPUs. All these registers can again be broadly classified into the following categories, based on their numerous usage:

i. General-purpose registers;
ii. Address registers;
iii. Data registers;
iv. Condition code registers.

Several design issues and alternative approaches have, however, been observed in relation to the classification and categorization of these registers and their specific usage, although each one has its own merits and drawbacks and there is, as such, no specific rule or perfect and complete solutions to these design issues. However, the modern trend is observed to be tilted in favour of using specialized registers. A new approach that finds advantages in the use of more registers (to the tune of hundreds or so) is implemented in some RISC (reduced instruction set computer) systems (discussed separately in Chapter 9). Incidentally, many processors, including those from ***Intel***, do not use condition codes at all. Instead, they use conditional branch instructions that indicate a specific comparison to be made, and act accordingly on the outcome of the comparison. Besides, they use some other mechanisms that serve, equally well, the purpose that condition codes provide. Thus, they find no need to use any such condition codes (registers).

The details of the functions and the responsibilities that are carried out by each of these registers are given in the website: http://routledge.com/9780367255732.

3.3.2 Control and Status Registers

There are various types of CPU registers available in most machines, used exclusively by the CPU to control its own operation. On a majority of machines, most of these registers are not accessible to the user, a few of them on some machines are found to be user-visible (program counter [PC] in DEC VAX system), and at certain points in time, some of them are visible to the machine instructions when executed under supervisor mode. However, there are *four key registers* in the processor that are essentially attached to instruction execution, which are described below.

Program Counter (PC): To execute a program, the CPU always keeps track of the address of the current instruction in successive memory location to be fetched, using a register known as the *program counter* (*PC*). After fetching the current instruction using the content of PC, the PC value is then automatically updated to point to the next instruction to be fetched in sequence until a branch or a jump instruction is encountered. The presence of a branch instruction in any form may modify the contents of the PC and may load a different value.

Instruction register (IR): The contents of the memory location being pointed to by the PC are then fetched by the control unit of the CPU. The contents of this memory location if interpreted as an instruction to be executed are loaded into the IR. Symbolically, this can be written as IR $\leftarrow$ [[PC]].

Memory address register (MAR): This register contains the address of a memory location the contents of which are to be fetched.

Memory buffer register (MBR): This register contains the word most recently read from memory or a word to be written to memory.

Besides, there are a number of other registers found in particular CPU designs that are used to contain control and status information. They are often known as the *program status word* (*PSW*) or *program control block* (*PCB*) and contain condition codes and other status information.

During execution, the instruction being pointed to by the PC is fetched and is loaded into the IR where it is decoded. Data is exchanged with memory using the MAR and MBR. User-accessible registers often exchange data with the MBR. In a bus-organised computer, the MAR is directly connected with the address bus and the MBR is directly connected with the data bus. Within the CPU, the ALU may have access to the MBR and other user-accessible registers, or alternatively, there may be *additional buffering facilities* at the boundary of ALU, serving as input and output registers (INPR/OUTR) for the ALU. These registers involve in the exchange of data with MBR and other user-accessible registers.

Last but not least, there is an issue related to register length that often made an impact on CPU design. Registers that are used to hold addresses must be long enough so that they can hold the largest address. Data registers also should be capable of holding values of most data types. Some machines exploit two contiguous registers to be used as one for holding double-length values (double precision), and also four contiguous registers to be used as one for holding extended length values (extended precision). However, not all registers are found in all the CPUs. Even the size of the registers is different in different machines. Also, the number of registers to be employed is not the same in all the CPUs. All these and similar aspects are entirely a designer's choice, depending on trade-off and other factors, including the operating system to be used and the ultimate objectives to be attained.

The processors of a somewhat different class of architecture, known as *RISC architecture* (explained in Chapter 9), have the abundant number of registers, and most of the instructions in the instruction set of those processors are thus register-oriented. Moreover, the number of registers and their internal organisation in this type of processors introduced by different manufacturers are also found varying over a wide range.

In fact, the increasing number of registers and their availability in various types in the processors, the growing number of functions they are to perform, and of course, their more efficient internal organisation actually set the stage for constantly improving processor performance.

3.3.3 Register Organisation in Microprocessor: IA-32/64 and MC68000

Most contemporary CPUs are realized today by the modern versions of powerful, versatile single-chip processors or *microprocessors* introduced by different vendors. But the two leading representative microprocessors in the industry are Intel X-86 series from **Intel** and MC 68000 series from **Motorola** with their constantly released new members, such as 8-, 16-, 32-, and 64-bit microprocessors in their own families. The register organisation of the microprocessors that came from these two families can be thus cited here as an example to give an overall idea as to how the register organisations are actually planned in microprocessor implementations, in general.

3.3.3.1 Motorola MC68000 Series

Although the initial product of the MC68000 series was essentially a hybrid between 16- and 32-bit architectures, subsequent releases were truly in 32-bit architecture. The 32-bit register set is partitioned into eight data registers and nine address registers, including one stack pointer (SP) register. There is also a 32-bit PC and a 16-bit status register (SR). The data registers although primarily used for data manipulation are also used for addressing similar to index registers, thus giving rise to a flavour of general-purpose activities. The 8-, 16-, and 32-bit-width data operations are allowed in these registers, which are determined by the respective operation codes given in the instruction. The address registers contain 32-bit addresses; two of these registers are also used as SPs, one for users and the other for the operating system, depending on the current execution mode. Both these registers are numbered A7, because only one will be used at a time. There are, however, no special-purpose registers.

The *instruction set* used in Motorola processor is a reasonably regular one. The available registers are, however, divided into two functional components, thereby enabling the processor to save one bit on each register specification that consequently helps to make the code somewhat shorter in length and more efficient. This approach appears to be an effective compromise between complete generality and code compaction.

3.3.3.2 Intel IA-32 Architecture

The **Intel X-86** uses a different approach in its register organisation, which is relatively tricky. Every register here is special-purpose, although some registers are also used as general-purpose or multipurpose. The multipurpose registers hold various data sizes (*bytes, words,* or *double word*) and are used for almost any purpose, as dictated by a program. The earlier 8086, 8088, and 80286 contain 16-bit architectures having a set of registers, which are also now a subset of registers provided in the full 32-bit internal architectures of 80386 to Core-2 microprocessors for maintaining downward compatibility with the earlier versions (Brey Barry B.). The register organisation of the member processors under IA-32 includes *eight 32-bit general-purpose registers,* such as EAX EBX, etc. used for all types of X-86 instructions, including indirect addressing. Some of these registers can also be used for other special purposes; i.e. the registers ECX, EDI, and ESI are implicit in the string instructions. The processor also has *six 16-bit segment registers,* including CS, SS, DS, ES, FS, and GS that contain segment selectors to identify specific segments containing instruction to be executed, the user-visible stack, and segments with other information for dedicated and implicit usages. Although this arrangement provides the benefit of compact encoding, it comes at the cost of reduced flexibility. Moreover, IA-32 provides a 32-bit register called *EIP* that contains the address of the current instruction (i.e. our usual **PC**), and a 32-bit register called *EFLAGS* that contains numerous *condition codes,* and *control bits,* such as *sign, overflow, carry,* and trap flag (TF), interrupt enable flag (IF), as well as various mode bits indicating the state of the processor. In addition, there exist *other registers,* such as **numeric, control, status,** and **tag word**, especially committed to the *floating-point unit.* Apart from all these, IA-32 employs *four control registers* (register **CR0 through CR3**) and a total of *eight 64-bit* **MMX registers**. In fact, the processor does not include specific MMX registers. Rather, the processor uses an aliasing technique that enables the eight existing floating-point registers (each of 80 bits, out of which 64 bits are used for mantissa and the rest 16 bits for exponent) to be used to work as an MMX register to store MMX operands only in the 64-bit mantissa portion of these registers. Thus, these eight registers actually serve a dual purpose. When used by an MMX instruction, these registers are referred to as *MM0* through *MM7.* To provide parallel operation (SIMD (single-instruction multiple data) approach, see Chapter 10) on standard multiple data lengths, four new packed data types are defined in MMX. These are explained in Section 3.9 (MMX data types) and Section 3.10.8.1 (MMX operations).

In order to access and specify the *global and local descriptor tables* located in the memory system (see Chapter 4), IA-32 contains a few registers that are not directly addressable by the software, known as *program-invisible registers.* Two such registers are GDTR (**global descriptor table register**) and IDTR (**interrupt descriptor table register**), which contain the base address of the descriptor table and its limit. One of the global descriptors is set up to address the local descriptor table which is accessed by the LDTR (**local descriptor table register**).

3.3.3.3 Intel IA-64 Architecture

The register organisation of the member processors under IA-64 contains many useful features, including most of the features of register organisation of IA-32, and some other types of registers of its own. However, we restrict ourselves at this point from entering into any further details on it. In fact, IA-64 (Core 2, Pentium 4, and onwards) provides **sixteen 64-bit general-purpose registers**, such as **RAX RBX** etc., including **eight additional 64-bit registers** (R8 through R15) used for all types of X-86 instructions. These additional registers are addressed as a byte, word (16 bits), doubleword (32 bits), or quadword (64 bits), but only the rightmost 8 bits is a byte. In addition, IA-64 provides a 64-bit register called *RIP* that contains the address of the current instruction (i.e. our usual PC), and also a 64-bit register called *RFLAGS* consisting of a set of 1-bit status and control flags that contains numerous condition codes and various mode bits indicating the state of the microprocessor, thereby controlling its operation. In the current architectural design and definition, the upper 32 bits of RFLAGS remain unused (Dulong, C.).

A brief detail of the register organisations with figures of both Motorola 68000 series and Intel IA-32 and IA-64 processors is given in the website: http://routledge.com/9780367255732.

3.4 Stack Organisation

One of the useful features that are included in the modern macro-architecture of CPU in most computers is a stack or *last-in–first-out* (LIFO) list. Block-structured high-level programming languages are normally implemented in such a way that when a procedure or function is exited, the storage it had been using for local variables is temporarily stored and then released, and the easiest way to achieve this goal is by using a *data structure called a stack.* A *stack* consists of data items (words, characters, bits, etc.) stored in consecutive order in a portion of a large memory or a *stack* that can be organised as a collection of a finite number of memory words or registers. Several operations are defined on stacks, and the most important ones are operations *for the insertion* of elements called *PUSH* (or push-down) and *for the deletion* of elements called *POP* (or pop-up, i.e. removing one item so that the stack pops up). The items are stored in such a manner that the item stored last is the first item retrieved (LIFO). Most of the time, the stack is partially filled with stack elements and the remainder is available for stack growth. Three addresses are thus needed to keep track of the status of the stack for proper operation, namely *SP, stack base,* and *stack limit,* and these are often stored in CPU registers.

Stack Pointer (SP) associated with each stack is a *register* or *memory word* that contains the address of the top of the stack (TOS). If an item is appended to or deleted from the stack, the pointer is incremented or decremented to contain the address of the new top of the stack (TOS). The number of elements in the stack, or *length* of the stack, is variable. If n is the number of bits available in SP, then the maximum number of locations that SP can point to is 2^n. This is the maximum size of the stack with this SP. ***Stack base*** contains the address of the bottom location in the reserved block. If the contents of the SP are equal to the contents of the stack base, the stack is empty. If an attempt is made to POP when the stack is empty, an error is reported. ***Stack limit*** contains the other end (upper end) of the reserved area for the stack. If the contents of the SP are equal to the contents

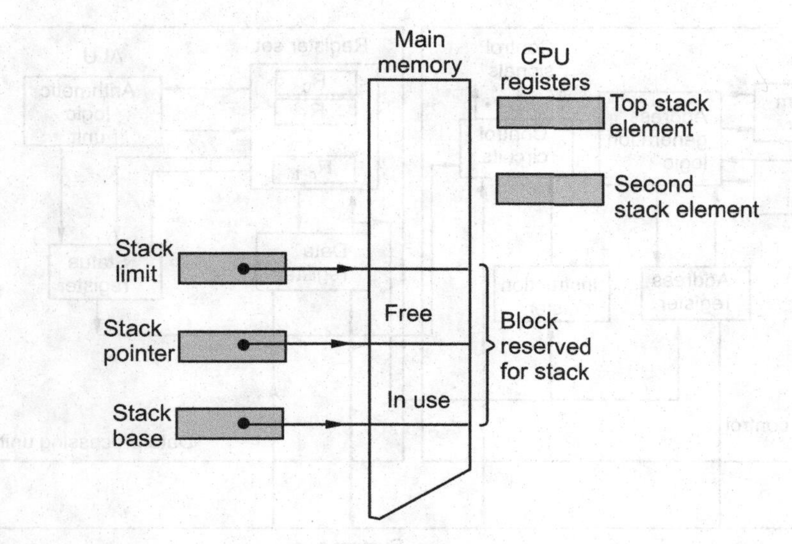

FIGURE 3.2
A typical stack organisation.

of the stack limit, the stack is full. If an attempt is made to PUSH when the block is fully utilized for the stack, an error is reported. Most computers do not provide hardware to check for stack *overflow* (full stack) or *underflow* (empty stack). The stack limits can be checked by using two processor registers: one to hold the upper limit (*stack limit*) and the other to hold the lower limit (*stack base*). After a push operation, SP is incremented and is compared with the upper-limit register, and after a pop operation, SP is decremented and is compared with the lower-limit register. Figure 3.2, however, depicts a block diagram showing all these elements. In some stack implementations, the top two stack elements are often stored in two separate registers to speed up stack operations, as shown in Figure 3.2. In this case, the SP is the third element of the stack. One of the distinct advantages of using stack, which is already implemented in stack-organised computers and also in other computers with this feature, is that it always provides relatively shorter machine instructions with no addresses (zero-address instruction) that summarily save both CPU time and necessary memory space for machine instructions – one of the vital criteria of a CPU design. For more details, see stack-organised computer, given in later section.

A brief detail of the stack organisation with an example showing the stack operations is provided with figure in the website: http://routledge.com/9780367255732.

3.5 Generalized Structure of CPU

Keeping this register organisation in view, an overall structure of a generalized CPU, in a slightly detailed form considering some typical basic elements of the ALU and control unit, is shown in Figure 3.3. Here, the program control unit and data-processing unit have separately illustrated with their respective individual processing components and their interconnections as well as the interconnections between these two units as a whole.

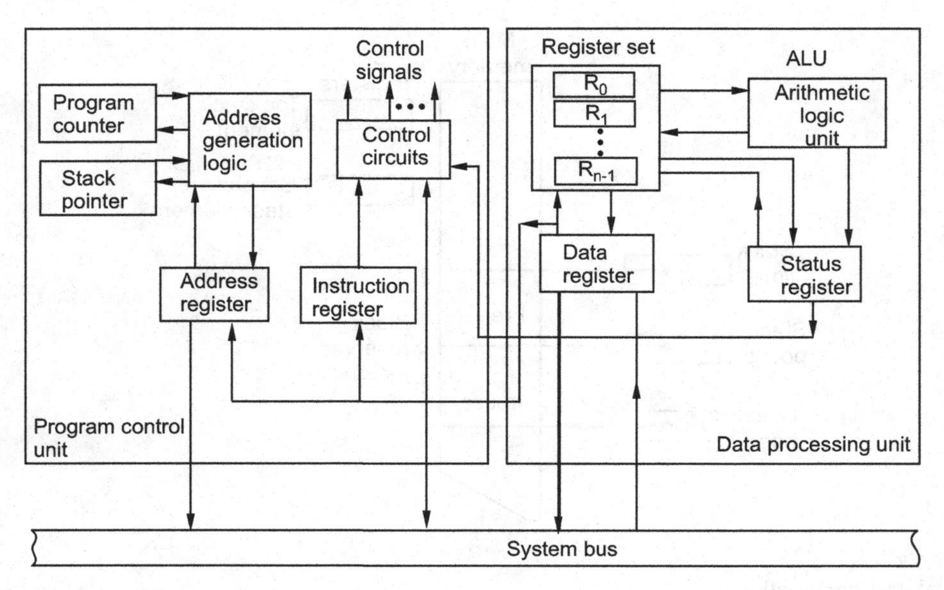

FIGURE 3.3
A schematic block diagram of a CPU with general-register organisation.

The interconnections path used between the components and also between these two units is called the *internal processor bus*, which is required to transfer data between various registers and the ALU during the actual computation or processing of data. It is interesting to observe the similarity between the internal structure of the computer as a whole and the internal structure of the CPU. In both the cases, there is a small collection of major resources (computer: CPU, I/O, memory; and CPU: ALU, control unit, registers) connected by data paths.

As shown in Figure 3.1, the CPU communicates directly with the main memory, which is a high-capacity, multichip random-access memory having relatively low speed than CPU via the system bus. This speed disparity between processor and main memory is truly an inherent problem till today, even with tremendous advancement in technological evolution and revolution. That is why, most computers nowadays use a comparatively high-speed, smaller-capacity cache memory placed in between the CPU and the main memory in order to reduce the repeated slower main memory visits required by the CPU. The CPU communicates directly with I/O devices, or I/O modules connected with I/O devices, in much the same way as it communicates with external memory via the system bus.

3.6 CPU Operation: Instruction Execution

The CPU executes each instruction in a series of small steps:

1. ***Fetch*** the current instruction pointed by the PC from memory into the IR;
2. Change the PC to point to the next instruction in memory;

3. ***Decode*** the instruction just fetched, to determine its type;
4. If the instruction uses data in memory, determine where they are;
5. Fetch the data, if any, into internal CPU registers Data Register (DR);
6. ***Execute*** the instruction;
7. Store the result in the appropriate place;
8. Go to step 1 to begin executing the next instruction.

This sequence of steps (micro-operations) is central to the operations of all types of CPUs and is frequently referred to as the ***fetch–decode–execute cycle*** or ***instruction cycle***. However, the exact sequence of events during an instruction cycle depends to a large extent on the design of the processor. Moreover, not every step in the instruction cycle is needed by each instruction for its execution. In fact, the sequence of steps to be followed for the execution of an instruction may vary, which entirely depends on the type of instruction being fetched. A check for pending interrupt requests is also usually included in the instruction cycle.

During an instruction cycle, the action of the CPU is defined by the sequence of micro-operations it executes. The time required by the CPU to execute the *shortest* well-defined micro-operation is the CPU ***cycle time*** or ***clock period*** T clock, which is a basic unit of time for measuring CPU actions. It should be noted that the number of CPU cycles required to completely process an instruction varies with the instruction type as well as to the extent to which the processing of an individual instructions can be overlapped. At this moment, we are strictly assuming that each instruction in sequence is fetched, decoded, and finally executed.

A brief detail of the instruction cycle along with a flow chart of the steps being followed is given in the website: http://routledge.com/9780367255732.

3.7 Instruction Set

Each operation of a CPU is determined by the instruction(s) it executes. These primitive instructions of the computer are referred to as ***machine instructions*** or ***computer instructions***. The collection of these different instructions that the CPU can execute is referred to as the CPU's ***instruction set*** that precisely categorizes a CPU and signifies its capability and versatility. This set should be *complete, efficient*, and above all, *easy to use*; fully satisfy the performance requirements of the computer; and be directly executed by the electronic circuits built in the form of *gates*.

3.7.1 Machine Instruction Elements

Each machine instruction specifies some particular operation, and hence, each instruction must contain sufficient information so that the CPU can execute the operation embedded in the instruction. The elements of a machine instruction are as follows:

- ***Operation code*** tells what action (operation) to be performed (e.g. ADD LOAD, etc.), and is specified by a unique binary code, known as *operation code* or *opcode*;
- ***Source operand*** are those that are input for the operation;

- ***Result operand*** The operation may produce a result after execution;
- ***Mode reference*** specifies the means (way) by which the operand can be accessed and obtained;
- ***Next instruction reference*** provides the CPU with the information of the next instruction to be fetched when the execution of the current instruction is completed.

Source operand and result operand, however, can be obtained in one of ***three*** places, namely *main memory* or *virtual memory*, *CPU register*, and *I/O device* module.

3.7.2 Instruction Formats and Design Criteria

Each (machine) instruction is divided into groups of bits called the *fields*, indicating the constituent elements forming the instruction. The layout of the instruction, or the way the instruction is expressed, is known as the *instruction format*. It is usually depicted in the form of a rectangular box symbolizing the bits of the instruction as they appear in memory words or in a control register. Three different types of instruction format exist: each must include an opcode (implicitly or explicitly), and zero, one, or more operands, as is shown in Figure 3.4.

Opcodes are generally expressed in *symbolic representation* (called a *mnemonic*). Each explicit operand is referenced using one of the addressing modes (to be discussed later), and the format must include implicitly or explicitly the addressing mode for each operand. Operands residing in *memory* are specified by their *memory address*. Operands residing in *processor registers* are specified with a register address, and lesser number of bits is required to address a register operand than an operand rȩsiding in memory. A *register address* in a general-purpose register computer is a binary number of k-bits that defines one of 2^k registers in the CPU. If a machine has 16 ($16 = 2^4$) registers, then 4 bits are required to address an individual register.

The design of an instruction format is a major issue and equally complex to realize due to the presence of many deciding factors, such as processor complexity, number of available processor registers, processor speed, memory characteristics and size, memory organisation, memory-transfer rate, bus structures and similar others , that need to be taken into account. The designer thus always negotiates with a lot of competing and conflicting factors while developing machine instructions and its formats, in which each factor appears to be equally significant. Therefore, it is the trade-off that will ultimately decide which factor will receive more importance to fit in the underlying design to achieve the ultimate target. Accordingly, a compromise may be needed on some factors, to strike a reasonable balance in between. It is thus found really critical to decide the effective instruction format length that can be efficiently utilized. In fact, an instruction set used in a computer will normally have a variety of instruction formats of different lengths to realize the ultimate objectives.

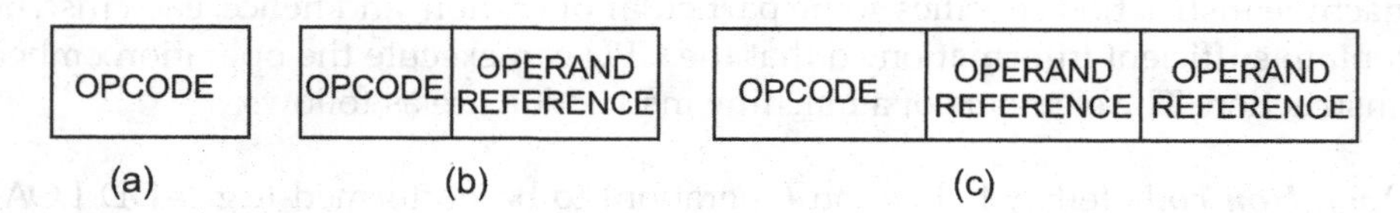

FIGURE 3.4
A simple instruction format with three different types: (a) zero operand (b) one operand (c) more operands.

A brief detail of instruction formats and design criteria is given in the website: http://routledge.com/9780367255732.

3.8 Types of Operands

Various operations in machine instructions are executed on different types of data. Usually, the type of a unit of data to be used is determined by the operation to be performed on it. The most important general categories of data used as operand are as follows:

i. Numbers;
ii. Characters;
iii. Logical data;
iv. Addresses;
v. String.

Apart from the above-mentioned data types, some machines also support dedicated specialized data types or data structures used for special types of operations.

Numbers are used as operands in all machine languages, and most computers commonly use *three types* of numerical data, namely (i) integer or fixed point, (ii) floating point, and (iii) decimal, represented in binary form (machine's internal form) in the defined instruction formats. **Character**-type operands contain textual data or strings of characters that need some form of internal representation using a variety of suitable codes (*ASCII, EBCDIC, BCD*, etc.) within the machine in the form of a sequence of fixed number of bits. Each character in this code is represented by a byte (8 bits) in which a unique pattern of 7 bits, i.e. a total of $2^7 = 128$ different characters comprising numeric characters (0–9), alphabetic characters (A–Z, and a–z),other special characters (+, *, &, @, etc.),and also some of the patterns of these 128 different types, are used to represent *control characters* for print control, communication procedures, and similar other purposes. The eighth bit in an 8-bit pattern of character representation is used as a *parity bit* for error detection or for other purposes. **Logical data** is a single unit of data, whether of numeric type or of character type, and is basically a set of bits. Various logical operations can be performed for manipulating individual bits (a word or other addressable unit) of such data, often referred to as *bit twiddling*. **Addresses** are unsigned data also used as operand in an instruction that do not participate directly in the operation indicated in the instruction; rather, it points to the location (i.e. indicating the address) where the actual value of the operand can be obtained. In many cases, some extra calculations must be computed to determine the actual address (location). Some machines even use different types of opcodes to indicate the operand as containing an address, and not data to be used directly in the operation. **String** is a set of characters of variable lengths used in many applications. Successive byte locations contain successive characters of the string. The beginning of the string is indicated by giving the address of the byte containing its first character. There are usually two ways to indicate the length (or end) of the string. A special control character representing the meaning "end of string" can be used as the last character in the string to signify the end of the string, or a separate memory word location or a processor register can be employed, which contains a number that indicates the length of the string in bytes.

3.9 Intel X-86 (IA-32 and IA-64) Data Types

Intel X-86 processors deal with numerous data types of different lengths as addressable units, such as 8 (bytes), 16 (word), 32 (doubleword), 64 (quadword), and 128 (double quad-word) bits in length using ASCII codes. All X-86 processors of Intel while storing data in memory always use the ***little-endian*** (for more details about endianness, see Chapter 4) convention of byte numbering in memory. Moreover, at the time of storing data of any type, of any length unit, there is no need to maintain the respective boundary alignments (i.e. storing of words does not require any alignment at even-numbered addresses; double-words need not be aligned at addresses evenly divisible by 4, and so on), which eventually facilitates efficient memory utilization and offers tremendous flexibility in organizing data structures. Apart from the general data types of different lengths as described earlier, the X-86 also provides an impressive array of *specific data types* that are identified and operated on by some particular instructions.

The numerical data types supported by X-86 for signed integers are represented in two's complement representations (see Chapter 7) and may be 16, 32, or 64 bits long. The floating-point data type, however, refers to a variety of types with different ranges of precision (see Chapter 7). All the numerical data types used in IA-32 and IA-64, however, conform to IEEE 754 standards. The characters used here are usually letters of the alphabet, decimal digits, punctuation marks, and so on, apart from other types of characters known as *special characters* and *control characters* used for specific purposes. The later versions of the Pentium line of processors virtually provide SIMD (see Chapter 10) operations especially for MMX applications using packed SIMD data types, a *derived data types* in which multiple operands are packed into a single-referenced memory item, and these multiple operands within an item are then executed in parallel by a single instruction that eventually gives a flavour of SIMD operations. Some of the commonly used packed data types of this form are as follows:

- **Packed byte and packed byte integer**: 8-bit bytes are packed into a 64-bit quad-word or 128-bit double quadword, interpreted as a bit field or as an integer.
- **Packed word and packed word integer**: 16-bit words are packed into a 64-bit quadword or 128-bit double quadword, interpreted as a bit field or as an integer.
- **Packed doubleword and packed doubleword integer**: 32-bit doublewords are packed into a 64-bit quadword or 128-bit double quadword, interpreted as a bit field or as an integer.
- **Packed quadword and packed quadword integer**: Two 64-bit quadwords are packed into a 128-bit double quadword, interpreted as a bit field or as an integer.
- **Packed single-precision floatingpoint and packed double-precision floating-point**: Four 32-bit floating-point or two 64-bit floating-point values are packed into a 128-bit double quadword, interpreted as a floating-point number.

3.10 Types of Instructions and Related Operations

Each instruction in the instruction set specifies some operation indicated by the *opcode,* included as one of the fields in the instruction. The number of different types of operations (expressed by opcodes) available in a machine varies widely from machine to machine, and this is considered as one of the important criteria of a machine design decision. However, we concentrate here only on some general types of operations that are found on almost all machines. The different types of commonly used operations that most CPUs perform can be typically categorized as:

i. Arithmetic;
ii. Logical;
iii. Shift operation;
iv. Data transfer;
v. I/O;
vi. Conversion;
vii. Transfer of control;
viii. System control.

3.10.1 Arithmetic

Basic arithmetic operations such as add, subtract, multiply, and divide are ALU functions; are mainly provided with operands having signed/unsigned integer (fixed-point) numbers, floating-point data, binary data, or decimal data; and are often applicable to *single-precision, double-precision,* or *extended-precision* data for floating-point numbers.

In case of ***accumulator-type organisation,*** an arithmetic addition may be defined as ADD X, where X is the address of the operand. The ADD instruction, in this case, results in the implicit accumulator operation, like AC ← AC + M [X], where M [X] is the content of memory word located at address X. This is an example of *one-operand* instruction.

In case of ***general register-type organisation,*** an arithmetic addition may be defined as ADD R1, R2; this actually denotes the operation: R1 ← R1 + R2; here, the destination operand (register R1) is the same as one of the source operands (register R1). This is an example of *two-operand* instruction. The same instruction can be written in another form using different source operands and destination operand like ADD R1, R2, R3; this denotes the operation R1 ← R2 + R3. This is an example of *three-operand* instruction.

In case of ***stack-organised CPU,*** an arithmetic addition may be simply defined as ADD. In stack-organised computers, operation-type instruction needs no address field, because here the operation is performed on the implicit two operands that are the top two elements of the stack. These two top elements will be popped off the stack: the addition operation will be performed and the sum will then be pushed into the TOS. All the operands are implied to be in the stack. This is an example of *zero-operand* instruction.

In addition to the usual arithmetic instructions for signed integer numbers, Intel IA-32 architecture incorporates instructions for integer multiply and divide, as well as instructions for operations on floating-point numbers.

For details, see the website: http://routledge.com/9780367255732.

3.10.2 Logical

Most of the machines have a variety of operations for manipulating individual bits of data (a *byte,* a *word,* or other *addressable unit*). These operations, known as *logical operations,* are based upon *Boolean operations.* Some of these operations listed below can be performed on Boolean or binary data.

A	B	NOT A	NOT B	A AND B	A OR B	A XOR B	A = B
0	0	1	1	0	0	0	1
0	1	1	0	0	1	1	0
1	0	0	1	0	1	1	0
1	1	0	0	1	1	0	1

Similar to arithmetic operations, all types of logical operations may also include data transfer operations before the actual action of ALU begins.

Some typical and most commonly used logical and bit-manipulation instructions are given in the website: http://routledge.com/9780367255732.

3.10.3 Shift Operation

In addition to bit-wise logical operations, most machines offer a variety of *shifting* and *rotating* functions. With a ***logical shift***, the bits of a word are shifted *left* or *right*. Whatever be the type of shift, on the one end the bit shifted out is lost, and on the other end a 0 is injected in for padding. Logical shift can be primarily used for isolating fields within a word. The 0s that are injected into a word displace undesirable information which is shifted out at the other end. The operation codes SHL and SHR are normally used to indicate logical shift-left and shift-right operations, respectively. Figure 3.5 illustrates such logical-shift operations for a better understanding.

The ***arithmetic shift*** operation treats the data as a signed integer, shifts the data to the left or right, and does not shift the sign bit. On an arithmetic right-shift (operation code SHRA),it is necessary that the sign bit in the leftmost position remains unchanged and the sign bit is normally replicated into the bit position to its right as depicted in Figure 3.6. With numbers represented in 2's complement notation, an arithmetic left-shift (operation code SHLA) or SHRA corresponds to multiplication or division by 2, respectively, provided there would be no occurrence of an overflow or underflow situation. The large IBM S/390 system provides this instruction, but many processors, such as Intel Itanium and IBM PowerPC, do not include this instruction.

The ***circular (rotate) shift*** preserves all of the bits being operated on. It circulates the bits stored in the register around the two ends without the loss of information, as shown

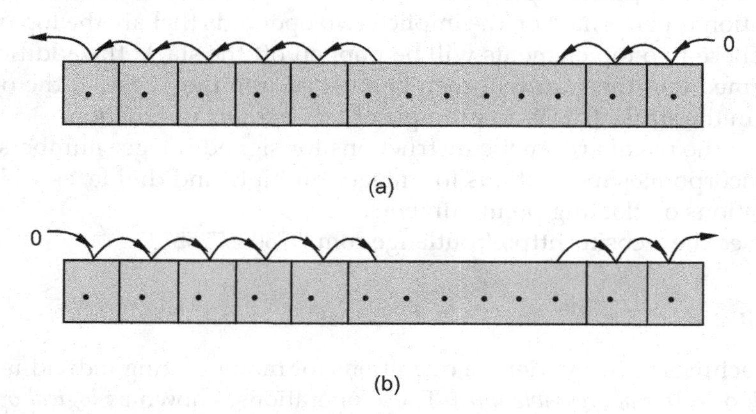

FIGURE 3.5
Logical shift operation. (a) Logical left shift and (b) logical right shift.

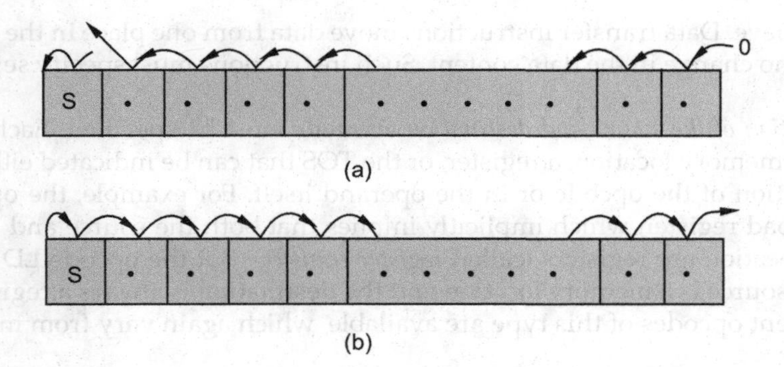

FIGURE 3.6
Arithmetic shift operation. (a) Arithmetic left shift and (b) arithmetic right shift.

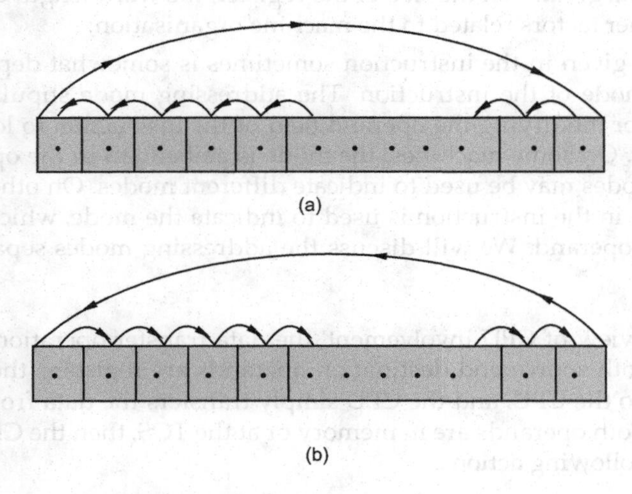

FIGURE 3.7
Rotate (circular) shift operation. (a) Left rotate and (b) right rotate.

in Figure 3.7. This is accomplished by connecting the serial output of the shift register to its serial input. Many useful actions employ the *shift rotate* operation (ROR for rotate right, and ROL for rotate left) as a part of their activities.

Similar to the arithmetic operation, all types of shift operations may involve data transfer operations before the commencement of actual ALU operation. Many machines have different types of the most commonly-used shift operations and their corresponding instructions. However, mnemonic names may differ on different machines for the same type of instructions.

A list of a few of these operations with examples is given in the website: http://routledge.com/9780367255732.

3.10.4 Data Transfer

The most fundamental and major machine instruction type available in any instruction set of a computer is *data transfer* instructions. In fact, the kind of data transfer instructions to be included in an instruction set indicates the kind of trade-offs that the designer intends to achieve. Data transfer instructions move data from one place in the computer to another with no change in the data content. Such instructions must specify several things:

i. The *location of the source and destination operands* must be specified. Each location can be a memory location, a register, or the TOS that can be indicated either in the specification of the opcode or in the operand itself. For example, the opcode LR means load register, which implicitly implies that both the source and the destination location are registers (called *register transfer*). But the opcode LD indicates that the source is a memory location and the destination is always a register. A lot of different opcodes of this type are available, which again vary from machine to machine;
ii. The *length of data* to be transferred must be specified. Again, this may be specified either in the specification of the opcode or in the operand itself. The length specification depends greatly on the size of the register, the word length of the memory, and some other factors related to the machine organisation;
iii. The operand given in the instruction sometimes is somewhat dependent on the addressing mode of the instruction. The addressing mode stipulates a rule for interpreting or modifying the operand field of the instruction to locate the operand to access. On some machines, the mode is embedded in the opcode itself, or different opcodes may be used to indicate different modes. On other machines, a separate field in the instruction is used to indicate the mode, which is then used to locate the operand. We will discuss the addressing modes separately in later sections.

From the point of view of CPU involvement, the data transfer operations seem to be the simplest type. If both source and destination operands are registers, then this operation is totally internal to the CPU, and the CPU simply transfers the data from one register to another. If one or both operands are in memory or at the TOS, then the CPU must perform some or all of the following actions:

- Calculate the actual memory address. Determine whether the address is in cache; if so, the operand is accessed. If not, issue the command to memory module;
- If the operand refers to virtual memory, translate the address from virtual address to actual main memory address and issue a command to memory module.

A list of the most commonly used data transfer instructions that are found in many machines is given below to give an idea of these types of instructions.

Operation	Mnemonic	Description
Load	LD	Transfer word from memory to processor.
Store	ST	Transfer word from processor to memory.
Move	MOV	Transfer word or block from source to destination.
Exchange	XCH	Swap contents of source and destination.
Clear	CLR	Transfer word of 0's to destination.
Set	SET	Transfer word of 1's to destination.
Push	PUSH	Transfer word from source to the TOS
Pop	POP	Transfer word from the TOS to destination.

3.10.5 Input/Output (I/O)

A variety of I/O instructions are available in different machines, depending mostly on the approaches taken in I/O organisation, such as *programmed I/O, DMA,* and above all, the availability of a *stand-alone separate I/O processor.* The I/O instructions ultimately transfer data between the processor registers and I/O modules or terminals. A list of the most commonly used I/O instructions that are found in many machines is given here for a clear understanding and ready reference.

Operation	Mnemonic	Description
Input (read)	IN	Transfer data from specified I/O port or device to destination, like memory, register.
Output (write)	OUT	Transfer data from specified source to I/O port or device.
Start I/O	SIO	Transfer instructions to I/O processor to initiate I/O operation.
Test I/O	TIO	Transfer status information from I/O system to specified destination.

3.10.6 Transfer of Control

In the normal course of execution, the instructions are fetched one after another from consecutive memory locations in sequence using the PC and then are executed by the CPU. However, in any program, there are a good number of instructions which when executed change the sequence of instruction execution by modifying the PC to contain the address of some other desired instruction located in memory. There exist, of course, certain substantial reasons as to why transfer-of-control operations are really required. The most common types of transfer-of-control operations found in the instruction set of many machines are as follows:

- Branch instruction;
- Skip instruction;
- Subroutine call instruction.

3.10.6.1 Branch Instructions

A branch instruction, sometimes called a *jump instruction,* has one of its operands indicating the address of the next instruction to be executed. Branch instructions are mainly of two categories: a *conditional branch* instruction, when the branch is taken (i.e. the PC is updated with the address specified in operand) only if a certain condition is met; otherwise, the next instruction in sequence is executed as usual (natural increment of PC). *Unconditional branch* instruction (BR or JMP) performs a jump to the target instruction without considering anything. A branch can be either to an instruction in the *forward direction* (i.e. an instruction with a higher address which means the downward direction of the program) or to an instruction in the *backward direction* (i.e. an instruction with a lower address which means the upward direction of the program).

Conditional branch is implemented using different strategies on different machines. One way is to provide an additional 1-bit or multiple-bit *condition code* that is set according to the result of some operations. As an example, a machine can implement an arithmetic operation (ADD, SUBTRACT, etc.) using a 2-bit *condition code* to be set with one of the four following values depending on the result of instruction execution: *0, positive, negative,* and *overflow.* This requires four different conditional instructions, namely:

BRZ Q	Branch to location Q if the result is zero.
BRP Q	Branch to location Q if the result is positive.
BRN Q	Branch to location Q if the result is negative.
BRO Q	Branch to location Q if overflow occurs.

The other approach also found in use is with a three-address instruction format. Here, both the comparison and the target instruction address are given in the same instruction. For example,

BRE R2, R5, X: Branch to X, if contents of R2 = contents of R5.

Different machines use different mnemonic names for same type of actions, and many machines even have varieties of conditional branch instructions.

3.10.6.2 Skip Instruction

Skip instruction is another common form of transfer-of-control instructions. The skip instruction does not require a destination address field; this includes an implied address. Typically, the skip implies that one instruction be skipped; thus, the implied address is the current content of PC plus the length of one instruction. Many machines have and use different types of skip instructions, and usually, SKP (skip) mnemonic is used. A typical example of this type of instruction is ISZ (*I*ncrement and *S*kip if *Z*ero).

ISZ R5	This means that R5 is incremented. If it is not zero, the next instruction by default will be executed. If it is zero, the next of next instruction will be executed.

3.10.6.3 Subroutine Call Instruction

One of the finest innovative approaches available in programming languages is the provision of *subroutine*, which is a self-contained sequence of instructions in the form of a computer program that performs a given computational task. It is incorporated into a larger main program, and may be *invoked* or *called* many times at various points in the main program to perform its specified function. Each time a subroutine is called (or invoked), a branch action is performed to arrive at the beginning of the subroutine to start executing its set of instructions. After the execution of the subroutine, a branch is again taken to come back (*return*) to the next instruction in the main program from where the call took place. Both of these are forms of branching instructions. This mechanism is illustrated in Figure 3.8.

The instruction used in a main program to transfer control to a subroutine is known by different names such as *call subroutine* (CALL), *go to subroutine* (GOSUB), *jump to subroutine, branch to subroutine, branch and save address,* or *branch and link* (*BL*). The return from the subroutine is accomplished usually by a RETURN (or RET) statement used as the last instruction of the subroutine. Different machines, however, use different mnemonic names for subroutine call instructions. A few of them are as follows:

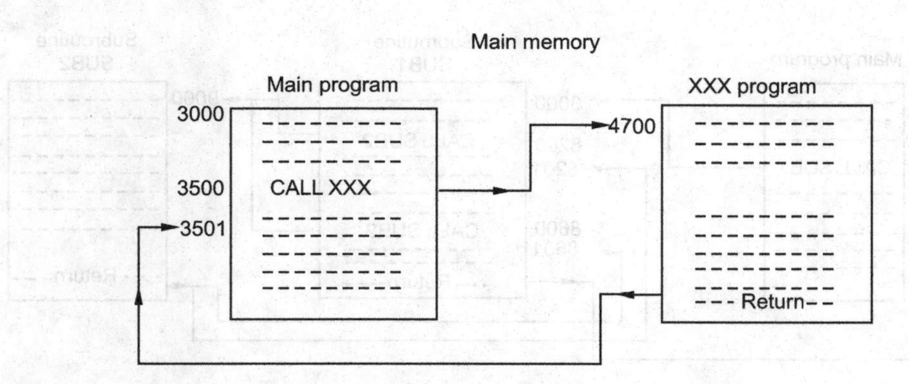

FIGURE 3.8
Subroutine call mechanism.

Operation	Mnemonic	Description
Call jump to subroutine	CALL/ENTER GOSUB	Place current program control information to known location (stack); jump to the subroutine specified address.
Return	RET/LEAVE	Replace (restore) contents of PC and other registers from the known location (stack).

In case of a subroutine call, several points are worthy to be mentioned:

i. Subroutine can be called many times;
ii. A subroutine can be called from more than one location in the main program;
iii. A subroutine call can appear in a subroutine itself. This is known as the *nested subroutine*. The *nesting* of subroutine can go to an arbitrary depth, of course, to a specified limit as already delineated;
iv. Each subroutine *call* is matched with a corresponding *return* in the called program.

The nested subroutine call mechanism is shown in Figure 3.9. The main program includes a call to subroutine SUB1 located at address 8000. When this call instruction is encountered and executed, the CPU suspends execution of the main program and begins execution of SUB1 by fetching the next instruction from location 8000. Within SUB1, there are two calls to SUB2 located at address 9000. In each case, when the call occurs, the execution of SUB1 is suspended and execution of SUB2 then begins. The RETURN statement in each subroutine causes the CPU to go back to the calling program and continue execution at the next instruction after the corresponding CALL instruction.

A subroutine call instruction in the main program consists of an operation code together with an address that specifies the beginning of the subroutine being called. This call instruction is executed performing in effect, mainly two operations:

i) The address of the next instruction of the call in the calling program available in the PC along with other relevant information is stored in a temporary location (stack) so that the subroutine knows where to return in the calling program after the completion of its own execution. This is called the *return address*;
ii) Program control is then transferred to the beginning of the subroutine.

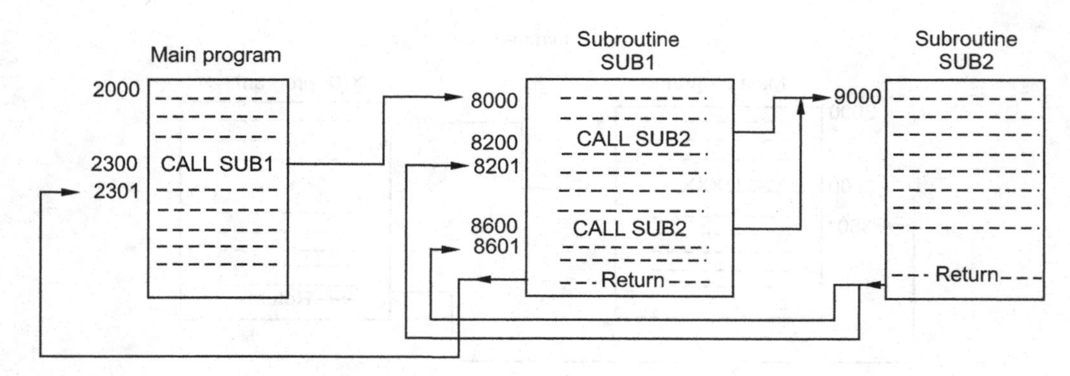

FIGURE 3.9
Nested subroutine call mechanism.

The last instruction of every subroutine is commonly called the *return from subroutine.* This instruction when executed transfers the return address (already stored) from the temporary location into the PC. Other relevant information that was already stored is also now reloaded from the temporary location to the respective place in the main program to enable the main program to continue its own execution. The CPU now reads the PC that causes a transfer of program control to the instruction next of the call instruction in the main program. Consequently, the execution of the main program is once again resumed.

Use of stack: This problematic situation of storing the return addresses can be avoided if different storage locations are employed for each use of the subroutine while another lighter-level (most recent call) use is still active. A more general and powerful approach for storing the return addresses is to use a ***memory stack***. When the CPU executes a call, it places the return address on the stack. When it executes a return, it uses the information located on the TOS. The current return address will always be at the TOS. The advantage of using a stack to store the return addresses is that when a number of subroutines are called in succession, all the respective return addresses can be pushed in sequence, one after another, onto the stack. The return instruction being used at the end of the subroutine causes the stack to pop the contents of the TOS, and this value is always the one that is then transferred to the PC. In this way, the return is always to the program which is most recently (i.e. the last one) called a *subroutine.* Figure 3.10 illustrates the use of a stack for the example as already depicted in Figure 3.9. A *subroutine call* is, however, implemented with the following machine instructions:

SP←SP–1: Decrement SP to point to next available space.

M[SP]←PC: Push content of PC on to the stack.

PC←Subroutine address: Transfer control to the subroutine.

If another subroutine is called by the current subroutine, the new return address is pushed onto the stack in the same way which will remain at the TOS and so on. The instruction (Return) that returns from the last called *subroutine* is implemented by way of machine instructions, such as:

PC←M[SP]: Pop stack (the top element of stack) and transfer to PC.

SP←SP + 1: Increment SP to access next item.

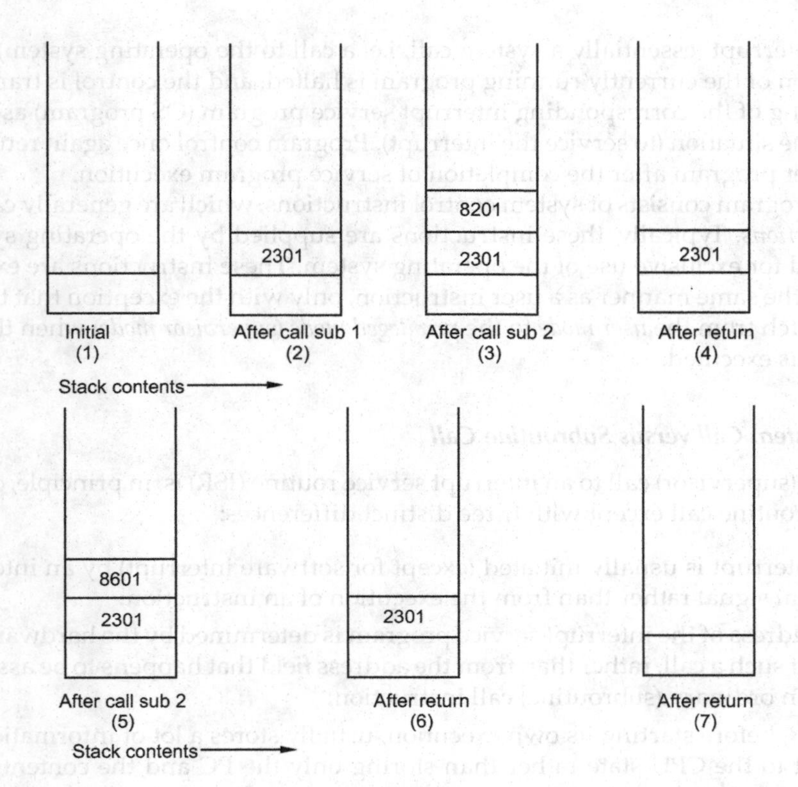

FIGURE 3.10
Use of stack to handle nested subroutine of Figure 3.9.

Apart from storing the return address, the stack implementation is also found to be a flexible approach, even in the event of *parameter passing* at the time of a call. When the CPU executes a call, it not only stores the return address onto the stack, but also stores in the stack the parameters issued by the calling procedure, to be passed to the called procedure. The called procedure is aware of this fact and can access the parameters from the stack. Upon return, the called procedure can also place the return parameters on the stack, *under* the return address. While a procedure is invoked (called), the entire set of parameters, including the contents of the register under use as well as the return address stored on a stack, are often collectively referred to as a ***stack frame***, and this is automatically handled by the hardware in one unit and the user need not be concerned in this regard.

In the Intel ***IA-32 architecture***, the processor stack data structure is used for the sake of convenience, to handle entry to and return from subroutines using some specified registers (ESP, EAX, EDI, etc.), and respective stack instructions to handle the stack.

A brief detail of different types of transfer of control instructions is given in the website: http://routledge.com/9780367255732.

3.10.7 System Control

A *user program* while currently running sometimes requires the assistance of a service program (operating system program), as a result of an external- or internal-generated request. This request is issued in the form of an ***interrupt***, which is essentially an unusual event.

When an interrupt (essentially a system call, i.e. a call to the operating system) is issued, the execution of the currently running program is halted, and the control is transferred to the beginning of the corresponding interrupt service program (OS program) as requested to resolve the situation (to service the interrupt). Program control once again returns to the original user program after the completion of service program execution.

Service program consists of system control instructions, which are generally called *privileged instructions.* Typically, these instructions are supplied by the operating system and are reserved for exclusive use of the operating system. These instructions are executed by the CPU in the same manner as a user instruction, only with the exception that the processor will switch from the *user mode* to the *privileged mode* (*supervisor mode*) when this type of instruction is executed.

3.10.7.1 System Call versus Subroutine Call

The system (supervisor) call to an interrupt service routine (ISR) is, in principle, quite similar to a subroutine call except with three distinct differences:

i. The interrupt is usually initiated (except for software interrupt) by an internal or external signal rather than from the execution of an instruction;
ii. The address of the interrupt service program is determined by the hardware at the time of such a call, rather than from the address field that happens to be associated with an ordinary (subroutine) call instruction;
iii. An ISR, before starting its own execution, usually stores a lot of information with regard to the CPU state rather than storing only the PC and the contents of the other registers.

The hardware procedure for servicing an interrupt is very similar to the procedure being used in handling subroutine call instructions. Here also, the stack approach is used for storing necessary information before jumping to the ISR. The additional information that is to be saved is the *PSW/PCB*, which keeps track of the CPU status at that moment. Hence, when the last instruction in the service program, *return from interrupt*, is executed, the stack is popped to retrieve the old PSW/PCB as well as the return address. This PSW/PCB is then transferred to the System Register (SR), and the return address to the PC. Thus, the CPU state is restored once again to user mode and the original program execution can now be resumed as if nothing has happened.

Different machines do have numerous types of system control operations depending on their organisation, especially the CPU organisation. The amount of information to be saved, and the way the control operation would be handled, is a critical issue which is primarily decided at the time of CPU design and subsequently in its implementation. Besides, a lot of other types of machine instructions such as HALT, WAIT, NOOP, etc. are also available in many machines. To what extent the various types of machine operations are to be included in the *instruction set* is the trade-off, and one of the major criteria in CPU design which again depends on the target that the CPU intends to attain.

3.10.8 Other Operations: IA-32 Instruction Set

Only a small set of common operations and the corresponding instructions to realize them that are available in a generic instruction set have been described. However, there may still be other useful operations and related instructions in the instruction set of any particular processor with suitable hardware to attain certain specific functional objectives. Here, we will describe a few more important instructions that are available in the instruction set of the **IA-32 architecture**, in particular.

Memory handling: The basic IA-32 instruction set contains some specialized instructions to realize memory segmentation activities related to memory handling. These are privileged instructions that can only be executed in supervisor mode from the OS level. They perform loading and reading local and global segment tables (called *descriptor tables*) as well as checking and modifying the privilege level of a segment.

3.10.8.1 MMX (Multimedia Extension) Operation

The MMX technology is used for multimedia applications that deal with an array of a large number of *pixel* (picture element), the smallest element of a video (digital) image that can be assigned an individual dot in a dot–matrix representation of a picture. Each pixel is represented by 8 bits, its colour component (red, green, and blue) comprises 8-bit integer data item, and the brightness is also an 8-bit integer data item. A pixel of an image thus may be represented by a 24-bit quantity. Multimedia applications, however, involve the manipulation of individual pixel, matrix multiplication and matrix convolution type of operations, and also operations on multiple numbers of pixels simultaneously. The same characteristics, however, also apply to sampled audio signals or speech processing, where a sequence of *signed digital numbers* represents the samples of a continuous analog audio signal taken at periodic intervals. All these features together require an SIMD kind of CPU architecture for needed processing of a multimedia application.

The **IA-32** MMX technology adds 57 new specific instructions known as *enhanced MMX instructions* to the instruction set of X-86 processors for performing multimedia tasks. MMX instructions operate on large arrays of small-packed audio and video data types, typically of 8 or 16 bits (see Section 3.8). Typical analog audio signal samples are of 16-bit units. Since the MMX registers are all 64 bits long, one can pack a total of 8 or 4 pixel values (each 8-bit or 16-bit integer) in a single register that can be used as a single-referenced memory item packed of these multiple operands (pixels) in the MMX instructions. These multiple operands packed within an item can then be loaded simultaneously and are then executed in parallel by a single instruction that eventually gives a SIMD operation, which is ideally suited for the integer pipeline of Pentium architecture. These fast parallel operations can yield an approximate speed-up of two to even eight times over comparable algorithms that do not use the MMX instructions (Atkins, 96). Each instruction typically takes a *single clock cycle* to execute. The operands on which these instructions are executed can be in memory or in the eight floating-point registers (see Section 3.2.1) referred to as *MM0* through *MM7*. With the introduction of **IA-64** architecture, Intel has further expanded this extension to include double quadword (128 bits) operands and floating-point operations. The additional new data types as included within the existing one are as follows:

- **Packed quadword**: Two 64-bit quadwords are packed into a 128-bit double quadword.
- **Packed single-precision floating point and packed double-precision floating-point**: Four 32-bit floating-point or two 64-bit floating-point values are packed into a double quadword, interpreted as a floating-point number.

In ARM (Acorn RISC Machines) RISC processor, apart from having different types of ADD instruction, there are a set of **parallel addition and subtraction instructions**, in which portions of two operands are operated on in parallel. For example, ADD16 instruction adds the top half-words of two registers to form the top half-word of the result, and adds the bottom half-words of the same two registers to form the bottom half-word of the result. This type of instruction is in extensive use in MMX applications.

A representative list (not an exhaustive one) of different types of instructions of IA-32 MMX instruction set is shown in the website: http://routledge.com/9780367255732 (Table 3.1).

The saturation arithmetic with an example showing its impact is depicted in the website: http://routledge.com/9780367255732.

A real-life example with MMX instruction is depicted in the website: http://routledge.com/9780367255732.

3.10.8.2 Streaming SIMD Extension (SSE)

Since MMX instructions use eight floating-point registers, it is hence not possible to execute an MMX instruction and a floating-point execution simultaneously. But many multimedia operations, like in video processing, require simultaneous operations involving floating-point numbers, and MMX instructions are then clearly disadvantageous. Moreover, since MMX instructions are executed using only the floating-point registers, a large number of processor clock cycles are unnecessarily consumed for context switching from the state of executing MMX instructions to the state of executing floating-point operations, and vice versa. This also includes, in addition, few cycles to initialize these registers as required, by executing additional instructions, for subsequent use. To get rid of these shortcomings, there was an urgent necessity to extend the MMX instruction set to include floating-point instructions. This extended instruction set called *streaming SIMD extensions or SSE instructions* has been used in Pentium III, and SSE instruction set has further been enhanced for use in Pentium 4.

The SSE instructions are SIMD instructions for single-precision floating-point numbers operated on *four* 32-bit floating points concurrently. To execute these instructions, a set of *eight* 128-bit *new registers*–named as *xmm0 through xmm7* – have been specifically defined for SSE, and each of them can hold *four* 32-bit single-precision floating-point numbers. Since different registers have been allocated, it is now possible to execute both fixed-point MMX instructions and floating-point operations simultaneously without unnecessarily consuming lot of useful cycles. These SSE instructions can also execute non-SIMD floating-point and SIMD floating-point instructions simultaneously. In memory streaming instruction executions, the data is pre-fetched into a specified level of the cache hierarchy, and prefetching this type of data into the L2 cache of Pentium III and Pentium 4 is certainly an effective way that summarily improves the memory system performance as a whole.

SSE instruction set has been further enhanced, keeping in view the added capabilities and the architectural changes incorporated within the **Pentium 4**. The innovative Pentium 4 NetBurst microarchitecture introduces Internet **SSE2** instructions, which are essentially an extension of existing SSE by adding 144 new instructions. This new instruction set, however, increases the accuracy of the 128-bit SIMD double-precision floating-point operations, supports new formats of packed data (new data types),and increases the speed of manipulation of 128-bit SIMD integer operations. With the introduction of SSE2, Intel has ultimately extended its SIMD capabilities that MMX technology and SSE technology together already delivered.

The next-generation 90-nm process-based Pentium 4 processor introduces the next version **SSE3** in 2004 when they released their next version of Pentium 4, the Prescot. The SSE3 instruction set includes 13 additional SIMD instructions over SSE2 that comprise five different types of instructions, namely *floating-point-to-integer conversion, complex arithmetic operations, video encoding, SIMD floating-point operations using array of structures format,* and *thread synchronization.* These additional instructions are mainly aimed towards enhancing 3D graphics, video, and multimedia applications, and some of them will be useful for improving thread synchronization. Summarily, the processor's capability is largely increased to speedily handle many important computations, including faster in-parallel floating-point computations required for 3D graphics applications, multimedia, and gaming.

3.11 Instruction Addressing Scheme

Different machines do have different CPU organisations with various types of machine instructions having different lengths (format) containing varying number of operands. An operand referenced in an instruction either contains the actual value of the operand or a reference to the address of the operand. To locate where the *operands* are, is accomplished by certain means called ***addressing***. So, addressing is a mechanism by which the corresponding operand is traced. On some level 3 machines (Figure 2.1), all instructions have the same length (format); on others, there may be two or three different lengths.

- ***Zero-addressing***

  A computer having instructions with no address field in the instruction format is called a *zero-address computer.* A stack-organised computer may have instructions that do not use any address field, such as ADD MUL, etc. Operation-type instruction in this category has implicit two operands that are the top two elements of the stack. These elements are popped off the stack, the operation will be performed, and the result will then be pushed into the TOS. For example, the instruction:

  ADD: It implies that TOS ← (A + B), where A and B are taken from the TOS, and the result is pushed on the TOS.

- ***One-Addressing***

  An instruction that uses only one operand (address field) in the instruction format is called a *one-address instruction*. One-address instructions generally use an *implied accumulator (AC) register* for all data manipulation. All operations are done mainly between the AC register and a memory operand. For example, to calculate A = B + C,

```
LOAD  B    :    AC ← M[B]
ADD   C    :    AC ← AC + M[C]
STORE A    :    M[A]   ←   AC
```

  where M [X], in general, implies the content of memory location having address X or PUSH A: This implies that TOS ← A.

  Other types of one-address instructions are also available in both *general-purpose register* machines and *stack-organised machines.*

- ***Two-Addressing***

  An instruction that uses two operands (address field) in the instruction format is called *two-address instruction.* In commercial computers, two-address instructions are the most commonly used instructions. Here, each address field can specify either a processor register or a memory word or a mix of these as well as literals. For example, the MOV instruction moves or transfers the operands to and from the memory and processor registers.

```
ADD    R1, A      :    R1 ← R1+M[A]
MOV    B, R1      :    M[B] ← R1
```

Other types of two-address instructions are also available in both *general-purpose register* machine and *accumulator-based machine.*

- ***Three-Addressing***

  An instruction that can use three operands (address field) in the instruction format is called a *three-address instruction.* Computers using three-address instruction formats can use each address field to specify a processor register, or a memory operand or a mix of these as well as literals. The *advantage* of the three-address format is that it results in short programs due to having three operands in one instruction when evaluating arithmetic expressions. The *disadvantage* is that the binary-coded instructions require too many bits to specify three addresses (longer length). For example,

```
ADD    R1, R2, R3          :    R1 ← R2 + R3
ADD    R1, X, Y            :    R1 ← M[X] + M[Y]
MUL    A, R1, R2           :    M[A] ← R1 × R2
```

  Other types of three-address instructions are only available in *general-purpose register* machines. Many mainframe systems, including the IBM 360/370/390 machines and their upper versions, all have varieties of three-address instructions.

3.12 Addressing Modes

Each machine instruction contains an opcode which is executed on the operands that sometimes have limited addressing capability while addressing over a wide range, particularly, to reference a large range of locations in main memory as well as in virtual memory. That is why, the operands need to be specified by way of using varieties of addressing techniques. The different ways by which the location of an operand is specified in an instruction are commonly referred to as *addressing modes.* The different types of machine instructions available in a CPU are also determined by these various addressing modes used with the operands and the related operations present in the machine instructions. The *addressing mode* actually specifies a rule for interpreting or modifying the address field of the instruction before the operand is actually accessed. The various types of addressing modes to be used in a CPU organisation are essentially a *trade-off* criterion and mainly serve the following purposes:

- To provide the user the programming versatility, addressing flexibility, and the number of memory references such as pointers to memory location, counter for loop control, indexing of data, program relocation, and similar other facilities;
- To reduce the number of bits in the address field of the instruction.

The most common types of addressing modes are as follows:

i. Immediate;
ii. Implied;
iii. Direct;
iv. Indirect;
v. Register;
vi. Register indirect;
vii. Displacement;
viii. Stack.

Most computer architectures provide more than one of these addressing modes. The decoding phase of an instruction cycle executed by the *control unit* determines the addressing mode of the instruction and thereby the location of the operand, apart from analysing the type of the operation to be performed.

The *addressing mode* of the instruction in some computers is specified with a distinct binary code of one or more bits, called the ***mode field*** (*mod* field in Motorola 68000 series), just like the operation code in the instruction format. The value in the mode field determines the type of addressing mode to be employed. Other computers use a single binary code that designates both the operation and the mode of the instruction. Instructions may be defined with a variety of addressing modes, and sometimes, two or more operands with their distinct addressing modes are present in one instruction. The addressing modes ultimately modify the address field of the instruction to generate the *effective address* (EA) of the operand that precisely signifies either a main memory address/virtual memory address or a register in the system. The actual mapping of a virtual address on to a corresponding physical address, which is not at all visible to the user, is one of the responsibilities of the memory management unit (MMU). However, a few of the most commonly used addressing modes will now be discussed.

- ***Immediate Mode***

  The operand value is actually present in the instruction itself rather than an address or other information describing where the operand is. Immediate-mode instructions have specific use, e.g., to initialize registers or a memory operand with a constant value expressed in 2's complement form. When such an operand is loaded into a data register or in a memory variable, the sign (leftmost) bit is extended to the left to the full-data word size. However, in some cases, the immediate binary value used is interpreted as an unsigned non-negative integer. For example, the instruction:

```
MOV  R1,  21H;
copies a byte-sized constant 21H into register R1. The operand value is
available within the instruction itself, and hence, no visit to any other
location, including memory, is required here.
```

Immediate addressing has the *advantage* of not requiring an extra memory reference to fetch the operand, thus saving some cycles of time in the instruction cycle. It has the *disadvantage* of restricting the operand to a number that can fit in an address field of an instruction, which, in most instruction sets, is small compared to the word length, which consequently limits their usefulness.

The *Intel CPUs* do not have an addressing mode for immediate operands. Instead, they have a large collection of distinct instructions in which one of the operands is immediate. The **MC** ***68000*** *series*, however, has an immediate addressing mode. IBM 360/370 computers use a single binary operation code that designates both immediate mode and the operation itself in the instruction. For example, the action of *Move Immediate* instruction (MVI):

```
MVI  2,  "5";    here, 5 will be moved to register 2, and no memory
reference is required.
```

- ***Implied Mode***

  The operands in this mode are implicitly specified in the definition of the instruction. For example, the instruction *complement accumulator* in an *accumulator-based machine* is an implied-mode instruction because the operand is in the accumulator register, which is implied in the definition of the instruction. In fact, in an *accumulator-based machine*, all register reference instructions that use an accumulator are implied-mode instructions. For example:

```
ADD  X ;    where X is the address of the operand. The ADD instruction in
this case results in the operation  AC ← AC + M[X],  and the result of
the addition will be in the accumulator. The accumulator AC is implied
to be involved in the operation although it is not mentioned in the
instruction. Zero-addressing instructions in a stack-organised computer
are essentially implied-mode instructions since the operands are always
implied to be on top of the stack.
```

- ***Direct Mode (Addressing)***

  In *direct addressing*, the address of the operand is given *directly* in the operand field of the instruction, which is the EA of the memory word where the operand is actually located. The operational view of this mode is depicted in Figure 3.11. For example, in a branch-type instruction, the address field specifies the actual branch address. Here, *Effective address* (*EA*) = *A*, the *given address*.

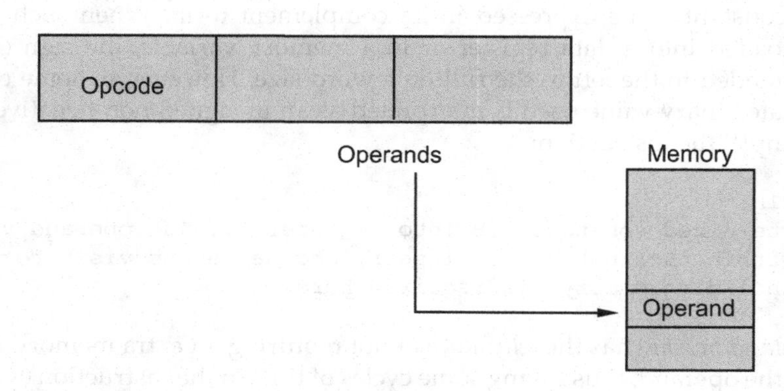

FIGURE 3.11
Direct mode addressing.

This technique is a very simple form of addressing and is found on almost all contemporary computer systems, including the advanced microprocessors of today. It requires only one memory reference to fetch each operand and no additional or special calculation to derive the address of the operand. The obvious limitation is that it covers only a limited address space since the length of the address field used is constrained by the instruction length. For example,

```
ADD  R1, X ;    where X is the address of the operand. The ADD
instruction in this case results in the operation: R1 ← R1 + M[X]; the
operand X resides in memory, and its actual address is precisely given in
the instruction. The operand can be accessed straightaway by only one
memory reference, and no other computation is required to locate X.
```

The ***Intel CPUs*** have direct addressing, but the directaddresses are too short to cover the entire address space. The ***MC 68000 series*** has *two forms* of *direct addressing*: one with 16-bit address (short form) and the other with a 32-bit address (long form). Memory addresses within the first 64 K (2^{16}) of memory can be referenced with the short form, whereas addresses above 64 K need the long form. These two forms are indicated by values in the 3-bit MOD field in the instruction format, and thus apply to all instructions with one or more operand fields.

- ***Register Direct Mode (Addressing)***

  In direct addressing, when the operands of an instruction refer to only a processor register, the instruction is said to be in the *register mode*. Thus,

```
Effective address (EA) = R, the specified register
```

For example, in the instruction:

```
MOV R1, R2 ;   the content of R2 is the operand, and the instruction
copies the word-sized content of register R2 into register R1.
```

In register addressing, the operand sometimes contains the register number that indicates one of the registers of the CPU. For example, if the content of a register address field in an instruction is 6, then register R6 is the intended address. The content of the register R6 is the value of the operand. This is illustrated in Figure 3.12. Typically, an address field that references register as operand will have from 3 to 5 bits, so that a particular register can be referenced from a total of 2^3 to 2^5, i.e. from 8 to 32 general-purpose registers. In fact, an operand that specifies k-bit field in an instruction can refer any one of 2^k registers.

The advantages of register addressing in machines designed with registers (general-purpose register machines) are as follows:

- Registers are faster in operation than the main memory;
- Only a small address field (a few bits) is needed to address a register in the instruction (length of the instruction is thus reduced);
- No time-consuming memory references (as these are slower in operation compared to CPU speed) are required.

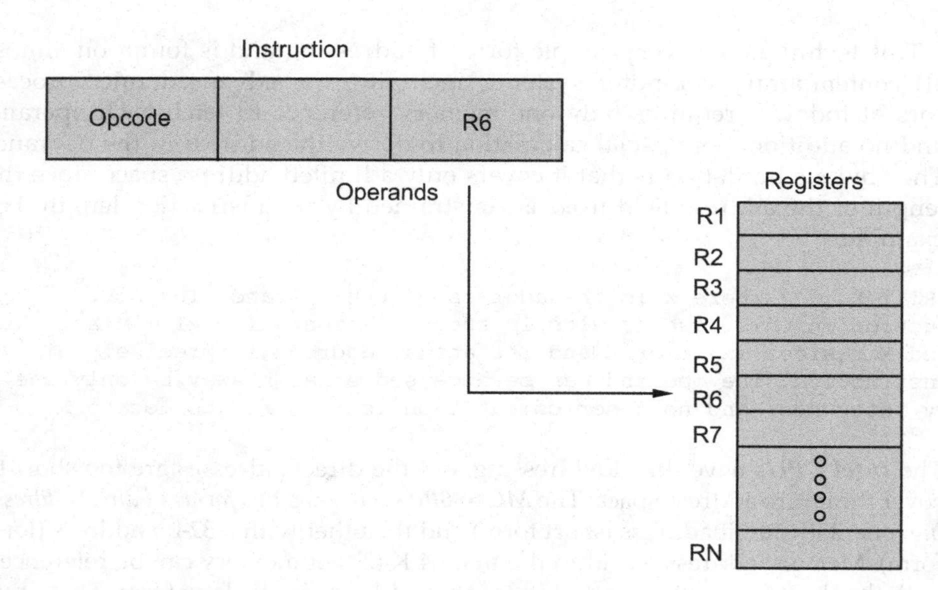

FIGURE 3.12
Register-direct addressing.

Since the number of available registers is limited (compared to the number of locations in main memory), their use in this mode makes sense only if they can be employed efficiently. If every operand is brought from main memory into a register, operated on once, and then returned to main memory once again, then a wasteful intermediate step has been added. Instead, if the operand in a register remains in use for multiple operations, then a real benefit can be obtained from register usage, and significant savings in operation time can then be achieved.

Most modern CPUs employ multiple general-purpose registers. It is the prerogative of the programmer, or the compiler designer, to decide which values should remain in the registers and which should be stored in the main memory, for efficient use of the available registers. For example,

ADD R1, R2, R3; where R1, R2, and R3 are the addresses of the operand. The ADD instruction, in this case, results in the operation:
R1 ← R2 + R3; all operands in this instruction are registers that require a few bits to refer to a register in the operand field of the instruction. The instruction length, in turn, becomes smaller. However, the contents of the registers are the desired values of the operands.

All modern large systems as well as the advanced microprocessors that came from both ***Intel*** and ***Motorola*** have a wide repertoire of instructions that take their operands from registers and, after execution, also leave their results in a register.

- ***Indirect Mode (Addressing)***

One of the limitations of direct addressing is that the length of the address field is usually less than the word length and is limited, and thus is unable to provide

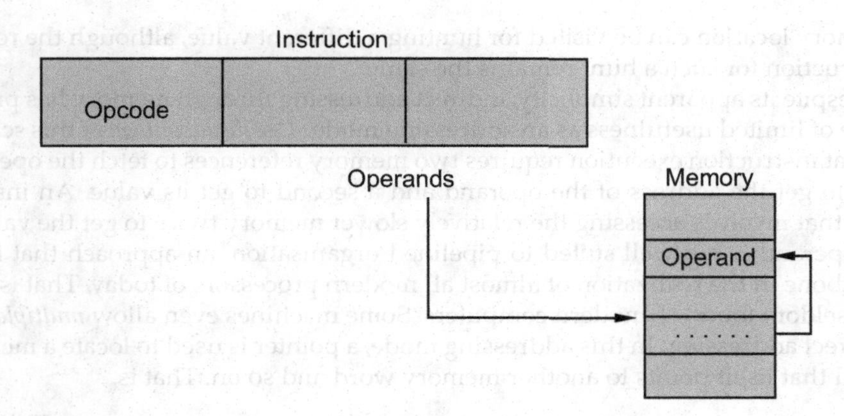

FIGURE 3.13
Indirect mode addressing.

the full address range. The indirect mode is a scheme in which the instruction does not provide the operand or its address explicitly. Instead, it provides information from which the memory address of the operand can be determined. In fact, the contents of the operand field in the instruction actually point to the memory word or register that contains not the operand but the address of the operand. The operational view of this mode is depicted in Figure 3.13.

An obvious advantage of this approach is that if the operand address field length in the instruction is l, then $l < N$, where N is the word length. This operand address in the instruction will now point to a memory location, and the content of this memory location is a full-word length N, which is now an address of actual target location. As a result, an address space of 2^N can now be accessed. The operand address in the instruction, however, specifies a register or a memory location which will now contain the ultimate full-word length address of the operand. The content of this memory location is fetched and is then used as an address to reach the destination location. The content of this destination location, as obtained, is the actual value of the desired operand. Here,

```
Effective address (EA)  =  Contents of (Address given as operand) = (A)
```

It is customary to denote *indirection* by placing the name of the register or the memory address given in the instruction in parentheses as is shown above. The parentheses here are to be interpreted as meaning *contents of*. For example,

```
MOV R1, [R2]; here the content of register R2 is not the operand, but the
address of the target location where the operand resides. This
instruction copies the word-sized data located to an address indicated by
the contents of R2 into the register  R1.
```

The register or memory location which is pointed to by the operand field (address) in the instruction that contains the address of an operand is called a *pointer*. Indirection and the use of pointers are important and powerful concepts in programming activities. By changing only the contents of a pointer, a different memory location can be visited for hunting a different value, although the related instruction for such a hunt remains the same.

Despite its apparent simplicity, indirect addressing through memory has proven to be of limited usefulness as an addressing mode. The *disadvantage* of this scheme is that instruction execution requires two memory references to fetch the operand: one to get the address of the operand and a second to get its value. An instruction that involves accessing the relatively slower memory twice to get the value of an operand is not well suited to pipelined organisation, an approach that is the backbone in the realization of almost all modern processors of today. That is why, it is seldom found in modern computers. Some machines even allow *multiple-level* indirect addressing. In this addressing mode, a pointer is used to locate a memory word that itself points to another memory word and so on. That is,

```
Effective Address(EA) = Contents of (...(Address given as operand))  = (...(A))
```

In this case, the targeted location of a full word has one bit used as an indirect flag (I). If the I bit is 0, then the content of the word is the EA. If the I bit is 1, then another level of indirect addressing is invoked. Without having much advantage, this approach has a serious drawback that three or more memory references may be required to fetch the ultimate operand.

The ***Intel processors*** have *indirect addressing* via register. Indirect addressing using a pointer in memory is not possible. The ***MC68000 series*** allows *indirect addressing* via the *address registers.* No other forms of indirect addressing are possible. The 68020 and 68030 allow *indirect addressing* via *memory, in several forms.* This additional feature is one of the main differences between 68000 and the newer CPUs.

Immediate, direct, indirect, and ***multiple-level indirect addressing,*** however, exhibit an interesting feature that is related to a certain *progression* in memory references:

- *Immediate addressing* requires *zero memory* references, because the operand is fetched along with the instruction;
- *Direct addressing* requires *one memory* reference to fetch the operand;
- *Indirect addressing* requires *two memory* references, one for the pointer and the other for the operand;
- *Multiple-level indirect addressing* requires at least *three memory* references, two or more for pointer and one for the operand.

Memory references in this context also include register references.

- ***Register Indirect Mode (Addressing)***

  As register addressing is analogous to direct addressing, so is register indirect addressing analogous to indirect addressing. Here, the instruction contains an operand specifying a register, and the contents of this register give the address of the operand in memory, rather than the value of the operand itself. A previous instruction should place the memory address of the operand in the register before using an instruction having this mode. As discussed earlier, the advantages as well as the shortcomings of this mode are essentially the same as for indirect addressing. Here also, the address field of the instruction uses only a few bits to

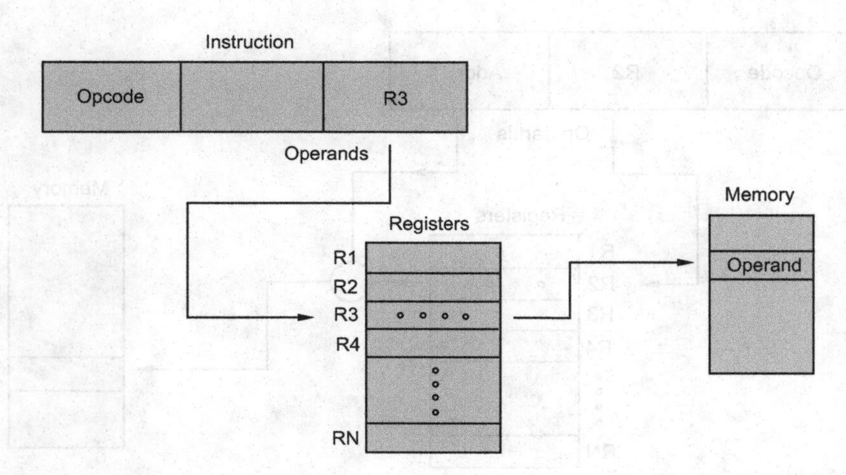

FIGURE 3.14
Register-indirect addressing.

select a register, the content of which is a full-word length, and can be used as an address of a full-word length to identify a memory location. This will, however, never be possible using a memory address directly in the instruction. Figure 3.14 illustrates the addressing mechanism used under this mode. Register indirect addressing, in addition, will always use one less memory reference than the usual indirect addressing.

Indirect addressing through registers is a powerful tool that is extensively used in the CPU organisation of all large systems as well as in the modern microprocessors. It can provide a high degree of flexibility. In fact, when the necessary absolute addressing is not obtainable, indirect addressing through registers makes it possible to access global variables by first loading the address of the operand in a specified register.

- ***Displacement Mode (Addressing)***

  This addressing mode essentially covers a broad category of a different type of addressing. In fact, this mode of addressing is virtually a *hybrid* (which combines the capabilities) of direct addressing and register indirect addressing techniques. It requires that the address field of the instruction be added to the contents of a specific register (PC, an index register, or a base register) to derive the EA of the operand. Figure 3.15 illustrates how the EA under this mode is derived.

  This approach, however, is known by a variety of names depending upon the context (type of the register) of its use. But the basic mechanism remains the same whatever be the context. This category is commonly referred to as *displacement addressing.*

  The EA of the operand in these modes is determined as: EA = Address part of instruction + Content of CPU register.

```
That is,  EA =  A + (R)
```

The instruction using this mode has two address fields: at least one of these is explicit. The value contained in one address field (value = A) is used directly

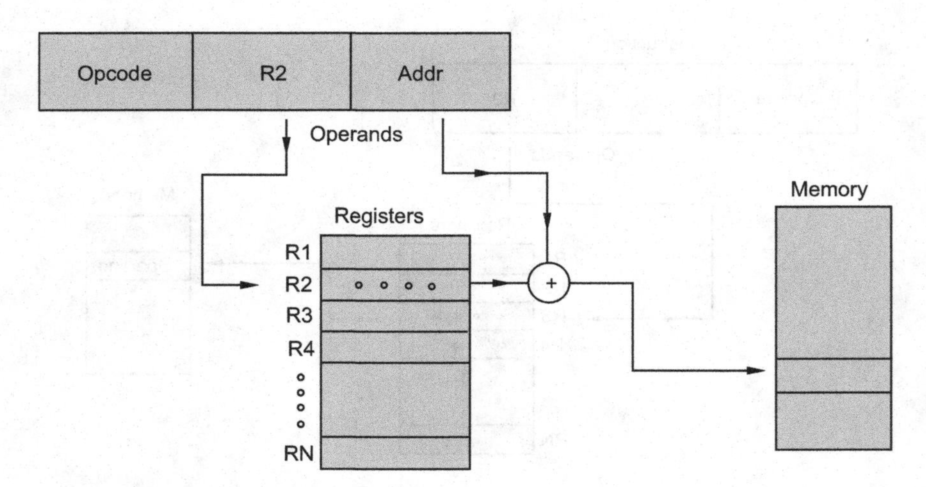

FIGURE 3.15
Displacement addressing.

(direct addressing). The other address field refers to a register or an implicit reference based on the opcode in use, whose contents are added to A to produce the EA (indirect addressing).

The most common forms of displacement addressing are as follows:

- **Indexed addressing;**
- **Base-register addressing;**
- **Relative addressing.**

- ***Indexed mode (addressing)***

  Many algorithms often require some operations to be performed on a sequence of data stored in consecutive memory locations. One approach to efficiently handle this situation is to use either a special register provided for this purpose or, more commonly, one or more registers from a set of general-purpose registers in the processor. In either case, it is referred to as an *index register*. Using this index register, a different type of addressing mode has been evolved, which is known as *indexed addressing mode*.

  In this mode, the address has two parts, which can be symbolically written as X(Ri), where one address field X references a specific main memory address (i.e. a constant value) and the other is the name of a register involved (known as *index register*) containing a positive displacement (offset) from that specified memory address. The contents of the index register are added to the address part of the instruction to obtain the EA. That is,

```
EA = X + [Ri]
```

Any operand in the consecutive location can be accessed with the same instruction by *incrementing / decrementing* the index register accordingly to facilitate access to consecutive operands. The register reference is sometimes explicit and sometimes implicit, depending on the strategy taken in the design of CPU. Some computers provide special instructions or addressing modes, or even special types of index registers that automatically increment or decrement themselves as part of the same instruction cycle. Automatic modification of an index register is called *autoindexing* that gives rise to ***autoincrement / autodecrement mode.*** In a byte-addressable memory, this mode is found useful in accessing successive bytes of some predefined list. But if successive words are needed to be accessed in a byte-addressable memory with, e.g., a 32-bit word length, the increment must be then 4 (32-bit = 4 × 8-bit). Computers that have autoincrement mode automatically increment the contents of the register by a value that corresponds to the size of the accessed operand. If an index-type instruction does not include an address field (X) in its format, the instruction then converts to the conventional *register indirect mode* of operation.

Besides the usual form of normal indexing, there exist two other possibilities: whether the indexing is to be performed before or after the indirection, namely ***postindexing*** and ***preindexing.*** The details of these are given in the website: http://routledge.com/9780367255732.

The **ARM architecture** contains both *postindexing* and *preindexing* in its load/store addressing (instruction) to provide a relatively rich set of addressing modes. What ARM refers a base register, it here acts here as an index register for both preindex and postindex addressing. These two modes can also, however, be used together to implement an important data structure called a *stack.*

Both the ***Intel*** and ***Motorola*** CPUs have a variety of different addressing modes that involve indexing. The Motorola 68000 series also has the facility of *autoindexing.*

- ***Base-Register Mode (Addressing)***

The primary aspects and the basic form of indexed addressing have been described. Several variations of this basic form exist that provide powerful mechanisms in accessing memory operands more efficiently in real-life programming situations. For example, a second register can be used in indexed mode addressing that will contain the offset X, in which case the indexed mode can be written as:

(R_i, R_k)

The second register is usually called the *base register.* The R_i register contains a displacement (usually an unsigned integer representation) from the base register given memory address. The EA is then derived as the sum of the contents of registers R_i and R_k. This form of indexed addressing known as *base-register addressing* consequently provides even more flexibility in defining and accessing operands, because both components present in the EA can then be made to be varied at will.

Yet, another variant of base-register addressing (an extended version of index mode) exists that uses two registers plus a constant. This form of addressing is commonly known as ***base-with-index and displacement (offset)*** addressing, which can be symbolically expressed as:

$X\ (R_i, R_j)$

In this case, the EA is derived as the sum of the constant X and the contents of registers R_i and R_j. This added flexibility is often found to be even more useful, particularly in the event of accessing multiple components inside each item in a record. In other words, this mode may be considered as virtually implementing a three-dimensional array.

Both ***Intel*** and ***Motorola*** CPUs have incorporated this powerful base-with-index and displacement addressing modes in their instruction set. In fact, all large systems, including the mainframe systems introduced by IBM, have also extensively used this form of addressing to make their instruction set even more versatile.

- ***Relative Mode (Addressing)***

  Another popular form of displacement addressing is commonly known as *relative addressing*. For relative addressing, or also called *PC-relative addressing*, the implicitly referenced register is often the PC. The contents of the PC are added to the address part of the instruction to derive the EA. Typically, the address field is treated as a *signed number* (either positive or negative) represented in 2's complement form for this operation. This EA indicates a memory location, which is relative to the address of the next instruction. For example, assume that the PC contains the number 500 at present. The instruction at location 500 is fetched from the memory and then the PC is automatically incremented by 1 to 501, to point to the next instruction. The just-fetched instruction has an address part which contains the number 19. The EA calculation for this relative address mode would then be $501+19 = 520$. This is 19 memory locations ahead from the address of the next instruction.

  Relative addressing exploits the concept of locality of reference. It is often used in branch-type instructions when branch address is in the close vicinity surrounding the instruction being executed. Under this scheme, a comparatively shorter address field is required in the instruction format since the relative address is specified with a reasonably smaller number of bits compared to the number of bits required to designate the entire memory address.

- ***Stack Mode (Addressing)***

  The organisation of stack has been already discussed in detail in Section 3.4. Since the SP is maintained in a register, so whenever a stack location in memory is referenced, the address of that location is in the SP (a register), and hence, the references to that stack location are a *register indirect address*. Based on this organisation, the stack mode of addressing is defined and derived, which is essentially a form of *implied addressing*. Operation-type instructions (machine instructions) neither need an address field nor include any memory reference, but implicitly operate on the TOS. Stacks are becoming quite common in microprocessors and also in the design of newer CPUs.

  None of the **Intel CPUs** have stack addressing, but they do have special instructions PUSH and POP to put items onto the stack and remove them, respectively. In contrast, the **Motorola 68000 CPUs** all have *stack addressing* using *autoindexing*.

  A brief detail of all addressing modes as discussed is given in the website: http://routledge.com/9780367255732.

3.13 Intel X-86 Addressing Modes: IA-32 and IA-64

The X-86 is equipped with a variety of these addressing modes as described, which aim to provide efficient execution of high-level language programs. But we will consider here only the generic addressing modes of the X-86 family of processors. The 64-bit operation of the **Pentium 4** and **Core 2** has, however, modified and extended a few of the usual aspects that are available in each of these modes. In this processor, a number of different types of registers (16 bits, 32 bits, and 64 bits) have been made available that are entrusted with specific responsibilities in providing different addressing modes. Figure 3.16 illustrates a representative generalized scheme of the address calculation mechanism involved in different types of addressing modes. There exist a set of *segment registers* that are employed to determine the current segment, which is the subject of present reference. Each segment register holds an index to the segment descriptor table, which contains the starting address of the corresponding segments. Associated with each such user-visible segment register is a *segment descriptor register* (also named in the same fashion in Figure 3.16 to indicate their correspondence), not programmer visible, which records the access rights for the segment, length (limit) of the segment, and the starting address of the segment. Altogether, six such segment registers are available, but which one is to be used for a particular reference depends on the context of execution and the instruction. In addition, there are also two other registers, namely the *base register* and the *index register*, that may be used to build an EA of the operand.

Intel provided all types of generic addressing modes as already discussed and also other useful modes of its own using different types and different lengths of dedicated and general-purpose registers along with other criteria befitting with its CPU organisation. Some of the most useful modes of addressing that are included are *immediate mode, register mode, register direct mode, register indirect mode, displacement mode, base with displacement (register relative) mode, base-plus-index mode, base relative-plus-index mode (base with index and displacement), scaled-index mode, scaled-index with displacement mode, based-scale index with displacement mode, relative mode, RIP-relative mode,* and many other useful ones.

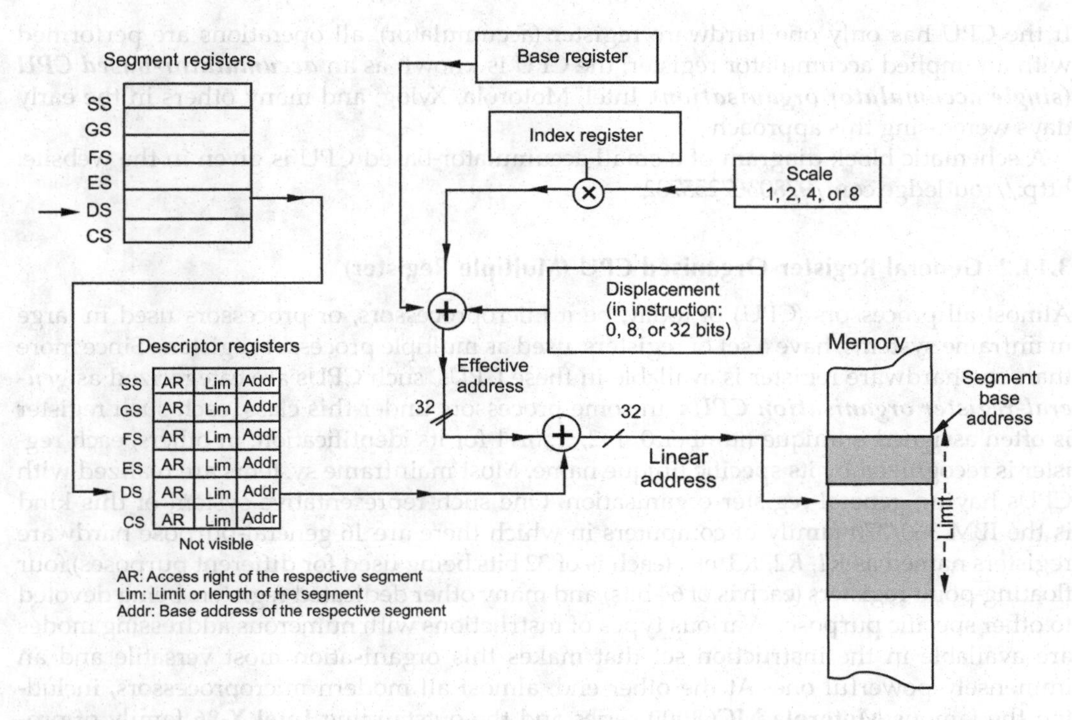

FIGURE 3.16
A schematic representation of addressing mode calculation mechanism (Intel processor).

All these different types of generic addressing modes that are available with the advanced X-86 processors, including Pentium and its subsequent upper versions, can also be described in a tabular form.

A brief detail of all these addressing modes, their working, and their usefulness with examples along with a tabular form of them (Table 3.1) is given in the website: http://routledge.com/9780367255732.

3.14 Register-Organised CPU

The normal approach of CPU design obeying Von Neumann concept includes one or more registers (both hardware and software) that support CPU operations to realize certain predefined objectives, and as such, the organisations of these registers play a decisive role in the design of the CPU. The number of registers available in a CPU, their types, the way they are being used while the CPU executes programs, and similar other features altogether determine a particular class of a specific type of CPU organisation, which can be broadly classified into two distinct categories, namely

Accumulator-based CPU (*single accumulator organisation*);
General register-organised CPU (*Multiple register*).

3.14.1 Accumulator-Based CPU (Single Accumulator Organisation)

If the CPU has only one hardware register (accumulator), all operations are performed with an implied accumulator register; the CPU is known as an ***accumulator-based CPU (single accumulator organisation)***. Intel, Motorola, Xylog, and many others in the early days were using this approach.

A schematic block diagram of a small accumulator-based CPU is given in the website: http://routledge.com/9780367255732.

3.14.2 General Register-Organised CPU (Multiple Register)

Almost all processors (CPU) of today, be it microprocessors, or processors used in large mainframe systems, have a set of registers, used as multiple processor registers. Since more than one hardware register is available in these CPUs, such CPUs are categorized as ***general-register organisation CPUs***. In some processors under this class, each such register is often assigned a unique number 0, 1, 2, ..., *n*–1 for its identification; in others, each register is recognized by its specific unique name. Most mainframe systems are realized with CPUs having general-register organisation. One such representative system of this kind is the IBM 360/370 family of computers in which there are 16 general-purpose hardware registers named as *R1*, *R2*, *R3*, etc. (each is of 32 bits being used for different purposes),four floating-point registers (each is of 64 bits),and many other dedicated registers being devoted to other specific purposes. Various types of instructions with numerous addressing modes are available in the instruction set that makes this organisation most versatile and an immensely powerful one. At the other end, almost all modern microprocessors, including the famous **Motorola MC68000** series and the outstanding **Intel X-86** family of processors, have all exploited general-register organisation (Section 3.3). A rough descriptive overview of the register organisation in the microprocessors of these two families has been already presented at the beginning of this chapter and also in the website: http://routledge.com/9780367255732. However, a schematic block diagram of such a general-register organisation of CPU is depicted in Figure 3.3.

Motorola, a strong competitor of Intel in this trade holding a share of nearly 50% or even more of the microprocessor market over many years with their marvellous products (MC 68000 series), finally decided to shift their focus from the then existing business line and started to concentrate more on the then a newer upcoming most promising area known as *mobile communication*. As a result, Motorola has sold its microprocessor division to a company, now called *Freescale Semiconductors, Inc.* Consequently, Intel – the only giant of earlier days – remains, still continues from its IA-32 architecture, and then exerts more efforts to develop more technology-enriched hardware to introduce IA-64 architecture, and today almost monopolizes and captures a major share of the marketplace of desktop and notebook systems, in addition to a limited share in the area of mid-range workstations. That is why today, Intel products in this trade have become the most dominant ones and deserve to be a subject matter worthy of discussions.

3.14.2.1 The Intel IA-32/IA-64 Architecture

The 32-bit Intel architecture, popularly known as *IA-32*, commenced from the introduction of Intel 80386 microprocessor, sometimes 1985 onwards, through progressively more powerful advanced processors up to the recent release of IA-64 Pentium and multicore series during a period of last odd 30 years, has implemented an evolution of the same basic instruction set architecture (IA-32)along with MMX, but with the continuous appropriate enhancement and modification befitting the underlying advanced form of more technology-enriched hardware. The different distinctive attributes that are found in IA-32, and later in IA-64 in the Pentium and multicore line of processors, have been discussed throughout this book at different places relevant to the respective topics under discussion at that point.

3.14.2.2 Register Organisation

The register organisation of the member processors under **Intel IA-32** and **IA-64** has been already described in Section 3.3.3.

3.15 Stack-Organised CPU (A Stack Processor)

Recall that the number of registers presents in a CPU and their internal organisation plays a decisive role in the design of register-based CPU, and that it also influences the corresponding machine instruction set design. A few alternatives of this approach were earlier explored: one of these gave rise to a different strategy in CPU design, commonly known as ***stack-organised CPU***. The main architectural feature of this new concept is that the instruction set will have implicit operands to be held only in a stack data structure, and only to those locating at the TOS, and results after required operations are also always returned to the TOS. All read and write operations are referred to *TOS* only. A stack memory thus replaces the *accumulator* and the *other CPU registers* used for temporary data storage. Moreover, here machine instructions having no operand addresses (zero-address instruction) are shorter in length, thereby saving both CPU execution time and necessary memory space – one of the vital criteria of a CPU design. The earlier series of large mainframe computers **B5500**, **B6500**, and **B6700** produced by Burroughs Corporation and later **HP3000** from Hewlett–Packard use this approach in their CPU design. A recent example of this type of CPU design has been implemented in the **SUN picoJava** microprocessor, for fast execution of compiled Java code.

The stack-organised CPU is used in computers with stack organisation (not stack orientation), which we have already discussed in detail in Section 3.4, including stack operations. It should be noted that stack-oriented operations are available in all CPUs, even those that do not belong to this class of stack-organised CPUs.

Stack is mostly located in memory, but its top few elements are held in hardware registers in processors to avoid frequent memory accesses, because all temporary storage locations are now part of the stack (memory). As a result, the stack access time will be drastically reduced since most accesses involve in only the top few elements which are in the registers, and therefore, only the register transfers are required within the processors. Different computer systems belonging to this category manufactured by different vendors use different number of registers to achieve their own target, which is essentially a trade-off criterion and a critical issue in this CPU design. The assigned memory for stack is partitioned into three segments: *program*, *data*, and *stack* separately, and a number of hardware registers are used as pointers to the program and data segments, as well as to handle and operate them. Stack computers have a variety of instructions which, when executed, perform operations on data that occupies the top few locations of the stack. The results thus generated are also left on the stack. During the execution, the data are moved between the stack and the memory. A computer organised around a stack offers several advantages, when compared to a multi register-organised CPU. Apart from many others, the major features are as follows:

1. Short instruction format because many instructions have no addresses;
2. Simplicity of the algorithm to convert the formulas, which in turn are easy to evaluate;
3. No need of complicated algorithms to optimize register usage.

None of the ***Intel CPUs*** have stack addressing, but they do have special instructions PUSH and POP to put items on the stack and remove them, respectively. In contrast, all of the ***Motorola 68000 series*** have stack addressing using *autoindexing.*

There are some special features of a generalized stack-organised CPU that emphasize certain elegant procedures these machines usually follow to execute the arithmetic expressions using *reverse Polish notation* (*RPN*).

3.15.1 Expression Evaluation and Reverse Polish Notation

A stack organisation is very effective for evaluating the arithmetic expressions. Under this organisation, the data-processing instructions do not need to contain addresses as they generally do in a conventional Von Neumann computer. For example, the ADD operation $a+b$ for a stack-organised machine is specified by the following sequence of instructions, and all these actions are hidden from the programmer, who does not have to worry about this at all:

PUSH *a*: Loads the memory operand *a* into TOS.
PUSH *b*: Loads the memory operand *b* into TOS above *a* causing *a*'s location to become TOS - 1.
ADD: The top two words of the stack are popped into the ALU where they are added. The sum is once again pushed back automatically into the TOS with no instruction needed and the SP is automatically adjusted.

Mathematical formulas are commonly expressed in what is known as *infix notation* where a binary operation appears between the operands (e.g. *u+v*). The arithmetic expressions can be represented in *prefix notation* where the operator is placed before the operands (e.g. *+uv*). This representation is often referred to as *Polish notation*. The *postfix notation*, referred to as *reverse Polish notation* (*RPN*), places the operator after the operands (e.g. *uv+*). It is to be noted that, regardless of the complexity of an expression, no *parentheses* are ever required while using *RPN*. This notation is ideal for evaluating arithmetic and other expressions on a computer which is stack-organised. The expression consists of *n* symbols, where each one is either an operand (variable or constant) or an operator. The procedure consists of first converting the arithmetic expression into its equivalent RPN using a suitable algorithm. The expression in RPN thus generated will then finally be evaluated by another algorithm using a stack.

A brief detail of a generic stack-organised CPU implementation with figure as well as the implementation of RPN with algorithms and solved examples is given in the website: http://routledge.com/9780367255732.

3.16 Stack-Organised Symbolic LISP Processor

A symbolic processor is a stack-organised machine that has been primarily developed for *artificial intelligence* (AI) applications. The machine architecture is divided into layers that allow the use of a pure stack model to design an overall simple instruction set, and the implementation is carried out with a simple stack-oriented machine. To make the operation faster, cache memory is used to implement the stack buffer and temporary memories (scratch-pad) to communicate with main memory. Most of the instructions are executed in one machine cycle following the dominant RISC philosophy. Integer instructions fetch operands from the TOS and place them into the stack buffer and in scratch-pad memory. Fixed-point additions are carried out in parallel. Floating-point operations are carried out by the tag processors. The system has been built to primarily execute Lisp instructions.

A brief detail of a generic LISP processor with figure is given in the website: http://routledge.com/9780367255732.

3.17 Summary

CPU is considered as the chief resource in the computer system, and consists of control unit, ALU, and a set of registers. The main task of the CPU is to fetch each active instruction from memory one after another, decode it, and then finally execute it. This sequence of actions known as an *instruction cycle* is considered as the central theme of any kind of CPU operation. The different types of registers and their organisations inside a CPU have been described with the illustration of both IA-32 and IA-64 architectures. The basic organisation of two different categories of CPU, based on the number of available registers, namely the *accumulator-based CPU* (a single hardware register) and the *general-purpose register CPU* (a set of general-purpose hardware registers) organised in many different ways, has been discussed, along with a completely different type of CPU organisation, known as *stack-organised CPU*, which has also been implemented in large commercial computers. A representative real-life stack-organised processor architecture (LISP processor) has been presented here for a clear understanding of this subject. The generic *instruction set* along with MMX instructions (*multimedia operations*) of both IA-32 and IA-64, which determines the types of the operations that a CPU can perform, to be decided at the time of CPU design, has also been explained. Numerous addressing schemes used in the instructions of a CPU that enormously influence the instruction execution time and also determine the performance of the CPU as a whole have been narrated. Generic addressing modes, including the important concepts of pointers, and indexed addressing employed in today's dominant Intel X-86 processors family, including both IA-32 and IA-64, have been presented as the real-life representative examples. In fact, this chapter introduces the basic structure and different types of organisation of a generic CPU with relevant topics, setting aside for the time being, its advanced forms such as pipelined/superscalar architecture and multicore architecture, which have been discussed in Chapter 8.

Exercises

3.1 Define register. "The number of registers present in a CPU and their internal organisation play a dominant role in the organisation of CPU" – justify. How many different types of CPU organisation are thus obtained based on the number of registers available in a CPU?

3.2 Give the name of the registers that are commonly available in all types of CPU organisation. Discuss their specific usage.

3.3 What do you mean by instruction cycle and machine cycle? Discuss the steps being followed in an instruction cycle that are common to all processors.

3.4 A microprocessor is clocked at a rate of 2 GHz.

   a. How long is a clock cycle?

   b. What is the duration of a particular type of machine instruction consisting of four clock cycles?

3.5 What do you mean by instruction set of a CPU? What are the issues that are to be addressed at the time of designing an instruction format?

3.6 What are different types of instruction formats that are commonly in use? Explain them briefly with suitable examples.

3.7 Define opcode. State with examples the different types of general operations that are found common on almost all machines.

3.8 A computer has a memory of 512K words of 32 bits each. The instruction format of this computer consists of four fields: an operation code field, a mode field to specify one of seven addressing modes, a register address field to specify one of 50 processor registers, and a memory address. Specify the instruction format indicating the number of bits in each field assuming an instruction takes one memory word. What is the maximum number of different operations that can be expressed by this instruction format?

3.9 Explain the mechanism and the steps being followed to execute a subroutine call. Why stack is found useful in handling subroutine call? What are the basic differences between a branch instruction, a subroutine call instruction, and a program interrupt? Discuss the differences being observed between software interrupt and subroutine call?

3.10 What are the implications of addressing scheme in the design of instruction set of a computer? Explain each type of addressing with examples.

3.11 The addressing modes is a trade-off criterion in CPU organisation – justify. What is the difference between an immediate, a direct, and an indirect addressing mode of instruction? How many references to memory are needed for each of such type of instruction to bring an operand in the processor register? What must the address field of an indexed addressing mode instruction be to make it the same as a register indirect mode instruction?

3.12 What are the preferred addressing modes for

i. Pointer variable

ii. Global variable

Explain why?

3.13 An instruction is stored in memory at location 400 with its address field at location 401.The address field has the value 500. A register R_1 contains the number 300. Evaluate the EA if the addressing mode of the instruction is (a) immediate, (b) direct, (c) register indirect, (d) indexed with R_1 as the index register, and (e) relative.

3.14 Explain the stack organisation of a computer system. Describe the basic constituents that are required to implement this type of organisation.

3.15 Write a program to evaluate the arithmetic expression:

$$Z = \frac{X+Y-(K \times U+V)-M}{E+F \times L}$$

a. Using an accumulator-type computer with one-address instructions.

b. Using a general register-type computer with two-address instructions.

c. Using a general register-type computer with three-address instructions.

d. Using a stack-organised computer with zero-address instructions.

3.16 Explain why a given arithmetic expression (infix) needs to be converted to RPN for effective use of a stack organisation. Convert the following expression into RPN, and show the evaluation procedure in the stack-organised CPU.

$$X \times Y + X \times (Y \times V + U \times W)$$

Suggested References

Brey, Barry B. *The Intel Microprocessors.* Upper Saddle River, NJ: Prentice-Hall, 2009.

Flynn, M. and Johnson, J. "On instruction sets and their formats." *IEEE Trans Comput*, March 1985.

Stallings, W. *Computer Organisation and Architecture*, Indian ed. Dorling Kindersley India Pvt. Ltd., 2010.

Hayes, J. P. *Computer Architecture and Organisation*, Int'l ed. WCB/McGraw-Hill, 1998.

4

Memory Organisation

LEARNING OBJECTIVES

- To study the memory system and its key characteristics.
- To underline the significance and usefulness of building up a well-defined memory hierarchy containing various types of memory components having the widest range of underlying technology, organisation, performance, and cost.
- To demonstrate the principles and working of Static random-access memory (SRAM), and Dynamic RAM (DRAM) memory cell.
- To explain the different approaches of RAM organisation for both SRAM and DRAM.
- To provide an introduction to advanced DRAM (SDRAM), DDR SDRAM and its different generations, RDRAM, and Cache DRAM (CDRAM).
- To enable a clear understanding of Read-Only Memory (ROM), their different varieties with numerous characteristics, organisations, and working principles.
- To introduce the reader to more advanced form of EEPROM:Flash memory, Flash Card, Flash drives, Pen drives.
- To study the different types of secondary memories including magnetic disks, tapes; and optical disks (CDs, DVDs), different types of commodity disks, and to introduce Redundant Array of Inexpensive Disks (RAID) approach along with its different levels.
- To explain the implementation mechanism of virtual memory and the strategies employed to make use of it profitably.
- To focus on cache memory, its design objectives, and related issues. Caches in multiprocessor systems and numerous ways to handle critical cache coherence problem.
- To provide an overview of Interleaved memory and its organisation, as also Associative memory and its different types of implementation.

The important breakthrough in the technological innovations in the electronic industry have been successfully implemented in the design and development of high density, high speed, and larger size memory system which plays a vital role, and became a dominant factor in building computers with relentlessly increased performance for succeeding generations. To fulfil the continuously increasing demand from constantly evolving high speed CPU, computer memory came out perhaps with the widest range of type, technology, organisation, performance, and cost to keep a computer system within an affordable limit. A typical computer system of today is thus found to be equipped with a diverse spectrum of memory subsystems maintaining a well-defined hierarchy; some are internal to the system (directly accessible by the CPU) and some are external (accessible by the CPU via an I/O module).This chapter, however, explains the architecture and organisation of different types of most commonly used memory subsystems. An attempt is made here to explore the salient features of different types of memory subsystems with their merits and drawbacks, as well as their limitations, and above all their organisations in contemporary representative computer systems.

4.1 Memory System Overview

The ultimate target of the computer is to process the information, and hence, a variety of devices are used to store this information which consist of instruction and data. Memory is a core component of computer subsystems (physical storage devices) which are employed for storage and subsequent retrieval of data and instructions required for computer operations. The *storage devices* along with the *methods* being used to control (or manage) the stored information constitute the ***memory system*** of the computer. Since the information stored in memory is involved in the processing executed by the processor, it is, hence, desirable that the processors must have fast, immediate, and uninterrupted access to the memory so that they can operate at their optimal speed. It is observed that a single memory using just one type of technology that can operate at speeds to keep up with the processor speeds is not viable. Even if it is feasible, it would be relatively costly, and may be usually not affordable. As a consequence, the typical computer system is equipped with a variety of memory units with very different physical characteristics, some of them are internal to the system (directly accessible by the CPU), and some are external (CPU can access via an I/O module) [Jacob, B]. In the following survey, memory's key characteristics and different aspects of organisation of memory subsystems have been studied.

4.1.1 Key Characteristics of the Memory System

A. **Location**: A small amount of working memory (a small set of high-speed *registers*) within the CPU (internal processor memory) is provided for temporary storage of instructions and data during execution. *Main memory,* also called primary memory, nowadays fabricated in the form of integrated circuits (ICs) is a relatively large fast memory used to store data and program during execution (internal to the processor). Its locations can be directly and promptly accessed by the CPU. *External memory,* also called auxiliary, secondary, or backing memory consisting of peripheral storage devices like magnetic disk (hard disk or removable disks),magnetic tapes, pen drive, CD-ROM etc. have much larger storage space but are slower in operation than main memory. Large data files, system programs, and others are usually stored in secondary memory permanently which are not continually required by the CPU. It is also used as an *overflow* memory when the capacity of main memory is exceeded. Information in secondary memory is accessed via I/O module which executes input–output programs that transfer the required information first to main memory and from there, CPU accesses that information.

B. **Storage Capacity**: Memory consists of a number of ***cells*** (or location). The base unit of memory that may contain a ***binary digit (0 or a 1)*** is called a ***bit***. An 8-bit cell is called a *byte*. One byte can store one character. Bytes are grouped into *words*. Common *word lengths* are 8, 16, and 32, and also 64 bits (1, 2, 4, and 8 bytes, respectively). However, the natural unit in memory organisation is usually ***word***. The size of the word is also different for different machines, and is often equal to the number of bits required to represent a number and also the instruction length. Many exceptions in this regard have also been observed. IBM used a word size of 32 bits, and that in BURROUGHS is of 48 bits, and CRAY-1 has a 64-bit word length. Each cell or location is identified by what is called its ***address*** by which information contained in that location can be referred and accessed. If a memory has n cells, it will have addresses *0* to n–1. All cells in a memory contain the same number of bits. If a cell consists of k bits, it can hold any one of 2^k different bit combinations.

C. **Addressable Unit**: In most systems, the addressable unit is the *word*, however, in some systems, it is a *byte*. If an address is of m-bit (contain m bits), the maximum number of cells (bytes) directly addressable is 2^m. If 1 byte can store one character, then an m-bit address can cover a memory which can store 2^m characters. If bytes are grouped into words and if a word consists of 4 bytes(i.e. 32bits), and if this word is taken as an addressable unit, then with the same m-bit address, a memory of size 2^m word, i.e. 2^{m+2} (as $4.2^m = 2^{m+2}$) byte can be addressed, i.e. the same length of address bit can address a memory of greater capacity, if it is accessed in terms of word unit. This is an important factor clubbed with other issues to be considered at the time of memory organisation.

Byte Ordering: Big-Endian and Little-Endian The bytes in a word can be numbered from left to right or right to left. When the bytes are numbered from left to right, i.e. when the lower-numbered bytes are used for the more significant bytes (the leftmost bytes) of the word, the numbering at the "big" (i.e. high order) end is called a ***big-endian*** used by the ***Motorola*** family. When the bytes are numbered using right to left, i.e. when the lower-numbered bytes are used for the less significant bytes (the rightmost bytes) of the word, i.e. the numbering at "little" (i.e. low order) end is called ***little-endian*** used by the ***Intel*** family. Both big-endian and little-endian assignments are used in commercial machines. In both cases, byte addresses 0, 4, 8, ... are taken as the addresses of *successive words* in the memory, and are the addresses used when specifying memory read and write operations for words. If computers stored integers, there would not be any problem with any type of ordering. But, if the data consists of a mixture of integers, character strings, and other types, then sending a record from big-endian to the little-endian one byte at a time, starting with byte 0 and ending with byte 4 (say) would cause a problem. It has reversed the order of the characters and integers in byte.

It is also necessary to specify the *labelling of bits* within a byte or a word. It is accepted as most natural ordering is similar to that of encoding of numerical data. The same ordering is also used for labelling bits within a byte, i.e. $b_7, b_6, \ldots, b_0$, from left to right. The reverse ordering, however, are also equally used in some computers.

D. **Unit of transfer:** In case of main memory, the number of bits read out of or written into main memory at a time is called *unit of transfer* which is generally equal to the unit of main memory as declared, but need not be always equal to a word or an addressable unit. For external memory, data are often transferred in much larger units (a chunk of information) and these are referred to as ***blocks***.

E. **Access method**: Among the different memory types, one of the important characteristics is the *method of accessing* units of data. Four types of such method (mode) may be distinguished. It is to be noted that access mode is a function of inherent characteristics of the technology being used and also on the organisation of the memory.

   i. **Sequential access (serial access)**: The unit of information being stored is called ***record***. Memory is organised to store set of records, and an access to these records can be made in a specific linear sequence. An ***addressing scheme*** is used to separate the records at the time of creation, and accordingly assists in the retrieval process. A shared read/write mechanism is used, and this must be moved from the current location to the target location. During this movement, all the intermediate records are passed and rejected till the desired record is obtained. Hence, under this scheme, the time to access an arbitrary record is highly variable. Magnetic tape units, cassette tapes mostly use sequential access.

   ii. **Direct access**: Under this scheme, the memory is organised into records or chunk of records called ***blocks***, and each individual records or blocks have a unique address based on mainly their physical location. At the time of access, the shared read–write mechanism uses the address to reach directly to a close vicinity, and then sequential movement is carried out until the final target location is reached. Here also, the access time varies depending mainly on the time required for sequential movement. All types of *Disk units* mostly use the direct access technique, also sometimes called *semi-random access*, but also use sequential access.

   iii. **Random access**: If the locations of the memory is accessed in any order, and access time is independent of the location being accessed, the memory is termed as *random-access* memory (RAM). Each addressable location in primary memory (RAM) has a unique, physically wired-in addressing mechanism. Any location in memory can be selected at random, and is directly addressable and accessible. The time to access a given location is independent of prior access or its position, and is constant for any location. Main memory (primary memory) systems are random access. Random-access memories based on semiconductor technology can also be used to construct serial-access memories, but the converse is not usually true.

   iv. **Associative**: This is, in principle, a random-access type of memory where a location is retrieved based on a portion of its *contents* rather than its address, and the related comparison for the desired match to reach the target location is done for all locations (words) simultaneously. This memory is thus also sometimes called *content addressable memory* (CAM). Here, each location has its own addressing mechanism like ordinary RAM, and hence, retrieval time is constant irrespective (independent) of its location or its prior access. *Cache* memory often employs associative access apart from the use of other techniques

F. **Performance view**: One of the most important aspects of memory device is its performance, apart from the consideration of its capacity. ***Three*** important parameters are used to determine the performance of memory device.

i. **Access time:** The rate at which the information can be read from or written into the memory is a measure to determine the performance of a memory. A convenient approach to measure performance is to calculate the *average time* required *to read* a fixed amount (an unit) of information, e.g. one word or one byte from the memory. This is termed as *read access* time or simply *access time*, and is denoted by t_A. The *write access* time is similarly defined, and is typically, equal to the read access time, but really not always. Access time primarily depends on the physical characteristics of the storage medium, and also to an extent on the type of access mechanism being used. For RAM, this is the time the memory takes to execute a *read* or *write* operation. It is usually calculated from the instant that a read/write is received by the memory unit to the instant at which all the requested information has been made available at the memory output terminals or have been stored at target location. For non-RAM (secondary storage), access time is the time to *position* the read–write mechanism at the target location. The *access rate* b_A of the memory is another widely used performance measure, and is defined as $1/t_A$, and is commonly measured in words per second.

ii. **Memory-cycle time (memory latency):** This concept is applicable to mainly RAM. This time consists of the access time plus any additional time required before a second access can start. This additional time may be required to regenerate data if they are read erroneously (not properly) or for transient to die out on signal lines. The access time t_A of a memory is defined as the time duration between the receipt of a read request by the memory and the delivery of the requested information to its targeted external devices. In case of dynamic memories and in destructive read—out (DRO), an additional time is required to carry out a restore (in case of DRO) and a refresh operation (for dynamic memories) before initiation of another memory access. This additional time plus the usual access time t_A is called t_C which is needed to *complete* any read or write operation in the memory, and is often loosely defined as ***cycle time*** of the memory or ***memory latency***. However, memory speed is measured both in terms of t_A and t_C when $t_A \neq t_C$. But, the overall computer system performance is measured mostly in terms of t_A and not on t_C, since t_A determines the length of time that a processor must wait after initiating a request for memory access. During remainder of t_C (i.e. the time equal to $t_C - t_A$), both the processor and memory can operate independently and simultaneously (in parallel). Still, the maximum amount of information that can be transferred to or from the memory every second is $1/t_C$, and this quantity called b_m, is the ***data-transfer rate*** or ***bandwidth***, and is measured in bits or words per second.

However, the effective bandwidth in a computer system (involving data transfers between the memory and the processor) is not determined solely by the speed of the memory, $1/t_C$; it also depends on the transfer capability of the links (bus) that connect the memory and the processor, typically the speed of the memory bus. Memory chips are usually designed to meet the speed requirements of popular buses. Thus, b_m cannot be increased at will. In fact, the dominating factor that intercepts and limits *memory bandwidth* (b_m) is the ***memory bus width*** w_m which is defined as the number of bits that can be transferred simultaneously (in parallel) over the memory bus, i.e. the number of wires present in the memory bus. It is generally the same as the size of the internal memory word, but not necessarily always. Thus, the bandwidth (b_m) clearly depends on the speed of access and transmission along a single wire ($1/t_C$), as well as on the number of bits that can be transferred simultaneously (in parallel) over the memory bus, i.e. the number of wires available in the memory bus. Thus, the bandwidth is the product of the rate at which data are transferred (and accessed), i.e. $1/t_C$ and the width of the data bus w_m, i.e. $b_m = w_m \times 1/t_C = w_m/t_C$ bits per second.

iii. **Transfer rate:** This is the rate at which data can be transferred to or from a memory unit. In case of RAM, this rate R is equal to 1/(cycle time). For non-random–access memory: $t_N = t_A + N/R$

Where t_N = average time to read or write N bits

t_A = average access time

R = transfer rate in bits per second (bps)

N = number of bits involved.

G. **Physical type:** Development in techniques for fabricating improved memory units are in progress, especially, a striking one is the achievement in semiconductor RAM ICs. Out of a variety of *physical types* of memory being employed today, the three most commonly used are:

a. Semiconductor memory using LSI, VLSI, and ULSI technology.
b. Magnetic surface memory used for disk and tape.
c. Optical sensitive surface memory in the form of optical disks which resembles magnetic disk in principle but using optical methods in read/write operation.

H. **Physical characteristics:** The information stored in memory using some physical processes is inherently unstable. As a result, the stored information may be lost over a period of time unless appropriate action is taken to preserve the data. This physical characteristics of data storage is expressed in terms of the attribute known as *performance of storage*. There are ***three*** important characteristics that can destroy information; *DRO, Dynamic storage,* and *Volatility*. The DRO is a phenomenon which exhibits that while a memory location is being read, the information stored in that location is destroyed during the reading process. Memories in which the reading process does not affect the stored data are said to have non-destructive readout (NDRO). Appropriate action is then taken at the time of such reading of DRO memories for storing the data afresh at that location to maintain the data unchanged even after reading. Secondly, certain memory devices have the property that the data stored in the form of electrical charges in the capacitors forming the memory tend to leak away over a period of time causing a loss of information stored in the memory. Some mechanism is thus needed to compensate this loss of charge, and thereby to restore the charges is called ***refreshing***. Memories which require periodic refreshing using certain strategy are called *dynamic memories* as opposed to *static memories* which require no refreshing. Finally, another physical process that can destroy the contents of a memory is the failure of its power supply. A *memory* is said to be ***volatile***, if the stored information decays naturally or is lost when electrical power is switched off. Most semiconductor memories are volatile. In a non-volatile memory, information once recorded (stored) remains without deterioration until deliberately changed. No electrical power is needed to retain information. Secondary memories including those of semiconductor types (ROM and variants of PROMs, such as pen drive, flash drives) are ***non-volatile.***

I. **Other attributes:** Several other attributes of memories that can critically affect not only the cost of a memory technology, but also increase the cost of systems' running overhead. One of such attributes is ***storage density*** which is a factor determining the physical size of a memory unit, and is generally measured in bits per unit area or per unit volume. An increase in physical size increases its space requirement and decreases its portability. ***Energy consumption*** by a memory unit is also a factor that significantly contributes to the running costs of a computer system. In addition, huge energy consumption indicates more heat generation which requires the use of expensive cooling instrument, and hence increases the overhead cost. Another factor is ***reliability*** which is measured by the mean time to failure (MTTF). In general, memories with no moving parts (semiconductor memory) have higher reliability than memories using mechanical motion like magnetic disk, magnetic tape, etc. Even memories that have no moving parts sometimes also suffer from reliability problems due to high-storage densities or high data-transfer rates. However, with the use of ***error-detecting*** and ***error-correcting*** codes, the reliability of any memory can be increased.

4.2 The Memory Hierarchy

To achieve optimal performance by a computer system and that too by a CPU, the memory system must be made comparable with the CPU speed so that while CPU executes instructions, it should not be kept waiting for a longer time for instructions or operands being fetched from memory. For RAM, the organisation, hence, is a key design issue. Computer memory perhaps exhibits the widest range of ***types, technology, organisation, performance,*** and also of ***cost.*** The cost must be reasonable compared with other components of the computer system. The design constraints on the memory of a computer can, however, be summed up considering a variety of technologies that are commonly used to implement memory systems.

Technology-wise, the ***three key characteristics*** of memory, namely, ***capacity, speed*** (access time), and ***cost*** are considered as the trade-off. From an application point of view, large capacity memory with low cost per bit is an essential requirement. But, to meet performance requirement, an expensive, relatively lower capacity memories with fast access time are also essential. Thus smaller, faster, and more expensive memories are supplemented by larger, slower, and cheaper memories. To accommodate this diverse requirement at the time of design, a compromise is made which dictates that not a single memory component or technology would be used, rather a ***memory hierarchy*** is to be employed for cost-effective performance. The ultimate target of creating a memory hierarchy with different types of memories at different levels is to support the CPU to work nearly at its optimal speed during the execution of programs, of course, obeying the trade-off parameters as already mentioned.

Since memory is constantly referenced by the processor, the principle of ***locality of reference*** is also considered as one of the key factors in the design of memory hierarchy. This means that the *percentage of access* to each succeeding lower level is substantially less

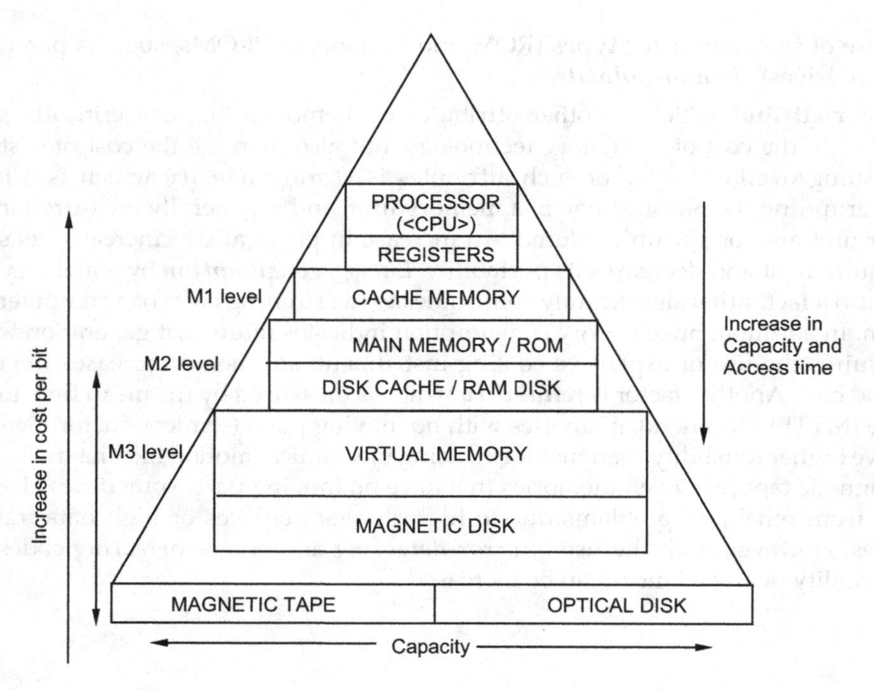

FIGURE 4.1
Contemporary memory hierarchy.

than that of the level above. A typical memory hierarchy is illustrated in Figure 4.1 based on the parameters:

i. *Access time*
ii. *Cost/bit*
iii. *Capacity*
iv. *Frequency of access by CPU*

If memory can be organised in this manner, and if the instructions and data can be distributed across the memory system, it is intuitively clear that this scheme will offer a given level of performance with a substantial reduction in overall costs.

Considering the hierarchy as shown, the fastest (*access time*), smallest, and most expensive type of memory consists of ***Registers***, internal to the processor. A processor nowadays typically contains a few dozen to hundreds of such registers. Now, with regard to *frequency of access*, the principle of ***locality of reference*** is followed here. This tells that the data most recently used is very likely to be accessed (referenced) again in the near future, and hence it is suggested to be kept in the fastest memory for improved performance. ***Cache memory*** fulfils this criteria which is of higher speed, smaller in size, but a costlier one. It is usually extended with main memory, and is not visible to the programmer, or indeed to the processor. It is placed nowadays inside the CPU (on-chip), in addition to, in between CPU and primary memory (off-chip), and is employed for the movement of data between main memory and processor register to improve performance. Typically, it is on-chip or off-chip CMOS, SRAM. ***Main memory*** is the principal internal memory system of the machine which has a unique address for each location.

It is generally of larger size, relatively slower (higher access time), and comparatively less costly. The distinguishing characteristics of these three types of memory just described are based on semiconductor technology. They are ***volatile***, and hence, must be provided with a constant power supply while in use. At the same level of main memory (RAM) in the memory hierarchy, the non-volatile **Read-Only Memory (ROM)** with all its variants, like **PROM**, **EPROM**, etc., are also present. Program and data are, however, stored more permanently on external non-volatile secondary mass storage devices, the most common are ***magnetic disk***, ***compact disk***, and ***tape***.

Disk is also used to provide an *extension* to main memory known as ***virtual memory***. In contrast, a portion of main memory can be used as a buffer to temporarily hold data that is supposed to be read from disk. Such a technique, sometimes referred to as a ***disk cache*** (RAM disk) improves the performance to a large extent. Disk cache, in general, is a purely software technique requiring a little bit additional hardware support, however.

A brief detail of this topic along with a table, listing the major characteristics of different types of memories is given in the following web site: http://routledge.com/9780367255732.

4.3 Semiconductor Main Memory

In earlier days, most computers had a common form of main memory (random-access storage) using ferromagnetic loops referred to as *cores*, and was thus commonly called *core memory*. To date the term persists to mean main memory, although all the internal memories (internal to the system and directly accessible by the CPU) of today are precisely semiconductor memories. The fundamental element of a semiconductor memory including main memory (RAM) is the basic storage cell (memory cell) which is primarily transistor circuits, made of both *Bipolar* and *MOS*, with MOS being the dominant circuit technology for realizing larger RAM chips. RAM is distinguished by one of its characteristics that it is possible both to read data from the memory and to write data into the memory quiet easily, using only the needed electrical signals. However, it is *volatile*; it requires constant power supply to remain active. If power is interrupted, then the entire contents of RAM are lost. Hence, RAM can be used only as temporary storage.

4.3.1 Random-Access Memory (RAM)

Random-access memories (RAMs) are semiconductor memories whose basic element is the memory *cell*. Each cell can be fabricated either with technology as used in SRAM (Static RAM) or in DRAM (Dynamic RAM). Each cell usually represents one bit, but a cell may contain a number of bits, and in that situation, all cells in the memory contain the same number of bits. These RAMs are characterized by the fact that every location can be accessed by its unique address independently. The *access time* and *cycle time* for every location are constant irrespective of position. Whatever electronic technologies are being used, all memory cells exhibit certain common properties, namely:

- All cells have two stable (or semi-stable) states, and these states can be used to represent binary 1 and 0.
- These cells are capable of being read to *sense* the state.
- They are capable of being written into (at least once) to *set* the state.

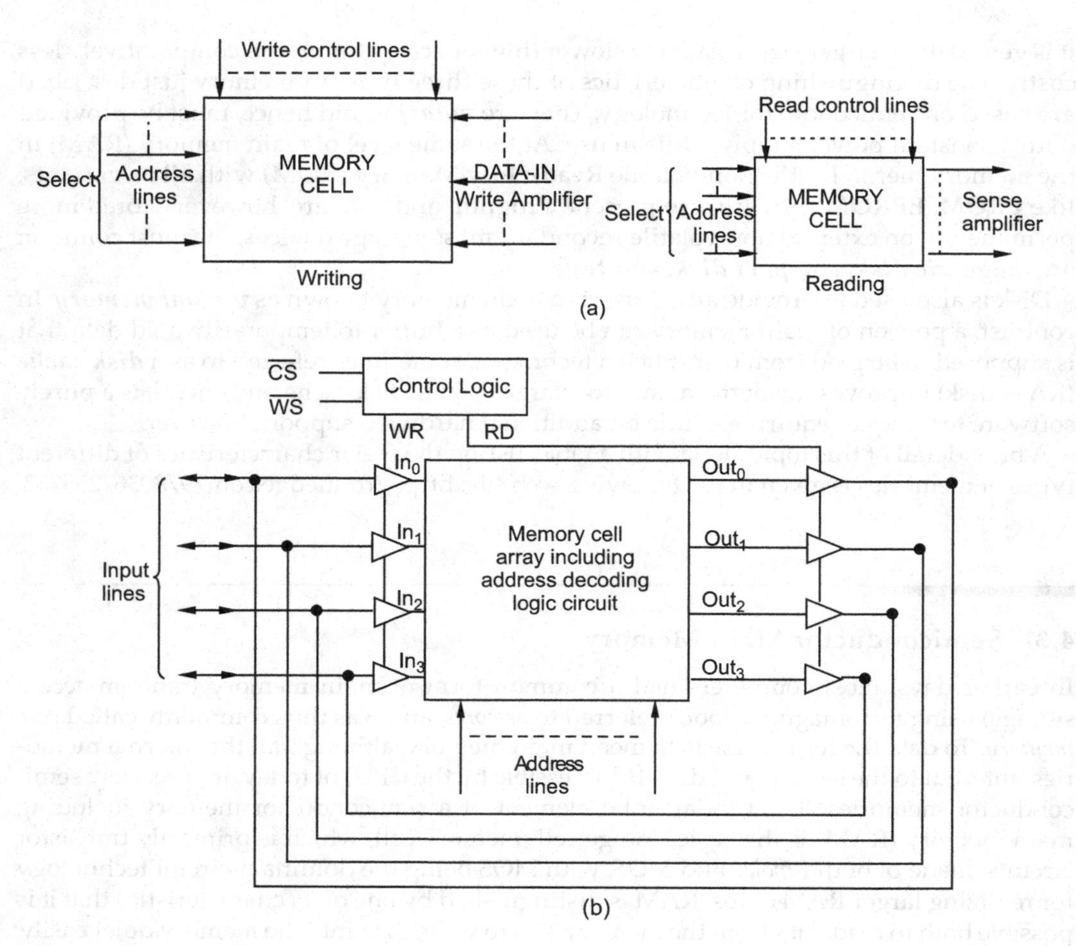

FIGURE 4.2
(a) Scheme of a memory cell operation (b) Logic circuit for a memory having common I/O lines.

Figure 4.2a illustrates the operation of an idealized model of a RAM cell with its external connections. Most commonly, the cell has ***three*** functional terminals, each one is capable of carrying an electrical signal.

i. The address lines are used to select the cell for either reading or writing. The number of address lines I of a memory unit is related to the number of locations m of the memory by the relation: $I = \log_2(m)$
ii. The control lines indicate either read or write operation. The related control signals are: (a) Chip Select or Chip Enable CS, (b) Write Enable WE (c) Output Enable (Read), or Output control OE
iii. A set of data lines is used for transferring data to and from the memory. For writing, these data lines provide an electrical signal that sets the state of the cell to 0 or 1. For reading, these data lines are used for the output of the cell's existing state detected by a sense amplifier. SRAM can be categorized into two types depending on the arrangement of data input/output lines.

a. Same set of data lines are used both for carrying data into the memory and to output data from the memory. For a common set of input/output lines, the data output lines are internally connected back, through tri-state buffers, to their corresponding data input lines as shown in Figure 4.2b.
b. Separate sets of data lines for incoming data and for outgoing data.

The details of internal organisation, functioning, timing, and the actual number of physical lines connected to a storage cell depend mostly on the specific integrated-circuit technology being used. Very often, one physical line has been found to have several functions, e.g. it may be used as both an address and a data line. Thus, the cell as shown in Figure 4.2 is not a physical one, rather it can be treated as a logical entity. However, we assume that individual cells can be selected for reading and writing operations.

4.3.2 Cell Organisation

Each storage cell of semiconductor memories (RAMs) is basically transistor circuits, containing one bit of information whose value may be either 0 or 1 at any point of time. A RAM chip precisely comprises such cells and a wide range of RAM chip capacity containing from a few million to even a few billion (giga=10^9) cells is now available using VLSI technology. For cell organisation, both *bipolar* and *MOS* transistor circuits are used. But, for producing larger capacity of RAM chip, MOS circuit technology is preferred. However, two distinctive basic categories of RAM (main memory) that are realized using two different technologies mainly exist. These RAMs are traditionally used in computers and are known as static RAM (SRAM) and dynamic RAM (DRAM).

4.3.3 Static RAM (SRAM)

A static RAM (SRAM) consists of circuits using the same logic components that are used in the processor, and is capable of retaining their state as long as power is applied. Traditional flip-flops are used to realize this memory that stores the binary information following conventional logic-gate configurations. Each static cell can store a single bit of either 0 or 1. Each cell generally requires six transistors for its realization using both bipolar and MOS technology.

Figure 4.3 schematically illustrates how a static RAM (SRAM) cell can be implemented. Two inverters are cross-connected to form a latch. The latch is connected to two bit lines of transistors T_1 and T_2. These transistors actually act as switches that can be opened or closed under control of word line. When the word line is at the ground level, the transistors are turned off, and the latch retains its state. For example, let us assume that the cell is in state 1, if the logic value at point C_1 is 1 and at point C_2 is 0. This state is maintained as long as the signal on the word line is at the ground level. The static RAM cell, however, can be realized using CMOS cell following the line as shown in Figure 4.3. The power supply voltage required for modern low-voltage versions of CMOS SRAMs is 3.3V, while it was 5V in older versions. SRAMs are said to be *volatile* memories since their contents are lost when power is not present.

Read Operation: For reading the state of the SRAM cell, the word line is activated by closing switches T_1 and T_2. If the cell is in state 1, the signal on bit line b is high and the signal on bit line b' is low. The opposite is true when the cell is in state 0. Thus, b and b' are complements of each other. Read/Write logic circuits attached at the end of the bit lines are used to monitor the state of b and b' and accordingly set the output.

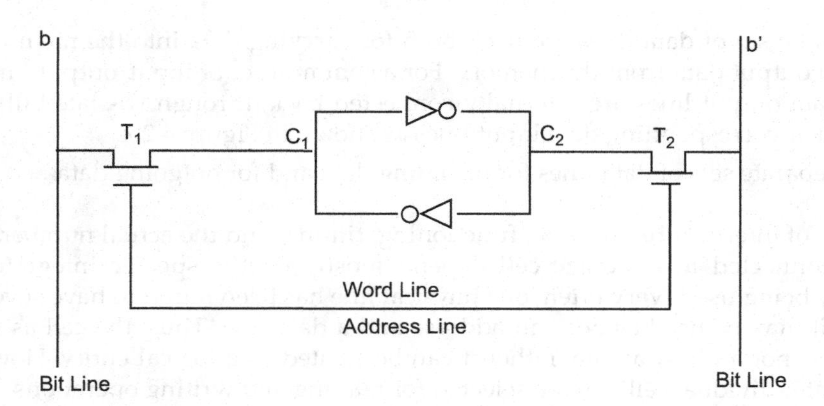

FIGURE 4.3
A static RAM cell.

Write Operation: For write operation, the state of the cell needs to be set, and this is done by placing the appropriate value on bit line *b* and its complements on *b'*, and then activating the word line. This consequently forces the cell into the corresponding state. The signals required on the bit lines are generated by the Read/Write logic circuits.

Two distinct ***advantages*** of CMOS SRAMs are observed. Firstly, they can operate with very low power consumption. Secondly, these SRAMs can be accessed very quickly. Access times are of just the order of a few nanoseconds as reported in commercially available chips. That is why, SRAMs are found to be extremely suitable for use in systems devoted to time-critical applications (e.g. real-time systems) where speed of execution is a major concern.

The RAM chips 6116, 6164 (or 6264), and 61256 (or 62256) are static RAMs and all are of CMOS type. These IC chips are run by single power supply +5V. The access time of these chips, on average, varies in the range of 100–120 ns. The outputs are tri-stated and TTL compatible. Although these memories are called CMOS RAMs, the memory cells use NMOS technology,and CMOS technology is used for the fabrication of the decoding, control, and internal interface circuits. Full CMOS RAMs are, however, also available for low-power applications.

The details of the realization of CMOS cell with figure are shown in the web site: http://routledge.com/9780367255732.

4.3.4 Dynamic RAM (DRAM)

Static RAMs are, no doubt, operationally fast, but they are realized at high cost and consume more space because several (six transistors) transistors are required to construct each cell. Designers were thus desperately looking for suitable methodology that could make less expensive RAMs using relatively simpler cells. Appropriate methods were ultimately found out to construct such simpler cells. Incidentally, such cells are inherently unable to retain their state indefinitely, rather have a tendency to change their own state dynamically, even when the power is on. Due to this attribute, the cell is commonly referred to as Dynamic RAMs or DRAMs [Jacob, B].

A dynamic RAM is made with cells that store data as charge on capacitors. The presence or absence of a specified amount of charge in a capacitor is interpreted as a binary 1 or 0. This capacitor is controlled by a transistor-switching circuit. Because capacitors have a natural tendency to discharge, the charge stored in a dynamic RAM cell tends to leak away with time, even with power constantly applied. Consequently, the information stored in it is thus in the way of being lost. Since the cell is usually required to store information for a much longer time, the amount of charge content lost from the cell by this time must be then periodically compensated (recharged) properly by restoring the capacitor charge to its original value. This process of restoration of charge in condenser must be carried out within a time duration before the charge of the capacitor decays to a level which is the minimum *threshold* for logic 1. This restoring process is known as ***refreshing*** the DRAM. Dynamic RAM thus requires periodic charge-refreshing mechanisms to maintain its stored data intact. As long as the charge in the capacitor stays above the threshold, no data is lost. Otherwise, if the capacitor is allowed to discharge too much and consequently the charge goes below the minimum threshold, logic 1 will eventually move to logic 0, thereby causing severe damage to the stored information. Most of the commercial DRAMs thus require refreshing the capacitor within a 2–4 ms interval.

It has been observed that about 3 % of the memory usage time only is spent on refreshing, and hence, refreshing has very little impact on the performance of the memory. It should be noted that while refreshing is going on, normal memory access cannot be performed. However, refresh cycles and normal memory access cycles can be interleaved and it has been accomplished in the advanced IC chip design.

A typical DRAM cell structure for an individual cell that stores one bit of information is depicted in Figure 4.4. The cell is constructed simply by a capacitor *C* and a transistor *T*. The transistor *T* here acts as a switch which is closed (allowing current to flow) if a voltage is applied to activate the address line and is open (no current flows) if no voltage is present on the address line. The address line is activated only when the *bit value* from this cell is to be read or written.

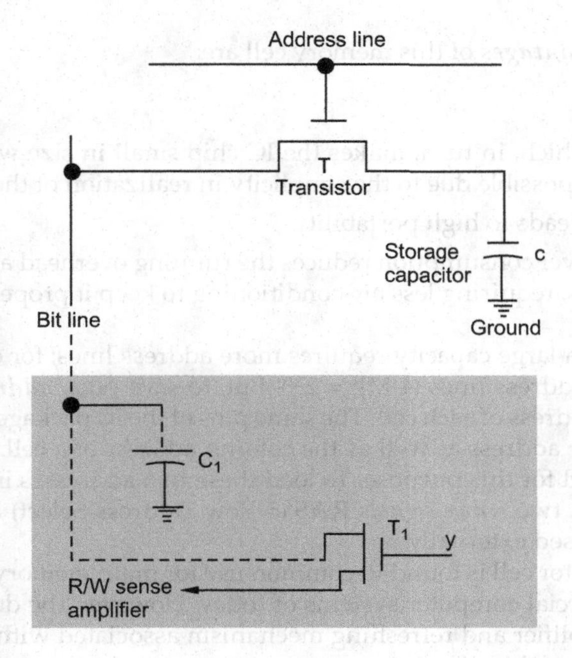

FIGURE 4.4
A scheme of typical DRAM memory cell structure.

Write Operation: For the write operation (i.e. to store information) in the cell, a voltage either high representing 1 or low representing 0 (i.e. an appropriate voltage signal) is applied to the bit line. A signal is then applied to the address line to turn the transistor T on. This causes a known amount of charge (bit value) to be transferred to the capacitor C to store. Charge is transferred to the capacitor C only if the data line is in the 1 state, and no charge is transferred, if the data line is in the 0 state.

Read Operation: For the read operation, the address line for the targeted cell is selected and then activated. As a result, the transistor T is turned on, and the charge already stored, if any, in the capacitor is transferred, and is fed out onto a bit line, and then to a sense amplifier for detection. The sense amplifier compares the capacitor voltage to a reference value (*threshold value*) that determines whether the cell contains logic 1 or logic 0. The read-out from the cell is *destructive* in nature (DRO) that discharges the capacitor C. This occurs because the selected storage capacitor C comes in parallel with the stray capacitor C_1 of the bit line. Usually, the value of C_1 is much larger than C because many such cells are connected to the same column line. Thus, the information being readout is now amplified and subsequently restored (rewritten again into the cell) in C for future use. This operation may, however, be combined with the periodic *refreshing* operations as required by dynamic memories of this type. The read operation now comes to an end. With its usual access circuitry, this memory cell also requires a few ***additional arrangements*** for keeping the provision of:

- Suitable mechanism for refreshing
- Necessary arrangement to recover the cell status due to destructive readout.
- DRAM is often bit-organised. Therefore, the read/write operation can be done for one bit at a time, and for this reason, a DRAM with a large capacity requires comparatively more address lines than its counterpart a Static RAM.

Still the ***distinct advantages*** of this memory cell are:

- Low cost per bit.
- Its small size which, in turn, makes the IC chip small in size with high cell (bit) density. This is possible due to the simplicity in realization of the DRAM cell.
- Smaller in size leads to high portability.
- Having low power consumption reduces the running overhead as well as less heat generation, thus, requiring less air-conditioning to keep it properly operative.

Since a DRAM with a large capacity requires more address lines; for example, a 4 Mb × 1 DRAM requires 22 address lines (4 Mb = 2^{22}). But, to save pins, ***address multiplexing*** is used to specify the address of each cell. The same pins of the IC package are thus *multiplexed* to input both the row address as well as the column address of a cell. Hence, appropriate decoders are required for this purpose. To load these two addresses into the row decoder and column decoder, two *strobe signals* **RAS'**(= Row Address Select) and **CAS'**(=Column Address Select) are used externally.

CMOS semiconductor cell is found in common use for main memory (primary memory) in almost all commercial computer systems of today. However, the detailed implementation of the sense amplifier and refreshing mechanism associated with its implementation are beyond the scope of this discussion.

4.3.4.1 Schemes For Refreshing DRAM

Normal decay of stored charge from the capacitor creates problems only when the information is read from the cell, because writing on the cell means putting some amount of charge to store in the capacitor. Thus, refreshing is mainly required to restore the charge level to its original value before any such reading is carried out. Refreshing is periodically done by cycling through the words once every 2–4 ms to restore the charge already decayed. This requires an extra control circuitry for refreshing mechanisms that are to be interleaved with normal memory access operations. Naturally, realization of dynamic RAM requires somewhat extra circuitry and additional mechanisms than those of their static counterparts. However, there are several types of refresh operations. Those are, namely:

i. RAS′ Only Refresh
ii. CAS′ Before RAS′ Refresh
iii. Distributed Refreshing
iv. Hidden Refresh
v. Burst Refresh

A brief detail of different types of refresh operations is given in the following web site: http://routledge.com/9780367255732.

4.3.5 SRAM vs DRAM: A Rough Comparison

- Since the transistor component used in SRAM is much faster in operation than the capacitor used in DRAM, the access time of SRAM is much lower, indicating that it is comparatively faster in operation than that of its counterpart DRAM. This emphasizes the fact that machines which are devoted to handle time-critical environment must use only SRAM, and not DRAM (e.g. aircraft control).
- Realization of one-bit memory cell of DRAM requires relatively less number of electronic components and small internal circuitry than those of its counterpart SRAM. Consequently, the room space needed by a DRAM cell is less than that of a SRAM cell. This makes the IC chip smaller in size with high cell (bit) density and consequently gives rise to a substantially large storage capacity in a single DRAM memory chip compared with SRAM memory chip of the same size. In other words, the packaging density of DRAM is significantly higher when compared with that of the corresponding SRAM.
- Due to the presence of relatively less number of electronic components and small internal circuitry in DRAM, the power consumed by DRAM is appreciably less when compared with compatible SRAM. Moreover, the principle used to realize SRAM employs transistors that usually consume more power than those of their counterpart DRAM which is realized based on the principle of using a capacitor to store the information, and thus results in reduced power consumption.
- DRAM requires more address lines, since it is often bit-organised, than its counterpart SRAM.
- DRAM needs an additional suitable mechanism, not required by SRAM, which is devoted to refreshing the DRAM.

- DRAM requires necessary additional arrangements to recover the cell status due to destructive read out (DRO), the SRAM never requires any such.
- SRAM is costlier than DRAM.
- The smaller size leads to the high portability of DRAM than that of SRAM.

4.3.6 RAM Organisation

Semiconductor memory ICs started in use on regular basis sometime in 1970s, and this memory comes in packaged chip. Each chip contains a number of memory cell that are arranged in a suitable manner.

We know that each cell has three functional terminals, each one is capable of carrying an electrical signal. Moreover, each such terminal of a cell must have a ***driver*** which acts as either an *amplifier* or a *transducer* of physical signal flowing through the related terminal. Thus a set of *address line drivers*, a set of *data line drivers*, various other *drivers*, *decoders*, and related *control circuits* are required for the proper functioning of the cells. All these are collectively referred to as the ***access circuitry*** of the memory unit. This access circuitry and other needed functional logics together have a significant effect on the trade-off parameters, such as, speed, capacity, cost, etc., and these parameters, in turn, have a critical impact on the organisation of memory cells on a chip. A general approach to minimizing the access circuitry cost in semiconductor memories is what is called *matrix* or *array organisation*. It has essentially two distinct features:

i. To simplify the connections between the cell and its allied access circuitry to form memory storage, the physical arrangement of cells demands rectangular arrays.
ii. Each cell C_i in the memory has a unique address A_i, and due to the rectangular arrangement of cells, this A_i becomes a *d-dimensional* vector $(A_{i1}, A_{i2}, \ldots, A_{id}) = A_i$. An address word A_i thus has d parts (d can be assumed as the column of the matrix) and each part is directed to a different address decoder and a different set of address drivers. A particular cell is selected by simultaneously activating all d parts of its address line. A memory unit with this kind of cell organisation is said to be a d-dimensional memory.

One-Dimensional (1D) Organisation: When $d = 1$, the array organisations are called *one-dimensional* or *1D* memories. Here, each cell containing 1bit is connected to one address line. The storage capacity of the unit is taken here as N bits consisting of N rows, and each word is of 1 bit (only one column). The access circuitry will select one of N rows from the address decoder, and each row requires one address driver (since, only one column) and hence, a total of N address drivers are required for the entire memory. A total of $\log_2 N$ address lines (width of address bus) supply the address of the word (row) to be selected. This arrangement is the simplest array organisation of cells to form the memory storage.

4.3.6.1 2D Organisation

For semiconductor memories, two-dimensional (2D and 2½D) organisational approaches have been used. In 2D organisation as shown in Figure 4.5, each row in the array is considered as a word, and the number of columns in a row are the number of bits forming the word. The array is organised into N words (rows), each word is of w bits. For example, a 4K-bit chip might contain 256 words (256 rows), each word is of 16bits (16 columns). All the elements of the array are interconnected both horizontally (row-wise) and vertically

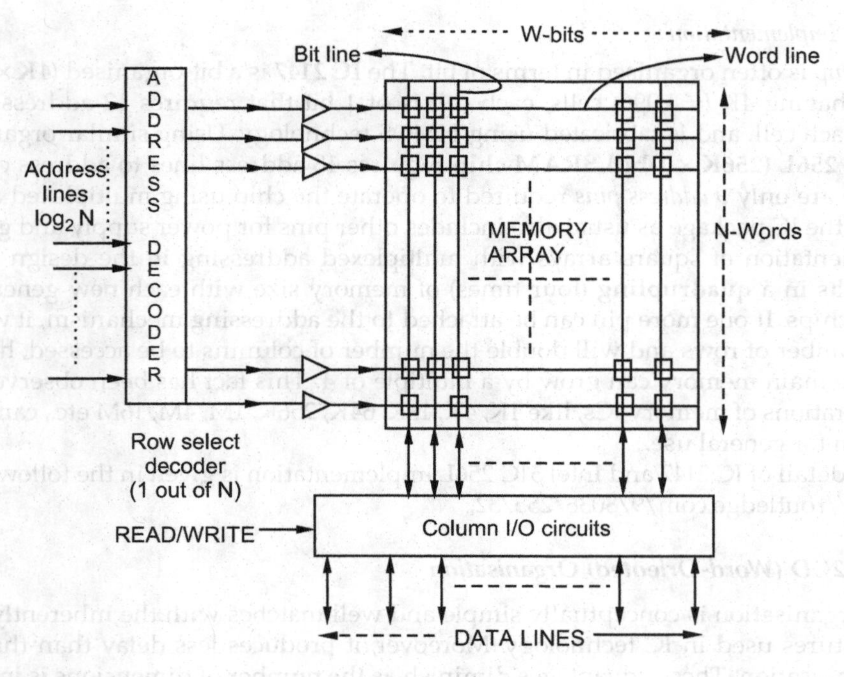

FIGURE 4.5
2D Memory organisation scheme.

(column-wise). Each horizontal line connects to all the *select terminals* of each cell in that row and each vertical line connects to all the *Data-In/Sense* terminal of each cell lying in that column.

At any point of time, an entire word, i.e. any row is selected by the signal on address lines. A total of $\log_2 N$ address lines are needed to address *N* words of memory. These address lines carry the signal as the address of the word to be selected. In our example, the memory has 256 words, i.e. there are 256 rows ($N = 256$), and hence, $\log_2 256 = 8$-bit address line is needed to select one of the 256 words. These address lines are fed into a decoder resident on a chip having $\log_2 N$ inputs and *N* outputs. When an address signal in the form of a bit pattern arrives at a decoder, it activates a signal output which corresponds to one of the word lines (rows) in the matrix and drives it. For example, an address input of 00001101 to the decoder causes the fourteen (14) word line (numbering the lines from 0 to $w - 1$) to be activated.

Data lines are simultaneously activated and used for the input/output of *B* bits (contents of the cell) of the corresponding row from and to a data buffer. Additional supporting circuitry such as *address line drivers, data line drivers* as well as *other drivers* are present on the chip for every cell to work. When the data from the cell is *read* (output), the content of the cell (value of each bit line) is passed through a sense amplifier and presented to the data line. When the data is *written* into the cell (input), the *bit driver* of each bit line is activated for a 1 or 0 according to the value of the corresponding data line. The number of columns in a row defines the word length of memory storage. Here, the word length is of *w*-bits, and hence, an entire word of *w*-bits can be accessed in each read or write cycle but the individual bit within a word is *not separately addressable*. A RAM of this type is sometimes referred to as an $N \times w$-bit ***word organised memory***.

4.3.6.1.1 *Implementation*

A *RAM chip* is often organised in terms of bit. The **IC 2147** is a bit-organised (**4K × 1**) static memory having 4K (= 4,096) cells, each cell is of 1 bit that requires 12 address lines to address each cell, and is fabricated using HMOS technology. Using similar organisation, **Intel 51C 256L (256K × 1-bit) SRAM chip** requires 18 address lines to address each cell. But, there are only *9 address pins* required to operate the chip using multiplexed addressing. Both the IC package as usual also includes other pins for power supply and ground.

Implementation of square arrays with multiplexed addressing in the design of RAM chip results in a **quadrupling** (four times) of memory size with each new generation of memory chips. If one more pin can be attached to the addressing mechanism, it will double the number of rows and will double the number of columns to be accessed, hence the size of the main memory can grow by a multiple of 4. This fact has been observed when new generations of memory ICs, like 1K, 4K, 16K, 64K, 256K, 1M, 4M, 16M etc., came out in succession for general use.

A brief detail of IC 2147 and Intel 51C 256L implementation is given in the following web site: http://routledge.com/9780367255732.

4.3.6.2 2½D (Word-Oriented) Organisation

The 2D organisation is conceptually simple and well matches with the inherently 2D circuit structures used in IC technology. Moreover, it produces less delay than that of the other organisation. These advantages diminish as the number of dimensions is increased. Thus values of D greater than 2 are rarely used. However, a number of *disadvantages* have been observed in 2D organisation, and those when modified give rise to another organisation known as 2½D organisation. The *major drawbacks* that are experienced in traditional 2D organisation are:

- The 2D physical structure consisting of large number of words (rows) for larger size of memory while having relatively only few bits per words (columns) as word length is limited for a number of reasons. This shows the dimension of the array is very long and narrow; hence demands an increased logic or circuit requirements.
- Moreover, each word line (row) must have one signal driver, and the number of driver increases as the number of words (rows) in the memory increases. Each such driver again must be connected to one logic gate of the decoder, and hence, logic circuit requirement increases as the number of words (memory size) increases.
- Each bit line requires one driver and one sense amplifier, and hence, the number increases as the word length (number of columns) of memory increases. This, in turn, requires a large number of external data lines, and in fact, the larger the word length, larger will be the number of external data lines requirement.
- Since, all the bits of a word are on the same word line and placed very closely, electromagnetic disturbances may occur, and multiple bits in a single word may then be affected, and hence, effective use of error-correction circuitry becomes difficult, and truly is impossible.

Most of these drawbacks have been adequately addressed, and successfully overcome with the introduction of 2½D organisation. In this organisation, the bits of a particular word are spread across multiple chips (wafer). The most common organisation is to permit only *1 bit of a given word on a chip and a number of such chips forms a package.* An integrated circuit (IC) is then mounted on a package that has pins for connections to the outside world.

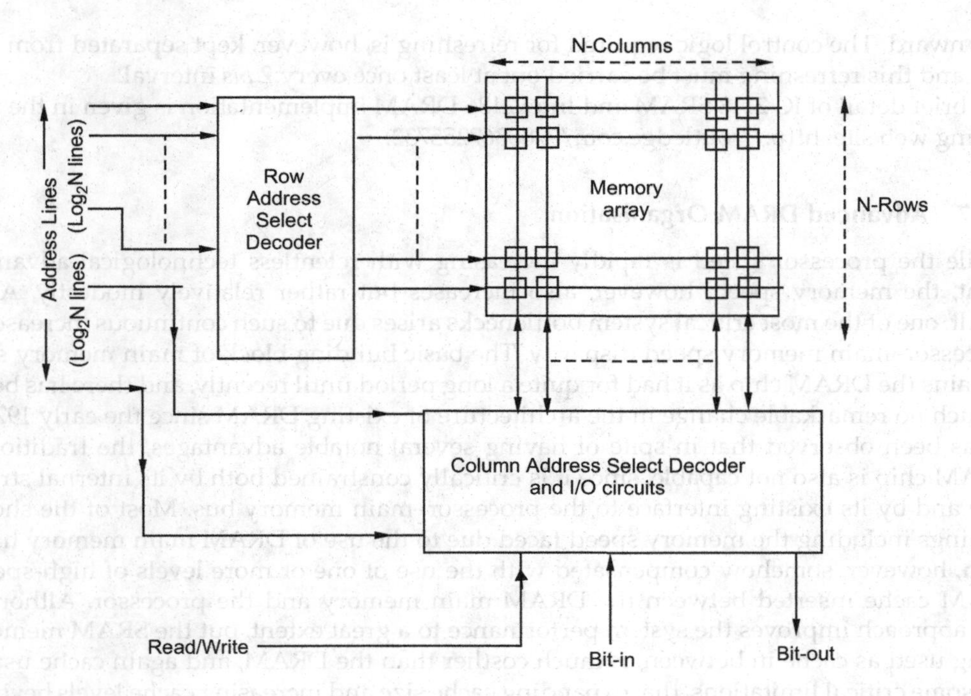

FIGURE 4.6
2½ D Memory organisation (one bit per chip).

In 2½D organisation as depicted in Figure 4.6, the chip as usual contains an array of bits. The array is typically a square one. As in 2D, here also the array itself functions in the same manner. That is, each cell is connected to a row line and a column line, and an entire row of cells is selected at any point of time whenever an operation is performed. The address lines, as before, carry the address of the word to be selected. This word address is split into two parts to select the bit of a particular word. One part of the address is presented to a decoder that selects one row, and the other part of the address is directed to another decoder connected to the column lines. This decoder activates one columns line and the cell situated at the intersection of this column and the already selected row is being read or written in the array. The other memory chips (wafer) in the package must be used simultaneously to supply the remaining bits of the desired word being addressed.

Finally, it would not be out of context to mention at this point that the term 2½D, however, gradually converted simply to the term 2D. The reason is that the original 2D organisation virtually faded away as the memory fabrication technology using the advanced form of modern VLSI technology steadily got matured with the passing of time, giving rise to single-chip high-density IC packaging.

4.3.6.2.1 Implementation

INTEL 2186 DRAM: This is a 64K-bit cells memory chip organised in a 2½D array in the form of 8K (2^{13}) 8-bit bytes (8K × 8 = 64K bit). The memory-cell array (8K = 2^{13}) has 13 input address bits, out of which 7 address bits select a particular R out of 128 (2^7 = 128) rows of this array. The remaining 6 address bits then select the particular column (out of 2^6 = 64 column) of this array. The intersection of this selected row and column is the target point where the eight columns of R that contain the desired word can be obtained going downward. The control logic required for refreshing is, however, kept separated from the cell, and this refreshing must be carried out at least once every *2 ms* interval.

A brief detail of IC 2148 SRAM and Intel 2186 DRAM implementation is given in the following web site: http://routledge.com/9780367255732.

4.3.7 Advanced DRAM Organisation

While the processor speed is rapidly increasing with relentless technological advancement, the memory speed, however, also increases but rather relatively modestly. As a result, one of the most critical system bottlenecks arises due to such continuous increase in processor–main memory speed disparity. The basic building block of main memory still remains the DRAM chip as it had for quite a long period until recently, and there has been as such no remarkable change in the architecture of existing DRAM since the early 1970s. It has been observed that in spite of having several notable advantages, the traditional DRAM chip is also not capable, since it is critically constrained both by its internal structure and by its existing interface to the processor–main memory bus. Most of the shortcomings including the memory speed faced due to the use of DRAM main memory have been, however, somehow compensated with the use of one or more levels of high-speed SRAM cache inserted between the DRAM main memory and the processor. Although this approach improves the system performance to a great extent, but the SRAM memory being used as cache in between, is much costlier than the DRAM, and again cache usage has some critical limitations, that expanding cache size and increasing cache levels beyond a certain point start to contribute diminishing returns.

Thus, once again it became essential to shift the focus to the actual problem rather than to somehow avoid or negotiate it indirectly, and that too, in the quest for some definite effective approaches to get rid of it. Consequently, a number of improvements in the basic DRAM architecture have been explored in recent years. Not all of them have been successfully implemented. However, the schemes that are currently most popular, and commonly in use are SDRAM, DDR-DRAM and its variants, and RDRAM. The CDRAM also cannot be kept set aside, and it is considered to be equally important.

4.3.7.1 SDRAMs (Synchronous DRAMs)

SDRAM is one of the most popular and widely used modified forms of DRAM, and unlike the traditional DRAM, which is asynchronous, SDRAM in operation is directly *synchronized* to an external clock signal to nullify the bad *wait state* associated with traditional DRAM. As a result, it can run at the highest possible speed of available processor–memory bus. The cell array in SDRAM is, however, organised similar to traditional asynchronous DRAMs, already discussed in the previous section. Since, this modified DRAM uses synchronous access, it is called synchronous DRAM or SDRAM. The principle that works behind is very simple. Here, the DRAM operates moving data in and out under the control of the system clock. While the processor or other master issues the request with related information as required by DRAM, it is then latched by the DRAM, and later it responds, only after performing all its own operations, consuming a predefined number of clock cycles. Meanwhile, the master remains free to do its own work while SDRAM is engaged in its own operations to process the request.

Figure 4.7 illustrates a schematic block design of an SDRAM. The address and data connections are buffered by means of two individual registers. The notable feature is that the output of each sense amplifier is connected to a latch. The SDRAM employs a

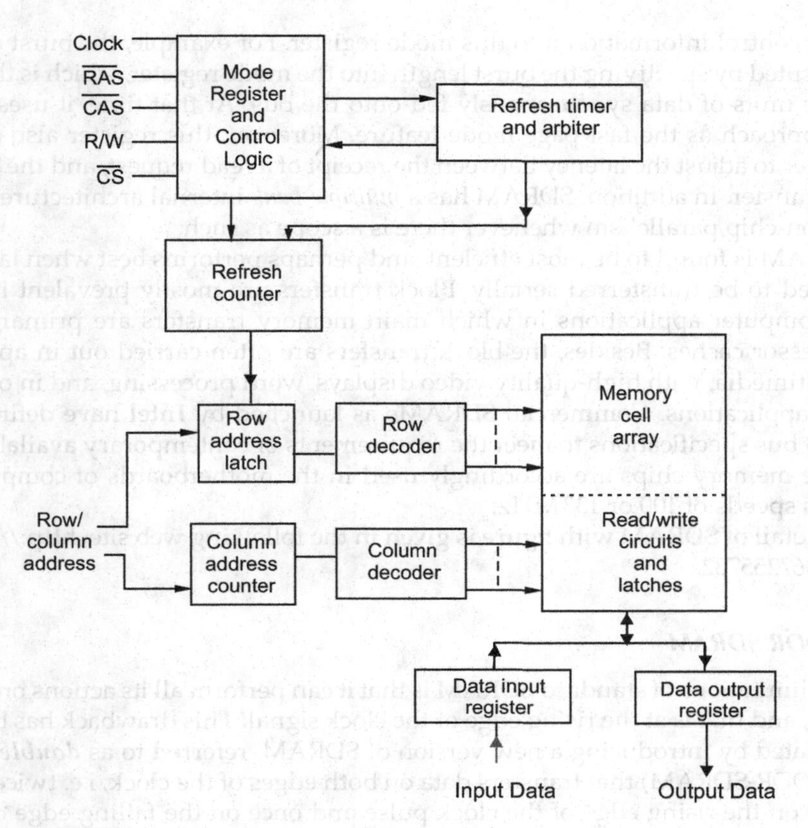

FIGURE 4.7
Schematic block diagram of a synchronized DRAM (SDRAM).

burst mode (already discussed in the previous section) to eliminate the address setup time, and row and column line pre-charge time after the first access. In burst mode, a series of data bits can be clocked out rapidly after the first bit has been accessed. This mode is effective when all the bits to be accessed are in sequence and in the same row of the array as the initial access. That is why, at the time of *read* operation, the contents of all the cells in the selected row (as per the supplied address) of SDRAM are to be loaded into the latches (buffer). Data available in the latches that correspond to only the selected columns (as per the supplied address) are then transferred into the data output registers as shown in Figure 4.7. During this operation, even if an access is made for the purpose of refreshing, the refreshing action will be simply carried out only on the contents of the cells, and not on the data located in the latches.

One of the salient features of the SDRAM is that it has an associated built-in *refresh circuitry.* A part of this circuitry is a *refresh counter* that provides the addresses of the rows being selected for refreshing. Normally, a typical SDRAM refreshes each of its rows in an interval of, at least, every 64 ms. Another key feature that differentiates SDRAM from traditional DRAM is the inclusion of a *mode register* and associated control logic attached with SDRAM as shown in Figure 4.7. This module provides a suitable mechanism to customize the SDRAM to suit specific system needs. It facilitates to realize various modes of operation that an SDRAM can usually perform. The particular mode is, however, selected by writing control information into this mode register. For example, the burst operations can be executed by specifying the burst length into the mode register, which is the number of separate units of data synchronously fed onto the bus. At that time, it uses the block transfer approach as the fast page mode feature. Moreover, this register also allows the programmer to adjust the latency between the receipt of a read request, and the beginning of a data transfer. In addition, SDRAM has a *multiple-bank* internal architecture that often improves on-chip parallelism whenever there is a scope as such.

The SDRAM is found to be most efficient, and perhaps performs best when large blocks of data need to be transferred serially. Block transfers are mostly prevalent in general-purpose computer applications in which main memory transfers are primarily to and from processor caches. Besides, the block transfers are often carried out in applications, like in multimedia, with high-quality video displays, word processing, and in other types of similar applications. Commercial SDRAMs as launched by **Intel** have defined **PC100** and **PC133** bus specifications to meet the requirements of contemporary available processors. These memory chips are accordingly used in the motherboards of computers with system bus speeds of 100 or 133 MHz.

A brief detail of SDRAM with figure is given in the following web site: http://routledge.com/9780367255732.

4.3.7.2 DDR SDRAM

One of the limitations of standard SDRAM is that it can perform all its actions once per bus clock cycle, and that is at the rising edge of the clock signal. This drawback has been, however, alleviated by introducing a new version of SDRAM, referred to as ***double data rate* SDRAM** (DDR-SDRAM) that transfers data on both edges of the clock, i.e. twice per clock cycle; once on the rising edge of the clock pulse and once on the falling edge of it. DDR-SDRAM accesses the cell array in the same way that traditional SDRAM does. The latency of these devices is also the same as that of standard SDRAMs. But, since they transfer data on both edges of the clock, their bandwidth is essentially doubled for long burst transfers. To make this possible to work with data at such a high rate, the cell array is organised in two banks. Each bank can be accessed separately. Consecutive words of a given block are stored in different banks. Such *interleaving* of words allows *simultaneous* access to two consecutive words located in two different banks that can now be transferred on successive edges of the clock. The concept of interleaving of memories, however, are discussed in Section 4.8.2 in more detail.

Continuous innovations with no indication of any let-up in the quest for further improved performance in existing DDR technology has been steadily progressing. As the technology is getting gradually matured, two generations of improvement to the DDR technology has evolved. The first one is the introduction of **DDR2** that increases the data-transfer rate by increasing the operational frequency of the RAM chip, and by increasing the prefetch buffer from 2 bits to 4 bits per chip. The prefetch buffer is essentially a memory cache located on the RAM chip that enables the RAM chip to preposition bits to be placed on the data bus as rapidly as possible. The next one is **DDR3**, introduced sometime in 2007, came out with some notable features as part of further improvement over the existing DDR2 including a higher clock rate, and an increase in the prefetch buffer size to 8 bits. Table 4.1 shows a rough comparison of the DDR generations in relation to some of their basic characteristics.

A brief detail of DDR read operation with timing diagram is given in the following web site: http://routledge.com/9780367255732.

TABLE 4.1
Comparison of Basic Characteristics of DDR Generations

Specification	Standard DDR (DDR1)	DDR2	DDR3
Voltage levels	2.5V	1.8V	1.5V
Prefetch buffer (bits)	2	4	8
Data transfer clock rate (MHz)	200–600	400–1,066	800–1,600
Front-side bus data rates (Mbps)	200, 266, 333, 400	400, 533, 677, 800	800,1,066, 1330,1600

4.3.7.3 Rambus DRAM (RDRAM)

As already mentioned, all DRAMs including the different generations of DDR chips use similar organisations for their cell arrays, and also access the cell array in a similar way. As a result, their latencies tend to be almost similar if the same components and same fabricating technology are employed in these chips. This *latency* as well as the *bandwidth* are often regarded as the *primary characteristics* of these DRAMs that mostly determine their performance as a whole. Again, the effective bandwidth as obtained in a computer system (involving data transfers between the memory and the processor) is not determined solely by the speed of the memory, but it also depends on the transfer capability, typically the speed of the memory bus. Memory chips are thus usually designed to meet the speed requirements of popular buses. The problem is that the bus speed can be attained only up to a certain limit and cannot be increased at will to improve the data-transfer rate. The only way to reasonably increase the data-transfer rate on a speed-limited bus is to increase the width of the bus by providing more data lines to widen the bus. From a design point of view, a reasonably wide bus is not only expensive, but needs a lot of space on the motherboard which is difficult to organise. An alternative approach that can negotiate this conflicting situation suggests the use of a comparatively narrow bus, but a relatively faster one in implementation. This approach was religiously used by **Rambus Inc.** to develop a proprietary design methodology known as ***Rambus.***

One of the key features of Rambus technology is its fast signalling method which is used to transfer information between chips. Instead of using signals that have usual voltage levels of either 0 or V_{supply} to represent the logic values, the signals here actually consist of much smaller voltage swings around a reference voltage, V_{ref}. This reference voltage is about 2V, and two logic values are represented by 0.3V swings above and below V_{ref}. This type of signalling is generally known as ***differential signalling.*** Small voltage swings, however, make it possible to have short transition times that, in turn, allow relatively higher transmission speed. Special techniques are thus here employed for the design of communication links (bus) to implement differential signalling, and special circuit interfaces are also designed to deal with these differential signals. All these together, however, ultimately put several constraints on making the bus wide. Rambus thus provides a complete specification for the design of such communication links called the ***Rambus channel.*** The earlier designs of Rambus, however, allowed for a clock frequency of 400 MHz, and while the data is transferred on both edges of each clock signal (PGT and NGT), the effective data-transfer rate then eventually attained 800 MHz. This design specified a channel that provided 9 data lines and a number of control and power supply lines. Eight of the data lines were intended for transferring a byte (8 bits) of data. The ninth data line can be used for other purposes: one such is parity checking. Subsequent enhancements specified additional channels. A two-channel Rambus, also known as **Direct RDRAM** has 18 data lines (16 actual data, and two parity) intended to transfer two bytes of data at a time, resulting in a signal rate of 800 Mbps on each data line. There are no separate address lines.

Rambus requires specially designed memory chips. These chips, however, use similar organisations for their cell arrays as found in standard DRAM technology. Multiple banks of cell arrays are interleaved to access more than one word at a time. Required circuitry needed by the cell arrays to interface to the Rambus channel is included on the chip. Such memory chips are known as ***Rambus DRAMs*** (RDRAMS) [CRISP,1997]. RDRAM chips are vertical packages, with all pins on one side. The chip exchanges data with the processor over 28 wires, no more than 12cm long. Presently, the bus can address up to 320 RDRAM chips, and is rated at 1.6 Gbps. This chip has been subsequently recognized, and later adopted by **Intel** for its ***Pentium*** and ***Itanium*** processors. Soon, it became the main competitor to SDRAM.

The special **RDRAM bus** delivers address and control information using an asynchronous block-oriented protocol. After an initial 480 ns access time, this produces the 1.6 Gbps data rate. What makes this speed possible is the bus itself, which defines impedances, clocking, and signals very precisely. Rather than being controlled by the explicit RAS, CAS, R/W, and CE signals as being used in traditional DRAM, an RDRAM gets a memory request over the high-speed bus. This request is communicated between the *master* (be it the processor or some other devices) and the RDRAM modules which serve as *slaves*, and it is carried out in terms of *packets* transmitted on the *data lines*. There are three types of packets: *request*, *acknowledge*, and *data*. A request packet issued by the master indicates the *type of operation* that is to be performed. The operation types include memory reads and writes, as well as reading and writing of various control registers present in the RDRAM chips. The request packet also contains the address of the desired memory location, and includes an 8-bit (byte) count that specifies the number of bytes needed to be transferred. When the master issues a request packet, the addressed slave (RDRAM) responds by returning a positive acknowledgement packet, if it can immediately satisfy the request. Otherwise, the slave indicates that it is *"busy"* by returning a negative acknowledgement packet, in which case, the master will try again. If the number of bits in a request packet exceeds the number of data lines, it then means that several clock cycles are needed to transmit the entire packet. Use of a narrow communication link is, however, compensated here by the availability of very high rate of transmission.

Figure 4.8 illustrates a schematic representation of the RDRAM consisting of a controller and a number of RDRAM modules connected via a common bus. The controller is located at one end of the configuration, and the far end of the bus is a parallel termination of the bus lines. As already mentioned, the bus includes 18 data lines cycling at twice the clock rate, thereby resulting in a signal rate of 800 Mbps on each data line. There is a separate set of 8 lines (Rambus channel, RC) used for addressing and controlling signals. There is also a clock signal that starts at the far end of the controller, propagates to the controller end, and then loops back. A RDRAM module sends data to the controller synchronously to the clock, to the master, and the controller sends data to the RDRAM synchronously with the clock signal in the opposite direction. The remaining bus lines include a reference voltage, ground, and power source.

RDRAM chips can be assembled into larger modules, similar to SIMMs and DIMMs (*see* next section: RAM module organisation). One such module called RIMM can hold up to 16 RDRAMs. Rambus technology, however, gradually matured with the passing of days and started to compete directly with the DDR SDRAM technology. Each one has its own several merits as well as certain drawbacks. Still, the one notable factor is that while DDR SDRAM technology is an open standard, RDRAM is a proprietary design of Rambus Inc.

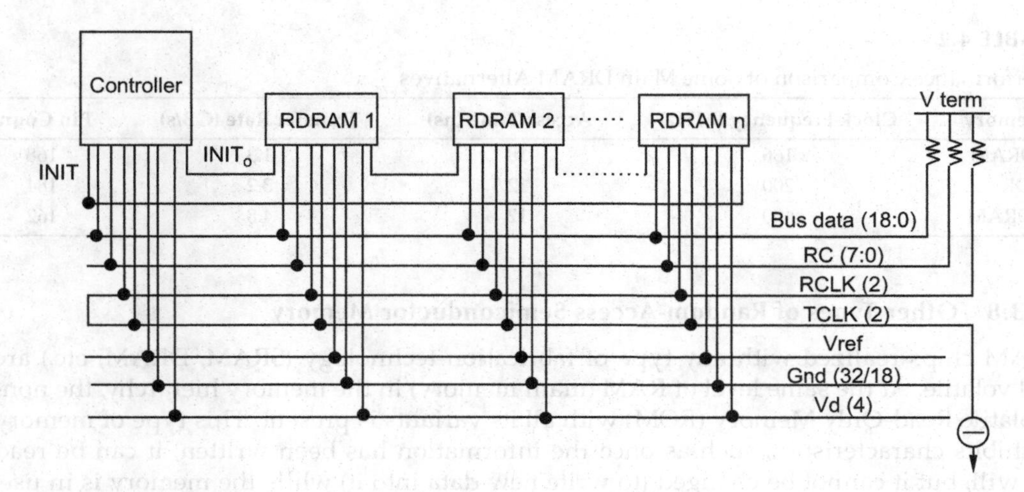

FIGURE 4.8
Rambus DRAM (RDRAM) structure.

for which the manufacturers engaged in fabricating RDRAM chips have to pay a royalty, an extra overhead added on the cost of the chips. However, from a performance point of view, even if it is found to be moderately adequate, the last word is certainly the price of the component that eventually often influences the final decision in regard to its usage.

4.3.7.4 Cache DRAM (CDRAM)

In recent years, a number of attempts have been made to further enhance the basic DRAM architecture. Some of them have been explored, a few of them have been implemented, and are presently in the market. CDRAM implementation is one such enhancement that received considerable attention. It is basically a combination of SRAM onto a DRAM in one single chip. A brief description of CDRAM is given here for an overall understanding. Cache DRAM (CDRAM) is essentially an integration of a small SRAM cache (16 Kb) onto a generic DRAM chip, developed and introduced by ***Mitsubishi Corporation***, Japan. The SRAM on the CDRAM can be used in many different ways. One such is that the SRAM portion of the memory present in CDRAM, by virtue of being faster in operation than its counterpart DRAM, can be used truly as a cache. This cache, as usual, consists of 64-bit lines. The cache mode of the CDRAM is, however, effective for ordinary random access to memory. The SRAM in CDRAM can also be used in another way, as a buffer to support the serial access to a block of data. As for example, to refresh a bit-mapped screen, the CDRAM can prefetch the data from the DRAM into the SRAM buffer. Subsequent accesses to the chip can then result in accesses entirely from the SRAM portion only, thereby making the usual memory accesses much faster.

A lot of schemes in relation to the extension of traditional DRAM architecture have been implemented to overcome its inherent constraints caused by its existing internal architecture as well as the limitations of its interface to the processor–memory bus. As a result, different categories of DRAM have been fabricated. But, only those schemes that currently dominate the market, and worthy of mention are essentially SDRAM, DDR–DRAM, and RDRAM. The performance of these products are, however, summarily compared in Table 4.2, taking into account mainly those basic parameters that have direct impacts on the performance of these RAMs.

TABLE 4.2

Performance Comparison of Some Main DRAM Alternatives

Memory	Clock Frequency (MHz)	Access Time (ns)	Transfer Rate (GB/s)	Pin Count
SDRAM	166	18	1.3	168
DDR	200	12.5	3.2	184
RDRAM	600	12	4.8	162

4.3.8 Other Types of Random-Access Semiconductor Memory

RAM chips realized with any type of fabrication technology (SRAM, DRAM, etc.) are all volatile. At the same level of RAM (main memory) in the memory hierarchy, the non-volatile Read-Only Memory (ROM) with all its variants is present. This type of memory exhibits characteristics, such as once the information has been written, it can be read at will, but it cannot be changed (to write new data into it) while the memory is in use, i.e., on-line alteration (if they at all can be altered) like RAM in the stored information is not possible. That is why, they are called read-only memories (ROMs). The basic type of ROM contains a permanent pattern of data that cannot be changed. However, there exist several variations of ROM with different characteristics, but all of them belong to the same family with non-volatile attributes, and as such they are commonly referred to as types of ROM. Such memories are found in extensive use in *embedded systems* and are also targeted to achieve certain specific objectives, and as such they seldom use disk storage devices. Their programs are usually stored in non-volatile semiconductor memory devices.

There are *five* basic ROM types:

i. *ROM*
ii. *PROM*
iii. *EPROM*
iv. *EEPROM*
v. *Flash Memory*

Each type has its own unique characteristics, which will be discussed in this section, but they are all different types of *memory* with *two* things in common:

- Data stored in these chips is non-volatile: it is not lost when power is removed.
- Data stored in these chips is either unchangeable or requires a *special operation* to change (unlike *RAM*, which can be changed as easily as it is being read).

However, each type is equally important, and each one by virtue of its own unique characteristics finds its own specific usage in particular application areas.

4.3.8.1 ROM (Read-Only Memory)

A ROM is a non-erasable storage device with permanently stored data which is injected at the time of its fabrication. While a ROM can be read at will, it is not possible to write new data into it. The presence of ROM offers the advantage that the data or program already stored in ROM means it is permanently available in main memory during run-time, and need not be loaded ever from a secondary device. A ROM is essentially a combinational circuit that consists of an array of semiconductor devices, such as *diodes, bipolar transistors* or *MOSFETS,* and is also equipped with a decoder and an encoder. With a ($2^n \times K$) bits capacity ROM, the 2^n number of locations can be addressed separately one at a time by the *n*-address inputs ($A_0, A_1, A_2, \ldots, A_{n-2}, A_{n-1}$) submitted to the decoder, and *K* denotes the number of bits at each location that contains stored data of any location available later on as the output. Thus, a ROM may be called a *multiple input/multiple output* combinational circuit.

Like any other integrated-circuit chip, ROM is created with the data actually wired-in to the chip as part of the fabrication process. This, in turn, invites two problems:

i. During fabrication, if one bit gets wrong, there is no way of error correction, and as a result, the whole batch of ROMs under fabrication must be discarded.
ii. The data insertion process into the ROM during its fabrication includes a relatively large fixed cost irrespective of whether a single copy or thousands of copies of a particular ROM are fabricated.

ROMs at Work: Similar to *RAM*, the ROM chips also contain a grid of columns and rows. But at the intersection of the columns and rows, ROM chips are fundamentally different from RAM chips, and it uses a diode or transistor to connect the columns and row lines to provide a value 1. For a value 0, the lines are not connected at all. A *diode* normally allows current to flow in only one direction and has a certain threshold, known as the ***forward breakover***, that determines how much current is required before the diode will pass it on. In silicon-based items such as processors and memory chips, the forward breakover voltage is approximately *0.6 V*. By taking advantage of this unique property of diode, a ROM chip can send a higher charge than the forward breakover down the appropriate column with the selected row grounded to connect at a specific cell. If a diode is present at that cell, the charge will be conducted through to the ground, and, under the binary system, the cell will be read as being "on" (a bit: 1). On the other hand, if there is no diode at that intersection to connect the column and row, the cell at that cross point stored a bit: 0. By changing the position of the diodes in the matrix, we can obtain different fixed outputs (essentially binary).

It can be seen from the working of a ROM chip that correct programming with perfect and complete data must be present when chip is fabricated (created). Creating the original error-free template for a ROM chip is often a laborious process, full of trial and error. But the benefits of ROM chips outweigh all its drawbacks. Once the desired template is completed, each actual chip production requires very little cost. They use very little power and are extremely reliable, and in the case of most small electronically controlled devices, it contains all the necessary fixed programming with related data that entirely controls the attached devices. Other potential applications of ROM include:

- Code conversion
- Character generator for CRT display from ASCII code generated by the depression of the key from the keyboard
- Function look-up table (table corresponding to the binary values of trigonometric, exponential, and other functions).

- In process-control system and in many other systems, often used in our daily life: an ideal example is the small chip used in the singing fish toy. This chip, about the size of a fingernail, contains the 30-second song clips written in ROM as well as the control codes to synchronize the motor to the music to swing the different parts of the toy's body in accordance with the rhythm of the song.

A brief detail of the implementation of ROM with figures is given in the following web site: http://routledge.com/9780367255732.

4.3.8.2 PROM (Programmable ROM)

Creating ROM chips totally from scratch is time-consuming and equally very expensive, if it is produced in a small scale. Mainly, for this reason, and also when only a small number of ROMs with a *particular memory content* is needed, developers found out a less expensive alternative type of ROM known as *programmable read-only memory (PROM).* Like the ROM, the PROM is also non-volatile. Blank PROM chips are available with a lesser cost, and may be coded (written into) only once by anyone with a special tool called a *programmer.* The writing process for the PROM is executed electrically, and may be performed by the supplier or customer at a later time than the original chip fabrication. Although PROM provides more flexibility and convenience than ROM, the ROM still remains attractive for cheaper high-volume production runs.

PROMs are fabricated by the manufacturers in which the active switching devices (like BJTs or MOSFETs) are placed *at each intersection* of rows and columns of the matrix, unlike *at the selected intersection* of rows and columns of the matrix as in ROM. A fusible link is associated with each active device in series for connecting a row and a column. For the BJT, this fusible link is placed at the emitter. In MOSFET, it is at the source. The fusible link (F_L) can be selectively blown out by the user at his own choice where the connections between a row and a column are not wanted. This blowing out of the F_L is done by passing sufficiently large current in pulsed form. The process of selective blowing of F_L is called the ***programming*** or ***burning*** the ROM. The ROMs in which these programming facilities are available are called ***Programmable ROM*** or ***PROM***.

Since all the cells have a fuse, the initial (blank) state of a PROM chip is all 1s. To change the value of a cell to 0, i.e. to store a *bit-0* at any cross-point of the memory matrix, a tool *"programmer"* is used that sends a specific amount of current to the cell to blow out the F_L so that the connection between the row and the column at that cross-point is disrupted permanently, thereby treating the cell at that position as 0. Still the active element used in the PROM is present there, and is of no use for that cross-point, but reserves a large silicon area. To store a *bit-1* at any cross-point of the memory matrix, it is required to keep the F_L intact at that cross-point so that by applying a proper address to the row and column decoders, the row line and column line can be connected through the active device present at the cross-point and the intact fusible link (F_L). Therefore, burning a PROM means selective programming for 1s. Once the PROM is burnt, the bit pattern in the memory matrix becomes fixed, i.e. programmed, and cannot be changed, but can be read over and over again. The PROMs are usually available in smaller sizes with TTL compatible output, Open-collector or Tri-state output. But nowadays, large capacity PROMs with electrically fusible *polysilicon* links have been developed for commercial use. Although PROMs can be programmed, and only once, they are more fragile than ROMs. A jolt of static electricity can easily cause even needed fuses in the PROM to burn out, thereby changing essential bits from 1 to 0, making the contents to become erroneous. However, blank PROMs are less costly, and are ideal for prototyping the devices.

Another variation of *read-only memory* is the ***read-mostly memory***, which is useful for many applications in which read operations are far more frequent than write operations, and that too also from a storage which should be non-volatile. There exist as such ***three*** common forms of read-mostly memory: *EPROM, EEPROM,* and *Flash memory.*

4.3.8.3 EPROM (Erasable PROM)

Working with ROMs and PROMs can be a wasteful business. Even though they are inexpensive per chip, the cost can add up overtime. To address this issue, a compromise between the performance of ROM and the erasability of RAM system is made, and that is by the use of *erasable and programmable* ROM or EPROM. The EPROM chip is *optically erasable*, and can be rewritten many times. The EPROM is erased *off-line* and it requires a special tool that emits a certain frequency of ultraviolet (UV) light. EPROM is read and written electrically like PROM, but it is more expensive. To program (write into) an EPROM, it is configured using a particular type of tool that provides voltage at specified levels depending on the type of EPROM used. Multiple update capability is its distinct advantage.

In an EPROM, similar grids of columns and rows are used. The cell at each intersection of a row and a column has *two* transistors. The two transistors are separated from each other by a *thin oxide layer* (10–20 nm deep). One of the transistors is known as the ***floating gate*** and the other as the ***control gate***. The floating gate's only link to the row (word line) is through the control gate. As long as this link is in place, the cell has a value of 1. To change the value to 0 requires a curious process called Fowler–Nordheim tunnelling. Tunnelling is used to alter the placement of electrons in the floating gate. An electrical charge, usually *10–13 V*, is applied to the floating gate. The charge comes from the column (bit-line), enters the floating gate, and drains to the ground.

This charge causes the floating-gate transistor to act like an electron gun. The excited electrons are pushed through, and trapped on the other side of the thin oxide layer, giving it a negative charge. These negatively charged electrons act as a *barrier* between the control gate and the floating gate. A device, called a *cell sensor*, monitors the level of the charge passing through the floating gate. If the flow through the gate is *greater than 50%* of the charge (threshold), it has a value of 1. When the charge passing through drops *below the 50%* of threshold, the value changes to 0. A blank EPROM has all of the gates fully open, giving each cell a value of 1.

Erasing Technique: To rewrite an EPROM, it should be *erased first*. The EPROM can be erased multiple times, and hence, allows rewriting of data virtually indefinitely. To erase it, a supply of a level of energy strong enough to break through the negative electrons blocking the floating gate is required. In a standard EPROM, this is best accomplished with UV light at a *frequency of 253.7* or with the use of mercury arc lamps emitting strong radiation at a wavelength of 2,537Å (4.9 ev). Because this particular frequency will not penetrate most plastics or glasses, each EPROM chip thus has a quartz window on top of it. The EPROM must be placed very close to the eraser's light source, within an inch or two, to work properly for a duration of 15–20 minutes. To erase an EPROM, a device called *EPROM* is used. An EPROM eraser is not selective, it will erase the entire EPROM. The EPROM to be erased must be first removed from the device in which it resides, and then the EPROM is placed under the UV light of the *EPROM eraser* for several minutes. An EPROM that is exposed for a long duration under the UV light can be over-erased. In such a case, the EPROM's floating gates are charged to the level that they are unable to hold the electrons at all.

EPROMs are rated in *k-bits* where *k* is equal to 1,024, and the EPROM number generally (but not always) reflects the size. A couple of examples are: the ***2716*** EPROM number ends in 16, and thus is *16Kbits* in size or16 × 1,024 or 16,384 bits. Now, 16,384 bits = 2,048 bytes or 2Kbytes. Thus, a 2716 is a 16K-bit EPROM, but is most often expressed as being 2Kbytes (2K × 8) in size. The chip ***IC 2732*** is **32K bits** having memory organisation 4K × 8. The chip ***IC 2764*** is a **64K bits** having a memory organisation of 8K × 8. Some EPROMs are word-wide or 16 bits (2 bytes) wide. These EPROMs are also rated in bits, such as ***27C1024*** is a 1,024K-bits, or 128K-bytes, or 64-word EPROM. Such EPROMs usually come in 40-pin packages, thereby providing extra pins that may be used otherwise.

Windowless package EPROM is One Time Programmable (**OTP**) and is less costly. ***Intel P2764*** (8K × 8) is one time programmable, and a highly reliable and fast device. This has been designed for high volume production. All the 8,192 locations (8K) can be programmed almost within 2/3 minutes.

EPROMs using UV erasing technique mainly have ***three*** drawbacks.

- Accidental erasure due to being exposed to the room lighting of fluorescent lamps and incandescent lamps, emitting a wavelength of about 3,000Å (= 4.1 ev), erase a device within 3 years against a normal life of 15 years. Direct sunlight can have the same effect within approximately 1 week. Manufacturers of this device provide opaque labels which are placed over the quartz windows to protect them from such unintentional erasure.
- They are expensive due to the special ceramic packing.
- They have to be removed from their circuit boards for erasing and reprogramming. For this third drawback, a special socket called ZIF (Zero Insertion Force) is often used with EPROMs so that they may be easily removed from and again reinstalled into the circuit board for the purpose of reprogramming, whenever needed.

Implementation: IC 2716 EPROM 2D organisation is typically used by most read-only (ROM), and read-mostly chips (EPROM, EEPROM) with all the bits of each word on one chip (wafer).

The implementation details of the 24-pin package **IC 2716** are briefly described in the web site: http://routledge.com/9780367255732.

4.3.8.4 EEPROMs (Electrically Erasable PROM)

Although EPROMs are a big step up from PROMs in terms of reusability, they still require dedicated equipment and a labour-intensive process to remove and reinstall them each time a change is necessary. Also, changes cannot be incrementally made to an EPROM; the whole chip must be erased before each upgrade. Electrically erasable programmable read-only memory (EEPROM) chips remove all these biggest drawbacks of EPROMs.

In EEPROMs:

- The chip does not have to be removed from the device for modification (rewritten).
- The entire chip does not require to be completely erased to change a specific portion of it. At any time, without erasing the prior contents, only the byte or bytes being addressed can be modified.
- To change or to modify the contents, it does not require any additional dedicated equipment.

The modified structure of the floating gate used in EEPROM is called ***Floating Gate Tunnel Oxide*** (FLOTOX). At each cross-point of a row and column of the cell matrix of the EEPROM, a flotox cell along with a normal MOSFET is used in series.

Erasing Technique: Instead of using UV light, it can be possible to return the electrons in the cells of an EEPROM to normal with the localized application of an electric field to each cell. This erases the targeted cells of the EEPROM, which can then be rewritten. The contents of EEPROMs can be changed even 1 byte at a time, which makes them versatile, but relatively slower in operation. Later on, hardware designers developed EEPROMs with the erasure region broken up into even *smaller "fields"* (instead of 1 byte) that could be modified individually without affecting the others. To alter the contents of a particular memory location, it copies the targeted entire field into an off-chip buffer memory, erasing the field, modifying the data in the buffer as required, and then rewriting it into the same field. This, however, requires considerable computer support to accomplish. In fact, EEPROM chips are too slow to be used in many products that require quick changes to the data already stored on the chip.

IC Chip ***2816*** is a 2K × 8 EEPROM with 24 pins. It is pin-to-pin compatible with ***2716*** EPROM as already discussed.

4.3.8.5 Flash Memory (Flash EEPROM)

The particular drawback (limitation) of the slower operational speed of EEPROM has been alleviated in *Flash memory,* a type of EEPROM that uses in-circuit wiring for erasing by applying an electrical field to the entire chip or to pre-determined sections of the chip called ***blocks.*** Flash memory was introduced sometime in 1984 combining the features of the high cell density of EPROM and the in-circuit erasing and programming facilities of EEPROM. A flash cell is based on a single transistor controlled by trapped charge, just like an EEPROM cell. In fact, the flash memory cell structure is very similar to that of EPROM, except for the use of a high quality and thin (10–20 nm deep) tunnelling oxide under the floating polysilicon gate. Programming means storing electrons in the floating gate. The cells are programmed by the *hot electron injection technique.*

While similar in many respects, there are also substantial differences between the flash and EEPROM devices. In EEPROM, it is possible to read and write the contents of a single cell (1 byte). In a flash device, it is possible to read the contents of a single cell, but it is only possible to write an entire block of cells. Consequently, flash memory works much faster than traditional EEPROMs due to writing data in an entire block of cells, usually 512 bytes in size, instead of 1 byte at a time. Of course, prior to writing, it copies the contents of the entire targeted field into an off-chip buffer memory, erasing the field, modifying the data in the buffer as required, and re-writing the contents of the buffer into the same field. To accomplish all these, it requires considerable computer support. However, PC-based EEPROM flash memory systems are often equipped with their own dedicated micro-processor system to carry out all these tasks. Flash devices, however, have a greater density, which leads to higher capacity, and consequently a lower cost per bit. They require single power supply voltage, and consume relatively less power during their operation. Flash drive is more or less a miniaturized version of this flash memory.

The chips ***27F64*** (8K × 8), **27F256** (32K × 8), and ***28F256*** (32K × 8) are a few examples of available flash memories that are commercially used. In traditional personal computer (PC) system, the chip marking "AMIBIOS" is the flash memory that shows a common use of this memory containing Binary Input Output System (BIOS) data.

Computers of today access modern flash memory systems very much like hard disk drives, where the controller system has full control over where information is actually stored. The low power consumption of flash memory makes it attractive for use, especially, in portable equipment implemented mostly by embedded systems that are driven by battery. Typical potential applications include cell phones, digital cameras, MP3 music players, hand-held computers, and many more, in which flash memory holds the essential software and needed data, thereby alleviating the need of a disk drive to be used. To meet the actual requirement, larger memory modules consisting of a number of such chips are thus often needed. Two commonly used implementation of such modules are of primary interest: *flash cards* and *flash drives.*

Flash Card: It is an unit which consists of a number of flash chips mounted on a card, thereby providing relatively larger memory space to the user. Such cards have a standard interface that enables them to be attached with a variety of products by way of simply plugging them into conveniently accessible slots available within the products. It is reported that Lexar introduced a Compact Flash (CF) card in 2000 with a USB connection, and a companion card read/write and USB cable that eliminated the need for a USB hub. Today, flash cards are commonly available in a variety of memory sizes, having typical sizes of 64, 128, 256 Mbytes, and even more.

Flash drives: Continuous improvement was on going with a target to obtain even more larger memory space, at least to the extent that could eventually be able to replace a standard hard disk drive. Consequently, these drives have been designed in a way that they would be fit in standard disk drive bays of any system. However, the storage capacity of these flash drives are significantly lower relative to the usual sizes of the hard disks at present. As the flash drives are semiconductor devices, they are relatively much faster in operation than the electro-mechanical hard disk drives. Moreover, for being an entirely solid state device, they naturally consumed much less power to carry out their operation in comparison with their counterpart hard disk drives. This attribute enables them to use in portable systems that are mostly battery-driven. In addition, they are also insensitive to vibration. But, the main drawbacks of flash drives are their much smaller capacity, and relatively higher cost per bit when compared with traditional hard disk drives.

4.3.8.6 USB Flash Drive: Pen Drive

A **USB flash drive** is a physically small and weighing less than 30gm, but a relatively high-volume data storage device that contains flash memory with an integrated Universal Serial Bus (USB) interface, and are typically removable and equally rewritable. It provides much faster data-transfer rate than relatively larger optical disk drives like CD-RW or DVD-RW drives (discussed in detail in later sections) with adequate storage capacity that eventually befits it for most of the applications frequently used on a regular basis. In addition, it is supported natively by almost all modern operating systems, and can be read by many other systems including current DVD players, and also by many mobile smartphones. It is interesting to note that nothing moves electro-mechanically in a flash drive; yet, the term *drive* persists. Because, computer systems *read* and *write* flash drive data in a similar manner using the same system commands as those for electro-mechanical disk drives, with the storage appearing to the computer operating system and user interfaces as just another drive. Flash drives are, however, very robust mechanically, and are more durable and reliable, because they have no moving components.

First Generation USB flash drives product was commercially introduced in 2000 under the brand name *"Thumb Drive"* by many different companies all over the world, and it had a storage capacity of 8 MB, more than five times the capacity of the then-common floppy disk. **Second Generation** flash drives introduced USB 2.0 connectivity having 480 Mbits/s (60 MB/s) transfer capacity, but the drives were still not able to make use of this full transfer capacity, because of technical limitations inherent in NAND flash. Although the fastest drives available at that time used a dual channel controller, they claimed to read at a rate of up to only 30 MB/s, and write at about half that speed which was considerably short of the transfer rate available from a current generation hard disk, or less than the maximum high speed USB throughput that could be obtainable. **Third Generation** flash drive was announced in late 2008 using USB 3.0 which offered radically improved data-transfer rates up to 5 Gbit/s (625 MB/s), although the befitting consumer devices were not available until the beginning of 2010. As of 2012, most USB 3.0 flash drives, like their predecessors, still were not able to utilize the full speed of the USB 3.0 interface due to the limitations of their memory controllers. However, single packaged devices with capacities of 64GB were readily available as of the end of 2011, and devices with 8GB capacity are quite economical. Storage capacities in this range are considered enough space for both the operating system software and some free space for the user's data. All USB 3.0 devices are, however, downward compatible with USB 2.0 ports.

Similar to generic hard drives, flash drive sectors are 512 bytes long, and the first sector can also contain a master boot record and a partition table. Most flash drives are supplied preformatted with a FAT32 file system, but can be reformatted also as usual to any file system supported by the host operating system. In **Windows Vista** and **Windows 7**, the **ReadyBoot** feature allows use of flash drives (from 4 GB in the case of Windows Vista) to augment operating system memory. **Windows 8** provided a feature called ***Windows To Go***, which allows users to carry a bootable copy of Windows 8 alongside their documents and programs in a USB drive. Most current PC firmware (BIOS) permits booting from a bootable USB drive, and such a configuration is known as a ***Live USB***.

Flash drive is very convenient for use often as a ***backup*** medium, mainly for the following reasons:

- This is simple for the end-user, and more likely to be done.
- The drive is physically small in size and convenient, and more likely to be carried off-site for safety.
- The drives are less fragile mechanically and magnetically than tapes.
- The capacity is often large enough for several backup copies of critical data.
- Flash drives are cheaper than many other traditional backup systems.

A brief detail showing the internal assembly of different components of the pen drive and their usage with the respective figure is given in the following web site: http://routledge.com/9780367255732.

4.3.9 RAM Module Organisation

A large memory is always a crying need, because it intuitively leads to a better performance, since more programs and data that are involved in the processing are now readily available in the memory during run-time, thereby reducing the frequent time-consuming visit to slower secondary memory for accessing the desired information. In the past, as the capacity of an available memory IC chip was not large enough to fulfil the constantly growing user requirements, a larger main memory was then built by organizing an array of memory IC chip directly on the *motherboard* (the main system printed-circuit board that contains the processor).

The details of this module organisation, its working, and addressing mechanisms are explained with figures in the following web site: http://routledge.com/9780367255732.

Modern present-day computers, even from a small personal computer to a typical workstation, usually have very large memories, to the order of several gigabytes. If such a memory is built up by using an array of even small-sized large-capacity DRAM chips directly on the *motherboard,* it will then occupy such an unacceptably large amount of space on the said board that is practically difficult to afford. Moreover, it is awkward and equally difficult to keep provision (slots) for future expansion of the memory, because space must be allocated, and associated wiring must be provided for realizing the additional maximum possible expected size.

All these limitations that stand as an obstacle on the way to realizing larger memory units within space constraints have led to the development in considering suitable handy packaging to produce a relatively larger memory known as **SIMM** (Single In-line Memory Module) and subsequently **DIMM** (Dual In-line Memory Module). Such a module is actually an assembly of several memory chips on a separate small board that can be directly plugged vertically into a single socket available on the motherboard. Different sizes of SIMMs and DIMMs are thus designed that use a socket of the same size .For example, 32M × 32, 64M × 32, and 128M × 32 bit DIMMs, all use the same 100-pin socket. Similarly, 8M × 64, 16M × 64, 32M × 64, and 64M × 72 DIMMs use a 168-pin socket. Such modules not only occupy a comparatively smaller amount of space on a motherboard, but they also allow easy expansion by simply making a replacement of the existing module, if the desired larger module uses the same socket as the smaller one, available already on the motherboard.

4.4 Serial-Access Memory: External Memory

Primary memories realized by contemporary semiconductor technologies are quite fast due to the directly accessible and independently addressable (random) features of each of its storage location. But, one of their major limitations is that they are volatile, i.e. permanent storing of information is not possible. Moreover, their storage capability is limited, mainly due to the cost consideration in relation to per bit of stored information, and hence, they cannot fulfil the storage-space requirements of the computer systems. Large storage requirements with permanent storing of most general-purpose computer systems are thus fulfilled with the use of serial-access memories (often called secondary/external memory). The most commonly used memories of this category are economically realized in the form of *magnetic disks, magnetic tapes, optical disks,* and nowadays *flash drives* (*pen drives*). These memories except flash drives (already discussed) will now be discussed in the subsequent sections one after another.

4.4.1 Characteristics

Most computers have *slower, cheaper,* and usually much *larger* serial memories to permanently hold bulk volume of data that are usually stored on a set of fixed paths or ***tracks***. Tracks are a concentric set of rings (or a set of parallel lines), and each track consists of a sequence of cells; each cell (storage location) is capable of storing 1 bit of information. Read–write circuitry is here *shared* among different storage locations. Every track has over it a number of fixed access points at which the information can be transferred to or from the track. Transfer of information to and from any track is accomplished by moving either the stored information, or the read–write unit (Head) or both. Positioning the read–write unit (Head) on the desired track is a direct one, and then the information being transferred to and from the accessed track is essentially serial. The larger size of serial memories realized by electromechanical equipment using magnetizable material to construct small simple storage cells is low-cost (inexpensive) and also portable, but they provide relatively long access time which is mostly due to the following factors:

i. The time taken to position the read–write unit (Head) on the desired track.
ii. The time elapsed due to relatively slow speed of the track movement.
iii. The time taken to transfer the data serially, and not in parallel to and from the memory.

Apart from the size to be accounted, the access time of serial memory is also considered as a critical parameter at the time of its selection. The major characteristics that create the differences among the various types of serial memories are [Mee, C., et al.]:

Head/Storage media movement: In a *movable-head system* (except a *fixed-head system*), there is only one read/write head which is mounted on an arm, and the arm, in turn, can be extended or retracted over the tracks.

Access Procedure: In a movable head system, the only read–write head is needed to move between different tracks. The *average time* required to move a head from one track to another is called the ***seek time*** T_S of the memory. Once the head is placed on the desired track, the targeted cell may or may not be just under the head. A certain amount of time is thus needed for the targeted cell to reach under the read–write head so that data transfer can begin. The average time needed for this movement of the cell is called the ***latency*** or sometimes called the ***rotational delay*** T_L of the memory. A substantial amount of time has thus been already consumed by way of *seek time* and *latency time* for only to reach the targeted information. So, once the destined word is found, it is not economical and even inefficient to access just one word per access. A chunk of words is, hence, normally accessed in place, and these words are usually grouped into a larger unit called ***block***. An entire block consisting of a group of words is then accessed using only one seek and one latency time instead of many.

Data Transfer Time: To compute an estimated average time required to transfer the entire information of a **sector** (consisting of several blocks) or a **block** (consisting of several words) in a serial-access memory, we assume for the sake of simplicity that the memory has closed rotating tracks (disk), and this time then essentially consists of *three* fundamental components:

a. First the head(s) is(are) to be placed on the right track (cylinder) containing target sector(s)consuming an **average** ***seek time***, say, T_S. Typical value of T_S is usually about 9 ms.

b. Head is now placed on the track. The start of the desired sector on the track may be just under the head, or may be at any position on the track. Hence, on an average, half the revolution is required to get the beginning of the target sector under the head. This time is the ***latency time*** T_L, or sometimes called ***average rotation*** which is,

$$T_L = 1/2 \times 1/r \ seconds = (1/2) \times (60 \times 1/rpm) \ minute$$

c. The read–write head is now positioned at the starting point of the requested sector(block). Transfer of data can now be started. The *rate* at which the data is transferred depends primarily on:
   - the speed at which the stored information is moving under the head. This is expressed in terms of revolution per minute (*rpm*). For example, the *rpm* of a rotating device is 7,200, it means that the device can accomplish 7,200 rotation per minute. This characteristic is often supplied by the vendor.
   - the number of bits to be transferred depends on the *storage density* along the track. The density is defined in bits per unit length.

The rate at which the information is transferred continuously to or from a track under any situation is called the ***data-transfer rate.*** If a track has a storage density of *k* bits per centimeter and moves at a velocity of *v* centimeter per sec past the head, then the number of bits being transferred per sec is *kv*. This *kv* is considered as the *data-transfer rate.* Suppose each track has an average capacity of *N* sectors (blocks), and each track is rotating at the speed of *r* revolutions per second, then; the **data-transfer rate** of the device is $r \times N$ sectors (blocks) per second.
So, the time required to transfer 1 sector (block) is = $1/(r \times N)$ second. In other words;

$$\text{Transfer time } (T_{\text{avg}}\text{transfer}) = 1/(r \times N) \text{ s} = 60 \times (1/\text{rpm}) \times 1/(\text{sector per track}) \text{ per minutes.}$$

Therefore, the **total average time** required to transfer the entire information from any specified target sector is:

$$T = T_S + T_L + T_{\text{avg}}\text{transfer}$$

Platters: A *single-platter* (a single circular platter) constructed of plastic, coated with a magnetizable material forms the disk. Data are recorded on circular tracks already created on the disk, and later retrieved from the disk via the head. During a read/write operation, the head is movable while the platter rotates beneath it. Floppy disks and CDs are of this type. Some disk drives accommodate *multiple platters* stacked vertically about an inch or less apart. Each one is fixed on a spindle at a definite distance along the length of the spindle. Multiple arms are provided in between the platters. Each arm has two heads; one for the lower surface of the upper platter, and the other is for the upper surface of the lower platter. In fact, each arm moves radially through inter-platter gap. All the arms now move together back and forth in a fixed linear path so that all heads move in unison to select a particular track or a set of tracks. The spindle is given a circular motion with the help of a rotating shaft attached to an electric motor to give a constant circular motion to the attached platter. The entire unit is known as a ***disk pack***.

Sides: If the magnetizable coating is applied to one side of the platter, the disk is known as a **single-sided** disk. Floppy disks of the early days, and nowadays CDs are of this type. For most disk (hard disks), the magnetizable coating is applied to both sides of the platter and are referred to as **double-sided**. Today's hard disk is of this type.

Portability: A *non-removable disk* permanently mounted in the disk drive is a sealed pack, and it is used as one unit. All the hard disks used in today's computers (except mainframes) are non-removable disks. A *removable disk* is mounted in a disk drive. The disk drive consists of an arm with the read/write head attached to it, a shaft that rotates the disk, and the entire electronic circuits required for transferring (input/output) binary data. The removable disk may be of single platter (CDs) or may be of multiple platter (disk pack used in a mainframe large system). The removable disk can be taken away manually from the disk drive and replaced with another disk on the same drive even when the computer is on. The advantage of this type of disk is that an unlimited amount of data is available with a limited number of disk systems. Furthermore, a disk may be moved from one computer system, and it can be mounted in another compatible system.

Head Mechanisms: It is an important parameter in electromechanically accessed magnetic memories that distinguishes between magnetic-disk and -tape memories, and also provides a clear classification of the disk into *three types*. The **Contact** type (magnetic tape) mechanism allows the head to come into physical contact with the surface of the magnetic medium during a read/write operation. In the **Fixed-gap** type, the read/write head is positioned a fixed distance apart above the magnetic surface, allowing an air gap. This very narrow space separates the head from a cell on the storage track of the surface, so that the action of the head can interact to transfer information between the head and the storage medium. In **Aerodynamic-gap** type (Winchester), the mechanism is used in a sealed drive assemblies that almost have no contaminants (dust, pollution, etc.). The head here in the shape of an aerodynamic foil rests lightly on the platter's surface (magnetic surface) when the disk is not in motion. When the disk comes into motion, the air pressure generated by the rotating disk is enough to make the foil rise above the surface to operate closer to the disk's surface but not in contact. It can then interact with the platters' surface to transfer information between the head and the storage medium. Nowadays, any aerodynamic head design used in a sealed-unit disk drive uses the term *"Winchester"*.

A brief detail of these topics with the respective figures and a solved problem based on these characteristics is given in the following web site: http://routledge.com/9780367255732.

4.4.2 Rotating Memory (Disk) Organisation

Serial access memories can be accessed in a number of ways. Let us assume that each word is stored along a single track in consecutive cells and that each access results in the transfer of a block consisting of n words, where $n \geq 1$. Now, the address of the information to be accessed is issued, and it is composed of *track address* which determines a particular track, and the *block address* which indicates the targeted block within the accessed track. This entire address is then applied to the address decoder to generate the respective track address and block address. The track address is then used to place the head on to the right track, but the targeted block cannot be accessed until it comes under the head. To determine its arrival, some type of *track position indicator* mechanism is required which constantly generates the address of the block that is currently passing the read–write head. The generated address is compared with the block address as produced by the address decoder. When the match occurs, the head is enabled, and data transfer begins between the storage track (or buffer) and the memory buffer register (MBR). When a complete block of information is totally transferred, the head is disabled.

Serial memories are constructed with various types of storage media and different types of access mechanisms, and special techniques are used to store all information only in tracks on the surface. Hence, all of them similar to magnetic disk, magnetic tape, optical disk, etc. follow this common strategy of organisation, but, of course, each is implemented differently as found to be deemed fit.

A brief detail of this topic with the respective figure is given in the following web site: http://routledge.com/9780367255732.

4.4.2.1 Read–Write Mechanism

In magnetic surface memories (or CDs), the tracks in a disk are in the form of concentric rings on a platter. The tracks in the tape are horizontal parallel lines along the length of the tape. Whatever be the formation of track, the read–write mechanism is *identical* in the disk and tape, irrespective of the type of storage media and the access mechanism being used. Each track has a number of cells, and each cell has two stable magnetic *states* defined by the *direction* or *magnitude* of magnetic flux (or magnetic intensity) in the cell that represents the logical 0 and 1 values. Electric currents are used for altering (writing) and sensing (reading) the *magnetic state* of the cell. An external read–write head formed with a ring of soft magnetic material around which there is a coil is used for this purpose. The read and write currents pass through the coils and produce a magnetic field. There is a gap between the two ends of the ring that permits the magnetic flux to pass into the space between the head and the cell to induce the respective fields in the cell on the track of the magnetic surface.

A brief detail of this topic with the respective figure is given in the following web site: http://routledge.com/9780367255732.

4.4.3 Device Controller: Rotating Memory

Operation of a rotating device, be it a disk or a magnetic tape, is controlled by a *device controller* circuit (as a part of the I/O module, see Chapter 5) which also provides an interface between the device and the bus that connects it to the rest of the computer system. The device controller uses suitable mechanisms including the DMA scheme to transfer data between the devices and the main memory. A device controller may be used to control more than one device of the same type, and is initiated by the operating system (OS) to access any device attached with it. The OS initiates the data transfers by issuing Read/Write requests to the controller with the necessary address information, such as *main memory address, disk address* (containing sector/track address and block address), *word count* (the number of words in the block to be transferred), and control information, such as, *seek, read, write,* and *error checking.* If the disk drive is connected to a bus that uses packetized transfers, then the controller must be equally equipped to handle such transfers. For example, a controller for a ***SCSI drive*** conforms to the SCSI bus protocol (described in Chapter 5, SCSI bus).

A brief detail of this topic with the respective figure is given in the following web site: http://routledge.com/9780367255732.

4.4.4 Magnetic Disk

Magnetic disks are the primary devices that build up the foundation of an external memory system virtually on all present-day computers. A magnetic disk is essentially a flat circular platter constructed of nonmagnetic metal or plastic with a thin coating of magnetizable material on both sides of its surface. More recently, glass has been introduced to make the platter that, by virtue, accrues a number of notable advantages. Each platter has two ***surfaces***, and each surface has up to several thousand ***tracks*** of the same width in the form of concentric rings. Data are recorded on the track, and later retrieved from it. Adjacent tracks are separated by gaps which prevent the interference of magnetic fields and serve some other purposes as well. The number of bits stored on each track are, however, the same. Hence, the density measured in bits per linear unit increases as one moves to the innermost track. Typically, each track is divided into several sectors (in the range of a few hundred sectors). Data are transferred to and from the disk in sectors/blocks. These sectors may be of either fixed or variable length. Adjacent sectors on each track are separated by *intra-track* gaps. Data in the form of a record is stored in sectors. Typical sector size may be 512 bytes. Each sector may contain more than one record or a record (large record size) may take more than one sector.

A disk drive is commonly referred to as a ***direct-access*** device. This means that to get to the desired sectors, it need not read all the sectors in between sequentially. It can access any track directly (randomly), and wait for the targeted sectors to come under the head, and then the action is carried out *sequentially* within the sector thus arrived at.

Most disks accommodate multiple ***platters*** stacked vertically one above the other separated by an inch or even less apart, and fixed to a common spindle (already described). This unit is known as a *disk pack* or commonly, nowadays called a *hard disk*. For example, a disk may contain five platters, i.e. $5 \times 2 = 10$ recording surface. In practice, the top-most and the bottom-most surface of a *removable disk* are not used for data-storing (recording) purpose. Hence, the available recording surface for *removable disk* here is $(10 - 2 = 8)$ 8 surface, and not 10. The disk capacity can be then computed as:

$$\begin{aligned}\text{Capacity} = &(\text{no. of bytes/sector}) \times (\text{average no. of sectors/track}) \\ &\times (\text{no. of tracks/surface}) \times (\text{no. of surface/platter}) \\ &\times (\text{no. of platters/disk})\end{aligned}$$

A typical disk configuration may be as follows:

512 bytes/sector, 350 sectors/track (average), 16,000 track/surface,
2 surface/platter, 5 platter/disk.

Disk manufacturers (vendors) often express the capacity of disk in units of gigabytes (GB), where 1 GB = 10^9 bytes.

A brief detail of this topic with the respective figure and a number of solved problems based on these characteristics is given in the following web site: http://routledge.com/9780367255732.

Disk Format: Each track on the disk has one *starting point*, and each sector on the track is identified by its *start* and *end mark*. These requirements are satisfied by using some control data (system's own data) recorded on the disk before the insertion of the user's data. These activities are performed at the time of ***formatting*** the disk carried by the operating system (OS), and hence, without formatting, the disk cannot be used. These extra data which are solely used by the disk drive (disk controller) and not accessible to the user, are written on the disk when it is formatted, and consume some amount of disk space. That is why, after formatting, the disk space available to the user is always less than the actual disk space declared by the vendor. Since the formatting of the disk is carried out by the OS, different OS thus creates sectors of different sizes, and different fields of different sizes in a sector to hold user data, and other control data needed for the disk controller to handle user's information. Thus, a disk formatted by an OS cannot be run under different OS. Windows uses FAT 32 format.

Representative disk formats with the respective figures, together with solved problems based on formatted disks are given in the following web site: http://routledge.com/9780367255732.

4.4.4.1 Commodity Disk Considerations

The cost and performance of a disk unit, apart from its storage capacity, mostly depend on its internal structure, and the *interface* attached with it being used to connect the disk to the rest of the system.

ATA/EIDE Disks: Standard disk interfaces employed in today's most widely used personal computers and their variants have, however, passed through numerous modifications from its very inception to match with the constantly emerging innovative technologies. Eventually an enhanced version was arrived at, which has become the standard known as *EIDE* (Enhanced Integrated Drive Electronics) or as *ATA* (Advanced Technology Attachment). Many disk manufacturers have introduced a range of disks with EIDE/ATA interfaces that can be directly connected to the most common PCI bus (discussed in Chapter 5). In fact, Intel's Pentium chip sets include a controller that allows EIDE/ATA disks to be connected directly to the motherboard. One of the distinct advantages of EIDE/ATA drives is their low price, mostly due to their use in the PC domain. One of their *main drawbacks* is that a separate controller is needed for each drive if two or more drives are to be used concurrently to improve performance.

SCSI Disks: Small Computer System Interface (SCSI) disks have a specific interface designed for the connection to a standard SCSI bus (discussed in detail in Chapter 5). This is a fast communication bus having its own specified protocol standard that allows the computer to connect to multiple devices that include CD-ROM/RW drives, tape drives, pen drives, printers, and scanners, apart from hard disk drives. These disks tend to be more expensive, but they yield better performance, mainly made possible by the advantages inherent to the SCSI bus in comparison with those of the PCI bus. That is why, these disks are found in extensive use in high-end servers. Concurrent accesses can be made to multiple disk drives because the drive's interface is actively connected to the SCSI bus, only when the drive is ready for a data-transfer operation. This is especially useful in applications where there is a large number of requests for small files, which is often the case in computers used as file servers.

4.4.4.2 RAID: Redundant Array of Inexpensive Disks

During the past decade, while processor speeds have enormously increased, main memory speeds have also improved but comparatively modestly, and the speed of disk storage devices have by far exhibited the smallest relative improvement in this regard as such. Of course, a spectacular improvement in the storage capacity of these devices has been observed. Still, the magnetic disks, the primary secondary devices, have several inherent drawbacks. The most important of them are:

- They have relatively slow data-transfer rates
- Their electromechanical construction is such that it essentially makes them prone to both transient and catastrophic failures.

Computer users thus constantly clamour for disks that can provide larger capacity, faster access to data, high data-transfer rate, and of course, higher reliability. Although high-performance devices tend to be normally expensive, yet it is possible to achieve a considerably high performance at a reasonable cost by using a number of relatively low-cost devices to be operated in parallel, as is found in a multiprocessor system. Multiple magnetic disk drives, thus similarly can be used to provide a high-performance secondary storage unit. This approach led to the development of arrangements of disks in an array that would operate independently and also in parallel. With multiple disks, separate I/O requests can be handled in parallel, as long as the data reside on separate disks. On the other hand, a single I/O request can also be executed in parallel if the block of data to be accessed is distributed across multiple disks.

The presence of multiple disks in the configuration opens a wide variety of ways of organisation of data as well as ways in which redundancy can also be added to improve reliability. At the same time, it is difficult to develop schemes for a multiple-disk database design that can be equally useable on a number of different hardware platforms and operating systems. Fortunately, industry has agreed on a standardized scheme to overcome this difficulty using *redundant array of independent disks* (RAID), which consists of universally accepted seven levels; from level zero through six, besides a few additional levels that have been proposed by some researchers. These levels do not indicate any hierarchical relationship, rather designate different architectures in a design that exhibits the following common characteristics:

1. RAID although consists of a set of physical disk drives, but when being driven by the operating system, it is recognized as a single logical disk drive.
2. Data are distributed in different fashion across multiple physical disk drives present in the array.
3. Redundant disk capacity is used to store parity information which enables recovery of data at the time of any transient or catastrophic failures of a disk.

The RAID technology distributes the data involved in an I/O operation across several disks, and performs the needed I/O operations on these disks in parallel. This feature consequently can provide fast access or a higher data-transfer rate, but it depends on the arrangement of the disks employed. The performance of any of the RAID levels, however, critically depends on the request patterns of the host system and on the layout of the data. High reliability is achieved by recording redundant information; however, the redundancy employed in the RAID arrangement is different by nature from that employed in conventional disk usage. A conventional disk uses a *cyclic redundancy checksum* (CRC) written at the end of each record for the sake of providing reliability, whereas redundancy techniques in a RAID employ extra disks to store redundant information so that original data can be recovered even when some disks fail. Recording of, and access to redundant information, however, does not consume any such extra I/O time, because both data and redundant information are recorded/accessed in parallel. Different RAID levels, however, essentially differ in the details of the characteristics as mentioned in (2) and (3) above, and the characteristic in (3) only is not supported by RAID 0 and RAID 1.

Seven different RAID schemes have been proposed that mainly differ in their implementations of redundancy techniques, and the disk striping arrangements have been employed. However, RAID level 0+1, and RAID level 1+0 are hybrid organisations based on RAID 0 and RAID 1 levels, but the RAID level 5 is the most *popular* RAID organisation.

However, the two main metrics, namely, (i) data-transfer capacity (rate), or the ability to move data, and (ii) I/O request rate or ability to satisfy I/O requests, mostly determine the performance differences among these levels.

A brief detail of seven different RAID schemes with their respective figures, and an example have been explained individually in the web site: http://routledge.com/9780367255732.

4.4.4.3 Disk Cache

The objective in the use of smaller and faster cache memory, also known as memory cache, interposed between main memory and processor, is to effectively minimize repeated relatively slower memory visits, and thereby to reduce the average access time by means of exploiting the principle of locality.

The same principle can be applied to disk devices to improve effective disk access time that consequently speeds up many operations related to disk memory, thereby improving the performance of the entire system as a whole. A *disk cache* is essentially a buffer (a dedicated area) in relatively faster main memory for some specified disk sectors. This disk cache usually contains a copy of some of the most recently used sectors of the file data and meta-data, like file map tables stored on the disk. When an I/O request is initiated on a file, the *file system* converts the offset of a data byte in the file into an address of a disk block along with an offset within that block. It then uses the generated address of this disk block to check whether the target block exists in the disk cache. If so, the request results in a cache hit, and is responded to via the disk cache. The net effect is an elimination of delays and the overhead associated with disk accessing, and a quick response via memory to the requesting process. Alternatively, if the requested block is not in the disk cache, a disk-cache miss occurs, and an I/O operation then involves in two copy operations; one copy is from the requested disk block to the disk cache in the disk, and the other is from the disk cache to the target memory location in the address space of the process that initiated the I/O operation. As a result, the entire I/O operation becomes expensive to complete. But, because of the principle of locality of reference, it is expected that when a block of data is fetched into the disk cache to meet any single I/O request, it is highly probable that the same block may be referenced in the near future once again. However, the design considerations of disk cache and its management are entirely software approaches which fall in the domain of operating system principles and design, and hence lie outside the scope of this discussion. Readers interested in this area, however, can consult the book by the same author [*Operating Systems*: P. Chakraborty].

4.4.5 Magnetic Tape

Magnetic tape was historically the first kind of most popular secondary memory in use for regular processing of data. Magnetic-tape memory, tape, and tape drive units are very similar to our domestic audio/video tape recorders but store binary digital information. The storage medium is a flexible *mylar tape* (plastic tape) coated with magnetic oxide. Information is generally stored in 9 parallel longitudinal tracks. A read–write head is used that can access 9 tracks simultaneously. Data are stored on tape one basic character (9-bit) at a time across the width of the head, and each character comprises 8 bits (one byte) and the remaining one bit is used as the parity-check bit. As with the disk, data are read and written in the tape in contiguous blocks called *physical records*. Adjacent blocks are separated by relatively large gaps, called *inter-block* gaps. Similarly, adjacent records within a block are separated by gaps, called *inter-record* gaps (IRGs). A tape drive is essentially a *sequential-access* device. This means that if the tape head is currently positioned at record 1, then to read or write physical record k, all the physical records, record 1 through $k - 1$ should be passed through one at a time. If the head is presently positioned beyond the target record, the tape must be rewound to bring it to the beginning of the desired record, and then start operating forward. Magnetic tapes are stored on reels of about 2,400 ft in length, and more so today as it provides a compact, inexpensive, and portable medium of storing large information of files. These tapes are also packaged in cartridges or cassettes which are analogous to audio-tape cassettes. It is still in wide use in spite of having the slowest speed but equally of lowest cost, and are generally used as back-up of large files apart from its frequent use to store regularly operated active files.

A brief detail of the magnetic tape with respective figures and its operation, along with their different generations, is given in the following web site: http://routledge.com/9780367255732.

4.5 Optical Memory: External Memory

The first successful non-erasable compact disk (CD) based on optical technology was introduced sometime in 1983 by Sony and Philips companies, designed to hold up to 75 minutes of data providing a total of about 3×10^9 bits (3 gigabits) of storage using digital representation of 16-bit samples of analog sound signals taken at a rate of 44,100 samples per second for achieving high-quality sound recording and reproduction. Subsequently, low-cost, higher-capacity, and read-only storage media suitable for computer usage referred to as CD-ROM have been developed. Over time, this technology, however, got continuously improved and gradually became highly mature, resulting in the introduction of a variety of optical disk system with different characteristics, almost at regular intervals. A brief summary of the different products of this category has been presented in Table 4.3. We will now examine all these varieties of popular optical disks one after the other.

4.5.1 Compact Disk (CD) Technology

Computer-usable CD-ROM, similar to the audio CD, has used an optical technology in which a sharply focused coherent laser light source is made incident on the surface of the rotating disk. The surface is arranged with physical indentations along the tracks of the disk. These indentations reflect the focused beam incident on them towards a photo-detector (optical sensor), which subsequently detects the stored binary patterns. Coherent light used here consists of synchronized waves of the same wavelength. If a coherent beam is combined with another beam of the same kind, and the two beams are in phase, then the intensity of the resultant beam will be high, giving a brighter light. Similarly, if the waves of the two beams are out of phase (180 degrees), they will cancel each other, thereby causing the resultant beam to be of feeble intensity. These characteristics are sensed by a photo-detector that eventually distinguishes a brighter spot (first case) from a relatively darker one (second case).

A cross-sectional view of a small portion of a CD-ROM is depicted in Figure 4.9. The bottom layer of a CD-ROM is normally a ***polycarbonate*** plastic that functions as a clear glass base. The surface of the plastic is programmed to record digital information (either music

TABLE 4.3

Numerous Optical Disks

CD: Compact disk of early days of 120mm in diameter was simply a non-erasable disk and capable of storing only digitized audio information of more than 60 minutes of uninterrupted audio playing time on one side of it.

CD-ROM: Compact disk Read-Only Memory is a non-erasable one-time recorded disk of 120mm diameter with standard speed and format used for storing computer-related data. A CD meant for 75 minutes can store information of about 680 Mbytes. The speed of the CD drive is expressed in terms of X (i.e. say, 40X) that influences the data-transfer rate.

CD-R: This compact disk called CD Recordable (**WORM**) allows to write *only once subsequently* after production (original writing). The disk controller of CD-R although is somewhat expensive than that of a CD-ROM, but allows to read the disk as usual. The resulting disk after writing can also be read by an ordinary CD-ROM drive.

CD-RW: This compact disk called CD Rewritable allows the user to erase and rewrite the disk multiple times. Compared with CD-ROM and CD-R, it is much better while considered for use as a secondary storage. In fact, it often competes even with the magnetic hard disk for many reasons including itscost-wise high storage capacity, easy removability, and high portability, and as such, they are found to be extremely suitable for archival storage of information.

DVD: This disk called the Digital Versatile Disk, a forerunner of Video CD (VCD), can store digitized, compressed representation of mostly video information, as well as bulk volume of other digital data, currently as much as seven times of a CD-ROM. The physical size of the DVD, however, is the same as the CD. DVDs of both 80mm and 120mm diameter are used, with a double-sided capacity of up to 17 Gbytes. The basic DVD is, however, a read-only, i.e. DVD-ROM, and is one-sided. The data-transfer rate in generic DVD is relatively much higher than the other optical disks because of its higher storage density.

DVD-R: This disk called the DVD Recordable is only one-sided, and can be written *only once subsequently* after production (similar to CD-R) using a DVD Recorder, and then can function as a DVD-ROM.

DVD-RW: This DVD Rewritable may be a two-layered and two-sided one, and is similar to a DVD-ROM, but the user is allowed to erase and rewrite the disk multiple times.

HD-DVD: This high-definition DVD (HD-DVD) using a disk format different from conventional DVDs provides even more storage capacity compared with traditional DVDs, thereby enabling to store high-definition videos. Higher storage capacity has been realized by way of achieving higher bit density, and that became possible by using a shorter wavelength laser beam of 405nm in the blue–violet range. The single-layer/single-sided HD DVD scheme can store 15 GB, and a dual layer HD DVD can store up to 30GB. However, HD-DVD was not able to survive for long and gradually went out of the market in the competition with Blu-ray DVD.

Blu-Ray DVD: It is also a high-definition DVD (HD-DVD), similar in many respects with HD-DVD, but uses a disk format different from conventional HD-DVD. It provides even more storage capacity compared with traditional HD-DVDs. In a single-layer/single-sided disk, the storage capacity is 25GB and a dual layer disk can hold up to 50GB. Blu-ray disk, however, had a competitive challenge from the existing HD-DVD, and eventually won in the format war of 2006–2008. Consequently, the Blu-ray scheme achieved market dominance, and became the de facto successor to the DVD format.

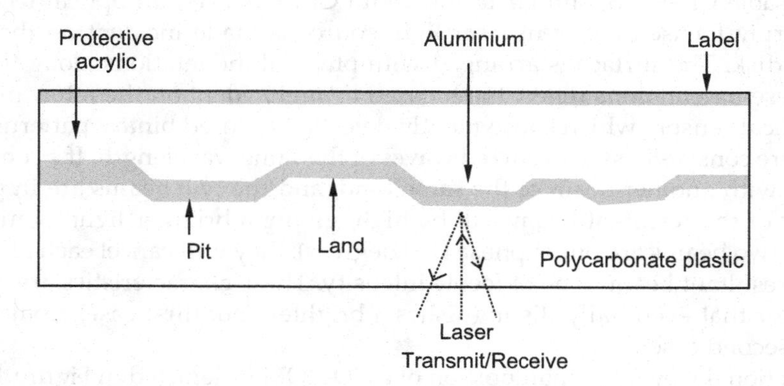

FIGURE 4.9
Cross-sectional view of a CD and its operation.

or computer data) by indenting (imprinting) the surface of the polycarbonate with a series of microscopic *pits*. The unindented parts (i.e. the areas between the pits) that remain on the surface are called *lands*. This indenting is first accomplished with a finely focused, high-intensity laser to create a programmed master disk. This master is then subsequently used to make a die to stamp out copies of this master onto other polycarbonates. The entire pitted surface of the copy is then coated with a thin layer of a highly reflective surface, usually aluminium or gold. This shiny sensitive surface is then protected against dust and scratches by a top coat of clear acrylic (lacquer). Finally, a label can be silkscreened to stamp onto the acrylic. The total thickness of the disk, however, comes to approximately 1.2 mm, most of which is contributed by the polycarbonate plastics. The other layers are extremely thin. Figure 4.10 exhibits a layered view of a CD-ROM.

Information is retrieved from a CD or CD-ROM by a low-powered laser that is housed in an optical disk player or in a CD-ROM drive unit. The laser source and the photo-detector are positioned below a polycarbonate disk. From the laser side, the pits actually appear as bumps with a convex surface with respect to the land. The disk is rotated with the help of a motor, and the beam that is emitted travels through the plastic, and is incident on the shining aluminium layer, and then reflects off the layer, and travels back towards the photo-detector.

When the disk rotates, the laser beam scans across the disk, and comes upon different situations. The intensity of the reflected laser light changes, resulting a low intensity when it falls on a pit having somewhat of a relatively convex surface, since the light here scatters. In contrast, when the laser beam falls on a *land* which is a relatively smooth surface, the reflected light does have a higher intensity. This change between pits and lands is detected by the photo-detector and is converted into a corresponding digital signal. A different situation arises when the beam moves through the edge where the pit changes to the land, i.e. a transition from a pit to a land and vice-versa. The beginning or end of a pit represents a 1; when no change in elevation occurs between intervals, a 0 is recorded. This pattern is, however, not a direct representation of the stored data. The CDs actually use a complex encoding scheme to represent error-free data. Each byte of data is represented by a 14-bit code, so that it can provide adequate error detection capability. (As the pits are very small, it is difficult to implement all of the pits perfectly, and as such, physical imperfections cannot be avoided that may cause to inject some errors in data while stored. For the audio/video CD, these errors in data have no perceptible impact on the reproduced sound or image. However, such errors are not acceptable in computer applications, and it is thus

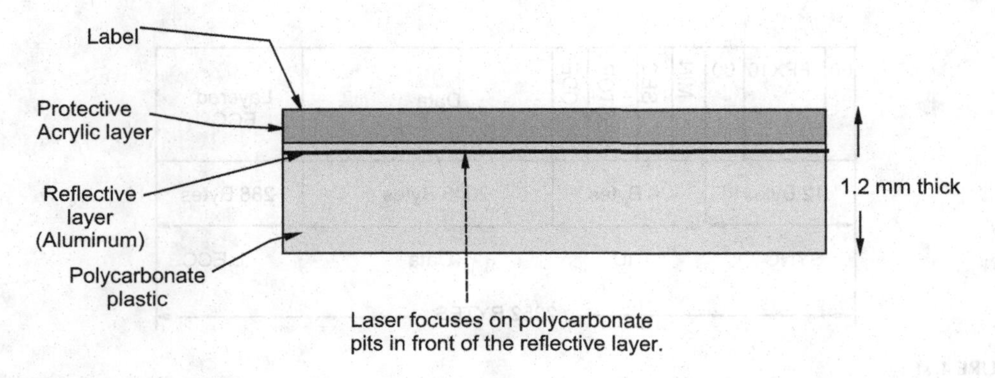

FIGURE 4.10
A layered view of CD-ROM (capacity 682 MB).

necessary to use additional bits to provide adequate error-checking and correcting capabilities to ensure the integrity of the stored data. CD used in computer applications are equipped with such capability). We restrict ourselves here from entering into any further discussion on the details of this code.

Unlike concentric tracks in the magnetic disk, the CD contains just one single physical spiral track, beginning near the centre and spiralling out to the outer edge of the disk, to realize greater capacity. But, it is customary to refer to each such circular path spanning 360 degrees as a separate track, which is analogous to the terminology used for the magnetic disk. The pits are, however, arranged along such tracks. The CD is 120 mm in diameter with a 15 mm hole in the centre. Data are stored on tracks that cover the area from a 25-mm radius to a 58-mm radius. The space between the tracks is 1.6 μm, or even less nowadays. Pits are 0.5 μm wide and 0.8–3 μm long. The *track density* is about 6,000 tracks/cm (more than 15,000 tracks on a disk) which is much higher than the density achievable in magnetic disks that usually ranges from 800 to about 2,400 tracks/cm.

4.5.2 CD-ROM (Compact Disk Read-Only Memory)

CD used in computer applications was historically called the CD-ROM, because after fabrication, its contents can only be read, similar to semiconductor ROM chips. Like audio CD, the CD-ROM also uses a spiral track, and the data are stored in the tracks in the form of blocks that are called *sectors*. All the sectors are of same size, and contains the same amount of information irrespective of whether it is located closer to the centre or away from the centre. Hence, the number of sectors per track is variable, and increases as we move towards the outer surface of the disk when the length of the track increases. These same-sized sectors are scanned by the laser beam at the *same rate* using a ***constant linear velocity*** (CLV). Sector-scanning at the same rate thus requires the *rotation of the disk at variable speed*, and the disk thus rotates more slowly for accesses near the outer edge to keep the sector scanning at the same rate, since the number of sectors increases in the outer track than those near the centre. The capacity of a track (the amount of data stored on a track) and the rotational delay both increases as one moves towards the outer edge of the disk.

Data stored in a sector (block) have several *different formats*. However, a representative ***block format*** of a CD-ROM is depicted here in Figure 4.11. The *beginning* of a block (*Sync field*) is marked by 1 byte of all 0s, followed by 10 bytes of all 1s, and then followed by 1 byte of all 0s. The *block address* is identified (indicated) by 4 bytes. One byte each for recording

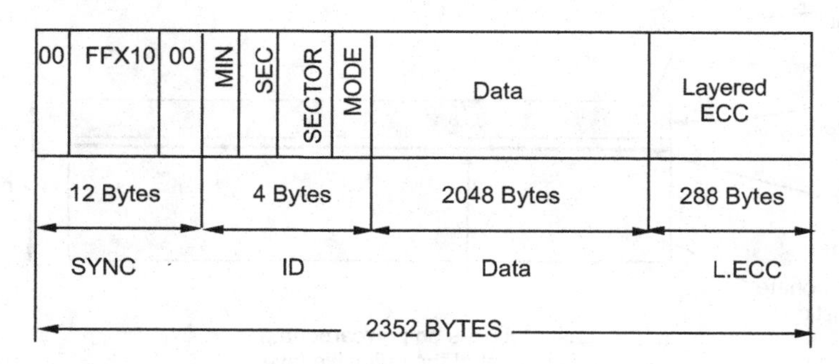

FIGURE 4.11
A representative CD-ROM block format.

minutes, seconds, sector and the fourth byte specifies a *mode* of data storage; e.g. Mode 0 indicates a blank data field, Mode 1 indicates that the last 288 bytes are used as error-detecting code, and Mode 2 indicates that the last field (288 bytes) is used as an additional *data field.* Therefore, in Mode 2, a block of CD-ROM stores 2,048 + 288 = 2,336 bytes of data. All these together are sometimes known as ***Header***.

The data field (2k = 2,048 bytes) is used for storing data, followed by 288 bytes of additional data which may be used either for error-correcting code or as an additional data field depending on the mode. The data is stored and retrieved sequentially along a spiral track with the use of CLV. In this disk, *random access is more difficult*, because locating a desired address involves many operations to carry out, such as:

- First, to move the head to place in the specified area.
- Then adjusting the rotation speed.
- Then reading the address.
- Then to make an access the specific sectors as desired.

Error detection and correction are done at more than one level. As already mentioned, use of a 14-bit code for each byte of stored information can correct single-bit errors. However, errors that occur in short bursts affecting several bits are subsequently detected and corrected using the error-checking bits present at the end of the sector (Figure 4.11).

The CD-ROM drives operate at a number of different rotational speeds due to the use of its constant linear velocity (CLV). However, the ***basic speed***, known as 1×, is 75 sectors per second, thereby providing a basic data rate of 75 × 2,048 = 153,600 bytes/sec (150 Kbyte/s approx.), using the Mode 1 format. With this speed and format, a CD-ROM based on standard designed for 75 minutes (75 × 60 = 45 × 10^2 sec.) of information has a data storage capacity of about 680 Mbyte. Note that the speed of the drive affects only the data-transfer rate, but not the storage capacity of the disk. Higher speed CD-ROM drives, such as, a 40X CD-ROM has a data-transfer rate that is 40 times higher than that of the 1X CD-ROM. It can be observed that this transfer rate (40 × 153,600 = 6 Mbytes/s approx.) is significantly less than the transfer rate usually obtained in magnetic hard disks, which are in the range of tens of megabytes per second. Moreover, as the length of the tracks is not the same across the disk, the amount of data stored per track is not constant. Hence, the ***addresses*** of the location of CLV disks are represented by units of minutes (0–59) and seconds (0–59) and blocks (0–74). This address information is placed at the beginning of each block.

However, the CD-ROM exhibits three ***distinct advantages*** when compared with the traditional magnetic hard disks, such as:

- Capacity of storing information is relatively much greater.
- Mass replication of data along with the disk is not costly, and can be done at ease. Magnetic disk, on other hand requires two disk drives at a time for such copying.
- Optical disk is reliable, long-lasting, removable, and safely portable allowing the disk to be used for archival storage. Most magnetic disks are non-removable and, if removable, not easily portable.

The CD-ROM also equally suffers from several critical ***disadvantages,*** when compared with magnetic hard disks. One such is its access time which is longer than that of a magnetic disk.

4.5.3 CD-Recordable: CD-R (WORM)

CD-R, the *Write Once Read Many* (WORM) disk was developed sometime in the late 1990s with a target, when only a few copies of compact disks (CDs) are required with new data which can be written (recorded) by a computer user *only once subsequently* after its production (original writing). The new disk can, however, be read at will as usual. In CD-R, the medium includes an *organic dye* layer which is activated by a high-intensity laser and is used to change reflectivity. A spiral track is implemented on the disk during the fabrication process. For the purpose of writing, a low-powered laser in a CD-R drive is used to create pits by burning the organic dye on the track. When a burn spot is heated beyond a critical temperature, enough heat is generated *to burst* the surface of the dye to make it opaque. Such spots reflect light of lesser intensity when subsequently read. The written data are stored permanently. The unused portion of a disk after writing once, however, can be further utilized for writing to record new sets of data at a later time (multi-session). The *Read operation* is carried out as usual by a laser in the CD-R that illuminates the disk surface, and the burst blisters provide higher contrast than that of their surrounding area, and thus are recognized quite easily. The disk drive of CD-R uses a laser beam of *modest intensity*, and is somewhat expensive than that of a CD-ROM, but the CD-R disk after writing can also be read by an ordinary CD-ROM drive. The CD-R optical disk thus facilitates to have a permanent copy of a bulk volume of user data, and hence, is attractive for archival storage of documents and files (written once).

4.5.4 CD-Rewritable (CD-RW): Erasable Optical Disk

The basic structure of CD-RW is similar to that of CD-R. But, with this type of optical disk, the data already recorded can be changed (overwritten) repeatedly as is the case with any magnetic disk. To fulfil this requirement, a number of proposals including *magneto optical mechanism* have been tried, but the only feasible technology as considered is based on a purely optical approach called ***phase change***. The *principle* behind this approach is as follows. Instead of using an organic dye in the recording layer as used in CD-R, the CD-RW here uses a material which is an alloy of *silver, indium, antimony,* and *tellurium*. This alloy exhibits an interesting and equally useful behaviour when it is heated and cooled. If it is heated above its melting point (about 500 degrees C) and then cooled down, it goes into an ***amorphous state*** in which it mostly absorbs the light incident on it, leaving only a feeble intensity of light for reflection. But, if it is heated only up to 200 degrees C, and this temperature is maintained for an extended period, the process of ***annealing*** takes place which ultimately leaves the alloy in a ***crystalline state***. While the alloy is in this state, it has a smooth surface which allows the incident light to pass through, and consequently provides a high intensity of light for reflection. A beam of laser light is, after all, used to change a material from one of its phases to the other.

If the *crystalline state* represents a *land* area, *pits* can now be created by heating the selected spot past the melting point. A reflective material is placed above the recording layer to reflect the incident light that travels back towards the photo-detector. The laser source and the photo-detector are positioned below the alloy film. At the time of *writing*, pits are created only at the selected spot with the help of the laser beam, the remaining area of the alloy film is left with a smooth surface. At the time of *reading*, the incident laser beam while encounters a smooth surface, it passes through, and then reflects back by the reflector with high intensity. If the incident laser beam encounters a pit, it is almost absorbed, and the reflected light thus obtained has a feeble intensity. These differences in the intensities of the reflected light are sensed by the photo-detector, by which it can recognize the stored data.

The ***erasing*** of stored data is accomplished by using the annealing process, which returns the alloy once again to a uniform *crystalline state*. That is why, unlike the magnetic disk in which erasing of the existing data is done automatically when new data is written over the existing one, here in CD-RW, erasing is to be carried out explicitly before new data is to be written.

The CD-RW drive is thus essentially a little bit different one that uses *three* different laser powers. The *highest power* is used to record the pits. The *middle power* is used to put the alloy into the *crystalline state*. It is often referred to as the ***"erase power"***. The *lowest power* is used to read the stored information. Some CD-RW drives operate in ZCLV, CAA, or CAV modes, but most work in the constant linear velocity (CLV) mode, and can also handle other types of CDs as usual. They can read CD-ROMs, and can read and write CD-Rs. They provide standard interconnection interfaces, such as SCSI and USB, for universal use.

CD-RW exhibits several distinct ***advantages.*** Modern CD-RWs now provide a truly low-cost, multiple-writes, easily removable, and highly portable storage medium that often competes even with magnetic hard disk. Presently, CD-RW drives are so fast that they are now almost commonly used for taking hard-disk backups on a regular basis. The introduction of CD-RW technology has eventually outweighed CD-R and made it less relevant. Last but not the least, a key advantage of the CD-RW is its higher reliability and longer life because the technology used for this optical disk has a tolerance that is much less fatal than that of high-capacity magnetic disks. Still, the primary ***disadvantage*** of this phase-change optical disk is that there is a limit on how many times a CD-RW disk can be rewritten. This is due to the fact that the material eventually and permanently loses its desirable properties after crossing a certain limit of repeated erasing and subsequent rewriting. Still, materials currently being used provide these processes to continue up to a theoretical limit of around 1,000 times.

4.5.5 Digital Versatile Disk (DVD)

The enormous success of the CD technology, and its wide acceptability in the user domain encouraged the designer to further go ahead in quest of even more storage capability that has eventually led to the emergence of high-capacity digital versatile disk (DVD) optical disk within a few years. It truly replaces the existing analog VHS video tape being played with a video cassette recorder (VCRs), and actually begins the digital era of the videos. While the physical size of a DVD (1.2-mm thick and 120 mm in diameter) is the same as that of a CD, but the storage capacity is currently as much as seven times of a CD-ROM. This huge storage capacity further facilitates to enrich the multimedia culture, thereby allowing more videos to incorporate into any software for regular use as well as software into videos for demonstration purposes.

Before the advent of DVD and Blu-ray, video-CD (VCD) introduced sometime in 1993 became the first format for distributing *digitally encoded* films on standard 120 mm optical disks (its predecessor, CD-video used analog video encoding). In the same year, two new optical disk storage formats were being developed. One was the *Multimedia Compact Disk* (MMCD), backed by Philips and Sony, and the other was the *Super Density* (SD) disk, supported by Toshiba, Time Warner, Matsushita, Hitachi, Mitsubishi Electric, Pioneer, Thomson, and JVC. However, the **DVD** disk storage format invented and developed by Philips, Sony, Panasonic, and Toshiba sometime in 1995 has achieved much larger storage by incorporating several distinct changes in the concept of existing CD design. Some notable ones are the following:

1. A laser diode red light with a wavelength of 635 μm (micron) is used instead of the infrared light laser used in CDs having a wavelength of 780 μm. This shorter wavelength makes it possible to focus the laser light on a relatively smaller spot. Consequently,
   - Bits are made packed more closely on a DVD.
   - Moreover, the spacing between the loops of a spiral track on a DVD is 0.74 μm, whereas in CD, it is 1.6 μm.
   - The minimum distance between pits along the spiral in a DVD is 0.4 μm, whereas in a CD, it is 0.834 μm.
   - The minimum width of the pits in a DVD is 0.4 μm, whereas in a CD, it is 0.5 μm.

   The ultimate result of these improvements is about a 7-fold increase in capacity, to about 4.7 GB (= 7 × 680 MB). This single-layered single-sided disk is defined in the standard as **DVD-5**. Further increases in capacity have been achieved by way of approaching towards making a two-layered and two-sided disks.
2. A ***double-layered disk*** makes use of two layers on which tracks are implemented on top of each other. Here, the pits of the second layer lands on the top of the first. The first layer is the clear base, as in CD disks. But, instead of using reflecting aluminium, the lands and pits of this layer are covered by a translucent material that acts as a semi-reflective layer. The surface of this material is then also programmed with indented pits to store data. A reflective material is placed on top of the second layer of pits and lands. The disk is read by adjusting the focus of the laser beam on the desired layers. The lasers in DVD drives can read each layer separately. When the beam is focused on the first layer, sufficient light is reflected by the translucent (semi-reflective) material to detect the stored binary patterns. When the beam is focused on the second layer, the light reflected by the reflective material corresponds to the information stored on this second layer. In both cases, the layer on which the beam is not focused reflects a much smaller intensity of light which is subsequently eliminated by the detector circuit considering it as noise. This technique almost doubles the capacity of the disk to about 8.5–8.7 GB, instead of the expected exact double 9.4 (2 × 4.7 = 9.4). This happens due to the lower reflectivity of the second layer which affects its storage capacity to a little extent that limits to achieve a straightaway full doubling. This disk is, however, called **DVD-9** in the standard.

Two single-sided disks can be put together to form a sandwich-like structure where the top disk is turned upside down. This can be done with two single-layered disk giving rise to a composite disk, known as ***double-sided single-layered disk*** providing a capacity of 9.4 GB (2 × 4.7 = 9.4). This disk is called **DVD-10** in the standard.

The above approach can be carried out with two double-layered disks giving rise to a ***double-sided double-layered*** disk that yields a total capacity up to 17 GB. This disk is specified as **DVD-18** in the standard. A layered view of a double-sided dual-layered DVD is depicted in Figure 4.12 (make a comparison with the corresponding CD layered-view as depicted in Figure 4.10). The notable characteristics and specifications of different types of

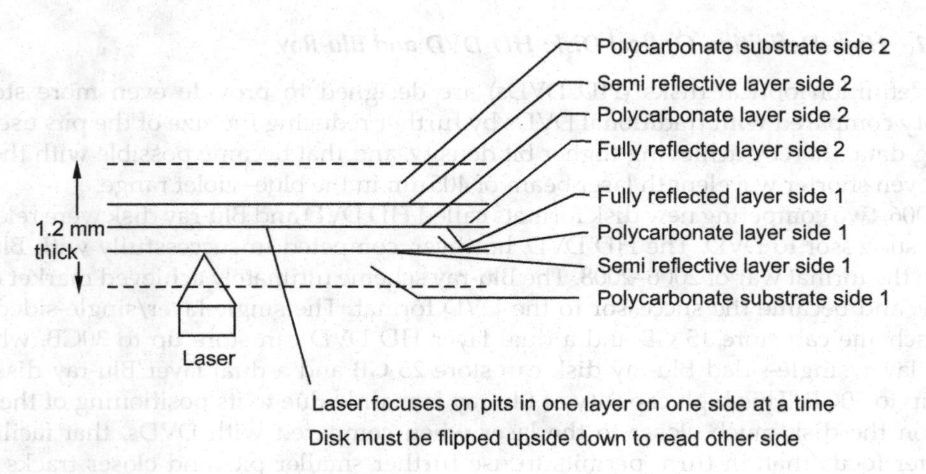

FIGURE 4.12
DVD-ROM double-side dual-layer capacity (17 GB).

TABLE 4.4
Specifications of DVD Disks

Read Mechanism	650 nm laser,10.5 Mbits/s (1×)
Write Mechanism	10.5 Mbits/s (1×)
Capacity	4.7 GB (single-sided, single-layer common) 8.5–8.7 GB (single-sided, double layer) 9.4 GB (double-sided, single layer) 17.08 GB (double-sided, double-layer rare)
Standard	DVD Forum's, DVD Books, and DVD+RW Alliance specifications

DVD disks are summarized in Table 4.4. However, the fundamental characteristics, such as, pit size, inter-spiral track gap, etc. are all same for all types of DVD disks, and, hence, are not included in this Table 4.4.

Writing speeds for the DVD were initially 1×, that is, 1,385 KB/s, in the first drives and media models. More recent models, at 18× or 20×, have 18 or 20 times of that speed. Note that for CD drives, 1× means 153.6 KB/s, about one-ninth as swift. The DVD drives provide an ***access time*** that is similar to that of CD drives. However, when the DVD disk rotates at the same speed, the ***data-transfer rate*** is much higher because of the higher density of pits available in DVD disks.

Pre-recorded DVDs are mass-produced using moulding machines that physically stamp data onto the DVD. Such disks, similar to CD-ROMs are known as *DVD-ROM*, because data can only be read and not written nor erased. Blank recordable DVD disks (DVD-R and DVD+R) can be recorded once, similar to CD-R, using a DVD recorder and then function as a DVD-ROM. Rewritable DVDs (DVD-RW, DVD+RW, and DVD-RAM), similar to CD-RW can be recorded and erased multiple times. The specifications of these disks are summarized in Table 4.4.

The DVD video format was first introduced by Toshiba in November 1996 in Japan, in the United States in March 1997 (test marketed), in Europe in October 1998, and in Australia in February 1999. In May 1997, the existing DVD Consortium was replaced by the DVD Forum, which is open to all other companies.

4.5.5.1 High-Definition Optical Disk: HD-DVD and Blu-Ray

High-definition optical disks (HD DVDs) are designed to provide even more storage capacity compared with traditional DVDs by further reducing the size of the pits used for storing data, thereby achieving higher bit density, and that became possible with the use of an even shorter wavelength laser beam of 405 nm in the blue–violet range.

In 2006, two competing new disk formats called HD DVD and Blu-ray disk were released as the successor to DVD. The HD DVD, however, competed unsuccessfully with Blu-ray disk in the format war of 2006–2008. The Blu-ray scheme ultimately achieved market dominance, and became the successor to the DVD format. The single-layer/single-sided HD DVD scheme can store 15 GB and a dual layer HD DVD can store up to 30GB, while a single-layer/single-sided Blu-ray disk can store 25 GB and a dual layer Blu-ray disk can hold up to 50GB. This high capacity of Blu-ray is mainly due to its positioning of the data layer on the disk much closer to the laser when compared with DVDs, that facilitates a tighter focus that, in turn, permits to use further smaller pits and closer tracks with lesser distortion. However, three categories of Blu-ray disk are available, namely, *read-only* (BD-ROM), *recordable-once* (BD-R), and *re-recordable* (BD-RE).

In spite of availability of higher-capacity disk with changed format, the standard DVD still remained dominant as of 2012, mainly due to the fact that the Blu-ray technology was not sufficiently matured to replace the existing culture. Moreover, *write* and *read* speeds of Blu-ray were relatively poor, and also the fact that the necessary hardware was still expensive, and not very readily available. That is why, a large majority of consumers were fully contended with the existing DVD practice, although the Blu-ray players and the now-defunct format HD DVD players were designed to be downward compatible, thereby allowing older DVDs to play at will with these devices, since the media are physically identical. Nevertheless, it was commonplace for major releases issued in "combo pack" format that included both a DVD and a Blu-ray disk (as well as, in many cases, a third disk with an authorized digital copy). Also, some multi-disk sets have used Blu-ray for the main feature, but DVDs for supplementary features (examples of this include the *Harry Potter* "Ultimate Edition" collections, the 2009 re-release of the 1967 *The Pioneer* TV series, and a 2007 collection related to Blade Runner). Another reason cited (as of 2012) for the slower transition to Blu-ray from the existing standard DVD was the necessity of, and confusion over *"firmware updates"* that needs an access permission while performing updates over the internet.

4.6 Virtual Memory

4.6.1 Background

In the early days of computers, memories were expensive and small, and a large program was then traditionally executed with the use of secondary memory, such as disk. A particular strategy was then used, called **Overlays**, in which a large program to be placed in the disk is divided by the developer into a number of small independent pieces (modules), such that each piece could be then fit into the available main memory. When the large program is executed, the overlay mechanism properly monitors the *to and fro* journey of each of these small modules from disk to main memory as and when needed and vice-versa, in place of fetching the entire large program in the small available memory. Here, the programmer has to do everything in relation to create and manage the overlays, and as such, they must have the required expertise about the overlay technique and its implementation, apart from having adequate skill to develop the algorithms to solve their own problems. The overlay technique could, however, somehow handle the more complicated time-sharing multiprogramming environment of those days, but that too in a limited way, and thus has been in wide use for this purpose over many years, and is still in heavy use for different purposes.

In 1961, a group of people at Manchester, UK, proposed a method for the automatic execution of the overlay process without any involvement and intervention of the programmers, and even no intimation to them about its internal working. This approach has been subsequently implanted in what is now called ***virtual memory***. This memory is *created* in a portion of the secondary memory by computer hardware, and casts an illusion to the user as if a large memory space, almost equal to the totality of secondary memory, is available to them for their use. It is physically a part of secondary memory, but behaves like a primary memory to the user, that is why, it lies somewhere in between primary memory and secondary memory, and hence, is legitimately called *virtual memory*. Virtual memory on its own automatically monitors the two levels of the memory hierarchy represented by main memory and secondary storage. But, it should, however, be noted that while virtual memory is created by the computer hardware, but is *managed* only by the operating system. That is why discussions over virtual memory appear both in the domain of computer architecture as well as in the operating system. It was first used in the computers during the 1960s, and by the early 1970s, virtual memory has been available on almost all computers. Even earlier microprocessors, including Intel 80386 and Motorola 68030 had a highly sophisticated virtual memory system. Without virtual memory, it would have been impossible to develop a time-sharing multi programmed computer system that is most common in any present-day environment.

4.6.2 Address Space

Each word in the physical memory is identified by a unique physical address and all such memory words in the main memory form a ***physical address space*** or ***memory space***. In systems with virtual memory, the address generated during the compilation of a program, and subsequently used by the system is called the ***virtual address,*** and the set of such addresses form a ***virtual address space*** or simply ***address space***. The users are informed that they have such total *address space* for their use. During execution, the virtual addresses are issued by the processor, but memory addresses only are required for processing. That is why, only currently executable programs or parts of the programs are brought from virtual memory into a smaller amount of physical memory which are then shared among those programs. The virtual addresses of the executing program while being issued must be then translated into their corresponding physical addresses of the locations where the said program (or part of the program) is placed in the main memory. However, the address translation mechanisms and management policies being used are often affected by the virtual memory model used, and by the organisation of the disk arrays, and certainly also of the main memory. Virtual memory approach also simplifies loading of the programs for execution, and permits necessary *relocation* of codes and data, allowing the same program to run in any location in physical memory with appropriate address mapping as discussed later.

4.6.3 Address Mapping

With virtual memory, the CPU generates *virtual addresses*, which are then translated by a combination of hardware and software tools to *physical addresses* that are used to access main memory. This process is called *memory mapping* or *address translation*. Let V be the virtual addresses generated by the CPU during the execution of a program. Let M be the set of physical addresses assigned and allocated to run this program. The automatic mechanism of a virtual memory system implements the following mapping:

$$f : V \rightarrow M\ U\{\phi\}$$

Since the physical memory location of any particular virtual address is dynamically (during run time) allocated and deallocated, the same virtual address may be allocated at different memory locations at different times during execution. Hence, mapping is a function of time. Thus, mapping can be elaborately defined as:

$f(v) = m$; when the virtual address v will be allocated a memory location $m \in M$

$= \phi$ when v is not found in M

when the virtual address v can be uniquely mapped into m by the function $f(v)$, there is a *memory hit* in M. When $f(v) = \phi$, this indicates that the referenced item addressed by v is not found at that time in memory, the situation is a *memory miss*. The performance of the virtual memory is highly influenced by the efficiency of this address translation process. In a multiprocessor system, the implementation of virtual memory is even more complex due to presence of additional problems like protection, coherence, and consistency, and hence, different types of address translation mechanisms are employed that mostly depend on the architecture of the computer systems.

4.6.4 Types of Virtual Memory

The virtual address space is divided into units (similar to modules used in the overlay approach) called ***pages***, and the corresponding unit of same size in the physical memory are called ***page frames*** or ***blocks***. Two types of virtual memory models are common. They are the following:

i. **Private virtual memory** which is used both in uniprocessor (single CPU) and in multiprocessor computer system. In case of a uniprocessor system, the entire virtual memory space is assigned to the single processor, no question arises with regard to being private or public. In case of a multiprocessor system, each processor is given a private virtual memory space. Virtual pages from different virtual spaces are mapped into the same physical memory. Here, the physical memory is shared by all processors.

ii. **Shared virtual memory,** in which different disjoint virtual address spaces for different processors are present in a single globally *shared virtual space*. An appropriate locking mechanism is required to safeguard each virtual space for making them to be mutually exclusive. Some areas of virtual space can also be shared by multiple processors at the same time.

A brief detail of this topic with appropriate figures is given in the following web site: http://routledge.com/9780367255732.

4.6.5 Address Translation Mechanisms

This mechanism involves the translation process of virtual addresses to physical addresses as shown in Figure 4.13. Various schemes for the translation process are in wide use. In any case, the translation process requires the use of *translation maps* or *mapping table* which can be realized in various ways. The mapping table may be stored

a. in a separate cache memory
b. in main memory
c. in associative memory.

For the case (a) above, an additional memory unit is required as well as one extra memory access time. For the case (b), the table itself takes space from main memory reducing the available area for the user to use. Moreover, two access to memory is required to provide data to CPU, in effect, the program will be then running at half speed.

Wherever the mapping table is stored, a mapping function is applied to the virtual address to generate a pointer to the mapping table. Both the main memory and virtual memory can be handled with either paged memory management or segmented memory management. But we confine our approach here only with paged memory management. Both the physical (main) memory and virtual memory are divided here into fixed-size

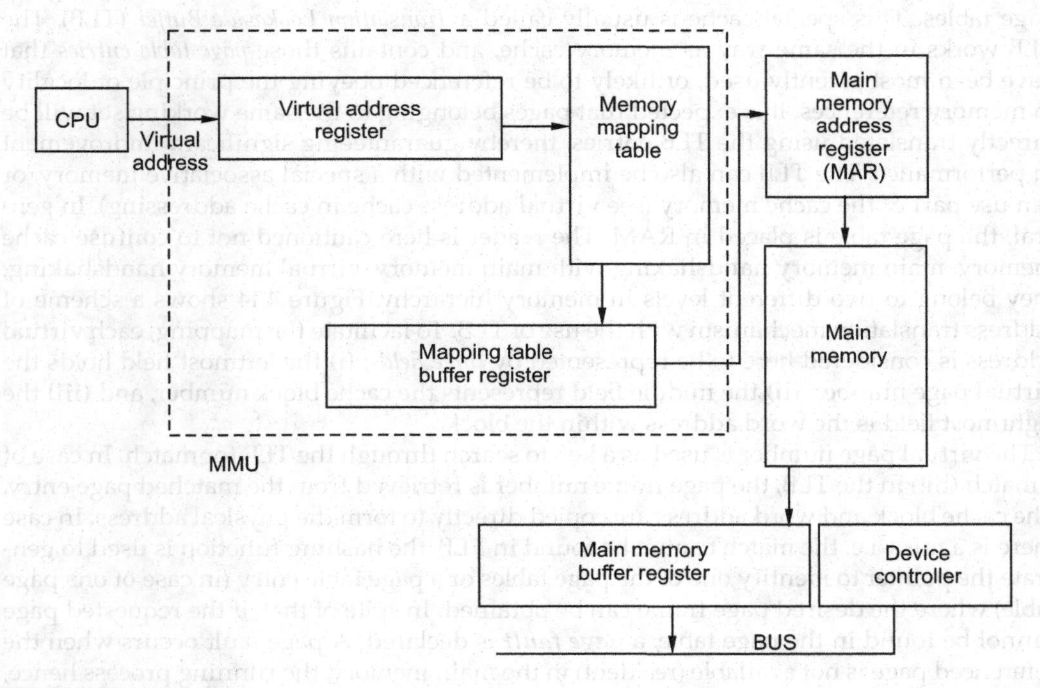

FIGURE 4.13
A scheme of mapping a virtual address to a physical address using memory map table.

units, called *page frames* and *pages*, respectively. Page tables (if more than one) are created, and are used as a mapping table to translate between pages and page frames. The page table entries are essentially containing address pairs (virtual page, page frame). As an analogy, $y = f(x)$ if x is the page, f is the page table, then y is the page frame. All exchange of information between virtual memory and main memory is carried out at the page level. As soon as an application program enters the execution state, the related information is injected into the page table. One page table can be created to accommodate many user processes or each process can use its own page table, and hence, the number of page tables (due to many processes being active) maintained in the main memory can be then very large.

Address translation mechanism and related virtual memory management functions are handled by a specialized unit called a *memory management unit* (*MMU*) which is nowadays inside the CPU, and is positioned between the effective-address generation unit (EAU) of the CPU and main memory as shown in Figure 4.13. MMU receives virtual addresses from the EAU, and converts them to the corresponding physical addresses for transmission to M or to memory cache, if it is present. The MMU also handles memory protection violation, swapping of data between main memory and secondary memory, and other exceptional situations that might happen.

4.6.6 Translation Lookaside Buffer (TLB)

Every virtual memory reference can cause two physical memory access: one to fetch the appropriate page table entry, and another one to fetch the desired data after receiving the information from page table. To overcome (reduce) this double memory access time, most virtual memory schemes make use of an additional special fast cache to store the original page tables. This special cache is usually called a *Translation Lookaside Buffer* (TLB), The TLB works in the same way as memory cache, and contains those *page table entries* that have been most recently used, or likely to be referenced obeying the principle of locality in memory references. It is expected that pages belonging to the same working set will be directly translated using the TLB entries, thereby guaranteeing significant improvement in performance. The TLB can also be implemented with a special associative memory, or can use part of the cache memory (see virtual address cache in cache addressing). In general, the page table is placed in RAM. The reader is here cautioned not to confuse cache memory–main memory handshaking with main memory–virtual memory handshaking; they belong to two different levels in memory hierarchy. Figure 4.14 shows a scheme of address translation mechanism with the use of TLB. To facilitate the mapping, each virtual address is considered here to be represented by *three fields*; **(i)** the leftmost field holds the virtual page number, **(ii)** the middle field represents the cache block number, and **(iii)** the rightmost field is the word address within the block.

The virtual page number is used as a key to search through the TLB for match. In case of a match (hit) in the TLB, the page frame number is retrieved from the matched page entry. The cache block and word address are copied directly to form the physical address. In case there is a miss, i.e. the match cannot be found in TLB, the hashing function is used to generate the pointer to identify one of the page tables or a page table entry (in case of one page table) where the desired page frame can be obtained. In spite of that, if the requested page cannot be found in the page table, a ***page fault*** is declared. A page fault occurs when the referenced page is not available (resident) in the main memory, the running process hence, is suspended. An interrupt (page-fault interrupt) is issued. A process switch is made to another ready-to-run process while the missing page as requested is transferred from the

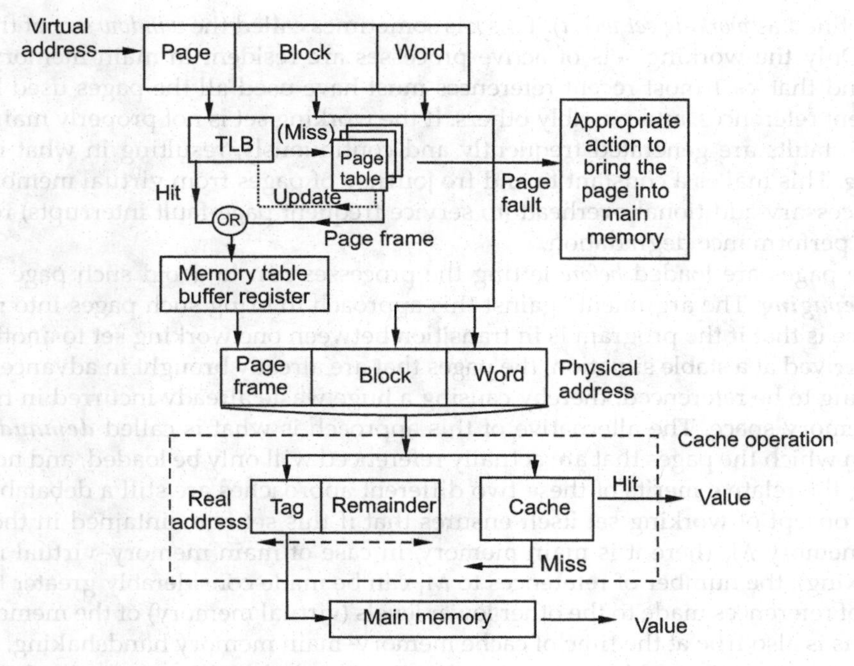

FIGURE 4.14
Translation lookaside buffer and cache operation.

disk unit to the physical (main) memory (interrupt servicing). The suspended process of address translation then resumes computation. It should be noted that both TLB entries and page table entries need to be dynamically updated to reflect the latest memory reference history, and that is maintained by using a few additional bits in the translation maps (in TLB, and memory page table).

The mapping scheme as described can be extended with ***multiple level page table*** (along with multiple page tables at each level). Multi-level paging requires multiple memory references to access a sequence of page tables to generate the desired physical address, and hence, consumes a relatively longer time, but can accommodate expanded memory space (larger memory), and can provide sophisticated protections of page accesses at different levels of the memory hierarchy to safeguard the users' interest. It is once again recalled that the virtual memory and its related aspects are created by the computer systems, but those are totally managed by the operating system using different policies, and hence are outside the domain of the organisation and architecture of the computer systems. Still, some topics related to their management are described here only for the sake of completeness of this discussion.

4.6.7 Working-Set Model

It has been observed that most programs do not reference their logical address space uniformly, rather randomly access only a small number of pages forming a set, and the changes in this set then occur slowly with time. The pages whose addresses are referenced during the time interval from $(t - t_1)$ to t denoted by $(t - t_1, t)$ constitute the *working set* $w(t, t_1)$. An alternative definition proposed by Denning is that at any instant time, t, there exists *a set consisting of all pages* used by the k most recent memory references.

This is defined as *working set w(k, t)*. This k is sometimes called the *window size* of the working set. Only the working sets of active processes are resident in main memory. It has been found that $k+1$ most recent references must have used all the pages used by the k most recent references, and possibly others. If the working set is not properly maintained, then page faults are generated frequently and continuously, resulting in what is called ***thrashing***. This makes a constant to and fro journey of pages from virtual memory causing unnecessary additional overhead (to service frequent page-fault interrupts) resulting in severe performance degradation.

If some pages are loaded *before* letting the processes run to avoid such page faults is called ***prepaging***. The argument against this approach to bring such pages into memory in advance is that if the program is in transition between one working set to another, and has not arrived at a stable situation, the pages that are already brought in advance, are not really going to be referenced, thereby causing a huge waste already incurred in time and costly memory space. The alternative of this approach is what is called ***demand paging*** scheme in which the pages that are actually referenced will only be loaded, and no others. However, the relative merits of these two different approaches are still a debatable issue. But, the concept of working set itself ensures that if this set is maintained in the fastest level of memory M_1 (here it is main memory, in case of main memory–virtual memory handshaking), the number of references to M_1 can be made considerably greater than the number of references made to the other lower levels (virtual memory) of the memory hierarchy. This is also true at the time of cache memory–main memory handshaking.

A brief detail of the working set and its implications in handling pages is given in the following web site: http://routledge.com/9780367255732.

4.6.8 Demand Paging Systems

Paged memory management uses various types of memory allocation policy to operate a virtual memory. Amongst them, demand paging system is one of the most often used policy that properly matches the paging scheme principle. In analogy to the well-known demand feeding algorithm for babies; when the baby cries, you feed it (as opposed to feeding at regular intervals of day). Similarly, in demand paging, pages are brought in memory only when an actual request is issued for a page, and not in advance, i.e. the pages are brought into main memory only upon demand. This policy allows only pages (instead of processes) to be transferred between the main memory and the swap device (virtual memory). The idea of demand paging nicely goes with the concept of working set which is formed only with the pages as being referenced on demand made by the active processes. The major advantage of this method is that it offers the flexibility to dynamically accommodate a larger number of processes within a limited size of physical memory in multiprogramming environment. This, in turn, will substantially increase the *system throughput* with less overhead (e.g. page fault), and decreases the memory traffic load. Pages of the process not under demand will not involve in this game. This demand paging policy was first implemented in UNIX BSD 4.0 release. Later, UNIX System V and many other modern operating systems also employed this demand-paging scheme in their memory management.

4.6.9 Page Replacement Principles

Since the total number of available page frame is much smaller than the number of pages, the frames will eventually be mostly fully occupied. In order to accommodate a new page to resolve the page fault situation, one of the resident pages must be replaced. Memory management policies, in fact, include the preemptive allocation and deallocation of page frames to active processes, and also replace page frames whenever required. *Page replacement* refers to the process in which a resident page in main memory (page frame) is replaced by a new page transferred from the virtual memory. This replacement operates on a per-process basis. Different page replacement policies, however, have been suggested. The ultimate goal of a page replacement policy is thus to maximize the hit ratio or equivalently, to minimize the number of possible page faults so that the effective memory-access time can be reduced. A good policy should also take into account the program locality property, and will after all target a page frame to remove whose absence in the memory would have the smallest adverse effect on the running programs. The policy to be chosen, however, is often affected by the page size, and by the number of available page frames. The effectiveness of a policy mostly depends on the program behaviour, memory traffic patterns encountered, as well as on the type of paging scheme being implemented.

When the removal of a page is ultimately required to make room for a new page to be brought in, the operating system will always try to choose a *clean page* to remove, i.e. a page frame that has not been changed, its disk copy is already up-to-date, no rewrite is, hence, necessary, and the new page will then simply overwrite the page frame to be evicted. But, if the page frame selected to be removed has been modified (*dirty page*) while in main memory, it must be written back to the disk to bring the disk copy to latest status, which thereby consumes additional time. Thus, an extra bit (dirty bit) is then needed to be attached with every page entry in the page table to indicate whether the page is *clean* or *dirty*, and this information is used to select a page at the time of page removal.

It is always desirable to maintain in memory a high ratio of clean pages to dirty pages to minimize the rewrite time at the next page fault. Some operating systems take a chance that whenever the disk is idle, a dirty page is selected by suitable prediction and is picked up, and it is then copied to the disk using DMA or data channels (IOP) while the CPU is busy with its other activities. Such copying never changes its memory contents. Writes to disk with the intention of making dirty pages clean are called ***sneaky writes***. It is still a debatable issue whether the sneaky write is really a profitable one if the administrative overhead attached with it is also taken into account.

Performance of a page removal algorithm is usually measured in terms of the *success function S*, the number of success, or the *failure function F*, the number of failures. If P is the number of page references in the page trace, then $S + F = P$. Another measure of performance is the *success frequency functions* $S = S/P$, or *failure frequency function* $f = F/P$, hence, $f = 1 - s$. The ***effectiveness*** of a page removal algorithm is measured in terms of how often the system will access virtual memory to fetch a page, it had replaced. If the replacement algorithm is a good one, the system would not have to go very often.

4.6.10 Page Replacement Policies

For the operating system, the best page to remove is intuitively the page that will never be needed again, or at least, not in the near future. Such algorithms MIN or ***Optimal replacement strategy,*** **OPT** have been unfortunately found to require some type of simulation run to get an advance knowledge of future page faults (interrupts). This simulation process is expensive to compute and, hence, OPT is not considered a practical replacement policy, and primarily has only theoretical importance. The other approach is to pick a random page to replace at each page fault. This is known as ***Random replacement.*** However, we will study here some useful page replacement policies specified in a demand paging memory management on real systems.

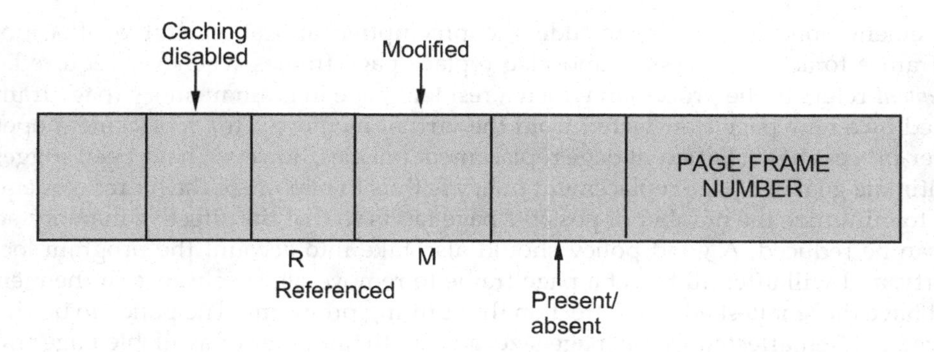

FIGURE 4.15
An entry in a page table using NRU.

4.6.10.1 Not Recently Used Page Replacement (NRU)

The operating system (OS) in most computers with virtual memory keeps track of status about pages whether used or unused with the help of the ***two status bits*** associated with each page in the page table. A typical page table entry is shown in Figure 4.15. Here, *"R"* bit is set when the page is referenced (Read or Written), and *"M"* bit is set when the page is written to (i.e. modified). It is important that these bits must be updated on every memory reference, so it is essential that they be set by the *hardware*. Once a bit is set to 1, it remains 1 until the OS resets it to 0 in software. If hardware does not have these bits, they can be simulated by *software* under the control of the OS.

"R" and *"M"* bits in a combined way can be used to build a simple effective paging algorithm. When a process is started, both these page bits initially are set to 0 by OS for all its pages. Periodically (e.g. on each clock interrupt), the R bit is cleared, to distinguish pages that have *not been referenced recently* from those that have been. When a page fault occurs, the operating system scans all the pages, and divides them into *four distinct categories* based on the current values of *R* and *M* bits.

Class 0:	Not referenced ($R = 0$),	Not modified ($M = 0$)
Class 1:	Not referenced ($R = 0$),	Modified ($M = 1$)
Class 2:	Referenced ($R = 1$),	not modified ($M = 0$)
Class 3:	Referenced ($R = 1$),	Modified ($M = 1$)

At first glance, although class 1 pages seem impossible, they occur when a class 3 page has its *R* bit cleared by a clock interrupt. Clock interrupts, however, do not clear the M bit, because this information is only needed to know whether the page has to be rewritten to disk or not, at the time of its removal.

The ***NRU*** algorithm removes a page at random from the ***lowest-numbered*** non-empty class. This algorithm implicitly indicates that it is always appropriate to remove a modified page that has not been referenced in at least one clock period (typically 20 ms) than a clean page that is in heavy use. The main attraction and ***distinct advantage*** of NRU is that it is easy to understand, efficient to implement, and offers a performance that, while certainly *not optimal, is often adequate.*

4.6.10.2 First-In–First-Out (FIFO)

Another *low-overhead* page removal algorithm is FIFO (First-In–First-Out) algorithm, in which a page selected for removal is the one which has been in the memory longest. Is FIFO, a good policy for the operating system? Probably not, as the pages that are replaced for being in memory longest times would be those that the system have been using for the highest times.

The columns in the page table are kept ordered so that the *"oldest"* page is always at the bottom, and the *"newest"* page is at the top. On a page fault, the page at the bottom is removed and the new page is added to the top of the list. In actual implementation of FIFO removal, the clock value indicated the time when a page was loaded and can be stored, and for removal, a search can be made for the *"oldest"* time. It is usually more efficient to maintain an ordered list, usually by means of pointers, so that the page to be removed can be selected without a lengthy search. However, FIFO has ***three key disadvantages***, namely;

1. If a page is frequently and continually used, it will eventually become the oldest, and will be removed; even though it would be needed once again immediately.
2. Some strange side effects can occur.
3. Others algorithms have been found to be more effective.

The most noted side effect, called the ***FIFO anomaly/Belady effect*** is that under certain circumstances, adding more physical memory can result in poorer performance, when one would expect a larger memory to result in a better one. The actual occurrence of page traces that result in this anomaly are, of course, very rare. Nevertheless, this unusual phenomenon coupled with the other objections as noted, has caused FIFO in its pure form be *dropped from favour*. However, pure FIFO has been modified in many different ways to make it more attractive to be modestly accepted. Of these, the schemes, namely, Second chance and Clock page replacement (Circular FIFO) are worthy of mention.

A brief detail of Second chance and Clock page replacement (Circular FIFO) is with respective figures in the following web site: http://routledge.com/9780367255732.

4.6.10.3 Least Recently Used Page (LRU)

One of the most popular page removal techniques among many ones with a good approximation to the optimal algorithm is called *Least Recently Used* (LRU). It selects the page for removal that has *not been referenced* for the longest time. FIFO, in contrast, removes the page that *has been in memory* for the longest time, regardless of how often and when it was referenced. LRU is based on the theory, that if a page is referenced, it is likely to be referenced again soon. Conversely, if it has not been referenced for a long time, it is unlikely to be needed in the near future. This approach is also the basis of the ***theory of locality***. According to this strategy, when a page fault occurs, throw out the page which is remained unused for the longest time. LRU policy, however avoids the replacement of frequently used blocks which can occur with FIFO.

An LRU page trace analysis as explained puts the most recently referenced page on the top of the M list. The least recently used page thus falls to the bottom of the M list, and is the candidate for removal. LRU has been found to possess many interesting theoretical properties, and functions in such a way that increasing memory size can never cause the number of page faults (page interrupts) to increase, in contrast to the FIFO anomaly.

Implementation of theoretically realizable LRU is not cheap. It is necessary to maintain a *linked list of all pages* in memory with the most recently used page at the front, and the least recently used page at the rear. The difficulty is that the list must be updated on every memory reference. Moreover, at the time of page replacement, finding a page in the list, deleting it, and then adjusting the link and placing the new one to the front is a very time-consuming operation. Either an expensive special hardware is needed, or it must be realized by a cheaper approximation in software.

On every instruction, switching and manipulating a linked list is prohibitively slow, even when implemented in hardware. However, there are other ways to implement LRU with special hardware. This method requires to equip the hardware with a *64-bit counter*, C, that is automatically incremented after each instruction. Furthermore, each page table entry must also have a field large enough to store the contents of this counter C. After each memory reference, the current value of C is stored in the page table entry for the page just referenced. When a page fault occurs, the operating system examines all the counters in the page table to find the *lowest one*. That page is the least recently used, and is selected to be a victim for replacement.

4.6.10.4 Performance Comparison

The effectiveness of a page removal algorithm to a large extent depends on the behaviour of the particular program being run. The performance of a page replacement algorithm depends mostly on *execution trace, address trace,* and finally on *page trace*. The best policy until now is the OPT algorithm. However, this policy cannot be implemented in practice, because it is not really possible to predict the future page reference of a program beforehand. The LRU algorithm is a popular policy being stood on a strong base of theory of locality. It often results in a high hit ratio. If the available main memory is more than the size of the working set, which is usual and very common, LRU then appears to perform very well. The Random policies and FIFO are very easy to implement with almost no demand of additional facility, but may perform badly because of violation of locality of program. The Second Chance and Circular FIFO attempt to optimize FIFO, and work within the frame of FIFO, and at best, can approximate the LRU, but cannot perform beyond it. The performance of NRU may be between the LRU and the FIFO, with the cost of some additional overhead in the bargain. It is efficient to implement, and offers a performance that, while certainly *not optimal*, it is *often adequate*. It is thus very difficult to declare any one as the winner, because there is no fixed superiority of any policy over the others. The reason is that the program behaviour, run-time status of the page frames, and many other similar parameters are to be taken here into consideration that surely influence the ultimate outcome as a whole.

A few examples on page replacement policies with solutions are given in the following web site: http://routledge.com/9780367255732.

4.6.11 Segmentation

Virtual memory system when implemented with a fixed-size page causes certain inconvenience with regard to program (both user and system) size, and the logical structure of the program to be executed. Actually, an application is essentially a collection of related program units. The units may vary in their size. Each unit contains closely related functions and/or data, and may obtain services from other units. The units are effectively a higher level of structuring an address space, and are, however, independent as far as compilation of the program is concerned, but are integrated at link, load, or run-time. Program developers really are not at all aware of how these units are mapped into a process's private address space during the execution. In fact, when a program runs, it appears that certain units of the program have almost filled up its allocated address space, while the other units of the program may have lots of room as unused space. One solution for this hassle may be to take some space from the unit having much room to release, and giving it to the unit which deserves it. This shuffling for the sake of smooth execution of the program is the additional headache of the programmer which is a nuisance at best, and a tedious unproductive work as well.

To relieve and to release the programmer from this burden of organizing the program by dividing into modules for proper memory utilization, a straightforward and extremely general solution is to *provide the machine* with many completely independent address spaces, called *segments.* In segmentation, a logical address space is organised into a collection of disjoint segments. Each segment constitutes a separate address space with a linear sequence of addresses, from *0* to some maximum. Different segments may, and generally do have different lengths which can even grow or shrink independently without affecting the others during execution (dynamic). From the program developers' point of view, memory consists of multiple disjoint address spaces or *segments,* each one is a single logical entity. They accordingly divide the programs and data into logical parts (called *segments*). Procedures, subroutines, array of data, tables, etc. are examples of segments. Segments may be generated by the programmer, or by the operating system. Users are not at all concerned with which code/data block (logical part) is mapped into which segment in the logical address space, and also the relative order of segment placement in the logical address space. Memory when managed with a strategy that allocates main memory by segments is called ***segmentation,*** and this memory management technique is called *segmented memory management.* During execution, when a segment is needed, but not currently available in the main memory, the entire segment is then fetched from secondary memory (virtual memory). It seems reasonable to maintain complete segments in the main memory as opposed to paging system which uses a part of it. A segment usually does not contain a mixture of different types of information like a procedure and an array, a stack and scalar variable, etc.

The address generated by a segmented program called *logical address* is converted into corresponding physical address by the memory management unit in a way similar to the virtual memory mapping concept. A complete program, and its data sets can be viewed as a collection of linked segments. The links indicate that a program segment may use or "call" another program or data segment. In a multitasking environment, the common programs that are being shared by other programs can be kept in one segment for the use of the others without having to produce multiple copies. An ideal example in this regard is a *shared library.*

4.6.11.1 Pure Segmentation

The segmentation and paging differs both in approach (strategy) as well as in implementation. Segments are user-oriented concepts, providing logical structures of programs and data in the virtual address space. Paging, on the other hand, concentrates on the management of physical memory. In a paged system, all pages are of fixed-size in contrast to the segments that are variable, indeed dynamic in size, and all page addresses form a linear address space within the virtual address space. The physical addresses assigned to the segments in memory are maintained in a memory map called a *segment table.* This table may itself be a relocatable segment.

The segmented memory is organised as a 2D address space. An address in this segmented memory has a two-part address, the prefix field called the *segment number* and the postfix field called the *offset* (address) within the segment. The offset targets the location within each segment form one dimension of the contiguous addresses. The segment numbers, not necessarily to be contiguous to each other, form the second dimension of the address space. When memory is segmented, and if one segment is evicted due to the need of the executing program, and a smaller segment is brought into that area, and if this process continues which is an obvious phenomenon during run-time, the memory will ultimately be divided into a number of chunks, some containing segments, and some containing holes. This phenomenon, popularly known as *external fragmentation* (also called checker boarding) wastes a good amount of memory in the holes, and is not at all desirable. This drawback although can be overcome by using the technique of *memory compaction*, but adds an expensive overhead from the administrative point of view. This, along with other inconveniences as already observed make pure segmentation ultimately drop from favour.

A brief detail of pure segmentation with respective figure is given the following web site: http://routledge.com/9780367255732.

4.6.11.2 Segmentation with Paging

When the segments are large, it may be inconvenient, and even sometimes impossible to accommodate them entirely in the allotted limited main memory. This leads to the idea of dividing each segment into pages, so that only those pages that are actually needed at any point of time have to be around. Paging and segmentation thus can be wilfully combined to make use of the advantages of both. Thus, virtual memory can be implemented with a type of *paged segments*. Within each segment, the addresses are divided into fixed-size pages. Each virtual address (logical address) is thus divided into *three* fields. The upper field is the *segment number*, the middle one is the *page number*, and the lower one is the *offset* within the page. The memory map then consists of a *segment table*, and a *set of page tables*, one set for each segment. The segment table contains for each segment an address pointing to the base of the corresponding page table. The page table is used in the usual way to determine the required physical address. Figure 4.16 shows a scheme of address translation of a virtual address to a main memory address when virtual memory is implemented under segmentation with paging.

The **distinct advantage** of breaking a segment into pages is that a contiguous region of main memory is not needed to store the entire segment. The segment broken into pages, and the corresponding page frames need not be contiguous, and hence, the various space allocation strategies like **best fit**, **first fit** can be dispensed with low overhead. Different trade-offs do exist among the size of the segment field, the page field, and the offset limit. This set places a limitation on the *number of segments* that can be declared by users within the permissible range offered by the operating system, the *segment size* (the number of pages within each segment), and the *page size*. All these three parameters (*number of segments, segment size*, and *page size*) although are determined and managed by the operating system, they are entirely dependent on, and have a bearing on the architecture of the system being used.

4.6.11.3 Paged Segmentation in Mainframe (IBM 370/XA)

The representative mainframe IBM 370 series uses a two-level virtual memory structure referring two levels as *segments* and *pages*. Different page sizes, and segment sizes have been used in this series, but the flagship 370/XA architecture uses page size of 4K bytes,

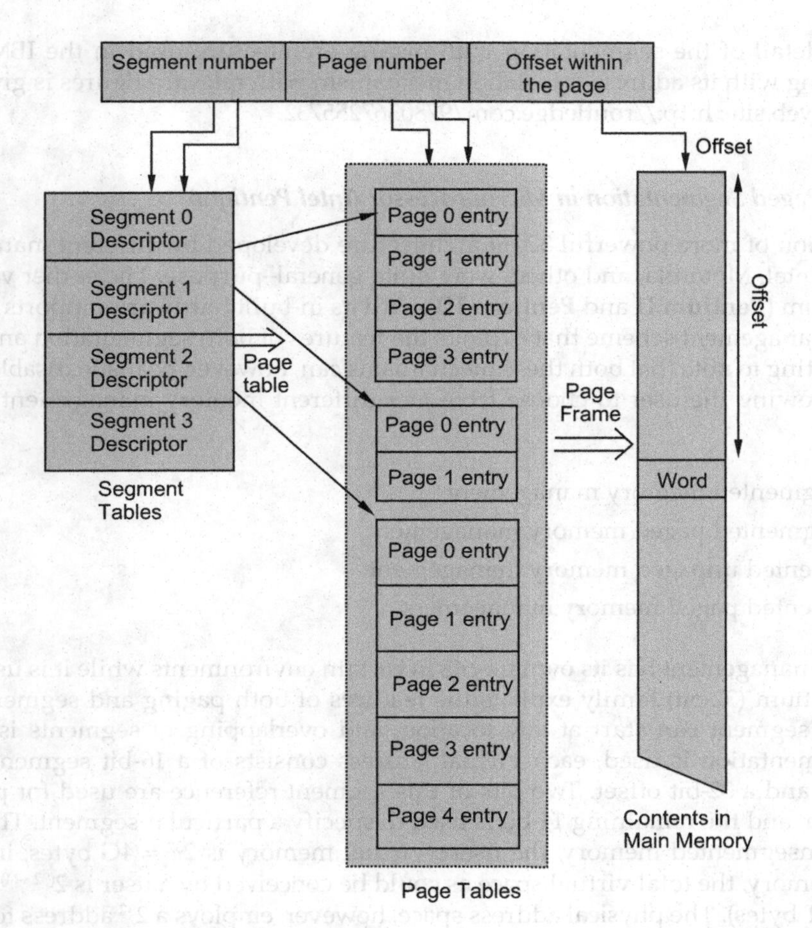

FIGURE 4.16
Paged segment memory management. Conversion of a two-dimensional virtual address into a physical main memory address.

and the segment size is 1M byte. The virtual address space is of 2 GB, thereby requiring a 31-bit (2 GB = 2^{31} bytes) virtual address to access. There is one *segment table* for *each* virtual address space. *Multiple virtual address spaces,* one per process is provided by the most powerful and popular multiple virtual storage (MVS) operating system used in 370 and higher versions of IBM systems, and also in the operating systems used in the DEC VAX systems. Each virtual address space is made up of the number of segments, and there is one entry in the segment table per segment. Each segment, as usual, has associated with it a page table containing entries, one entry per page in the segment. Thus, each 31-bit virtual address consists of the *byte index* specifying one of 4K bytes (2^{12}) within a page, the *page index* specifying one of 256 pages (2^8) within a segment, the *segment index* identifies one of 1024 (2^{10}) user-visible segments, and the bit-0 is used for other purposes.

Use of *Translation Lookaside Buffer* (TLB) and an innovative page replacement technique are two additional salient features of system/370 memory management. The exact implementation differs from one model to another model, and again depends on the architecture of the systems and the type of operating systems being used.

A brief detail of the segmentation with paging architecture used in the IBM 370/XA system along with its address translation mechanism with relevant figures is given in the following web site: http://routledge.com/9780367255732.

4.6.11.4 Paged Segmentation in Microprocessor (Intel Pentium)

The evolution of more powerful 32-bit architecture developed by different manufactures including Intel, Motorola, and others were quite general–purpose. The earlier versions of Intel Pentium (**Pentium II** and **Pentium III**) with its in-built hardware supports provide a memory management scheme that exploits the features of both segmentation and paging. It is interesting to note that both these mechanisms can, however, be made disabled at will, thereby allowing the user to choose from *four* different memory management schemes, namely,

i. Unsegmented memory management
ii. Unsegmented paged memory management
iii. Segmented unpaged memory management
iv. Segmented paged memory management

Each such management has its own merits in certain environments while it is used.

Intel Pentium (32-bit) family exploits the features of both paging and segmentation in which the segment can start at any location, and overlapping of segments is allowed. When segmentation is used, each virtual address consists of a 16-bit segment number (reference) and a 32-bit offset. Two bits of this segment reference are used for protection mechanism, and the remaining 14-bit is used to specify a particular segment. Thus, while with the unsegmented memory, the user's virtual memory is 2^{32} = 4G bytes, in the segmented memory, the total virtual space as could be conceived by a user is $2^{(32+14)} = 2^{46} = 64$ terabytes (T bytes). The physical address space, however, employs a 2^{32} address for a maximum of 4G bytes. If the segment size is itself 4G bytes, equal to the size of the entire physical memory, it indicates that only one segment is in the main memory, i.e. the segmentation is essentially disabled. Virtual address space is divided into two parts. One-half of the virtual address space ($2^{46} \div 2 = 2^{45}$K bytes) is global, shared by all processes, the remainder is local, and is distinct (mutually exclusive) for each process.

A virtual address consisting of a 16-bit segment selector, and a 32-bit offset is submitted for address translation mechanism as depicted in Figure 4.17. This 16-bit segment selector consists of a number of fields, one of these is a segment number/index field which points to a 32-bit entry in the segment table. The contents of this 32-bit entry is then used for paging mechanism which is actually a two-level table lookup operation. The first level is a page directory requiring 10 bits which contains 1,024 (= 2^{10}) entries. This splits the linear memory space (2^{32} = 4G byte) into 1,024 (2^{10}) page groups, each one is 4M byte in length ($2^{32} \div 2^{10} = 2^{22}$ = 4 Mbytes). Each such page group (i.e. each entry in the page directory) contains an address that corresponds to the base address of its own page table. Each page table contains up to 1,024 (= 2^{10}) entries. The content of each such entry in the page table corresponds to the base address of a single page, each page size is 4 Kbytes ($2^{22} \div 2^{10} = 2^{12}$ = 4K bytes).

Memory management has the option of using one page directory for all processes, or one page directory for each process, or some other combination of the two. The page directory for the current task is always in main memory. However, page tables may be in virtual memory.

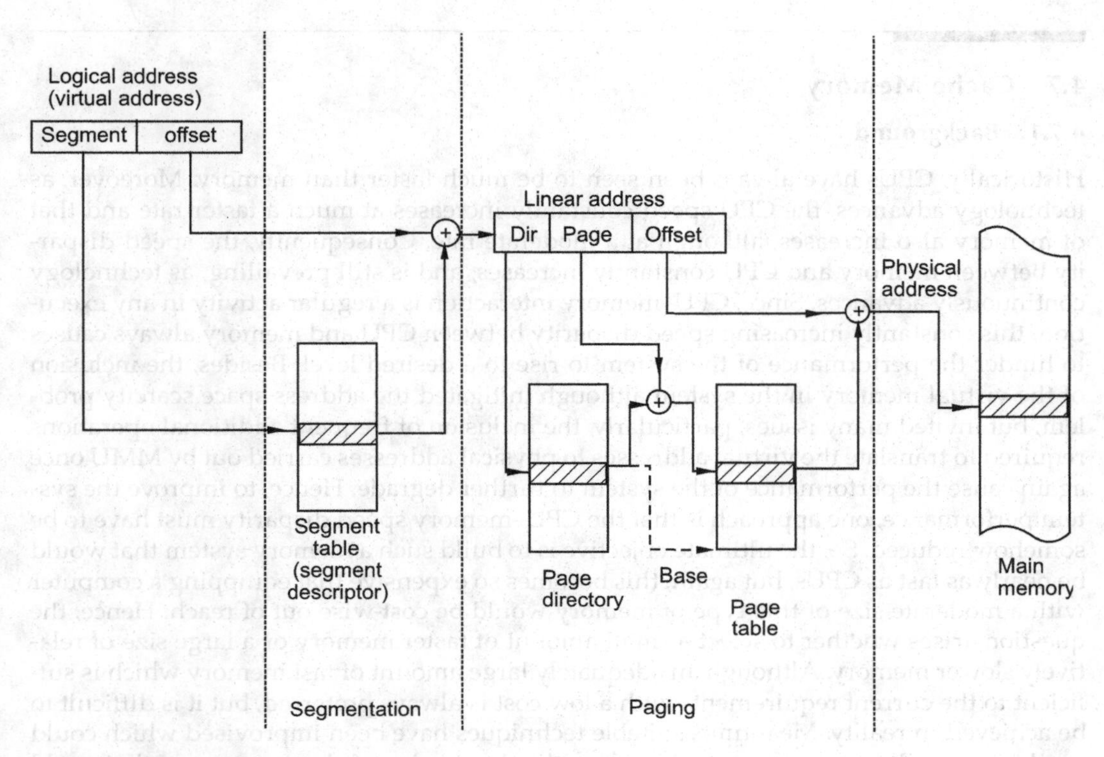

FIGURE 4.17
Memory address translation mechanisms in Pentium.

Pentium processor provides a new enhancement with provision of two different page sizes to use. The PSE (page size extension) bit found in Page Directory Entry permits the OS programmer to define a page as either 4K bytes or 4M bytes in size. When the 4M byte page size is used, there is only one level of table lookup for accessing pages. When the MMU accesses the page directory, it finds that the page directory entry has its PSE bit set to 1. The high-order 10 bits (i.e. bits 22–31) of the 32-bit address then defines the base address of a 4M byte page in memory. The remaining 22 bits out of 32-bit address is the offset within this 4M byte (2^{22}) page, and is then used to access any location within this page. The use of 4M byte page size, however, significantly reduces the memory management overhead (the time consumed to access a particular location) as well as the huge storage requirement for maintaining page directory and page table when a large main memory is used.

As usual, translation lookaside buffer (TLB) is used, that can hold 32 most recently used page table entries, to directly access the physical location at the time of address translation process. Each time the page directory is changed, this buffer is made cleared. In addition, four control registers are used to handle regular paging and page-fault situation when the cache miss occurs. For the sake of clarity, the translation lookaside buffer and memory cache mechanisms are not included in Figure 4.17. Interested readers can consult the book of the same author [*Computer Architecture and Organisation*, First Edition, P. Chakraborty].

A brief detail of the segmentation with paging mechanism used in Intel Pentium system exhibiting different tables, different formats of each entry, and their implications used in different tables with related figures, along with the address translation mechanism with relevant figures, is given in the following web site: http://routledge.com/9780367255732.

4.7 Cache Memory

4.7.1 Background

Historically, CPUs have always been seen to be much faster than memory. Moreover, as technology advances, the CPU speed constantly increases at much a faster rate and that of memory also increases, although at a moderate rate. Consequently, the speed disparity between memory and CPU constantly increases, and is still prevailing, as technology continuously advances. Since, CPU–memory interaction is a regular activity in any execution, this constantly increasing speed disparity between CPU and memory always causes to hinder the performance of the system to rise to a desired level. Besides, the inclusion of the virtual memory in the system although mitigated the address space scarcity problem, but invited many issues, particularly, the inclusion of frequent additional operations required to translate the virtual addresses to physical addresses carried out by MMU once again cause the performance of the system to further degrade. Hence, to improve the system performance, one approach is that the CPU–memory speed disparity must have to be somehow reduced. So, the ultimate objective is to build such a memory system that would be nearly as fast as CPUs, but again, this becomes so expensive that equipping a computer with a moderate size of this type of memory would be cost-wise out of reach. Hence, the question arises whether to select a small amount of faster memory or a large size of relatively slower memory. Although an adequately large amount of fast memory which is sufficient to the current requirements with a low cost is always preferred, but it is difficult to be achieved in reality. Meantime, suitable techniques have been improvised which could combine a smaller size of faster memory with a larger size of slower memory that could attain a speed nearly that of the fast memory, and at the same time could satisfy almost the same capacity as of a large memory at a moderately affordable price. This small, fast, and comparatively costlier memory is historically called ***cache*** (from the French word "cache", meaning to hide) [Jacob, B].

4.7.2 Objective

It has also been observed that CPU while executing a program does not really access memories completely at random, rather the accesses are found to remain confined within the domain of recently used areas. This is true when instructions of a program fetched from consecutive locations in memory are sequentially executed, except for jumps and procedure calls. Similarly, program loops or subroutines, matrix manipulation program, procedures for handling tables, iterative procedures, array of numbers, and similar others tend to be localized and repeatedly refer to a certain portion of common memory locations. The result of all these observations together highlights that there is a tendency in which most of the program references are localized to a small fraction of the total memory for both instructions and data. This fact that the memory references over any short interval of time tend to use only a small fraction of total memory is called the ***principle of locality,*** which actually forms the basis of all caching systems. The remainder of memory is accessed relatively infrequently.

Locality of reference is found mainly in data accesses, and not as strongly as in code accesses. *Three types* of locality have been observed. ***Temporal locality*** states that recently accessed *items* (instructions or data) are likely to be accessed very soon. For example, *subroutines, iterative loops, process stacks,* etc. will be accessed again and again within a short period. This locality thus entails in *nearness in time* that causes access to be remained confined within the domain of recently used areas. ***Spatial locality*** says that items whose *addresses* are near one another tend to be referenced close together in time. For example, operations on arrays or tables involve accesses of a certain memory block in the address space. There is a high probability that the other data in the same block will be needed soon. Spatial locality, however, implies *nearness in space/distance*. Figure 4.18 shows the temporal locality and spatial locality for a series of memory references made over time that corresponds to the given sequence of addresses. ***Sequential locality*** is observed in the execution of instructions of a program that normally follows a sequential order (in-order execution) unless branch instructions cause the control to go out-of-order (out-of-order execution). In ordinary programs, the ratio of in-order executions to out-of-order executions is roughly estimated to be 5 to 1. These sequential instructions reside normally in a chunk of memory close to one another. This chunk is accessed again and again due to the in-order execution of the sequential instructions in the program giving rise to what is called *sequential locality*. Both the temporal locality and the spatial locality as found during program execution, however, have a strong impact on the overall execution, and hence, these localities can be then wilfully exploited by the efficient use of the cache in the memory hierarchy to improve the performance of the system as a whole.

The temporal locality phenomenon can be advantageously exploited in a way by bringing any information (instruction or data) in the cache whenever it is referenced. If adequate space is not available in the cache, then some of the not-recently used existing ones can be evicted from the cache to make room for the new ones. Similarly, the spatial locality phenomenon can be favourably exploited in a way by bringing a block of contiguous information (instruction or data) in the *cache block* (often called *cache line*) whenever it is referenced, and not just the requested one only.

Thus, the general approach would be that the active portions of the program and data are always attempted to be placed in a fast small cache memory so that most of the frequent slower main memory visits made by the processor can be avoided during execution. As a result, the average memory access time will then be reduced, resulting in the fast execution of the program, and thereby improving the performance of the entire system as a whole. Although the cache is only a *small fraction of the total size* of main memory, a *large fraction of memory requests* will be found in the fast cache memory because of the locality-of-reference property of the executing programs. In multiple-processor systems with multiple CPU and input–output processor (IOP), each processor can function efficiently due to its own separate private cache for faster independent operations.

Summarily, a cache is essentially a small, fast, and costly memory placed in between a processor and main memory that contains *a copy of recently used portions of main memory*, and acts as a buffer creating a two-level internal memory. The cache is then considered as the fastest component in the memory hierarchy, and approaches nearly the speed of CPU components. The term *"cache"* is commonly reserved for a *general-purpose buffer memory*

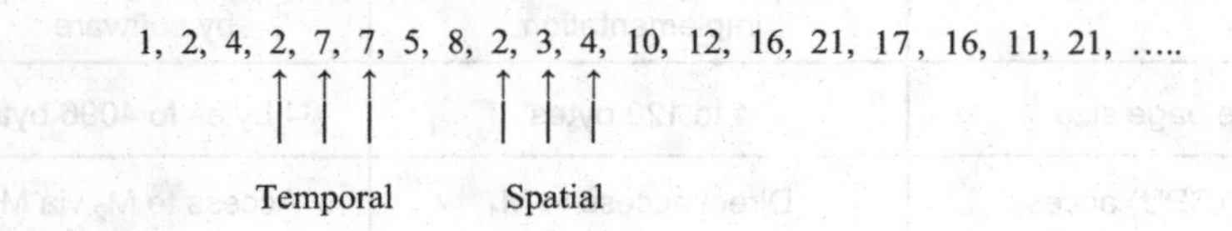

FIGURE 4.18
An illustration of different locality of references.

designed to store instructions and data associated with the execution of all types of programs. Cache, however, was first commercially introduced on the IBM 360/85 system in 1968. The inclusion of this element into the memory hierarchy has dramatically improved the overall performance of the entire system.

4.7.3 Hierarchical View

It is a common practice to build a three-level memory hierarchy system taking two largely independent two-level hierarchies. One of these two-level hierarchies is cache C_1 and main memory M_1, i.e. (C_1, M_1), and the mechanism used in (C_1, M_1) is part of the computer architecture, implemented totally in hardware, and typically not visible even to the operating system which is the master of the system (that is why, it is called *cache*, meaning *hidden*). This two level is especially organised as fully internal memory system for fast operations. The other two-level is main memory M_1 and secondary memory M_2, i.e. (M_1, M_2) which may be organised as a virtual memory and disk cache systems to fulfil other targets. Historically, cache memory appeared on the heels of virtual memory, and therefore, many of the techniques already available in virtual memory–main memory (M_1, M_2) management are used almost in the same line in the management of cache memory–main memory (C_1, M_1) organisation. Still, there are some major differences between (C_1, M_1) organisation and (M_1, M_2) organisation which are summarized in Figure 4.19.

Since the cache memory is faster in access than that of main memory by a factor of 5–10, it is thus organised as a supporting memory module to the main memory, and hence, (C_1, M_1) lies in a higher memory hierarchy, and also operates at a higher speed than (M_1, M_2). A common arrangement of the CPU, cache, main memory, and secondary memory is illustrated in Figure 4.20. With the relentless progress of the VLSI semiconductor memory technology, use of caches indeed have become economically viable, and it then started to be used in various forms with fast access time as one of its basic characteristics. In segmentation and paging scheme, the *Translation Lookaside Buffers* ***(TLB)*** are special-purpose ***address caches*** designed to store frequently accessed segment or page tables.

4.7.4 Principles

When the CPU needs to access memory, the cache is examined, if the requested word is found in cache, it is read from the fast cache memory. If the word addressed by the CPU is not found in the cache, the main memory is accessed to read the word. A block of words varying from one word to about 16 words adjacent to the one just accessed is then

Characteristics	Cache-main memory (C_1, M_1)	Main-secondary memory (M_1, M_2)
Ratios of typical access time	(5–10)/1	1000/1
Memory management system	Special hardwae for implementation	Implemented mostly by software
Average page size	4 to 128 bytes	64 bytes to 4096 bytes
Processor (CPU) access	Direct access to M_1	Access to M_2 via M_1

FIGURE 4.19
Major parametric difference between cache–main memory and main–secondary memory hierarchies.

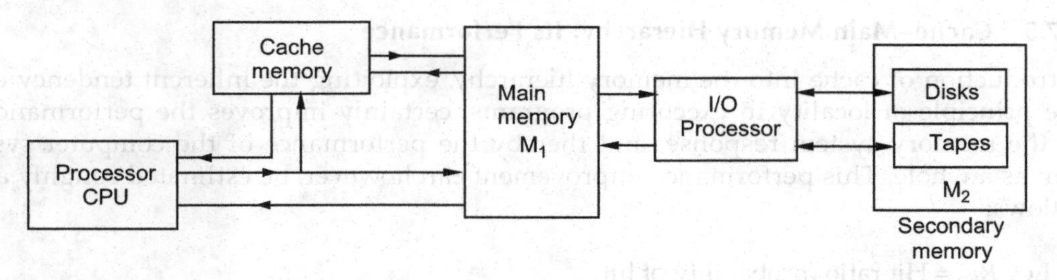

FIGURE 4.20
Data flow in a system with CPU, memory, secondary memory and an off-chip cache.

transferred from main memory to cache memory. In this way, some data are always transferred to cache beforehand so that future reference to main memory can get the required words automatically in the fast cache memory. But, if the word addressed by the CPU is not found even in the main memory, the virtual memory is accessed to read the word. A page of virtual memory containing the word just accessed is then transferred from virtual memory to a page frame in the main memory, and from there to cache memory in the way already explained. If a word is referenced k times in a short interval, it is expected that the computer will first need 1 reference to slower memory, and then remaining $k - 1$ references to fast cache memory. The larger the value of k, the better would be the overall performance.

The performance of cache memory depends on whether the memory references issued by the processor have been found in cache or not. When the CPU refers to memory and finds the words in cache, it is said to produce a ***hit***. If the word is not found in cache, it is mostly in main memory, and it then treats as a ***miss***. The ratio of the number of hits divided by the total number of CPU references to memory (hits plus misses) is called the ***hit ratio*** (h) or ***probability of hit***. Following our small example of the last paragraph, we can then write:

$$\text{hit ratio } (h) = (k-1)/k$$

and then the

$$\text{miss ratio} = (1-h)$$

Hit ratios of 0.9 and above have been often observed by running representative programs. This high hit ratio, on the other hand, ensures the validity of the *principle of locality*. With these definitions, the average (mean) access time can be calculated as follows: if c is the cache access time, m, the main memory access time, and h, the hit ratio then,

$$\text{average access time} = c + (1-h)\cdot m$$

as $h \to 1$, i.e. all references to memory are found in cache, the average access time then approaches c. On the other hand, as $h \to 0$, a memory visit is needed every time, so the access time approaches $c+m$, here, the time c is required *to check* the cache (unsuccessful), and then a time m to do the memory reference. On some systems, the memory references can be started in parallel with the cache search to shorten the time consumption, so that if a cache miss occurs, the memory cycle has already been started. However, this strategy requires different treatments, which are actually handled by the respective *microprograms*.

4.7.5 Cache–Main Memory Hierarchy: Its Performance

Introduction of cache into the memory hierarchy, exploiting the inherent tendency of the principle of locality in executing programs, certainly improves the performance of the memory system response, and thereby the performance of the computer system as a whole. This performance improvement can however, be estimated roughly as follows:

Let, R_{hit} = Hit ratio (probability of hit).

R_{miss} = Miss ratio (probability of miss) at the cache level, and particularly, holds for any level.

T_{hit} = Time to complete the memory reference activity, when hit occurs at any given level in the memory hierarchy.

T_{miss} = Time to complete the memory reference activity, when miss occurs at any level, and have to go down to the next level in the memory hierarchy.

Thus, the average time required to complete any memory reference irrespective of hit or miss (or taking into account the probability of both hit and miss) would be:

$$T_{avg} = R_{hit} \times T_{hit} + R_{miss} \times T_{miss}$$

However, R_{hit} should be always 100% for the lowest level (bottom-most level) in the memory hierarchy.

A number of problems with solutions related to this topic are given in the following web site: http://routledge.com/9780367255732.

4.7.6 Cache Design

Cache is controlled by the Memory Management Unit (MMU) and also by cache controller, and is transparent to the operating system and also to the programmer. Cache stores a set of frequently referenced main memory contents M(A_i) into cache blocks (cache pages) where A_i is the memory block address. Each cache block is a sub-block of some main-memory page. The contents of the entire cache array are thus *copies* of a set of small non-contiguous main memory blocks attached with addresses. When an address is sent from the CPU at the beginning of a read or write memory access cycle, the relevant part of address (or the address itself) is compared by the cache with all its currently stored addresses. If there is a match, i.e. a *cache hit*, the cache then accesses the word M(A) corresponding to the address A just matched. It completes the appropriate *read* or *write* operation within the memory cycle. If the address A is not matched with any of the stored addresses in cache, a *cache miss* occurs. Usually, the cache then initiates a sequence of one or more read cycles to copy the required memory block P(A) into the cache containing the desired item M(A). If necessary, the cache controller first saves in main memory the existing contents of any cache page frame P′(A) it selects by any standard replacement algorithm to make a room for P(A), if the cache is found to be full. The size of the cache page is often designed in such a way that an entire page can be fetched in one main-memory cycle. The actual cache–main memory transfer procedures vary from cache to cache, and read/write operations are generally handled in a different way.

4.7.7 Cache Design Issues: Different Elements

Cache design and cache organisation involve many of the issues as have already been found in other memory module organisation. Although there are various types of cache implementations, a few basic design elements serve to classify and differentiate cache architectures. The primary elements which have a strong influence in cache design are:

a. Cache size
b. Cache block size
c. Mapping functions
   Fully associative
   direct
   Set associative
   Sector associative
d. Cache initialization
e. Write policy
   Write through
   Write back
   Write once
f. Replacement Algorithm
   Least-Recently Used (LRU)
   First-In First-Out (FIFO)
   Least-Frequently-Used (LFU)
   Random Replacement

Out of all of the aforementioned issues, the two most important issues that are specifically peculiar to cache design are:

i. The way in which main-memory information is transferred (*mapped*) into cache memory.
ii. When a *cache–write operation* will modify (change) the contents of a cache, how the corresponding main memory location will be updated.

4.7.7.1 Cache Size

The size of the cache at the time of its design is a compromise between two different opposite approaches:(i) the size should be *small* enough so that cost-wise it should be very close to that of main memory alone. Moreover, the small size permits it to be accommodated within a limited available chip and board area;(ii) the size should be *large* enough so that most of the memory references could be available in the cache, and hence, the overall average access time would be as close to that of the cache alone; the ultimate motive of using a cache. On the contrary, the larger the size, the larger will be the number of gates involved in addressing the cache which will eventually make the cache slightly slower in operation than the small ones, even when built with the same IC technology, and placed in the same chip, and circuit board.

With the continuous advancement in electronic technology, the logic density has enormously increased. Consequently, it has created a massive impact on the overall cache design including the size of caches which has been found to have radically changed over the past decade. Moreover, a new concept has been emerging in the usage of multiple caches, and that too at different levels, providing a comparatively larger size of cache in totality which has become a de facto current standard norm. In addition, it has become possible to have a cache on the same chip as the processor; the ***on-chip*** cache, and that too again at different levels within the CPU chip. All these aspects, and their implications in actual implementations have been separately discussed in later sections. Summarily, it can be said since the cache operation and its performance are very sensitive to the nature of the applications, it is not only difficult, but nearly impossible to conclude what the optimum size of cache should be.

4.7.7.2 Block Size

The choice of the block size of cache is one of the deciding factors in cache performance. Larger block size could store not only the desired word when it is retrieved and placed into the cache, but some number of adjacent words also. As a result, the hit ratio will then automatically increase because of the *principle of locality* (spatial locality). But, larger blocks reduce the *number of blocks* that can fit into a cache of fixed size. As a result, a small number of blocks results in frequent replacement of data shortly after it is fetched, increasing unnecessarily the overhead of cache operation. On the other hand, while a smaller block size increases the number of blocks available in fixed-size cache, the hit ratio cannot be appreciably improved. Smaller block size can only store a few units of information which is not enough to meet all the near-future references that could arrive according to the *principle of locality* (spatial locality). As the block size increases from small to larger sizes, the hit ratio will at first instance increase, and as the block becomes even larger, the hit ratio will then begin to decrease because each additional word newly fetched is farther from the requested word, and therefore less likely to be needed in the near future. In fact, the relationship between cache size, cache block size, and hit ratio is very complex to derive, and no such convincing optimum value has been found yet. It is more or less depending on the characteristics of a particular application. However, a reasonable approach close to optimum is a size of about 4–8 addressable units (bytes, or words).

4.7.7.3 Mapping Schemes

Four types of mapping schemes (as already mentioned) are of practical interest. Two of them are of fundamentally different cache organisation, along with a third form which is a *hybrid* of the first two, and the fourth one is the extension of the third form. Each scheme has its own merits as well as drawbacks. However, the ultimate performance depends mainly on cache-access patterns, cache organisation, and management policy used, apart from other similar considerations.

Cache–Main Memory Handshaking The transfer of information from main memory to cache memory is conducted in units of *cache blocks* on *cache lines*. Blocks in caches are called *block frames* which are denoted as C_i for $i = 1,2, \ldots,m$. The corresponding *memory blocks* are denoted as B_j for $j = 1,2, \ldots,n$. It is legitimately assumed that $n \gg m$ and

n = total number of blocks in memory = 2^s (say)

m = total number of block frames in caches = 2^r (say)

where

s = number of bits required to address a main memory block.

and r = number of bits required to address a cache memory block.

In an word-organised memory system, let us assume, for the sake of simplicity that each block or block frame consists of b words where $b = 2^w$ (e.g. block size (b) = 8 words = 2^3 words, $w = 3$).

Then the cache consists of (cache size) $m \times b = 2^r \times 2^w = 2^{r+w}$ words.

The main memory has (memory size) $n \times b = 2^s \times 2^w = 2^{s+w}$ words and, the length of the address to access this memory requires $(s + w)$ bits.

Let us assume that the entire cache is divided into v sets and let $v = 2^t$ sets. If k is the number of block frames in each set, then $k = m/v = 2^{r-t}$ [m is the total number of block frames in the entire cache].

4.7.7.3.1 *Fully Associative Cache*

When a processor presents an address A to the cache, the ultimate target is to quickly compare the address A to *all* the addresses stored in the cache to find a *hit* or a *miss*. Serial scanning all the cache entries is unacceptably slow. The fastest technique is thus to implement the entire cache array as a *single associative* or *content-addressable memory* (CAM). The associative memory stores both the address and contents (data) of the memory word. This permits any location in cache to store any word from main memory. Hence, this cache organisation offers the most flexibility in mapping cache blocks. But, pure associative memories are extremely expensive, so it is just feasible to use them only in moderate size, such as those used in microprocessor-based computer systems.

When a memory address A is presented to the CAM, it then compares the address A to all addresses currently stored in the cache in parallel (simultaneously). As illustrated in Figure 4.21, each block of main memory can be placed in *any* of the available block frames *at random*. To maintain this flexibility an s-bit tag (block number of memory address in high-order bits) is needed in each cache block.

Each cache block frame contains one memory block of data and its address (memory block number+word address within the block) along with a bit (valid) indicating whether the block frame is currently in use or not. When the computer is switched on, or is reset, all the valid bits are set to 0 to indicate that no cache entries are valid. When a program is completed or is terminated, the valid bit of all the block frames used by the program is set to 0 to identify those frames to be now available for use. When an address A is sent by CPU to cache, this address A will be simultaneously compared with all addresses currently stored in the cache. If it indicates a match, i.e. a *hit*, the cache responds to the request directly by either reading or writing the data portion of the cache, and the valid bit is set to 1. But, if the comparison fails, i.e. the requested address A is not currently available to cache, a *miss* occurs and the cache then forwards the requested address A to the main memory. The main memory responds similar to the usual CPU-memory request by executing a read or write cycle, and then sends back a cache page containing the data and the requested address A to cache to store it for future use. If the cache-array is found full, then an entry

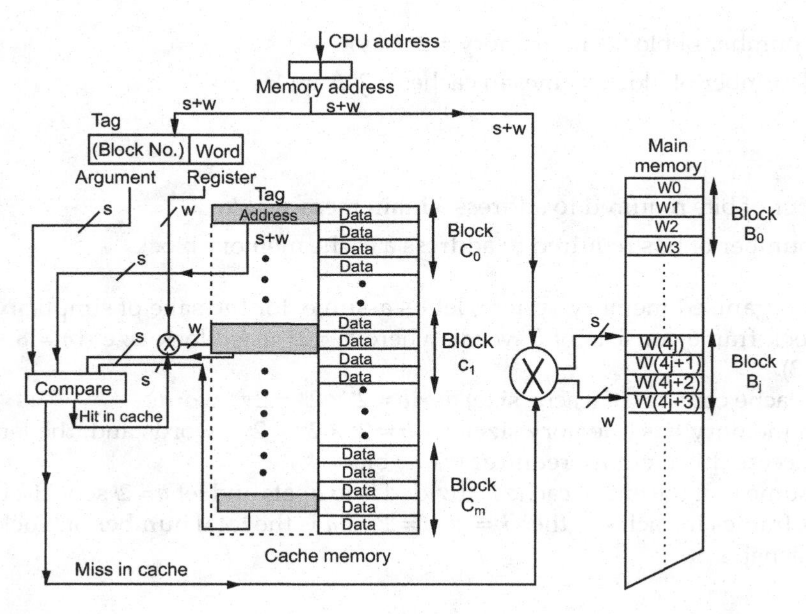

FIGURE 4.21
Associative mapping and search scheme with all block tags.

is selected for eviction to make room for this new one. Commonly, the least-recently used (LRU) replacement policy is implemented using special hardware that constantly monitors cache block frame usage. In practice, the selection as to which one would be thrown out has to be decided very quickly (of the order of nanoseconds). However, many machines including VAX usually pick a block frame at random without spending much time for any algorithm to decide.

It is to be noted that making a cache as part of the main memory (off-chip) as used here, rather than making it part of the CPU (on-chip), it takes more time for the CPU to access the cache. Moreover, it does not reduce the traffic on the CPU-memory bus. This aspect once again receives a considerable importance when multiple processors share a common memory system. As a result, memory-based caches (off-chip) of this type are much less common than processor-based (on-chip) caches.

The name ***fully associative cache*** is derived from the fact that an *m-way associative search* (m is the total number of blocks in cache, and a search all over such blocks at a time) requires the memory block number (tag) to be compared with all block frame tags in the cache. This is illustrated by a scheme as shown in Figure 4.22 using a four-way mapping with a *fully associative search*. Here, out of n memory blocks, 16 possible blocks are mapped into the 16 cache block frame. Since 16 (= 2^4) block frame is searched simultaneously, the block address (tag) of main memory is contained in cache in 4 bits. Although this scheme offers the highest flexibility in implementing block replacement policies to attain a higher hit ratio, it demands at the same time a higher implementation cost for the cache to be a fully associative one.

The distinct **advantage** of the fully associative cache is that it usually has fewer misses. Also, there is no *conflict misses,* and therefore a higher hit rate, in general, can be obtained. Moreover, fully associative property can be exploited in a better way to implement a relatively speedier block replacement policy which can be realized with the help of a

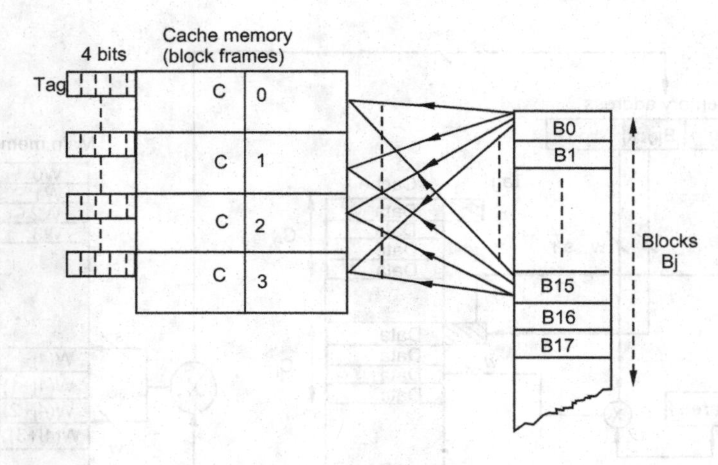

FIGURE 4.22
Representative scheme of fully associative cache. Every block is mapped to any of the four block frames identified by the tag.

microprogram in a special hardware for simultaneous comparison. The major drawback, of course, is the costly hardware required for such an expensive search.

A few problems with solutions on this topic is given in the following web site: http://routledge.com/9780367255732.

4.7.7.3.2 Direct-Mapping Cache

Associative memories used as cache memories with associative mapping offers the most flexibility in mapping cache blocks using special hardware that can compare a given block number to every entry in cache simultaneously. This added hardware makes the associative cache expensive compared with RAMs. Thus, an attempt has been made to explore the possibility of using a *RAM for the cache.* The direct-mapping cache is the outcome of such an investigation.

In the general case, there are $m = 2^r$ block frames in the cache memory, and $n = 2^s$ blocks in the main memory. This cache organisation is based on the principle that $n/m = 2^{s-r}$ number of memory blocks, separated by equal distances in memory to be directly mapped to one unique block frame in the cache. Since the number of blocks in cache is m, the mapping is defined using a modulo-m function. Block B_j is mapped to block frame C_i:

$$B_j \rightarrow C_i \text{ if } i = j \text{ (modulo } m)$$

This implies that there is always a unique block frame C_i that each B_j can be loaded into. Although the direct mapping is very rigid, it is also one of the simplest cache organisation techniques. Figure 4.23 illustrates direct mapping assuming the block (b) contains four words, i.e. $b = 2^w = 2^2$; $w = 2$.

When the CPU generates a memory request, the memory address is divided into three fields:

i. The lower w bits (here, $w = 2$ bits) specify the *word offset* (indicate the unique word within the block) within each block.
ii. The next higher s bits specify the *block address* in the main memory.
iii. The leftmost $(s - r)$ bits specify the tag to be matched in the comparison process.

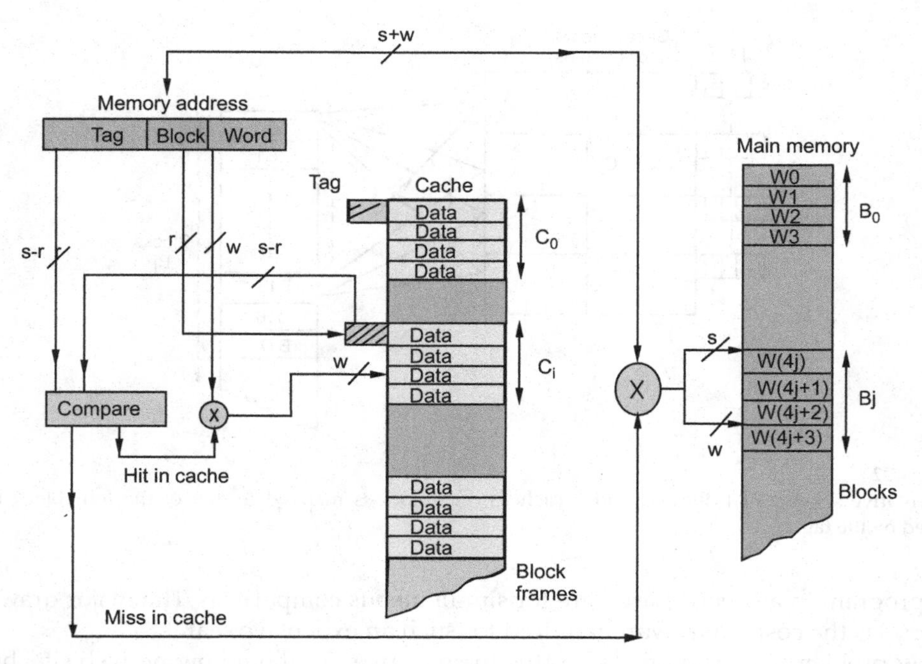

FIGURE 4.23
Direct mapping cache memory addressing.

At the time of comparison, the block field (r-bits) is used to implement the modulo-m placement where $m = 2^r$. Once the block frame C_i is uniquely identified with the use of block field (r), the tag associated with the already identified block frame is compared with the tag in the issued memory address. When the tag in the block frame matches with the tag in the addressed memory block, a cache *hit* occurs, otherwise it is a cache *miss*. When cache hit occurs, the word offset (w) is used to identify the desired word within the addressed block frame. When a miss occurs, the entire memory address ($s + w$) is used to access the desired main memory location. The leftmost s bits locate the addressed block, and the lower w bits locate the particular word within the block.

Certain ***distinct advantages*** are obtained from direct-mapped cache. It requires less area in its realization as only one comparator is needed, and only a few tag bits are required for its implementation. Moreover, this mapping scheme does not need to store the entire addresses of items in the cache as is the case in the associative mapping scheme. In addition, its operation is relatively fast, because it can return the contents of cache almost in parallel with search if a hit occurs.

However, one of the ***severe drawbacks*** of the direct-mapping scheme is that the cache hit ratio drops sharply, if two or more frequently used blocks happen to map onto the same region in the cache. Although this possibility appears to be fortunately less by the fact that such blocks are relatively far apart in the logical memory address space, but it may, however, degrade the overall performance of the cache to a considerable extent. For this reason, direct-mapped caches tend to use a larger cache size with more block frames to avoid contention.

Implementation Example Although the electronic technology, particularly the integrated-circuit (IC) technology at the early days was not so mature as of today, but still, due to having a number of distinct advantages offered by a direct-mapped cache as mentioned above, the giant manufacturers were naturally inclined to implement this scheme into their most popular large mainframe systems of those days. Both IBM and DEC thus implemented direct-mapping cache in their most successful large systems IBM 370/158 and DEC VAX/8800.

A few problems with solutions on this topic are given in the following web site: http://routledge.com/9780367255732.

4.7.7.3.3 Set-Associative Cache

In spite of several distinct advantages that direct-mapping cache has than its counterpart associative cache, the severe drawback is the fact that multiple memory blocks (modulo-*m* apart in main memory, m = number of block frames in cache memory) map onto the same cache block frame can cause problems. If many such referenced memory blocks are there that happen to map on the same cache slot, the cache performance will then certainly degrade. The ultimate goal, however, is then to alleviate this drawback, and thereby to improve the cache performance as a whole.

The way to get out of this difficulty is to expand the direct-mapped cache to have *more than one entry* per cache block frame to avoid this collision. This design, however, offers a compromise that exhibits the strengths of both the direct-mapping and fully associative approaches while reducing their disadvantages. In fact, a direct-mapped cache with multiple entries per cache block frame is called a *set-associative cache,* and it is illustrated in Figure 4.24.

Let m block frame in cache be divided into v sets. If k is the number of block frame per set (set size), then $v = m/k$. Each set in v is identified by a d-bit *set number* where $2^d = v$. The cache block frame tags are now reduced to $s - d$ bits. This approach is illustrated in Figure 4.25.

At the time of cache search, the set is identified with a technique in the line of direct-mapping, and once a particular set is found, *k-way* associative search is carried out within the set to reach the target block using the associative mapping technique. This means that the *tag* needs to be compared with the k tags within the identified set. For this reason, this technique is legitimately called a hybrid of the direct-mapping and associative-mapping techniques.

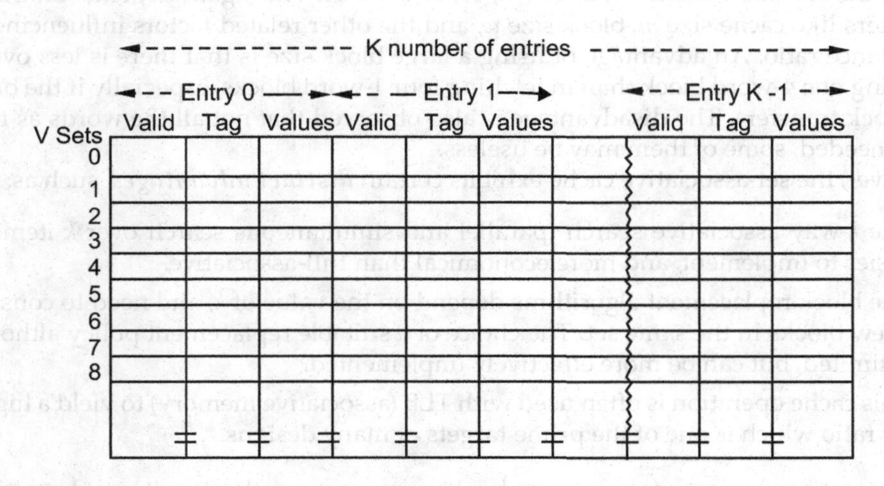

FIGURE 4.24
A set-associative cache with k number of entries per set.

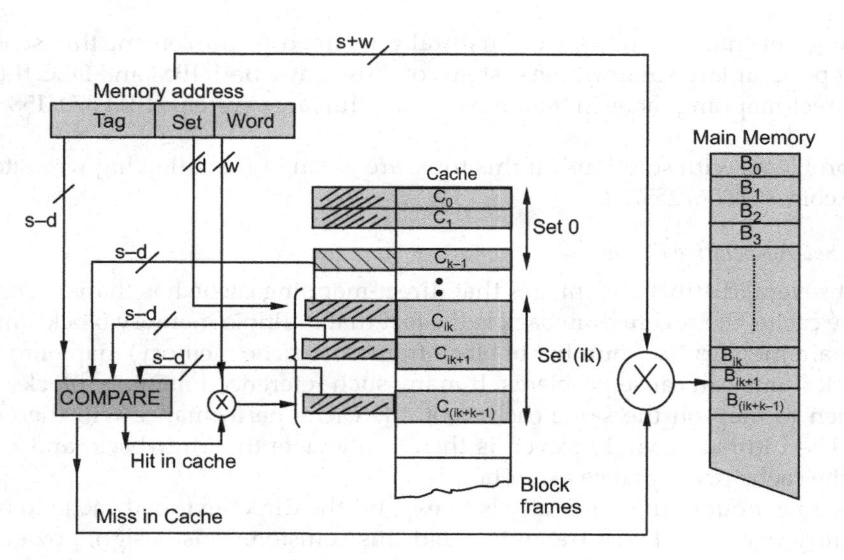

FIGURE 4.25
An associative (*k*-way) search over *k* cache blocks per set.

When $v = 1$, i.e. the entire cache is considered to be a single set. This means:

$$v = 1 = 2^0 = 2^d, \text{ i.e., } d = 0$$

and hence $s - d$ bits becomes $s - 0 = s$ bits which participate in the associative search, indicating a fully associative mapping.

If $k = 1$, then $v = m/k$ means $v = m$, i.e. the number of set is equal to the number of cache block frames. This means one entry per block frame which is essentially a pure direct-mapped cache.

Design Considerations: The set size k, and the number of set v in fixed cache size m are related by $m = v \times k$. Since, m is constant, v and k are related inversely with each other. In real life, the set size k is chosen as 2, 4, 8, 16, or even 64. This again depends on the other parameters like cache size m, block size w, and the other related factors influencing cost/performance ratio. An advantage of using a large block size is that there is less overhead in fetching one 4-word block than in fetching four 1-word blocks, especially if the bus executes block transfers. The disadvantage is also observed that not all the words as fetched may be needed, some of them may be useless.

However, the set-associative cache exhibits certain ***distinct advantages***, such as:

i. The k-way associative search (parallel and simultaneous search over k items) is easier to implement, and more economical than full-associative.

ii. The block replacement algorithms depend on the value of k, and need to consider a few blocks in the same set. The choice of a suitable replacement policy although is limited, but can be more effectively implemented.

iii. This cache operation is often used with TLB (associative memory) to yield a higher hit ratio which is one of the prime targets of many designs.

Implementation: The set-associative caches have more strengths than its weaknesses. That is why they are considered the most popular cache design built into many commercial computers. Different manufacturers use different sizes of caches with different parametric values [set size (k), number of sets (v), etc.], and block sizes to attain their own target.

A tabular representation of the set-associative cache organisation of a few representative popular systems and a few problems with solutions on this topic are given in the following web site: http://routledge.com/9780367255732.

4.7.7.3.4 Sector Mapping Cache

This scheme is one of the oldest techniques using the idea to partition both the cache and main memory in fixed-size *sectors*. So, there is a one-to-one mapping possible between each memory sector and available cache sector frames. Fully associative search is applied on sector frames using sector number as key. This scheme, however, offers a design alternative to set-associative caches.

Since the CPU issues memory requests in terms of blocks and not for sectors, the sector tag can be filtered out from the issued memory address, and can then be compared with all sector-tags in cache using a fully associative search. If the outcome is a cache hit, the matched sector frame is found, the block field is then used to locate the target block within that sector frame. If there is a cache miss, only the missing block is fetched from main memory, and must be placed in the appropriate block frame within a destined sector frame. Figure 4.26 illustrates a sector mapping with a sector size of 4 blocks.

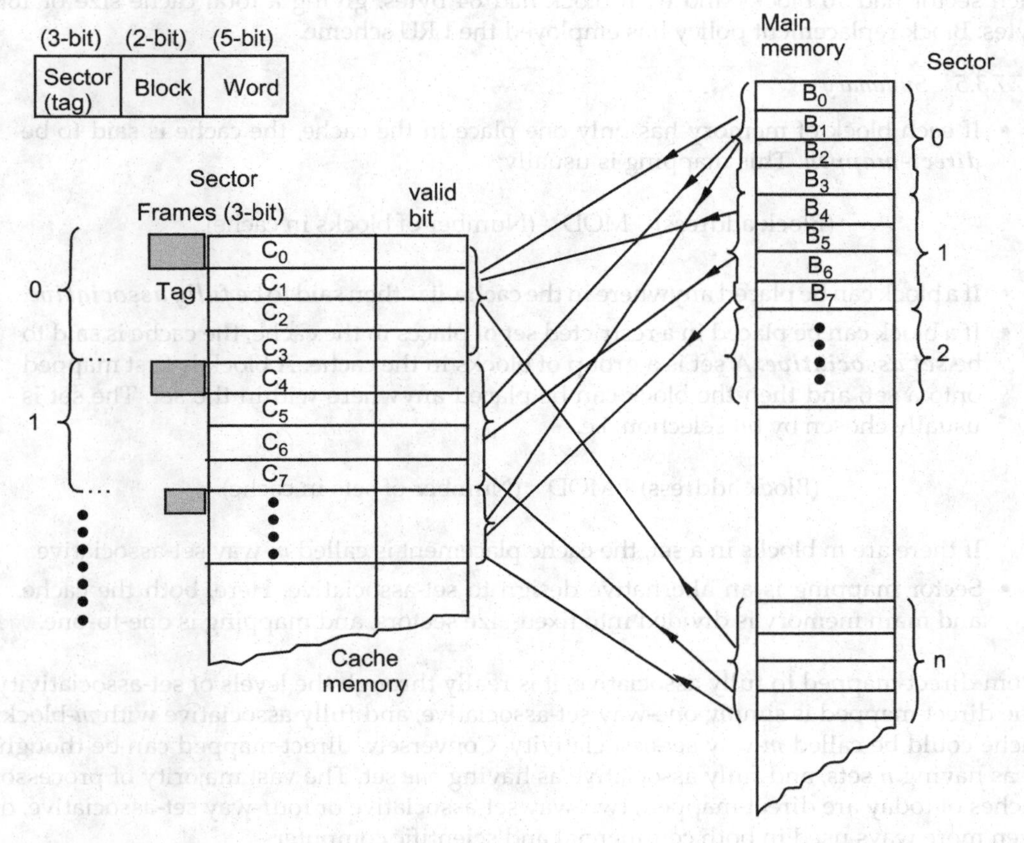

FIGURE 4.26
A scheme of sector mapping cache organisation.

As with the other mapping scheme, a valid bit here also is usually attached to each block frame to indicate whether it is valid or invalid. The other block frames in the same sector are marked invalid to indicate the validity of recently replaced block frame. Multiple valid bits, however, can be used to record the states of other blocks. Sector mapping cache offers some ***advantages***, a few of them are:

- Flexibility in implementing various block replacement algorithms. Different replacement policies when implemented yield different design approaches.
- Fully associative search across a limited number of sector tags is also found to be reasonable and economical.
- Sector partitioning, as a whole, favours in grouping cache frames while partitioning at both ends.
- Each memory sector can be mapped to any of the sector frames with full associativity at the sector level.

With appropriate choice in design, the cost/performance ratio of sector-mapping caches is very close to that of set-associative design, and is even found to be superior when it conforms with the system's architecture. This scheme was first implemented in the then most popular and versatile IBM 360/85 model where cache memory was formed with 16 sectors, each sector had 16 blocks and each block had 64 bytes, giving a total cache size of 16K bytes. Block replacement policy has employed the LRU scheme.

4.7.7.3.5 Summary

- If each block of memory has only one place in the cache, the cache is said to be ***direct-mapped***. This mapping is usually;

  (Block address) MOD (Number of blocks in cache).

- If a block can be placed anywhere in the cache, it is then said to be ***fully associative***.
- If a block can be placed in a restricted set of places in the cache, the cache is said to be ***set associative***. A set is a group of blocks in the cache. A block is first mapped onto a set, and then the block can be placed anywhere within the set. The set is usually chosen by bit selection, i.e.

  (Block address) MOD (Number of sets in cache).

  If there are m blocks in a set, the cache placement is called *m*-way set-associative.
- Sector mapping is an alternative design to set-associative. Here, both the cache and main memory is divided into fixed-size sectors, and mapping is one-to-one.

From direct-mapped to fully associative, it is really through the levels of set-associativity. The direct-mapped is simply one-way set-associative, and fully associative with *m*-blocks cache could be called *m*-way set-associativity. Conversely, direct-mapped can be thought of as having *n* sets, and fully associative as having *one* set. The vast majority of processor caches of today are direct-mapped, two-way set-associative or four-way set-associative, or even more ways used in both commercial and scientific computers.

4.7.7.4 Cache Initialization

Cache initialization is one of the important aspects in cache organisation and its operation. The cache is initialized when the computer is switched on, or the computer is reset, or when the main memory is loaded with a set of programs from secondary memory. When a program is completed or terminated, the portion of the cache the program was using is initialized. After initialization, the cache is considered to be empty or available for use, but, in fact, it contains some non-valid data. So, some means should be used to indicate whether the cache entry is valid for use, or non-valid to be overwritten, if required. This is accomplished with the addition of a *valid bit* to every cache item.

When the cache is initialized, all the valid bits are set to 0. When a word is loaded into cache from main memory, its corresponding valid bit is set to 1, and remains set unless the cache word has to be initialized again. When the valid bit of a word is 0, this means that the corresponding word can be used for new entry from main memory, and will be written over the existing non-valid data. A word in cache having valid bit 1 can only be replaced by another word coming from main memory. Cache miss occurs when the comparison fails to find the item from only those having valid bit 1. Thus, valid bit plays an important role in cache operation, and thus is employed for cache initialization.

4.7.7.5 Writing into Cache

Another important issue of cache design and its organisation is how to handle the write operation. When the CPU issues a request, and it is available in cache during a read operation, the transfer simply takes place between cache and CPU; the main memory is not involved in this transfer activity. However, if the operation is a write, two strategies are commonly used by the system.

4.7.7.5.1 *Write-Through (Store Through)*

The most commonly used simple procedure is to update the cache entries for every write operation, with corresponding memory entries being updated immediately in parallel. This strategy is called the *write-through* method. This method is not only time-consuming, but sometimes even a single entry is found repeatedly updated both at cache end and at memory end. The updated entry in the main memory in that situation seems to be redundant, so long it is residing in cache. But, the advantage of this approach is that it ensures that the cache entries are always the same as the corresponding memory entries. This characteristic becomes really important in the computer system which uses direct memory access (DMA) transfer policy for I/O operation. It then ensures that the data residing in main memory are valid, and most recently updated, so that an I/O device while communicating with memory through DMA would receive the most currently updated data at all times.

4.7.7.5.2 *Write-Back (Copy Back)*

The second strategy is called the *write-back* method. In this method, the cache entry is only updated during a write operation, and does not update the corresponding memory location. Each entry in cache is thus required a bit to be used as a flag telling whether the cache entry has been modified since it was loaded into the cache. When the entry is evicted from the cache to make room for others, it is copied then into main memory only if the flag bit tells it to do so. The advantage of the write-back method is that while a word resides in the cache, it may be updated several times, and all the requests from the CPU for the word are readily satisfied by the cache, it does not matter at all whether the copy in main memory is updated or out of date. But, only when the item is removed from the cache, an accurate latest copy available in cache need be then rewritten into main memory for future use. Since memory references is not encountered very frequently, this approach looks lucrative from a performance point of view.

Both *write-back* and *write-through* have their own strengths. With write-back, writes occur at the speed of cache memory, and multiple writes within a block require only one write to the main memory. Since, some writes do not go to memory, write-back uses less memory bandwidth, making write-back lucrative and attractive in multiprocessor environments. Write-through is easier to implement than write-back, and has the advantage that the next lower level (main memory) always has the most current copy of data. This is an important aspect for I/O, and as multiprocessors are fickle, they want write-back for *on-chip* caches to reduce the memory traffic, and write-through for *off-chip* caches for consistency of data with lower-levels (main memory) of the memory hierarchy. With write-through, the CPU must wait for write stalls, but that can be minimized with the use of a write buffer which allows the processor (CPU) to continue as soon as the data is written to the buffer, thereby letting processor execution and memory updating from the buffer to overlap henceforth.

4.7.7.6 Replacement Policy (Algorithm)

Since the size of the cache is fixed, when it is full and or otherwise a new block is required to be brought into the cache, one of the existing blocks must be replaced to make room for the new one. For direct-mapping, there is no choice, only one possible slot is to be replaced for any particular addressed block. For the fully associative and set associative, an appropriate replacement algorithm is needed. Since cache memory appeared after the introduction of the virtual memory, most of the replacement algorithms used by the memory management (a part of OS) to handle main memory–virtual memory handshaking are also equally applicable to handle cache memory–main memory handshaking. As cache is not visible to OS, that is why, such a befitting algorithm must be implemented in hardware (in cache controller circuit) to work, and that also achieves high speed in turn. A number of algorithms are available for this purpose, out of which *four* are very common.

First-In-First-Out (FIFO) is one of the easiest approach to implement. It selects that block in the set to replace which has been in the cache the longest. It is implemented with the standard round-robin or circular-buffer technique.

Least-Recently Used (LRU) is a corollary of the principle of locality. It is based on the assumption that more recently used memory locations are more likely to be referenced, then the best candidate for disposal is the least recently used block. In this way, it reduces the chance of throwing out information that may be once again needed very soon. Hence, it selects to replace that block in the set which has been in the cache the longest with no reference to it. To implement it, accesses to blocks are recorded. Each slot includes a USE bit. When a slot is referenced, its USE bit is set to 1, and the USE bit of the other slot in that set is set to 0. At the time of replacement, this USE bit is consulted, the slot whose USE bit is 0 is used to select as victim for replacement.

Least-Frequently Used (LFU) is another possibility of replacement policy. It selects to replace that block in the set which has experienced the fewest references until that time. LFU could be implemented by associating a counter with each slot.

Random replacement is used to spread allocation uniformly. This technique is not based on any usage, a slot is picked at random from among the candidate slot. Some systems generate pseudo-random block numbers to get reproducible behaviour, which is particularly useful when debugging hardware. Random replacement, however provides only slightly inferior performance when compared with an algorithm based on usage.

4.7.8 Multiple-Level Caches

The cache itself can be implemented at one or multiple *levels*, depending on its speed and application requirements. As IC chip design and fabrication technology relentlessly progressed over the days, it became possible to put very fast caches straightaway within the processor chip (*on-chip*), and that too, nowadays with *different levels*. Since, the *on-chip* cache (*multiple level*) is connected with the processor, via the short and speedy data paths internal to the processor and not via an external bus, the on-chip cache accesses will be appreciably faster that will definitely speed up execution times, and thereby increase the overall system performance. Moreover, when the *on-chip* cache is in operation (*hit*), it does not use the external bus, and consequently the load on external bus not only gets reduced during this period, but the bus would be then free, and can carry out its own other on-going important activities by this time.

Usually, the on-chip cache (even with multiple levels within the processor chip) is relatively quite small due to only a limited area being available within a CPU chip as well as to abide by certain design metrics. This raises demands to consider additional external caches (off-chip) on the processor board to be included in contemporary designs. One of the main reasons behind such an inclusion is that in the event of *on-chip* cache misses, the *off-chip* cache which is relatively larger than the *on-chip* cache and comparatively faster than the main memory could be immediately consulted for a quick retrieval of missing information rather than to go directly to main memory which is much slower than *off-chip* cache. As a result, the overall system performance would be found to be definitely improved. Such organisation of caches is known as *two-level* cache design: the *first-level* (L1) internal cache (primary) is within the CPU chip (*on-chip*), and the *second-level* (L2) or secondary cache is on the CPU board (*off-chip*). Nowadays, the *on-chip* cache itself consists of multiple levels, say, L1 and L2, and even L3, and in that situation, the *off-chip* cache belongs to level L4.

Figure 4.27 illustrates a scheme of this arrangement. If an L2 SRAM (off-chip) cache having a speed comparable to bus speed is used, then the very often missing information from L1 (on-chip) can possibly be obtained and retrieved immediately from this faster L2, rather than to get it by visiting the relatively slower DDR or DRAM main memory. Although the inclusion of L2 (off-chip) cache in the memory hierarchy offers a potential improvement in the overall system performance, this again depends mostly on the hit rates in both L1 and L2 caches that, in turn, are determined by different elements of cache design (Section 4.7.7). Despite having these advantages, incorporation of multilevel caches in the system architecture, however, invites a lot of complications in handling many of the design issues relating to caches.

In most of the contemporary cache designs, L2 (off-chip) cache does not use the system bus as the path for transfer between the L2 cache and the processor, rather uses a separate data path for fast processing that, on other hand, reduces the load on the system bus. Moreover, as the VLSI technology rapidly advances, and as logic density (less than 0.20 μm of today, as compared with thickness of hair which is about 75 μm) has remarkably increased, the space requirement of processor components within CPU continued to decrease, and as a result, more and more on-chip area now became available that once again has inspired the designer to fabricate L2 cache within the processor. Most microprocessors of today have thus moved the L2 cache onto the processor chip, and added

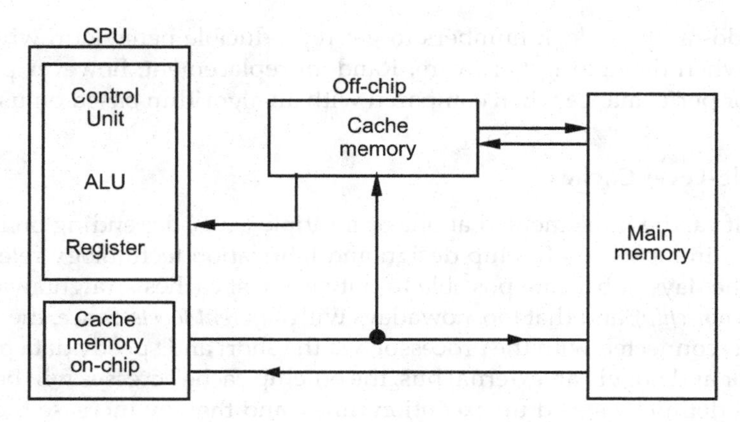

FIGURE 4.27
Multiple level (two-level) caches: on-chip and off-chip.

an external L3 cache on the processor board. ***Intel Pentium III***, in particular, uses this specific design approach. As usual, the L3 cache also initially uses a separate external bus. More recently, most microprocessors have included the L3 cache within the processor (*on-chip*) that is likely to provide further improvement in system performance. ***Intel Pentium 4*** (described later) follows this approach.

During execution, if a word just addressed cannot be found in the on-chip (L1) cache, a request is sent to the next level (L2) cache. If that fails, the next L3 off-chip cache is checked. If it also fails, the main memory is then ultimately consulted.

The *first-level* generally uses the Write-Through (WT) policy, and the *second-level* uses the Write-Back (WB) policy when they are used as data cache (D-cache). The first-level cache is always a subset of the second-level cache, and the second-level cache is also a subset of the third-level cache, and so on. Caches when used in two or more levels require appropriate measures at the hardware level to ensure information consistency.

The first-level cache can be small enough to match the clock cycle time of the fast CPU, while the second-, and onwards levels can be large enough to capture many accesses that would otherwise go to main memory, thereby trying to reduce the effective cache misses.

The average number of cache misses at the on-chip level is known as *Local miss*, and those in the off-chip level is known as *Global miss*. The local miss rate is large because the on-chip-level cache skims the cream of the memory accesses, and that is why, the *global miss rate* is a more useful measure; it indicates what fraction of the memory accesses that leave the CPU to go all the way to memory.

4.7.9 Unified Cache and Split Cache

It has been observed that whenever a cache is addressed, the address is ultimately either of an instruction of a currently executing program, or of a data used by the running program. Initially, the *on-chip* single cache was used to store both instructions and data that are recently referenced. More recently, it has become a common practice to split the cache to separately handle individual access to instructions and data from different caches. When a cache is used only to store instructions but not data, it is called *Instruction cache* or *Instruction lookaside buffer*, in short, *I-Cache*. The advantage of restricting a cache to have only instructions is that, unlike data, instructions do not change, so the contents of an instruction cache need never be written back to main memory. When a cache is used only to store data, it is called *data cache* or simply *D-cache*, the contents of which undergo frequent changes, and require to be written back to main memory to keep the memory contents updated. For this reason, the data is sometimes put into different caches for better performance.

When both instructions and data are put into the same cache, it is called ***unified cache design***. Unified cache, being one cache unit to store both instructions and data, exhibits a few potential advantages which are related to effective cache design, and its less-complicated implementation. In contrast, when instructions and data are stored in different caches for execution conveniences, it is called ***split (mixed) cache design***. This approach has been first commercially implemented in ***Motorola 68030***, and later in a much modified form in Motorola 68040, and also in ***Intel Pentium processor***. Instruction caches, however, have lower miss rates than data caches. Separating instructions and data within caches reduces misses due to the decrease in conflicts between instruction blocks and data blocks, but this split also fixes the cache space committed to each type. The total cache size, however, will remain the same whether the unified design or split design is implemented. Besides, in split cache design, there can be *separate buses* and *ports* for both D-cache and I-cache as is found in ***Motorola 68040***, which could easily double the effective bandwidth between the memory hierarchy and the CPU. Separate caches also provide the opportunity to optimize each cache design separately: different capacities, block sizes, and associativity that may lead to yield more improved performance. This two design approaches are, however, equally acceptable in commercial computers due to their underlying individual merits and strengths.

But, with the advent of modern advanced electronic technology, the processor architecture has been radically changed from conventional non-pipeline to *scalar pipeline*, and thereby to *superscalar design* to exploit extensive *instruction-level parallelism* for the sake of even more performance improvement. As a result, this insisted to persuade more towards split cache than unified cache design. The recent trend is, thus, to have separate dedicated instruction cache in such architecture that can be used for prefetching predicted future instruction to effectively realize parallel instruction execution (instruction-level parallelism) more efficiently. A sufficiently large size of instruction cache is, therefore, most essential and inevitable, particularly, in superscalar design in which the degree of instruction parallelism can be then greatly increased.

Moreover, if the data and instructions of running programs can be kept separated in different caches, then these caches can service the different stages of the pipeline simultaneously without stalling the stages temporarily. Otherwise, there could be a cache contention between different stages of a pipeline that may severely degrade the overall performance of the pipeline. There still remains other reasons, apart from all these considerations and explanations, that constantly encourage the designers to be more inclined in favour of using split cache design in contemporary system architectures. With the passing of days, the cache sizes (both unified and split caches) at different levels also have been remarkably improved, and different cache sizes are now used at different levels, consisting of both unified as well as split caches in the implementation of a variety of representative contemporary processors.

A table showing the sizes of multiple-level unified and split cache design of some representative contemporary processors is given in the following web site: http://routledge.com/9780367255732.

4.7.9.1 Implementation: PENTIUM Cache Organisation

The split cache design culture has been indulged, fully exploited, and probably best implemented in all the members of the Pentium family of processors, from initial versions of ***Pentium*** (1993), core version ***P5*** which was equipped with an on-chip L1 split cache of 16 Kbytes (8 Kbytes instruction cache and 8 Kbytes data cache). The ***Pentium Pro Processor*** (1995), core version ***P6*** contained similar (like P5) specification of L1 cache (8 Kbytes instruction cache and 8 Kbytes data cache) and additionally a 256 Kbytes off-chip L2 cache operated at a speed of 60 MHz or 66 MHz. ***Pentium II*** (1997) represents a new direction for Intel. It included an in-built off-chip L2 cache mounted on a small circuit board (instead of placing it on the motherboard) on which the Pentium processor itself has been placed. The size of this L2 cache is 512 Kbytes with an 8-way set-associative, and a cache block size of 128 bytes. ***Pentium III*** was, however, launched with an added off-chip L3 cache in which the size of the on-chip L1 cache was 512 Kbytes.

The initial version of ***Pentium 4*** processor (2000) had an L1 cache size still remains 32Kbytes, with future versions possibly containing a L1 cache of size 64 Kbytes. The ***Pentium 4 Extreme Edition*** contains a 2 Mbytes L2 cache, the Pentium 4e contains a 1 Mbytes L2 cache, whereas the ***Core 2*** contains a 2 Mbytes or 4 Mbytes L2 cache. Both the L2 and L3 caches in Pentium are 8-way set-associative with a cache line size of 128 bytes.

Placement of caches at different levels in their respective positions within the architecture of Pentium 4 is extremely critical due to the reason that this architecture attempts to accommodate many different conflicting factors with a compromise for the sake of performance improvement. A schematic block diagram of the Pentium 4 cache organisation, considering mainly the placement of three types of caches with a simplified view, is shown in Figure 4.28. Within the processor core, only the four major components that interact with these three different levels of caches are shown. Each such component in turn, may comprise one to more number of pipeline stages (see Chapter 8) however. Here L1 cache is a split cache, but both L2 and L3 are unified caches.

The **Fetch/Decode unit** is responsible to sequentially fetch program instructions from the L2 cache. It then decodes each instructions into a series of micro-operations which are then stored in the L1 instruction cache. The **out-of-order execution logic unit (scheduler)** performs scheduling on queues of micro-operation instructions contained in the L1 instruction cache. It may schedule the execution of micro-operations in a different order other than the order found in the fetch/decode unit (i.e. in the order they were fetched), depending on the availability of related data and other resources. Sometimes, this unit employs the *predictive approach* in the execution of micro-operations that may be found to be deemed fit at that moment to negotiate the near-future situations. The **Execution unit** carries out the micro-operations taking the required data fetched from the L1 data cache, and stores the result temporarily into registers. The **Memory-related subsystem unit** carries out the functions related to fetching of information from memory whenever needed. This unit comprises L2 and L3 caches, and the system bus which is used to access memory and I/O subsystems. In the event of L1 and L2 cache misses, the main memory is consulted, and the information is brought into L3, and from there to L2 and L1, whenever needed. The data cache here employs a **write-back** policy. Data is written to main memory only when it is removed from the cache, and an update happens. Besides, the Pentium 4 processor can also be dynamically configured to accommodate **write-through** caching.

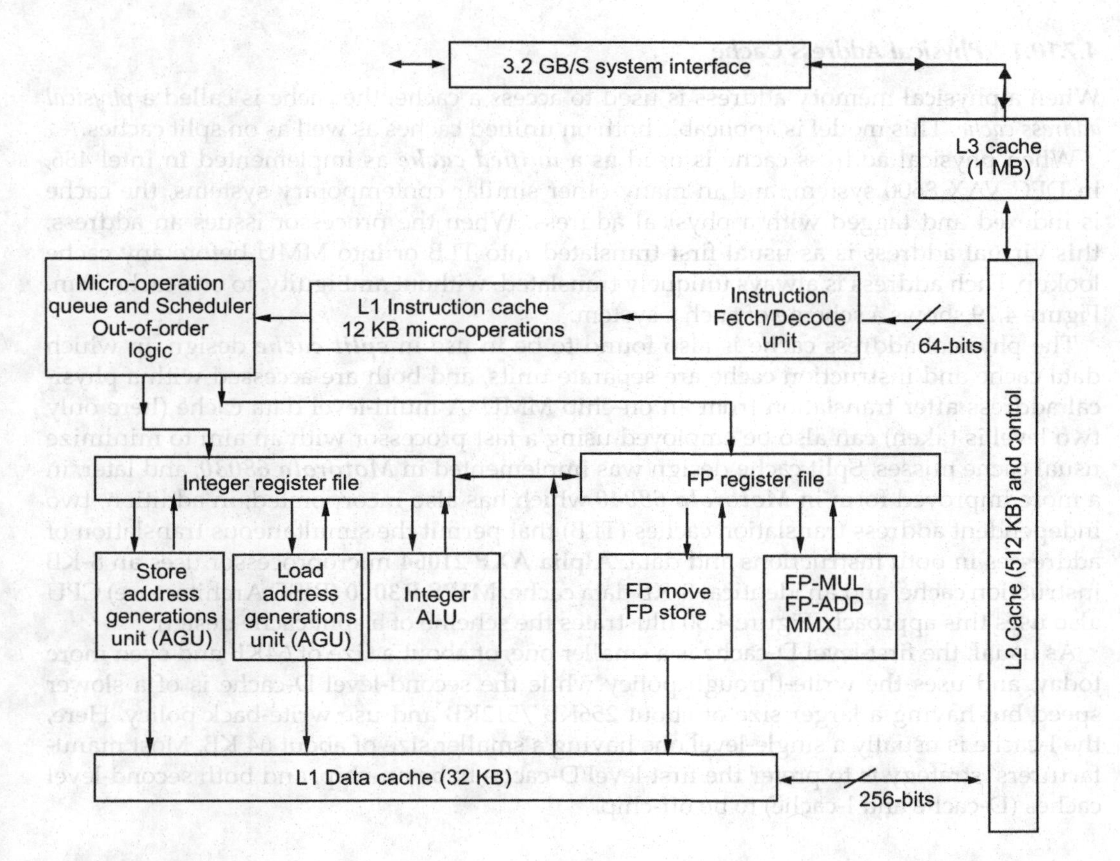

FIGURE 4.28
Block diagram of split cache design in Pentium 4.

4.7.9.2 Motorola RISC MPC7450 Cache Organisation

The **MPC7450** is a **Superscalar processor** using a relatively deeper pipeline up to seven stages. The superscalar performance is realized by way of issuing up to *four instructions* per clock cycle to its functional units. The cache design here is really peculiar and most scintillating, both on-chip and off-chip caches are used. The Level 1 (L1) on-chip cache is a split one consisting of a separate 32 KB for instruction and data, and uses the 8-way set-associative technique. The Level 2 (L2) cache is also on-chip, but a unified one having a size of 256 KB, also uses the 8-way set-associative technique. The off-chip Level 3 (L3) cache can be of a size of 1MB or 2MB, and is accessed over a 64-bit bus. Since, both L1 and L2 are on-chip, L1–L2 handshaking takes place at the rate of processor speed using a data path of 256-bit width.

4.7.10 Cache Addressing

CPU while accessing a cache can address the cache either by a physical address or by a virtual address irrespective of whether it is a unified cache or a split cache. This leads to *two* different models in cache design.

4.7.10.1 Physical Address Cache

When a physical memory address is used to access a cache, the cache is called a *physical address cache*. This model is applicable both on unified caches as well as on split caches.

When physical address cache is used as a ***unified cache*** as implemented in Intel 486, in DEC VAX 8600 system, and in many other similar contemporary systems, the cache is indexed and tagged with a physical address. When the processor issues an address, this virtual address is as usual first translated into TLB or into MMU before any cache lookup. Each address is always uniquely translated, without ambiguity, to one cache item. Figure 4.29 shows a scheme of such a system.

The physical address cache is also found to be in use in ***split cache*** design, in which data cache and instruction cache are separate units, and both are accessed with a physical address after translation from an on-chip MMU. A multi-level data cache (here only two level is taken) can also be employed using a fast processor with an aim to minimize usual cache misses. Split cache design was implemented in ***Motorola 68030***, and later, in a more improved form in ***Motorola 68040*** which has also incorporated, in addition, two independent address translation caches (TLB) that permit the simultaneous translation of addresses in both instructions and data. **Alpha AXP 21064** microprocessor uses an 8-KB instruction cache, and an identical 8-KB data cache. **MIPS R3000** (RISC Architecture) CPU also uses this approach. Figure 4.30 illustrates the scheme of a split cache design.

As usual, the first-level D-cache is a smaller one, of about a size of 64KB and even more today, and uses the write-through policy while the second-level D-cache is of a slower speed but having a larger size of about 256KB /512KB and use write-back policy. Here, the I-cache is usually a single-level one having a smaller size of about 64 KB. Most manufacturers' strategy is to prefer the first-level D-cache to be on-chip, and both second-level caches (D-cache and I-cache) to be off-chip.

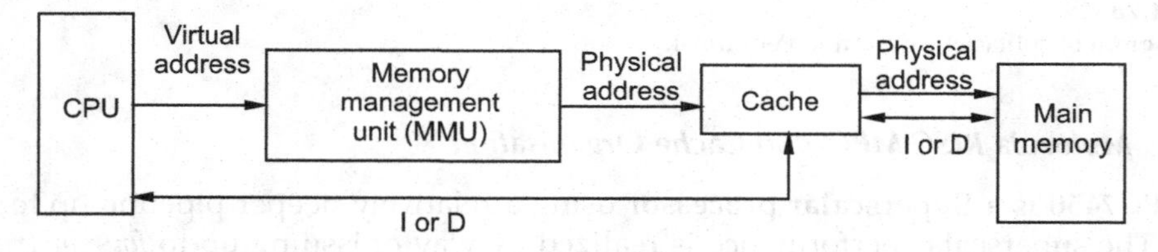

FIGURE 4.29
Unified cache design when accessed by physical address.

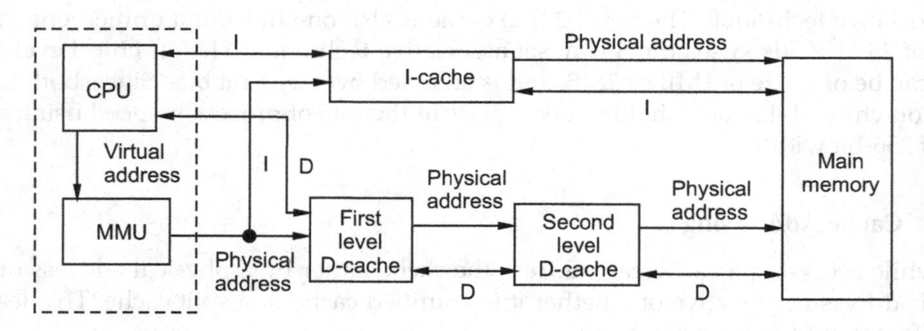

FIGURE 4.30
Split cache design when accessed by physical address.

The advantages of physical address caches lie in its design simplicity which requires little intervention of the operating system. As it is accessed with physical addresses, no problem arises in accessing, since the cache has the same index/tag. Cache *flushing* is not required, if proper bus watching is provided on system bus while servicing the requests from CPU or DMA–I/O. One of the drawbacks of this design is the slowdown of cache access due to waiting for the completion of the address translation mechanism executed by MMU/TLB.

4.7.10.2 Virtual Address Cache

One of the major shortcomings of the physical address cache is its slowdown in accessing the cache until the MMU/TLB completes address translation. This drawback has been alleviated in the virtual address cache which is indexed and tagged with *virtual address*, hence it is called *virtual address cache*. Figure 4.31 shows an example of this scheme for a unified cache. The virtual address will go to the cache and MMU simultaneously. Translation of virtual address to physical address is done by MMU parallel with cache lookup operations. Hence, cache lookup is not delayed at all, rather the cache access efficiency is increased due to overlapping with the MMU address translation process. The MMU translation would yield physical main memory address which is saved for later use by the cache for write-back, if required. Since hits are much more common than misses, virtual addressing effectively eliminates address translation time in case of a cache hit.

The serious ***drawback*** associated with a virtual address cache is ***aliasing*** or ***synonyms*** problem which means that different logically addressed data is appeared to have the same index/tag in the cache. This is severe, particularly when multiple applications (processes) are executed in a system, and if two or more processes attempt to access the same physical cache location. There are various ways to solve this problem, and each system using virtual address cache has solved this problem from its own viewpoint. For example, UNIX has solved this problem with *periodically flushing* the cache or completely flushing the cache with each application context switch and handled it from its kernel level. SUN system has solved it using a *Process Identifier Tag* (PID) or a *Physical Address Tag* to each line of the cache to identify the application (process) that this address refers to. Whatever the approach be considered, the ultimate objective is to enhance the cache performance avoiding complexity as much as one can.

Figure 4.32 illustrates a scheme of virtual address cache used in split cache organisation. Exploiting virtual address cache design, ***Intel i860***, a RISC processor has used split caches

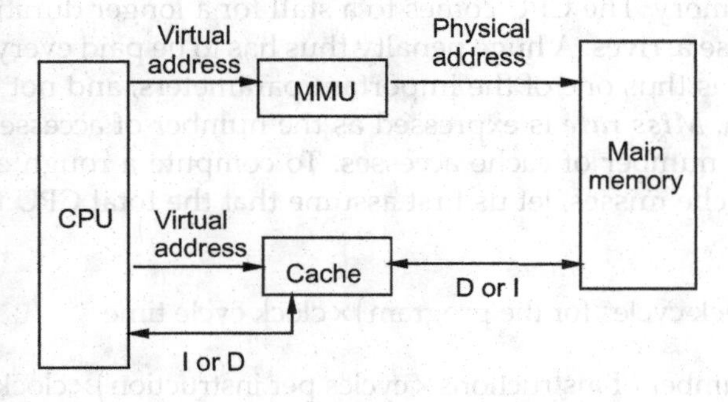

FIGURE 4.31
Virtual address access in a unified cache.

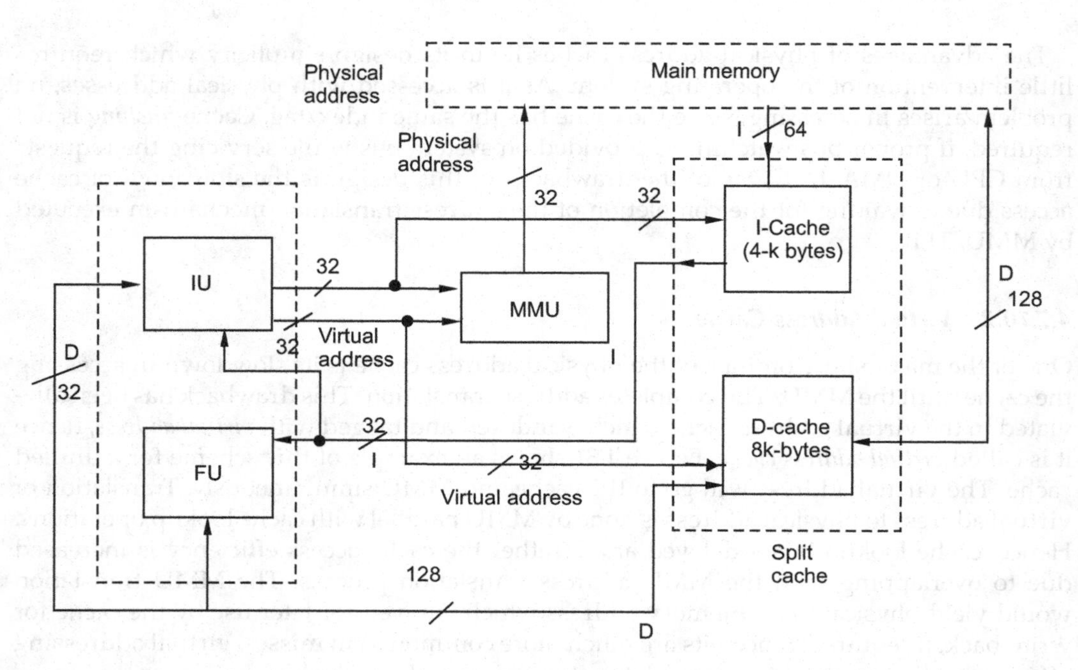

FIGURE 4.32
Virtual address accessed in a split cache.

for data and instructions separately. The Data cache (D-cache) is 8K-bytes and Instruction cache (I-cache) is 4K-bytes, and both are of 32-bytes block size. Set-associative cache organisation (2-way) is implemented with 128 sets in D-cache, and 64 sets in the I-cache. The instruction length is 32-bits. Virtual addresses are also of 32-bits wide generated in the integer unit (IU), the physical address is, however, also 32-bits long.

4.7.11 Miss Rate and Miss Penalty

In practice, the designer's target is to achieve as high a hit ratio as possible at any level in the memory hierarchy. For a cache miss, it is called *block miss,* since the unit of transfer is *block,* and in case of main memory, it is called *page fault* because *page* is the unit of transfer. Every time a miss occurs, the next higher level of memory is consulted. This causes a substantial amount of additional time to be spent in accessing the next higher level of relatively slower memory. The CPU comes to a stall for a longer duration in such a situation until the response arrives. A huge penalty thus has to be paid every time when misses occur. The miss rate is thus one of the important parameters, and not the only one at the time of cache design. ***Miss rate*** is expressed as the number of accesses that have missed divided by the total number of cache accesses. To compute a rough estimate of penalty being paid due to cache misses, let us first assume that the total CPU time required for a program to execute is;

$$\text{CPU time} = (\text{CPU clock cycles for the program}) \times \text{clock cycle time}$$

$$= (\text{Total number of instructions} \times \text{cycles per instruction}) \times \text{clock cycle time}$$

$$= (\text{IC} \times \text{CPI}) \times \text{clock cycle time} \quad (4.1)$$

where, IC = Instruction Count = Total number of instructions executed, and CPI = Cycles Per Instruction.

If, for example, the clock cycle time (clock period) is 2 ns, then clock rate is 1/2 ns = 500 MHz.

In fact, CPU time must include not only the CPU clock cycles that CPU itself requires only to execute the instructions, but also the number of cycles during which the CPU is stalled, waiting for the response from memory which accesses the needed operand to provide, and this can be referred to as *memory stall cycles*. Hence,

$$\text{CPU time} = (\text{CPU clock cycles} + \text{memory stall cycles}) \times \text{clock cycle time} \quad (4.2)$$

In case of a cache hit, the CPU clock cycles is hardly affected, if at all affected, its impact is negligible, but when cache miss occurs, the CPU is stalled, and memory stall cycles increase due to the needed visit to a slower main memory. The number of such memory stall cycles depends on both the number of misses, and the number of extra cycles needed due to every miss (extra overhead), commonly known as ***miss penalty***. Thus,

$$\text{Memory stall cycles} = \text{Number of misses} \times \text{Miss penalty}$$

$$\text{Now, Number of misses} = \text{Total number of instructions} \times \text{Misses per instruction}$$

$$= \text{IC} \times \text{Misses per instruction}$$

$$= \text{IC} \times \text{Memory references per instruction} \times \text{Miss rate}$$

and the miss penalty can now be calculated with the following formula:

$$\text{Miss penalty} = \frac{\text{Memory access time}}{\text{Clock cycle}}$$

For example, if the memory access time of a computer system is 200 ns, and the clock rate is 500 MHz [i.e. clock cycle = 1/clock rate = 1/(500 × 10^6) s = 2 ns].

Then, Miss penalty = 200 ns/2 ns = 100 cycles

It is to be noted that *miss rate* and *miss penalty* are two different parameters creating impact on memory stall cycles, and thereby determining the CPU execution time. In case of cache hit, the memory stall cycles is taken as zero, and there is no penalty, otherwise it is computed separately, and this penalty is included in the equation of CPU execution time calculation as mentioned above.

4.7.11.1 Types of Cache Misses and Reduction Techniques

It is observed that in spite of having large size caches with adequate block size, and even using optimal replacement algorithm to keep the contents of caches always updated, modified, and relevant to all CPU requests, cache miss cannot be avoided, and is becoming a regular, natural, and normal event. However, for every cache miss, a huge penalty has to be paid, and hence, the ultimate objective of cache organisation is thus aimed to reduce the number of such cache misses as much as one can. Cache misses can occur due to numerous reasons. Let us now classify all possible cache misses that can happen into ***three*** distinct categories:

i. When the execution just begins, an attempt to access any cache block for the first time leads to a cache miss because the needed information cannot be in the cache block, and the required block is to be then brought into the cache to negotiate the situation. This is ***compulsory***, inevitable, and cannot be avoided. These type of misses are commonly called ***cold start misses*** or even sometimes called *first reference misses.*

ii. Cache misses also occur since the cache cannot contain within its limited size, all the blocks referenced by the CPU all the time during execution. This is known as ***capacity miss*** which may occur because the blocks just thrown out of the cache to make room for other recently referenced entries are once again immediately demanded. Even use of the most optimal (efficient) replacement algorithm and effective cache organisation cannot always avoid this situation.

iii. Cache misses can also happen due to the replacement strategy being followed. If the block mapping strategy is set-associative or direct mapped, conflict arises between blocks having the same index but different sets. Too many such blocks that have already arrived need to be mapped to their sets, but the room is limited. ***Conflict misses*** will occur because one such block needs to be thrown out to make room for a similar class of blocks, but unfortunately may be referenced within a short interval. These are called ***collision misses*** or *interference misses.*

Numerous techniques have been devised to reduce the number of cache misses, thereby reducing the miss rate. Unfortunately, many such techniques that reduce miss rates also increase *hit time* or *miss penalty.* However, the simplest classical way to reduce miss rate is to ***increase the block size.*** Larger block sizes will reduce compulsory misses, and also exploit the advantage of spatial locality. On the other hand, larger block sizes reduce the number of blocks in the limited size cache, thereby increasing collision misses, and even capacity misses, if the cache is small. All these collectively result in an increase in miss penalty that outweighs the gain obtained from a decrease in miss rate, and hence, a rigid implementation of this approach has been dropped from favour. ***Higher associativity*** is another classical technique that improves miss rates but that too at the expense of increased hit time resulting in an increase in miss penalty. It appears to be almost fully associative (minimal miss rate), if the size of available cache is larger. Other techniques that have been later developed to reduce miss rate with minimum impact in the miss penalty include:

- Use of an *additional* small, fully associative cache between a cache and its refill path to negotiate the situation of primary cache miss.
- Additional hardware facility (extra cache) to provide prefetching of instructions and data. This means that while the caches continue to supply the instructions and data from their storage, the prefetched instructions are being fetched in parallel and arrive at cache simultaneously. Such type of nimble cache is called ***non blocking cache*** or *lookup-free cache.*
- Use of *optimizing compiler* that enables to carry out smooth prefetching by way of appropriate reordering the instruction codes during compilation without affecting

the program semantics. It has been observed from the outcome of many experiments using different types, and sizes of caches that this results in the notable improvement of spatial and temporal locality of the data, that summarily reduces the miss rate.

4.7.11.2 Miss Penalty and Reduction Techniques

From the day one of cache usage, it has been observed from numerous experiments with various types of caches as well as the cache performance formula reveal that the miss penalty is equally a dominant factor, and that its improvements could even yield a better result than that can be obtained by simply improving the miss rate. Moreover, the advancement of electronic technology and its current trends have been continuously improving the speed of processors, even faster than moderately improved main memory, resulting in miss penalties becoming more costly, and that too gradually increasing over time. Numerous attempts thus have been made to reduce this miss penalties, but most of the techniques improvised in this regard have impacts on CPU. The only technique that sets the CPU aside, and relieves it in this regard has been realized with the use of ***second-level caches***.

It has also been observed that the use of *faster cache* to match with the CPU speed alone cannot be the only solution to reduce the performance gap between processors and main memory. Use of *larger cache* can also be deployed to overcome this problem by way of splitting it over two levels between processor and main memory. Adding another level of cache in the hierarchy between original cache and memory is straightforward. While the faster first-level cache can be small enough having clock cycle time nearly matched with that of the fast CPU, the second-level cache may be a comparatively slower one, but large enough to arrest most of the accesses that would otherwise go to memory, thereby reducing the effective miss penalty. Although, the use of *second-level cache* may be an useful approach, it may invite some complications while performance analysis is being carried out.

The access mechanism being followed by *two-level cache* is: each CPU request will be first intercepted by *first-level cache* (L1), and the cache will be searched. If it is a hit, the CPU request will be then promptly serviced, and the time taken to service may be called *Hit time of L1* (HT_{L1}). If it is a miss, it will be passed to the *second-level cache* for appropriate action. The miss penalty for L1 is thus to be paid. The second-level cache will now be searched. If it is a hit, the CPU request will be then serviced, and the time taken to service it may be called *Hit time of L2* (HT_{L2}). If it is a miss, it will be then passed to the main memory for appropriate action. The miss penalty for L2 is then to be paid. It is to be noted that the second-level miss rate is measured on the leftover of first-level cache. Let the Miss rate and the Miss penalty for L1 be denoted by MR_{L1} and MP_{L1} respectively. Similarly, they are MR_{L2} and MP_{L2} respectively for L2. MR_{L2} is sometime called *local miss rate* which is the number of misses in this cache divided by the total number of accesses to this cache. Another term *global miss rate* is defined as the number of misses in this cache divided by the total number of memory accesses generated by the CPU. *Global miss rate* of L2 is $MR_{L1} \times MR_{L2}$. It is expected that the local miss rate is always large because the first-level cache L1 skims the cream of all the memory accesses, and hence, the global miss rate is becoming a more useful measure. It identifies the fraction of memory accesses that forces the CPU to visit all the way to memory. Let us now estimate an *average memory access* time while such a two-level cache L1 and L2 referred as first-level and second-level cache respectively is used.

$$\text{Average memory access time} = HT_{L1} + (MR_{L1} \times MP_{L1}) \text{ and}$$

$$MP_{L1} = HT_{L2} + (MR_{L2} \times MP_{L2})$$

So,

$$\text{Average memory access time} = HT_{L1} + MR_{L1} \times (HT_{L2} + MR_{L2} \times MP_{L2})$$

4.7.12 Caches in Multiprocessor

The emerging dominance of the microprocessors blessed by the radical developments in VLSI technology in the beginning of the 1980s motivated many designers to create small-scale multiprocessors where several processors were on the motherboard, shared a single physical memory built around a single shared bus. In the mid-1990s, with large caches, the bus, and the single memory can satisfy the memory demands of a multiprocessor built with a small number of processors. In this architecture, each processor as shown in Figure 4.33 has been given uniform access to this single shared-memory, and hence, these machines are sometimes called to have ***uniform memory access (UMA)*** architecture. This type of UMA architecture consisting of a ***centralized shared memory*** connected with a small number of processors (CPU and IOP) and supported by large caches which significantly reduce the bus bandwidth requirements, is currently by far the most popular organisation, and is extremely cost-effective. Another approach consists of multiple processors with ***physically distributed memory.*** To support *larger processor counts*, memory must be distributed among the processors, with each processor having a part of memory of its own to support the bandwidth demands. An illustration of such an architecture is shown in Figure 4.34.

Whatever is the memory–module design, typically, all processing elements (CPU and I/O processors), main-memory modules (shared or distributed), and I/O devices are attached directly with the ***single shared bus*** that eventually results in potential communication bottleneck, leading to contention and delay whenever two or more units try to access the main memory simultaneously. This severely limits the number of processors

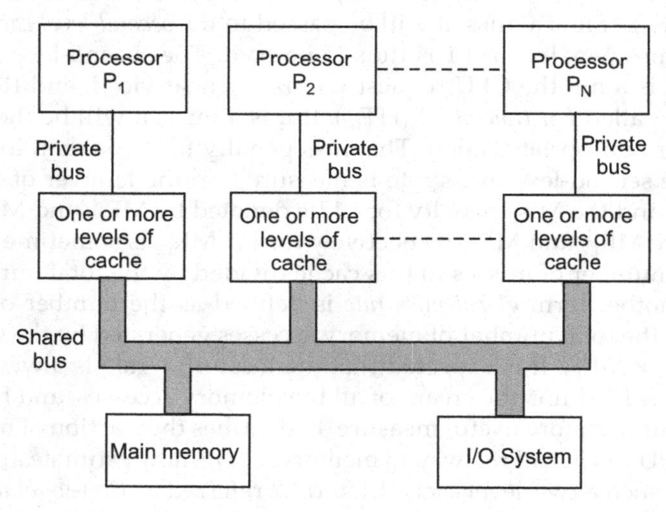

FIGURE 4.33
Basic structure of a centralized shared-memory multiprocessor.

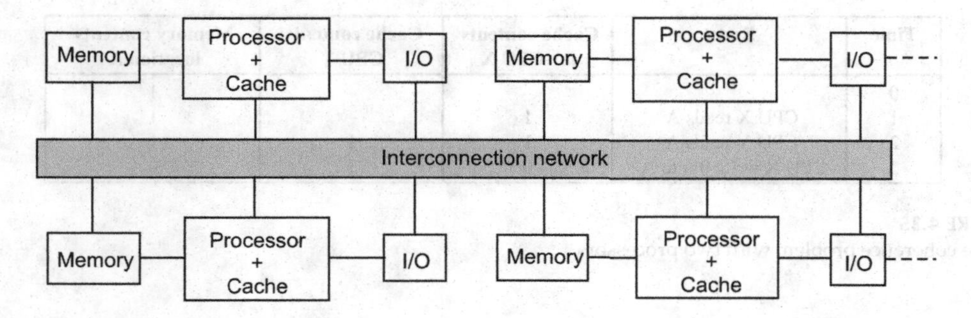

FIGURE 4.34
Basic architecture of a distributed memory machine containing individual nodes (= processor + some memory + typically some I/O and interface to interconnection network).

that can be included in such a system without an unacceptable degradation in performance. To improve the performance considerably, each CPU is, however, provided with a ***cache*** or a ***local memory*** connected by a dedicated local bus which can even include local IO devices [Lee, R. L., et al.]. Local memory supports caching of both *private* and *shared data.* All these local buses are, in turn, connected with the main shared system bus.

When a *private item* is cached, its location is simply migrated to the cache, thereby reducing the average access time and much of the routine memory traffic being taken away from the main shared bus, that summarily then releases the shared bus to perform its other important functions. But, when *shared data* are cached, the shared value can be replicated in multiple caches attached with multiple processors. This replication often facilitates to achieve a reduction in contention for the shared system bus that may be used for shared data items. But, these shared data might be independently operated by multiple processors individually at their own end, likely generating different values (inconsistent values) of the same item by different processors almost at the same time. Thus, caching of shared data, on the other hand, causes to introduce a new critically typical problem, historically called ***cache coherence*** problem or *cache inconsistency.*

4.7.13 Cache Coherence

In shared-bus multiprocessor system using shared memory, the introduction of caches with/without local memories to each processor is a compulsion to reduce the average access time in each processor, and to reduce the contention for the shared system bus. Thus, each processor has its own private cache which allows both data and instructions to be accessed by the respective processor without using the shared system bus. Each processor then views the memory through its own individual cache. If a multiprocessor has an independent cache in each processor, there is a possibility that two or more caches containing different values of the same location (variable) at the same time. This is generally referred to as the ***cache coherence*** or ***multi-cache consistency problem.*** Figure 4.35 illustrates the problem, and shows how two different processors can have two different values for the same location.

In Figure 4.35, a single memory location (*A*) is read and written by two processors X and Y. Initially, we assume that neither cache contains the variable *A* and that *A* has the value 1 in main memory. We also assume a write-through cache, however, a write-back cache has similar complications, and sometimes, even something more. After the value of *A* has been written by X, both X's cache and memory location *A* has the value 1, but Y's cache does

Time	Event	Cache contents for CPU X	Cache contents CPU Y	Memory contents location A
0		–	–	1
1	CPU X reads A	1	–	1
2	CPU Y reads A	1	1	1
3	CPU X write 0 into A	0	1	0

FIGURE 4.35
Cache coherence problem with two processors.

not. After time interval 3, a different version of the same location (A) is found in different caches. The change made by X is not known to Y, and Y had the old *stale data* violating the semantics (ethics) of shared memory (when one process changes a word, subsequent reads by other processes must return the new value). This inconsistency in the value of the same location in different caches is called the ***cache coherence*** problem. Without an appropriate solution, caching, hence, cannot be used, and bus-oriented multiprocessors would then be limited to only two or three CPUs. A mechanism thus needed that allows each cache to be informed about changes to *shared information* stored in any other caches to ensure cache consistency.

4.7.14 Reasons of Coherence Problem

The entire set of information used in any program can be classified as *shared versus not-shared*, and also *read-only versus writable data* that are not shared, i.e. data that are private to a process, cannot cause cache inconsistencies, because they will never appear in other caches simultaneously. Similarly, data that are read-only (like instructions) cannot cause any problems because they are never changed, no questions at all arise about different versions. Then, the category that causes all the trouble is ***shared and writable data.*** Whatever write policy be adopted, the problem remains same. A write-through maintains consistency between memory and the originating cache, but the other caches in the system remain inconsistent, since they still hold the old values. In a write-back policy, main memory is not updated at the time of any change in a particular cache. The copies in main memory, and the other caches are then found to be inconsistent. Memory, however, will ultimately be updated when the modified data in the cache are copied back to memory.

Another concern to this inconsistency problem arises in systems having direct memory access (DMA) activity in conjunction with an IOP (I/O channel) connected to the system bus. The DMA may modify (input) locations in main memory that may also reside in cache without updating the cache. The DMA may also read (output) a memory location that is even present in cache and not currently updated, but will be updated later only at the time of write back. In both the situations, inconsistent data will be passed. This indicates that not the caches alone, but *IOP should also be considered as a party* while addressing cache coherence problem [Lilja, D. J.].

4.7.15 Cache Coherence Problem: Solution Methodologies

In a coherent multiprocessor, the caches provide both *migration* and *replication* of shared, writable data. Coherent caches provide migration, since a data item can be moved (migrated) to a local cache, and is used there in a transparent fashion, this obviously reduces the latency to access a shared data item which is remotely allocated. Coherent caches also support replication of shared data that is being simultaneously read, since the caches make a copy of the data item in the local cache, thereby reduces both latency of access and contention for a read shared data item. Supporting this migration and replication is critical to performance in accessing shared data. Various schemes thus have been proposed to solve this cache coherence problem in shared/distributed memory multiprocessor systems.

The problem can be solved by hardware and/or software means. Hardware-only solutions, by virtue, have the advantage of higher speed and program transparency, but are quite expensive. Software-based solutions to ensure cache consistency require the ability to tag information (data) at the very beginning of program compilation as either cacheable (whether can be kept in a cache) or non-cacheable. We briefly discuss some of these schemes.

4.7.15.1 No Private Cache

Since the presence of private cache to each processor with shared writable data is the root of this problem, a simple scheme is not to permit any private cache for processor in place to provide a shared cache memory associated with main memory. Every data accessed by any processor is then to be done through the shared cache. Hence, there will be no inconsistency in data. But, this method will neither reduce the latency to access a shared data item nor minimize the bus contention. It is at the same time disobeying the principle of *closeness* of CPU to cache. Thus, rather than trying to solve the problem, it can at best be an attempt called a scheme of avoidance.

4.7.15.2 Software Solution

Private cache to each processor in a multiprocessor system is a must, and there is as such no alternative from a performance point of view. Then, the problem indicates the category that causes all the trouble is only *shared, writable data.* One approach may then be to allow only *non-shared* and *read-only* data to be stored in caches. Such items are called *cacheable,* and the remaining *shared writable data* are to be operated only in main memory, and no cache can migrate or replicate these items in their private region. These items are called *non-cacheable.* When a program is compiled, the compiler must tag data either *cacheable* or *non-cacheable,* and the system hardware must ensure that only *cacheable* data are to be stored in caches. Cache coherence can then be implemented by a *write-through* policy that requires a processor to mark a shared cache item (non-cacheable) *K* as *invalid*, or to be *deallocated,* whenever the processor writes into *K*. When the processor references *K* again, it insists to access main memory, thereby always getting the most recent version of *K*. Invalidation, however, is not a good approach, because it may sometimes enforce the removal of needed data from the cache, thus reducing its hit ratio, and thereby increasing the main memory traffic. Moreover, this method, however, eventually imposes some restriction on the type of data to be stored in caches, and invites an extra software overhead leading to a situation of some degree in degradation in system performance.

A *slight variation* in this approach is a scheme that permits *writable data* to exist in *at least* one cache. This method generates a *centralized global table* at the time of compilation. The status of all memory blocks is stored in the central global table. Each block is identified either Read-Only (RO) or Read-Write (RW). All caches can have copies of blocks identified as RO. Only one cache can have a copy of an RW block. Thus, if the data are updated in the cache with an RW block, the other caches will not be at all affected, since they do not have a copy of this block. Inconsistency in data in different caches will then no longer exist.

Another scheme that provides a solution is based on collecting all the shared, writable data together and putting them in a separate segment, page, or part of the address space. The compiler will take this responsibility, and will put all shared writable data in a pre-defined area in virtual memory. During run time, caching would then be *turned off* for addresses lying in this pre-defined area, and all references would be negotiated directly from memory and be written back to memory. It would never be cached, no matter how often they are used. Addresses being referred beyond this jurisdiction would be cached as usual. Implementation of this software solution demands that the cache hardware should be equipped with the ability to disable caching for a range of addresses, and hence, requires some additional hardware assistance.

4.7.15.3 Hardware-Only Solution

While the *software approach* is offering a formidable solution to eliminate the cache coherence problem, but it is so achieved at the cost of notable *degradation in performance*, mainly due to:

- disabling caching for shared, writable data,
- increasing in main-memory traffic,
- reduction in hit ratio, and for other similar reasons.

For applications, where heavy use of shared writable data is inevitable as in the case of some database systems, this solution is found to be not only inadequate, but fails to provide at least a reasonable return. We, hence, need to look at a comparatively speedy hardware solution with program transparency by introducing a protocol to maintain coherent caches (to remove inconsistency). The set of rules implemented by the CPUs, caches, and main memory for preventing different versions of the same memory block contents from appearing in multiple caches is called the ***cache consistency protocols***. These protocols also maintain coherence for multiple processors [Archibald, J., et al.]. A key to implementing this protocol is to *track the state of any sharing of a physical memory data block*. There are ***two classes*** of protocols in use, which exploit different techniques to track the sharing status:

- **Directory based**: The sharing status of a block in physical memory is kept just in one location, called the *directory*. Distributed shared-memory (DSM) architecture (NUMA) uses this technique [Agarwal, A., et al.].
- **Snooping**: *Every cache* that has a copy of the data item from a block of physical memory also *has a copy of sharing status of the block, and no centralized state* like directory is kept. The cache controllers are specially designed to *monitor* or *snoop* on the bus to determine whether or not they have a copy of a block that is requested on the bus from CPUs and IOPs. These devices are called *snooping caches* or sometimes *snoopy caches.*

A simple method based on snoopy cache protocol is to adopt a *write-through* policy using the procedure as explained. All the snoopy controllers watch the bus for *memory write operations.* When a word in any cache is modified by writing into it, the corresponding memory location is accordingly updated by virtue of write-through. All the local snoopy controllers in other caches check their entry individually to determine if they have a copy of the item that has been just modified. If a copy exists in any remote cache, the location is marked *invalid*. As all caches snoop on every bus writes, whenever an item is modified in any cache, the main memory is accordingly updated, and it is removed (invalidated) from all other caches. At a later time, when a processor accesses the invalid item from its own cache, the response is equivalent to a cache miss, the corresponding main memory location is consulted, the updated item from main memory is transferred to the cache in question. No inconsistent versions, hence, are present in any cache, and the coherent problem is thereby arrested [Vernon, M. K., et al.].

Snooping protocols generally are used in systems with caches attached to a single shared memory. These protocols can use a *pre-existing physical connection:* the bus to memory, to interrogate the status of the caches. Particularly, small-scale multiprocessors built with microprocessors having single-shared memory makes this protocol most popular.

Many variations of these different protocols have been proposed, explored, implemented, and analysed. We will briefly discuss here two basic approaches to the *snoopy protocol*. For the sake of simplicity in delivery, we are intentionally avoiding here to include the mechanisms involving multiple-level (L1 and L2) caches as well as not considering systems built on distributed multiprocessors. In any case, this would not, however, add any such new principles.

4.7.15.3.1 Write-Invalidate Protocol: MESI Protocol

The write-invalidate approach is probably the most widely used one in commercial multiprocessor system built with microprocessors, such as ***Pentium 4*** and ***PowerPC***. With this protocol, there can be multiple readers, but only one writer at any point of time. As usual, a line may be shared among several caches for reading purposes. But, when one of the caches wants to perform a write to the line, it first issues a notice that invalidates this specific line in the other caches making the line exclusive to the writing cache. Once the line becomes exclusive, the owning processor can then easily execute the local writes in no time until some other processor requires the same line.

When this protocol is used, it marks the state of every cache line (using two extra bits in the cache tag) as ***Modified, Exclusive, Shared***, and ***Invalid***. That is why, this protocol is sometimes called MESI protocol. This method ensures that a processor has *exclusive access* to a data item before it writes that item. This style of protocol is called a *write invalidate protocol* because it invalidates other copies on a write. It is by far the most common protocol, both for snooping and for directory schemes. Exclusive access ensures that no other readable or writable copies of an item exist when the write occurs; all other cached copies of the item are invalidated. Figure 4.36 shows an example that ensures coherence by way of adopting an invalidation protocol.

On a snooping bus for a single cache block *A* with *write-back* caches are attached two processors X and Y. Since the write requires exclusive access (here processor X), any copy

Event	Bus activity	Contents of CPU X's cache	Contents of CPU Y's cache	Contents of memory location A
				0
CPU X reads A	Cache miss for A	0	–	0
CPU Y reads A	Cache miss for A	0	0	0
CPU X writes 1 to A	Invalidation for A	1	–	0
CPU Y reads A	Cache miss for A	1	1	1

FIGURE 4.36
Invalidation protocol using snooping bus with write-back caches.

held by the reading processor (here, processor Y) must be invalidated. Thus, when the read by the processor Y occurs, it misses in the cache and is forced to fetch a new copy of data from main memory. At this moment, CPU X responds with the value from its cache cancelling the response from memory as requested by CPU Y. Both the contents of Y's cache and memory contents of *A* (write-back) are updated. This is illustrated in line 4 of Figure 4.36. Writing processor thus has exclusive access preventing any other processor from being able to write simultaneously. If two processors do compete to write the same data simultaneously, one of them wins the race, causing the other processor's copy in cache to be invalidated. Meanwhile, if this other processor wants to complete its write, it must obtain a new copy of the data with the updated value. In other words, this protocol enforces ***write serialization***. This scheme has been used in several IBM 360/370 series of computers, and also especially in the larger model IBM 3033.

4.7.15.3.2 Write-Broadcast Protocol: Write-Update

The alternative to a *write-invalidate* protocol is a *write-update* protocol. With this protocol, there can be multiple readers as well as multiple writers also. When a processor wishes to update a shared line in its local cache, the word to be updated is informed and distributed to all others, and caches attached with other processors containing that line would then update it.

With this protocol, all the cache copies of a data item are to be *updated* as soon as the item is written (modified). The processor executing the write operation in its own cache will ***broadcast*** to all caches in the system (and to main memory) via the shared bus. Every cache then examines its assigned addresses to see, if the broadcast item is present in its own. If it is, the item in question will then be updated in cache, otherwise the broadcast item will be simply ignored. In this way, the consistency of shared data in all caches is maintained. To keep the bandwidth requirements of this protocol under control, it is useful to track whether or not the item in cache to be broadcast is shared; that is, whether contained in other caches. Another disadvantage of this technique is that every cache write operation, which is very frequent, forces all other caches to check the broadcast data, making the caches then unavailable for their own normal processing. In the decade, since these protocols were developed, write invalidate protocol, however, has ultimately emerged as the winner for the vast majority of designs. Figure 4.37 shows an example of a Write Broadcast protocol in operation.

Neither of these two approaches, however, seems to be superior to the other under all circumstances. Performance greatly depends on the number of local caches, and the pattern of memory reads and writes. In a bus-based multiprocessor, since the bus and memory bandwidth is usually the most demanded commodity, *invalidation* has become the

Event	Bus activity	Contents of CPU X's cache	Contents of CPU Y's cache	Contents of memory location A
				0
CPU X reads A	Cache miss for A	0	–	0
CPU Y reads A	Cache miss for A	0	0	0
CPU X writes 1 to A	Write broadcast of A	1	1	1
CPU Y reads A		1	1	1

FIGURE 4.37
Broadcast protocol using snooping bus with write-back caches.

protocol of choice for most implementation. Since, the current trends are towards increasing processor performance and related increase in bandwidth, update schemes (Broadcast protocol) are expected to be used very infrequently, and at this juncture, the invalidate scheme naturally gets the edge over broadcast protocol. Nevertheless, some systems implement ***adaptive protocols*** that employ both *write-invalidate* and *write-update* mechanisms.

4.7.16 Two-Level Memory Performance: Cost Consideration

Let us now look at some of the parameters which are relevant for an *assessment of cost* to be incurred in a two-level memory organisation. The two-level memory module may, however, be: **(i)** Cache memory–Main memory or **(ii)** Main memory–Virtual memory. Whatever be the memory module, to find an estimate to express the average cost of this memory module while accessing an item, we must consider not only the size of these two levels of memory M_1 and M_2, but also the cost per bit of these two individual memory systems to find an average cost of these two-level memory modules [Przybylski, S. A.]. We assume,

T_S = Average access time (system)
T_1 = Access time of M_1 (e.g., cache, disk cache etc.)
T_2 = Access time of M_2 (e.g., main memory, virtual memory (disk), etc.)
H = Hit ratio (fraction of time, reference is found in M_1 or the probability of finding a given reference in M_1)
C_S = Average cost per bit for the combined two-level memory
C_1 = Average cost per bit of upper-level memory M_1
C_2 = Average cost per bit of lower-level memory M_2
S_1 = size of memory M_1
S_2 = size of memory M_2

Then

$$C_S = \frac{C_1S_1 + C_2S_2}{S_1 + S_2}$$

Our target is to attain $C_S \approx C_2$. Given that $C_1 \gg C_2$, this requires $S_1 \ll S_2$.

From the perspective of access time, the use of a two-level memory is to obtain a significant performance improvement, we thus need to have T_S approximately equal to T_1 ($T_S \approx T_1$). Given that T_1 is much less than T_2 ($T_1 \ll T_2$), and hence, a hit ratio of close to 1 is demanded.

It is natural that one would like M_1 to be as small to hold down cost, and at the same time would like it to be large enough to improve the hit ratio, and thereby to enhance the performance. The objective would be then to find out a suitable size of M_1 that could satisfy both the requirements to a reasonable extent. This, in turn, demand the answers of a number of questions. A few of those are:

i. What value of hit ratio is expected to satisfy a desired level of performance?
ii. What size of M_1 will ensure the required hit ratio as thought of?
iii. Whether this size of M_1 as perceived will match the related cost to be incurred?

To get this, let us define a quantity T_1/T_S, which is referred to as the ***access efficiency***. It is a measure of how close average access time (T_S) is to M_1 access time (T_1). From Section 4.7.4, we have

$$T_S = T_1 + (1-H) \times T_2$$

and hence,

$$\frac{T_1}{T_S} = \frac{1}{1+(1-H)\times T_2/T_1}$$

This shows that *access efficiency* is a function of hit ratio H with a quantity T_2/T_1 as parameter. For Cache memory–Main memory, cache access time is about five to ten times faster than main memory access time, (i.e. T_2/T_1 is 5–10) and in case of Main memory–Virtual memory, main memory access time is about 1,000 times faster than disk access time (i.e. T_2/T_1 is about 1,000). Thus, a hit ratio in the range of 0.8–0.9 is expected to be required to attain the desired level of performance.

The ultimate aim is thus to attain the desired hit ratio so that the performance goal can be nearly met. This hit ratio again depends on a number of parameters including the design of two-level memory (cache design also), the replacement algorithm, the nature of the software being executed, and many more. All these parameters are so selected so that a greater degree of locality can be realized. The degree of locality to a large extent is determined by the relative memory size (S_1/S_2) which is indicated in Figure 4.38. Now, if $S_1/S_2 = 1$, i.e. if M_1 is same as M_2 in size, then the hit ratio will be 1.0; all of the items in M_2 are always stored also in M_1. Now suppose that there is *no locality*, i.e. all the references are completely at random, then the hit ratio should be a linear function of relative memory size S_1/S_2 as shown in Figure 4.38. If the locality is appreciable, a high value of hit ratio can be achieved even with small values of relative memory size S_1/S_2, i.e. with the small size of upper-level memory S_1. This is indicated by the lines "moderate locality" and the "strong

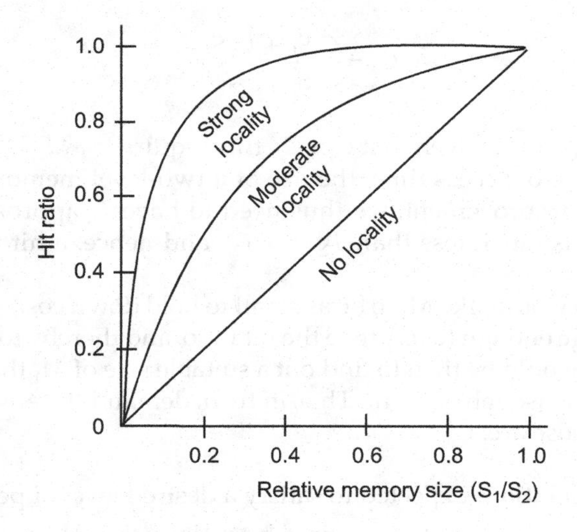

FIGURE 4.38
Hit ratio as a function of relative memory size.

locality" in Figure 4.38, where a comparatively lower value of relative memory size (S_1/S_2) can yield a higher hit ratio.

Numerous studies have revealed the fact that rather small cache sizes can offer a higher hit ratio above 0.8, regardless of the size of the main memory. Usually, a cache size of 128–256K or even 512 words (on-chip), and off-chip of a multiple megabyte is generally found adequate, whereas the main memory is now typically in the multiple gigabytes. In the cases of virtual memory and disk cache, this observation still remains the same. This relative size of the two memories not only satisfies the users' requirements from the performance point of view, but also satisfies designers' targets to cost implications. If a relatively small size of upper-level memory can achieve a good performance, then it is natural that the cost per bit of the two levels of memory will tend to approach that of the cheaper lower-level memory.

4.7.17 Memory Hierarchy Design: Size and Cost Consideration

Let us assume the design of a three-level memory hierarchy with the current standard specifications for memory characteristics. The target is to attain a roughly effective memory-access time $t = 10.00$ μs as the design goal with a cache hit ratio of about $h_1 = 0.98$, and a hit ratio in the main memory $h_2 = 0.9$. Also, the total cost of the memory hierarchy is budgeted with some upper-bound. It can be shown with suitable computations that increasing of main memory size and a corresponding decrease in disk capacity for the sake of staying within the budgeted cost will not at all affect the cache hit ratio. Moreover, the effective memory-access time can once again be reduced without disturbing the budget, if the main memory and virtual memory management issues (like page replacement, page size, etc.) are properly addressed and resolved with care.

A brief detail of this topic supported by a table with a rough estimation is given in the web site: http://routledge.com/9780367255732.

4.8 Interleaved Memory Organisation

4.8.1 Background

Main memory is centrally located, since it is both the destination of input and the source for output, and satisfies the demands of CPU requests and also serves as the I/O interface at the same time. Performance measures of main memory emphasize both *latency* and *bandwidth* parameters. Amdahl's law warns us what will happen if we concentrate only on one parameter to speed up the computation while placing less importance on and/or ignoring the other parameters. So, both the latency and bandwidth improvement should be considered at the same time with equal importance. *Memory latency* is traditionally the primary concern of cache usage, and is minimized by using various types of cache organisation. While *memory bandwidth* is the primary concern of and critical to Processor and I/O, the memory design goal in this regard is targeted to broaden the *effective memory bandwidth* so that more and more memory words can be accessed per unit time. The ultimate aim is, however, to match the memory bandwidth with the processor bandwidth and with the bandwidth of the bus to which the memory is attached. Benefit in bandwidth improvement using caches can also be obtained by increasing each cache block size, but that may not be a cost-effective approach.

Innovative organisations of main memory are, hence, needed. ***Memory interleaving*** is such a new organisation aimed more at the improvement of *memory bandwidth* than at reduction of *latency* with almost no extra cost. Increasing *width* is one way to improve bandwidth, but another benefit of this is to extract the potential parallelism by way of having multiple memory in a memory system as demanded by pipeline processor and vector processor for their optimal performance. These processors often require simultaneous access to memory from their two or more sources (stages). An instruction pipeline sometimes accesses memory to fetch anew instruction, and at the same time to get an operand from two different segments(stages) of the processor for another executing instruction lying in the pipeline. An arithmetic pipeline in a similar way also requires two or more operands at the same time before entering the execution stage of the pipeline. This simultaneous accesses require two memory buses. It can be avoided, if the memory can be divided into a number of modules connected to a common memory address bus and data buses.

4.8.2 Memory Interleaving

Here, the main memory is constructed with multiple modules. Memory chips can be organised in banks to read or write multiple words at a time rather than a single word. These memory modules are connected to a system bus or a switching network to which other resources such as processors or I/O devices are connected to communicate with memory modules. A memory array is thus formed with these memory modules when each memory module has its own address register and data register (like MAR and MBR).

Each bank is often of one-word width so that the width of the bus and cache does not need any change, and each bank returns one word per cycle. Sending different addresses to several banks simultaneously permits them to operate at the same time so that multiple words can be accessed in parallel, or even, at least, in a pipelined fashion.

Consider a memory system formed with memory interleaving having $m = 2^x$ memory modules, with each module containing $w = 2^y$ words. The total capacity in the memory system is $m \times w = 2^{x+y}$ words. These words are assigned with usual linear addresses. Linear addresses could be assigned in different ways giving rise to different memory organisation, both for *random access* and for *block access*. Block-access at consecutive addresses is sometimes needed for fetching a sequence of instructions or for accessing a set of data linearly ordered. This size of the block may correspond to the size of the cache block, or to that of several cache blocks (cache lines). The memory organisation should also consider this aspect of block-access on contiguous words at the time of its design.

The number of modules (banks) present in a memory system is called ***Interleaving factor*** or ***degree of interleaving***. *Word-oriented* interleaving optimizes sequential memory access, and is equally well-matched to catch read miss as the words in a block of cache are read sequentially. *Write-back* caches make sequential write to yield more efficiency from word-interleaved memory.

4.8.3 Types of Interleaving

Low-order X-bits of the memory address is used to identify the target memory module (bank). The high-order Y-bits of the said memory address are the word address (displacement or offset) of the target location within each module. The same address can be applied to all memory modules simultaneously. Such type of arrangement of modules to support memory addressing is called **low-order interleaving**. Figure 4.39 illustrates the scheme of this interleaving. In this arrangement, contiguous memory locations are

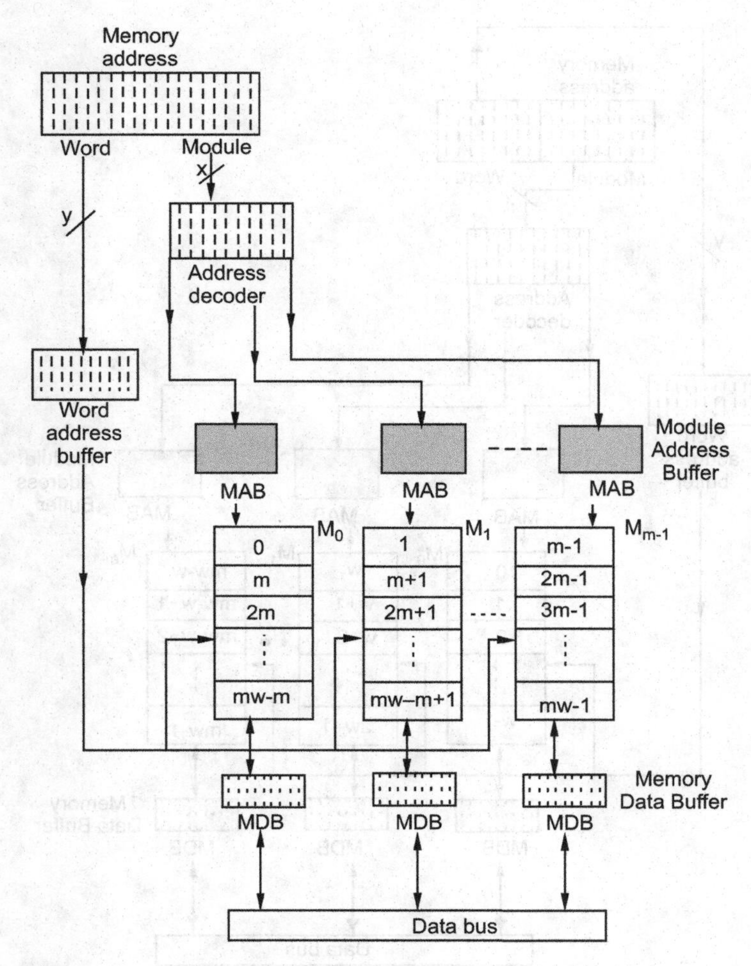

FIGURE 4.39
Low-order *m*-way interleaving word-address scheme with $m = 2^x$ modules, each module having $w = 2^y$ words.

spread across the *m*-modules horizontally. Low-order interleaving facilitates block-access in a pipelined fashion.

When failure is detected in one module, the remaining modules cannot be used. The fault isolation cannot be carried out in this low-ordered organisation. A module failure in this arrangement may paralyse the entire memory bank. That is why, this type of organisation is not *fault tolerant*.

High-order X-bits of the memory address is used as the module address and the low-order Y-bits as the target word address within each module. Contiguous memory locations are hence assigned to the same module. Such type of arrangement of modules to support memory addressing is called **high-order interleaving**. Figure 4.40 shows such an arrangement. Only one word is accessed from each module in each memory cycle. Block access of contiguous locations thus cannot be obtained from high-order interleaving.

On the other hand, since sequential addresses are assigned to each module in this organisation, it is easier to handle module failure. When one module failure is detected in a memory bank of *m*-memory modules, the faulty memory module is isolated, the remaining

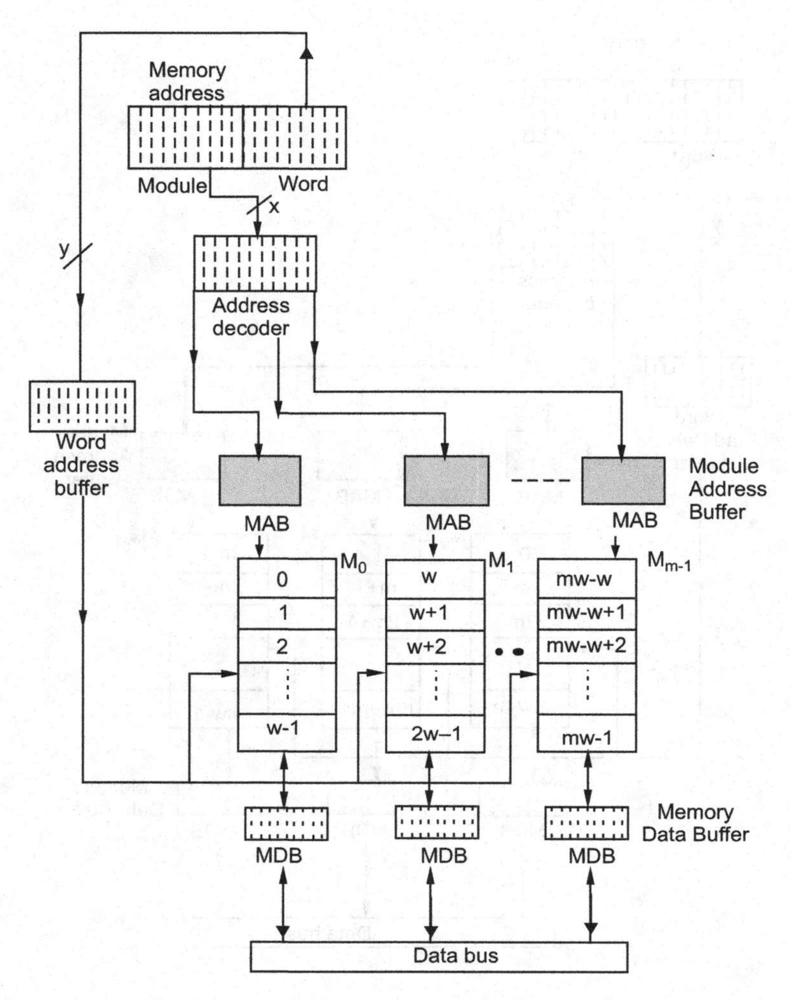

FIGURE 4.40
High order m-way interleaving word-address scheme with $m = 2^x$ modules, each module having $w = 2^y$ words.

modules can still be used by opening another window in the address space. Fault tolerance is a salient feature of this organisation apart from having other distinct disadvantages.

4.8.4 Interleaving in Motorola 68040

The Motorola 68000 series was introduced sometime in the late 1970s. This was one of the first chips to use 32-bits (4 bytes) for addresses to access, in principle, a memory having 2^{32} different locations (words). The memory organisation of the Motorola 68040 processor, however, served to reduce the overall access time in DRAM. The 68040 performs ***burst accesses*** to read or write 16 bytes of data in 4 adjacent long words between its caches and memory in a single bus transaction. The interleaved memory configuration is designed to speed up 68040 burst accesses by as much as 30% (of course, the actual speed up depends on DRAM access time and system clock speed). The four long words of a burst access are spread across two physical modules of DRAM; the individual access over each module can

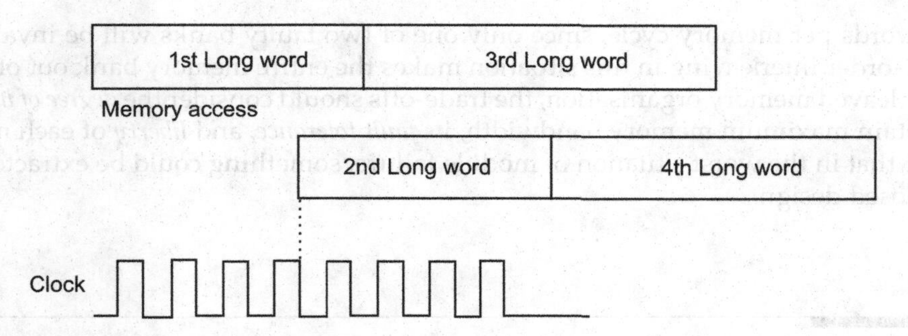

FIGURE 4.41
Interleaved burst access timing.

be overlapped to reduce overall access time, and to hide part or all of the memory access delay. This is illustrated in Figure 4.41.

4.8.5 Conclusion

Some aspects of this memory organisation encourage low-order interleaving, other aspects indicate a clean sweep in favour of high-order interleaving. However, high-order and low-order interleaving can again be combined to yield many different interleaved memory organisations. These different types, however, offer normally a better bandwidth and that too, even in the case of module failure. One of such representative organisations is shown in Figure 4.42 using a four-way low-order interleaving for a clear understanding of this hybrid organisation. Here, low-order interleaving is organised in each of two memory banks.

The advantage of this arrangement is that in case of module failure, this organisation of the two-bank four-way design as shown in Figure 4.42 will still offer a reduced bandwidth

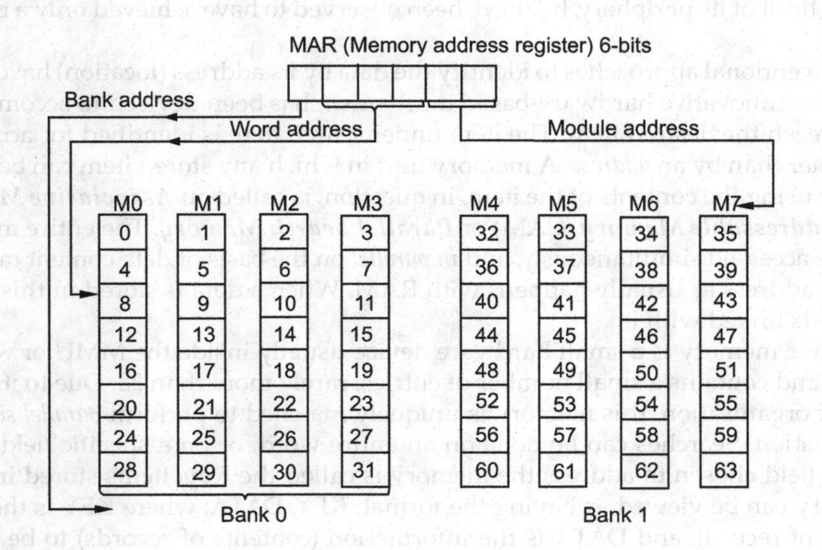

FIGURE 4.42
Four-way interleaving within each memory bank; 1-bit to address one of two banks; 2-bit to address one of 4 modules within each bank; 3-bit to address any of 8 words in any module.

to four words per memory cycle, since only one of two faulty banks will be invalid. The pure low-order interleaving in this situation makes the entire memory bank out of use.

In interleaved memory organisation, the trade-offs should consider the *degree of interleaving* to obtain maximum memory bandwidth, its *fault tolerance*, and *liberty* of each memory banks so that in the worst situation of module failure, something could be extracted from the proposed design.

4.9 Associative Memory Organisation

4.9.1 Background

As technology constantly progresses, CPU-main memory speed disparity continuously increases, thereby creating a severe bottleneck at the CPU's end. Inclusion of cache memory in the memory hierarchy primarily solves this problem of latency. Main memory when organised in an interleaved fashion reduces mostly the bandwidth problem. But, supply of data to CPU requires, *firstly* to reach the particular target data from a list of it, and that too as fast as possible. Thus, searching of target data to reach it before it is supplied is an inherent process both in user-oriented as well as in system-oriented applications. The traditional search procedure usually follows a strategy that chooses a sequence of addresses, reads the memory content at each address, and then compares the information read with the item being searched until a match occurs, or the sequence of addresses thus searched comes to an end with no match. Under this scheme, the total time required to access the desired data depends on the number of accesses to memory which again depends on the location of the target item, the organisation of this list of items, and the efficiency of the search algorithm being employed as well. Many different techniques have been proposed in this regard, a few of them have been devised to optimize this approach of searching within the limit of its periphery, but have been observed to have achieved only a minimum at best.

Thus conventional approaches to identify the data by its address (location) have been left away, and an innovative hardware-based mechanism has been devised to accomplish fast search to reach the desired item. The item under search here is identified for access by its *content* rather than by an *address.* A memory unit in which any stored item can be accessed directly by using the contents of the item, in question, is called an ***Associative Memory*** or ***Content Addressable Memory*** (CAM) or ***Parallel Search Memory***. The entire memory of this type is accessed simultaneously, and *in parallel* on the basis of data content rather than by specific address as usually happens with RAM. When a data is stored in this memory, no address is linked with it.

Associative memory is a small hardware device usually inside the MMU or within the CPU chip, and contains a small number of entries, rarely more than 32. Due to its particular form of organisation, this memory is uniquely matched to perform *parallel searches* by data association. Searches can be done on an entire word, or on a specific field within a word. The field chosen to address the memory is called the ***Key***. Items stored in associative memory can be viewed as having the format: KEY, DATA; where KEY is the address (a subfield of record), and DATA is the information (contents of records) to be accessed. Applications where the search time is assumed to be very critical and must be very short are the ideal situation for associative memory to use.

Associative memory is generally used to contain the data which are heavily in use in the form of a table during the execution of a process. Sometimes, it is used to contain a small fraction of the page table entries which are frequently demanded, and thus accelerates the mapping process of virtual addresses to physical addresses without going through the entire page table available in RAM. It is, hence, sometimes called *Translation Lookaside Buffer*(TLB). The MIPS R2000, a RISC machine that has eventually taken the associative memory idea to its limit. Here, the CPU contains a 64-entry associative memory on the CPU chip, each entry in this memory is of 64-bit, holds the virtual page number and other related information. When the CPU generates a virtual address during execution, this memory's entries are used for the purpose of faster address translation. It is to be noted that the addressing of associative memory is performed with the contents of one of the fields (here, virtual page number) in each row of the table.

The basic difference between associative memory and RAM is that associative memory is *content-addressable* allowing parallel access of multiple memory words, whereas the RAM must be accessed by specifying the word addresses. The inherent parallelism in associative memory has a great impact on the architecture of ***associative processors,*** a special class of SIMD array processors which are updated with associative memories.

An associative memory is more expensive than a RAM of the same size, because each cell must have storage capability as well as logic circuits for matching its content with the supplied argument. Some additional circuit like ***Select Circuit*** is also included in the hardware mechanism to provide other services as will be discussed later.

The ***major advantage*** of associative memory over the RAM is its capability of performing parallel search and comparison operations, which are needed in many important applications, such as table look-up, information storage and retrieval of rapidly changing databases, radar-signal tracking, execution of image processing, and real-time artificial intelligence computation.

The ***major disadvantage*** of associative memory is its much increased hardware cost. Currently, associative memory is much expensive than RAM, even though both are built with integrated circuitry. However, with the rapid advent of VLSI technology, the price gap between this type of memories is gradually reducing.

4.9.2 Implementation

4.9.2.1 Word-Organised Associative Memory

It consists of a memory array of m words, each is of n bits, and also the related logic for this m words with n bits per word. In this organisation, several registers are employed to accomplish different responsibilities. Figure 4.43 shows the structure of this type of a simple associative memory. The functions that different registers perform are as follows:

Input register (I): The input register I holds the input. This means that it holds the data to be written into the associative memory, or the data to be searched for. At any instant, it holds one word of data, i.e. a string of n bits of the memory. Consequently, the length of the input register is n-bit.

Mask register (M): Each unit of stored information is a fixed-length word (record).Any subfield of the word may be chosen as the key. The mask register provides a mask for choosing a particular field or ***key*** (i.e. the key to be searched) in the input register's word. The maximum length of this register is n bits, because this register has to hold a portion of the word or all bits of the word to be searched. For example, consider an inventory file containing various items, where each item is a record in the file. Each such record contains

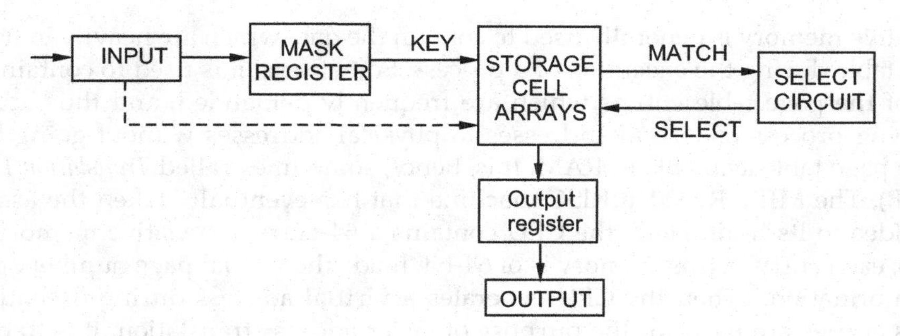

FIGURE 4.43
Block structure of a simple associative memory.

several fields, such as product code, type, product name, description, etc. Any field can be chosen as a *key* for searching an item over this file. At this moment, assume that the *"productcode"* field will be used here as the *key* for searching an item over this file. This desired key is specified by the mask register whose contents identify the bit positions (not necessarily be adjacent) in the record that define the key field.

To illustrate the searching mechanism involved with associative memory, let the input register and mask register contain the following information:

$$\text{Input register (I)} = 1011\quad 1001$$

$$\text{Mask register (M)} = 1111\quad 0000$$

and the file contains three records (words) to be searched, for example, are given as:

$$\text{Words } 1 = 1001\quad 1101$$

$$\text{Words } 2 = 1101\quad 0101$$

$$\text{Words } 3 = 1011\quad 1100\quad \Rightarrow \text{matched}$$

The contents of the mask register indicates four 1s in its leftmost four bits. This signifies that four leftmost bits of input register will be used as key for searching. The contents of the key field as obtained from the input register is then 1011, and this string of bits will be searched over all the records present in the file. Only that (or those) record (s) will be selected as match which contains 1011 in its leftmost four bits (i.e. at the corresponding position as given in the mask register) irrespective of the contents of the other fields in the record (s). It is found that there is a match for word 3 only, and not with others.

The current key is compared simultaneously with all stored words, those that match the key will emit a match signal, which enters a *select circuit*. This circuit does have a select register S of *m* bits, one for each *memory word*. If matches are found after comparing input data in I register with key field in M register, then the corresponding bits in the select register (S) are set. The select circuit enables the data field to be accessed (the back arrow from select circuit to S).

If several entries have the same key (i.e. more than one match), then the select circuit determines which data field to be read out; it may, for example, read out all matched entries in some pre-determined order. *Since, all words in the memory (storage cell arrays) are*

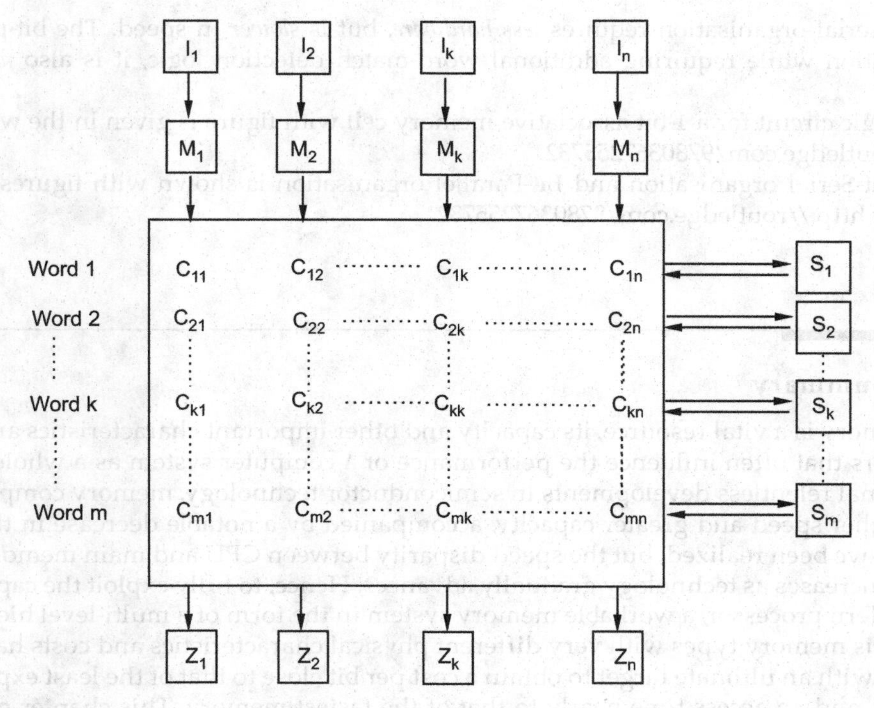

FIGURE 4.44
A representative block diagram of an associative memory of size $m \times n$.

required to compare their keys simultaneously, each must have its own match circuit. The match and select circuits make associative memories much more complex and expensive than conventional memories (RAM). The arrangement of the memory array and four external registers linked with the associative memory system are depicted in Figure 4.44.

In practice, most associative memories have the capability of word-parallel operations, that is, all words in the associative memory array are involved in the parallel search operation. This differs radically from the *word serial* operations encountered in *RAM*s.

Based on how *bit slices* are involved in the operation, there are mainly ***two different*** associative memory organisation:

(i) Bit-parallel organisation, (ii) Bit-serial organisation.

i. ***Bit-parallel organisation:*** Under this scheme, the comparison process is performed in a *parallel-by-word* and *parallel-by-bit* fashion. All bit-slices which are not masked off by the masking pattern are involved in the comparison process. Essentially the entire array of cells is involved in a search operation *P*arallel *E*lement *P*rocessing *E*nsemble (PEPE). **Burroughs** Corporation has employed associative memory with bit-parallel organisation.

ii. ***Bit-serial organisation:*** This memory organisation operates with one bit slice at a time across all the words. The particular bit slice is selected by an extra logic and control unit. The bit cell read out will be used in subsequent bit-slice operations. The ***associative processor STARAN*** (Goodyear Aerospace) has the bit-serial memory organisation.

The bit-serial organisation requires *less hardware*, but is *slower* in speed. The bit-parallel organisation while requiring additional word-match detection logic, it is also *faster* in speed.

The logic circuit for a 1-bit associative memory cell with figure is given in the web site: http://routledge.com/9780367255732.

The Bit-Serial organisation and Bit-Parallel organisation is shown with figures in the web site: http://routledge.com/9780367255732.

4.10 Summary

The memory is a vital resource, its capacity and other important characteristics are critical factors that often influence the performance of a computer system as a whole. With sensational relentless developments in semiconductor technology, memory components with higher speed and greater capacity, accompanied by a notable decrease in the cost per bit have been realized, but the speed disparity between CPU and main memory constantly increases as technology gradually advances. Hence, to fully exploit the capability of a modern processor, a workable memory system in the form of a multi-level hierarchy of various memory types with very different physical characteristics and costs has been built up with an ultimate target to obtain a cost per bit close to that of the least expensive memory, and an access time nearly to that of the fastest memory. This chapter primarily presented the pioneering technological and most important organisational details of needed memory design that are contemplated in most of the contemporary advanced computing system.

Main memory (RAM) cell realization using dominant technologies in the form of Static RAM (SRAM) and Dynamic RAM (DRAM), and subsequent organisation of these cells in 2D arrays to form a full-fledged RAM chip have been described. Subsequent developments of DRAM with constant improvement in fabrication technology eventually made it possible to realize DDR, DDR1, DDR2, and also DDR3 to meet the ever increasing demand of main memory space and speed. Introduction of non-volatile ROM and its different variants located at the same level of RAM in the memory hierarchy have supplemented the primary memory support by way of providing different application activities. Still, the main memory space scarcity problem remains as the computing environment gradually changes, and this crisis has been ultimately mitigated by way of including a relatively larger virtual memory (a portion of the disk, but behaves like a primary memory) in the memory system as a standard feature. But, this requires an additional time-consuming hardware-supported mechanism built within the processor for translating the issued virtual-memory addresses to actual physical addresses (primary memory address) needed during run-time. As a result, the memory response time further increases that summarily degrades the existing overall system performance. To minimize this degradation in performance as well as to compensate the constantly increasing speed disparity between CPU and main memory due to relentless advancement of electronic technology, one or more intermediate faster, smaller, and comparatively costlier memories known as *cache memory* are injected in the memory system in between CPU and main memory to reduce the effective memory access time as experienced by the processor. Nowadays, almost all modern systems including more powerful modern microprocessors, not only use large size caches, but incorporates multi-level on-chip caches (L1, L2, and even L3) apart from using large off-chip cache on board. Splitting of caches and efficient use of them with effective balancing of other design parameters including appropriate mapping scheme have been able to substantially minimize the undesirable effects of memory latency. Caches, when used in multiple-processor architectures provide them enormous processing capabilities to handle many different classes of constantly emerging numerous application areas. But, at the same time, they invite a critical issue known as *cache inconsistency* or *cache coherent* problem. Appropriate mechanisms (mostly by means of hardware-based approach now a days) are thus employed to get rid of this serious problem to enjoy the ultimate benefit of cache usage. All these aspects as mentioned have been adequately addressed and described with respective figures to complete the discussion convincingly.

Secondary storage, in the form of traditional *magnetic disks* and *tapes,* along with relatively modern devices such as *optical disks* (CDs and DVDs) provides the largest storage capacity with least possible cost in the memory hierarchy. Nowadays, extensive use of portable secondary devices, such as flash card, flash drives, pen drives, etc. which are relatively smaller in size, but provide moderately adequate storage space with reduced cost, have virtually become a strong competitors of hard disks. The RAID approach organizing many inexpensive disk drives in an array arrangement provides a high-performance large secondary storage unit in which each disk could operate independently, and also in parallel with others.

Interleaved memory is an innovative organisations of main memory that can extract potential parallelism by creating many main memory units in a memory system, aiming primarily to improve *memory bandwidth,* more than to reduce *memory latency,* with almost no extra cost. *Associative memory* or sometimes known as *Content Addressable memory* (CAM) or *Parallel Search Memory* using a pioneering hardware-based mechanism has been devised to supplement main memory, mainly to accomplish fast access to the desired item by its content rather than by an address as conventional approaches do. The entire memory of this type is accessed simultaneously, and in parallel on the basis of data content (used as address) rather than by specific address as is done usually with RAM. When a data is stored in this memory, no address is linked with it.

Exercises

4.1 What is meant by memory hierarchy? Why is such an hierarchy required? What is the basis on which such an hierarchy pyramid had been developed? Draw a suitable memory hierarchy block diagram with the parameters you have considered.

4.2 Discuss big-endian and little-endian methods used in byte ordering. What is the difference between direct access and random access?

4.3 Discuss the parameters that determine the performance of a memory system.

4.4 Describe with a suitable diagram the storage structure of a bipolar storage cell and explain the reading and writing operations on the cell.

4.5 Explain with a diagram the reading and writing operations of a basic static MOS cell.

4.6 "In spite of requiring additional arrangement for its operation, memory system realized using DRAM cell is still more advantageous than its counterpart SRAM cell". Justify the statement with your comment.

4.7 Why does a DRAM cell need refreshing? State briefly the different schemes being used for refreshing DRAM.

4.8 How many128 × 16 RAM chips are needed to construct a memory capacity of 4,096 words (16 bit word)? How many lines of the address bus must be used to access a memory of 4,096 words. How many of these lines will be common to all chips? For chip select, how many lines must be decoded?

4.9 Using four numbers of 128 × 8 RAM modules, design the following memory organisations: (i) 512 × 8, (ii) 256 × 16, (iii) 128 × 32

4.10 Describe with diagram the two-dimensional (2D or 2½D) organisation of semiconductor memories. Show with a diagram one such representative IC chip.

4.11 What is the difference between SDRAM and DRAM? What are the different types of ROM? Discuss their merits and drawbacks.

4.12 Describe the technique being used in erasing EPROM. What are the drawbacks of this method?

4.13 What is virtual memory? Why is it called so? What are the advantages obtained from a virtual memory system?

4.14 What is meant by logical address space and physical address space? Explain the mechanism with an example how logical address is converted into physical address (address translation mechanism)?

4.15 It is observed that with the advent of more advanced technology, the chip can be developed with increasing capacity of memory storage, and hence the virtual memory is no longer needed and can be dropped from future computers. Comment with reasons in favour of and against this thought.

4.16 A two-level memory system has eight virtual pages on a disk to be mapped into four-page frames in the main memory. Assume the page frames are initially empty. During execution, a certain program generated the following page trace:1, 2, 0, 2, 1, 7, 7, 6, 0, 1, 2, 0, 3, 0, 4, 5, 1, 5, 2, 4, 5, 6, 7, 6, 2, 4, 7, 3, 2, 4, 4

a. Show the successive virtual pages residing in the four-page frames with respect to the above page trace using the circular FIFO page replacement policy. Compare the hit ratio in the main memory.

b. Repeat part(a) for the LRU page replacement policy. Compute the hit ratio in the main memory.

c. Compare the hit ratio in parts (a) and (b) and comment on the effectiveness of using the circular FIFO policy to approximate the LRU policy with respect to this particular page trace.

4.17 "The page replacement policy to be adopted has a severe impact on the performance of the system." Discuss. What is the working set? How working set model affects the main memory hit ratio?

4.18 What are the different page replacement policies that are commonly used? Write two page replacement policies that are used in a virtual memory system.

4.19 Consider a two-level memory, M_1 and M_2. Let the hit ratio of M_1 be h. Let c_1 and c_2 be the costs per kilobyte, s_1 and s_2 the memory capacities, and t_1 and t_2 the access times, respectively.

a. Under what conditions will the average cost of the entire memory system approach c_2?

b. What is the effective memory-access time t_a of this hierarchy?

c. Let $r = t_2/t_1$ be the speed ratio of these two memories. Let $E = t_1/t_a$ be the *access efficiency* of the memory system. Express E in terms of r and h.

d. What is the required hit ratio h to make $E > 0.95$ if $r = 100$?

4.20 Explain the need of auxiliary memory devices. How are the serial access memories organised? In what way are they different from main memory? Give the major differences being observed between tape drive and magnetic disk.

4.21 Define the following for a disk system:

T_S = seek time;

R = rotation speed of the disk, in revolutions per second

N = number of bits per sector

C = capacity of a track, in bits

T_A = time to access a sector; Build up a formula for T_A as a function of these other parameters.

4.22 A disk pack has15 surfaces. Storage area on each surface has an inner diameter of 20 cm and an outer diameter of 30 cm. Maximum storage density on each track is 2,000 bits/cm, and minimum spacing between the tracks is 0.25 mm. What would be the storage capacity of the pack? What would be the data-transfer rate in bytes/sec at a rotational speed of 3,600 rpm?

4.23 How long does it take to read a disk with 800 cylinders, each containing five tracks of 32 sectors each? First, all the sectors of track 0 are to be read starting at sector 0, then all the sectors of track 1 starting at sector 0, and so on. The rotation time is 20 ms, and a seek takes10 ms between adjacent cylinders and 50 ms for the worst case. Assume switching between tracks of a cylinder can be done instantaneously.

4.24 Calculate how much disk space (in sectors, tracks, and surfaces) will be required to store 5,000 logical records, each is of 120byte, if the disk is a fixed sector of 512 bytes/sector, with 100 sectors per track, 120 tracks per surface, and 8 usable surfaces. Assume that records cannot span 2 sectors and ignore any file header records and track indexes, etc.

4.25 Higher blocking factor allows more records to be accommodated in a tape reel and also takes lesser amount of time to process the tape, whereas using comparatively lower blocking factor, lesser number of records can be stored in the same tape with more time to process the tape. Explain why? In a practical situation, still we prefer reasonably lower blocking factor. Explain.

4.26 What is the transfer rate of a 9-track magnetic tape unit whose tape speed is 120 inches per second and whose tape density is 1,600 linear bits per inch (bpi)?

4.27 A9-track magnetic tape reel of length 2,400 ft having a tape density of 1,600 linear bits per inch is rotating at a speed of 200 inches per second past the recording head. Data on tape are organised in physical records, where each physical record contains a fixed number of user-defined records call *logical records*. The inter record gap is 0.8 inch.

   a. How long will it take to read a full tape of 240-byte logical records having a blocking factor 10 (each physical record contains 10 logical records)?
   b. What will be the time taken if the blocking factor is 30?
   c. How many logical records will the tape hold with each of the above blocking factors?
   d. What is the effective over all transfer rate for each of the above two blocking factors?
   e. What is the capacity of the tape?

4.28 From the manual, it is found that a commercial magneto-optical disk drive (CD-ROM) has the following specifications:

Formatted storage capacity of unit with 1,024-byte sectors	650 GB
Formatted storage capacity of unit with 512-byte sectors	650 GB
Read data-transfer rate with 1,024-byte sectors	0.87 MB/s
Read data-transfer rate with 512-byte sectors	0.79 MB/s
Write data-transfer rate with 1,024-byte sectors	0.29 MB/s
Write data-transfer rate with 512-byte sectors	0.26 MB/s

   a. It is observed that larger (1,024 byte) sector provides greater storage capacity and higher data-transfer rates than the smaller (512 bytes) sector. Explain the reasons.
   b. The larger sector size appears to have all the advantages, so why is the smaller size ever used?
   c. Why does the writing operation take more time than reading?

4.29 What are cache memories? Many computers use cache block size in the range of 32–128 bytes. What would be the major advantages and disadvantages being observed in making the size of cache blocks larger or smaller? Explain with the help of a neat sketch the operation of a cache indicating the write-back and write-through schemes.

4.30 Compare the relative merits of the four cache memory organisations:

   a. Fully associative cache
   b. Direct-mapping cache
   c. Set-associative cache
   d. Sector-mapping cache

4.31 Discuss the advantages and disadvantages in using a common cache or separate caches for instructions and data. Explain the support from data paths, MMU and TLB, and memory bandwidth in the two cache architectures.

"Presence of multi-level caches in the memory organisation definitely enhances the performance." Define multi-level cache and justify the statement.

4.32 The cache design is associated with the following terms. Explain each one with relative merits and demerits, if appropriate:

a. Factors affecting cache hit ratios

b. Write-through versus write-back caches

c. Private cache versus shared cache

d. Cacheable versus non-cacheable data

e. Cache flushing policies

4.34 Consider a two-level memory hierarchy, a cache (M_1) and memory (M_2) with the following characteristics:

M_1: 32K words, 50 ns access time

M_2: 1M words, 400 ns access time

Assume eight-word cache blocks and a set size of 256 words with set-associative mapping.

a. Show the mapping between M_1 and M_2.

b. Calculate the effective memory-access time with a cache hit ratio of $h = 0.95$.

4.35 The memory unit of a computer system is 128K × 16, and has a cache memory of 2K words. The cache memory uses direct mapping with a block size of four words.

a. How many bits are there in the tag, index, block, and word fields of the address format?

b. How many bits are there in each word of cache, and how are they divided into functions? Include a valid bit in the format.

c. How many blocks can the cache accommodate?

4.36 The access time of a cache memory is 90 ns and that of main memory is 900 ns. It is estimated that 80% of the memory requests are for read and the remaining 20% for write. The hit ratio for read accesses only is 0.9. A write-through procedure is used.

a. Considering only memory-read cycles, what is the average access time of the system?

b. When both read and write requests are considered, what would be the average access time of the system?

c. What is the hit ratio considering the write cycles also?

4.37 A hierarchical cache–main memory subsystem has the following specifications:

a. Cache access time is 50 ns.

b. Main memory access time is 550 ns.

c. 80% of memory requests are for read.

d. Hit ratio is 0.9 for read access.

e. Write-through scheme is employed.

Estimate:

a. average access time of the system considering only memory-read cycle.

b. average access time of the system both for read and write request.

4.38 A four-way set associative cache memory uses blocks of four words. The cache can accommodate a total of 2,048 words from the main memory. The main memory size is 256 × 32.

   i. Construct the cache memory with all pertinent information needed to formulate it

   ii. What is the size of the cache memory?

4.39 What is cache miss? State and explain the different categories of cache miss. Explain the reasons behind such cache misses. Discuss three techniques to reduce the miss rate.

4.40 What is meant by cache coherence problem? What are the reasons that create such problem? What are the different schemes being used to solve cache coherence problem?

4.41 What is meant by interleaving of memory? What are the advantages that can be extracted from this arrangement? What is meant by degree of interleaving? Discuss briefly the different types of interleaving and their relative merits and drawbacks.

4.42 Derive the logic of one cell of an associative memory. Where do you find the use of associative memory advantageous? Why are large associative memories rarely used?

Suggested References and Websites

Archibald, J. and Baer, J. L. "Cache coherence protocols: Evaluation using a multiprocessor simulation model." *ACM Trans Prog Comput Syst*, vol. 4, pp. 273–298, November 1986.

Crisp, R. "Direct RAMBUS Technology: The new main memory standard." *IEEE Micro*, November/December 1997.

Hennessy, J. L. and Patterson, D. A. *Computer Architecture: A Quantitative Approach*. San Mateo, CA: Morgan Kaufmann, 1990.

Jacob, B., Ng, S. and Wang, D. *Memory Systems: Cache, DRAM, Disk*. Boston, MA: Morgan Kaufmann, 2008.

Przybylski, S. A. *Cache and Memory Hierarchy Design*, 2nd ed. San Mateo, CA: Morgan Kaufmann, 1990.

Smith, A. "Cache memories." *ACM Computing Surveys*, September 1992.

The RAM Guide: Provides an overview of RAM technology as well as a number of useful links.

Optical Storage Technology Association: Provides an abundant source of information about the technology used in optical storing, and vendors as well as extensive list of relevant links.

5

Input–Output Organisation

LEARNING OBJECTIVES

- To provide an overview of the principles of input–output (I/O) system organisation and the related design of I/O modules.
- To enable a clear understanding of interrupts and their types, and the hardware and software needed to support them.
- To detail the different types of I/O operations, including direct memory access (DMA) for high-speed devices and I/O processors (IOPs) for large systems.
- To facilitate an understanding of buses, their different types (PCI (*peripheral component interconnect*), SCSI (small computer system interface), USB (Universal Serial Bus)), and distinctive features along with the corresponding bus designs.
- To detail the features of FireWire, an advanced high-speed and high-performance serial bus for digital audio and video pieces of equipment.
- To introduce the reader to InfiniBand, an advanced high-speed communication link for modern distributed systems, cluster architectures, and high-end servers.
- To discuss ports and their different types, including serial, parallel, and USB ports.

The synergy between computers and communications is at the heart of today's information technology revolution. Apart from the CPU (central processing unit) and a variety of memories constituting the internal memory system, the third key resource of a modern computer system is a set of various I/O modules attached with numerous peripheral devices that can now be interfaced to the system bus or central control to communicate with the main system for migration of information from the outside world into the computer or from the computer to the outside world. Besides, most computers also have a wired or wireless I/O connection to the Internet to communicate with the external world. Thus, I/O modules, on the one hand, work in tune with the CPU and main memory, and on the other hand, they establish a connection with the external peripheral devices (mostly electro-mechanical) to execute physical I/O operations. The continuous development in the design of I/O modules and its allied interfaces makes it more and more intelligent, and consequently enhances the performance of the computer system to which it is attached. Large computer systems nowadays have such powerful I/O modules that an I/O module can be itself treated as a stand-alone full-fledged computer system. This chapter presents an overview of the I/O organisation detailing different types of I/O modules and their interfaces with the main computer systems, including the buses and ports, and their structures and functions, and above all, the mechanisms employed to control these devices to achieve certain predefined goals.

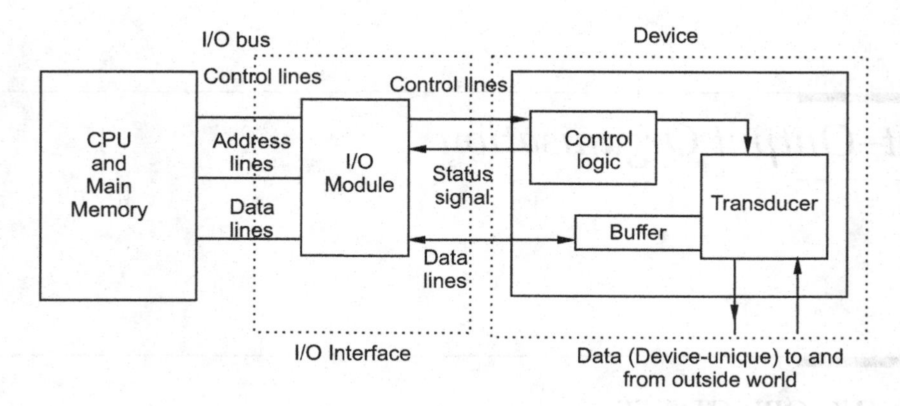

FIGURE 5.1
Block diagram of an external device along with its I/O interface.

5.1 Input–Output System

The *I/O system* performs the task of transferring information between main memory or the CPU and the outside world. Figure 5.1 shows a model of an I/O system, which includes an I/O module that works as a mediator to communicate between the high-speed CPU or memory and the much slow-speed electro-mechanical devices of different types connected by the well-defined links (buses). This link is used to communicate control, status, and physical data between the I/O modules and the external devices, as shown in Figure 5.1, for realizing desired I/O operation. Each of these different types of devices as connected with the system must have a control logic of its own that will manage the operation of the device as directed by the I/O module. The device should also have necessary electronic circuits in the form of a transducer to convert the I/O data into appropriate forms for execution. Usually, a small *buffer* is associated with the transducer to temporarily hold the data during the data transfer operation to support smooth coordination between the slower I/O devices and the relatively faster memory or CPU.

Nowadays, the I/O organisation is considered as one of the key parameters to characterize a particular generation of computers, which in turn determines the class and category of the computer and the devices attached to it. In fact, the differences between a small (micro, mini) and a large (supermini, mainframe, supercomputer) system, besides other characteristics, mostly depend on the capability and capacity of its attached I/O module, the amount of hardware available to communicate with its peripherals, and also the number of peripherals that could be hooked up with it.

5.2 I/O Module: I/O Interface

The primary responsibility of the I/O module and its associated interfaces is to *resolve the differences* (mainly due to the devices with various characteristics having different types of operations with numerous data formats and codes along with the different disparities in the data transfer rate with the CPU and memory) that exist between the central computer system and the different types of peripheral devices attached with it. To perform the responsibilities as described, a set of useful functions is executed by the I/O module (I/O interface) at the time of data transfer that are two fold in nature: *CPU–I/O module handshaking* and *I/O module device negotiation*. CPU–I/O communication is mostly carried out via memory which is located centrally (Figure 5.1), supporting the CPU, on the one hand, and communicating with the I/O module, on the other hand. Fundamentally, there exist *many different methods*, consisting of various arrangements of different sets of lines (buses) that connect the I/O modules with the memory in a system. Of course, each of these methods has its own performance merits and drawbacks, and the related cost implications also vary over a wide range.

A brief detail of I/O module, its different functions, and its communication with memory is given in the website: http://routledge.com/9780367255732.

5.2.1 I/O Module Design

The ultimate target of I/O module design is to what extent it can relieve and release the CPU from the tedious burden of I/O activities (non-intelligent activity), and hence, this design varies considerably depending on the amount of intelligence to be injected into an I/O module so that how much burden of an entire I/O operation it can bear independently using the different types of external peripherals attached with it. The I/O logic within this module interacts with the CPU (or rest of the computer system) hiding the details of the devices via a set of control lines (e.g. system bus lines), as shown in Figure 5.2. These lines are also used by the CPU to communicate with I/O modules in the form of issuing commands. Other control lines attached with the I/O logic are used to control and drive the peripherals. In addition, every I/O module is also equipped with a set of addresses; each address in the set is the address (identification) of the peripheral attached to the module.

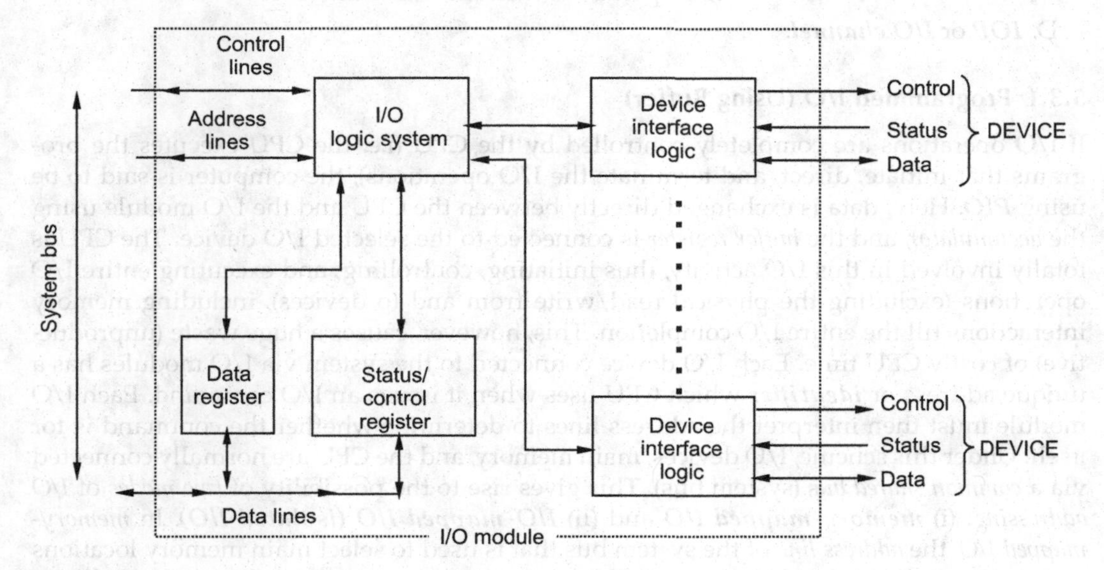

FIGURE 5.2
Schematic I/O module between CPU (rest of the system) and devices.

If an I/O module is quite primitive and needs a detailed control from the CPU during I/O operation, this type of I/O module is usually referred to as an ***I/O controller*** or ***device controller,*** which again may be of various types depending on their intelligence and capabilities, and to the extent they require assistance from the CPU and its involvement in the execution of I/O operation. If the I/O module is equipped to handle most of the detailed I/O processing burdens on its own setting the CPU aside, relieving the CPU totally from the headache of I/O operations, presenting to the CPU a high-level interface, then this type of I/O module is generally referred to as an ***I/O channel*** or ***I/O processor.***

A brief detail of I/O module design is given in the website: http://routledge.com/9780367255732.

5.3 Types of I/O Operations: Definitions and Differences

I/O organisations (I/O systems) are usually distinguished by the extent to which the CPU is involved in the execution of I/O operations and is relieved from I/O activities. The CPU merely executes the I/O instructions and may accept the data temporarily, but the ultimate transfer of information to and from external devices involves the memory unit. That is why, the data transfer between the central computer and I/O devices may be handled in a variety of ways, mainly, either between an *I/O device and the CPU* or between an *I/O device and main memory*. This gives rise to ***four*** different I/O schemes, which are in current use:

A. ***Programmed I/O (PIO)***;

B. ***Interrupt-driven I/O***;

C. ***DMA***;

and, the ultimate target is to find out a technique to completely free the CPU from any involvement in I/O operations with the introduction of

D. ***IOP*** or ***I/O channel.***

5.3.1 Programmed I/O (Using Buffer)

If I/O operations are completely controlled by the CPU (i.e. the CPU executes the programs that initiate, direct, and terminate the I/O operations), the computer is said to be using ***PIO***. Here, data is exchanged directly between the CPU and the I/O module using the *accumulator*, and the *buffer register* is connected to the selected I/O device. The CPU is totally involved in this I/O activity, thus initiating, controlling, and executing entire I/O operations (excluding the physical read/write from and to devices), including memory interactions till the entire I/O completion. This, however, causes a huge waste (unproductive) of costly CPU time. Each I/O device connected to the system via I/O modules has a unique address or ***identifier*** which CPU uses when it issues an I/O command. Each I/O module must then interpret the address lines to determine whether the command is for itself. Under this scheme, I/O devices, main memory, and the CPU are normally connected via a *common shared bus* (system bus). This gives rise to the possibility of *two modes* of *I/O addressing*: (i) ***memory-mapped I/O*** and (ii) ***I/O-mapped I/O (isolated I/O).*** In *memory-mapped I/O*, the *address line* of the system bus that is used to select main memory locations can also be used to select *I/O devices*. Each junction between the system bus and I/O device is called an ***I/O port*** and is assigned a unique address. The I/O port includes a *data buffer* *register,* which is here a part of main memory address space (Motorola 68000 series). That is why, it is called *memory-mapped I/O.* A memory reference instruction that causes data to be fetched from or stored at address X automatically becomes an I/O instruction if X indicates the address of an I/O port. For most CPUs of this type, a relatively large set of different instructions is available for referencing memory. An *advantage* of memory-mapped I/O is that this large repertoire of instructions can also be used for I/O activities, thereby enabling more efficient programming. The *second strategy* in organisation is sometimes called ***I/O-mapped I/O (or isolated I/O),*** where the memory and I/O address space are kept totally separate. Here, the I/O port includes a *data buffer register* which is associated with the I/O module (devices), and memory reference instructions do not affect the I/O devices, but separate I/O instructions are required to activate the I/O. Consequently, an I/O device and a main memory location may have the same address. If I/O-mapped I/O (isolated I/O) strategy is used, there are only a few I/O instructions. This scheme is used in INTEL 8085 (using 8255A chip) and 8086 series of processors.

More details of PIO, and a comparison between the two modes, are given in the website: http://routledge.com/9780367255732.

5.3.2 Interrupt-Driven I/O

The primary *disadvantage* of PIO is that the CPU is totally involved in the slow I/O operation, and spends most of its time remaining idle called ***busy waiting.*** The way to get rid of *busy waiting* is to have the CPU issue an I/O command to an I/O module as usual to start the I/O device, and tell the I/O module to generate an ***interrupt*** when it is done. While the I/O module starts working, the CPU is no longer involved in this activity and would proceed to do its some other useful work. The I/O module will interrupt the CPU at the right time to request service when it is ready to exchange data with the CPU. The CPU would then again get involved, leaving its own ongoing processing, and execute the data transfer as usual, and after then would go back to resume its former processing. This approach known as *interrupt-driven I/O* can be accomplished by setting the INTERRUPT ENABLE bit in a device register; the software would expect that the hardware will give it a signal when the I/O module would request.

Although interrupt-driven I/O is more efficient, and is a big step forward from PIO due to eliminating needless waiting of the CPU, it is still far from optimal. The problem is that every time an interrupt occurs, it requires the CPU to get involved for every word (character) of data that goes from memory to I/O module or from I/O module to memory, and interrupt processing then becomes expensive. However, the Intel **8259A** chip supports interrupt-driven I/O that can handle up to 8 I/O modules, and a cascade arrangement can extend it to handle even up to 64 modules.

A brief detail of interrupt-driven I/O and various forms of interrupt implementations are given in the website: http://routledge.com/9780367255732.

5.3.2.1 Interrupt-Driven I/O: Design Issues

In implementing interrupt-driven I/O, ***two aspects*** in the design are to be considered.

a. There will almost invariably be multiple I/O modules connected to the system. How the CPU would determine which device (source) issued the interrupt;

b. If multiple interrupts have occurred, how the CPU would decide the order of their processing (priority).

These two aspects are logically linked with each other because multiple interrupts issued by multiple devices may be attached with single I/O module or may be attached with multiple I/O modules. The first task of the interrupt system is to identify the source of the interrupt. There is also a high probability that several sources request interrupt services simultaneously, or while an interrupt routine is running, a second I/O device wants to generate its interrupt. The solutions consist of techniques commonly in use, which fall broadly into ***four*** general categories:

i. Multiple interrupt lines (independent requesting);
ii. Software poll;
iii. Daisy Chaining (vectored, hardware poll);
iv. Bus arbitration (vectored).

5.3.2.1.1 Multiple Interrupt Lines

Providing *multiple interrupt lines* between the CPU and I/O modules is probably the most straightforward approach towards the solution of this problem. This method also corresponds to ***independent requesting*** of interrupts issued from devices. This scheme actually uses separate BUS REQUEST and BUS GRANT lines for each unit sharing the bus. The bus control unit immediately identifies all requesting units individually and is able to respond very rapidly to requests for bus access. Priority is also determined by the bus control unit, and may be programmable. While this method has some distinct advantages, it also suffers from severe drawbacks. The ***PDP-11*** Unibus system has implemented this technique (bus control) by *combining* this independent requesting with a daisy chaining (hardware) approach (discussed next).

A brief detail of the different design approaches of interrupt-driven I/O is given in the website: http://routledge.com/9780367255732.

5.3.2.1.2 Priority and Level: Its Determination

The categories **(ii), (iii),** and **(iv)** as mentioned above are significant when several devices (masters and/or slaves) connected to a *shared bus* (common bus) would simultaneously issue interrupts, requesting access to CPU at the same time. The problem of selecting one I/O device to service, from many such devices that have generated interrupts, bears a strong resemblance to the bus arbitration process for bus control. In fact, these categories belong to essentially different individual-selection mechanisms that determine appropriate selection considering the priority of each individual interrupt being issued, while negotiating such concurrent competing requests. *Priority* of simultaneous interrupts can be realized by the techniques implemented by *software, hardware,* or a *combination of both.*

5.3.2.1.2.1 Polling (Software Approach) The software method to identify the highest-priority interrupt is realized by a polling procedure. In this method, there is one common branch address for all interrupts. The program that takes care of interrupts begins at the branch address and polls the interrupt sources in the order of their *priority.* The highest-priority source is tested first, and if its interrupt signal is on, control branches to a service routine for this source; otherwise, the next-lower-priority source is tested, and so on. The particular serviced routine thus reached belongs to the highest-priority device among all devices that interrupted the computer at that instant. As depicted in Figure 5.3, this method uses a set of lines called *poll count lines* which are directly connected to all units in parallel on the

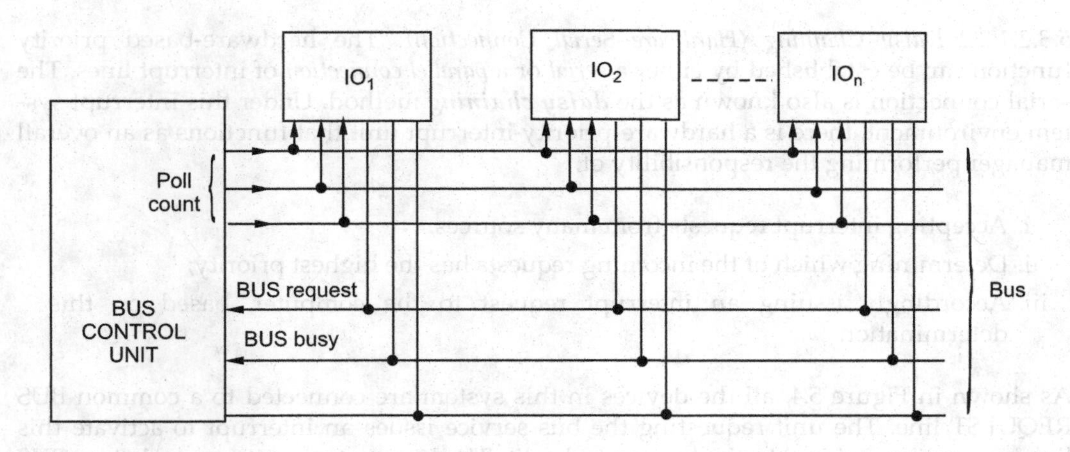

FIGURE 5.3
Polling priority interrupt.

bus. When a unit requests access to the bus, it puts a signal on the common BUS REQUEST line. In response to the received signal, the bus controller proceeds to *generate a sequence of numbers on the poll count lines*. These numbers, which can be thought of as device addresses, are compared by each unit with the unique address already assigned to that unit. When a requesting device I_J finds that its address matches the number on the poll count lines, it activates the BUS BUSY. The bus controller *responds by terminating the polling process* and I_J connects to the bus. The *priority* of a unit, however, is clearly determined by the *position of its address* in the polling sequence, which can be altered under program control, and hence does not depend on the physical position of the device unit on the bus.

Advantages

i. The sequence of numbers generated by the bus controller to match the unique unit address is normally programmable (the poll count lines are connected to a programmable register); hence, selection priority can be altered under program control;
ii. A failure in one unit need not affect any of the other units.

Disadvantages

i. The cost of more control lines (*k* poll count lines instead of one BUS GRANT line) to achieve flexibility (independence) over the connected units;
ii. The *number of units* that can share the bus is limited by the addressing capability of the poll count lines;
iii. If there are many interrupts, the time required to poll them may exceed the time permitted to service the I/O device. In this situation, a hardware-priority-interrupt unit can be used to speed up the operation.

Intel 8259A is one such programmable interrupt controller being used with the respective CPUs that alone manages the interrupts with priorities issued by the numerous devices.

5.3.2.1.2.2 Daisy Chaining (Hardware–Serial Connection) The hardware-based priority function can be established by either a *serial* or a *parallel connection* of interrupt lines. The serial connection is also known as the ***daisy chaining*** method. Under this interrupt system environment, there is a hardware-priority-interrupt unit that functions as an overall manager performing the responsibility of:

i. Accepting interrupt requests from many sources;
ii. Determining which of the incoming requests has the highest priority;
iii. Accordingly issuing an interrupt request to the computer, based on this determination.

As shown in Figure 5.4, all the devices in this system are connected to a common BUS REQUEST line. The unit requesting the bus service issues an interrupt to activate this line by sending a signal to the bus control unit. The bus control unit responds to a BUS REQUEST signal only if BUS BUSY is inactive. This response takes the form of a signal placed on the BUS GRANT line. On receiving the BUS GRANT signal, the requesting unit enables its physical bus connections and activates BUS BUSY for the duration of its new bus activity. Since each interrupt source (device) has its own interrupt vector, it can then directly access its own interrupt service routine. As the BUS GRANT line is connected serially from unit to unit, if two units simultaneously request bus access, the one closest to the bus control unit receives the BUS GRANT signal first and gains access to the bus. *Selection priority* is therefore completely determined *by the order in which the units are linked* together (chained) over the BUS GRANT line. The device with the highest priority is placed in the first position (closest to the bus control unit), followed by lower-priority devices up to the device with the lowest priority, which is placed last in the chain.

Advantages

i. Daisy Chaining is much faster than any software control and requires very few control lines employing a very simple arbitration algorithm;
ii. It can be used essentially with an unlimited number of devices.

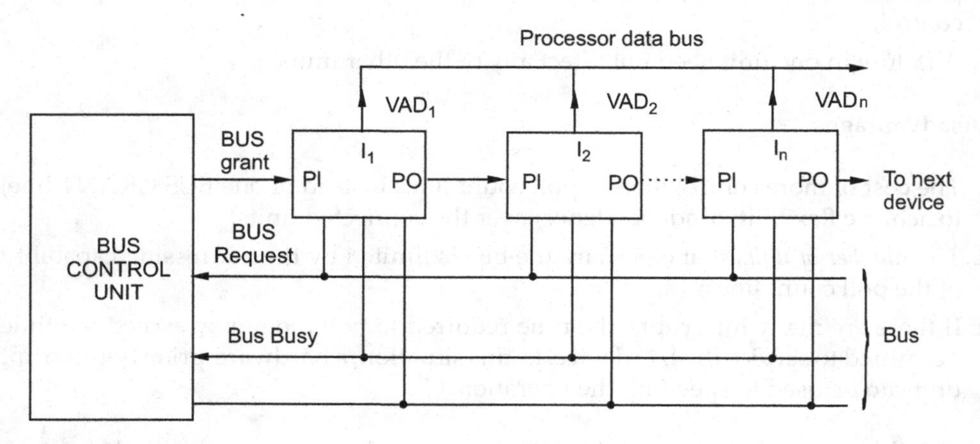

FIGURE 5.4
Daisy chain priority interrupt.

Disadvantages

i. Since priority is wired-in, the priority of each unit cannot be changed under program control, hence offering less flexibility;

ii. If bus requests are generated at a sufficiently high rate, a high-priority device can lock out a low-priority device;

iii. There is a susceptibility to failures involving the BUS GRANT line and its associated circuitry. If the first device is unable to propagate the BUS GRANT signal, then all other devices cannot gain access to the bus.

5.3.2.1.2.3 Bus Arbitration Technique This technique makes use of *vectored interrupts*. With bus arbitration, an I/O module by any means must *first* gain control of the bus (by raising the interrupt request line) before it can issue the interrupt. Thus, only one module can raise the line at a time. When the CPU detects the interrupt, it can easily identify the source and responds on the interrupt acknowledge line. The requesting module then places its vector (address of the service routine) on the data lines.

A brief detail of the different design approaches of interrupt-driven I/O as discussed is given in the website: http://routledge.com/9780367255732.

5.3.3 Direct Memory Access (DMA) I/O

5.3.3.1 Introduction

The problem still remains; interrupts at different stages are required: more for PIO and relatively less for highly efficient interrupt-driven I/O, and processing an interrupt done only by CPU is, however, expensive. In addition, for both the cases, all data transfers also involve CPU and must be routed only through the CPU. As a result, the costly CPU time is badly wasted that causes an adverse impact on CPU activity, and also limits the I/O transfer rate. A befitting mechanism is thus needed that would target to ultimately relieve and release the CPU to a large extent from this hazardous time-consuming I/O-related activities, and that can be accomplished by simply letting the peripheral devices themselves to directly manage the memory buses without the intervention of the CPU. This would definitely improve not only the speed of data transfer but also the overall performance of the system, as the CPU is now released to carry out its own useful work.

5.3.3.2 Definition

With the development of hardware technology, the I/O device (or its controller) can be equipped with the *ability to directly transfer a block of information* to or from main memory *without CPU's* intervention. This requires that the I/O device (or its controller) is capable of generating memory addresses and transferring data to or from the system bus: i.e. it then must be a bus master. The CPU is still responsible for *initiating* each block transfer, and the I/O device controller can then carry out the physical data transfer without further program execution by the CPU. The CPU and I/O controller interact only when the CPU must yield control of the system bus to the I/O controller in response to a request issued from the controller. This request is in the form of an ***interrupt***, to be serviced by the CPU. The CPU is now involved to process the interrupt. After servicing the interrupt, the CPU is no longer there, and it can then go back to resume the execution of its previously ongoing executing program. The access of the bus is now transferred and is now rest with the controller (requesting device) which then completes the required number of cycles for the data transfer, and then again hands over the control of the bus back to CPU. This type of I/O capability is called ***direct memory access (DMA)***. For large volumes of data transfer, DMA technique is found much faster than if it is carried out by the CPU, and is observed to be adequately efficient. To implement DMA and interrupt facilities, most computers may then require the system's I/O interface to contain *special DMA* and *interrupt control units*.

5.3.3.3 Essential Features

Figure 5.5 shows a block diagram indicating how the DMA mechanism works.

- The I/O device is connected to the system bus via a special interface circuit called a *DMA controller*;
- The DMA controller contains a data buffer register ***IODR*** (input–output data register) to temporarily store the data, just like in the case of PIO. In addition, there is an *address register* ***IOAR*** (input–output address register), and a *data count register* ***DC***;
- The IOAR is used to *store* the address of the next word to be transferred. It is automatically incremented after each word transfer;
- The DATA count register ***DC*** *stores* the number of words that remain to be transferred. It is automatically decremented after each transfer and tested for zero. When the data count reaches zero, the DMA transfer halts;
- These registers allow the DMA controller to transfer data to and from a contiguous region of main memory.

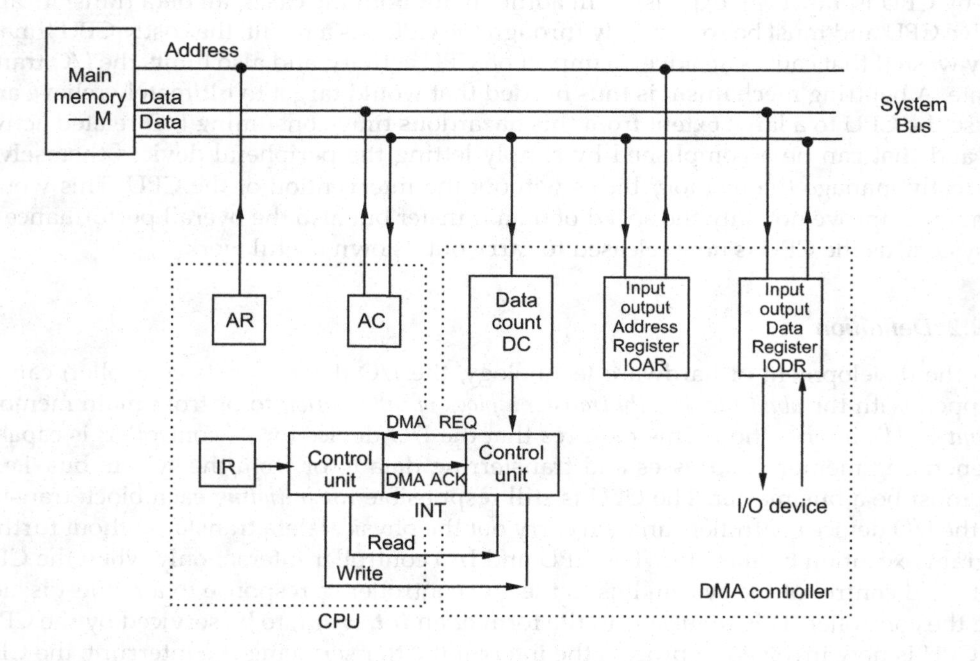

FIGURE 5.5
Schematic diagram of DMA I/O.

The controller is normally provided with an interrupt capability, and in this situation, it sends interrupts to the CPU to signal the beginning and end of the data transfer. The logic necessary to control the DMA activities can easily be placed in a single integrated circuit. DMA controllers are available that can supervise DMA transfers involving several I/O devices, each with a different *priority* of access to the system bus.

5.3.3.4 Processing Details

When the CPU intends to receive or send (read or write) a block of data, it issues a request to the DMA module with certain information, and the *DMA transfer operation* then proceeds as follows:

- The read or write control line between the processor and the DMA module is used by the CPU to intimate the DMA whether a read or write is requested;
- The identification (address) of the I/O device to be involved is intimated by the CPU to the DMA, and is communicated on the data line;
- The CPU executes two I/O instructions, which load the DMA registers: the IOAR with the base address of the main-memory region to be used in the data transfer and the DC with the number of words to be transferred to or from this region;
- When the DMA controller is ready to transmit or receive data, it activates the DMA REQUEST line to the CPU. The CPU then activates DMA ACKNOWLEDGE and there after relinquishes control of the data and address lines, and gets back its own work. The CPU now waits for the next DMA breakpoint. In fact, DMA REQUEST and DMA ACKNOWLEDGE are essentially BUS REQUEST and BUS GRANT lines for the system bus. Simultaneous DMA requests from several DMA controllers, if any, can be resolved by using one of the bus-priority control techniques;
- The DMA controller now transfers data directly to or from the main memory. After a word is transferred, IOAR and DC are incremented and decremented, respectively;
- If the DC is decremented to zero, the DMA controller again relinquishes control of the system bus. It may also send an interrupt signal to the CPU, and the CPU responds either by halting the I/O device or by initiating a new DMA transfer;
- If the DC is not decremented to *zero*, but the I/O device is not ready to send or receive the next batch of data, the DMA controller returns control to the CPU by releasing the system bus and deactivating the DMA REQUEST line. The CPU responds by deactivating DMA ACKNOWLEDGE and resumes normal operation.

Intel 8257 chip supports four DMA channels, by which four peripheral devices can independently request for DMA data transfer at a time. The DMA controller has 8-bit internal data buffer, a read/write unit, a control unit, a priority-resolving unit along with a set of registers. **Intel 8237** critically differs architecturally from 8257 and provides a better performance compared to 8257. It is an advanced programmable DMA controller capable of transferring a byte or a bulk of data between system memory and peripherals in either direction. Memory-to-memory data transfer facility is also available in this chip. This DMA controller can be interfaced to the processor family 80x86 with DRAM memory. Similar to 8257, the 8237 also supports four independent DMA channels (numbered 0, 1, 2, and 3), which may be expanded to any number, by cascading more number of 8237.

Here, each channel can be programmed independently, and any one of the channels may be made individually active at any point of time. But the distinctive feature of this chip is that it provides many programmable controls and dynamic reconfigurability attributes, which eventually enhance the data transfer rate of the system remarkably.

Many CPUs like those of the **MC 68000** series have no internal mechanism for resolving *multiple DMA requests*; this must be done by external logic. The DMA controller **68450 chip** contains ***four copies*** of the basic DMA controller logic that enables the 68450 to carry out a sequence of DMA block transfers *without reference* to the CPU. When the current data count reaches zero, a DMA channel that has been programmed for chained DMA transfer (as mentioned in Bullet 4) fetches the new value of DC and IOAR from a memory region (MR) that stores a set of DC–IOAR pairs. A special memory address register in every DMA channel holds the base address of MR.

DMA is subsequently accepted as a standard approach, commonly used in all ***personal computers, minicomputers,*** and ***mainframes*** for carrying out I/O activities.

5.3.3.5 Different Transfer Types

Under DMA control, data can be transferred in one of the several following ways:

1. **DMA block transfer**: This type transfers a sequence of data word of arbitrary length in a *single continuous burst* when the DMA controller is the master of the system bus. Block DMA supports the maximum I/O data transmission rate, but it may cause the CPU to remain inactive for relatively longer periods. Auto-initialization may be programmed in this mode. This DMA mode is particularly required by secondary memory devices like magnetic disk drives where data transmission cannot be stopped or slowed down without loss of data, and block transfers are the norm.
2. **Cycle stealing**: This approach allows the DMA controller to use the system bus interspersed with CPU bus transactions while transferring long blocks of I/O data by a sequence of DMA bus transactions. During these cycles, the CPU will have to wait to get the control of the bus because DMA always has a higher bus priority than the CPU, as I/O devices cannot tolerate delays. The process of taking bus cycles away from the CPU by a DMA controller, or by way of forcing the processor to temporarily suspend its operation, is called ***cycle stealing***. Cycle stealing not only reduces the maximum I/O transfer rate, but also reduces the interference by the DMA controller in the CPU's activities. It is possible to completely eliminate the interference by designing the DMA interface so that the bus cycles are to be stolen only when the CPU is not actually using the system bus. This is known as *transparent DMA*. Thus, by varying the degrees of overlap between CPU and DMA operations, it is possible to accommodate many I/O devices having different data-transfer characteristics.
3. **Demand transfer**: In this mode, the device continues transfers until DC (count) is reached zero, or an external condition (end of process) is detected, or DMA REQUEST signal goes inactive.
4. **Cascading**: In this mode, more than one 8237 can be connected level-wise together to the host 8237 to provide more than four DMA channels. The priorities of the DMA requests, however, may be preserved at each level.

5.3.3.6 Implementation Mechanisms: Different Approaches

The DMA mechanisms can be implemented in a variety of ways.

1. Here, the DMA module and the I/O devices individually share the system bus with the CPU and memory as shown in Figure 5.6. The DMA module is acting here as a surrogate (a substitute) for the CPU. The DMA controller uses PIO to exchange data between memory and an I/O device. Like the CPU-controlled PIO, this approach also requires two bus cycles for each transfer of a word. This configuration, while it may be looked inexpensive, is clearly inefficient also;
2. This drawback of consuming more bus cycles at the time of data transfer can be reduced substantially if the DMA and I/O functions are integrated. This means that there is a separate path between the DMA module and one or more I/O modules that does not include system bus. This is shown in Figure 5.7. Here, the DMA logic may be a part of an I/O module or may be a completely separate module that controls one or more I/O modules;
3. The approach already mentioned in (2) can be modified one step further by connecting I/O modules to the DMA module using an I/O bus. The transfer of data between the DMA and I/O modules can then take place off the system bus and the system bus will be used by the DMA module only at the time of exchanging data with the memory. This approach will reduce the number of I/O interfaces in the DMA module to one, and at the same time offers an easily expandable configuration. Figure 5.8 shows a schematic design of this approach.

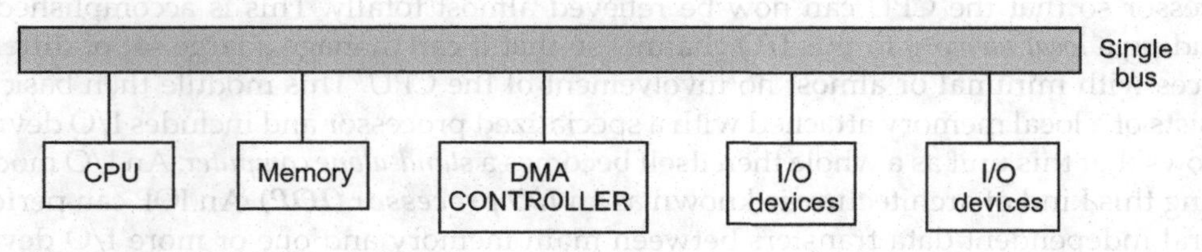

FIGURE 5.6
Single bus: DMA detached from 1/0.

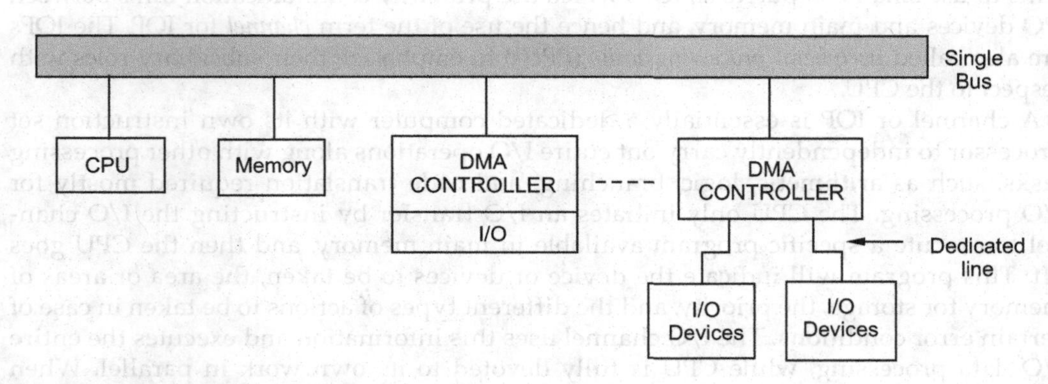

FIGURE 5.7
Single bus: DMA-I/O integrated.

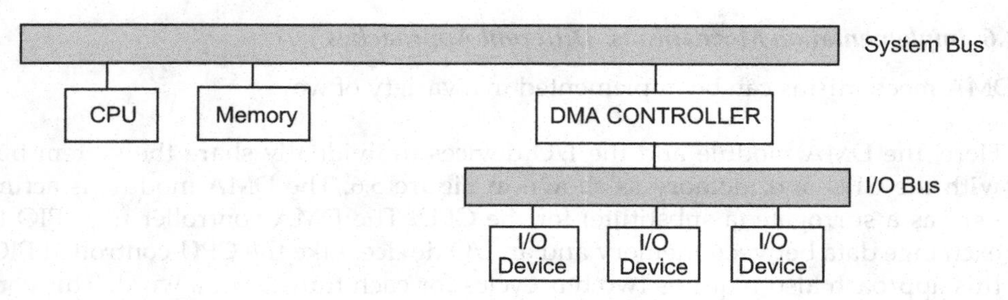

FIGURE 5.8
DMA-I/O with separate I/O bus.

5.3.4 I/O Processor (I/O Channels)

5.3.4.1 Introduction

While the introduction of a ***DMA*** controller in the I/O nodule is a radical breakthrough, it is after all not able to totally freeing the CPU. Moreover, DMA sometimes uses many bus cycles at a time, as in the case of a disk I/O, and during these cycles, the CPU will have to wait for bus access (as DMA always has a higher bus priority) that summarily restricts the performance to attain the desired level. Further development is thus targeted in quest of an enhanced I/O module so that this modified I/O module could control the entire I/O operation on its own, setting the CPU totally aside. This type of I/O module that almost fully relieves the CPU from the burden of I/O execution is often referred to as an ***I/O channel***. The final and ultimate approach is then to convert this I/O channel to a full-fledged processor so that the CPU can now be relieved almost totally. This is accomplished by including a *local memory* to this I/O channel so that it can manage a large set of different devices with minimal or almost no involvement of the CPU. This module then basically consists of a local memory attached with a specialized processor and includes I/O devices. It shows that this unit as a whole then itself becomes a *stand-alone computer*. An I/O module having this kind of architecture is known as an ***I/O processor (IOP)***. An IOP can perform several independent data transfers between main memory and one or more I/O devices without recourse to the CPU. Usually, an IOP is connected to the devices it controls, by a separate bus system called the ***I/O bus*** or ***I/O interface***. It is not uncommon for larger systems to use small computers as IOPs which are primarily communication links between I/O devices and main memory, and hence the use of the term *channel* for IOP. The IOPs are also called *peripheral processing units* (*PPUs*) to emphasize their subsidiary roles with respect to the CPU.

A channel or IOP is essentially a dedicated computer with its own instruction set processor to independently carry out entire I/O operations along with other processing tasks, such as arithmetic, logic, branching, and code translation required mostly for I/O processing. The CPU only initiates an I/O transfer by instructing the I/O channel to execute a specific program available in main memory, and then the CPU goes off. This program will indicate the device or devices to be taken, the area or areas of memory for storage, the priority, and the different types of actions to be taken in case of certain error conditions. The I/O channel uses this information and executes the entire I/O data processing, while CPU is fully devoted to its own work in parallel. When I/O activity is over, the channel interrupts the CPU and sends only all related necessary information. Traditionally, the use of I/O channels has been observed to be associated with mainframe or large-scale system attached with a large number of peripherals (disks and tape storage devices), which are used simultaneously by many different users in multitasking as well as in on-line transaction processing (OLTP) environment handling bulk volume of data. As the development of chip-based microprocessors has dramatically progressed, the use of I/O channels has now extended to minicomputers and even to microcomputers. However, the fully developed I/O channel is best studied on the mainframe system, and possibly the best-known example in this regard is the flagship IBM/370 system.

5.3.4.2 I/O Channel

The IOP in the IBM 370 system is commonly called a *channel.* A typical computer system may have a number of channels, and each channel may be attached to one or more similar or different types of I/O devices through I/O modules. Three types of channels are in common use: a **selector channel,** a **multiplexor channel,** and a **block multiplexor channel** (a hybrid of features of both the multiplexor and selector channels). The interface being used from an I/O module (channel) to a device (i.e. a device controller along with a device) is either a *serial* or a *parallel.* Although a *parallel interface* is traditionally a common choice for high-speed devices, but with the emergence of next-generation advanced *high-speed serial interfaces,* parallel interfaces have eventually lost their inherent importance to a considerable extent, and hence, are found to be much less common. However, the I/O channel is best implemented in the mainframe system, and possibly the well-known example in this regard is the flagship IBM/370 system.

A brief detail of I/O channel along with its different types, and its implementation in IBM/370 system, is given with figure in the website: http://routledge.com/9780367255732.

5.3.4.3 I/O Processor (IOP) And Its Organisation

The I/O channel has been finally promoted to a full-fledged IOP using a mechanism (already described in the "Introduction", Section 5.3.4) to make the CPU almost totally free from any I/O activities. The handshaking between the CPU and IOP at the time of establishing communications may take different forms depending mostly on the particular configuration of the computer being used. However, the memory unit in most cases acts as a mediator providing *message centre* (input–output communication region (IOCR)) facilities where each processor leaves some information for the IOP to follow. This is one form of *indirect handshaking.* The *direct handshaking* between CPU and IOP is generally done through dedicated control lines. Standard DMA or bus grant/acknowledge lines are also used for arbitration of the system bus between these two processors. However, Figure 5.9 illustrates here a schematic block diagram of a representative system containing an IOP. A sequence of steps is then required to be followed at the time of CPU–IOP interaction and communication for needed information exchange.

During IOP operation, the CPU is free and executes its own tasks with other programs. There may be a situation when the CPU and the IOP both compete with each other to get simultaneous memory access, and hence, the IOP is often restricted to have only a limited number of devices so that the number of memory accesses can be minimized. In the case of operation of a slower device, this may even lead to a situation of memory-access saturation, since I/O operations use DMA, and the CPU may then have to wait during this transfer, which may cause a notable degradation in CPU performance.

All the **Intel CPUs** have explicit I/O instructions to *read* or *write* bytes, words, or longs. These instructions specify the I/O port number desired, either directly as a field with the

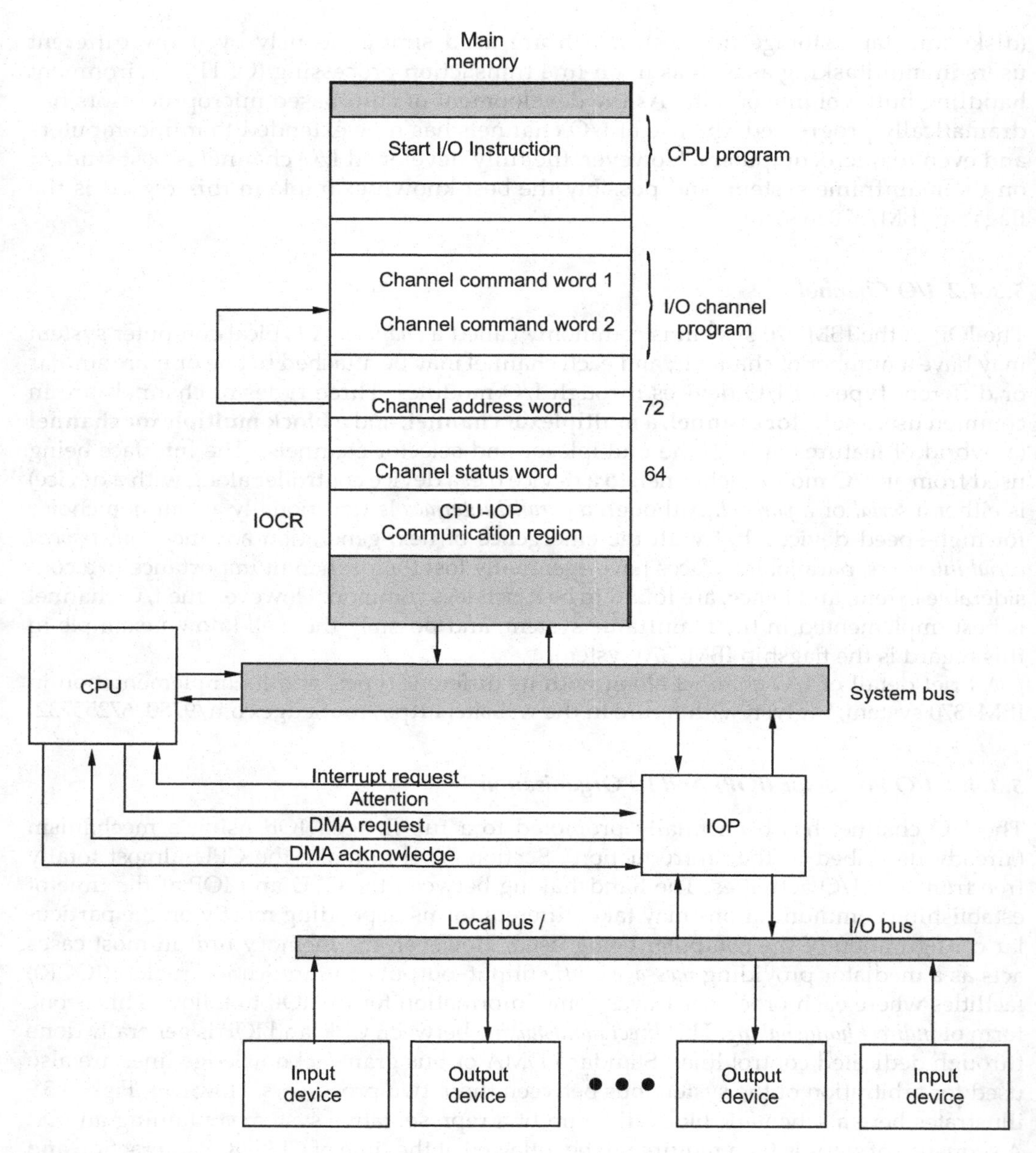

FIGURE 5.9
Block diagram of a representative system containing an IOP.

instruction or indirectly using the register DX to hold it. In addition, of course, DMA chips (Intel 8257/8237) are frequently used to relieve the CPU from handling I/O burden. None of the **Motorola chips** have I/O instructions. It is expected that the I/O device register will be addressed via memory mapping. Here too, DMA is widely used.

A brief detail of IOP along with its working is given with figure in the website: http://routledge.com/9780367255732.

5.4 Bus, Bus System and Bus Design

Three fundamental resources of a computer, namely the CPU, memory, and I/O, are interconnected via a collection of parallel wires called a *bus*. A bus is actually a communication medium connecting two or more resources. A key characteristic of a bus is that it is a *shared* transmission medium, on which multiple devices can be connected, and a signal issued by any one of these attached devices is available for reception by all other devices attached to the bus. A bus may be of a single line, and a sequence of binary digits (bits) can then be transmitted *serially* one bit at a time across this single line. A bus may also consist of multiple lines, and these bus lines can be used in *parallel* to transmit the same binary digits all at a time. For example, a data of 8 bits can be transmitted over an 8-bus line simultaneously. However, a number of varied types of buses were in widespread use from very early days introduced by different vendors, being employed by different systems built with a diverse spectrum of chips. A few notable ones from early days are IBM PC/AT bus, EISA (Intel 80386), Multibus (Intel), VME bus (Motorola 68000 series), Omnibus (PDP-8), Unibus (PDP/11 series), PCI bus, SCSI bus, USB, etc. Standardization in this area seems very unlikely. Recent developments in modern bus technologies, however, give rise to the introduction of much faster and more versatile buses, such as *FireWire serial bus* and *InfiniBand* serial bus.

5.4.1 Bus Structure

A bus that connects major computer components such as CPU, memory, and I/O is called a *system bus*. It usually consists of, typically, 50–100 parallel copper lines providing different functions etched onto the motherboard, with connections spaced at regular intervals for plugging in memory, I/O, and other add-on cards. Although there are many different bus designs, on any bus, these lines can be classified into *three basic functional groups*: **data lines, address lines**, and **control lines**. In addition, there may be power distribution lines that supply power to the attached modules. Each of these groups, however, carries various types of signals indicating respective operations to be performed.

5.4.2 Bus Arbitration

Normally, in a computer, the CPU is the *bus master* having control of the bus most of the time. In reality, an I/O module may become bus master while reading or writing directly to memory. A co-processor may also need to become bus master at a certain point of time. Since only one unit at a time is allowed to gain control of the bus (to be a bus master), it is thus essential to have some mechanism in place, to prevent the chaos at the time of simultaneous access of the bus by different resources. This mechanism is known as *bus arbitration*.

5.4.3 Bus Protocol

At the time of designing a resource (e.g. a processor), designers have the liberty to use any kind of bus they want *inside* the chip, but the situation is different when a third party designs a circuit board in which the said resource will be placed on the system bus. So, they must abide by certain *well-defined rules* concerning bus operation, and all the characteristics of the devices that would be attached to it, as well as other areas relating to bus usage. These rules are called ***bus protocol***.

5.4.4 Bus Design Parameters

Although a variety of different bus implementation techniques exist, there are a few basic parameters or design elements that serve to classify and differentiate buses. The key elements to be considered are as follows:

a. **Type**: This element may be (i) *dedicated* and (ii) *multiplexed* (time multiplexing);
b. **Method of arbitration**: This element may be (i) centralized and (ii) distributed;
c. **Timing**: This element may be (i) synchronous and (ii) asynchronous;
d. **Bus width**: This element may be (i) address bus and (ii) data bus;
e. **Data transfer type**: This element may be (i) read, (ii) write, (iii) read–modify–write, (iv) read-after-write, and (v) block.

5.4.5 Bus Interfacing: Tri-State Devices

A bus line represents a logic path with a potentially very large fan-in and fan-out having many devices of different types connected to it. One of these devices is *active* (*master*) at any instant and can *initiate* bus transfers, whereas others are *passive* (*slaves*) and *wait* for requests. When the CPU orders a disk controller to read or write a block, the CPU is acting as a *master* and the disk controller is acting as a *slave*. However, later on, the disk controller may act as a *master* when it commands the memory (*slave*) to accept the words it is reading from the disk drives. Several combinations of such master/slave relationships may feature with resources. Some typical configurations are shown in Table 5.1. It is interesting to note that *in no situation can memory ever be found to be a bus master.*

The binary signals that computer resources often yield are mostly not strong enough to power a bus, especially, if the bus is relatively long or has many devices on it. For this reason, most bus masters are connected to the bus by a chip called a ***bus driver***, which is essentially a ***digital amplifier***. Similarly, most slaves are connected to the bus by a ***bus receiver***. For devices that can act as both master and slave, a combined chip called a *bus transceiver* is used. These bus interface chips are often **tri-state devices** to allow them to float (disconnect) when they are not needed, or are hooked up in a somewhat different way called ***open collector***, which provides the desired effect. Tri-state devices are those which can output 0, 1, or none of these, z (open circuit). When two or more devices on an open collector line assert the line at the same time, the result is the Boolean OR of all the signals. This arrangement is often called ***wired-OR***. On *most buses, some of the lines are tri-state and others which need the wired-OR property are open collector.* If two or more resources want to become bus master at the same time, the *bus arbitration* is required to prevent this chaos by way of promptly making or breaking the connection of the competing resources with the

TABLE 5.1

Different Combinations of Master/Slave Configuration

Master	Slave	Example
CPU	Memory	Fetching instruction and data
CPU	I/O	Initiating data transfer
CPU	Co-processor	Handling of floating-point instructions
I/O	Memory	DMA
Co-processor	Memory	Fetching operands

bus using a *tri-state device* (*non-inverting buffer*) when it is required. Tri-state devices when used in the design of shared buses exhibit many distinct advantages, and all these have been already discussed in detail in Chapter 2.

A brief detail of bus organisation using tri-state buffer is given with figure in the website: http://routledge.com/9780367255732.

5.4.6 Some Representative Bus Systems of Early Days

Some notable and worthy bus systems introduced by reputed giant vendors of early days with their own specifications are IBM PC/AT Bus (for PC with Intel 80286), IBM PS/2, The EISA (for PC with Intel 80386 and onwards), Intel MULTIBUS (Multibus I and Multibus II), Motorola VME bus, DEC PDP UNIBUS, DEC VAX SBI, and MASSBUS.

A more detailed description of each of these is, however, given in the website: http://routledge.com/9780367255732.

5.4.7 PCI (Peripheral Component Interconnect): Local Bus

New approaches in the bus design gradually evolved, with the emergence of ***PCI***, to make full use of the constantly emerging advanced technology, and also to fulfil the ever-increasing needs of the computing environment. It was developed by Intel (1992) for the ***Pentium processor*** as a truly processor-independent, low-cost, and high-performance local bus with a bus speed up to 133MHz for interconnecting chips, expansion boards, and processor/memory subsystems, thereby replacing earlier bus architectures such as EISA,VL, and Micro Channel. Devices connected to the PCI bus appear to the processor as if they were directly connected to the processor bus, and they are assigned addresses in the memory address space of the processor. PCI provides a general-purpose set of functions, and the design was able to accommodate emerging high-speed devices (like disks) as well as to support both single and specialized needs of multiple-processor-based systems having a variety of microprocessor-based configurations. PCI was later adopted as an industry standard administered by the *PCI Special Interest Group* ("PCI SIG"), extending its definition to also characterize it a standard expansion bus interface connector for add-in boards, and as such, it is summarily still maintaining its position as an industry standard even after about two decades since its inception.

A schematic diagram of a typical use of the PCI bus with its interconnection to different resources in a single-processor system is shown in Figure 5.10a. Here, a combined DRAM controller and ***bridge*** to the PCI bus provides a tight coupling with the processor that can deliver data at high speeds. The bridge acts as a *data buffer* that negotiates the speed disparity between the PCI bus and the processor's I/O capability. In a *multiprocessor system*, the main system bus supports processors, caches, main memory, and the PCI bridges. These PCI bridges may be connected with one or more PCI configurations on the other side, as shown in Figure 5.10b. The presence of the bridge not only keeps the PCI bus independent of the processor speed as usual, but also provides a rapid transmission of data. In addition, a ***PCI-to-PCI bridge*** mechanism has been defined by PCI SIG where bridges are ASICs (application-specific integrated circuits) that electrically isolate two PCI buses while allowing bus transfers to be forwarded from one bus to another. Each bridge device has a *primary* PCI bus and a *secondary* PCI bus. Multiple bridge devices may be cascaded to create a system with many PCI buses. In some processors, like **Compaq Alpha**, the PCI processor bridge circuit is built on the processor chip itself, thereby further simplifying system design and packaging.

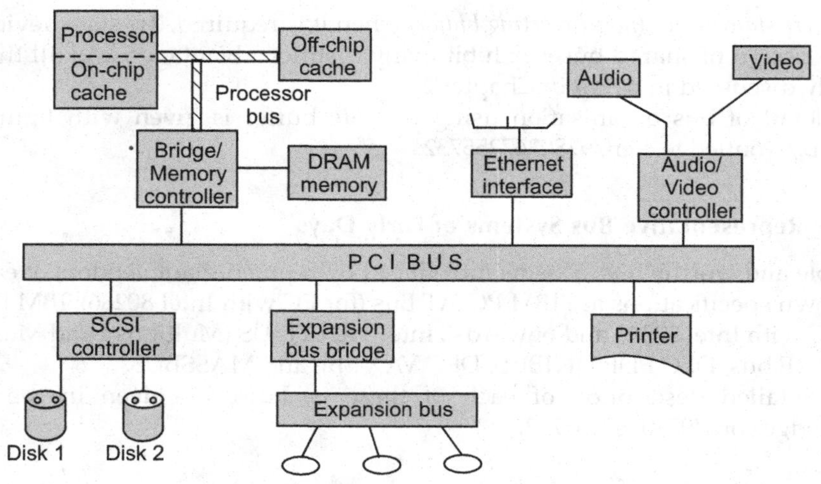

(a) Use of PCI bus in a typical standalone desktop system

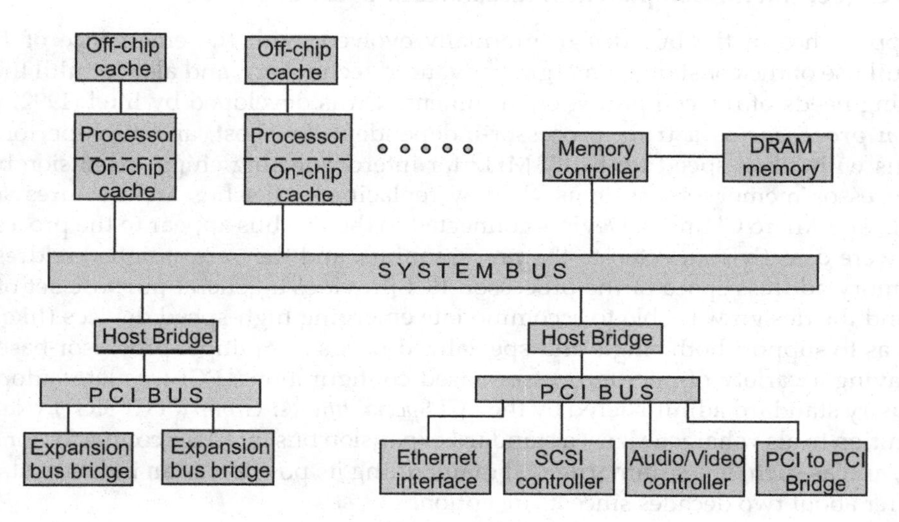

(b) Use of PCI bus in a typical Server system

FIGURE 5.10
A schematic representation of PCI configuration: (a) use of PCI bus in a typical stand-alone desktop system and (b) use of PCI bus in a typical server system.

One of the *salient features* that the PCI pioneered is its *plug-and-play* capability for connecting I/O devices. When a new device needs to be added to the system, it requires the device interface board only to be plugged in to one of the slots connected with the bus provided in the motherboard. The software then entirely bears the rest of the responsibilities for its operation. PCI, however, requires very few chips to configure it as a 32-bit or a 64-bit bus, and easily supports other buses attached with it. PCI interface is available over a wide range of I/O devices and is in use in systems based on many other processor families, including **SUN**.

Some of the key activities relating to the major characteristics of PCI bus are described one after another in brief in the website: http://routledge.com/9780367255732.

5.4.8 SCSI (Small Computer System Interface) BUS

SCSI (pronounced "scuzzy") has its origin based on an open system interface called *SASI* (*Shugart Associates System Interface*) developed in 1981 by **Shugart Associates** in collaboration with **NCR Corporation**, and then was modified and subsequently ratified by ANSI to announce it as ***SCSI***. It is a *fast communication bus* that allows the computer to connect multiple devices, which may include *hard drives, CD-ROM/RW drives, scanners, printers, tape drives* etc.. While its earlier versions can transfer 8-bit data (narrow bus) and then 16-bit data (wide bus) at a time, the data transfer rate in its recent version is intended to be up to 640 megabytes/s or even more, and yet a higher rate is still anticipated. The maximum transfer rate again depends on the length of the cable being used and also on the number of devices being connected. The maximum capacity of the bus is 8 devices for a narrow bus and 16 devices for a wide bus. There are ***three*** basic specifications of SCSI, namely SCSI-1, SCSI-2, and then SCSI-3, which itself is again varied on the specifications of *SCSI parallel interface* (SPI) being used by the SCSI devices while communicating with each other. SCSI has the *plug-and-play* capability with *three* primary components, namely *the controller, the device*, and *the cable*. The controllers are connected to the SCSI bus using *daisy chaining* (*fixed priority*). A ribbon cable is used inside the computer for SCSI devices to attach to the SCSI controller. This cable has a single connector at each end and may have one or more connectors along its length to attach other devices in daisy chaining. Each device with built-in *adapter* typically has two SCSI connectors: one is used to connect to the previous device and the other is used to connect to the next device, thereby forming a chain that can be thought of as a tiny *local area network* (*LAN*) in which the SCSI controller looks like a *network router*, the adapter is comparable to the *Ethernet* card, and each SCSI device is like a computer on this network. As usual, SCSI bus is observed to have several distinct *advantages* and also suffers from certain potential *drawbacks*.

A brief detail of these different types of SCSI and its operations is, however, given in the website: http://routledge.com/9780367255732.

5.4.9 Universal Serial Bus (USB)

USB – an outcome of a collaborative effort of several computer and communications companies, including Hewlett–Packard, Intel, Compaq, Microsoft, Philips, Lucent, and Nortel Networks, introduced by USB Implementers Forum (USB-IF) in 1995/1996 with **USB 1.0** – is a simple and low-cost mechanism to connect almost all types of devices to the computer system. The USB has rapidly gained a wide acceptance in the market place, and with its numerous attractive features and capability, it immediately becomes the ultimate interconnection method of choice for most computer devices. Many small-sized versions of these connectors are also introduced especially for embedded products. Recent USB ports are used for many other activities, thereby ultimately putting an end to all hassles faced by the user while attempting to connect devices to a computer system either on *serial ports* or on *parallel ports*. The USB gives a single, standardized, easy-to-use way through a "plug-and-play" mode of operation to dynamically connect up to **127 devices** to a computer online, and also for device-to-device communications. As a result, almost every peripheral, including network adapters, is now made with USB facility, or comes with a USB version. Many USB devices come with their own built-in cable with an "A" connection on it, or have a socket on it that accepts a USB "B" connector. By using different connectors on the upstream ("A" connectors) and downstream ("B" connectors) end, it is impossible to ever get confused.

With USB, the computer acts as the **host** and up to *127 devices* can be connected to the host, either directly or by way of USB hubs. The **USB 2** interface has a maximum data rate of up to **480 Mbps**, and the **USB 3** interface has an even larger data transfer rate and the maximum data transfer rate is up to **5 Gbps**. A USB cable consists of **four wires**: two wires are used for power (+5 V and ground) and the other two wires are used to carry data. On the power wires, the computer can supply up to 500 milliamps of power at 5 V, and the low-power devices (like mice) can draw their power directly from the bus. High-power devices (such as printers and scanners), however, have their own power supplies and draw minimal power from the bus. Hubs can have their own power supplies to provide power to devices connected to the hub. USB devices are **hot-swappable (plug-and-play)**, auto-detected by operating system (OS), and can be connected and disconnected at any point of time on-line. Many USB devices can be put to **sleep** by the host computer when the computer enters a power-saving mode. In addition, all audio/video devices (microphones, speakers, and web cam), either externally connected or built-in, are used in modern computers as standard input devices. They require an analog-to-digital (A/D) converter and a digital-to-analog (D/A) converter while transferring data as input and output, respectively. A sampling process is thus needed inside the system for both input and output that yields a continuous stream of digitized samples. Such a data stream is called *isochronous*, which means that successive events (representing 4 bytes of data) are separated by equal periods of time. USB supports this type of transfer of such isochronous data, thereby yielding a data rate of about 1.4 megabits/sec. Different signalling schemes, however, are used for different speeds of transmission, but for high-speed links, differential transmission is used.

USB is a low-cost ***serial bus***, free from any invasion of *data skew*, in which *clock and data information* can be encoded together and transmitted as a single signal, thereby providing a high data transfer bandwidth by using a high clock frequency. As such, it offers *three types* of bit rates ranging from **1.5 Mbps** to **5 Gbps** (see "USB features") to accommodate the various needs of different devices attached with this bus.

When a device (or USB root hub) is attached to a host computer, it is connected to the system bus (I/O bus) with an arbitrarily assigned a local 7-bit address for its identification by USB host software, and is then recorded for its own use. When a device is *taken off* (i.e. disconnected or powered off), all identification information is then deleted from its own record. The host (computer) always polls each of its USB port and hub periodically to update its status information in order to learn, if by this time, any other new devices may have been added or disconnected. USB software provides bi-directional communication links (channels), called *pipes*, between the application software and the I/O devices. The *pipe* actually connects an I/O device to its specific device driver, and the connection is established when a device is plugged and is recognized by the USB software. The information being transferred over USB is arranged in the form of packets, and each one consists of one or more bytes of information arranged by one or more fields containing different kinds of information.

USB 2.0, a ***high-speed USB***, an upgrade of USB 1.1 (an advanced version of USB 1.0) introduced in 1998, was released in April 2000 that provides additional bandwidth for multimedia and storage applications with a data transmission speed 40 times faster than USB 1.1. In 2002, USB 2.0 was revised to provide three speed modes (1.5, 12, and 480 megabits per second) that support low-bandwidth devices, such as keyboards and mice, as well as high-bandwidth ones, such as high-resolution webcams, scanners, printers, and high-capacity storage systems. As a result, it has supported the PC industry in forging ahead with the development of next-generation PCs, and user applications with improved functionality and increased productivity. It, however, enables the user to run multiple PC applications at once or several high-performance peripherals simultaneously.

USB 3.0, similar to its predecessor USB 2.0, offers significantly improved data transfer rates compared to all its ancestors. Even though it was announced in late 2008, consumer devices were not available to totally make use of their potentials until the beginning of 2010. The USB 3.0 interface specifies the transfer rates up to 5 Gbit/s (625 MB/s) or 596 Mbit/s, compared to USB 2.0's 480 Mbit/s (60 MB/s). All USB 3.0 devices are downward compatible with USB 2.0 ports and support full-duplex operation. Computers with USB 3.0 ports are becoming very popular and common. The USB 3.0 port expansion cards are available to upgrade older systems, and many newer motherboards feature two or more USB 3.0 jacks. Even though the USB 3.0 interface allows extremely high data transfer speeds, as of 2011, many peripherals, including most USB 3.0 flash drives, have not been able to utilize the full speed of the USB 3.0 interface due to the limitations inherent in their memory controllers.

A brief detail of almost everything about USB is given in the website: http://routledge.com/9780367255732.

5.4.10 FireWire Serial Bus

Both the electronic technology and the computer architectures with allied resources are continuously advancing towards the next generation providing enormous processor speeds of an order of gigahertz with comparable memory speed, and even small storage devices are now capable of holding several gigabytes. Application domain also dramatically progressed with a regular inclusion of audio/video and images to support constantly increasing user density of diverse disciplines. Meanwhile, computers are continuously getting even smaller in size, but with added demand for more I/O support. Mobile telephones, handheld pocket-sized computers, and other tiny gadgets really not only provide very little room for connectors to be installed, but also place soaring demand for high data rates so as to handle audio/video and images. Continuous research in this area thus went on, which ultimately culminated in the introduction of a high-performance serial bus, commonly known as *FireWire* that conforms to an IEEE standard 1394. Although FireWire was named and initiated by Apple in 1986, its development was coordinated by the IEEE P1394 Working Group, and major contributions were also made by engineers from Texas Instruments, Sony, Digital Equipment Corporation (DEC), IBM, and INMOS/SGS Thomson (now known as *ST Microelectronics*). This 1394 interface is also known by the brands **i.LINK** implemented by Sony, and **Lynx** introduced by Texas Instruments.

FireWire is a serial bus that fully supports both *isochronous real-time* data transfer and *asynchronous* applications as well. It is widely accepted because it fulfils its ultimate intention to replace parallel SCSI in many applications, owing to its high speed, a simplified and more adaptable cabling system using lesser number of wires with less wider and less-expensive connectors (space and cost) for being serial, ease of implementation using adequately increasing cable length with almost no electrical noise, and above all, being relatively less error-prone with reasonably lower implementation costs. In fact, it is now well accepted not only in the computer arena, but also finds its wide usage in consumer electronics products such as digital cameras; DVD players/recorders; and televisions for audio, video, and images, including text, that are sent and received from and to digitized sources. FireWire is comparable with USB and is sometimes preferred over the more common USB for its greater effective speed and power distribution capabilities, although often these two technologies are considered together. Benchmarks carried with Apple Mac OS show that the sustained data transfer rates are higher for FireWire than for USB 2.0, but lower than for USB 3.0, and are observed to be more varied on Microsoft Windows. However, the relatively expensive hardware needed to implement FireWire has proved it to be a failure in displacing USB mostly in low-end mass-market computer peripherals, where product cost is a major constraint. That is why, USB in this domain is still reigning with more market share.

Technical specifications: FireWire can handle various types of devices having numerous characteristics using a daisy chain topology (configuration) and also a tree structure allowing up to a total of 63 such devices connected off a single port (as opposed to the parallel SCSI's electrical bus topology). Various types of devices such as mouse, external disk drives, laser printer, and LAN hook-ups using different ports can now be replaced with only this single connector. It allows peer-to-peer device communication, like communication between a scanner and a printer, to take place without using system memory or the CPU. FireWire is what is known as ***hot plugging***, and provides automatic configuration and identification to set device IDs with no human intervention, and that also irrespective of the relative location of devices. All FireWire devices are identified by an IEEEUI-64 unique identifier, in addition to the well-known codes indicating the type of device and the protocols it supports. In a tree-structured topology, one of the nodes is elected as root node and always has the highest id, and the ids are assigned in an order of equivalently traversing the tree depth first and post-order during the execution of self-id process which happens after each bus resets. In FireWire configuration, there is as such no termination, and the system automatically performs a configuration function to assign addresses (ids). FireWire also supports multiple hosts per bus. A simple FireWire configuration as exhibited in Figure 5.11 confirms the truth of this statement.

FireWire is capable of safely operating vital systems providing both ***asynchronous*** and ***isochronous*** modes of transfer, and judiciously arbitrates bus control in allocating bandwidth to the devices, when multiple devices interact with the bus. In isochronous mode, variable size of data in a sequence of fixed-sized packets is transmitted at regular intervals using simplified addressing with no need for acknowledgement. It is important especially for critical devices, including control of the rudder, mouse operations, and data from temperature/pressure sensors in an aircraft that require continuous, guaranteed uninterrupted bandwidth. That is why, this mode, however, is always privileged in the bus compared to its co-owner competitor, asynchronous transfer. For example, 80% of the bus time in IEEE 1394 is usually reserved for isochronous cycles, leaving asynchronous data to use only a maximum of 20% of the bus time.

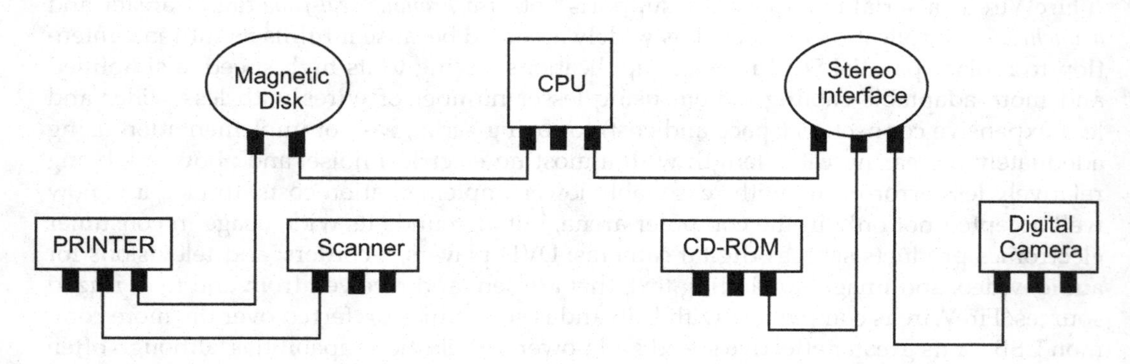

FIGURE 5.11
A representative configuration of a single FireWire.

The most common implementation of FireWire uses copper cable, which can be up to 4.5 m (15 ft) long and is more flexible than the most parallel SCSI cables. In its *six-conductor* or *nine-conductor* variations, it can supply up to 45 W of power per port at up to 30 V, allowing moderate-consumption devices to operate without a separate power supply. Last but not least, up to 1022 FireWire can be interconnected using bridges, thereby enabling a system to support as many peripherals as are usually needed.

Protocols: Unlike the other bus systems, FireWire standard specifies a set of ***three layers of protocols*** organised as stack to standardize the way in which the communication over the serial bus between the host and the devices can take place. They are *physical layer, link layer,* and *transmission layer.*

Physical layer converts binary data into corresponding electrical signals for numerous physical media with the defined data rates from 25 to 3200 Mbps as reported, and also provides an ***arbitration mechanism*** that decides which node gains control of the bus and at what time to transmit data. Each arbitration round lasts about 125 μs. The mechanism being used is of two forms; each form ensures that only one device will transmit data at any point of time. The simplest form is based on the *tree-structured* arrangement of the nodes (devices) in which during the round, the root node (device nearest the processor) sends a cycle start packet. All nodes requiring data transfer respond, with the closest node winning. After the node is finished, the remaining nodes take turns in order. This goes on until all the devices have used their portion of the 125 μs. The other form is a special case that follows a *daisy chaining* (fixed priority) *mechanism.* Both these mechanisms are again supplemented by two additional functions: *fairness arbitration* (FCFS; first come first serve) and *urgent arbitration* (priority-driven). With isochronous transfers, high-priority isochronous nodes, however, are permitted to control the bus for duration to the extent of even 75%–80% of the available bus time.

Link layer defines and describes the transmission of data in the form of packets using ***asynchronous*** and ***isochronous*** modes. In the event of *asynchronous*, a variable amount of data and data of several bytes related to the transmission layer (to be discussed next) is transferred as a packet to an explicit address, and a related acknowledgement is returned. In case of *isochronous* transmission, a variable amount of data is transferred in a sequence of fixed-size packets transmitted at regular intervals using simplified addressing with no need for acknowledgement. For devices that deal with data in a regular fashion, like digital audio/video, isochronous transmission is found very conducive, and ensures that data can be delivered within a specified time interval with a guaranteed data rate.

Transmission layer defines a request–response protocol that hides all the underlying lower-layer details of FireWire from applications which reside at a relatively higher level.

We stop at this point and will not enter into any further details on this subject which is at present outside the scope of this discussion. Interested readers, however, can go through (Anderson,'98) and (Wickelgren,'97) for more information and a better understanding of this topic.

5.4.11 InfiniBand

InfiniBand is essentially a *high-speed link* having especially typical I/O specifications for data flow among processors and intelligent I/O devices with large storage configurations, mainly for use in *cluster system architecture* (discussed in detail in Chapter 10) in which a number of computer systems are to be connected together, which ultimately provides a *single-system image.* This product is eventually an outcome of the merger of two competing projects aimed at the high-end server market: ***future I/O*** (backed by HP, Cisco, Compaq, and IBM) and ***next-generation I/O*** (developed by Intel, Microsoft, and Sun, and supported by a number of other companies). InfiniBand was originally envisioned as a comprehensive interface only for *networking of storage area* with low latency, high bandwidth, low-overhead interconnect for commercial datacentres, although it might perhaps only connect servers and storage to each other, while leaving more local connections to other protocols and standards like PCI.

InfiniBand is basically a ***switched fabric*** communication link using *switch-based architecture* that enables servers, remote storage, and other network devices to be attached in a central fabric of switches and links. It can connect up to 64000 servers, storage systems, and networking devices with high-performance computing, which can be considered good enough for creating a gigantic computing environment. Its distinctive features include *high throughput, low latency, scalability, quality of service,* and *failover.* InfiniBand host channel adapter (HCA) and Network switches (Figure 5.12) are manufactured by Mellanox and Intel (acquired Qlogic's InfiniBand business in January 2012).

The InfiniBand has been observed to possess several ***distinct advantages*** over other traditional I/O interconnects, including the most popular PCI. While each of the others requires its own specific I/O interface for interconnection, InfiniBand does not need to have any basic I/O interface hardware inside the host (server) system. With InfiniBand, networking of storage, remote storage, and interconnections between servers are accomplished by attaching all devices to a central fabric of switches and links. Figure 5.12 illustrates a representative InfiniBand architecture with its key elements. Permitting the server (host) to have removed the I/O interface hardware from its chassis makes it more flexible and easily scalable, as independent nodes may be now added at will as and when needed. In this way, it helps to attain greater server density and enough room now is available inside the server (host) that can be used otherwise for its own useful purposes. Another *key feature* of InfiniBand is that it reduces the location constraint of the device with respect to the CPU, and that the other interconnects, including PCI, often impose. In PCI, the devices must be located within a range of the order of centimetres from the CPU; the

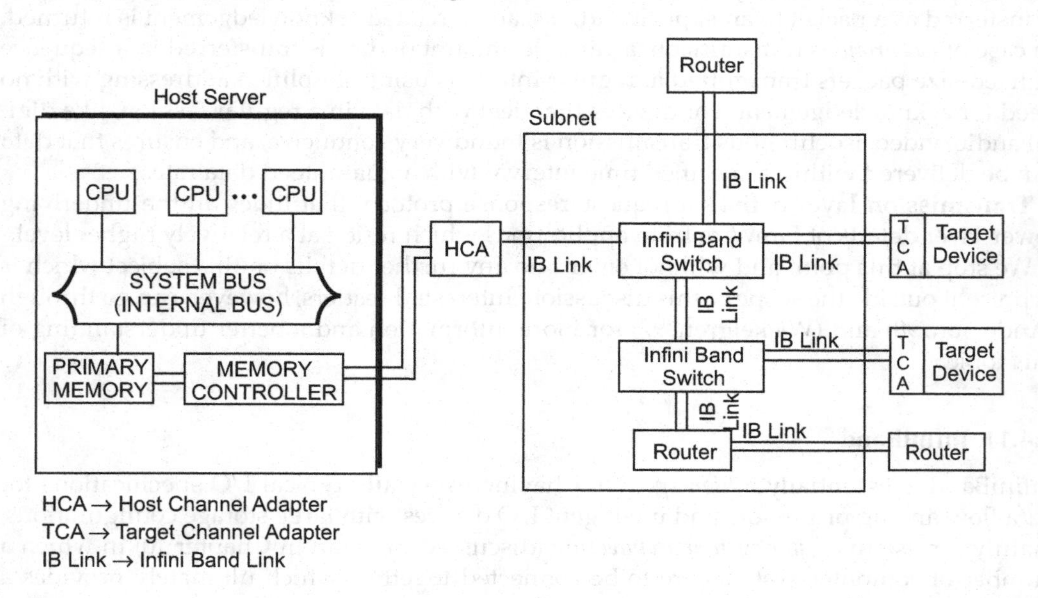

FIGURE 5.12
A representative arrangement of InfiniBand switch fabric.

InfiniBand channel design enables the I/O devices to be placed up to 17 m away from the host (server) using copper wire, up to 30 m using *multimode optical fibre,* and up to 10 km with *single-mode optical fibre.* Yet, the data transmission rate being achieved is as high as 30 Gbps (against up to 4 Gbps with PCI).

Basic components: The basic components that constitute the core architecture of InfiniBand, as shown in Figure 5.12, are HCA, *target channel adapter* (TCA), *InfiniBand switch,* and *InfiniBand links* (IB-Link), *subnets,* and *routers.*

HCA connects the server to an InfiniBand switch. Each server contains a *HCA* and is attached to the server at a memory controller with access to the system bus and controls traffic between the processor and memory, and between HCA and memory. It exploits DMA mechanism to directly read and write memory. *TCA* is used to connect storage systems, routers, and other peripherals to an InfiniBand switch. Each peripheral has a *TCA,* which is an intelligent device that handles all I/O functions without any intervention of the server's processor. All transmissions begin or end at a *channel adapter. InfiniBand switch* enables the establishment of point-to-point physical connections to numerous devices and transfers traffic from one link to another desired endpoint. Servers and devices communicate through their adapters, via the switch. The switch is an intelligent device that discovers all TCAs and HCAs in the fabric with the help of a control protocol and assigns logical addresses to each. This is done without interrupting or involving the server's processor. *Link* (an IB-Link) is a *serial link* that exists between a switch and a channel adapter (HCA and TCA), or between two switches to connect them. *Subnet* is essentially a domain consisting of one or more interconnected switches and the links that connect other devices to those switches. Although Figure 5.12 shows a simple subnet containing only two switches, more complex subnets may be required when a large number of devices are needed to be interconnected. Subnets provide *broadcast* and *multicast* transmissions that are confined within the domain of the subnet. *Router* connects multiple InfiniBand subnets, or connects an InfiniBand switch to a network, such as LAN, wide area network (WAN), or storage area network (SAN).

InfiniBand latency: The single data rate (SDR) switch chips have a latency of 200 ns, *double data rate* (DDR) switch chips have a latency of 140 ns , and *quad data rate* (QDR) switch chips have a latency of 100 ns . The end-to-end latency range, as reported, spans from 1.07 μs MPI latency (Mellanox ConnectX QDR HCAs) to 1.29 μs MPI latency (Qlogic InfiniPath HCAs) to 2.6 μs (Mellanox InfiniHost DDR III HCAs). InfiniBand also provides RDMA (remote direct memory access) capabilities for low CPU overhead with a latency, as reported, is less than 1 microsecond (Mellanox ConnectX HCAs).

InfiniBand protocols: InfiniBand standard specifies a layered protocol architecture that consists of a set of ***four layers of protocols,*** namely *physical layer, link layer, network layer,* and *transport layer,* to standardize the communication between the hosts and the devices.

Physical layer: Its specification defines and describes the physical media, including copper and optical fibre, as well as three types of *link speeds,* namely 1X, 4X, and 12X, that provide unidirectional signal transmission rates. For each type of these link speeds, this layer also specifies *five data rates,* namely SDR, DDR, QDR, *fourteen data rate* (FDR), and *enhanced data rate* (EDR). For example, A 12X QDR link carries 120 Gbit/s raw, or 96 Gbit/s of useful data, and is typically used for *cluster architecture* and *supercomputer* interconnects and also for inter-switch connections. The InfiniBand future roadmap also envisages to realize "HDR" (high data rate), and subsequently "NDR" (next data rate) within a few years, but these data rates are not yet tied to any specific speeds.

Link layer: This layer defines and describes the basic structure of the data packet to be transmitted containing data (or a portion of the total data), plus an addressing scheme and control information that assigns a unique link address to every device attached in a subnet. This layer also includes the logic to set up temporary connections (logical channel), and controls data through switches from source to target within a subnet. Each physical link can provide up to 16 such logical channels. This packet also contains an error detection code to provide reliability. Each such packet is of up to 4 KB, and all such relevant packets are taken together to form a *message*.

Network layer: It essentially *routes* packets between different InfiniBand subnets.

Transport layer: Apart from performing its primary responsibilities, it also provides a *reliability* mechanism for transfer of packets from source to destination across one or more subnets.

Applications: InfiniBand has been adopted in enterprise datacentres, e.g., *Oracle Exadata and Exalogic Machines* (use Sun computing hardware), financial sectors, *cloud computing* (an InfiniBand-based system won the best of VM world for cloud computing) and many more. **IBM** released a cluster of IBM System p servers (POWER 6/7) that communicate with one other, and also with a shared-disk cluster database *DB2 pureScale* using an InfiniBand interconnect. *Scale-out network storage* manufacturers increasingly adopted InfiniBand as a primary cluster interconnect for modern NAS designs, like IBM SONAS. A number of the TOP 500 supercomputers have used InfiniBand, including the former reigning fastest supercomputer, the *IBM Roadrunner.*

A brief detail of InfiniBand structure along with its applications is given in the website: http://routledge.com/9780367255732.

5.5 PORT and Its Different Types

The port is a predefined junction on the bus where a resource or an additional circuit can be placed. A computer port is an addressable location with a specific address by which it is accessed. Usually, a resource (peripheral) is made connected with the port by using a specific cable, one end of which is connected with the resource (peripheral) and the other end with a connector is to be plugged into the predefined port. In some cases, the peripheral circuits are hard-wired to the port. Software of the peripheral device controls and monitors the port circuits by reading and writing to the port's address. There are various types of ports available, but we will restrict our discussion in brief only to:

i. Serial port;
ii. Parallel port;
iii. USB port.

5.5.1 Serial Port

Serial port has been an integral part of most computers from its inception, and the name *serial* comes from the fact that a serial port *serializes* data, taking a byte (8 bits) of data, and transmits these 8 bits *one bit at a time*. The *advantage* is that a serial port uses less-wide (needs only one wire to transmit) and less-expensive cables with no additional shielding mechanism to prevent any electrical interference between the wires, and also inexpensive connectors with lesser number of pins to bend or break. Also, synchronization between wires is not required here, and as such, it is free from this problem that gets even worse with increased cable length. The *disadvantage* is that it takes (8 times) longer time to transmit the data than it would be if there were 8 wires. Serial ports use a well-defined protocol and a standard connector of 9 pins or 25 pins, in which each pin is assigned for different purposes to establish the desired communication. Serial ports, also called **COMmunication (COM)** ports, are ***bidirectional*** that allows each device to receive and to transmit data, either using the same pins that would limit communication to **half-duplex** (one direction), or using different pins that would allow communication to **full-duplex**, in which information can travel in both directions at the same time. A special controller chip (Intel's 8251A, the **universal synchronous asynchronous receiver/transmitter**, **USART**) can be used which is then attached to the system by plugging RS-232 interface cards into the bus. Although many of the newer systems have discarded the serial port completely in favour of *USB* connections, still few computers provide serial ports for some other reasons. All contemporary OSs in use today still support serial ports.

A brief detail of the standard serial port is shown with figure in the website: http://routledge.com/9780367255732.

5.5.2 Parallel Port

Most of the computers (also PCs) have a ***parallel port*** as a commonly used interface to mainly connect a printer and also other devices to the computer, in spite of USB being increasingly popular. With a parallel port, a character of 8 bits (1 byte) is sent at a time in ***parallel*** to each other. A typical parallel port, however, is capable of sending 50–100 kilobytes of data per second. Although parallel ports were originally specified as unidirectional, IBM offered a new bidirectional parallel port design, known as ***standard parallel port*** (*SPP*) with the introduction of the *PS/2* in 1987. In 1991, Intel, Xircom, and Zenith created the ***enhanced parallel port*** (EPP) with data transmission rate ranging from 500 kilobytes to 2 megabytes per second. It was targeted specifically for *storage devices* (*non-printer devices*) that needed the highest possible transfer rate. In 1992, ***Microsoft*** and ***Hewlett–Packard*** jointly announced a specification called ***extended capabilities port*** (ECP). While EPP was geared towards other devices to attach, ECP was mainly designed to provide improved speed and functionality for printers. In 1994, the IEEE 1284 standard that included both EPP and ECP specifications for parallel port devices was released.

A brief detail of the parallel port is shown with figure in the website: http://routledge.com/9780367255732.

5.5.3 USB Port

USB port, though not very old, is increasingly popular for being built up with advanced technology to offer a faster and more flexible interface for connecting nearly all devices to the computers, thereby almost replacing both serial and parallel ports completely. It requires almost nothing as such for configuring the hardware or software, and is very different from the legacy (serial and parallel) interfaces it is replacing. While accommodating various types of devices, it may use any of ***three speeds***, namely *low speed, full speed, or full and high speeds*. While in operation, there are ***four transfer types***, namely *control transfer, interrupt transfer, isochronous transfer,* and *bulk transfer.* USB ports differ from many other ports because all ports, each with its own connector and cable, with a host controller here on the bus share a single data path to the host (see USB bus, described earlier). Here, all ports (devices) share the available time, but only one port (device), or the host, is operative at a time. A single host may also support multiple USB host controller; however, each will do so with its own bus. Other types of interfaces that support multiple ports sharing a single data path include IEEE-1394 and SCSI. In contrast, with the traditional serial interface, each port is *independent* from the other ports, and can *independently send and receive data* simultaneously with the other ports. For more details, see USB bus already discussed earlier.

A brief detail of the USB port is shown with figure in the website: http://routledge.com/9780367255732.

5.6 Summary

I/O system organisation nowadays is of immense importance, and it critically influences the performance of the system as a whole. The organisations of I/O systems are distinguished by the extent of CPU involvement in I/O operations. Four basic approaches to I/O transfers are of common interest. The simplest technique is PIO in which the CPU is totally involved and performs all the necessary control functions of an I/O operation. The second approach is interrupt-driven I/O in which CPU is partly relieved. The third approach is based on providing I/O devices with *DMA* in which data transfers can be implemented independently without CPU involvement. Maximum speed and total independence in I/O are, however, achieved by providing I/O channels, and then IOPs, which are capable of executing their own programs to manage the entire I/O operations.

Buses are used as the primary means of interconnecting different resources of a computer system. Fundamental characteristics and different useful parameters of a generic bus system have been explained. Three popular interconnection standards, namely PCI, SCSI, and USB, have been briefly described with their individual strengths and drawbacks. A most-favoured high-speed, high-performance *serial bus*, FireWire, having high-speed serial bus interface standards, especially required for digital audio and video pieces of equipment, has been described. Easy availability of immensely powerful multiple-processor servers with reduced cost now requires typical I/O specifications having low latency, high bandwidth, high throughput, and low-overhead interconnect that might perhaps only connect servers and storage to one other for building up large commercial datacentres. InfiniBand, a *switched fabric* communication link using *switch-based architecture*, satisfies all these requirements and also aims to fulfil the constantly increasing demand that additionally includes quality of service, failover, and also scalability.

Port is used to attach a peripheral device or an additional circuit with the existing system. Popular computer ports such as serial, parallel, and USB are of common importance. They usually exhibit different approaches that meet the needs of various devices and the related interconnecting buses, and also reflect the increasing importance of *plug-and-play* features that ultimately make it possible to summarily realize users' ease and handiness.

Exercises

5.1 The transfer rate between a CPU and its attached memory is higher in order of magnitude than the mechanical I/O transfer rate. How can this imbalance cause inefficiencies?

5.2 What is the function of an I/O interface? What are the responsibilities being performed by I/O modules? What are the criteria that must be fulfilled at the time of designing an I/O module?

5.3 What are the various modes of data transfer between a computer and peripherals? Explain.

5.4 Differentiate between I/O-mapped I/O (isolated I/O) and memory-mapped I/O. What are their relative advantages and disadvantages?

5.5 What do you mean by busy waiting of CPU? Define interrupt. How do interrupts work? Discuss how an interrupt is processed.

5.6 Distinguish between vectored and non-vectored interrupts.

5.7 What is UART (universal asynchronous receiver transmitter)? What is the basic advantage of using interrupted initiated data transfer over transfer under program control without an interrupt?

5.8 Assume a computer without priority interrupt hardware. Any one of many sources can issue the interrupt that interrupted the computer, and any interrupt request results in storing the return address and branching to a common interrupt service routine. Explain how a priority can be established in the interrupt service program. (Hint: Section 5.3.2.3 Multiple Interrupts and Section 5.3.2.5)

5.9 Discuss the role of hardware and software in relation to the actions they take at the time of servicing an interrupt.

5.10 What is DMA mode of data transfer? When is this mode used? Describe the functions of different internal registers used in a DMA controller. How does cycle stealing DMA differ from burst-mode (block) DMA? Why does DMA always enjoy higher priority than CPU while both attempt to communicate with memory?

5.11 A computer uses DMA to read from the disk. The disk has 64 sectors per track, and each sector is of 512 bytes. The disk rotation time is 16 ms. The bus is 16 bits wide, and bus transfers take 500 ns each. The average CPU instruction requires two bus cycles. How much is the CPU slowed down by the actions of DMA?

5.12 Differentiate between polled I/O and interrupt-driven I/O. What are the vectored interrupts? How they are used in implementing hardware interrupts?

5.13 In the daisy chain priority interrupt as shown in Figure 5.6, device I_1 requests an interrupt after device I_2 has sent an interrupt request to the CPU but before the CPU responds with the interrupt acknowledge. Explain with a diagram what will happen.

5.14 What are the advantages of a parallel priority interrupt. Show with a real-life example how interrupt-driven I/O is implemented with an interrupt controller.

5.15 How are multiple DMA requests handled in the implementation of a DMA controller? Give suitable examples. "DMA data transfer may sometimes lead to cache coherence" – justify and explain.

5.16 Why do I/O devices place the interrupt vector on the bus? Would it be possible to store the information in a table in memory instead?

5.17 "IOP is virtually a stand-alone computer"– justify the statement. What are the advantages of using an IOP? How is the IOP organised when attached with a main computer?

5.18 Define each of the following I/O control methods: PIO, DMA controllers, and IOPs. What are the advantages and disadvantages of each method with respect to program-design complexity, I/O bandwidth, interface hardware costs, and system throughput?

5.19 Discuss the role played by the bus when the computer is in operation. Explain the structure of a generalized bus system giving its main constituents. Why is the single-shared bus so widely used as an interconnection medium in both sequential and parallel computers? What are its major disadvantages?

5.20 What do you mean by bus arbitration? Why is bus protocol required? Discuss the key elements (parameters) that are to be considered at the time of designing a bus.

5.21 Analyse the three bus-arbitration methods: daisy chaining, polling, and independent requesting with respect to communication reliability in the event of hardware failures.

5.22 Define each of the following terms in the context of bus design: master unit, handshaking, lock signal, wait state, tri-state, and bus transceiver. (Hint: Section 5.4.6)

5.23 Why do most modern PCs maintain a hierarchy in the buses? Discuss the hierarchy and the responsibilities being handled by each bus in the bus hierarchy of the PCI bus (see the PCI bus given in the website to answer).

5.24 Why is the SCSI bus so popular in today's small systems? State and explain its salient features and distinct advantages. "SCSI is not treated as only a simple bus, but it is called a *SCSI network*"– justify with reasons.

5.25 What is USB? Discuss its distinctive features that put USB in the forefront of the list of bus systems being used in the small computer systems.

5.26 What is FireWire? What are the salient features that made it more acceptable in the common user market?

5.27 Discuss in brief the protocols that are employed in FireWire serial bus transmission mechanisms.

5.28 "Evolution of InfiniBand is a big step towards a notable development in the information transmission mechanisms": justify the statement in the light of distinguished features that this technology possesses.

5.29 Describe with a diagram the basic components with their interactions that constitute the core architecture of InfiniBand.

5.30 Describe in brief the major features of the protocols that are employed in InfiniBand technology.

5.31 What do you mean by a port? What is the relationship between a port and a bus? What are the differences between a serial port and a parallel port? Discuss the major features that have been observed in SPP, EPP, and ECP.

5.32 Describe with a diagram in brief the basic components and the major features of a USB port (hub) (see the websites to answer).

Suggested References and Websites

Abbot, D. *PCI Bus Demystified*. New York: Elsevier, 2004.

Anderson, D. *Fire Wire System Architecture*. Reading, MA: Addison-Wesley, 1998.

Gustavson, D. B., "Computer buses — A tutorial." *IEEE Micro*, vol. 4, pp. 7–22, August 1984.

Kagan, M. "InfiniBand: Thinking outside the box design." *Communication System Design*, September 2001.

Shanley, T. *InfiniBand Network Architecture*. Reading, MA: Addison-Wesley, 2003.

1394 Trade Association: Includes technical information and vendor pointers on FireWire.

InfiniBand description, available at www.csdmag.com

PCI Local Bus Specification, available at www.pcisig.com/developers.

PCI Pointers: Links to PCI vendors for additional information.

T10 Home Page: A technical committee on National Committee on Information Technology Standards working on lower-level interfaces, including small computer system interface (SCSI).

Universal Serial Bus Specification: available at www.usb.org/developers.

6

Control Unit: Design and Operation

LEARNING OBJECTIVES

- To explain the need for a control unit to manage the operations of resources, including the CPU (central processing unit) operations at the time of program execution.
- To study the design issues of a traditional control unit.
- To discuss the control unit logic and different types of control unit signals.
- To describe the design of a generic control unit module with its constituent components.
- To introduce control unit implementation: hardwired approach and microprogrammed approach.
- To study the issues in microinstruction design: horizontal and vertical microprogramming.
- To explain the concept of nanoprogramming and its implications.

CPU performs numerous functions, including the executions of instructions carried out by its *data processing part* (Chapter 3). To provide these functions, the various elements of the CPU are monitored and controlled by a unit, historically known as the *control unit* of CPU (the instruction set processor). In fact, the control unit located inside the CPU as one of its constituent elements administers, monitors, and coordinates all the resources in the system, including the various elements of the CPU itself for providing numerous functions and their subsequent operations.

Traditionally, when the design and the architecture of the computer, including its functional structure (both the data processing part and the control part), are decided and finalized, it is then permanently implemented by making suitable circuits of hardware components such as gates, flip–flops, and other digital circuits. It cannot be then often changed, and any modification in it, if is needed, would then require extensive hardware alterations that are not only very difficult, but equally expensive to realize, and also time-consuming. When the control functions are realized in this manner, by way of assembling hardware components with a predefined logic and design, the generated control unit is known as a *hardwired control*.

With the ongoing revolutionary development in the area of electronic technology, the architecture and organisation of the *processors*, and also of the computer systems, have gradually evolved in steps over the years aiming at continuous increase in performance. Consequently, more flexibility is then needed to keep these continuous upgrades consistent so that higher versions with the same architecture (family concept) would include everything of the recent olds and something more to provide growth potential, and also to permit old programs to run straightaway on new machines without much reprogramming for a user installation to make it viable economically and technically. This flexibility is achieved when the control unit is designed with a different approach, called *microprogramming*, and the control units are then known as *microprogrammed control units*. The principle of microprogramming is really an elegant and systemic method for controlling the sequence of operations in a digital computer system.

These two different concepts and approaches in control unit design give birth to two different schools (categories) in defining and describing computer architecture: **RISC (reduced instruction set computer)** architecture and **CISC (complex instruction set computer)** architecture. RISC architecture will be discussed in detail in Chapter 9. This chapter is mainly targeted towards explaining separately the structure, function, and behaviour of both types of control units that are, by and large, common to all types of computers of their respective categories.

6.1 Introduction

CPU executes each instruction of the program in a series of small steps, called the *instruction cycle*, in which each such step is once again required to be broken down further to a series of smaller steps (microinstructions), and each such smaller step is fundamental and very simple, and when executed by the CPU, it is referred to as a *micro-operation* that accomplishes very little task. Figure 6.1 illustrates this relationship between the various steps we have

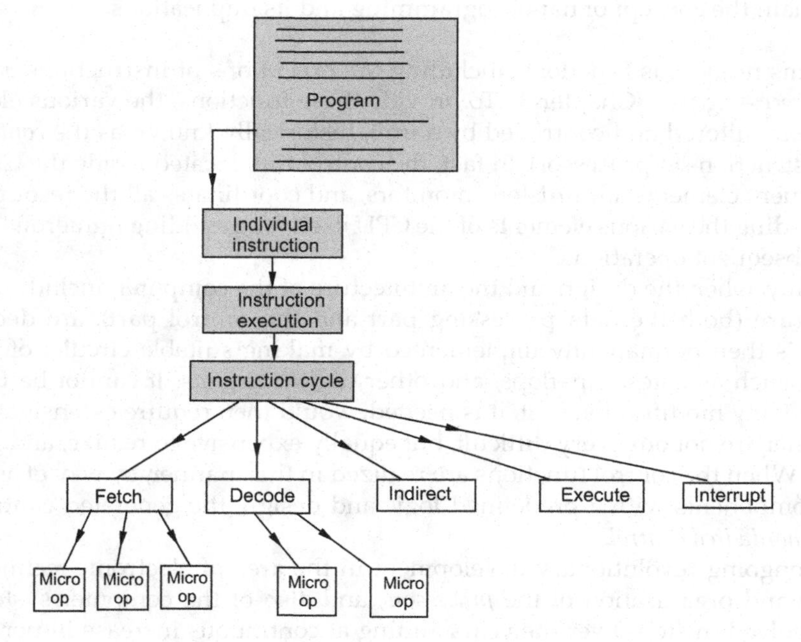

FIGURE 6.1
Flow of program execution with constituent elements.

just mentioned. Each micro-operation or a set of micro-operations is obtained by the execution of one microinstruction. The number of different types of micro-operations available in a given system is finite and depends on the design of the system. The function of the control unit is to initiate (not execute) particular sequences of such available micro-operations by issuing specific control signals to the data path. Various types of control signals are usually generated in the *control unit* by hardware (or by *instructions* in the control unit) using conventional logic design techniques. These control signals enter the data path at specific "control points" where they select the particular function to be performed at *specific times* and route the data through the respective parts of the data path unit. Different control signals specify different control functions, and each such control signal is a binary variable. When this variable is in one binary state, the corresponding micro-operation is executed, and when it is in the opposite binary state, there is *no change* in the state of the corresponding functional component. The active state of a control variable may be either the 1 state or the 0 state depending on the design or the application. Let us now examine the micro-operations with an example as given below, to gain an understanding as to how the events of any instruction cycle can be described in terms of a sequence of such micro-operations.

6.2 Micro-Operations: Fetch Cycle

Let us now look at the fetch cycle (one of the steps in the instruction cycle) and describe the sequence of events (smaller steps) that are to be carried out for the execution of this cycle, and its related impact on the CPU registers. The control unit here is intimately involved, monitoring the entire execution by way of **sequencing** the events and **timing** (when and how long) the data path. This timing in the basic computer is realized using a *master clock generator,* which emits regularly spaced clock pulses. The steps to be executed now *in sequence* using successive time units (T_1, T_2, T_3, T_4) are: the processor specify the address of the memory location of the information to be fetched using the program counter (PC) and move that address to its memory address register (MAR) at the instant T_1, i.e.

```
T1: MAR ← (PC); and send a control signal on the memory bus for Read
operation.
T2: Wait; for the requested Read operation to be completed.
```

The requested data is now available and is to be stored into the memory buffer register (MBR), and at the same time, in parallel, PC is incremented by 1 to get it ready to point the next instruction to be executed (since these two operations do not interfere with each other and create no conflicts, they can be operated in parallel), i.e.

```
T3: MBR ← Memory;  PC ← PC + 1
```

The next step is to move the content of the MBR to the instruction register (IR) and frees up all the registers (MAR and MBR) for its next use.

```
T4: IR ← (MBR)
```

This grouping of micro-operations could have been done in any possible manner to operate each step in a single time unit, if there is no direct conflict, yet the action here at T_2, in practice, may require one or more clock cycles, depending mostly on the speed of the resource being addressed. In a similar manner, the other constituent cycles (decode, execute, indirect operand fetch, interrupt, etc.) in an instruction cycle can also be likely obtained as a series of required micro-operations. In this way, the sequence of all elementary micro-operations required to execute one complete instruction of a program can then be generated and executed in an appropriate sequence to get the desired result.

6.3 Design Issues

The control unit primarily monitors mainly the operations within the CPU and also of the entire system. The design of the control unit for a basic non-pipelined CPU must explicitly define the events that the control unit must cause to happen. Thus, the functional characteristics in the design of a control unit naturally depend on the following parameters:

1. **Basic components available in a CPU**: The basic elements are the registers, ALU (arithmetic and logic unit), internal interconnection paths, external connection paths, and control unit.
2. **Micro-operations that the CPU must perform**: The appropriate micro-operation(s) from the set of various types of micro-operations available in the CPU is (are) required to be executed by the different elements of the CPU under the supervision of the control unit to execute an instruction.
3. **Specific functions that the control unit must perform at appropriate times so that the respective micro-operations can be executed**: To perform this responsibility, the control unit performs two basic tasks, namely the **sequencing**, which means that in which sequence these micro-operations are to be executed, and the **timing** (when and how long) of the different active elements of the CPU to be involved in the execution to get the desired result. In order to perform these functional activities, the control unit requires certain *input* which will be internally processed by the control unit itself using its own control logic, and subsequently will produce such an *output* in the form of control signals so that the intended functions on the CPU (and also on the system) elements can be realized.

A schematic generalized approach of a generic control unit module is illustrated in Figure 6.2, indicating the different types of inputs it requires and the various types of corresponding outputs it generates. The inputs are described in the following.

IR: It contains the current instruction to be executed, including the opcode which is used by the control unit to determine (decoding) what types of micro-operations are to be performed during the execution cycle.

Clock: The timing of CPU operations is synchronized by a *master clock*, which generates regularly spaced clock pulses when each such clock pulse defines a specific time unit. The control unit causes one micro-operation (or a set of simultaneous micro-operations) to be executed at *specific times* within each of such clock pulses. The time interval between two successive clock pulses is sometimes referred to as the ***clock cycle time*** or ***processor cycle time***. The number of clock cycles required

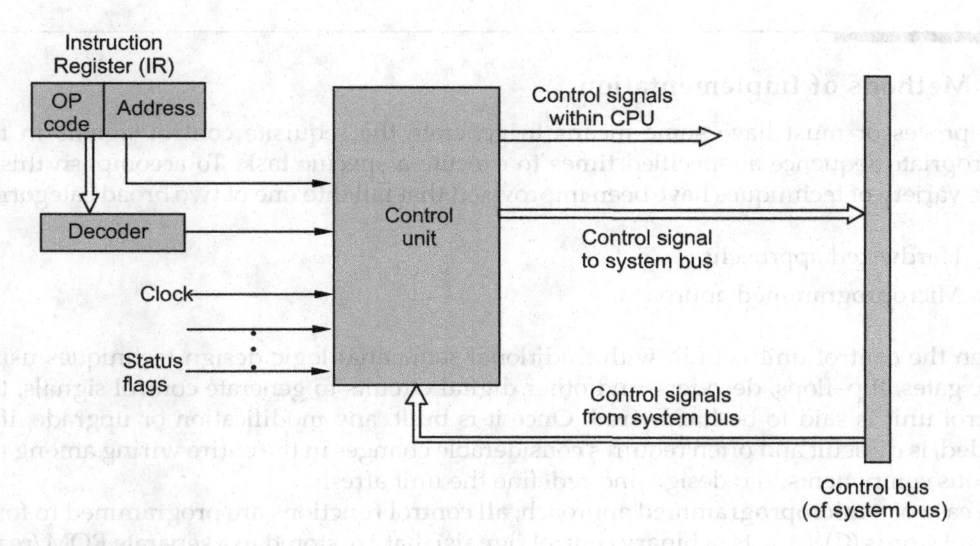

FIGURE 6.2
Schematic design of a control unit.

for the execution of a given instruction is fixed, but varies from one instruction to another.

Status flags: The control unit requires the *status control signal* to determine the present status of the data path and the outcome of the previous data path operations.

Control signals from the control bus: The control unit expects control signals from the control bus to get the status of the external resources such as memory and I/O modules. The control unit, after processing the input signals, routes appropriate *output signals* to specific control points at specific times to initiate the intended micro-operations.

Control signals within the CPU: There are two types of control signals within the CPU: one type activates the operations for moving data between internally connected registers, and the other type initiates specific ALU functions.

Control signals external to the CPU: These control signals are required to monitor the operation of the resources located outside the CPU.

Let us now examine how the control unit monitors the fetch cycle operation as explained in the last section. The first step in this cycle is to transfer the contents of the PC to the MAR, i.e. MAR←(PC). After that, the control unit causes other events to happen by sending the following control signals: (i) a control signal that causes the contents of the MAR to be placed onto the address bus; (ii) a control signal for memory read (memory–read control signal) is sent to the control bus; and (iii) a control signal that allows the contents of the data bus (data already read from memory and on the data bus) to be stored in the MBR, i.e. MBR←Memory, and in parallel, a sequence of control signals is required that adds 1 to the contents of the PC and stores the result of addition back to the PC. After the completion of these steps, different types of similar sequences of appropriate control signals are required for executing the operation between MBR and IR, the remaining portion of the fetch cycle. Other steps of an instruction cycle are executed in the same fashion.

6.4 Methods of Implementation

The processor must have some means to generate the requisite control signals in the appropriate sequence at specified times to execute a specific task. To accomplish this, a wide variety of techniques have been improvised that fall into one of two broad categories:

- Hardwired approach;
- Microprogrammed approach.

When the control unit is built with traditional sequential logic design techniques using logic gates, flip–flops, decoders, and other digital circuits to generate control signals, the control unit is said to be **hardwired**. Once it is built, any modification or upgrade, if is needed, is difficult and often requires considerable changes in the entire wiring among the various components to redesign and redefine the unit afresh.

In case of a **microprogrammed** approach, all control functions are programmed to form control words (CWs) (sets of binary control signals) that are stored in a separate ROM (read-only memory) within the CPU called the *control memory* (*CM*). All the control functions that are to be activated simultaneously can also be grouped together to form CWs. From the CM, these CWs are fetched one at a time, and the individual control fields within the word are then routed to various control units to activate their appropriate circuits. These circuits activated in specific sequence actually perform the required tasks as desired.

Both these two approaches have been explained in detail in the subsequent sections. However, to project these two different approaches briefly, Figure 6.3 illustrates the general structures of generic hardwired and microprogrammed control units.

6.4.1 Hardwired Control

In this approach, the control unit is viewed as a group of *fixed* sequential logic circuits or a finite state machine (and/or otherwise a combinatorial circuit) that interprets instructions, and subsequently generates and transmits certain sequences of appropriate control signals considering specific states and externally supplied instructions. Since the unit is composed of fixed circuits, it has relatively little flexibility, and moreover, the complexity

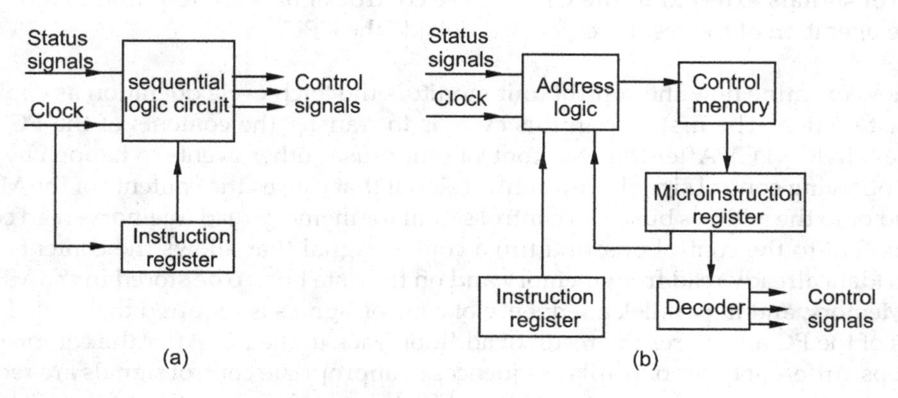

FIGURE 6.3
A representative general structure of two approaches of control unit design.

of the instruction set it can implement is limited. However, the ultimate objective of this design is to minimize the number of digital components used in the circuit and to realize an optimal fast mode of operation.

Numerous techniques in the design of these control units have been proposed to attain different goals involving various trade-offs, mainly the amount of hardware to be used and the speed of operation to be realized, and of course, the cost of the design process. But no one single method, including the **one-hot method** (concentrates mainly on simplifying the design process of the control unit for easy maintenance), can yield an acceptable circuit design at a reasonable cost. The most logically acceptable design thus may be to construct several independent different sequential circuits that can be later linked together in a specific manner to realize the desired control unit. Whatever the techniques being adopted, the **key inputs** are mainly the *IR, the clock, the status flags, and control bus signals.* Figure 6.4 illustrates an overview of a hardwired control unit.

Inputs: The IR provides the encoded opcode which is converted by the instruction decoder in order to produce a single output consisting of a particular combination of control signals. For variable-length opcodes, the control unit design is, however, assumed to be more complex. The clock issues a repetitive sequence of pulses that are used to measure the duration of micro-operations, including the time required to propagate the signals along data paths and other related circuitry. A sequence counter (SC) thus may be used to keep track of the timing signals. After each clock pulse, SC is incremented by 1 so that the timing signals can go through a sequence T_0, T_1, T_2, and so on. At the end of the instruction cycle, the control unit must feed back to the counter to re-initialize it at T_1. The other two inputs are status flags (condition codes) and control bus signals. Each individual bit in these inputs has some specific meaning and is directly fed into the encoder. All these inputs as shown in Figure 6.4 are then sent to the encoder of the control unit that eventually generates the ultimate individual control signals which flow to the targeted unit for the required operation.

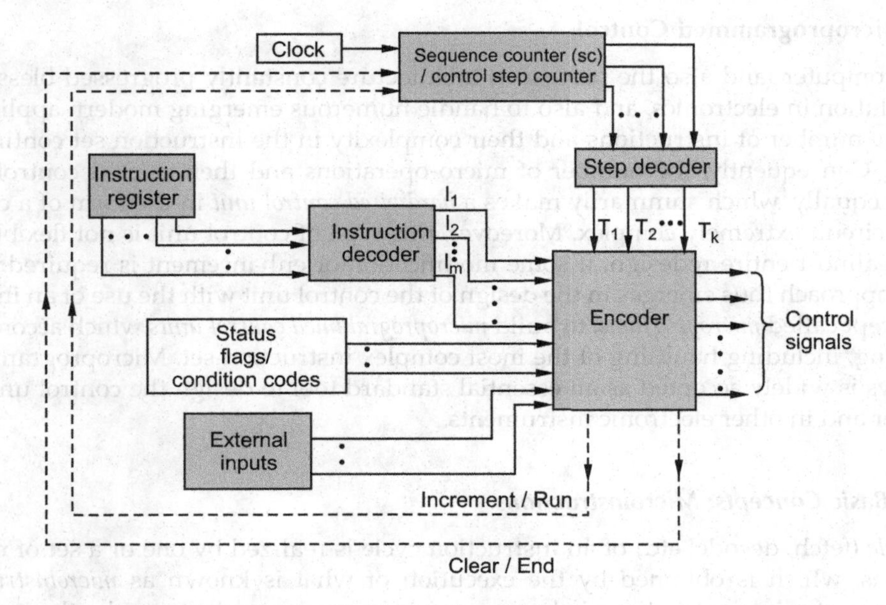

FIGURE 6.4
Hardwired control unit organisation with separate encoding/decoding functions.

6.4.1.1 Control Unit Logic

Different types of control signals are generated by the control unit to initiate different micro-operations which are then executed by the CPU to accomplish its various tasks. For example, the micro-operation required *to read the data from the external bus to MBR*, when memory, after reading, sends information to the CPU. This happens in the *fetch cycle* when the instruction is read by memory and sends to CPU, and also at the *operand-fetch cycle* when the value of operand is read by memory and sends to CPU. Considering these two situations, Boolean expressions are then formulated for the required control signals that can initiate these micro-operations. This case, as an example, actually exhibits the related logic to be required, and is then to be built in hardwired control unit.

A brief detail of the logic of this specific case and its formulation is given in the website: http://routledge.com/9780367255732.

6.4.1.2 Control Signals in Accumulator-Based CPU

CPU organisations are of various types and so are the control units. The control unit design for a basic, non-pipelined CPU mainly differs in degree and complexity, depending on the types of components used in the design of the CPU itself, and the category to which the CPU belongs. This again determines the type of the instruction set that the CPU must have in order to attain its objectives. These instructions in the instruction set when executed by the available hardware require the relevant control actions (micro-operations) to be initiated by the respective control signals. These signals are, however, issued from the control unit at specified times.

An example of a control unit with the control signals exploited in a simple accumulator-based CPU, using only a few of the design issues, is given in the website: http://routledge.com/9780367255732.

6.4.2 Microprogrammed Control

As the computer (and also the processor) architecture constantly progressed blessed by the revolution in electronics, and also to handle numerous emerging modern application areas, the number of instructions and their complexity in the instruction set continually increase. Consequently, the number of micro-operations and their related control lines increase equally, which summarily makes a *hardwired control unit* in the form of a combinatorial circuit extremely complex. Moreover, this type of control unit is not flexible and requires almost entire redesign, if some modification or enhancement is required. A far simpler approach thus emerges in the design of the control unit with the use of an innovative concept called *microprograms*, to build *microprogrammed control units* which accomplish everything, including handling of the most complex instruction set. Microprogramming nowadays is widely accepted as an essential standard tool to design the control unit of a computer and in other electronic instruments.

6.4.2.1 Basic Concepts: Microinstructions

Each *cycle* (fetch, decode, etc.) of an instruction cycle is realized by one or a set of micro-operations, which is obtained by the execution of what is known as *microinstruction*. A sequence of microinstructions is known as a *microprogram*, which can be thus viewed as an equivalent to a subcycle to be processed. The microprogram is sometimes called *firmware* due to its position somewhere in between hardware and software. It is relatively easy to design and simpler to implement in firmware than its hardware counterpart.

Each micro-operation for its execution requires a set of control signals generated from the control unit that flow through different control lines, and these signals are in either *on* or *off* state represented by a binary digit, either 1 or 0, respectively. Thus, the states of all the control lines required for each micro-operation consist of a set of respective binary digits that ultimately form a word known as *control word* (*CW*) or *microinstruction*. Each such CW with a particular bit pattern (1 and 0) when executed gives rise to one or a set of micro-operations. Different CWs required for different types of needed micro-operations can be stored in a special memory known as *control memory* (*CM*) or *control store* available inside the CPU chip. Each CW stored in CM has a unique address for its access. A sequence (set) of CWs corresponding to the control sequence of a machine instruction constitutes the **microroutine** for that instruction. Figure 6.5 shows an example describing how the different microroutines required for each cycle (fetch, execute, etc.) are organised in the CM.

While control unit provides different CWs for different micro-operations to be executed, its other task is to *sequencing micro-operations*, which means to find out the next micro-operation to be executed after the completion of the currently executing micro-operation. This can be realized in two ways: one approach is that a part of the currently executing microinstruction may include a field specifying the address of the next microinstruction in the CM to be executed, and the other way is to use a microprogram counter (µPC), similar to PC, to keep track of successive microinstructions to be read from the CM. This sequencing is again often influenced by the presence of various conditions such as branching, indirect cycle, external inputs, and status of condition codes, which may summarily cause a total departure from the natural manner of sequencing (successive microinstructions),and imposes to follow a particular path out of many alternatives to realize the befitting courses of actions. Thus, one of the ways is to add an additional address field to each CW in order to indicate the location in the CM where the next CW (micro-operation) is to be found for

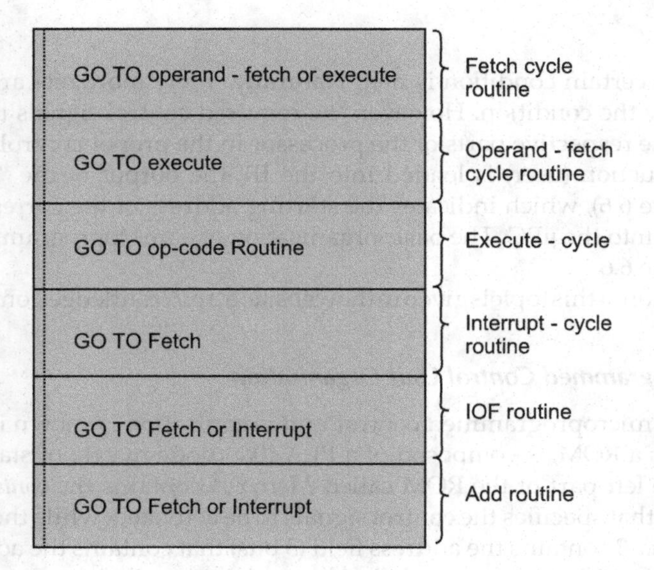

FIGURE 6.5
Various microroutines in CM.

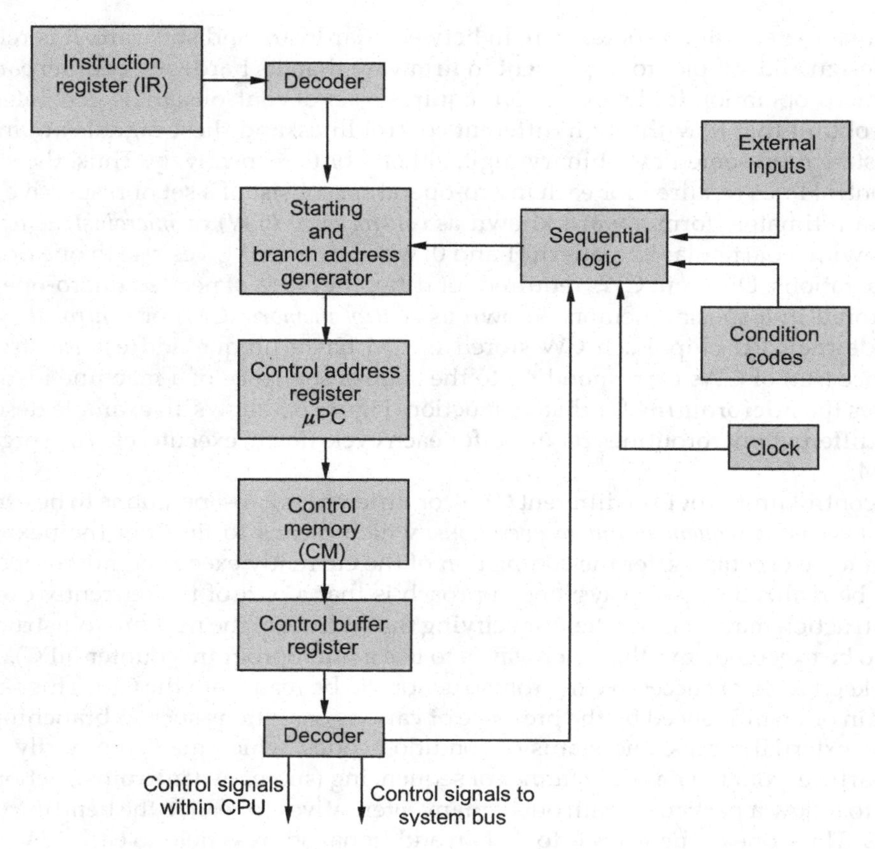

FIGURE 6.6
Basic organisation and functioning of microprogrammed control unit.

its execution, if a certain condition is met. Naturally, a few more bits are also needed in the CW to specify the condition. However, the required control signals ultimately would be delivered to the respective parts of the processor in the proper control sequence. Every time a new instruction (user) is loaded into the IR, the output of the "starting address generator" (Figure 6.6), which indicates the starting address of the corresponding micro-routine is loaded into the μPC. The basic organisation of a microprogrammed control unit is shown in Figure 6.6.

A brief description of this topic is given in the website: http://routledge.com/9780367255732.

6.4.2.2 Microprogrammed Control Unit Organisation

A simple form of microprogrammed control unit organisation is shown in Figure 6.7. The CM, organised as a ROM, is composed of a PLA-like diode matrix of standard gates and multiplexers. The left part of the ROM called *Matrix A* contains the *control fields* of every microinstruction that specifies the control signals to be activated, while the right part of the ROM called *Matrix B* contains the address field (3 bits) that contains the address of the next microinstruction in CM to be executed. Each row of the matrix A is one microinstruction.

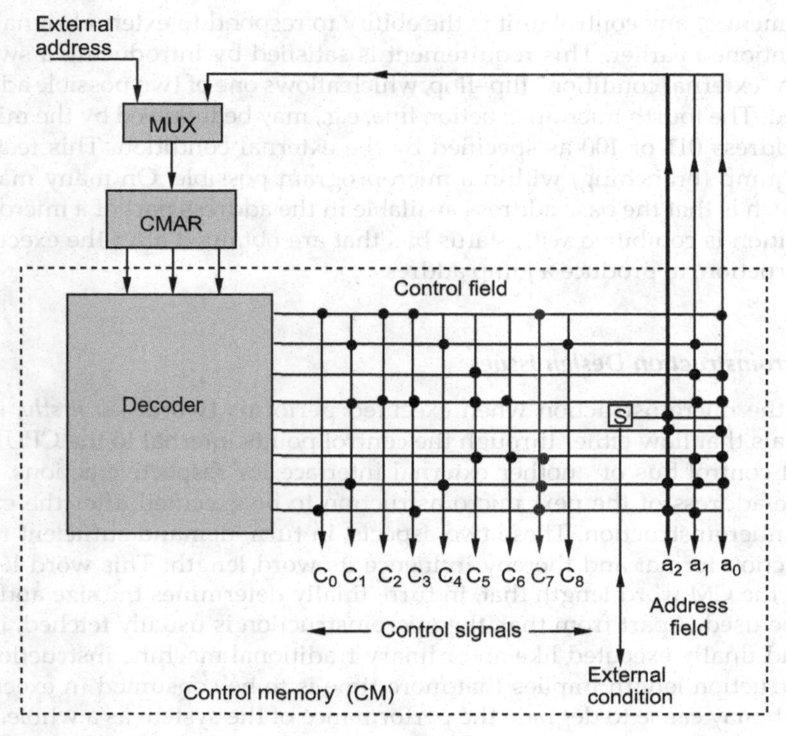

FIGURE 6.7
Basic structure of a microprogrammed control unit.

According to the inventor of microprogramming, Maurice V Wilkes, each bit k_i of a control field corresponds to a distinct control line c_i (vertical line of matrix A). When $k_i = 1$ in the microinstruction, c_i is activated; otherwise, c_i remains inactive.

The rows of CM represent the microinstructions; the columns of CM represent either control lines or address lines. A register called the *control memory address register* (*CMAR*) stores the address of the current microinstructions. The address is decoded, causing one of the horizontal lines (i.e. 1 × 8) of the diode matrix to become active. All vertical lines connected by a black dot (denoting the presence of a diode) to any given horizontal line are activated when the horizontal lines become active. When the top horizontal line, e.g., of the CM, which represents the microinstruction with address 000 (top line 000, next line 001, etc.) is selected, control lines C_0, C_2, C_3, C_7 are activated. At the same time, the address field contents (001) are sent to the CMAR where they are stored and used for the address of the next microinstruction to be executed.

In a typical microprogrammed CPU, each machine instruction is executed by (or interpreted by) a specific microprogram stored in the CM. The machine instruction opcode after suitable encoding provides the starting address for its microprogram which is then loaded in CMAR. The CMAR may, however, be also loaded from an external source that typically provides the starting address of a microprogram stored in the CM. The CMAR is usually loaded from the address field of a microinstruction, which generally contains the base address of its successors. This base address is typically OR-ed with a few status bits to produce the final address of the next instruction to be executed.

A requirement of any control unit is the ability to respond to external signals or conditions as mentioned earlier. This requirement is satisfied by introducing a switch *S* controlled by an "external condition" flip–flop, which allows one of two possible address fields to be selected. The fourth microinstruction line, e.g., may be followed by the microinstruction with address 011 or 100 as specified by the external condition. This feature makes conditional jump (branching) within a microprogram possible. On many machines, the other approach is that the base address available in the address part of a microinstruction under execution is combined with status bits that are obtained after the execution of the current instruction, to produce a jump address.

6.4.2.3 Microinstruction Design Issues

In essence, the microinstruction when executed performs two tasks: *firstly*, it generates control signals that flow either through the control points internal to the CPU or through the external control bus or another external interface for respective actions. *Secondly*, it provides the address of the next microinstruction to be executed after the execution of the current microinstruction. These two aspects, in turn, demand sufficient room in the microinstruction format and thereby influence its word length. This word length again depends on the CM word length that, in turn, finally determines the size and the cost of the CM to be used. Apart from that, the microinstruction is usually fetched, interpreted, decoded, and finally executed like an ordinary traditional machine instruction. Increase in microinstruction length implies that more time is to be consumed in executing these stages, which may cause to degrade the performance of the system as a whole. Hence, the length of the microinstruction is a vital aspect that needs to be carefully decided, and the decision to be taken in this regard, however, depends on a number of issues; the most crucial and dominating factors to be considered are as follows:

i. Since the bits of a microinstruction specify a micro-operation, what would be the maximum number of bits that can be kept in the microinstruction so as to realize many simultaneous desired micro-operations? In other words, what would be the *degree of parallelism* intended at the micro-operation level?
ii. Whether each bit of a microinstruction will directly involve in generating a control signal or whether control information will be encoded using a lesser number of bits in the instruction format, and later will be decoded at runtime to generate the required control signals. This encoding approach, however, can minimize the microinstruction length.
iii. Whether one bit in the address part of a microinstruction will directly produce one bit of the next address, or more complex address sequencing using fewer bits will be employed so that some bits in the instruction format can be saved to reduce its length.

While addressing these issues, microinstructions can be classified in a variety of ways. Distinctions that are commonly made, especially on the format of microinstructions, include the following:

- Horizontal/vertical;
- Unpacked/packed;
- Indirect/direct encoding;
- Hard/soft microprogramming.

6.4.2.3.1 Parallelism

Microinstruction format can be categorized as follows:

a. If each microinstruction specifies only a *single micro-operation* similar to conventional machine instruction, it is known as the ***vertical microinstruction***. Naturally, these types of microinstructions are relatively simple, short in size, need little effort to construct, easy to modify if required, and above all, not much difficult to maintain. The CM required to accommodate this type of microinstruction does have a limited size in word length, which, in turn, substantially reduces the cost of the CM. However, on the other hand, many such microinstructions are then required to execute a given operation, taking a longer time to complete. This leads to an appreciable degradation in the expected level of CPU performance. Figure 6.8 illustrates a representative of this type used in the IBM 370/145 model that consists of 4 bytes, when the leftmost byte is an opcode that specifies the micro-operation to be performed. The next two bytes specify operands, which, in most cases, are the addresses of CPU registers. The rightmost byte provides information used to construct the address of the next instruction to be executed;
b. If a single microinstruction can specify *many micro-operations* that can be executed *in parallel* at any instant with no conflict, giving rise to an essence of **parallelism**, it is known as the ***horizontal microinstruction***. The *parallelism* as obtained can be implemented in a number of ways. One such method to realize this parallelism is to use a longer microinstruction in which each bit or a group of consecutive bits straightaway specify one micro-operation or the operand, and many such micro-operations and operands can then be embedded in one microinstruction. Figure 6.9 shows a microinstruction of this type used in the IBM 360/50 machine. It consists of 90 bits being partitioned into a number of separate fields of various sizes supporting various purposes. A total of 21 such fields are used as control fields, each specifying a particular micro-operation, and the remaining fields are used to generate the address of the next microinstruction to be executed as well as to detect errors by way of parity bits. The implementation of this type of design, however, requires much expertise in the area of related hardware, because every control signal needs to be individually managed by the microprogrammer.

In spite of having a lot of merits, this scheme, however, suffers from several **serious drawbacks**. The control signals generated from different control fields are not all useful, and also cannot operate in parallel, and hence will never be used. This causes a huge wastage in microinstruction bits that, in turn, unnecessarily consumes costly CM space, and hence, this approach has been ultimately dropped from favour.

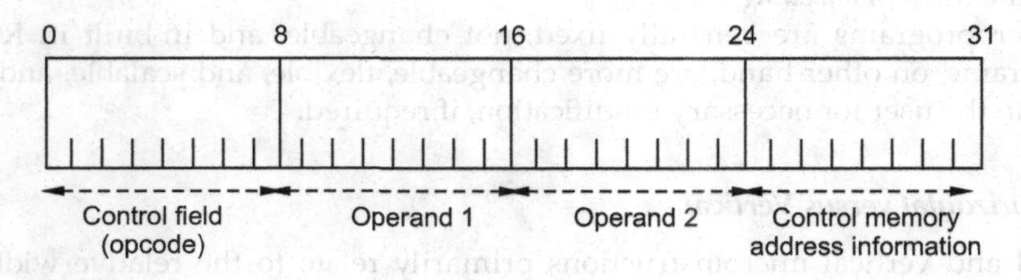

FIGURE 6.8
Vertical microinstruction format used in IBM 370/145 system (single micro-operation).

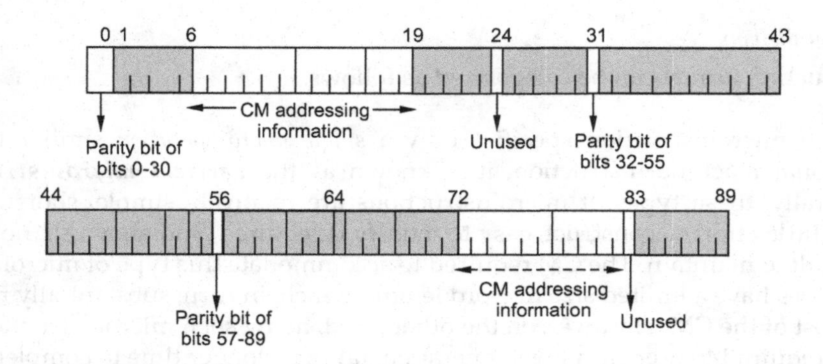

FIGURE 6.9
Microinstruction format specifying many micro-operations (shaded areas are control fields) used in IBM 360/50 system.

6.4.2.3.2 Encoding

Let N be the total number of control signals needed by a control unit for its operation, and this requires N possible distinct combinations of a certain number of bits, which is equal to $\log_2 N$. Unfortunately, this form of encoding in practice cannot be put to use due to many reasons, such as encoding with lesser number of bits than usually required demand a corresponding extra and more complex decoding logic during runtime, which makes the control logic module more intricate and eventually takes more time to operate than is the norm. However, in spite of all these hindrances, the entire proposition in favour of encoding can still be made workable with some compromises, by allowing a few more bits than $\log_2 N$ to include in the encoding scheme. After all, encoding has been used to primarily reduce the CM width and is done in such a way that the microprogrammer can view the CPU as an aggregate function or resource at a higher level, and not going deep into the details of it. Since it is not necessary for the microprogrammer to individually monitor each and every control signal under the encoded scheme, the workload of microprogram development then becomes relatively less and comparatively easy to handle.

Thus, the microinstruction format of the control signal portion falls in a broad spectrum, which can also be treated as a gamut of design strategies. Microinstruction formats are commonly divided into two types forming a pair, and three such different pairs exist, namely unpacked/packed, hard/soft, and horizontal/vertical, where each pair describes the same thing but emphasizes different design characteristics.

Packing indicates a means by which the minimum number of microinstruction bits can be identified to specify a particular control task. As the bits become more *packed,* a given number of bits can contain more needed information. Thus, *packing,* in other words, also implies some form of *encoding.*

Hard microprograms are generally fixed, not changeable, and in-built in ROM. *Soft* microprograms, on other hand, are more changeable, flexible, and scalable, and are also accessible to the user for necessary modification, if required.

6.4.2.4 Horizontal versus Vertical

Horizontal and vertical microinstructions primarily relate to the relative width of the microinstructions. A rule of thumb says that vertical microinstructions have lengths lying in the range of 16–40 bits and horizontal microinstructions have lengths lying in the range of 40–100 bits. The other distinguishing characteristics in these two styles of microinstructions are as follows:

Horizontal	Vertical
• Many bits in the instruction, and unencoded.	• Few bits in the instruction, and highly encoded.
• High degree of parallelism.	• Limited ability in parallelism.
• Little or almost no control logic.	• More complex control logic.
• Fast execution.	• Relatively slow execution.
• Needed detailed view of hardware.	• Needed aggregated view of hardware
• Difficult to program.	• Easy to program.
• Optimizes performance.	• Optimizes programming.

Horizontal microinstruction sets look very much like machine instruction sets. Vertical microinstruction, on the other hand, permits to construct a variety of different microprograms using the same microinstruction set. Vertical microinstructions can be broadly viewed and roughly estimated, similar to RISC instructions (not microinstructions, since RISC architecture is against the microprogramming approach), both in the small amount of parallelism they specify and in their limited size to ensure a single-cycle execution style.

6.4.2.5 Encoding Schemes

While the unencoded format used in horizontal microinstruction has many advantages, it suffers from some inherent severe drawbacks that completely outweigh its distinct advantages. Hence, a low degree of encoding is often used to overcome these and many other drawbacks. When the encoding scheme is adopted, the microinstruction, as viewed, consists of a set of fields; each field contains a code and not control lines. If a microinstruction has N such fields, each field after decoding causes one or more control lines, which, in turn, activate one or more control signals, and all such control signals from *N* different fields would then result simultaneously. Figure 6.10 illustrates a scheme of such encoding.

Microinstruction usually consists of several fields of different categories. Now the obvious question arises as to what types of fields would then be taken as constituents so that an encoded microinstruction can be safely formed. Strategy-wise, two approaches can be considered. When the encoding of fields is based on the functions executed within the machine, and the fields are designated and categorized by the function type, it is called *functional encoding*. For example, one such function within the machine is, say, transferring data to the system bus. This function can be performed by various sources; one field can be designated for this purpose, and each code in that field will then specify a different source. The other approach is *resource encoding* which views the machine as a collection of a set of independent resources, such as memory, I/O, ALU, control bus, etc., and then assigns one field to each such resource.

Another view of encoding is whether it is *direct or indirect*. The type of encoding that we have considered till now is *direct encoding*. With *indirect encoding*, the control line of one field after decoding is used to determine the interpretation of another field, as shown in Figure 6.10d. For example, consider an ALU which is capable of performing eight different types of arithmetic operations and eight different shift operations. A 1-bit field could be used to decide whether it is an arithmetic operation or a shift operation. Based on this finding, another 3-bit field would then indicate the specific operation required. It is obvious that this technique requires two levels of decoding, which makes the control logic much more complex, and the entire process will then be even slower.

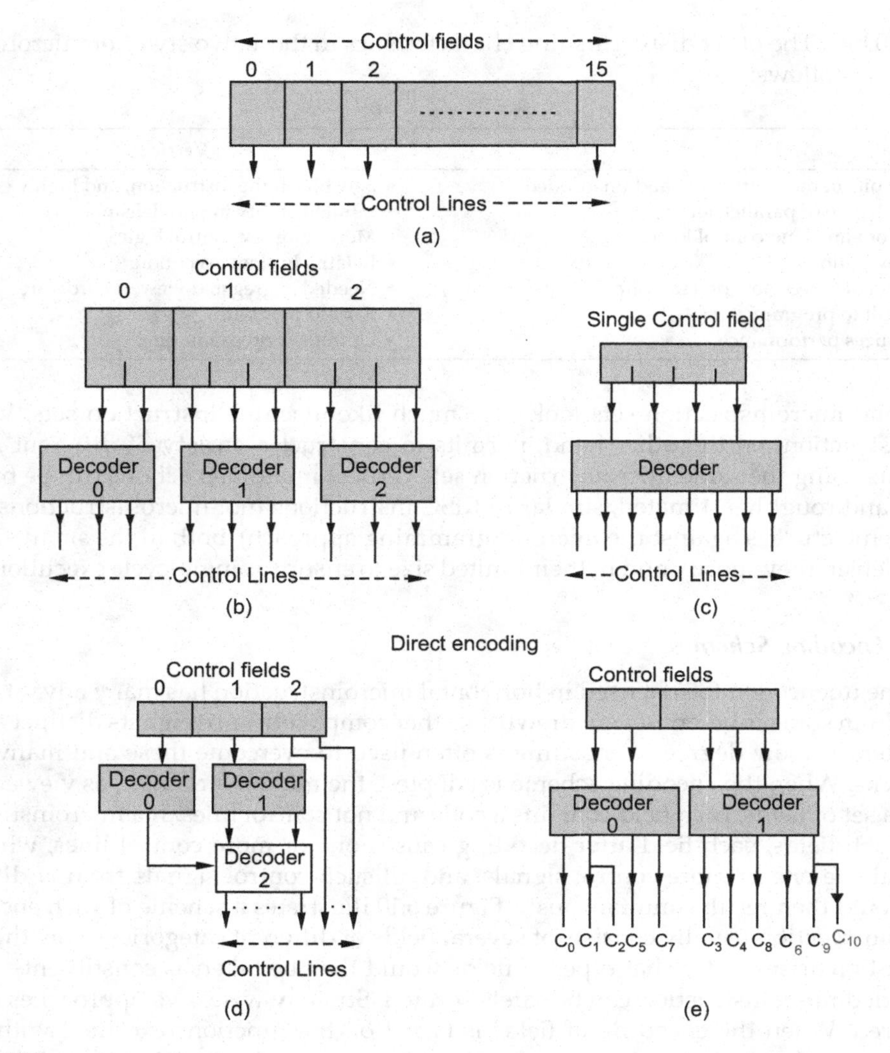

FIGURE 6.10
Different control field formats: (a) no encoding, (b) some encoding, (c) complete encoding, (d) indirect encoding, and (e) maximum parallelism with minimum number of bits.

6.4.2.6 Addressing Schemes

One part of each microinstruction in its basic design is devoted to storing the address of the next instruction to be executed. In case of normal sequencing of microinstructions, this approach in specifying explicit addresses is absolutely fine to facilitate fast operation, since no time is spent in address generation for the next instruction. But at the same time, it indulges a waste in CM space. To avoid this wastage and to make the microinstruction length shorter, the address field can be totally eliminated from all microinstructions, except the branch instructions, by introducing a μPC, as the primary source of the next microinstruction. The μPC plays a role similar to a PC at the normal instruction level, and sometimes, it can be used as a CMAR when the microinstruction is being fetched from CM.

The conditional branching feature, which cannot be ignored, may be implemented by including two possible next addresses in the microinstruction format itself without using the μPC. This approach is simple to implement and easy to design; however, it leads to a waste of CM space by making the microinstruction longer. However, there are other lucrative ways as well, to overcome this drawback.

One such approach is to use a status signal generated by the data path component for condition testing at the time of appropriate branching. Since several such conditions exist, it is desirable to have an additional condition-select code subfield in the microinstruction format to incorporate all possible conditions that can happen. As the branch address is available in the body of the microinstruction itself, this condition code is tested and the corresponding branch address is loaded into the CMAR to reach the target instruction. CM space, while providing these branch addresses in a microinstruction, can even be minimized by storing some low-order bits of the address instead of the complete address. This technique, known as *short jump,* however, restricts the branching within the CM to a limited range. A lot of other alternative methods are also in use, and each method has its own merits as well as shortcomings.

6.4.2.7 Emulation

In a microprogrammed CPU, each machine instruction during execution is not substituted by its corresponding microroutine as happens in the case of compilation. Rather, the instruction is interpreted, and the corresponding microroutine, which acts as a real-time interpreter for the instruction, is executed. This process is known as *emulation,* and the set of microprograms that interpret a particular instruction set or machine language L of a CPU is called an *emulator* for L. A microprogrammed computer M_1 having machine language L_1 can be made to execute programs written in a different machine language L_2 of a very similar computer M_2, by placing an additional emulator for L_2 in the CM of M_1 keeping its own emulator of L_1 intact. M_1 in that situation is said to be capable of *emulating* M_2. Thus, the major benefit of microprogramming is that it permits more than one model of computers to have the same instruction set, and more than one instruction set per computer. An example of the former benefit is the IBM system 360/370 family of computers. An example of the latter benefit is the compatibility feature available on many modern computers as just explained with M_1 and M_2.

Emulation permits the replacement of obsolete equipment with a comparatively more up-to-date similar type of machine. If the new computer can fully emulate the old one, then almost no change or minimum change in software is required to run the programs of the old machine on the new one. Emulation ensures easy transitions from old to new machines with minimum effort even with machines having totally different architecture.

6.4.2.8 Merits and Drawbacks

As a design activity, microprogramming can be compared with assembly language programming. However, the microprogrammer does require a more detailed knowledge of the processor hardware than the assembly language programmer. Symbolic languages, similar to assembly languages, are normally used to write microprograms; these are referred to as *microassembly languages.* A **microassembler** is then required to translate such microprograms into executable programs that can be *stored in* CM in the fast ROM inside the CPU.

Microprogramming, while implemented in the design of a control unit, makes the microarchitecture straightforward enough. The decoders and sequencing logic unit of a microprogrammed unit are very simple pieces of logic in comparison with its counterpart, the *hardwired control unit*. A microprogrammed unit is found to be both cheaper and less error-prone, and above all, more flexible for incorporating necessary modifications and enhancements, if required. Microprogramming has overcome, to a large extent, the limitations of functional components used in a system.

The *major drawbacks* of a microprogrammed control unit are that the microinstructions, like normal machine instructions, are fetched and decoded, and thus consume an extra amount of time, which makes the entire system somewhat slower than its counterpart, the *hardwired control unit* of comparable technology. To present the user with a highly versatile ability, microprograms get larger and more complicated, their interpreter gets bigger and slower, and more room is required in chips to accommodate them. This again affects the design of the chips which is basically made of silicon transistors having relatively slower switching times.

Despite all these shortcomings, microprogramming is still considered a dominant technique for implementing control units in modern micro, mini, supermini, and large mainframe processors from giant manufacturers.

6.4.2.9 Application Areas

The first microprogrammed control computer was announced and launched by IBM in its System 360 family in April 1964. The advantages of microprogramming were compelling enough for IBM to make this move. Since then, microprogramming has become an increasingly popular vehicle for a variety of applications, especially for implementing the control unit of a CPU. Till then, many improvements in the proposed design have been constantly made to enhance the implementation techniques of the microprogramming approach and its effective optimization. With passing days, it has been also observed that the application areas have become increasingly varied and widespread. The major areas of current applications where microprogramming can be considered to be effectively exploited are as follows:

- Realization of microprogrammed computers;
- Emulation;
- Operating system support;
- High-level language support;
- Microcontrollers;
- Microdiagnostics;
- User-oriented tailoring.

A brief detail of this topic is given in the website: http://routledge.com/9780367255732.

6.4.3 Nanoprogramming

Nanoprogramming is a befitting approach in situations when many microinstructions occur several times in a microprogram. The concept of nanoprogramming was first introduced in the QM-1 computer designed sometimes in the 1970s by Nanodata Corporation.

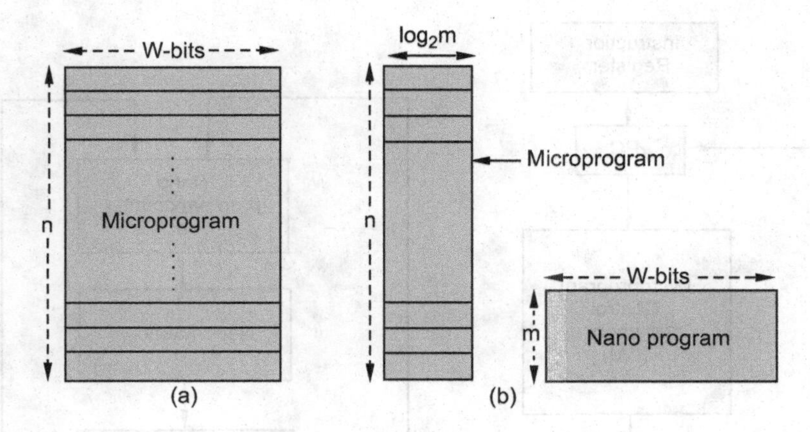

FIGURE 6.11
CM models: (a) conventional microprogram – one level; (b) corresponding nanoprogram – two levels.

Subsequently, this approach has been successfully exploited in many areas, including in the versatile Motorola 68000 series of microprocessors.

Figure 6.11 illustrates the concept of nanoprogramming. Part (a) shows a microprogram having n microinstructions, each W bits wide. A total of nW bits of CM is needed to store this microprogram. Assume that a careful study of this microprogram revealed that only m different microinstructions were actually in heavy use, with $m \ll n$. A special m-word W-bit nano-memory could be used to store m such unique microinstructions. Each microinstruction in the original program in CM could then be replaced by the address of the nano-memory word containing the desired microinstruction. Since the nano-memory contains only m words, the CM in question will only need instructions, each being $\log_2 m$ bits wide (the address of microinstruction held in nano-memory). This is illustrated in part (b) of the figure. Each microinstruction held in nano-memory is wide and allows for parallel operations within the logic unit. Thus, the use of CM is analogous to vertical microprogramming, since a relatively large number of small instructions are involved. The use of nano-memory, on the other hand, is analogous to horizontal microprogramming providing needed parallel operations.

The microprogram is executed as follows. The desired word is first fetched from the CM and is then placed in the microinstruction register. It is then used to point to the targeted nano-memory word to get the required microinstruction, which is then fetched and placed in the nanoinstruction register. The bits of this register are then used to control the gates for one cycle. At the end of the cycle, the next word is fetched from CM and the process is repeated. This is illustrated in Figure 6.12.

The CM space is not wasted since only those microinstructions that require the use of nano-memory will be pointing to it. Moreover, if two microinstructions in CM require identical nano-memory instructions, then each microinstruction can point to the same nano-memory instruction.

The **primary objective** of this approach is to save costly CM space, but at the cost of considerable slower execution. A machine with this two-level CM arrangement will run slower than the original one because the fetch cycle will now require two memory references: one to the CM and one to the nano-memory. These two fetches can never be overlapped to reduce time consumption.

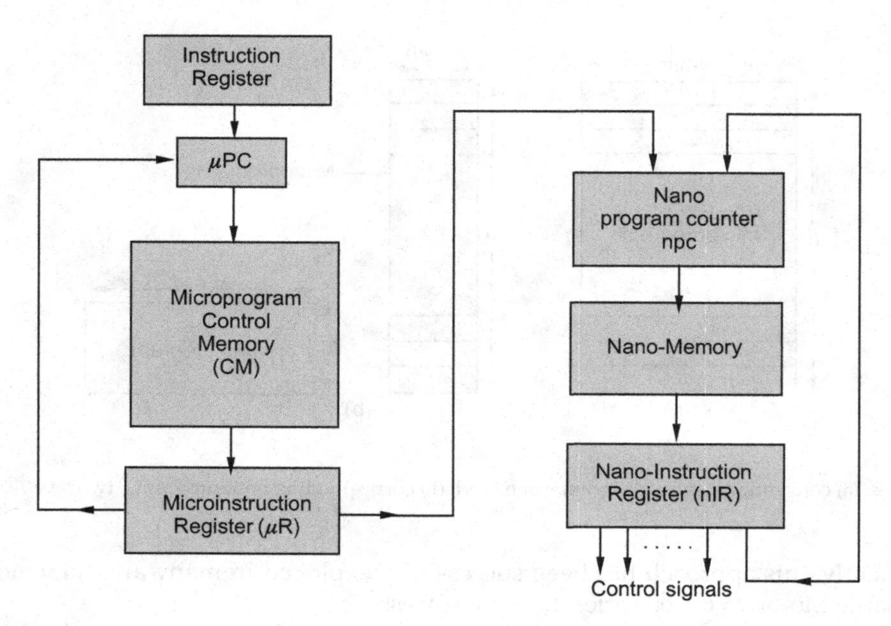

FIGURE 6.12
Two-level CM organisation for nanoprogramming.

Example

Suppose an original microprogram is of 4096 × 100 bits, but consists of only 128 different microinstructions. A nano-memory of 128 × 100 bits will be sufficient to hold all such microinstructions. The CM then becomes only of 4096 × 7 bits ($128 = 2^7$, 7 bits are required to uniquely define an address in nano-memory), and each word in CM will be pointing to a nanoinstruction. Hence,

Without using nanoprogramming, the required CM size is:

$$4,096 \times 100 \text{ bits.}$$

Using nanoprogramming, the total amount of CM size is

$$\left[4,096 \times 7(\text{control memory}) + 128 \times 100(\text{nano-memory})\right]$$

Thus, the memory saving in this example is:

$$4,096 \times 100 - \left[(4,096 \times 7) + (128 \times 100)\right] = 368,128 \text{ bits.}$$

Optimization in nanoprogramming, however, could further be made possible. Recall that nanoprogramming is best suited and is found to be most effective when the same microinstructions are heavily used. If two microinstructions that were not really the same but appear to be almost the same could be treated as being the same, then the microprogram would contain fewer distinct microinstructions, each with a higher frequency of usage. Optimization is now carried out with a slight variation from the basic idea. Words in the nano-memory may be parameterized. For example, assume that two microinstructions may differ only in the field indicating which register is to be gated onto some bus.

By putting the register number in the CM instead of in the nano-memory (the register field in the nanoinstruction contains zero), the two microinstructions can both point to the same word in nano-memory. When the word is fetched and put into the microinstruction register, the register field is taken from the CM instead of nano-memory. Obviously, this type of operation demands special hardware, which is again not very complex to realize. But this approach, in turn, will definitely increase the width of the CM, and hence, it is difficult to decide whether it is a profitable proposition or not. The type of microinstruction set in use, however, will have the last word in this context.

6.5 Summary

The design of a digital system, especially a CPU, consists of two typical individual parts: the control unit and the data processing unit. The function of a control unit is to issue control signals to the data processing unit at specified times that selects and sequences the desired data processing operations. Control units can be implemented by way of two distinctive fundamental approaches: hardwired control and microprogrammed control, and both these approaches with their salient features and implementation details have been discussed in the text with their individual strengths and drawbacks. Microinstructions are often interpreted by nanoinstructions which directly control the hardware. The primary objective of the nanoprogramming approach is to save costly CM space, but at the cost of indulging considerably slower execution due to a two-level CM arrangement that is operated in sequence and can never be overlapped. Nanoprogramming is best suited and is found to be most effective when the same microinstructions in CM are found in heavy use.

Exercises

6.1 Give the scheme of a generalized approach of a control unit module indicating the different types of inputs it requires and the corresponding various types of outputs it generates.

6.2 Describe the implementation technique of a hardwired control unit showing the inputs to be required and the purpose of using these inputs in this implementation.

6.3 Show the micro-operations and related control signals arbitrarily (in the same manner as shown in Figure 6.2 in the website: http://routledge.com/9780367255732) for the following instructions:

i. Load accumulator

ii. Store accumulator

iii. Add to accumulator

iv. Complement accumulator

v. Jump

6.4 Explain the difference between hardwired control and microprogrammed control. Define:
   i. Micro-operation
   ii. Microinstruction
   iii. Microprogram
6.5 Explain with an arbitrary example the microinstruction format for the CM and show with a diagram how mapping is done from an instruction code to microinstruction format. Is it possible to design a microprocessor without a microprogram? Are all microprogrammed computers also microprocessors?
6.6 What is CM and CW? Explain the organisation of CM. How control unit is implemented using CM?
6.7 Assume that a simple CPU has four major phases to its instruction cycle: fetch, indirect, execute, and interrupt. Two 1-bit flags designate the current phase in a hardwired implementation.
   a. Why these flags are needed?
   b. Why they are not needed in a microprogrammed control unit?
6.8 What are the factors required to be considered at the time of designing microinstructions?
6.9 Describe with a diagram the implementation technique used in the organisation of a microprogrammed control unit.
6.10 What do you mean by horizontal microprogramming and vertical microprogramming? Compare these two approaches of microprogramming. Explain with an example a horizontal microinstruction format and a vertical microinstruction format.
6.11 "Microprogramming technique is an innovative approach". Discuss its merits and drawbacks. What are the areas where this approach can be profitably used? Explain one such area along with the implementation of this technique.
6.12 What do you mean by nanoprogramming? Explain with an example how nanoprogramming improves the performance of microprogramming environment.

Suggested References

Segee, B. and Field, J. *Microprogramming and Computer Architecture*. New York: Wiley, 1991.

Carter, J. *Microprocessor Architecture and Microprogramming*. Upper Saddle River, NJ: Prentice-Hall, 1996.

Vassiliadis, S., Wrong, S., and Cotofana, S. "Microcode processing: Positioning and directions." *IEEEMicro*, vol. 23, no. 4, pp. 21–30, July–August 2003.

7

Arithmetic and Logic Unit Organisation

LEARNING OBJECTIVES

- To describe the different types of number systems, their numerical representations, and the relationships with each other.
- To enunciate in detail
  - Representation of decimal fixed-point and floating-point numbers in the binary system;
  - Floating-point representation with its normalized form, including IEEE standards;
  - Floating-point arithmetic and implementation of the floating-point unit;
  - Addition and subtraction of signed numbers with overflow consideration.
- To explain the basic structure of an ALU (arithmetic and logic unit) with the required hardware elements.
- To describe different types of basic adders and subtracters with the overflow design principle and practical high-speed adders, including carry-lookahead adder (CLA) and carry-save adder (CSA).
- To explain Booth's algorithm for multiplication of signed and unsigned numbers.
- To study the methods for division of signed and unsigned numbers using machine-based algorithms.

Every computer has machine-level instructions that perform numerous arithmetic and logic operations on data operands, both integer and real (floating-point) numbers, as well as characters and character strings given in the program to accomplish various tasks. To understand how these operations are being carried out, one should first know how numbers and characters are represented in a computer, and how they are then manipulated by appropriate algorithms for basic arithmetic and logic operations. The most natural way to represent a number in a computer system is straightaway by a string of bits, called a *binary number*, and a text character similarly can also be represented by a distinct string of bits called a *character code*.

7.1 Numerical Representations: Number Systems

The numerical values of various quantities used in every sphere of life are basically represented in two ways, namely **analog** and **digital**. But we are, at present, concerned only with the issues related to digital number systems. However, number can also be expressed in many different ways using different numbering system: the most common ones are the *decimal, binary, octal,* and *hexadecimal* systems, yet the *decimal system* is the most familiar one because of its everyday use. Moreover, conversions between each of these number systems are most regular and are often carried out to accomplish various tasks.

7.1.1 Decimal System

The decimal system is composed of *ten* different numerals or symbols or digits, such as 0, 1, 2, 3, 4, 5, 6, 7, 8, and 9, and as such, the *base* or *radix(r)* of this decimal system is 10. It is a *positional-value* system in which the value of a digit depends on its position. The various positions of the digits in the number relative to the decimal point carry different weights that can be expressed as powers (both positive and negative) of 10, the radix. The presence of decimal point, however, actually separates the positive power of 10 from its negative powers. However, the value of a number in decimal system is then calculated by summing up all multiplied values, individually of each digit with an integer power of 10 (the radix). *In general, any number is simply the sum of the products of each digit value and its positional value.*

7.1.2 Binary System

Unfortunately, the decimal number system seems to be not suitable to lend itself to convenient implementation in digital systems. Almost every digital system uses the *binary* (base-2) *number system* as the basic number system for its operations. Other number systems, however, are often used to interpret or represent binary quantities for the convenience of the users who work with these digital systems. In the binary system, there are only two symbols or possible digit values, 0 and 1 (don't confuse this 0 and 1 with the decimal 0 and 1), and hence, the *radix* or the *base* of this system is 2, and the ***binary digit*** is often abbreviated to the term ***bit***. Even with such a relatively small base, this base-2 system can be used at ease to represent any quantity that can be represented in decimal or in other number systems. The binary system is also a *positional-value system,* wherein each binary digit has its own *value* or *weight* expressed as a power of 2 (the *base* or *radix*).

All types of conversions from binary to decimal and decimal to binary, including their fractional representations, are given in the website: http://routledge.com/9780367255732.

7.1.3 Hexadecimal and Octal System

The hexadecimal (in short **hex**) number system uses 16 possible digit symbols, digit 0–9 plus the letters A, B, C, D, E, and F, and hence, it has the base 16. The digit positions as usual are weighted as powers of 16. The octal system uses eight possible digits, namely 0, 1, 2, 3, 4, 5, 6, and 7, and hence, it implies base 8. The digit positions here are thus weighted as powers of 8. Similar to the decimal system, here also the value of a number in both the systems is calculated by summing up all multiplied values, individually of each digit with an integer power of their respective base (radix, *r*). The relationship between hexadecimal, octal, decimal, and binary number systems is simple. Since the base of the hexadecimal system is 16 (= 2^4), and that of an octal system is 8 (= 2^3), each hexadecimal digit represents a group of *four* binary digits and each octal digit represents a group of *three* binary digits. It is important to note that hex digits A through F are equivalent to the decimal values 10–15.

All types of conversions from one to another of all these number systems, including their fractional representations, are given in the website: http://routledge.com/9780367255732.

7.1.3.1 Merits of Hex and Octal Systems

Hex and octal systems are often used in a digital system as one kind of a *shorthand* way to conveniently represent strings of longer sequence of binary digits (bits) in situations when human involvement is associated. Since, in today's computer environment, bit strings as long as 64 bits are not very uncommon, it is more convenient and less error-prone to write the binary numbers in hex or octal, and it is then relatively easy to convert back and forth between binary and either hex or octal. The conversion can be executed relatively quickly without going through any intermediate computations that make one to realize the effectiveness as well as the usefulness of this tool in digital systems.

7.1.4 BCD (Binary-Coded Decimal) Code

When numbers, letters, or words are represented by a special group of symbols, it is said to be *encoded*, and the group of symbols is simply called a *code*. All digital systems use some form of binary numbers for their internal operation, but the external world is still decimal in nature. This requires frequent conversions between decimal and binary systems that become quite complicated and even long for large numbers. For this reason, a means of *encoding decimal numbers* that combines some features of both the decimal and the binary systems is used in certain situations. If each digit of a decimal number is represented straightaway by its binary equivalent, the result is a code, called *binary-coded-decimal* (hereafter abbreviated as *BCD*). Since a decimal digit can be as large as 9, four bits are required to code each decimal digit, and only the four bit numbers from 0000 through 1001 are used (the binary code of 9 is 1001_2). In other words, only the 10 combinations out of the 16 possible 4-bit binary combinations are used here as code groups. As a result, BCD code always requires more bits than the straight equivalent binary code of a decimal number, and is therefore somewhat *inefficient*. To illustrate the BCD code, take a decimal number 235, in which each decimal digit is changed to its corresponding binary equivalent to express it in BCD code, and side by side, its straight binary equivalent is also shown.

2	3	5	(decimal)	
↓	↓	↓		
0010	0011	0101	(BCD)	12 bits
1 1 1 0 1 0 1 1			(binary)	8 bits

It is important to note that BCD is not another number system such as decimal, binary, octal, and hexadecimal. In fact, it is precisely the decimal system with each decimal digit converted to its binary equivalent, without considering the number as an entity; whereas, the binary representation of a decimal number takes the *complete decimal number* by value and represents it in binary. However, the distinct ***advantage*** of the BCD code lies in the relative ease of converting it to and from decimal, and is especially significant from a hardware point of view, because in a digital system, it is the logic circuits that perform the conversions to and from decimal.

7.1.5 Gray Code

When multiple input conditions in digital circuit are continuously changing almost at the same time at very high speeds, the situation may be misinterpreted, leading to an erroneous outcome. For example, with a 3-bit number, when the number 3 (binary 011) changes to 4 (binary 100), all three bits individually must change their states at the same time. In order to reduce the likelihood of a digital circuit misinterpreting a changing input, and also to facilitate such changes faster, the ***Gray code*** has been evolved as a way to represent a sequence of numbers. The unique aspect of the Gray code is that only one bit ever changes between two successive numbers in the sequence.

The mechanisms and its related circuits to convert a binary code to its corresponding Gray code, and vice versa, and also a table of 3-bit binary and corresponding Gray code values are shown in the website: http://routledge.com/9780367255732.

7.2 Number Representations: Binary Systems

Any arbitrary number can thus be represented with just the binary digits 0 and 1. In general, if an n-bit sequence of binary digits $b_{n-1}b_{n-2}\ldots b_1b_0$ is given, where $b_k = 0$ or 1 for $0 \le k \le n-1$, then this sequence can represent unsigned integer values V in the range 0 to 2^{n-1}, where

$$V = b_{n-1} \times 2^{n-1} + b_{n-2} \times 2^{n-2} + \cdots + b_1 \times 2^1 + b_0 \times 2^0$$

$$\sum_{k=0}^{n-1} 2^k b_k$$

For non-negative integers, this representation is simple and straightforward. For example, with a 6-bit sequence ($n = 6$), it is possible to represent the numbers from 0 to $2^6 - 1$, i.e. from 0 to 63, as (0 = 000000, 1 = 000001, ..., 63 = 111111). Similarly, the negative integers need to be properly represented along with the positive integers. In fact, out of several available conventions, the following three systems are of common interest to represent such numbers:

- **Sign-magnitude;**
- **1's (one's) complement;**
- **2's (two's) complement.**

7.2.1 Sign-Magnitude Representation

While positive values have identical representations in all systems, negative values have different representations. However, all of them involve treating the most significant (leftmost) bit being reserved to represent the sign of the number (known as *sign bit*), and the remaining ($n - 1$) bits in the n-bit number indicate their magnitude. If the sign bit is 0, the number is positive. Negative values are represented by changing the most significant bit from 0 to 1 in bit sequence of the corresponding positive value. The simplest form of representation that employs a sign bit is known as *sign-magnitude representation* in which the most significant (leftmost) bit represents the sign of the number, and the rightmost $(n - 1)$ bits of an n-bit word hold the magnitude of the integer. Sign-magnitude representation, however, suffers from several critical drawbacks, including two different representations of 0 (e.g. $+ 0_{10} = 000000$, and $- 0_{10} = 100000$), and as such summarily being dropped from favour while integer arithmetic is implemented in the hardware (ALU) of the computer.

7.2.2 1's (One's) Complement Representation

This representation also makes use of the leftmost bit of the binary number to represent the sign. But it differs from the sign-magnitude representation in the way in which the other bits are interpreted. Here, the negative value of a number is obtained by simply complementing each bit of the corresponding positive number, including the sign bit. As a result, all positive integers in this representation have the leftmost bit always equal to 0, and all negative integers necessarily have the leftmost bit equal to 1. While this representation is more or less consistent for all types of integer arithmetic carried out by the hardware (ALU) of the computer, similar to sign-magnitude, this representation also suffers from the same drawback: the digit 0 here also has two different representations ($+ 0_{10} = 000000$, and $- 0_{10} = 111111$).

7.2.3 2's (Two's) Complement Representation

In order to alleviate the limitations and drawbacks of the above-mentioned two methods, the most common scheme to represent negative numbers is 2's (two's) complement representation in which the most significant (leftmost) bit is also used here as sign bit. But it differs from the other two methods in the way in which the other bits of the integer are interpreted here. The *2's complement* of a positive number is obtained by taking the Boolean complement of each bit of the corresponding positive number (as is done in 1's complement), and then adding 1 to the resulting bit pattern viewing it as an unsigned integer. Most important is that the number 0 is identified as *positive*, and therefore has a 0 sign bit and a magnitude of 0s in all remaining bits. For a n-bit *positive* number, the sign bit , b_{n-1}, is always zero, and the remaining bits in the number represent the magnitude of the number in the same manner as usual. Therefore, the range of positive integers that can be represented is from 0 to $2^{n-1} - 1$ (the largest integer is $2^{n-1} - 1$, with the sign bit zero, and all of the magnitude bits here are 1). Any large number beyond this range would then require more bits to represent. For a n-bit negative number, the sign bit, b_{n-1}, is 1, and the remaining $n - 1$ bits together can take on any one of the 2^{n-1} values. Therefore, the range of the negative integers that can be represented with n bits is from -1 to -2^{n-1}.

It is always attempted to assign the weight to the bit of negative integer in such a way that arithmetic operation can be handled straightaway. In *unsigned integer* representation, to compute the value of an integer from the bit representation, the weight of the most significant bit is $+2^{n-1}$. For a representation with a *sign bit,* it turns out that the desired arithmetic properties can be achieved, if the weight of the most significant bit is made -2^{n-1}. This is the convention used in 2's (two's) complement representation, which finally gives the following expression to symbolize negative numbers:

$$\text{2's (Two's) Complement of } V = -2^{n-1} b_{n-1} + \Sigma 2^k b_k$$

This equation is ultimately able to define the ***2's (two's) complement representation*** for both positive and negative n-bit integers. For positive numbers, the sign bit $b_{n-1} = 0$, which makes the term $-2^{n-1}b_{n-1} = 0$, and the above equation then reduces to the usual definition of a non-negative integer. For negative numbers, the sign bit $b_{n-1} = 1$, which makes the first term to -2^{n-1} which is to be then added with summation term: i.e. 2^{n-1} is to be subtracted from the summation term to yield a negative integer. The 2's (two's) complement system is the most efficient method for performing important arithmetic operations, addition and subtraction, although it appears to be a somewhat unnatural representation from the human point of view. Still, it is almost universally accepted for use as the processor representation for all types of integers.

7.2.3.1 Conversion: Decimal to 2's Complement and Vice Versa

The nature of the 2's (two's) complement representation can be usefully exhibited with the help of a graphical device, like a *value box,* in which the value on the far right of the box is 1 (i.e. 2^0), and each succeeding position to the left is double in value (i.e. the power of 2 is incremented by 1), until the leftmost position which is negated (Figure 7.1a). It is clear from Figure 7.1a that the highest negative 2'scomplement number that can be represented is -2^{n-1}. If any of the bits other than the sign bit is 1, it adds a positive amount to the number.

Abnormality: Although the 2's complement operation is always valid to derive the negation of any signed integers, there are two unusual cases that should be taken into account. The first one is that the 2's complement (i.e. negation) of number 0 expressed in 8-bit representation gives rise to a carry 1 in the ninth bit position. If it is ignored, the correct result is obtained, which means that the negation of 0 is 0 as it should be. *Ignoring* this carry-out is, however, a natural approach in order to obtain the correct result. The **second** unusual case seems to be even more critical. If a number is taken having a bit pattern of 1 followed by $n-1$ zeros, the 2's complement operation of this number (i.e. negation) will revert back the same number (including the sign) which should not be. The number 10000000 (2's complement representation of –128) when are made 2's complemented, the same number (including the sign) is once again obtained, which is definitely an abnormality.

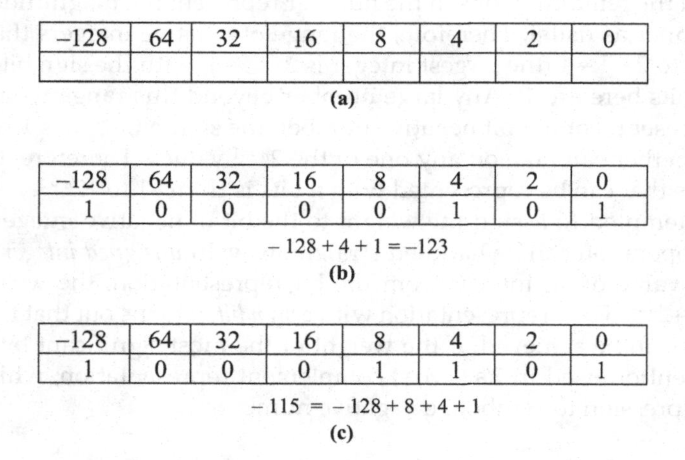

–128 64 32 16 8 4 2 0
(a)
–128 64 32 16 8 4 2 0
1 0 0 0 0 1 0 1
– 128 + 4 + 1 = –123
(b)
–128 64 32 16 8 4 2 0
1 0 0 0 1 1 0 1
– 115 = – 128 + 8 + 4 + 1
(c)

FIGURE 7.1
Value box representation for conversion between 2's complement binary and decimal. (a) An eight position 2's (two's) complement value box, (b) conversion of binary 2's complement value 10000101 to decimal, and (c) conversion of decimal –115 to binary 2's complement value.

Conversion between different bit lengths: It is often necessary to represent a number of an n-bit integer to another form of m-bits where $m \geq n$. The prescribed rule for *converting 2's complement n-bit integers to m-bit* integers is to move the sign bit to the new leftmost position and fill all other extended bits with copies of the sign bit. For positive numbers, fill in extended bits with zeros, and for negative numbers, fill in those with ones. This action is usually called ***sign extension***.

A brief detail with explanation of all these topics is given in the website: http://routledge.com/9780367255732.

7.3 Addition and Subtraction: Signed Numbers

Addition of two numbers in 2's complement representation proceeds simply as if the two numbers were unsigned integers. To *add* two numbers, add their n-bit representations, ignoring the *carry-out* bit from the most significant bit position. The sum will be the algebraically correct value in the 2's complement notation as long as the answer lies in the range of -2^{n-1} through $+2^{n-1} - 1$. Addition of two negative numbers sometimes requires discarding the carry bit, and the sum value is then 2's complemented to obtain the correct result in the usual form. Still, there may be some instances in any addition process when the result may be larger than can be held in the n-bit length being used. This condition is called *overflow*, which will be discussed in the next section.

Subtraction of two numbers in 2's complement representation can be easily carried out using the rule which tells that to *subtract* two numbers, A and B, i.e. to perform $A - B$, form the 2's complement (negation) of B, and then add it to A. Here, B is the *subtrahend* and A is the *minuend*. As usual, the result will be algebraically correct if the answer lies in the range of -2^{n-1} through $+2^{n-1} - 1$ in the 2's complement representation system. When the answers do not fall into the representable range, it is said that arithmetic *overflow* has occurred, which will be discussed in the next section.

7.4 Overflow: Integer Arithmetic

In the 2's complement number system, n bits can represent the values that must lie within the range -2^{n-1} to $+2^{n-1} - 1$. For example, with 4 bits, the range of numbers that can be represented is −8 through +7. When the result of an arithmetic operation falls outside the representable range, it is said that an *arithmetic overflow* has occurred. When an overflow occurs, the ALU within the processor must signal this event so that no further attempt would be taken henceforth to use the result.

With n-bit unsigned numbers, overflow is indicated by an output carry bit $c_{n-1} = 1$. For example, adding the (8-bit) unsigned numbers $X = 230_{10} = 11100110$ and $Y = 40_{10} = 00101000$ using an adder yields $Z = X + Y = 11100110 + 00101000 = 00001110$ with a carry bit = 1 from c_{n-1}, i.e. from c_7. Now the result $Z = 00001110$ corresponds to 14_{10}, which is $230_{10} + 40_{10}$ (modulo $2^8 = 256$) and is the result of this addition that "wraps around" when the largest number $2^n - 1$, in this case $11111111 = 255_{10}$, is exceeded. On appending c_7 (= 1) to Z, we get

$c_7\,Z = 100001110 = 270_{10} = 256_{10} + 14_{10}$, which is the sum in ordinary (modulo infinity) arithmetic. Unsigned arithmetic operations are often viewed as modulo–2^n operations only, and overflow is not explicitly detected. This happens when computing memory addresses in a computer system, for instance, when addresses simply wraparound to 0 after the highest address is reached.

However, this does not work when signed numbers are added (or are involved in arithmetic operation). For example, with 4-bit signed numbers, if we go to add numbers +6 (0110) and +5 (0101), the output in 4-bit will be 1011, which is the code of–5, an incorrect result. Here, the carry-out signal from the most significant bit (here, it is 4th bit) position is 0. Similarly, if we try to add–5 (1011) and –4 (1100), we get the output in 4-bit as 0111, which is equal to +7, another incorrect result, and in this case, the carry-out signal from the most significant bit (here it is 4th bit) position is 1. Thus, overflow is caused by carry (not always, see Figure 7.2a), but the implication of carry and overflow is not the same. Usual carry can be ignored, and the result is still correct. Unusual carry results in overflow, and the answer of an arithmetic operation then becomes incorrect. *Overflow may occur if both summands have the same sign*, as shown in Figure 7.2. Clearly, the addition of numbers with different signs cannot cause overflow. This leads to the following conclusions:

1. Overflow can occur only when two numbers are added that have the same sign;
2. The carry-out signal from the most significant bit position is not a sufficient indicator to designate the overflow when adding signed numbers.

Instead, a simple way to detect overflow is to examine the sign of the two summands A and B and the sign of the result. When both the operands A and B have the same sign, and the result is not the same as the sign of A and B, an overflow then occurs. In summary, the prescribed rule is:

Overflow rule: When two numbers are *added*, and they are either both positive or both negative, then overflow occurs, if and only if the result has the opposite sign.

ADDITION

(a)		(b)	
0100	(+4)	1010	(−6)
+ 0101	(+5)	+ 1001	(−7)
1001	Overflow	10011 ↑	Overflow
The result is −7			

SUBTRACTION

(c)				(d)			
0110	(+6)	0110	(+6)	1010	(−6)	1010	(−6)
− 1010	(−6) ➔	+ 0110	(+6)	− 0100	(+4) ➔	+ 1100	(−4)
		1100	Overflow			10110	
The result is −4						Overflow ↑	

FIGURE 7.2
Representation of overflow with signed numbers.

Carries 0 1	Carries 1 0
+ 83 = 0 1 0 1 0 0 1 1	−95 = 1 0 1 0 0 0 0 1
+ 62 = 0 0 1 1 1 1 1 0	− 79 = 1 0 1 1 0 0 0 1
+145 = 1 0 0 1 0 0 0 1	−174 = 0 1 0 1 0 0 1 0
(a)	(b)

FIGURE 7.3
Representation of overflow with signed numbers.

Alternative approach: Overflow can be identified and detected using a different mechanism, but, of course, keeping intact the fundamental requirements as mentioned above in conclusions (1) and (2). This means that overflow may occur if the two numbers being added are of the same sign. Numbers having different sign when added will not give rise to any overflow, because the result after addition will produce a result which will be always smaller than the larger of the two original numbers. To explain and exhibit the overflow, consider the following examples of 8-bit addition, as shown in Figure 7.3, with 8-bit signed integers in 2's complement form in which the most significant bit in each binary number is the sign bit.

Clearly, the results as obtained in both the cases are incorrect. The 8-bit result as obtained in (a) should have been positive but has a negative sign bit. Similarly, the 8-bit result as obtained in (b) should have been negative but has a positive sign bit. However, if the *carry-out* from the sign bit position is treated as the sign (it is 0 in (a), and it is 1 in (b) as shown by the underline at the top of the figure) of the result, then the 9-bit answer thus obtained in each case is found to be correct. Since the 9-bit result cannot be accommodated in 8-bit storage, it is declared that overflow has occurred.

A simple way to detect overflow in this approach is to examine the *carry into* the sign bit position (i.e. C_{n-1} at a_{n-1} bit position of the integer, which is 1 in example (a)) and the *carry out* from the sign bit position (i.e. C_n at an assumed bit position a_n which is at the left of the sign bit a_{n-1} of the integer, which is 0 in example (a)). If these two carries are both 0 or are both 1, then there is no overflow. If these two bits are different (i.e. the carries jointly are either 01 or 10), then an overflow exists. In other words, the overflow occurs if the Boolean expression $C_n \oplus C_{n-1}$ is 1 (i.e. true).

7.5 Characters

Similar to the representation of numbers as is carried out by the computer, it must be able to handle non-numeric text information consisting of alphabetic as well as special characters. Set of characters represented by computers can be letters of the alphabet, decimal digits (treated as characters, and not their values), punctuation marks, special characters, and similar other characters. They are represented by specific prescribed codes that are 8-bits (one byte) long. One of the most widely used such standard code is the American Standards Committee on Information Interchange, in short ***ASCII code,*** in which each character is represented as a 7-bit code. Another standard code, especially used by IBM and others, is ***EBCDIC code*** (Extended Binary Coded Decimal Interchange Code) in which eight bits are used to denote a character.

7.6 Arithmetic and Logic Unit (ALU)

Most computer operations, including arithmetic and logical operations on data, are executed in the ALU within the processor at the machine instruction level. The ALU is many times faster than any other resources connected to a computer system. To facilitate ALU to work, all other components of the computer system, such as memory, registers, I/O (input/output) units, control unit, and buses, participate in the working of ALU operation, mainly to supply data into the ALU for it to process, and subsequently to take the results back out, if required. Operands are brought for specific operation and are temporarily stored in the highest-speed storage elements called *registers* connected with the ALU within the processor. Each register can store one word (32 bits or 64 bits) of data. A block representation of an ALU along with its interconnections with the rest of the processor is depicted in Figure 7.4. The ALU, in fact, is a combinational circuit so that most of its operations, including entire register transfer operations from the source registers through the ALU, and finally into the destination register, can be performed mostly in one clock pulse period. The control unit actually provides the necessary appropriate signals to monitor and control the entire operation of the ALU as well as the movement of data (operands) to and from the ALU.

Compared with arithmetic operations, logic operations are relatively simple to implement using combinational circuitry. They require only independent Boolean operations on individual bit positions of the operands, whereas carry/borrow lateral signals are required in arithmetic operations.

7.7 Fixed-Point Arithmetic

Four basic arithmetic operations for fixed-point as well as for floating-point numbers are *addition, subtraction, multiplication,* and *division.* The arithmetic algorithms and the related logic circuits needed to implement these arithmetic operations are the main focus of this section and onwards. It has been shown that two *n-bit signed numbers* expressed in 2's complement representation can be added/subtracted using *n*-bit binary addition/subtraction,

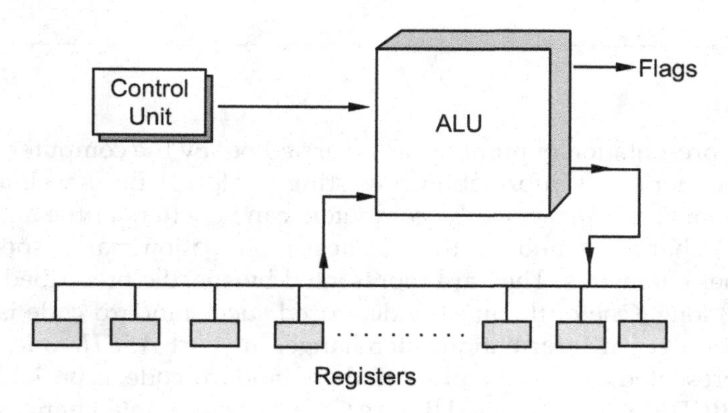

FIGURE 7.4
Representative block diagram of ALU with inputs and outputs.

treating the sign bit the same as the other bits. In other words, a logic circuit that is designed to add/subtract *unsigned* binary numbers can also be used equally well to add/subtract *signed* numbers in 2's complement form. The time needed by an ALU to perform an arithmetic operation often influences the performance of the processor. Addition/subtraction in this regard takes relatively lesser time than multiplication and division which require more complex circuitry than either addition or subtraction operation. That is why, some modern techniques are used in today's advanced computers that can perform all arithmetic operations at a comparatively higher speed.

7.7.1 Addition and Subtraction

Addition and subtraction for fixed-point numbers are the most fundamental operations that are found in the instruction set of every computer since inception. The elementary scheme to accomplish addition and subtraction is shown in the form of a basic circuit in Figure 7.5, which consists of a binary adder as central element that executes addition/subtraction operation with two *unsigned integers* sent from two registers X and Y, and then produces the result with an overflow indication. The result after computation may be stored in one of these registers or in a third register (depending on the design). The overflow, if happened, is usually stored in a 1-bit flag (for 0 = no overflow, 1 = overflow). For subtraction, the subtrahend, i.e. here the content of register Y, is passed through a 2's complementer so that the 2's complement of its content can be presented to the adder. Appropriate signals from the control unit are here required to ensure whether the contents of register Y will need to go through the complementer or not, which again depends on whether the operation is subtraction or addition, decoded by the control unit in advance. To implement all these, the required hardware can, however, be designed in many different ways that involve various trade-offs between operating speed and related hardware cost.

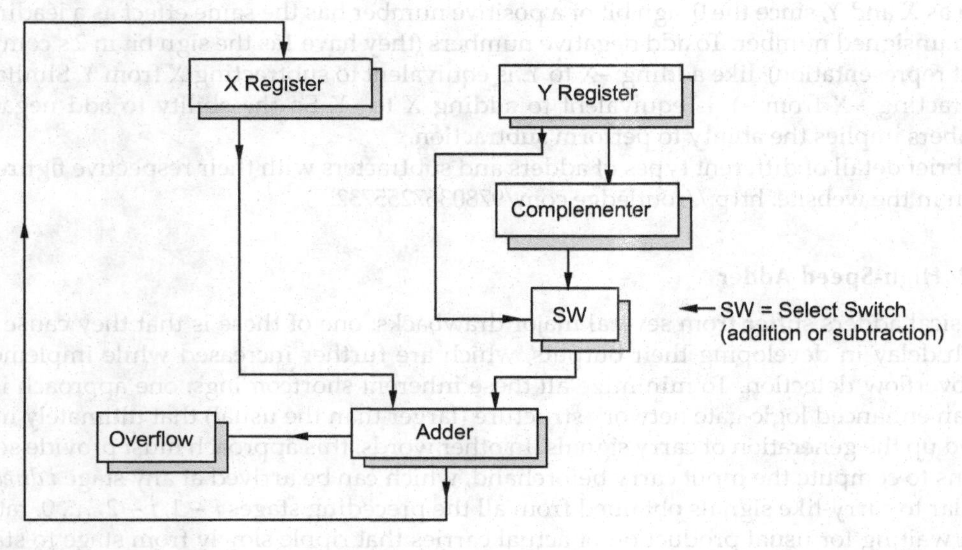

FIGURE 7.5
Schematic block diagram with hardware components for addition and subtraction.

7.7.1.1 Basic Adders

In general, the addition of two *n-bit numbers* X and Y is performed by subdividing the numbers into stages X_i and Y_i of length n_i where $n > n_i > 1$. X_i and Y_i are added separately, and the resulting partial sums are then combined to form the overall sum. The formation of this sum, however, involves assimilation of carry bits generated by the partial additions. However, a 1-bit adder (bit-by-bit addition), also called a ***full adder***, can be directly implemented in various ways that can be modified even to work as a full subtracter, since subtraction of a number (*Y*) from another number (*X*) can be defined as:

$$X - Y = X + (\text{two's complement of } Y) = X + \bar{Y} + 1$$

Thus, the subtraction can also be done using a full adder.

Parallel adder: Circuits that can add all bits of two *n-bit numbers* at a time, together with an external carry-in signal c_{in}, in one clock cycle are called *parallel adders* or simply *n*-bit adders. The simplest such adder can be formed by connecting *n* numbers of 1-bit full adder giving rise to a cascade configuration of *n-bit* full adder. As each 1-bit adder stage may supply a carry-bit to the stage on its left, all such carry signals or ripples must then propagate through this (cascade) adder from right to left, giving the name of this configuration an *n-bit ripple-carry adder or carry-propagation adder* (*CPA*). In the worst case, a carry signal can ripple through all *n* stages of the adder. The input carry signal c_{in} is normally set to 0 for addition. The maximum propagation delay of an *n*-bit ripple-carry adder is *nd* (*d* is the delay of a full-adder stage), which essentially determines its operating speed in synchronous circuit design. This adder is, however, relatively expensive in terms of hardware cost than a serial adder, and the cost simply increases linearly with *n*, the word size of the numbers to be added.

7.7.1.2 Subtracters

The *n*-bit adders can operate addition correctly on both unsigned and positive numbers such as X and Y, since the 0 sign bit of a positive number has the same effect as a leading 0 in an unsigned number. To add negative numbers (they have 1as the sign bit in 2's complement representation), like adding $-X$ to *Y*, is equivalent to subtracting *X* from *Y*. Similarly, subtracting $-X$ from $-Y$ is equivalent to adding *X* to $-Y$. So the ability to add negative numbers implies the ability to perform subtraction.

A brief detail of different types of adders and subtracters with their respective figures is given in the website: http://routledge.com/9780367255732.

7.7.2 High-Speed Adder

Classical adders suffer from several major drawbacks: one of these is that they cause too much delay in developing their outputs, which are further increased while implementing overflow detection. To minimize all these inherent shortcomings, one approach is to use an enhanced logic-gate network structure (larger than the usual) that ultimately must speed up the generation of carry signals. In other words, this approach must provide some means to compute the input carry beforehand, which can be arrived at any stage *i directly*, similar to carry-like signals obtained from all the preceding stages $i - 1, i - 2, \ldots, 0$, rather than waiting for usual production of actual carries that ripple slowly from stage to stage. Adders that use this principle are called ***carry-lookahead adders***.

7.7.2.1 Carry-Lookahead Adder (CLA)

The basic principle lying behind an *n*-bit CLA is that it is formed from *n* stages, each of which is basically a full adder, but modified by replacing its carry output line c_i at stage *i* by two auxiliary signals called g_i (*generate*) and p_i (*propagate*), respectively. Let z_i be the sum, and c_i be the carry-out of two 1-bit numbers x_i and y_i at stage *i*. Then, by suitable mathematical calculation, it can be shown that the generate function g_i and the propagate function p_i at stage *i* can be defined only in terms of x_i and y_i at the stage *i* without depending on the value of the carry of its just previous stage c_{i-1}. Moreover, all g_i and p_i functions for all *i*'s in an *n*-bit adder can be formed *independently* and *in parallel* in one logic-gate delay after the X and Y vectors are applied to the inputs of the adder. Similarly, the output function z_i at the stage *i* can also be defined in terms of x_i and y_i at that stage *i*.

Although a substantial improvement is thus achieved by this design in reducing critical gate delays, the number of gates, however, grows almost in proportion to n^2 as *n* increases. In contrast, the number of gates in a two-level adder of the sum-of-products type grows exponentially with *n*, while the number of gates in a ripple-carry adder grows linearly with *n*. Moreover, the complexity of the carry-generation logic embedded in the CLA, including its gate count, its maximum fan-in, and its maximum fan-out, increases steadily with the increase in *n*. Moreover, it is found that the last AND gate and the OR gate require a fan-in of $i + 2$ in generating c_i. For c_3 in the 4-bit adder, a fan-in of 5 is thus required. This is not only almost about the limit of practical gates, but consideration of cost in practice is also a matter of concern that often limits *n* in a single CLA module not to exceed four or so.

More details of CLA with mathematical calculations and corresponding figures are given in the website: http://routledge.com/9780367255732.

7.7.2.2 Adder Expansion

CLA, no doubt, is relatively enriched with several distinct advantages compared to all of its contemporary counterparts. But the inherent shortcomings as already explained encompassed with its embedded carry-lookahead logic often limit *n* in the addition/subtraction operation of *n*-bit integers to keep *n* bound within four or so. That is why, this adder design cannot be directly extended to longer operand (increased value of *n*) sizes.

One possible solution in this regard may be to improvise a method by combining ripple-carry propagation and carry-lookahead approaches together to let carry signals be handled in a reasonably efficient way. This approach, however, can be easily exploited to design larger adders of this kind needed to execute add instructions with longer operand sizes, say up to 64 bits. If we replace *n* number of 1-bit (full) adder stages in the *n*-bit ripple-carry design with *n* number of *k*-bit CLAs, we obtain an *nk*-bit adder. For example, four 4-bit adders, each one of a 4-bit carry-lookahead circuit, can be then connected in ripple-carry design to realize a relatively larger 16-bit adder. Similar approach can be employed with eight 4-bit CLAs to be connected to form a 32-bit adder.

The detail of this approach with related figure is given in the website: http://routledge.com/9780367255732.

7.7.2.3 Carry-Save Adder (CSA)

The inherent drawbacks of CLA, or of any derived adder based on this principle, often limit *n* in the addition/subtraction operation of *n*-bit integer that keeps *n* bound within a small range. In practice, arithmetic operation often requires longer operand (increased value of n) and, at the same time, the addition of several operands at a time to realize the final result (as happens with multiplication operation that requires the addition of several partial products as summands to arrive at the final result). Moreover, simultaneous addition of multiple operands is almost a regular feature in any computation, apart from the event of parallel processing. Hence, more powerful fast adders are required for negotiating such situation that can add many operands at a time instead of only two. A technique thus devised by which it is possible to accomplish high-speed multioperand addition is known as ***carry-save addition***, and the respective adder is called ***carry-save adder*** (*CSA*) that eventually speeds up the addition process. To illustrate the strength and effectiveness of such an adder, consider the following example with decimal integers as shown in Figure 7.6.

In this example, four decimal numbers are taken for addition. Firstly, all the digits in the unit place (first column) are added that produce a sum of 9 and a carry digit of 1 written in the second line shown by arrow (←) shifted left by one position. Similarly, all the digits in the tens place are added that produce a sum of 0 and a carry digit of 2 which is written in a similar way like the addition of digits in the unit place. Likewise, all the digits in the hundreds place are added that produce a sum of 8 and a carry digit of 1 which is written in a similar way as mentioned above. These summations (column-wise) can be then executed in parallel to produce a sum vector of 809 and a carry vector of 121, since no propagation of carry is required from the addition of digits in units place to the addition process of digits in tens place, and similarly from the tens place to the hundreds place, and so on. When the addition of all digits present in the operands is over, thereby producing a sum vector and a carry vector, the sum vector and the shifted carry vector are then simply added in the conventional way using either CPA or CLA that produces the final result.

Assume that the CSA here takes three binary numbers I, J, and K, and produces two outputs, namely the *sum vector* S and the *carry vector* C. The *sum vector* S and the *carry vector* C, however, can be obtained using the following expressions:

$$S = I \oplus J \oplus K$$

and

$$C = IJ + JK + IK$$

Here, all logical operations are performed bit-wise.

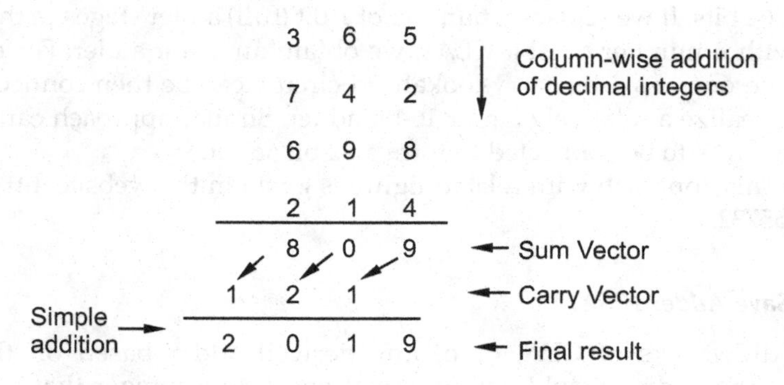

FIGURE 7.6
An illustration of carry-save addition technique with decimal numbers.

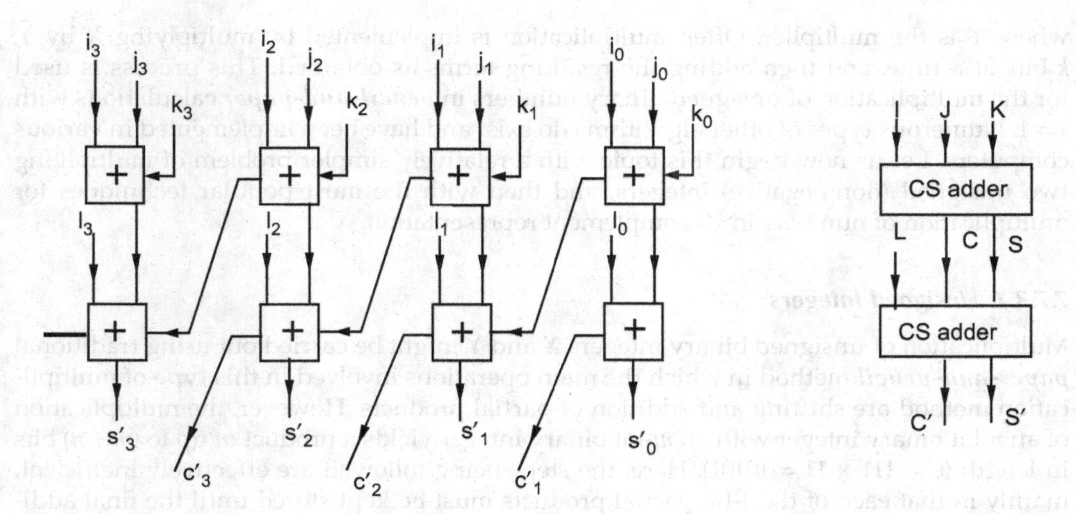

FIGURE 7.7
A representative scheme of a two-stage CSA.

The final arithmetic sum of three inputs, i.e. Z = I + J + K, is obtained by adding the two generated outputs S and C, i.e. Z = I + J + K = C + S, using a CPA or CLA.

In general, an *n*-bit CSA consists of *n* disjoint full adders. As shown in Figure 7.7, its input is three *n*-bit numbers to be added, while the output consists of the *n* sum bits forming a word S, and the *n* carry bits forming a word C. Unlike the adders discussed so far, there is no carry propagation within the individual adders. The outputs S and C can be fed into another *n*-bit CSA where, as shown in Figure 7.7, they can be added to a fourth *n*-bit number *L*. It is to be noted that the carry connections are shifted to the left to correspond to normal carry propagation. In this way, *m* numbers can be added using a tree-like network of CSAs to produce a result of the form (S, C). However, to obtain the final sum, S and C must be added by a conventional adder (CPA or CLA) with usual carry propagation.

Carry-save addition mechanism appears to be highly conducive and well suited to pipelined implementation. Multistage CSA circuit can now be built up in the form of a **Wallace tree** (see Chapter 8) to handle these groups (all groups in one stage) as just explained, and can be then implemented as a sequence of substages within the multiplier circuit built in the *execute stage* in a pipelined architecture. The type of operands used can also, however, be modified easily to handle *floating-point numbers.* The input significands (*mantissa*) are then processed in a fixed-point multiplier pipeline. The *exponents* are combined by a separate fixed-point adder, and a *normalization* circuit can be then included at the appropriate position in the *execute* stage within the pipeline.

7.7.3 Multiplication

Fixed-point multiplication is relatively a complex operation compared to addition/subtraction whether implemented in hardware or software. It requires comparatively more hardware, and as a result, it is not usually included in the instruction set of smaller processors. Multiplication is normally there implemented by a simple but slow method of repeated addition. To compute $X \times Y$ is only to add the multiplicand X to itself *Y* times, where Y is the multiplier. Often multiplication is implemented by multiplying X by Y, k bits at a time, and then adding the resulting terms as obtained. This process is used for the multiplication of unsigned binary numbers in ***pencil-and-paper*** calculations with $k = 1$. Numerous types of other algorithms do exist and have been implemented in various computers. Let us now begin this topic with a relatively simpler problem of multiplying two unsigned (non-negative) integers, and then with the most popular techniques for multiplication of numbers in 2's complement representation.

7.7.3.1 Unsigned Integers

Multiplication of unsigned binary integers X and Y might be carried out using traditional ***paper-and-pencil*** method in which the main operations involved in this type of multiplication method are shifting and addition of partial products. However, the multiplication of an n-bit binary integer with an m-bit binary integer yields a product of up to $(m + n)$ bits in length (e.g. $111 \times 11 = 10101$). Here, the steps being followed are effectively inefficient, mainly in that each of the 1-bit partial products must be kept stored until the final addition step is completed. To make it more efficient, firstly, it is possible to perform a running addition on the partial products rather than waiting until the end. This means that it is desirable to add each partial product term as it is generated to the sum of the preceding terms to form the current *partial product*. This eliminates the need for storage of all the partial products; of course, fewer additional registers are needed to accomplish this. Secondly, it is feasible to save some amount of time on the generation of partial products; for each 1 on the multiplier, an add operation and a shift operation are required, but for each 0 in the multiplier, only a shift is required.

Shift-and-add algorithm: Following the above-mentioned strategy, multiplication can now be performed using an adder circuitry in the ALU, and carrying out a number of sequential steps. The circuit performs multiplication by using a single n-bit adder n times to implement the spatial addition needed in this multiplication. At the start, the multiplicand and multiplier are loaded into two registers (say, X and Y), respectively. A third register (say, the Z register) is also needed and is initially set to 0. There is also a 1-bit register (say, C register) that holds a potential carry bit generated from addition. Initially, this register is set to 0.

The multiplication operation is carried out in the following way. Control logic reads the bits of the multiplier ($Y_{n-1}, Y_{n-2}, \ldots Y_0$) starting from Y_0 one at a time. If $Y_0 = 1$, then the multiplicand (X) is added to the Z register, and the result is stored in the Z register, with the C bit being used to hold carry. Then, all of the bits of the C, Z, and Y registers are shifted one bit to the right so that C bit goes into Z_{n-1}, Z_0 goes into Y_{n-1}, and Y_0 is lost. If $Y_0 = 0$, then no addition is performed, and only the right-shift is carried out. This process is repeated for each bit of the original multiplier. Each cycle is ended with C, Z, and Y being shifted right one bit position to allow for growth of the partial product as the multiplier is shifted out of register Y. Because of this shifting, multiplier bit Y_i appears at the LSB (least significant bit) position of Y to generate the Add/No-add signal at the correct time, starting with Y_0 during the first cycle, Y_1 during the second cycle, and so on. In this way, all the multiplier bits after being used are discarded by the right-shift operation. Note that the carry-out from the adder held in C register is the leftmost bit of the partial product $(i + 1)$, and it must be shifted right with the contents of Z and Y. After n cycles (end of multiplication), the high-order half of the product is held in register Z, and the low-order half is in register Y, giving the resulting product of $2n$-bits.

A brief detail of this topic with required hardware and also an appropriate flowchart of this scheme as well as an example showing its working are given in the website: http://routledge.com/9780367255732.

7.7.3.2 Signed-Magnitude Numbers

The multiplication of *signed-magnitude* numbers requires a straightforward extension of the unsigned case as already discussed above. The magnitude part of the product $P = X \times Y$ is computed as usual by the shift-and-add multiplication algorithm, and the sign p_s of product P is computed separately from the sign of X and Y as follows: $p_s := x_s \oplus y_s$. The signed-magnitude multiplication can be, however, then implemented on the same lines using the same sequential method as already described, after including the computation of the sign of the product result separately in the existing circuit.

7.7.3.3 Signed-Operand Multiplication

Multiplication of 2's complement signed operands is somewhat different and exhibits some peculiarities, especially when the operands are negative: e.g. if we consider, multiplying unsigned integers 11 (1011) by 13 (1101) to yield a product 143 (10001111). If we interpret these two numbers in 2's complement representation, we have –5 (1011) multiplied by –3 (1101), giving –113 (10001111). This example reveals that straightforward multiplication will not work if both the multiplicand and the multiplier are negative. In fact, it can be shown by other examples that straightforward multiplication will also not work if either the multiplicand or the multiplier is negative. More precisely, it can be said that the multiplication process must then treat positive and negative operands differently.

To resolve this dilemma, there exist a number of ways. One of these would be to convert both the multiplier and multiplicand to positive numbers, perform the multiplication, and then take the two's complement of the result if and only if the sign of the two original numbers differs. In order to avoid all such complications, particularly the final transformation step as mentioned above, one simpler technique that is most commonly used is Booth's algorithm. This approach additionally has the merit of potentially speeding up the multiplication process, relative to a more straightforward approach.

7.7.3.3.1 Booth's Algorithm

One of the most elegant and widely used schemes for two's complement multiplication was proposed by Andrew D. Booth sometimes in 1951. Booth's algorithm is simple to understand and easy to implement that employs both addition and subtraction, and it treats positive and negative operands uniformly, with as such no special actions required for negative numbers separately. It generates a $2n$-bit product while uniformly treating both positive and negative 2's complement n-bit operands. Moreover, one of the distinct advantages of this algorithm is that it can be readily extended in various ways to speed up the multiplication process. A version of this algorithm has been found to have been implemented in multiply instruction of reputed RISC (reduced instruction set computer) processor ARM6.

Booth's algorithm can be described as follows. Here too, the multiplicand and the multiplier are placed in two registers X and Y, respectively. There is also a 1-bit register placed logically to the right of the least significant bit (Y_0) of the Y register and designated as Y_{-1}. The results of the multiplication as usual will finally appear in the Z and Y register.

At the beginning, the registers Z and Y_{-1} are initialized to 0. As before, control logic here also starts scanning from the rightmost bit of the multiplier onwards, one at a time. Now, as each bit of multiplier is examined, the bit to its right is also examined. The comparison between these two bits is then made, which indicates the following:

- If the two bits are the same (i.e. the current bit under scan and the one just on its right are the same ($1 \leftarrow 1$ or $0 \leftarrow 0$)), then all of the bits of the Z, Y, and Y_{-1} registers are simply shifted to the right 1 bit;
- But if these two bits differ, i.e.
    - if it is $1 \leftarrow 0$, meaning that the current bit under scan is 1 and the one just on its right is 0, then the multiplicand X is *subtracted* from Z register, and then the registers Z and Y are shifted to the right 1 bit.
    - if it is $0 \leftarrow 1$, meaning that the current bit under scan is 0 and the one just on its right is 1, then the multiplicand X is *added* to the Z register, and the registers Z and Y are then shifted to the right 1 bit.

In all the cases, the right shift is such that the leftmost bit of Z, namely Z_{n-1}, not only shifted into Z_{n-2}, but also remains in Z_{n-1}. This is required to preserve the sign of the number in the result of the multiplication to be obtained in Z and Y. This action is commonly known as the *arithmetic shift* since it preserves the sign bit.

It is important to note the significance of using the 1-bit register Y_{-1}. Now to start the multiplication process, we always assume that there is a virtual bit designated as Y_{-1} on the right of the rightmost bit of the multiplier that is always taken as 0, since initialized with 0. As the right shift is always given in all the cases, Y_0 always holds the current bit of the multiplier to be scanned, and Y_{-1} consequently always holds the bit which is just to the right of the currently scanning bit of the multiplier. Thus, after shift (completion of one cycle), it is the pair (Y_0, Y_{-1}) that is always to be taken into consideration to determine the type of the actions (add, subtract, shift) to be taken.

Figure 7.8 shows an example detailing the sequence of steps mentioned in Booth's algorithm for the multiplication of 6 by 5. The result of this multiplication is finally obtained in Z and Y, i.e. 0 0 0 11 1 1 0, which is equal to 30 as usual.

Booth's algorithm can also be used equally well, directly for negative multipliers, as shown in Figure 7.9, for the multiplication of 6 by –5. The result of this multiplication is finally obtained in Z and Y, i.e. 1 1 1 00 0 1 0, which is equal to –30 as usual. In fact, by using

Z	Y (= 5)	Y_{-1}	X (= 6)	
0 0 0 0	0 1 0 1	0	0 1 1 0	← Initial values
1 0 1 0	0 1 0 1	0	0 1 1 0	(Z ← Z – X) First cycle
1 1 0 1	0 0 1 0	1	0 1 1 0	(Shift)
0 0 1 1	0 0 1 0	1	0 1 1 0	(Z ← Z + X) Second cycle
0 0 0 1	1 0 0 1	0	0 1 1 0	(Shift)
1 0 1 1	1 0 0 1	0	0 1 1 0	(Z ← Z– X) Third cycle
1 1 0 1	1 1 0 0	1	0 1 1 0	(Shift)
0 0 1 1	1 1 0 0	1	0 1 1 0	(Z ← Z + X) Fourth cycle
0 0 0 1	1 1 1 0	0	0 1 1 0	(Shift)

FIGURE 7.8
Multiplication example using Booth's algorithm (6 × 5 in Z and Y).

Z	Y (= – 5)	Y_{-1}	X (= 6)		
0 0 0 0	1 0 1 1	0	0 1 1 0	← Initial values	
1 0 1 0	1 0 1 1	0	0 1 1 0	(Z ← Z – X)	First cycle
1 1 0 1	0 1 0 1	1	0 1 1 0	(Shift)	
1 1 1 0	1 0 1 0	1	0 1 1 0	(Shift)	Second cycle
0 1 0 0	1 0 1 0	1	0 1 1 0	(Z ← Z + X)	Third cycle
0 0 1 0	0 1 0 1	0	0 1 1 0	(Shift)	
1 1 0 0	0 1 0 1	0	0 1 1 0	(Z ← Z – X)	Fourth cycle
1 1 1 0	0 0 1 0	1	0 1 1 0	(Shift)	

FIGURE 7.9
Multiplication example using Booth's algorithm (the product of 6 × –5 in *Z* and *Y*).

other examples, it can be shown that Booth's algorithm works equally well with any combination of positive and negative numbers, making no distinction between them as such.

Working of Booth's algorithm with both negative multipliers and negative multiplicands is shown in the website: http://routledge.com/9780367255732.

7.7.3.3.2 Alternative Method

Figure 7.10 illustrates the normal method and Booth's algorithm with a more compact manner using a different procedure from that of the examples just discussed. Figure 7.10a shows a normal multiplication scheme with a 4-bit signed operand –5 as multiplicand, multiplied by +7, the multiplier, to get the 8-bit product, –35. The sign extension of the multiplicand in the partial product is shown with an underline. Thus, the hardware as used for unsigned binary integer multiplication (already shown in the website) can also be used here for negative multiplicands if it provides for sign extension of the partial products.

Figure 7.10b shows a different multiplication procedure using Booth's algorithm as already defined, taking the same example of a 4-bit signed operand –5 (multiplicand) multiplied by +7 (multiplier) to get the 8-bit product, –35. The multiplier bit here is *recoded* (bit-pair recoding) when it is scanned from right to left following the original rules as already described above in Booth's algorithms, but essentially with a very little redefinition used for this type of multiplication scheme. It is to be remembered that there is always an implied 0 that lies to the right of the least significant bit (i.e. the rightmost bit) of the multiplier. Now, the scheme is as follows:

- When moving from 0 to 1 (1 ← 0), the current bit 1 of the multiplier will be replaced by –1;
- When moving from 1 to 0 (0 ← 1), the current bit 0 of the multiplier will be replaced by +1;

(a)		(b)	
1 0 1 1	(Multiplicand = – 5)	1 0 1 1	(= –5)
0 1 1 1	(Multiplier = + 7)	+ 1 0 0 –1	(= + 7)
1 1 1 1 1 0 1 1		0 0 0 0 0 1 0 1	
1 1 1 1 0 1 1	⟸ sign extension is shown by underline ⟹	0 0 0 0 0 0 0	
1 1 1 0 1 1		0 0 0 0 0 0	
0 0 0 0 0		1 1 0 1 1	
1 1 0 1 1 1 0 1	(= – 35)	1 1 0 1 1 1 0 1	(= – 35)

FIGURE 7.10
Multiplication example (–5 × 7) using different method to implement Booth's algorithm: (a) normal multiplication scheme and (b) Booth's multiplication scheme.

- When moving either from 1 to 1 or from 0 to 0 (i.e. the current bit under scan and its immediately preceding right bit of the multiplier are the same ($1 \leftarrow 1$ or $0 \leftarrow 0$)), the current bit of the multiplier, be it 1 or 0, will be replaced by 0.

In Figure 7.10b, it really occurs (least significant bit of the multiplier is 1). That is why, the least significant bit of the multiplier here is recoded as –1 according to the above rule (since the original 1 in the multiplier changes from 1 to –1 due to the implied 0 lying to its right). Thus, the recoded multiplier in this example becomes +100 –1 when the original multiplier was 0111. Here also, the sign extension of the multiplicand in the partial product is shown with an underline (Figure 7.10b).

Following the same method as already depicted in Figure 7.10b, it can be shown by suitable example that Booth's algorithm can also be equally applicable in multiplication with any combination of positive and negative numbers in a similar manner, making no distinction between them as such.

Some distinct features (remarks) of Booth's algorithm are given in the website: http://routledge.com/9780367255732.

7.7.3.4 Fast Multiplication: Carry-Save Addition

In multiplication, the final result is ultimately produced by adding the partial products (called *summands*) as generated row-wise during the multiplication process due to bit-by-bit multiplication of the multiplier with the multiplicand, and is placed in rows after rows with necessary left shifts. When these rows are added column-wise to obtain the final result, the superior carry-save addition technique (already explained in the previous section) using a CSA can be effectively employed here to make the multiplication operation even faster.

In multiplication with longer operands, many summands are produced that need to be added to produce the final result. A more significant reduction in computation time can then be regularly achieved by way of grouping the summands in threes, and carry-save addition can now be performed on each of these groups in parallel to generate a set of S and C vectors in one full-adder delay. Next, the set of the S and C vectors thus obtained can again be grouped into threes, and carry-save addition once again is performed on them, generating a further set of S and C vectors in one more full-adder delay. This process, however, will be continued until there are only two vectors remaining. They can be then simply added by a conventional ripple-carry adder (CPA) or by a CLA (already explained in the previous section) in a usual way to produce the final desired product.

7.7.4 Division

Division operation is comparatively more complex than multiplication operation. If two unsigned integers, a divisor V and a dividend D, are given, then after division, a third number Q, the quotient, is obtained, such that $Q \times V$ equals to or is very close to D. If R is considered as remainder which is required to be less than V, i.e. $0 \leq R < V$, we can then write $D = Q \times V + R$. This relationship $D = Q \times V$ suggests that there exists a close correspondence between division and multiplication, specifically the dividend (D), quotient (Q), and divisor (V), as shown in $D = Q \times V$, corresponds to the product, multiplicand, and multiplier, respectively. This correspondence means that similar algorithms and circuits can be used for both multiplication and also for division. In multiplication, the shifted multiplier is added to the multiplicand to form the product. In division, the shifted divisor is subtracted from the dividend to form the quotient. As multiplication often ends with a double-length product, division begins with dividend, often extended up to double length. Due to these similarities, division is essentially based on the same general principles as positive number multiplication, by relating the way the multiplication operation is done manually to the way it is done in the logic circuit. For integer division, the basis for the related algorithm is similar as before, the pencil-and-paper approach, and the operation involves repetitive shifting, and addition or subtraction.

7.7.4.1 Unsigned Integers

The traditional ***pencil-and-paper approach*** used in the division of unsigned decimal numbers can be equally implemented in a similar manner in the division of binary numbers, with the exception that the *divisor, dividend, quotient,* and *remainder* here all are bits of 0 and 1.

7.7.4.1.1 Machine-Based Algorithm

Following the traditional method, division can be performed by a machine algorithm using an adder circuitry in the ALU along with the execution of a number of required steps. At the start, the *n*-bit unsigned *divisor* is placed in the *M-register* where it remains throughout the division process, and the *n*-bit unsigned *dividend* in the *Q-register.* A third register, the *A register,* is also needed and is initially set to 0.

7.7.4.1.2 Restoring Division

At each step, the *A* and *Q* registers *together* are shifted 1 bit to the left. *M* is subtracted from *A* to determine whether *A* divides the partial remainder. If the most significant bit of the result of subtraction is 1, then the result is a negative number (<0), which indicates that *A* is smaller than *M*, and consequently, q_0 gets a 0 bit and *M* must be added back to *A* to restore *A* to its previous value. Otherwise, if the most significant bit of the result of subtraction is 0, then the result is a positive number (>0), which indicates that *A* is greater than *M*, and the division holds. Consequently, q_0 gets a 1 bit and the result of the subtraction is treated as a partial remainder. The count is then decremented, and the division process continues *n* steps. At the end, the quotient is in *Q* register, and the remainder is in the *A* register. Since, whenever $q_0 = 0$, the partial remainder (*A*) is added to the divisor (*M*) to restore the previous value of *A*, that is why, this straightforward technique is sometimes called *restoring division.* In fact, whenever the result of the subtraction is negative, a restoring addition is performed.

The algorithm just described assumes that the divisor *V* and the dividend *D* are positive and that too $|V| < |D|$. If $|V| = |D|$, then the quotient $Q = 1$ and the remainder $R = 0$. But if $|V| > |D|$, then $Q = 0$ and the remainder $R = D$. An algorithm can now be built up, summarizing all these actions as mentioned which can be described as follows:

1. An *n*-bit positive divisor is loaded into register M. Load the dividend D into [A, Q] registers. The dividend must be expressed as a 2*n*-bit positive number.

   For example, the 4-bit dividend (D = 13) 1101 becomes 00001101, and load it into [A, Q] registers as shown in the example (Figure 7.11);
2. Shift A, Q left 1-bit position;
3. Perform A ← A − M, i.e. [A ← A + (−M)]. This operation subtracts the divisor M from the contents of A.

4. a. If the result is non-negative (i.e. the most significant bit of A = 0), then set $q_0 \leftarrow 1$.
   b. If the result is negative (i.e. the most significant bit of A = 1), then set $q_0 \leftarrow 0$ and restore the previous value of A (i.e. $A \leftarrow A + M$).
5. Repeat steps 2 through 4 as many times as there are bit positions in Q (*n* times);
6. Finally, the remainder is available in A, and the quotient is in Q.

Figure 7.11 depicts a logic circuit arrangement that implements *restoring division*. It is amazing to note its similarity with the structure as implemented for multiplication that was shown in shift-and-add algorithm for multiplication of unsigned integers given in the website. At the beginning of the operation, the *M* register holds an *n*-bit positive divisor throughout, and an *n*-bit positive dividend is loaded into register *Q*. In each step, the 2*n*-bit shift register [*A*, *Q*] is shifted 1 bit to the left. The position thus vacated at the right-most end of the *Q* register can be used to store the quotient bit thus generated at that step. In this way, when the division process terminates, *Q* contains the quotient, whereas *A* contains the (shifted) remainder. It is clear that a division instruction when executed using these sequential steps by related hardware takes much more time to execute than a multiply instruction. Several techniques, however, have been subsequently improvised with the use of the emerging faster hardware components to speed up this division process.

The required subtraction is facilitated considering $M| < |Q|$, and by using 2's complement arithmetic as shown with the use of $-M$, which is added to *A* to give the effect of $A - M$. The extra bit at the left end of both *A* and *M* (sign-extended) accommodates the sign bit during subtractions. That is why, an $(n + 1)$-bit adder is used here instead of *n*-bit adder which was used in case of multiplication. Left shift of 1 bit as usual is given on [*A*, *Q*]. An example is depicted in Figure 7.12 showing the steps of the division of 4-bit numbers being processed by the circuit as shown in Figure 7.11.

The two 4-bit numbers; the divisor M (m_3, m_2, m_1, m_0) is 0100 (= 4) and the extra bit of M is m_4 (= 0, shown in Figure 7.12), and the dividend Q (q_3, q_2, q_1, q_0) is 1101 (= 13) are taken for division. Initially, the register A (a_3, a_2, a_1, a_0) and the extra bit of A (i.e. a_4) are set to 0. During division, whenever the value of A is restored, it is done by adding the current

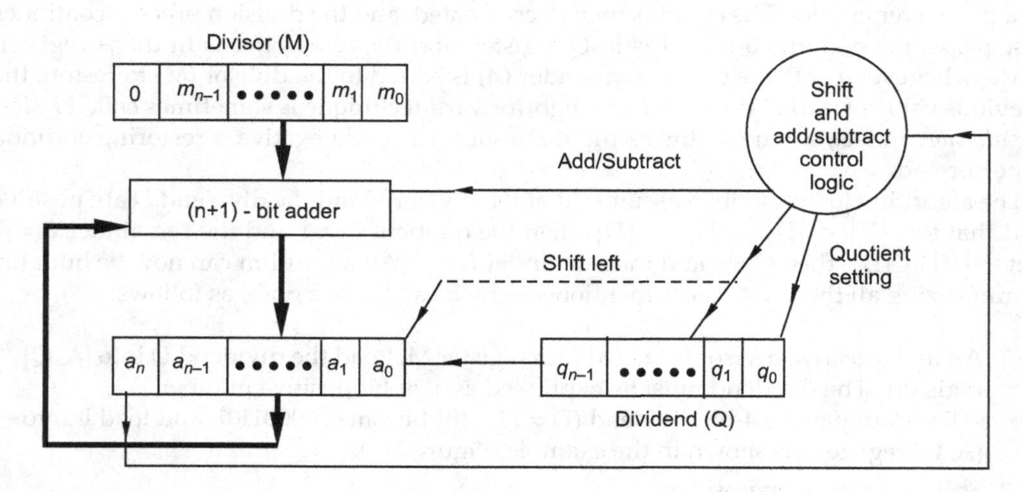

FIGURE 7.11
Schematic logic circuit diagram for unsigned binary division consisting of register configuration (restoring division).

A	Q(q_3 q_2 q_1 q_0)	M	–M	
00000	1101	00100	11100	Initial values
00001	101	00100	11100	Shift left
11100				–M = 2's Complement of M
11101				Subtract, A ← A –M,[=A + (–M)]
00001	1010			Restore A, i.e. A ← A +M, set q_0 = 0
00011	010			Shift left
11100				–M=2's Complement of M
11111				Subtract, A ← A –M, [=A+(–M)]
00011	0100			Restore A, i.e. A ← A +M, set q_0 = 0
00110	100			Shift left
11100				–M=2's Complement of M
00010				Subtract, A ← A –M, [=A+(–M)]
00010	1001			set q_0 = 1
00101	001			Shift left
11100				–M=2's Complement of M
00001				Subtract, A ← A –M, [=A+(–M)]
00001	0011			set q_0 = 1
Remainder	Quotient			

FIGURE 7.12
Example of restoring two's complement division (division of 13 by 4) using the logic circuit of Figure 7.11.

value of A (which is $A - M$) with M. After the division is complete, the n-bit result of this division is:

Quotient in Q (q_3, q_2, q_1, q_0) = 0011 (= 3), and the remainder in A = 0001 (= 1)

An appropriate flowchart of this scheme is shown in the website: http://routledge.com/9780367255732.

7.7.4.1.3 Non-restoring Division

One of the critical drawbacks of *restoring division algorithm* lies in its need for restoring the value of A in the event of an unsuccessful subtraction (sign bit of partial remainder is 1 after subtraction, i.e. negative). Thus, an attempt can be made to improve this restoring division algorithm so that the unpleasant action of restoring A after an unsuccessful subtraction can be avoided.

Recall the restoring division algorithm, and consider the sequence of operations that takes place after the subtraction operation. If A is positive, we shift it left 1 bit and then subtract M. This denotes that shifting A left 1 bit means we are getting $2A$, and then subtracting M signifies that we are actually performing $2A - M$. Now, if A is negative, we restore A by performing $A + M$, and then we shift it left 1bit and subtract M. This implies that actually $(A + M)$ is shifted left 1 bit, indicating that we are performing $2(A + M)$, i.e. $2A + 2M$, and then we are subtracting M, denoting we are really performing $[(2A + 2M) - M)]$; i.e., it actually becomes $2A + M$. This reveals that when A is negative, no restoration of A is required; instead simply shift A left 1 bit to obtain $2A$, and then add it with M to yield the required $2A + M$. As usual, the quotient bit q_0 is accordingly set to 1 when A is positive, and q_0 is set to 0 when A is negative. Summarizing all these actions as described, it is now possible to develop the following algorithm for non-restoring division.

Step 1: Execute the following ***n*** times:

i. If the sign of A is 0 (positive), shift A, Q left 1 bit position, and subtract M *from* A; otherwise, shift A, Q left 1 bit position and add M to A;
ii. Now, if the sign of A is 0 (positive), set q_0 to 1; otherwise, set q_0 to 0.

Step 2: If the sign of A is 1, add M to A.

In fact, Step 2 is additionally needed to leave the proper positive remainder in A at the end of the n cycles of Step 1.

An appropriate flowchart of this scheme is shown in the website: http://routledge.com/9780367255732.

The logic circuitry as shown in Figure 7.11 can also be used just as well to perform this algorithm. It must be noted that restore operations are no longer needed here and that exactly one Add or Subtract operation is performed per cycle. Figure 7.13 shows how the division example as already shown in Figure 7.12 is executed here by using a non-restoring division algorithm.

No simple algorithms exist for directly performing division on *signed operands* that are comparable to the algorithms for signed multiplication. However, the **division with *negative operands*** can be dealt in a similar manner with some preprocessing. In this regard, a simple shortcut method may be improvised, which can be described as follows:

Recall that the division operation can be recognized by defining it as: $D = Q \times V + R$.

A	Q(q_3 q_2 q_1 q_0)	M	–M	
00000	1101	00100	11100	Initial values
00001	101	00100	11100	Shift left
11100				–M = 2's Complement of M
11101				Subtract, A ← A –M, [=A + (–M)]
	1010			set q_0 = 0
11011	010			Shift left
00100				M= 00100
11111				Add M to A, i.e. A ← A + M
	0100			set q_0 = 0
11110	100			Shift left
00100				M= 00100
00010				Add M to A, i.e. A ← A + M,
	1001			set q_0 = 1
00101	001			Shift left
11100				–M=2's Complement of M
00001				Subtract, A ← A –M, [=A+(–M)]
	0011			set q_0 = 1
⟵ Remainder ⟶	⟵ Quotient ⟶			

Step 2 of above algorithm here is not required because the sign bit of A is 0 which is positive, otherwise if the sign bit of A becomes 1 at the end (i.e. negative), Step 2 of the algorithm is required, i.e. addition of M with A is required at last to obtain positive remainder at A.

FIGURE 7.13
Example of non-restoring two's complement division (division of 13 by 4) using the non-restoring division algorithm. $D = 13$, $V = 4$: $\rightarrow Q = 3$, $R = 1$.

If we now consider all possible combinations of signs of D and V of the previous example (Figure 7.13), we arrive at the followings:

$D = 13$	$V = 4$	→	$Q = 3$	$R = 1$
$D = 13$	$V = -4$	→	$Q = -3$	$R = 1$
$D = -13$	$V = 4$	→	$Q = -3$	$R = -1$
$D = -13$	$V = -4$	→	$Q = 3$	$R = -1$

It is interesting to note (from Figure 7.13) that 13 ÷ (–4) and (–13) ÷ 4 produces different remainders. Moreover, it is observed from the above that the *magnitude* (absolute value) of Q and R in all the cases are remained unaffected, irrespective of the signs of D and V, and that the *signs* of Q and R can also be easily derived from the signs of D and V. It is indicated from the above combinations that sign (R) = sign (D) always, whatever be the sign of V, and sign (Q) = sign (D) × sign (V). Hence, one way to perform 2's (two's) complement division is to convert the operands into unsigned values before division, and after carrying out the division in the usual direct method, the signs are to be accounted for at the end by the complementation, as required. The restoring division algorithm, however, chooses this approach in its working whenever division with signed operands is performed.

7.8 Floating-Point Representation

Till now, we have considered only binary integers which are essentially fixed-point numbers, both unsigned and signed. This means that there exists an implied binary point which lies at the right end of the number. With a fixed-point notation, it is thus possible to represent a range of positive and negative integers on both sides of integer 0. With the same format, however, it is also possible to represent fractional numbers as well by simply assuming that the binary point is just to the right of the sign bit. In the 2's complementary system, consider an n-bit binary fraction B as:

$$B = b_0\, b_{-1} \cdot b_{-2} \ldots b_{-(n-1)}$$

The signed value F of this fraction can be represented as:

$$F(B) = -b_0 \times 2^0 + b_{-1} \times 2^{-1} + b_{-2} \times 2^{-2} + \cdots + b_{-(n-1)} \times 2^{-(n-1)}$$

where F lies in the range of: $-1 \le F \le 1 - 2^{-(n-1)}$

However, this approach with fixed-point notation has certain limitations. Neither very large numbers nor very small fractions that are often involved as parameters in scientific computations can be represented. One way to get out of it is to exploit scientific notations that we often use with decimal numbers. For example, the number 245,000,000,000 can be represented as 2.45×10^{11}. Similarly, a decimal fraction like 0.0000000000245 can be represented as 2.45×10^{-11}. In effect, what is done here is to dynamically slide the decimal point to a convenient location, and using the appropriate *exponent* of 10 to keep track of the decimal point. This means that the position of the decimal point is variable and is adjusted according to the need as computation proceeds. In such a case, the decimal point is said to *float*, and the numbers, hence, are called the ***floating-point numbers***. This distinguishes them from the fixed-point numbers, whose decimal point is always in the same position.

This approach, however, enables to represent a range of very large and very small numbers with the use of only a few digits.

As the position of the decimal point in the floating-point number is changeable, it must be explicitly mentioned in the floating-point representation. For example, decimal numbers often used in scientific notation may be written as 7.035×10^{21}, -5.427×10^{27}, 3.529×10^{-18}, -8.045×10^{-25}, and so on. These numbers are said to be offered to four *significant digits*. The ***scale factors*** (such as 10^{27} and 10^{-25}) indicate the position of the decimal point with respect to the significant digits. The power appearing over the base, like 27 in 10^{27}, is called the ***exponent***. By convention, when the decimal point is placed to the right of the first (nonzero) significant digit, the number is said to be ***normalized***. It is assumed that the floating-point numbers are stored in normal form only; hence, the final result of each floating-point arithmetic operation should be normalized. It is to be noted that in the decimal system, the base, 10, in the scale factor is fixed and implicit, and hence does not need to appear explicitly in the machine representation of a floating-point number. Thus, only the sign, the significant digits (significand) S, and the exponent E in the scale factor are needed to constitute the representation. The same approach can be taken with binary numbers in which the base is 2 which is implicit, appears in the scale factor, and the significant digits (significand) S and exponent E are constituted with only 0s and 1s. Thus, a floating-point number N, be it decimal or binary, can be represented in the form by its sign, a string of significant digits (significand) S, also commonly called the ***mantissa***, and an *exponent* E to an implied base B for the scale factor, as $N = \pm S \times B^{\pm E}$. Here, the radix point in S is assumed to the right of the most significant bit of the significand S. That is why, there is only one bit to the left of the radix point. To explain the principles used in representing the binary floating-point numbers, let us consider an example as shown in Figure 7.14 in which a typical floating-point format of 32 bits is exhibited, which is a standard computer word length.

The leftmost bit stores the ***sign*** of the number (0 for positive and 1 for negative). The ***exponent*** value is stored in the next 8 bits to an implied base 2 that provides a scale factor of a reasonable range. The representation used in describing *signed* exponents is known as the ***biased representation***. A fixed value called the *bias*, which typically equals $(2^{k-1} - 1)$, where k is the number of bits used in signed binary exponent, is added to the given exponent to derive an *unsigned integer* which is to be stored as an exponent in the exponent field. In this case, the 8-bit exponent field yields the unsigned numbers 0 through 255. With a bias of $(2^{k-1} - 1)$ (i.e. $(2^{8-1} - 1) = 127$), the value actually stored in the exponent field is an unsigned integer $E' = E + 127$, where E is the given signed exponent in the number. This approach is also called ***excess-127*** format. Thus, E' lies in the range of $0 \le E' \le 255$. However, the end values of this range, 0 and 255, are used to represent *special values*. That is why, the range of E' for normal values lies in the range of $1 \le E' \le 254$. This means the actual original exponent E lies in the range of $-126 \le E \le 127$. This excess-x representation for exponents, however, enables an efficient comparison of the relative sizes of two floating-point numbers. Finally, the remaining 23 bits of the 32-bit format are used to store the ***significand*** (or sometimes loosely called the *mantissa*).

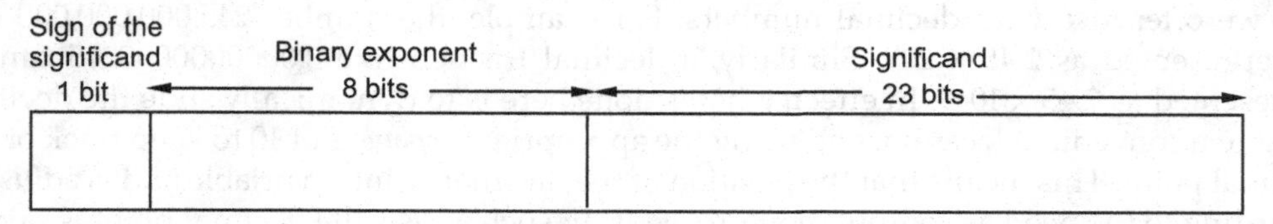

FIGURE 7.14
Representation of typical 32-bit floating-point format.

7.8.1 Normalized Form

A particular floating-point number can be expressed in many different ways. For example, all the numbers, such as 0.1010×2^7, 1010×2^3, and 0.10100×2^8, are equivalent, in which the significand is expressed in binary form and the exponent is expressed in decimal for the sake of more clarity in the explanation. As already mentioned, in order to conveniently express and simplify operations on floating-point numbers, it is always required that they be in ***normalized form***. A binary number in normalized form is one in which the most significant bit of the significand is 1 (nonzero), and by convention, there is only one bit to the left of the radix point. Thus, a normalized nonzero number can be expressed in the form:

$$\pm 1.\, x \cdot x \cdot x \cdot x \ldots\ldots\ldots\ldots x \times 2^{\pm E}$$

where x is a binary digit, either 0 or 1. As the most significant bit in the normalized form is always 1, it is not at all necessary to store this bit and as such not explicitly represented; rather, it is implicit, and it is assumed to appear just to the immediate left of the binary point. This leading implicit "1" in the magnitude part (significand) of a normalized sign-magnitude number never needs any space to actually store, since it can always be inserted by the arithmetic circuits that process the numbers. Consequently, in the IEEE 754 format, the complete significand (mantissa) is actually 1. M where "1" is the *hidden* leading bit that is not stored with the number.

Hence, the 23 bits stored in the significand field actually represent the *fractional part* of the significand (mantissa), i.e. the bits to the right of the binary point. In essence, this 23-bit field is used to actually store a 24-bit significand. Figure 7.15 shows a machine representation of a 32-bit floating-point binary number in the normalized form, almost on the same lines as Figure 7.14. It is apparent that if a number is not normalized, it can be always put in normalized form by shifting the radix point to the right of the leftmost "1" bit and adjusting the exponent accordingly.

To summarize, the binary numbers expressed in floating-point representation in normalized form have the following specifications:

- The sign of the number is to be stored in the most significant bit (first bit) of the word;
- The first bit of the actual significand (mantissa) is always 1 and need not be stored in the significand field. Only the fractional part of the significand is stored in the significand field;
- The *bias* value that typically equals $(2^{k-1} - 1)$, where k is the number of bits used to express the signed binary exponent, is to be added to the given exponent to derive an *unsigned integer* which is to be stored in the exponent field.

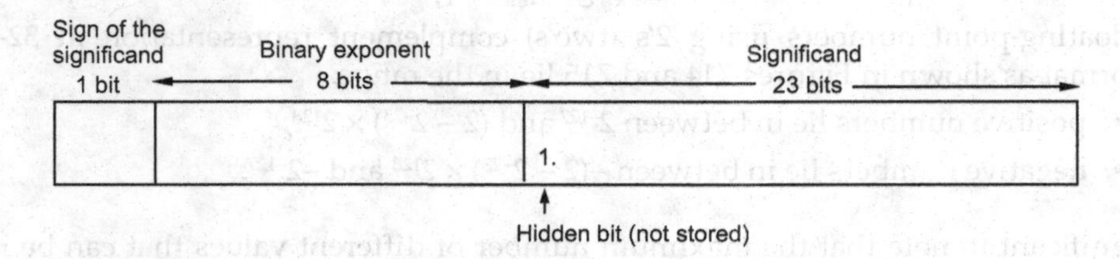

FIGURE 7.15
Representation of typical 32-bit binary floating-point format in normalized form.

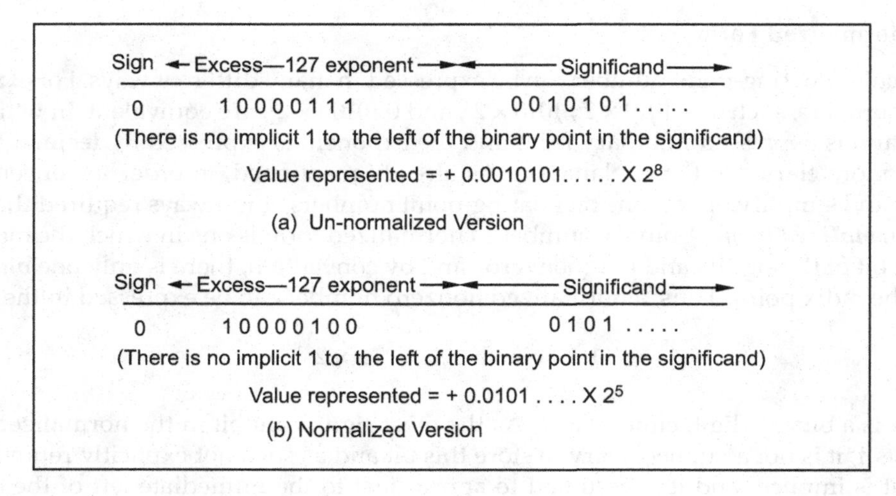

FIGURE 7.16
An example of machine representation of unnormalized and normalized formats of binary floating-point number.

Thus, a 32-bit floating-point number that conforms to the IEEE 754 standard using significand (mantissa) expressed in 1. *M*, and exponent representation as 8-bit excess-127 code (as mentioned before) for any real number *N*, could be described by the formula:

$$N = (-1)^S \, 2^{E-127} \, (1.M)$$

Figure 7.16 shows an example consisting of an unnormalized value $0.0010101\ldots \times 2^8$, and its normalized version $1.0101\ldots\ldots \times 2^5$. Since the scale factor is in the form 2^i, shifting the radix point right or left in the significand by one bit position is compensated by a decrease or an increase of value 1, respectively, in the exponent.

More examples with worked-out solutions relating to this topic are given in the website: http://routledge.com/9780367255732.

7.8.2 Range and Precision

The ranges of integers as well as floating-point numbers that can be represented by a 32-bit format are pictorially depicted in Figure 7.17. It can be seen that:

- Using 2's (two's) complement representation, all of the *integers* ranging from -2^{31} to $2^{31} - 1$ can be represented in 32-bit length, giving a total of 2^{32} different numbers;
- Floating-point numbers using 2's (two's) complement representation in 32-bit format as shown in Figures 7.14 and 7.15 lie in the range:
  - positive numbers lie in between 2^{-127} and $(2 - 2^{-23}) \times 2^{128}$,
  - negative numbers lie in between $-(2 - 2^{-23}) \times 2^{128}$ and -2^{-127}.

It is significant to note that the maximum number of different values that can be represented with 32 bits using floating-point notation is still 2^{32}. These ranges of numbers as explained above are pictorially depicted in Figure 7.17. Integers can be represented with

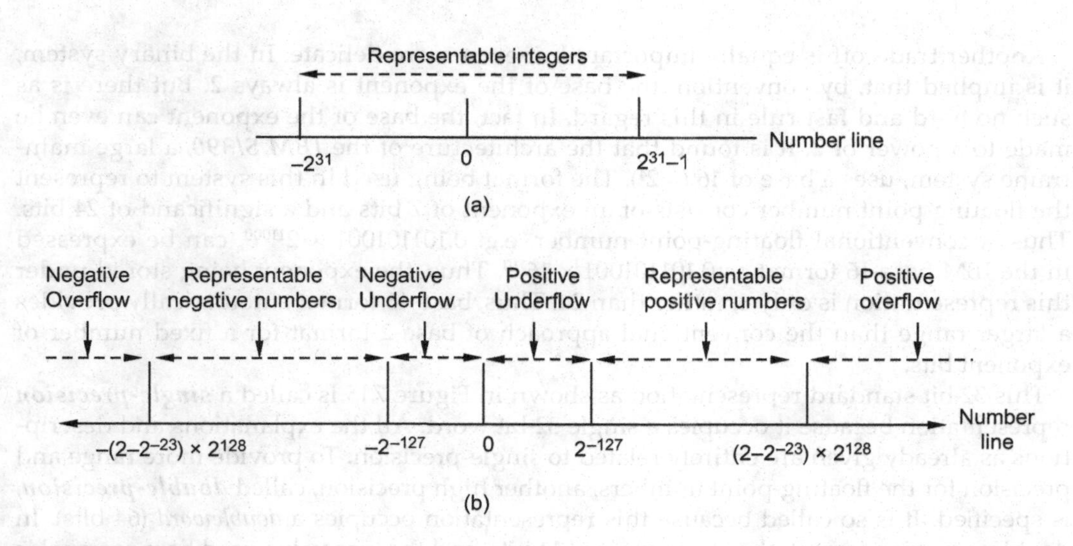

FIGURE 7.17
Representable numbers in 32-bit formats. (a) Two's complement integers (single format, 32-bit) and (b) floating-point numbers (single format, 32-bit formats).

exact values. Floating-point numbers, on the other hand, are often represented in practice mostly approximately, sometimes with truncation or rounding-off to the nearest value, depending on the format being prescribed.

It can be seen from Figure 7.17b that in the region on the extreme left of the number line, there is *negative overflow*, where negative numbers less than $-(2 - 2^{-23}) \times 2^{128}$ belong. Similarly, on the extreme right of the number line, there is *positive overflow* region, where positive numbers greater than $(2 - 2^{-23}) \times 2^{128}$ belong. When the numbers lie in one of these two regions, it gives rise to what is called an ***arithmetic exception***. Negative numbers greater than -2^{-127} but less than 0 lie in the ***negative underflow region***. The number 0 stands in the middle of the prescribed number line; all other regions are located on both sides of it. Just on the right of 0 where positive numbers greater than 0 but less than 2^{-127} reside is the region called ***positive underflow***. All the numbers lie in the domain constituted by the *negative underflow* and *positive underflow* regions including 0 (Figure 7.17) are expressible numbers (denormalized numbers, discussed later).

The exponent present in a floating-point number determines the ***range*** (span) of the numbers that can be expressed in a prescribed format. Use of more exponent bits in the format implies a larger exponent, which, in turn, enables to achieve a greater range of expressible floating-point numbers. The significand determines whether two nearly-equal numbers can be exactly expressed individually, and distinctly within the confinement of the given format. Thus, the significand relates to the ***precision*** of the numbers that are involved in computation of large numbers. If the number of bits in the significand is increased, then more and more almost-equal numbers can be expressed distinctly giving higher density of these numbers. But still, only a fixed number of different values can be expressed when the format is fixed. Now, the most vital question that arises is whether the designer of the computer system will give more importance to the *exponent* providing more bits to enlarge it to extend the range, or offer more bits to the significand to increase the resolution of the precision of a floating-point number when it is to be dealt with; it entirely depends on the trade-off of the related design. However, the ultimate target is to realize a reasonable balance in between.

Another trade-off is equally important but even more delicate. In the binary system, it is implied that, by convention, the base of the exponent is always 2. But there is as such no hard and fast rule in this regard. In fact, the base of the exponent can even be made to a power of 2. It is found that the architecture of the ***IBM S/390***, a large mainframe system, uses a base of 16 (= 2^4). The format being used in this system to represent the floating-point number consists of an exponent of 7 bits and a significand of 24 bits. Thus, a conventional floating-point number, e.g. $0.101101001 \times 2^{11000}$, can be expressed in the IBM base-16 format as $0.101101001 \times 16^{110}$. Thus, the exponent being stored under this representation is only 6, rather than 24. Thus, base-16 format automatically provides a larger range than the conventional approach of base-2 format for a fixed number of exponent bits.

This 32-bit standard representation as shown in Figure 7.15 is called a ***single-precision*** representation because it occupies a single 32-bit word. All the explanations and descriptions as already given are entirely related to single-precision. To provide more range and precision for the floating-point numbers, another high precision, called ***double-precision,*** is specified. It is so called because this representation occupies a *doubleword* (64 bits). In double-precision format, the exponent is of 11 bits, and the bias value used here as usual is $(2^{k-1} - 1)$, i.e. $(2^{11-1} - 1)$ or 1023. Therefore,11-bit excess-1023 exponent E' (= E + 1023) has the range $1 \le E' \le 2046$ for normalized values, with 0 and 2047 used to indicate *special values,* as before. The significand, however, is of 52 bits, which provides a precision equivalent to about 10 decimal digits.

7.8.3 IEEE Standard: Binary Floating-Point Representation

In order to maintain uniformity in the representation of floating-point numbers implemented in different processors introduced by different vendors, a standardization in this regard was ultimately defined in **IEEE Standard 754** sometimes in 1985 for both single-precision (32bit) and double-precision (64bit). It is now used on almost all contemporary processors as well as on numeric co-processors, including ***Intel IA-32*** and ***IA-64*** architectures. This standard, in addition, also defines two ***extended formats;*** *single-extended* and *double-extended,* whose exact format differs from one implementation to another, and truly is implementation specific. The extended formats essentially provide additional bits to both exponent (extended range) and the significand (extended precision) to make them even longer than those found in the usual double-precision format. As the extended format has a larger range and greater precision, it is extremely useful and is often used to hold the intermediate results of any ongoing computation that may not be accommodated in the single-precision or in the double-precision format without rounding-off/truncation. The effect of doing repeated rounding-off/truncation in the computations at intermediate stages may shift the finally obtained value from its actual value by a reasonable extent that may eventually cause infusion of significant error in the final result. Extended format, in this situation, provides sufficient room to accommodate the intermediate results without performing any rounding-off/truncation that facilitates minimizing the errors occurred in the final result due to repeated rounding-off/truncation needed at intermediate stages. However, each of these four formats, namely single format, single-extended format, double format, and double-extended format, has their own distinct features and salient characteristics.

A brief detail of this topic with related figure and table is given in the website: http://routledge.com/9780367255732.

7.8.4 Exceptions and Special Values

Exception is defined as a peculiar situation that arises when the result of a computation falls outside the prescribed domain of expressible range (both positive and negative) of normal numbers. It happens if any of the followings is found while performing operations: *underflow, overflow, divide by zero, inexact,* and *invalid.* The first three have already been mentioned. ***Inexact*** is defined as a situation when the result of the computation requires rounding in order to be represented in one of the normal formats. An ***invalid*** exception occurs if operations, such as 0/0 or $\sqrt{-1}$, are attempted. When an exception occurs, the processor must then set the corresponding exception flag, and the related interrupt is issued. Control is then passed to the respective interrupt service routine (ISR), which may be system-defined or even user-defined. Alternatively, the application program itself can examine the situation, and test on its own for the occurrence of such exceptions as necessary, and then decide what to do and how to proceed. Usually, when exception occurs, the results are set to **special values.**

As already mentioned, not all bit patterns as represented under IEEE formats are interpreted in the usual way. Instead, some bit patterns are used for representing some **special values.** The extreme exponent values of all zeros (indicating the exponent value 0), and all ones (the exponent value 255 of the excess-127 in single format and 2047 of the excess-1023 in double format) are used to represent the special values. The following classes of numbers, however, are special:

- When $E' = 0$, and the mantissa fraction $M = 0$, the value exact 0 is represented. It is useful to have a specific representation of an exact value of 0;
- When $E' = 255$ (single format) or 2047 (double format), and the mantissa fraction $M = 0$, the value ∞ is represented, where ∞ is the result of dividing a normal number by zero. It is also useful to have a representation of infinity. This leaves it up to the user to decide whether to treat it as an overflow or as an error condition or to carry the value ∞ and proceed as if nothing has happened;
- The sign bit is still part of these representations, so there are $\pm$ 0 and $\pm\ \infty$ representations;
- When $E' = 0$, and the mantissa fraction $M \neq 0$, *denormal numbers* are represented. Their value is $\pm\ 0.M \times 2^{-126}$. Therefore, they are smaller than the smallest normal number. There is no implied 1 to the left of the binary point; i.e., the bit to the left of the binary point is zero, and the actual exponent E is –126 or –1022, and M is any nonzero 23-bit (single-format) or 52-bit (double-format) fraction. The ultimate target of introducing denormal numbers is to allow *gradual underflow,* providing an extension of the range of normal representable numbers, which is sometimes useful in dealing with very small numbers in certain situations;
- When $E' = 255$ (single format) or 2047 (double format), and the mantissa fraction $M \neq 0$, the value represented is called ***not a number (NaN).*** A NaN is the result of performing an invalid operation, such as 0/0 or $\sqrt{-1}$.

Therefore, when the values of the exponent lie in the range of 1 through 254 (for single format), and 1 through 2046 (for double format), normalized nonzero floating-point numbers are represented. Since the *exponent* is *biased,* the range of actual exponent is –126 through +127 (for single format), and –1022 through +1023 (for double format). As usual, a normalized number requires a 1 in the bit to the left of the binary point. As this bit is implied, it offers an effective 24-bit (for single format) or 53-bit (for double format) significand, usually called the *fraction* in the IEEE standard.

7.8.5 Floating-Point Representation: Merits and Drawbacks

Floating-point representation format is used to express numbers; it exhibits certain distinct merits and also some drawbacks at the same time. A few of its ***merits*** are as follows:

- Floating-point representation provides the relatively easy method to express numbers of any type and of any size;
- Many different standards are used to represent the floating-point numbers;
- Hardware circuits implemented to handle floating-point numbers can also handle fixed-point numbers equally well, but the converse is not true;
- Relatively very large numbers can be expressed under this representation within the confinement of comparatively limited hardware facilities;
- Both range and precision of numbers obtained by this representation, especially before and after any arithmetic operation, are adequate to negotiate any type of critical situation during execution.

Some of the major ***drawbacks*** that are observed in this representation are mainly as follows:

- This representation is, however, found to be relatively complex while in use, because for any number to be represented, it uses two fields, exponent and significand (mantissa);
- Each of these two fields has its own individual rules to be abided by, and certain other rules to be followed in conjunction, defined by the respective standards;
- Specific hardware circuits are required to handle numbers under this representation;
- This representation is mostly implementation-specific.

7.9 Floating-Point Arithmetic

Let X and Y be two floating-point numbers, be expressed as (X_S, X_E) and (Y_S, Y_E), respectively. Therefore, the numerical value of X is $X_S \times B^{X_E}$ and that of Y is $Y_S \times B^{Y_E}$. To explain this, some realistic assumptions in this respect are needed to be made, which are as follows:

- X_S is an n_S-bit two's complement or sign-magnitude binary fraction;
- X_E is an n_E-bit integer in excess 2^{n_E-1} code, implying an exponent bias of 2^{n_E-1};
- B is the base which here is equal to 2.

The above assumptions also hold good for Y. In addition, it is also assumed that the floating-point numbers are stored in their normalized form only, and thus, the final result of each floating-point operation should also be normalized.

TABLE 7.1

Rules for Basic Operations Involving Floating-Point Operands

Operation	Respective Rules
Add/Subtract	1. Checking of zeros for both the numbers *X* and *Y*. 2. To equalize the exponent of the two input numbers *X* and *Y*, choose the number with the smaller exponent and shift its significand right a number of steps equal to the difference in exponents of two numbers, *X* and *Y*. 3. Set the exponent of the number after shift equal to the largest exponent. 4. Perform addition/subtraction on the significands and determine the sign of the result. 5. Normalize the result value, if necessary.
Multiply	In multiplication and division operations, no alignment of mantissas is needed. 1. Add the exponents and subtract 127. 2. Multiply the significands, and determine the sign of the result. 3. Normalize the resulting value, if necessary.
Divide	1. Subtract the exponents and add127. 2. Divide the mantissas, and determine the sign of the result. 3. Normalize the resulting value, if necessary. *In multiply and divide rules, 127 is added or subtracted. This is due to using the excess-127 notation for exponents.

Basic operations: General methods for floating-point addition, subtraction, multiplication, and division are given in Table 7.1. For addition and subtraction, it is necessary to ensure that both operands have the same exponent value. This may require shifting the radix point on one of the operands to realize alignment. Multiplication and division are relatively simple, because the significands (mantissas) and exponents can be processed independently. The floating-point operations normally produce the usual expected expressible results, but at times, they may give rise to one of these situations, such as:

i. *Significand underflow* is observed while aligning significands that digits may flow off the right end of the significand. To cope with this situation, some form of rounding-off is required, to be explained later in section;
ii. *significand overflow* occurs when the addition of two significands of the same sign may result in a carry-out of the most significant bit. This can be fixed by realignment, which will be explained later;
iii. *Exponent underflow* happens when a negative exponent becomes less than the minimum possible exponent value (e.g. –145 is less than –127) in the prescribed format. This means that the number is too small to be represented, and thus may be considered to be equal to 0; and
iv. *exponent overflow* happens when a positive exponent exceeds the maximum possible exponent value defined in the prescribed format. This may be designated in some systems as +∞ or –∞.

7.9.1 Addition and Subtraction

Floating-point addition and subtraction are relatively complex since the exponents of the two input operands must be made equal before the corresponding significands can be added or subtracted. Following the floating-point format as already described, the two operands must be placed in the respective registers within the ALU to execute the required operation. The floating-point includes an implicit bit in the significand, but that bit must be made explicit at the time of executing the operation. The procedures being followed to perform addition and subtraction, however, are explained in Table 7.1.

During addition/subtraction, if the signs of two numbers are the same, there also exists the possibility of significand overflow, the rectification of which, in turn, may invite exponent overflow. Whatever be it is, the appropriate actions would then be taken with suitable intimation, and possibly the operation is to be halted, and the subsequent needful actions are then required. After addition/subtraction, the result may be required to be normalized, which may invite exponent underflow. Again, suitable actions should be taken to resolve the situation.

A typical flowchart for performing addition/subtraction incorporating all the activities as mentioned in Table 1along with a solved example is given in the website: http://routledge.com/9780367255732.

7.9.1.1 Implementation: Floating-Point Unit

A floating-point arithmetic unit can be built up by connecting two loosely coupled fixed-point arithmetic circuits, one to be used as an exponent unit and the other as a significand (mantissa) unit. As the significand unit is required to perform all four basic arithmetic operations on the significands, a conventional fixed-point arithmetic circuit (already described earlier) can be used for this purpose. The exponent unit, however, is implemented by a relatively simpler circuit, capable of only adding, subtracting, and comparing exponents of the input operands. Comparison of exponents can be made by a comparator or by subtracting the exponents. With this idea, a schematic structure of a floating-point unit can be built up on the lines of the illustration shown in Figure 7.18.

The exponents of the input operands are loaded in registers $E1$ and $E2$, which are connected to an adder that computes $E1 + E2$. The comparison of exponents required for addition and subtraction is made by computing $E1 - E2$ (i.e. $E1 + (-E2)$, essentially is an addition) and placing the result in a counter E. The larger exponent is then determined from the sign of E. The bit-shift of one of the significands (mantissas) required before the addition/subtraction of the significands can be controlled by E. The magnitude of E is sequentially decremented to zero. After each such decrement, the corresponding significand located in the significand unit is shifted one-digit position. After the needed alignment of the respective significand

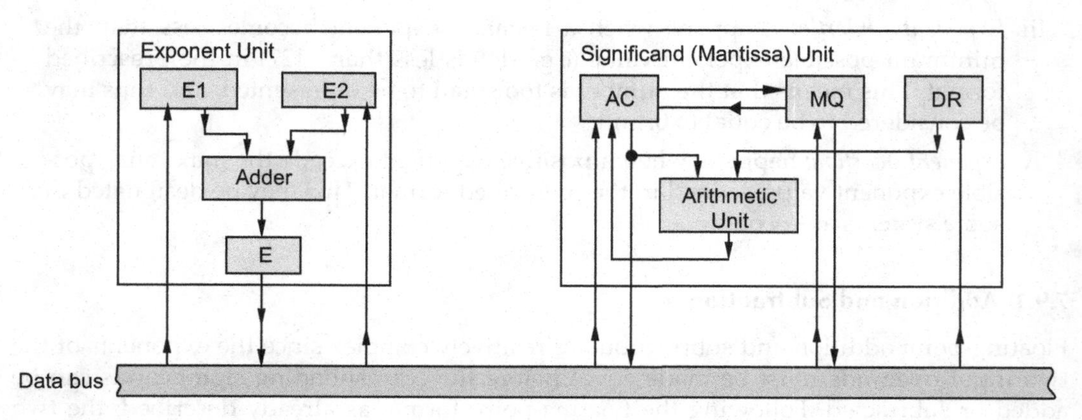

FIGURE 7.18
Schematic block diagram of a floating-point arithmetic unit.

(equalizing the exponent, i.e. when E becomes 0, of the two input numbers X and Y), they are processed in the usual manner depending on the type of arithmetic operation being required. The exponent of the result is also computed and is placed in E.

All the computers have the fixed-point arithmetic instructions as well as the floating-point instructions; it is, hence, always desirable to have a single unit within the ALU to execute both these types of instructions. But, as the sophisticated, faster, and also cheaper electronic technology is now readily available in abundance, it is almost common nowadays in most of the computer systems to incorporate separate units: one dedicated for fixed-point integer (FXU) and another for floating-point arithmetic operations (FPU). Separation of these two individual units located within the architecture of ALU facilitates the execution of fixed-point and floating-point instructions to continue in parallel.

7.9.2 Multiplication and Division

Multiplication and division are relatively simpler and somewhat easier than addition and subtraction, in that no alignment of significand (equalization of the exponents) is needed. As usual, the input operands here are represented in 2's (two's) complementary form.

In **multiplication**, if either operand is 0, the result is automatically declared as 0. The next step is to add the exponents. If the exponents are stored in biased form, the sum of the exponents would then contain double the bias value. Hence, the bias value must be subtracted from the sum. The result may sometimes give rise to a situation of exponent overflow or underflow which must be intimated with the termination of the process. However, if the exponent of the product (result) lies within the specified range, the next step is to multiply the significands of the input operands, taking into account their signs, as is done for integer multiplication (already described earlier). The product (result) will be double the length of the multiplier or multiplicand, which one is larger. The extra bits may be lost due to rounding-off the result. After obtaining the product, the result as usual needs to be normalized, and rounded-off, if required. The action of normalization may sometimes lead to a situation of exponent underflow. Appropriate actions should then be taken to resolve the situation.

Division is performed almost on the same lines as multiplication. Here too, the testing of 0 is to be carried out first. If the divisor is 0, an error is to be declared, or the result may be set to infinity, as per the guidelines of the particular implementation. But, for having a dividend of 0, the final result will be 0. The next step is to subtract the divisor exponent from the dividend exponent. This subtraction removes the bias, which must be added back in, but this addition may result in exponent overflow. However, appropriate tests are then made to inspect exponent underflow or overflow, if any, and a befitting test report can then be accordingly issued. The next action is to divide the significands of the two input operands. Finally, the result as obtained will go through the usual process of normalization, and rounding, if needed.

Two typical flowcharts for separately performing multiplication and division incorporating all the activities as described in Table 1, respectively, are shown in the website: http://routledge.com/9780367255732.

7.9.2.1 Implementation: Floating-Point Multiplication

A multiplier circuit can be implemented using a multistage CSA circuit (already described earlier). This circuit is popularly known as a *Wallace tree* after the name of its inventor (Wallace 1964). The inputs to the adder tree are n terms of the form $M_i = x_i\ Y\ 2^k$. Here, M_i

represents the multiplicand Y multiplied by the ith multiplier bit weighted by the appropriate power of 2. Suppose M_i is $2n$-bit long, and that a full double-length product is required, the desired product P is $\sum_{i=0}^{n-1} M_i$. This sum is computed by the CSA tree that produces a $2n$-bit sum and a $2n$-bit carry word. The final carry assimilation is then usually performed by a fast adder, a CLA, for instance, with normal internal carry propagation.

A brief detail of this topic along with a befitting figure is given in the website: http://routledge.com/9780367255732.

7.10 Precision Considerations: Guard Bits

Several important aspects are associated with the implementation of floating-point operations and their subsequent representations. Although the significands (mantissas) of initial operands and final results are limited to a specified number of bits (e.g. 24 bits for single format, including the implicit leading 1), it is still often necessary to have some extra bits to accommodate the results of the execution of intermediate steps of any type of arithmetic operation. Fortunately, the ALU registers that hold the exponent and significand of each operand prior and after a floating-point operation are always greater in length than the length of the significand plus an implied bit. The register thus automatically provides additional bits to the operands as well as to the final results, and these types of extra bits that are retained are often called the ***guard bits***, which help to realize the maximum accuracy in the final results.

7.10.1 Truncation

At the time of generating any result, it is often required to remove the guard bits from the extended significand by appropriately chopping it off to bring it to a specified length that simply approximates the longer version. This type of act making no changes in the retained bits is commonly known as ***truncation***.

Truncation and its significant impact on the final result are, however, given in the website: http://routledge.com/9780367255732.

7.10.2 Rounding: IEEE Standard

Rounding is essentially a variant of truncation that also disposes guard bits (extra bits) when represented in a specific format. Different types of rounding, including **Von Neumann rounding**, regular *rounding* in general, and also the *rounding* specified in the IEEE 754 floating-point standards, as the default mode for truncation are mostly common. All other *truncation methods* specified in IEEE standards are referred here as *rounding modes* with a list of *four* alternative approaches, namely *round to nearest*, *round to* $+\infty$, *round to* $-\infty$, and *round towards zero* are significant.

When the guard bits are not present or are removed by *truncation* (and *rounding*) at each intermediate step of computation, the amount of error ultimately crept into the final result may be then appreciably high. Therefore, the way that the *guard bits* and *truncation* (*rounding*) are to be used as specified in the IEEE floating-point standard is to enforce a maximum within half a unit in the LSB position of the final result. In general, this requires a rounding scheme in which only *three guard bits* are to be carried along with the needed operations during the computation of each intermediate step. The *first two* of these three bits are the two most significant bits of the section of the significand to be removed. The *third bit* is the logical-OR of all the bits beyond these first two bits in the full representation of the significand. From an implementation point of view, this bit is relatively easy to maintain during the intermediate computational steps to be performed. Initially, this bit is set to 0. If a 1 is shifted out through this position, the bit becomes 1 and retains that value. That is why, this bit is sometimes called the ***sticky bit***.

Details of different types of rounding as stated above, including those as mentioned *truncation methods* specified in IEEE standards, are given in the website: http://routledge.com/9780367255732.

7.10.3 Infinity; NaNs; and Denormalized Numbers: IEEE Standards

IEEE 754 has not only defined and described various methods to be adopted for different rounding modes as already explained, but also formulated many other aspects, including the procedures, to be followed for the floating-point arithmetic so that whatever be the hardware platform used for execution, a uniform and predictable result can always be obtained. However, the main focus is, at present, on *three* such important aspects related to floating-point arithmetic introduced by IEEE standards, namely *infinity, NaNs,* and *denormalized numbers.*

Infinity: Real arithmetic considers infinity as a limiting case that always produces the infinity values in certain situations abided by the following:

$-\infty <$ (any finite number) $< +\infty$

Any arithmetic operation involving infinity, excepting some special cases as will be discussed later, precisely yields the obvious results such as:

If x is any finitely expressible number, then

$x + (+\infty) = +\infty$	$(+\infty) + (+\infty) = +\infty$
$x + (-\infty) = -\infty$	$(-\infty) + (-\infty) = -\infty$
$x - (+\infty) = -\infty$	$(-\infty) - (+\infty) = -\infty$
$x - (-\infty) = +\infty$	$(+\infty) - (-\infty) = +\infty$
$x \times (+\infty) = +\infty$	$x \div (+\infty) = +0$

The other cases in this category involving ∞ are essentially NaNs.

NaN: A *NaN* is essentially a special value encoded in floating-point format, which is generated as the result when an invalid operation is performed. NaNs are of *two types* in general: (i) **signalling NaNs** and (ii) **quiet NaNs**.

A *signalling NaN* conveys (signals) an exception whenever an invalid operation involving an operand is attempted. Signalling NaN is accompanied with such values that lie beyond the domain prescribed by IEEE standards.

A *quiet NaN*, on the other hand, smoothly propagates through almost every arithmetic operation without issuing any intimation of an exception, e.g., the operation $(-\infty) - (-\infty)$, $0 \times \infty$, $0 \div 0$, etc. As a *quiet NaN* moves lurking, it somehow appears to be hazardous because it may give rise to a situation that may sometimes be fatal.

Although IEEE 754 Standard provides the same general format for both types of NaNs, these two kinds of NaNs are precisely implementation-specific in the way they will be represented, so that they can be uniquely identified by the system to appropriately handle numerous exception conditions.

Denormalized numbers: The normalization process is to be executed compulsorily to generate normalized numbers in any floating-point arithmetic operation following IEEE 754 Standard. However, if only normalized numbers are used, then there exists a reasonable gap between the smallest normalized number and 0 (Figure 7.17). In the case of single format (32-bit) under IEEE 754, there are 2^{23} representable numbers in each interval, and the smallest representable positive number is 2^{-126}. If the denormalized numbers can be included in this format, an additional $2^{23} - 1$ numbers could then be uniformly added between 0 and 2^{-126}.

Denormalized representation has an exponent 0, and in the fractional part f, there is no assumed leading 1 before the binary point. In case of denormalized numbers, the exponent field is 1-bias, instead of 0-bias, where bias = $(2^{k-1} - 1)$, k is the number of bits in the exponent field. Therefore, the value of a denormalized positive number is $f \times 2^{-126}$. For example,

The largest denormalized 32-bit (single-precision) number is:

$$0.1111......1 \times 2^{-126} = \left(1 - 2^{-23}\right) \times 2^{-126}$$

The smallest denormalized 32-bit (single-precision) number is

$$0.000......1 \times 2^{-126} = 2^{-23} \times 2^{-126} = 2^{-149}$$

Denormalized numbers are, therefore, equally useful and hence are included in this standard to mainly handle the cases of ***exponent underflow***. When the *exponent* of the result becomes too small (a large negative exponent), the result needs to be denormalized to get out of the situation by simply right shifting the fraction (significand) and incrementing the exponent accordingly for each such shift until the exponent comes within a representable range.

The inclusion of denormalized numbers in the IEEE 754 Standard prevents the density of representable numbers to increase as one approaches from the point of smallest representable normalized number towards 0. Thus, the use of denormalized number precisely helps to smoothen the said density to be mostly uniform in the said domain, and that is why, it is sometimes referred to as ***gradual underflow***. In effect, it fills the gap reducing the width between the smallest representable nonzero number and zero, and minimizes the effect of exponent underflow to such a level that is almost comparable to normalized numbers with rounding-off.

7.11 Summary of Floating-Point Numbers

The significance of the usual bit patterns in the IEEE 754 Standard formats and its interpretations, including some unusual bit patterns, to represent special values have been already described. The extreme exponent values of all zeros (0) and all ones (255 in single format and 2047 in double format), however, define special values that consist of many different types, as already explained.

Value = $(-1)^{-S} \times M \times 2^E$, where S = sign, M = significand (mantissa), and E = exponent

Bias = $2^{k-1} - 1$, where k = number of bits in exponent field
For 32-bit (single-precision), bias = $2^{8-1} - 1 = 127$.
For 64-bit (double-precision), bias = $2^{11-1} - 1 = 1023$.
Normalized: Significand field has implied (hidden) leading 1. Exponent field contains at least one "1". E = unsigned value of exponent field with appropriate bias.
Denormalized: Significand field has implied leading 0. All exponent field bits are equal to 0. E = 1–bias.
Special cases: *NaN*, *infinity*, and *denormalized* (already described).

7.12 Summary

Numerous arithmetical and logical (non-numerical) operations on various types of operands, including fixed-point and floating-point numbers, are carried out by the data processing part of a CPU, a major constituent of which is an ALU. Most of the modern processors nowadays incorporate numerous types of instructions in their instruction set to enable ALU to carry out all these numerous operations, and in many cases also accompanied by the required hardware to process floating-point instructions as well. Computer arithmetic circuit designs, however, presently exhibit several interesting well-developed logic designs, including high-performance adder designs, and sophisticated design of both multiplication and division units using Booth algorithm and restoring/non-restoring division algorithms, respectively. Floating-point and other complex operations are implemented by an autonomous execution unit within the CPU or by a supporting co-processor which is a program-transparent extension to the CPU. A floating-point processor is typically composed of *a pair of* fixed-point ALUs: one to process exponents and the other to process mantissas. Special circuits are yet needed for normalization, and also for exponent comparison and mantissa alignment in the case of floating-point addition and subtraction. The floating-point number representation standard proposed by IEEE has been described, and a set of rules under this specification for performing all four basic arithmetic operations has been given, including the options of special values and exceptions.

Exercises

7.1 What range of decimal values can be represented by a four-digit hex number? Convert 3259_{10} to hexadecimal, and then from hexadecimal to binary.

7.2 How many bits are required to count up to decimal 1 million? Convert 874_{10} to octal, and then from octal to binary.

7.3 A small process-control computer uses hexadecimal codes to representits16-bit memory addresses.

a. How many hex digits are required?

b. What is the range of addresses in hex?

c. How many memory locations are there?

7.4 Convert the decimal number 927.45_{10} to an equivalent binary number.

7.5 What is an advantage of encoding a decimal number in BCD rather than in straight binary? What is a disadvantage?

7.6 Represent the decimal value 195 by its straight binary equivalent. Then encode the same decimal number using BCD code. Convert the binary number 0101 to its equivalent Gray code.

7.7 Explain with reasons the minimum and maximum value of integers that can be represented by an *n*-bit data using

a. signed-magnitude method

b. signed 1's complement method

c. signed 2's complement method

7.8 Represent the number –21 in 8-bit format using

a. signed-magnitude method

b. signed 1's complement method

c. signed 2's complement method

7.9 Explain the principle of a CLA with an appropriate diagram considering binary numbers of 4 bits. State the merits of this adder as well as its drawbacks.

7.10 Show how to extend the 16-bit design of Figure 7.18 to a 64-bit adder using the same two component types: a 4-bit adder module and a 4-bit carry-lookahead generator.

7.11 Write down the Boolean expression for overflow condition when adding or subtracting two binary numbers expressed in two's complement. [For answer, see Section 7.3, Alternative approach.]

7.12 Show that the logic expression $c_n \oplus c_{n-1}$ is a correct indicator of overflow in the addition of 2's complement integers.

7.13 "CSA is called a *fast adder*". Explain with an example why it is so called. Describe the principle and its operation with a schematic diagram.

7.14 Multiply each of the following pairs of signed 2's complement numbers using the Booth's algorithm, assume that X is the multiplicand and Y is the multiplier.

i. $X = 110101$ and $Y = 011011$

ii. $X = 010111$ and $Y = 110110$

iii. $X = +14$ and $Y = -13$

7.15 Use Booth algorithm to multiply –28 (multiplicand) by –12 (multiplier), where each number is represented using 6 bits.

7.16 Describe the situation of worst case and the best case when Booth's algorithm is used.

7.17 With restoring method used in two's complement integer division algorithm, the value in the A register must be restored following unsuccessful subtraction. A slightly more complex approach, known as *non-restoring*, avoids the unnecessary subtraction and addition. Derive an algorithm for the latter approach.

7.18 Divide –121 by 13 in binary two's complement notation, using 8-bit words. Use both restoring and non-restoring division approaches.

7.19 How the non-restoring division algorithm can be derived from restoring division algorithm?

7.20 What is meant by floating-point representation of a number system? Why it is so called? How a binary number N is represented in a floating-point notation?

7.21 What are the four essential elements of a number in the floating-point representation?

7.22 A floating-point number system uses 16 bits for representing a number. The most significant bit is the sign bit. The least significant nine bits represent the significand (mantissa), and the remaining six bits represent the exponent. Assume that the numbers are stored in the normalized format with as usual one hidden bit

a. Give the representation of -2762.5×10^{-2}.

b. Compute the value represented by 1 001010 011000000.

7.23 What is the benefit obtained by using biased representation for the exponent portion of a floating-point number?

7.24 What would be the bias value for

a. A base-8 exponent ($B = 8$) in a 5-bit field?

b. A base-16 exponent ($B = 16$) in a 6-bit field?

7.25 A 32-bit number can represent a maximum of 2^{32} different numbers. How many different numbers can be represented in the IEEE754 single-precision 32-bit format? Explain.

7.26 Represent the following decimal numbers in IEEE 754 single-precision format:

a. –7

b. –1.75

c. 389

d. 245. 625

e. 1/16

f. –1/32

7.27 The following numbers use the IEEE 32-bit floating-point format. What is the equivalent decimal value?

a. 0101 0101 0110 0000 0000 0000 0000 0000

b. 1100 0011 0110 0000 0000 0000 0000 0000

c. 0011 1111 1010 1100 1000 0000 0000 0000

7.28 Compute the content of mantissa field when –26.75 is to be stored with the value of M (mantissa) interpreted as $(-1)^s\ 2^{E-31} 1.\ M$, where mantissa M is of 8 bits, exponent is of 6 bits, and 1 bit is used for sign. [Hint: $(26.75)_{10} = 11010.11 = 1.101011 \times 2^4$, $M = 10101100$]

7.29 Consider a floating-point format with 8 bits for the biased exponent and 23 bits for the significand (mantissa). Show the bit pattern for the following numbers in this format:

a. –549

b. 0.645

7.30 Show how the following floating-point additions are performed in which significands are truncated to 4 decimal digits.

a. $6.487 \times 10^2 + 5.693 \times 10^2$

b. $8.546 \times 10^2 + 7.425 \times 10^{-2}$

Show the results also in normalized form.

7.31 Show how the following floating-point calculations are made in which significands are truncated to 4 decimal digits.

a. $8.748 \times 10^{-3} - 6.593 \times 10^{-3}$

b. $7.756 \times 10^{-3} - 2.259 \times 10^{-1}$

Show the results also in normalized form.

7.32 Show how the following floating-point computations are performed in which significands are truncated to 4 decimal digits.

a. $(6.432 \times 10^2) \times (2.154 \times 10^0)$

b. $(7.756 \times 10^3) \times (2.259 \times 10^2)$

Show the results also in normalized form.

7.33 State and explain the significance and implications of *guard bits*.

7.34 Which of the following truncation technique has the smallest unbiased rounding error and why?

i. Chopping

ii. Rounding

iii. Von Neumann rounding

iv. Both (ii) and (iii)

7.35 If $X = 1.586$, find the relative error if X is truncated to 1.58, and if it is rounded to 1.59.

7.36 In computer-based calculations, one of the critical errors occur when two nearly equal numbers are subtracted. Assume $X = 0.23186$ and $Y = 0.23143$. The computer truncates all values to four decimal digits. $X' = 0.2318$ and $Y' = 0.2314$

a. What are the relative errors for X' and Y'?

b. What is the relative errors for $Z' = X' - Y'$?

7.37 Explain how *NaN* and *infinity* are represented in IEEE754 standard.

Suggested References and Websites

Hamacher, C., Vranesic, Z. G., and Zaky, S. G. *Computer Organisation*, 5th ed., Int'l. ed. McGraw-Hill Higher Education, 2002.

Hayes, J. P. *Computer Architecture and Organisation*, Int'l ed. WCB/McGraw-Hill, 1998.

Mano, M. *Logic and Computer Design Fundamentals*. Upper Saddle River, NJ: Prentice-Hall, 2004.

IEEE 754: The IEEE 754 documents, related publications and papers, and a useful set of links related to computer arithmetic.

Part II

High-End Processor Organisation

Computer architecture is a joint venture of its hardware organisation and its supporting available software/programming facilities. The performance of a computer system is the result of the coordinated effort of its hardware resources and the system/application software. The hardware core is formed with processors, memory, peripheral devices, and the interconnecting buses. Software programs interface with the hardware thus implemented and extract the potential of the hardware to greatly enhance the speed, performance, and the portability of user programs while running on different machine architectures. After the release of Von Neumann's architectural design concept, computers were built with the fundamental resources in the form of a sequential machine executing scalar data. This has since then gone through a series of evolutionary changes to improve and enhance the performance but could not cross a limit due to its design limitations as well as the *mandatory obligation of sequential execution of instructions in programs.* Computer performance has since then been increasing and steadily approaching its physical limit by the use of available faster hardware technologies and improved design in processor architecture, following the conventional path.

Many important areas involving the execution of numerous computational problems still remain beyond the capabilities of the fastest contemporary machines, even after increasing the capacity and improving the speed of the resources. One way to handle this issue is to exploit functional parallelism. Functional parallelism can be realised in two ways. One possibility is to build computers using multiple functional units – perhaps hundreds of low-cost processors (or processing elements) and their allied circuits that can work in parallel on common tasks. This is known as ***processor-level parallelism.*** Another possibility is to speed up the *single processor* by arranging the hardware so that more than one operation can be performed at the same time. This is called ***instruction-level parallelism*** that ensures the increase in the number of operations performed per unit time, leading to a substantial increase (speedup) in performance although no single instruction is executed in less than its predefined allotted time. This approach, in other words, encourages the practice of pipelining at various processing levels.

8

Pipeline Architecture

LEARNING OBJECTIVES

- To explain the pipeline approach in CPU architecture that provides for concurrent execution of many machine instructions (instruction-level parallelism (ILP)) for performance improvement.
- To formulate appropriate methods for pipeline control and collision-free scheduling of instructions.
- To study the different types of hazards which create hindrances in the smooth operation of pipelined processors.
- To describe numerous design approaches that combat the various types of these inherent hazards to reduce their ill effects.
- To detail the features of superpipeline architecture, and superscalar architecture of processor along with the numerous related issues, for further performance improvement.
- To describe in brief very long instruction word (VLIW) architecture and explicitly parallel instruction computer (EPIC) architecture in processors.
- To facilitate an understanding of multithreading as a means of even finer hardware thread-level parallelism (TLP) in processor design, applicable in scalar, superscalar, and VLIW processors.
- To illustrate the features of multicore architecture: its definition, design issues, and effective multicore organisation.
- To describe realization of multicore organisation with hardware multithreading architecture to yield perhaps the highest possible processor performance.

8.1 Pipeline Concept

The basic concept of a pipeline is very simple. A pipeline is similar to an assembly-line operation used in manufacturing plants. Henry Ford invented the assembly-line in the early 1890s to build all cars in stages. For example, in an automobile assembly-line, there are many steps being followed to build a car. Each step (e.g., preparing the chassis, installing the engine, adding the body, etc.) contributes something to the car production. Each step operates in parallel with the other steps operating on a different car. This means that while the second group of workers is just installing the engine on an already prepared chassis of one car done by the first group, the third group is adding the body on another car having the chassis and engine fitting completed. At the same time, the first and second groups are engaged with their own work on another new car assembly. As a result, it is possible to have new cars being rolled out of the assembly line in quick succession. It has been observed that some ideas have stood the test of time and have an immense enduring quality that can be applied equally well in many different ways in diverse environments. Incidentally, this assembly-line idea, in particular, has been implemented in designing a processor in the form of a pipeline (Kogge, P. M.).

8.2 Pipeline Approach: Instruction-Level Parallelism

The processor while executing a program follows an instruction cycle (fetch–decode–execute) for executing each instruction in the program, one after another. The pipeline technique splits up this sequential process (fetch–decode–execute cycle) of instruction execution into suboperations. Each subprocess is then executed in a special dedicated segment (*stage*) that operates concurrently with all other segments inline executing different suboperations on different instructions. Thus, pipelining is essentially an *implementation technique* whereby the execution of multiple instructions can be overlapped. Thus, the pipeline approach gives rise to an essence of parallelism, but only at the instruction level, and is thus legitimately called ***virtual parallelism***.

A *linear pipeline* is visualised as a collection (cascade) of processing segments; each segment in the pipeline completes a part of an instruction execution in a way similar to how the task is partitioned. The result obtained from the computation in each segment is then passed to the next segment in the pipeline. Instructions enter into the pipeline at one end, progress through the stages (segments), and usually exit at the other end but not necessarily just as the assembling of cars would go in an assembly line. The pipelines being employed in the design of processors, however, may be of various types, as we will see in later sections. But whatever be the type, in this *pipeline architecture*, only one instruction is always issued to the pipeline at every clock cycle. That is why this pipeline is sometimes referred to as *scalar pipeline*.

A *characteristic* of pipelines is that several different computations can be in progress in different segments with different instructions at the same time. The overlapping of computation is made possible by associating a *register* (buffer) with each segment in the pipeline. The register provides isolation between (adjacent) segments so that each can operate on distinct data simultaneously.

8.3 Implementation

To demonstrate the *principle* of pipelining, we use for the sake of simplicity, an instruction that can be implemented in at the most five clock cycles. The *five clock cycles* are described as follows. This principle of pipelining can be applied to even more complex instruction sets (CISC–like), such as RISC relatives, although the resulting pipelines would then naturally be more complex.

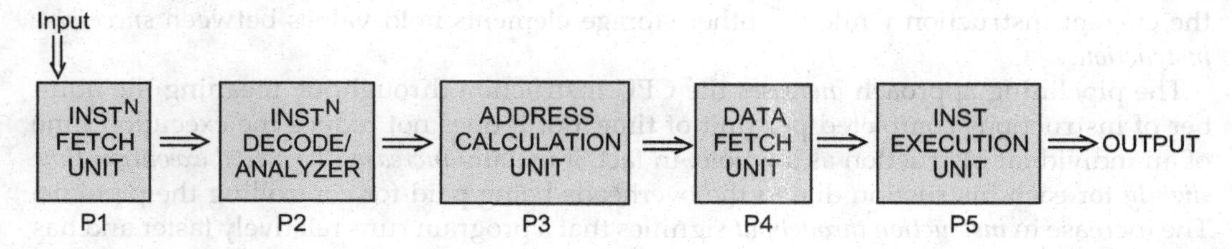

FIGURE 8.1
A five-segment pipeline.

A CPU as shown in Figure 8.1 would then comprise of five *processing units* P1–P5, in which each such unit is assumed to take one cycle to finish its execution (task); then the stages of execution look like the following:

1. *Instruction Fetch Cycle (P1)*: IR ← MEM (PC), PC ← PC + 1

   The content of the PC is the address of the instruction to be fetched from memory into the instruction register (IR). The content of the PC will then be incremented, and the new content of the PC will hold the address of the next sequential instruction needed in subsequent clock cycles.
2. *Instruction Decode/Analysis/Register Fetch Cycle (P2)*

   The instruction thus fetched and available in IR is decoded, and register (IR) is read by two temporary registers (operands) which would be used in later clock cycles. Decoding is done in parallel with reading registers, and this is possible because these fields are at a fixed location in the instruction format. This technique is usually known as *fixed-field decoding.*
3. *Effective Address Calculation Cycle (P3)*

   The address of the operands is now computed for all types of instructions (register–register, register-immediate, branch instruction, etc.), and the effective address thus obtained is placed into ALU output register.
4. *Memory Access/Data Fetch/Branch Completion Cycle (P4)*

   The address of the operand thus obtained from the preceding stage (cycle) is used to access the memory, if needed. In case of *load, store,* and *branch* instructions, data either returns from memory and is placed in the LMD (Load Memory Data)/ MBR (Memory Buffer Register) register or is written into the memory. In case of branch instruction, the PC is replaced with the branch target address in the ALU output register or the PC remains as it is, with already incremented (step 1), targeting the next sequential instruction.
5. *Instruction Execution/Write Back Cycle (P5)*

   The instruction will now be executed, and the result will be in the register file whether it comes from the memory system which is in LMD or from the ALU which is in ALU output register.

The above descriptions (P1–P5) show how an instruction flows through the datapath. At the end of each clock cycle, every value computed during that clock cycle and required for a later clock cycle (whether for the current instruction or for the next instruction) is written into a storage device, which may be memory, a general-purpose register, the PC, or a temporary register. The temporary registers hold values between *clock cycles* for the current instruction while the other storage elements hold values between *successive instructions.*

The pipelining approach *increases* the CPU instruction throughput, meaning the number of instructions completed per unit of time, but it does not reduce the execution time of an individual instruction as a whole. In fact, it usually *increases the actual execution time slightly* for each instruction due to the overheads being paid for controlling the pipeline. The increase in *instruction throughput* signifies that a program runs relatively faster and has lower total execution time, even though no single instruction runs faster.

8.4 Linear and Nonlinear (Static and Dynamic) Pipelines

Pipelining can be broadly classified into *two* distinct categories.

a. The traditional ***linear pipelines*** are ***static*** pipelines because they are built to perform only one predefined fixed function at specific times in a forward direction from one stage to next stage (Figure 8.1). In a static pipeline, repeated evaluations of the same function are performed with different data for a specified period of time. The operation of a static pipeline can only be changed after the pipeline has been *drained*. A pipeline is said to be drained when the last input data leaves the pipeline. The performance of static pipelines is severely degraded when the operations change frequently, since this requires the pipeline to be *drained* and *refilled* each time.

b. ***Dynamic*** pipelines are ***nonlinear pipelines*** that can perform more than one operation at a time. This pipeline has the provision to be reconfigured to execute variable functions at different times. This requires *feed forward* (bypass) and *feedback* (for repeated use of a stage) connections in addition to the streamlined connections between the stages. Figure 8.2 illustrates an example of a three-stage dynamic pipeline of an execution stage that performs subtraction and multiplication on different data at the same time. To perform multiplication, the input data must go through stages 1, 2, and 3; and to perform subtraction, the data only needs to go through stages 1 and 3. Therefore,while the first stage of the subtraction process can be performed on an input data D_1 at stage 1, the last stage of the multiplication process is performed at stage 3 on a different input data D_2. It is thus essential to ensure that the inputs D_1 and D_2 do not reach the stage 3 at the same time. In fact, the mechanism that controls data which is to be fed to this type of pipeline is much more complex than its counterpart – static pipelines.

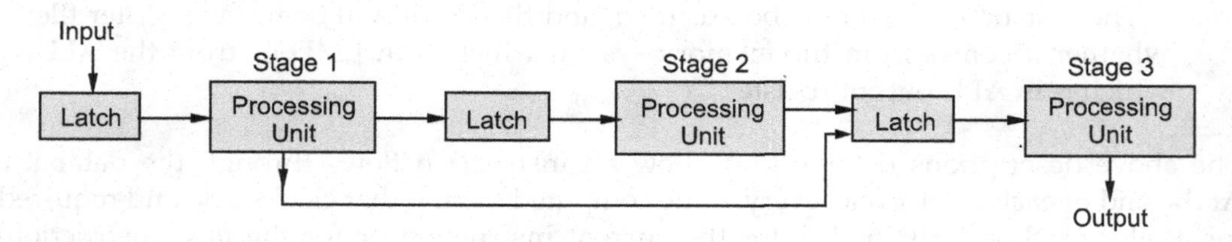

FIGURE 8.2
A three-stage dynamic pipeline of execution stage.

A given function (instruction) can easily be *partitioned* into a sequence of linearly ordered subfunctions in a *static pipeline*. However, partitioning the function in a *dynamic pipeline* becomes quite complicated because the pipeline stages are often interconnected with additional loops apart from having normal streamlined connections.

The ***scheduling*** of successive subfunctions in a linear pipeline is straightforward, but in dynamic pipelines having *feed forward* and *feedback* connections, this becomes a nontrivial task. Due to the presence of these additional connections, the output from a dynamic pipeline does not necessarily come out from the last stage of the pipeline as it generally happens with a linear pipeline. Dynamic pipelining has the *advantage* of evaluating different functions by the *same pipeline* using different *dataflow* patterns.

8.4.1 Linear Pipeline: Asynchronous and Synchronous Models

Depending on the *control of data flow* through different stages along the pipeline, the linear pipelines are broadly classified into ***two*** categories:

Asynchronous Model: Here,the data flow between the adjacent stages is controlled by a handshaking protocol. When stage S_i is ready to transmit, it sends a *ready signal* to stage S_{i+1}. The stage S_{i+1} receives the incoming data and then sends back an *acknowledgement signal* to S_i. Multicomputers exploit this idea in designing communication channels while using *pipelined wormhole routing* in message passing. The throughput rate in this method is *variable* since different stages have different amounts of delay.

Synchronous Model: Here, delays between stages are made to be approximately *equal* in all stages. To achieve this, clocked latches are used as interfaces between stages. The latches are made with *master–slave* flip-flops which can isolate inputs from outputs. Upon arrival of a clock pulse, all latches transfer data to the next stage *simultaneously*.

The ***pipeline frequency*** is defined as the inverse of the *clock period*, i.e. $f = 1/t$. If one result is expected to come out of the pipeline *per cycle*, f represents the *maximum throughput* of the pipeline. Depending on the *initiation rate* (initiation of an input data refers to the time when the data enters the first stage of the pipeline) of successive task entering the pipeline, the *actual throughput* of the pipeline may be *lower* than f. This is because more than one clock cycle (caused by the presence of delays) has elapsed between successive task initiations.

An example with solution and asynchronous and synchronous models of linear pipeline aregiven in the website: http://routledge.com/9780367255732.

8.4.2 Characteristics and Behaviour: Space-Time

The behaviour of a linear pipeline can be illustrated with a space-time diagram which shows the segment utilisation as a function of time. Figure 8.3 shows six instructions (tasks) I_1 through I_6 that are executed in *five segments*.

Initially, I_1 is handled by segment 1. After the first clock cycle, I_1 will be passed to segment 2 while segment 1 is busy with I_2. Continuing in this manner, the first task I_1 is completed after the fifth clock cycle. From then on, the pipeline completes a task in every clock cycle, no matter how many segments (stages) there are in the pipeline system. Once the pipeline is full, it takes only one clock period to yield an output for successive tasks already in the pipe.

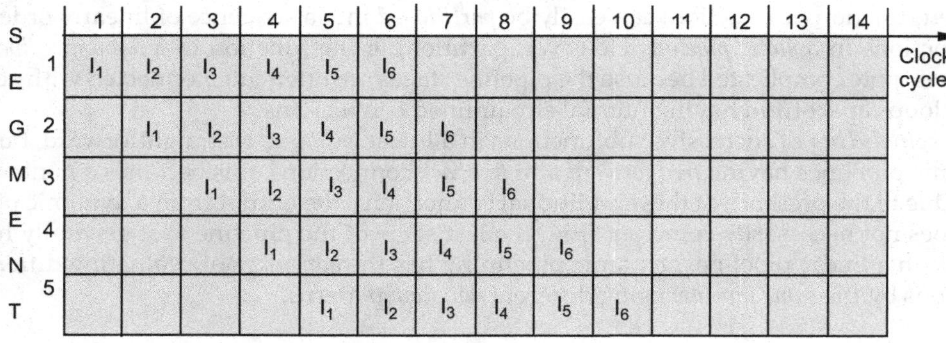

Segment	1	2	3	4	5	6	7	8	9	10	11	12	13	14
1	I_1	I_2	I_3	I_4	I_5	I_6								
2		I_1	I_2	I_3	I_4	I_5	I_6							
3			I_1	I_2	I_3	I_4	I_5	I_6						
4				I_1	I_2	I_3	I_4	I_5	I_6					
5					I_1	I_2	I_3	I_4	I_5	I_6				

Clock cycle →

FIGURE 8.3
Space-time diagram for a pipeline.

8.4.3 Speed-Up, Efficiency and Throughput

Consider a pipeline consisting of k segments, which is used to execute n instructions (tasks). The first instruction (task) I_1 requires a time equal to $k \times 1 = k$ clock cycles (one clock cycle time may be of any duration) to complete its operation, since there are k segments in the pipe. The remaining $(n-1)$ instructions (tasks) emerge from the pipe at the rate of one instruction (task) per clock cycle, and they will be completed after a time equal to $(n-1)$ clock cycles. Therefore, the total time taken for n instructions (tasks) using a ***k*-segment** pipeline will be $k + (n-1)$ clock cycles.

If one clock cycle time = ω, called ***clock period***, then total time required with this pipeline is

$$T_k = \{k + (n-1)\}\omega$$

Now, consider a nonpipeline unit that performs the same operation and takes a time equal to t_n, where $t_n = k\omega$, (k is the number of stages and ω is the time for each stage). This time is needed to complete each task. Hence, the total time required for n tasks is $T_L = n \times t_n = nk\omega$.

The ***speed-up*** factor for a k-stage pipeline over a nequivalent nonpipelined processor is defined as

$$S = T_L/T_k = nk\omega/\{k+(n-1)\}\omega = nk/\{k+(n-1)\} = nk/\{n+(k-1)\}$$

As the number of task increases, n becomes much larger than $k-1$ and $n+(k-1)$ approaches the value of n. Under this condition, the speed-up becomes (from the above equation)

$$S \approx nk/n \approx k \quad (8.1)$$

This shows that the theoretical maximum speed-up that a pipeline can provide is k, where k is the number of segments (stages) in the pipeline. The maximum speed-up is very difficult to achieve because of *data dependencies* between successive tasks (instructions), *program branches, interrupts*, and other similar factors.

The ***reasons*** why the pipeline cannot operate at its maximum theoretical rate are as follows:

a. Different segments may take *different times* to complete their own suboperations.
b. The clock cycle must be chosen to equal the time delay of the segment with the maximum propagation time. This again causes all other segments to waste time while waiting for the next clock to arrive.
c. It is not correct to assume that a nonpipeline circuit has the same time delay as an equivalent pipelined circuit.
d. Many of the intermediate registers which are required in a pipeline circuit will not be needed in a single-unit nonpipeline circuit.

The speed-up factor $S = nk/\{k+(n-1)\}$ is a function of n, the number of operations or instructions, and k, the number of subdivided pipeline stages. As depicted in Figure 8.4, for small values of n, the speed-up can be very poor. The smallest value of S is 1 when $n = 1$. From Eq. 8.1, the larger the number of pipeline stages k, the higher would be the potential speed-up performance.

When, $n = 64$ and $k = 8$, $S = 7.1$, but, when $n = 64$ and $k = 4$, $S = 3.7$ [≈4].

In fact, the value of k depends on the maximum number of stages into which instruction processing can be efficiently broken down. This number, in turn, depends on the complexity of the instruction set, the nature of the internal datapath of the CPU, and the organisation of the external main memory. However, the number of pipeline stages (k) also cannot be increased indefinitely due to some *practical constraints* which are primarily *cost, control, complexity, circuit implementation,* and *packaging limitations*. Furthermore, the stream length n also affects the speed-up. However, the longer the stream, the better the speed-up would be in a pipeline.

The *finest level* of pipelining is called ***micropipelining***, with a subdivision of pipeline stages at the *logic gate level* (*lowest level*). However, in practice, much of the pipelining is staged at the *functional level*, with $2 \leq k \leq 15$. Very few pipelined processors are designed to exceed ten stages in real computers. On the other hand, the *coarse level* for pipeline stages can be implemented at the *processor level*, called ***macropipelining***. The number of stages being used in a pipeline often depends on the trade-off between the performance and cost. The optimal choice for such a number, however, can be determined by obtaining the peak value of a ***performance/cost ratio*** (PCR), discussed in website: http://routledge.com/9780367255732.

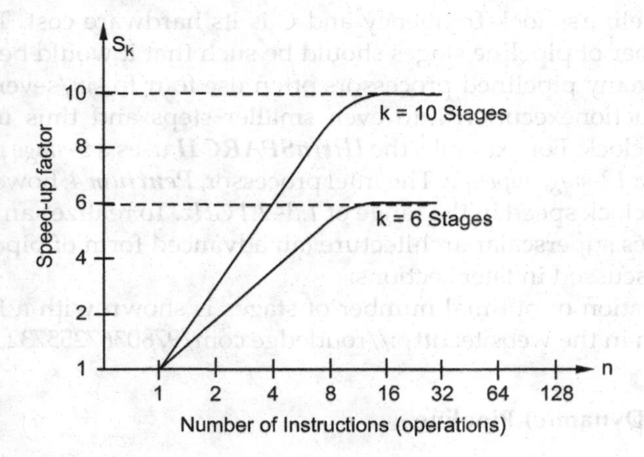

FIGURE 8.4
Speed-up factor as a function of number of operations.

Efficiency and Throughput: The *efficiency*, E_k of a linear k-stage pipeline is the ratio of speed-up factor and number of stages, which is k here. Thus,

$$E_K = S/k = nk/k\ [k+(n-1)] = n/[k+(n-1)] \tag{8.2}$$

Obviously, E_k approaches 1 when $n \to \infty$, and lower bound of $E_k = 1/k$ when $n = 1$.

Pipeline efficiency indicates the accumulated rate of utilisation of all stages present in the pipeline. The ***utilisation*** of a pipeline is defined as the percentage of time that the stages of the pipeline are used over a sufficiently long period of time. A pipeline is said to be utilised 100% of the time when every stage is utilised during each clock cycle.

The ***throughput*** H_K, also called ***bandwidth*** of a pipeline, is defined as *the average number of tasks (operations) performed per unit of time at each stage.* If t is the clock cycle time of a k-stage pipeline, then

$$H_K = n/[k+(n-1)] \times t = f \cdot n/[k+(n-1)] = f \cdot E_k, (f = 1/t = \text{frequency})$$

The *maximum throughput* $H_K = f$ when $E_k = 1$ as $n \to \infty$.

The pipeline *efficiency* and pipeline *throughput* are related to each other. In some situations, they are in close conformity; i.e. higher throughput accompanies higher efficiency. In other situations, they may have a contrary conclusion.

Optimal Number of Stages: Intuitively, the larger the number of pipeline stages (k), the higher the potential speed-up performance. However, in practice, most pipelining is staged at the *functional level* with $2 \le k \le 15$. In fact, most pipelined processors use the number of stages that seldom exceeds 10. Inclusion of each new stage S_i introduces some new hardware cost, the cost of latch (buffer register) between successive stages, and the cost of associated control logic. That is why, the optimal choice of the number of pipeline stages would be such that it should be able to maximize the performance of the pipeline and that the cost of doing so should also be acceptable. A pipeline performance/cost ratio (PCR) has been defined by Larson as

$$\text{PCR} = \frac{\text{Maximum throughput}}{\text{Pipeline cost}} = \frac{\text{Pipeline's clock frequency}}{\text{Pipeline cost}} = \frac{f}{C} \tag{8.3}$$

Where f is the pipeline's clock frequency and C is its hardware cost. Thus, the ultimate choice of the number of pipeline stages should be such that it would be able to maximise this PCR. In fact, many pipelined processors often use four to six/seven stages normally. Others split instructionexecution into even smaller steps and thus use more pipeline stages and a faster clock. For example, the ***UltraSPARC II*** uses a *9-stage pipeline* and ***Intel's Pentium Pro*** uses a *12-stage pipeline*. The Intel processor, ***Pentium 4***, however, has a *20-stage pipeline* and uses a clock speed in the range of *1.3–2.1 GHz*. To realizes an even faster operation, Pentium 4 uses superscalar architecture (an advanced form of pipeline architecture) which has been discussed in later sections.

A detail computation of optimal number of stages is shown with a figure, and solved examples are given in the website: http://routledge.com/9780367255732.

8.4.4 Nonlinear (Dynamic) Pipeline

A dynamic pipeline has *feedforward* and *feedback* connections along with the usual streamlined connections. That is why this pipeline structure is sometimes called a *nonlinear pipeline.*

The *scheduling* of successive subfunctions in a dynamic pipeline thus becomes a nontrivial task due to the presence of these additional connections. The output from a dynamic pipeline does not necessarily come out from the last stage of the pipeline as it generally happens with a linear pipeline. Dynamic pipelining has the *advantage* of evaluating different functions by the *same pipeline* using different *data flow* patterns as already explained.

8.4.4.1 Reservation Table

A reservation table is a pictorial representation of the space-time flow of data through the different stages in a pipeline for the evaluation of one function. With a given pipeline configuration, the evaluations of different functions can generate multiple reservation tables.

A pipeline reservation table shows the stages of a pipeline that are in use for a particular function. Each *stage* of a pipeline is represented by a row in the reservation table, where each row, in turn, is broken into columns, usually one per clock cycle. The number of *columns* (also called *evaluation time*) indicates the total number of time units required for the pipeline to evaluate a particular function. To indicate that a stage S_k is in use at time t_i, an X is placed at the intersection of the row and column in the table corresponding to that stage and time unit. For example, to evaluate the input function A as shown in Figure 8.5, the data here must go through the stages 1, 2, 3, 1, and 2 progressively at times t_0, t_1, t_2, t_3, and t_4, respectively, indicated by *X*'s at the respective positions to yield the desired result. Multiple checkmarks (*X*'s) in a row means repeated usage of the same stage in different cycles. Contiguous checkmarks in a row indicate prolonged usage of a stage over more than one cycle. Multiple checkmarks in a column imply that multiple stages are being used in parallel during a particular clock cycle. The *evaluation time* of a function is determined by the number of columns it uses. The reservation table, for a *static pipeline*, is trivial because the dataflow follows a linear streamline. In a *dynamic pipeline*, since a nonlinear pattern is followed, more than one function can be performed at a time, and the reservation table becomes more interesting and reasonably complicated.

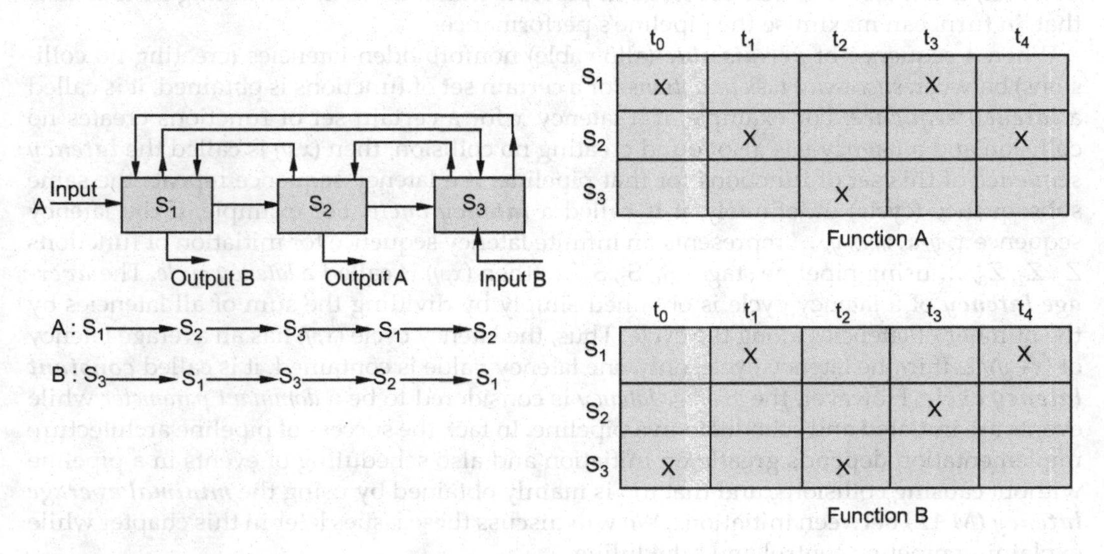

	t_0	t_1	t_2	t_3	t_4
S_1	X			X	
S_2		X			X
S_3			X		

Function A

	t_0	t_1	t_2	t_3	t_4
S_1		X			X
S_2				X	
S_3	X		X		

Function B

FIGURE 8.5
A three-stage dynamic pipeline with feedback and feedforward connections for two different functions and their corresponding reservation tables.

If two functions A and B are given for a *static pipeline*, a *single* reservation table can specify both the functions at the same time. But in case of a *dynamic pipeline*, more than one reservation table may be required to be specified; the evaluation of the function A and that of the function B correspond here to two separate reservation tables as shown in Figure 8.5. The function A requires five clock cycles to be evaluated, and the function B also requires five clock cycles as shown in Figure 8.5. The evaluation time taken by two functions in the pipeline may or may not be the same.

8.4.4.2 Latency and Collision

Initiation of an input data refers to the time when the data enters the first stage of the pipeline. The *duration of time* measured in terms of time units (clock cycles) between two successive initiations of a pipeline is the ***latency*** between them. A latency of *j* means that a delay of *j* time units or that the two successive initiations are separated by *j* clock cycles. Two or more initiations that use the same *pipeline stage* at a particular instant of time will create a ***collision***. A collision will occur *at any stage* if two pieces of input data are initiated with a latency *equal to the distance between two X's* in the corresponding row of the reservation table.

For example, the table in Figure 8.5 for function A has two X's in the first row (S_1) with a distance 3. Therefore,if a second piece of data is initiated in the pipeline 3 units of time after the first,a collision will occur in stage 1. So, latency 3 will create a collision in stage 1 for this function. Other latencies similarly may create collisions in the other stages. Latencies that cause collisions are called ***forbidden latencies***; others known as ***nonforbidden latencies***, creating no collisions as such, are *permissible*. Collisions indicate *resource*(stage) *conflicts* in the pipeline that must be avoided at the time of initiating a sequence of pipeline instructions, since they often reduce the performance gained by the pipeline from its ideal speed-up. The fundamental requirement for avoiding collision is simply to delay initiations of new instructions by time periods *not appearing* in the forbidden list. Collision avoidance, however, is a major and delicate issue in pipeline initiation than scheduling instructions that, in turn, can maximise the pipeline's performance.

When a sequence of ***permissible*** (allowable) nonforbidden latencies (creating no collisions) between *successive task initiations* for a certain set of functions is obtained, it is called a ***latency sequence***. For example, if a latency x for a certain set of functions creates no collision and a latency y is also found creating no collision, then (x,y) is called the ***latency sequence*** of this set of functions for that pipeline. If a latency sequence repeats the same subsequence (cycle) indefinitely, it is called a ***latency cycle***. For example, if the latency sequence $x, y, x, y, x, y, \ldots$ represents an infinite latency sequence for initiation of functions $Z_1, Z_2, Z_3, \ldots$ using pipeline stages $S_1, S_2, S_3, \ldots$, then (x,y) is called a *latency cycle*. The ***average latency*** of a latency cycle is obtained simply by dividing the sum of all latencies by the number of latencies along the cycle. Thus, the latency cycle (x,y) has an average latency of $(x+ y)/2$. If in the latency cycle, only one latency value is contained, it is called ***constant latency cycle***. However, the *average latency* is considered to be a *dominant parameter* while events are initiated and scheduled in a pipeline. In fact, the success of pipeline architecture implementation depends greatly on initiation and also scheduling of events in a pipeline without causing collisions, and that too is mainly obtained by using the ***minimal average latency (MAL)*** between initiations. We will discuss these issues later in this chapter while explaining pipeline control and scheduling.

A detailed explanation of latency and collision with figures is given in the website: http://routledge.com/9780367255732.

8.5 Areas of a Pipeline

Any instruction can be decomposed into a sequence of suboperations of about the same complexity, and each such suboperation can then be executed at the respective stage of the pipeline during a predefined cycle time. In this way, a stream of instructions, hence, can be executed by a pipeline in an overlapped manner. There are *three* major areas of computer design where the pipeline organisation is found to be beneficial. Thus, the pipelines are usually divided into the following main classes: (i) instruction pipelines, (ii) arithmetic pipelines, and (iii) memory-access pipelines.

An ***instruction pipeline*** operates on a stream of instructions by overlapping and decomposing the *fetch, decode,* and *execute* phases of the instruction cycle. It was used first in IBM 7030 in some form and later in a few other computers of the decade of the '60s. It once again re-emerged in the 1980s and was massively used in RISC machines, as one of the major contributors towards achieving RISC's high performance. It has also been widely used in many high-end mainframe CISC machines and in contemporary microprocessor-based small systems like Intel 80X86 series as well as in Motorola 68000 series.

An ***arithmetic pipeline*** divides an arithmetic operation into suboperations for execution in the different segments of a pipeline.

A ***memory-access pipeline*** is operated using separate instructions and data caches to minimise delays that are caused due to the non availability of precise information needed in instruction execution within specified intervals.

A pipeline in each of these classes as mentioned above can again be designed in two categories: ***static*** or ***dynamic***. The implementation technique for each of these classes is obviously different for each category.

8.5.1 Instruction Pipeline

An instruction pipeline reads consecutive instructions from memory by one segment (*fetch* segment), while previous instructions already fetched are being executed in other segments. This causes the instruction *fetch, decode,* and *execute* phases of an instruction cycle to overlap and perform simultaneous operations on several different instructions at different segments giving rise to an increase in the performance of a processor. An m-stage instruction pipeline can thus overlap the processing of a maximum of m instructions.

A typical instruction execution under pipeline architecture consists of a sequence of operations, including *instruction fetch, decode, calculate operands' addresses, fetch operands, execute,* and *write back operand* phases. These phases are ideal for overlapped execution on a linear pipeline. Each such phase may require one or more clock cycles to execute, depending on the type of the instruction and the design of the processor/memory architecture. The design of the instruction pipeline will be most efficient if the instruction cycle is divided into segments of equal duration. The time that each phase takes to complete its function depends to a large extent on the instruction and the way it is executed. Let us now consider the following decomposition of instruction processing.

1. **Fetch Instruction (FI)**: Read the next expected instruction into a buffer (prefetch buffer) from a cache memory (or main memory).
2. **Decode Instruction (DI)/Register Fetch**: Determine the opcode and the operand specifiers. Access the register file to read the registers. The output of the general-purpose registers is then put into temporary registers.

3. **Calculate Operand Address (CO):** Calculate the effective address of each source operand. This includes direct addressing, displacement, register–indirect, indirect, or other different forms of address calculation.
4. **Fetch Operand/Memory Access (FO):** Access memory, if needed, to fetch each operand. For a load instruction, data returns from memory and is placed in the LMD (Load Memory Data) register. If it is a store, then data from register is written into memory. In either case, the operand address as computed in the prior cycle is used. But operands in registers need not be fetched.
5. **Execute Instruction (EI):** The ALU performs the indicated operation on the operands obtained in the prior cycle and stores the result, if any, in the specified destination operand location. This involves execution of Memory reference, Register–Register ALU instruction, Register Immediate ALU instruction as well as Branch instruction.
6. **Write Back (Operand) (WB):** Write the result into the register file or store the result in memory.

As an illustration, let us assume a six-stage pipeline having equal duration at each stage as described above. It can be shown that this pipelined architecture can reduce the execution time for a sample of 8 instructions from 48 time units (for nonpipelined architecture, $8 \times 6 = 48$) to 13 ($= 1 \times 6 + (8 - 1) \times 1$) time units. It is interesting to note as shown in Figure 8.6 that each instruction passes through all six stages in the pipeline, which will not be the case for certain instructions. A *load* instruction, for example, does not require WB stage. However, for the sake of simplicity, it is assumed that the hardware is so designed that each instruction requires all six stages, and all these stages can be performed in parallel.

The flowchart of an instruction pipeline considering all the six stages as already explained is illustrated with figure in the website: http://routledge.com/9780367255732.

8.5.1.1 Limitations

Certain factors often cause to limit the performance enhancement. In fact, there are several intricacies involved in the pipeline, creating such hindrances that oppose the operation of instruction pipeline to attain its optimal rate. Some of them that are notable are as follows:

- If all the stages of a pipeline are not of *equal duration*, there will be some *waiting* involved at various pipeline stages.
- If the workload assigned to a stage in the pipeline is too much, the time taken to complete the operation at that stage may be unacceptably long. This relatively extra time spent at one stage will inevitably create a bottleneck in the pipeline system. The pipelined operation in this situation is said to have been *stalled*. To remove this bottleneck, that is, the source of congestion, one solution may be to further subdivide the affected stage to reduce the workload. Another solution may be to incorporate multiple copies of the affected stage into the pipeline.
- Some types of instructions do not require all the stages, and hence some stages are skipped. A register-mode instruction, for example, *load, Add Register (AR)*, etc., does not need an effective address calculation (CO) of the operand.

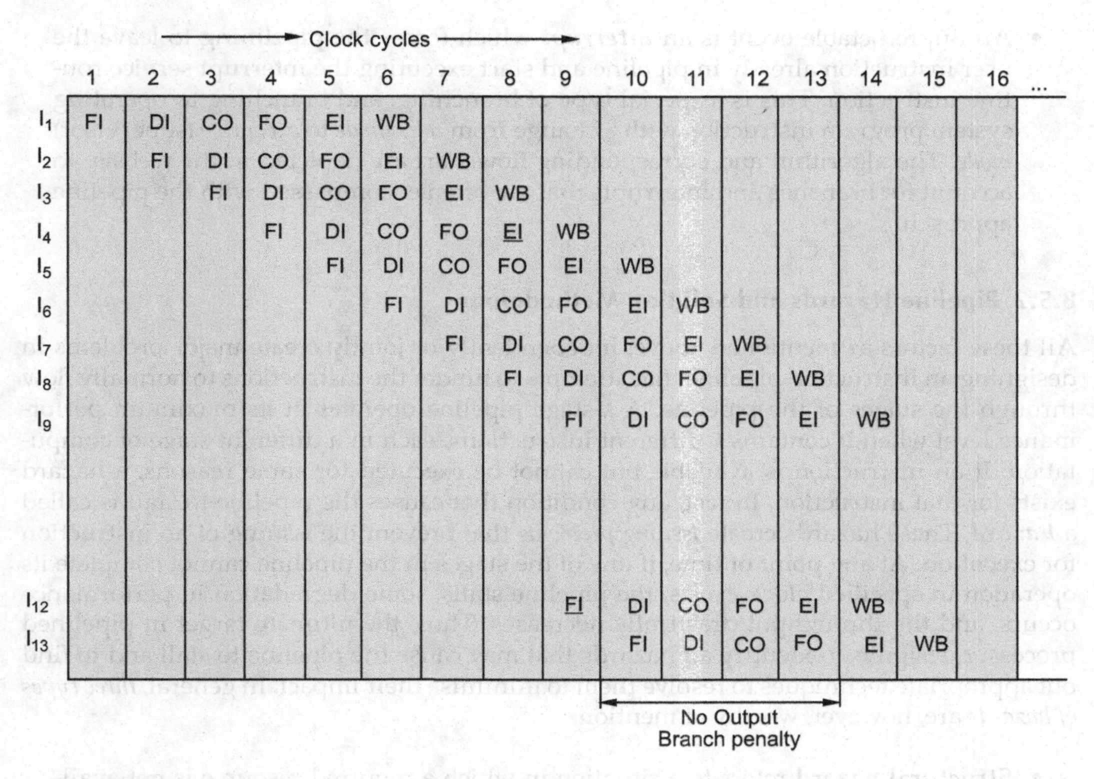

FIGURE 8.6
Limitation of an instruction pipeline due to the effect of a conditional branch.

- *Resource conflicts*: This is caused when two or more stages/segments (FI, FO, WB segments) involve a memory access at the same time for different instructions in the pipeline. This causes one segment to wait until another has completed its operation with the memory. Hence, there will be some *waiting* involved at various pipeline stages.
- Another difficulty is the *fetching problem* that arises due to the presence of the *conditional branch instruction*. This instruction can cause to invalidate several instructions previously fetched and already under operation in prior stages. Figure 8.6 depicts the effects of conditional branch. We assume here that instruction 4 is a conditional branch to instruction 12. Until instruction 4 is executed, there is no way of knowing which instruction will come next. The pipeline in this example, as usual, simply loads the next instruction (instruction 5) and onwards in sequence and proceeds through the stages. The branch is only being taken, if it is determined in the eighth time unit. At this point, the pipeline must be cleared of instructions (instructions 5–8) that are not useful. Killing(draining) off the instructions currently in the pipeline, sometimes called ***squashing***, is required, and a fresh start (refilling) is needed. During time unit 9, the new branch target instruction enters the pipeline. No instructions are completed during time units 10–13. This happens due to the fact that the system could not anticipate the branch, which leads to a situation commonly known as ***performance penalty (Branch penalty)***.

- An unpredictable event is an ***interrupt*** which forces the pipelining to leave the user instruction already in pipeline and start executing the interrupt service routine instruction. This is a special type of branching, and branching to operating system program instruction with a change from *user mode* to *executive* (supervisor) *mode*. The algorithm and corresponding flowchart are depicted in the website to account for branches and interrupts that are often encompassed with the pipeline approach.

8.5.2 Pipeline Hazards and Solution Methodology

All these factors as mentioned above, independently or jointly create major problems in designing an instruction pipeline that attempts to hinder the instructions to normally flow through the stages of the pipeline. A *k*-stage pipeline operates at its maximum performance level when it contains *k* different instructions, each in a different stage of computation. If an instruction is available but cannot be executed for some reasons, a hazard exists for that instruction. In fact, any condition that causes the pipeline to stall is called a ***hazard***. These hazards create *issuing problems* that prevent the issuing of an instruction for execution. At any point of time, if one of the stages in the pipeline cannot complete its operation in specified clock cycles, the pipeline stalls, some degradation in performance occurs, and the throughput drastically decreases. Thus, the ultimate target in pipelined processor design is to identify all hazards that may cause the pipeline to stall and to find out appropriate techniques to resolve them to minimise their impact. In general, *three types of hazards* are, however, worthy of mention.

- **Structural hazard** refers to a situation in which a required resource is not available (or is busy) for executing an instruction.
- **Data hazard** refers to a situation in which there exists a data dependency (operand conflict) among the operands being processed with a prior instruction. Under this situation, either the source or the destination operands of an instruction are not available at the specified time in the pipeline. This leads to cause certain delay in some operations, and the pipeline eventually comes to a stall.
- **Control hazard** refers to a situation in which an instruction, such as a conditional or unconditional jump, subroutine call, interrupt, and other program-control instructions, causes a change in the program flow. As a result, the steady stream of sequential instruction flow to the execution units is interrupted, and the net effect is that the pipeline ultimately comes to a stall, and squashing the pipeline is immediately required.

8.5.2.1 Structural Hazard and Solution Approaches

A structural hazard usually occurs mainly as a result of resource conflicts between instructions. One type of structural hazard that may occur is due to:

i. The inherent design of the execution units. If an execution unit that requires more than one clock cycle (such as multiplication operation that consumes longer time) is not by itself fully pipelined or is not replicated (to have multiple identical copies of the stage), then a sequence of instructions that uses this unit cannot be subsequently (one instruction per clock cycle as per pipeline design) issued for execution. This pipeline stage ultimately appears to be the bottleneck in the instruction flow.

ii. Each row in the reservation table corresponds to each stage in the pipeline. There may be multiple checkmarks (*X*'s) in a row indicating repeated usage of the same stage in different cycle. Too many checkmarks in a row in the reservation table implies heavy loading on that particular stage which may lead to cause a bottleneck in the pipeline. To avoid this situation, *multiple copies* (multiple functional units) of the accused stage are normally used in pipelined processor design. All such multiple execution units are supposed to operate in parallel provided the dependencies inside the unit are resolved.

iii. Another type of structural hazard that may occur is due to the design of the register files. If a register file does not have multiple write (read) ports, multiple writes (reads) to (from) registers cannot be performed at the same time (simultaneously). For example, the instruction pipeline under certain situations might want to perform register writes in a clock cycle by two different instructions I_1 and I_2. Even though the instructions and their data are all available, the pipeline is stalled because of one hardware resource, the register file, which has only one write port and therefore cannot handle two write operations at the same time. This situation can, however, be *resolved* by using register files with multiple input/output ports.

iv. One of the most common cases in which the structural hazard may arise is access to memory. One instruction in the pipeline may need to access memory as part of its WB stage operation, while another instruction in the pipeline is performing its FO stage operation. If instructions and data reside in the same cache (memory) unit, then only one operation can proceed, and the other operation has to wait till the current operation in action is completed, thereby summarily causing a delay. *Memory-access conflicts* are sometimes resolved by:

   a. Having caches to store the desired value, or FO and WB stages may be null.
   b. Using two memory buses, one each for accessing the instruction module and data module respectively, in separate caches for simultaneous operation. **MIPS R4000**, a RISC computer, uses this approach.
   c. Splitting the instruction fetch and data fetch from memory across two distinct stages (segments) in the *superpipeline*. MIPS R4000 system uses this approach out of its CPU's eight-stage pipeline design.

8.5.2.2 Data Hazard (Data Dependency) and Solution Approaches

In a nonpipelined processor, the instructions are executed in sequence in the same order as in the program, and only after the execution of any instruction is completed, does the execution of the next instruction begin. However, in a pipelined processor, this may not be true; here the instruction executions are precisely overlapped. As a result, any instruction in the pipeline may be started and even completed before its previous instruction is completed. This overlapping of instruction executions may invite a problem at the time of execution of *data-dependent instructions*. If an instruction at any stage of the pipeline is asking for an operand which is yet to be generated or supplied by one of its previous instructions in the pipeline but is not ready to provide, the execution of the current instruction under discussion will be delayed, and the pipeline will eventually be stalled. Such a situation is due to what is called a *data hazard*. A data hazard is any condition in which either the

	1	2	3	4	5	6	7	8	9
I_1	FI	ID	CO	FO	EI	WB			
I_2		FI	ID	CO	FO	EI	WB		

FIGURE 8.7
Instruction I_2 having data dependency on I_1.

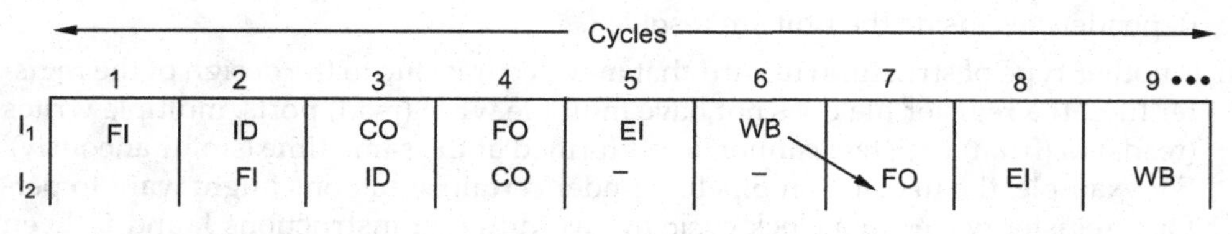

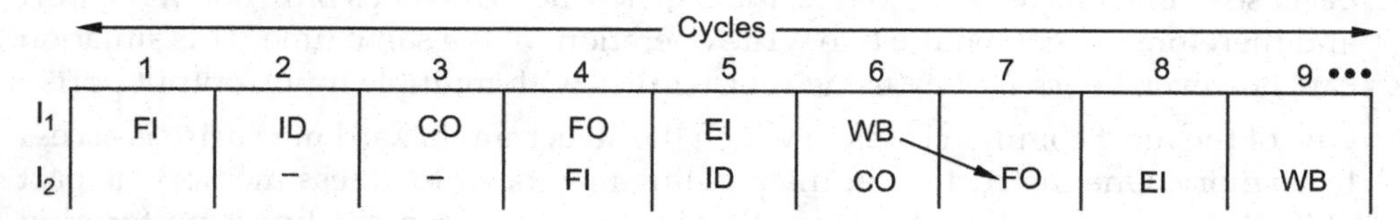

Cycles
1
2
3
4
5
6
7
8
9 •••
I_1
FI
ID
CO
FO
EI
WB
I_2
–
–
FI
ID
CO
FO
EI
WB
FI stage is delayed by two clock cycles

FIGURE 8.8
Two ways of executing data-dependent instructions.

source or the destination operands of an instruction are not available at the specified time in the pipeline. The net effect is that some operation has to be delayed, and the pipeline comes to a stall. For example, consider two consecutive instructions as follows.

```
I1: Add R1,R2,R3 →   R1 = R2 + R3
I2: Add R4,R1,R5 →   R4 = R1 + R5
```

In Figure 8.7, the instruction I_2 has a data dependency on I_1 because it uses the result of I_1 (i.e. the contents of register R_1) as input data. If the instructions were sent through a pipeline in the normal manner, I_2 would be in the FO stage before I_1 passed through the WB stage. This will result in a scenario where I_2 would fetch the old contents of R_1 for computing the current value for R_4, leading to an invalid result. To obtain a valid result, I_2 must not enter the FO stage until I_1 has completed its WB stage. This implies that the execution of I_2 will be delayed for two cycles, or in other words, the instruction I_2 is said to be *stalled* for two cycles. This situation is illustrated in Figure 8.8. When an instruction is stalled, all the instructions that are positioned after the *stalled* instruction will also be stalled. However, all the instructions before the stalled instruction in the pipeline remain unaffected and can continue their execution as usual.

8.5.2.2.1 Solution Approaches

Solution based on delaying execution to resolve data dependency can be accomplished in two ways:

a. Certain data dependencies can be resolved using a *static scheduling* scheme which is supported either by hardware or by software using an optimizing compiler. This scheme is used in the pipeline design where the execution order of the instructions as given in the program is preserved.

One way to resolve the data dependency by using a ***hardware mechanism*** is to delay the FO or FI stages of I_2 for two clock cycles. This is depicted in Figure 8.8. This delay is inserted by way of adding an extra hardware component called a ***pipeline interlock***. A *pipeline interlock* detects the dependency and appropriately delays the dependent instructions until the conflict is resolved. Another way to *minimise* the data hazard stalls to solve the data dependency is to use the **Operand Forwarding** approach. The data hazard is due to the instruction I_2 (Figure 8.8) waiting for data to be written (WB) in the register file by I_1. However, these data are available at the output of the ALU once the execute stage (EI) of I_1 is completed, and hence I_2 need not to wait for the completion of WB stage of I_1. Thus, the delay can be reduced or possibly eliminated, if we arrange for the result of instruction I_1 to be forwarded directly for use in step FO of I_2. The data forwarding mechanism is shown in Figure 8.9 with the arrow connection lines, and the corresponding pipeline stages are also indicated with an arrow in the stages. The two multiplexers connected at the inputs to the ALU permit the data on the destination bus to be selected instead of allowing the original contents to enter naturally into the ALU (EI stage) from its previous stages. Thus, the execution of I_2 proceeds with little or almost no interruption.

b. A ***software mechanism*** is to let the compiler to solve the data dependency by way of ***optimizing the compiler*** itself. A compiler or a postprocessor is thus used to increase the separation between the interlocked instructions that create the data dependencies. During compilation, the compiler detects such dependency between data and instructions. It then rearranges the instructions so that the dependency is not hazardous to the system, and at the same time, it will not affect the program semantics (logics). If it is not possible to rearrange the instructions, NOP (no operation) instructions are inserted in between to create delays. A processor

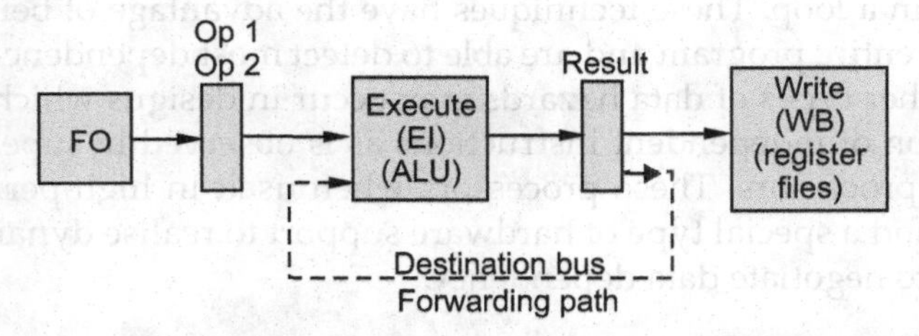

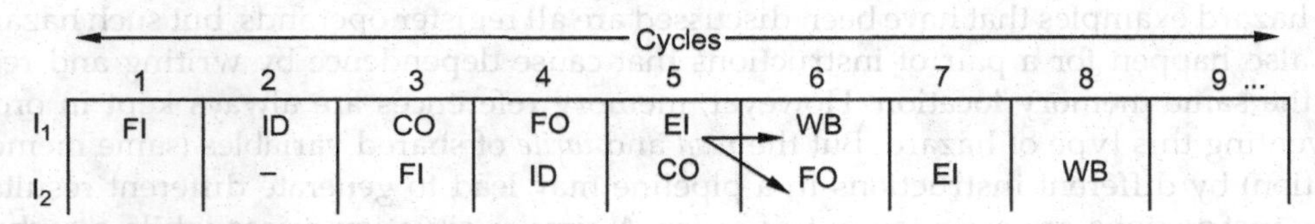

FIGURE 8.9
Operand forwarding in a pipelined register and reduction in delay.

I_1 : Add $R_1, R_2, R_3 \rightarrow R_1 = R_2 + R_3$

I_2 : Add $R_5, R_1, R_4 \rightarrow R_5 = R_1 + R_4$

I_3 : Add $R_6, R_6, R_7 \rightarrow R_6 = R_6 + R_7$

I_4 : Add $R_9, R_8, R_7 \rightarrow R_9 = R_8 + R_7$

Cycles

	1	2	3	4	5	6	7	8	9
I_1	FI	ID	CO	FO	EI	WB			
I_3		FI	ID	CO	FO	EI	WB		
I_4			FI	ID	CO	FO	EI	WB	
I_2				FI	ID	CO	FO	EI	WB

FIGURE 8.10
Rearranging the order of instruction execution.

architecture that has several hardware implementations offering different features may not need this insertion of NOP instruction. However, insertion of NOP instruction while solving the data dependencies, but leads to create larger code size, and as such, more time is taken to complete execution that would eventually lead to a considerable performance degradation. For example, consider the four instructions as shown in Figure 8.10; I_2 is dependent on I_1. Hence, these four instructions may be reordered so that I_3 and I_4 which are not dependent on I_1 and I_2 can be inserted between I_1 and I_2.

Most of today's low-end pipelined computers exploit some form of the static scheduling scheme using optimised compilers. This approach is adequately flexible to realise and comparatively cheaper to implement. Many different static scheduling techniques to check dependency have been developed to exploit parallelism in a loop. These techniques have the advantage of being able to look ahead at the entire program and are able to detect most dependencies beforehand. However, other types of data hazards may occur in designs which allow concurrent execution of independent instructions as is observed in superpipelined and superscalar processors. These processors when used in high-performance computers demand a special type of hardware support to realise dynamic scheduling at run time to negotiate data dependency.

8.5.2.2.2 Data Hazard Classification

The hazard examples that have been discussed are all register operands, but such hazards can also happen for a pair of instructions that cause dependence by writing and reading the same memory location. However, memory references are always kept in order, preventing this type of hazard. But the *read* and *write* of shared variables (same memory location) by different instructions in a pipeline may lead to generate different results if these instructions are executed out of order. A similar situation arises while a cache is accessed. Cache misses could often cause the memory references to get out of order if the processor is allowed to continue working on later instructions while an earlier instruction that missed the cache was accessing memory. The entire pipeline is then forced to stall on a cache miss, effectively allowing the instruction that suffered from such a miss to run for multiple clock cycles to access memory.

a. However, data hazards of this type may thus be classified as one of *three types*, depending on the order of *read* and *write* accesses executed by the instructions. The hazard names denote the *execution ordering* of the instructions that must be maintained to generate a valid result; otherwise an invalid result might occur. The possible data hazards are: (i) *RAW* (*read after write*), also called an *antidependency*; (ii) *WAR* (*write after read*); (iii) *WAW* (*write after write*), also referred to as an *output dependency*; and (iv) *RAR* (*read after read*), but this case is not a hazard, because both are read operations that never change the contents of the common element. However, it is implied that if the order of completion of instruction execution in the program is kept preserved, WAR and WAW types of hazards cannot happen.
b. There are other data dependencies that are either impossible or mostly hard to detect at the time of compilation, or cannot even be detected at compile time (static approach). For example, it is not always possible at the time of compilation to determine the actual memory addresses of load and store instructions that can be used to resolve a possible dependency between them. The actual memory addresses, however, are only known during runtime, and thus *dynamic checking* can be profitably utilised to determine the dependencies between instructions that are either really impossible or hard to detect at the time of compilation. However, dynamic checking at runtime with limited hardware support cannot always offer sufficient lookahead provision that can be used to efficiently exploit all the parallelisms available in a loop. That is why, in practice, a combined static–dynamic dependency-checking methodology is often used to take advantage of both the approaches.

A brief description of all these types of hazards is provided with examples in the website: http://routledge.com/9780367255732.

8.5.2.2.3 Dynamic Scheduling

Dynamic dependency checking and subsequent *dynamic scheduling* at runtime can be realised by two of the most commonly used techniques: ***Tomasulo's register-tagging*** approach and the ***scoreboarding*** scheme. The basic concept behind these methods is to use a mechanism that can identify the availability of operands and functional units in successive computations during runtime.

The key concepts in *Tomasulo's method* are the inclusion of *reservation stations* consisting of a series of functional units equipped with a set of input registers (source 1 and source 2), the innovative addition of *common data bus* (CDB), and the useful introduction of a simple *tagging scheme*. The reservation stations provide the wait for operands and hence relieve the functional units from such a headache. The CDB reduces the delay by supplying the result of an operation directly from the output of the functional unit to the reservation stations. The tagging scheme preserves dependencies between successive operations while encouraging concurrency. The real essence of the Tomasulo's method is to encourage the concurrent execution of instructions by using extra hardware facilities. The compiler still has enough options to substantially influence the *degree of concurrency* realised already by the hardware mechanism. That is why, in practice, a combination of hardware and software techniques is often used to increase concurrency for enhanced performance.

The *scoreboarding method* employs a number of tables to keep track of the status of the execution. The *instruction status table* dictates whether or not an instruction would be issued for execution. If the instruction is ultimately issued, the table prompts the stage to which the instruction belongs, and, if necessary, delays it until the forthcoming predictable hazards are removed. The independent instructions, however, are allowed to be executed out of order. The *functional unit status table* indicates whether or not the unit is busy. For a busy unit, the table also indicates the destination register and the availability of the source registers. A source register for a unit is said to be available if it does not appear as a destination for any other unit. The *destination register status table* is used to solve data hazards between instructions. It has an entry for each register; the entry indicates the active functional unit that will write to this register. Each time an instruction is issued for execution, the instruction's destination register is marked busy and remains busy till the instruction execution is completed. When a new instruction is considered for execution, its operands are checked to ensure that there will be no register conflicts with prior instructions still under execution.

Dynamic instruction scheduling requires extra hardware to work, and thus, was implemented in the past only in high-end processors used in mainframes and supercomputers. Most microprocessors in those days were simply contended with the use of static scheduling because they had only limited hardware support. Due to the advent of superior electronics technology, the scenario has gradually changed. Almost all RISC processors and also Pentium (CISC) series have religiously and effectively implemented pipelined architectures, and these scalar and superscalar processors have been built with adequate hardware support to implement *dynamic scheduling* at runtime. Modern microprocessors like Motorola MC 88000 (RISC) and Intel Pentium series (CISC) have successfully implemented the scoreboarding scheme by way of using *forwarding logic* and *register tagging* (see RISC Architecture, Motorola processor).

Brief descriptions of Tomasulo's method and scoreboarding method with respective figures are given in the website: http://routledge.com/9780367255732.

8.5.2.3 Control Hazard

In any computer program using any programming language, there is normally a need for some kind of statement that allows the flow of control to be something other than sequential. Instructions that do this are called *branches.* Whenever a branch instruction is executed, the action after the branch instruction is called a *branch taken* that fetches a non-sequential or remote instruction, called a *branch target.* The address of the branch target is then to be loaded into the program counter (PC), which will invalidate all the instructions that are either in the pipeline or prefetched in the buffer. This invalidation leads to *draining,* and subsequent *refilling* of the pipeline afresh from the branch target onwards is carried out whenever a branch is taken. As a result, the performance of the pipeline is adversely affected since the address of the next instruction to be executed is not exactly known until the instruction that causes branching (transfer of control out of sequence) is executed. The net effect is that the throughput of the pipeline is drastically reduced, when compared to that of a sequential processor. The clock cycles lost due to draining of partly executed instructions lying at different stages in the pipeline and subsequent refilling of the pipeline to start afresh is often referred to as the ***branch penalty.*** It is to be noted that the presence of a branch instruction does not cause the pipeline to automatically drain and begin refilling. However, a branch, if it is not taken, allows the normal uninterrupted sequential flow of instructions as usual to continue in the pipeline to be executed.

Only when a branch is taken does the problem arise. This summarily indicates that the presence of the branch instruction, if not handled properly, is one of the critical problems in designing and operating an instruction pipeline (Dubey, P., et al.).

Branch instruction, in general, can be classified into *three* groups:

i. Unconditional branch (simple branch, e.g., *go to, subroutine call, interrupt servicing, return*, etc.)
ii. Conditional branch (e.g., *if–then–else*).
iii. Loop branch (e.g., *for loop*).

An *unconditional branch* always alters the sequential program flow, and accordingly sets a new target address in the program counter, and the execution flow then starts from the target address called a *target path* onwards, rather than incrementing the program counter by 1 to point to the next sequential instruction as is normally the case. A *conditional branch* sets a new target address in the program counter only when a certain condition, usually based on a condition code, is satisfied. It then selects a path of instructions that starts from the target address called a *target path*. If the condition is not satisfied, the path starts from the next sequential instruction of the branch instruction as usual, and is called a *sequential path*. A *loop branch* in a loop statement usually jumps back to the beginning of the loop and executes the loop either a fixed or variable number of times depending on certain situation (data-dependent).

8.5.2.3.1 Unconditional Branches and Solution Approaches

To demonstrate the operation of a pipeline while executing an unconditional branch instruction, let us consider Figure 8.11 which shows a program segment consisting of a sequence of instructions I_1–I_3 stored at successive memory locations being executed in a five-stage pipeline. Assume that the instruction I_2 is an unconditional branch instruction and the corresponding branch target instruction is I_K. In clock cycle 4, the decoding operation for I_3 is in progress while the branch instruction I_2 computes its target address. In clock

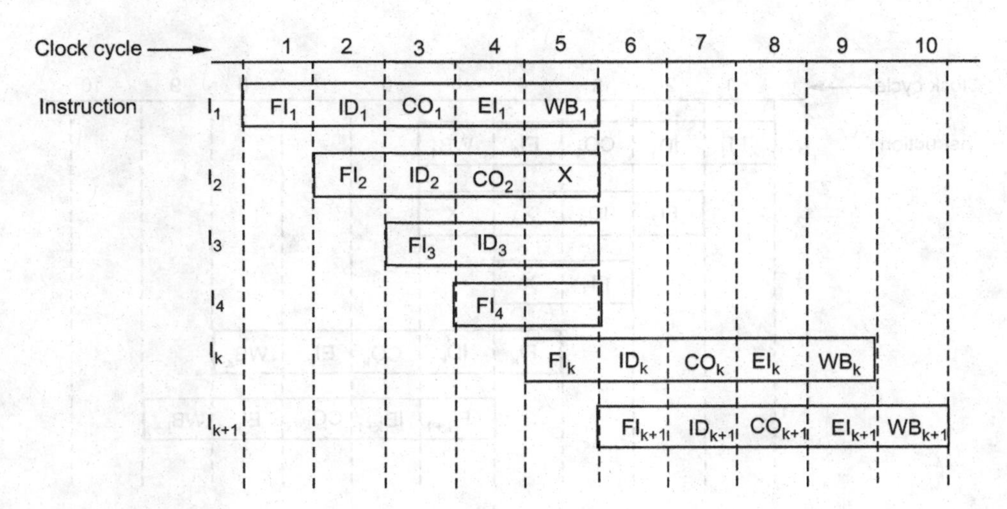

FIGURE 8.11
Idle cycles caused by branch instruction [address computed at the CO stage].

cycle 5, the processor must discard I_3 and I_4 which have been partly processed and clear the pipeline for all the instructions that appeared after the branch instructions and already in different pipeline stages. The processor then fetches the branch target instruction I_K. The Execute stage is instructed to do nothing during that clock period. The ultimate effect is that the entire pipeline then comes to a stall. This clearing of pipeline stages by draining out the partly executed instructions from all the stages appearing after the branch instruction and also idle staying of units (as shown X and blank in the Figure 8.11) causes a significant degradation in pipeline performance which is often referred to as ***branch penalty***. For a deeper pipeline (pipeline having a higher number of stages), the branch penalty paid may be even higher.

One feasible approach aimed at reducing the branch penalty is to compute the branch address *earlier* in the pipeline as shown in Figure 8.12 where target address has been computed in the decoding stage (ID), rather than at a later stage (CO stage) as carried out above. The instruction I_4 has not even entered the pipeline, the number of pipeline stages to be cleared is decreased, and the branch penalty to be paid will be definitely reduced. Typically, the fetch unit has dedicated hardware to *identify a branch instruction* and *compute the branch target address* as quickly as possible after an instruction is fetched. This approach thus substantially reduces the branch penalty, even for a deeper pipeline.

Another approach being used to reduce the branch penalty is to employ a sophisticated fetch unit that can fetch instructions before they are needed and put them in a queue which can store several instructions. An additional separate unit, called the *dispatch unit*, takes instructions from the front of the queue, performs decoding functions, and finally issues them to the next stage in the pipeline for execution as shown in Figure 8.13.

To make this scheme effectively operative, the fetch unit must be equipped with sufficient decoding and processing capabilities to identify and execute branch instructions. The fetch unit always tries to maintain the queue full most of the time to ensure an adequate supply of instructions for processing. This approach reduces the impact of occasional delays due to fetching of instructions while the pipeline is facing a branch or a cache miss, as the dispatch unit still continues to issue instructions from the instruction queue. Conversely, when the pipeline stalls, for example, due to a data hazard, the dispatch unit is not in a position to issue instructions from the instruction queue. The fetch unit, however, at that

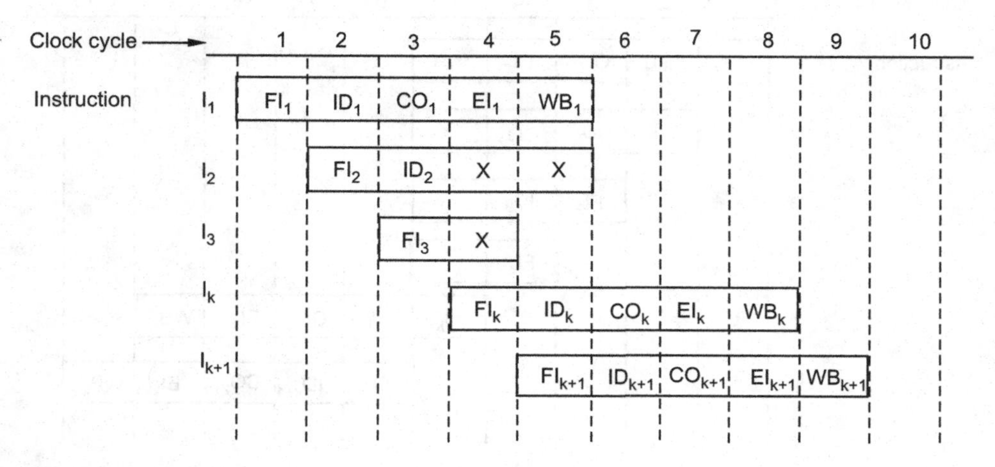

FIGURE 8.12
Idle cycles caused by branch instruction [address computed at the earlier stage (DI stage)].

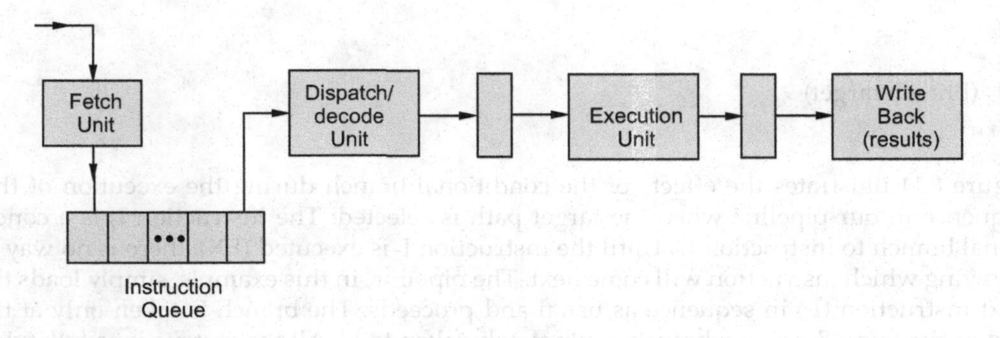

FIGURE 8.13
Instruction queue included in the hardware-pipelined organisation.

time still continues to fetch instructions and add them to the instruction queue to keep it full. Since the fetch unit is capable of identifying and executing branch instructions, this operation is carried out by the fetch unit concurrently with the execution of other instructions. This technique is sometimes referred to as ***branch folding.*** When the branch folding occurs due to the appearance of a branch instruction, the instruction queue must have at least one instruction other than the branch instruction to avoid a pipeline stall. That is why, in many processors, the bus width between the fetch unit and the instruction cache is such that it allows reading more than one instruction in each clock cycle.

The instruction queue has a similar contribution while dealing with ***cache misses.*** When a cache miss occurs, the fetch unit is then engaged to read the desired cache block from the main memory or from a secondary cache leaving its primary task of instruction fetching. As a result, the pipeline may be stalled. But in such a situation, the dispatch unit removes this stall by sending instructions continually for execution as long as the instruction queue is not empty. When the fetch operation is once again resumed, the instruction queue is refilled. Thus, in the event of a cache miss, while it has the support of a nonempty instruction queue, has almost no impact on the rate of instruction execution.

8.5.2.3.2 Conditional Branches and Solution Approaches

A *conditional branch* when being taken, breaks off the execution of normal sequential flow of instruction stream, requiring a new target address to be set in the program counter. As a result, it causes the pipeline to stall, since a new path of instructions is selected other than the usual flow of the next sequential instruction that starts from the target address. We will now examine the effect of branch instructions on the overall performance of the pipeline.

8.5.2.3.2.1 Effect of Branching Among all the preceding branch types already mentioned, conditional branches are possibly the hardest to handle. Appearance of a conditional branch instruction can invalidate several instructions already partly processed. A similar unpredictable event is the occurrence of an interrupt. However, for example, consider the execution of a sequence of instructions including a conditional branch instruction as given below.

I_1
I_2 (conditional branch to I_K)
I_3
I_4

I_5
⋮
I_K (branch target)
I_{K+1}

Figure 8.14 illustrates the effects of the conditional branch during the execution of this sequence in our pipeline when the target path is selected. The instruction I_2 is a conditional branch to instruction I_K. Until the instruction I_2 is executed (EX), there is no way of knowing which instruction will come next. The pipeline, in this example, simply loads the next instruction (I_3) in sequence as usual and proceeds. The branch is taken only at the end of time unit 5, when a branch is ultimately taken to I_K. All the instructions following the branch in the pipeline would then be useless and have to be flushed, thereby wasting a number of useful cycles already consumed. At time unit 6, the instruction I_K ultimately enters the pipeline. No instructions are completed during time units 7–9; this is the *performance penalty* (branch penalty) incurred because the branch could not be anticipated. The number of pipeline cycles wasted (ill effects) between a branch taken and refilling of its branch target is called the ***delay slot***; let it be denoted by d. In general, $0 \le d \le k-1$, where k is the number of pipeline stages.

The impact of such ill effects due to branching in this type of pipeline is computed and is given in the website: http://routledge.com/9780367255732.

8.5.2.3.2.2 Different Techniques The above analysis implies that the pipeline performance is adversely affected due to the presence of branch instruction, and the degradation seems to be appreciable when the instruction stream is sufficiently long. To reduce the negative effects of conditional branching on processor performance, several techniques have been proposed. Some of the better known techniques are as follows:

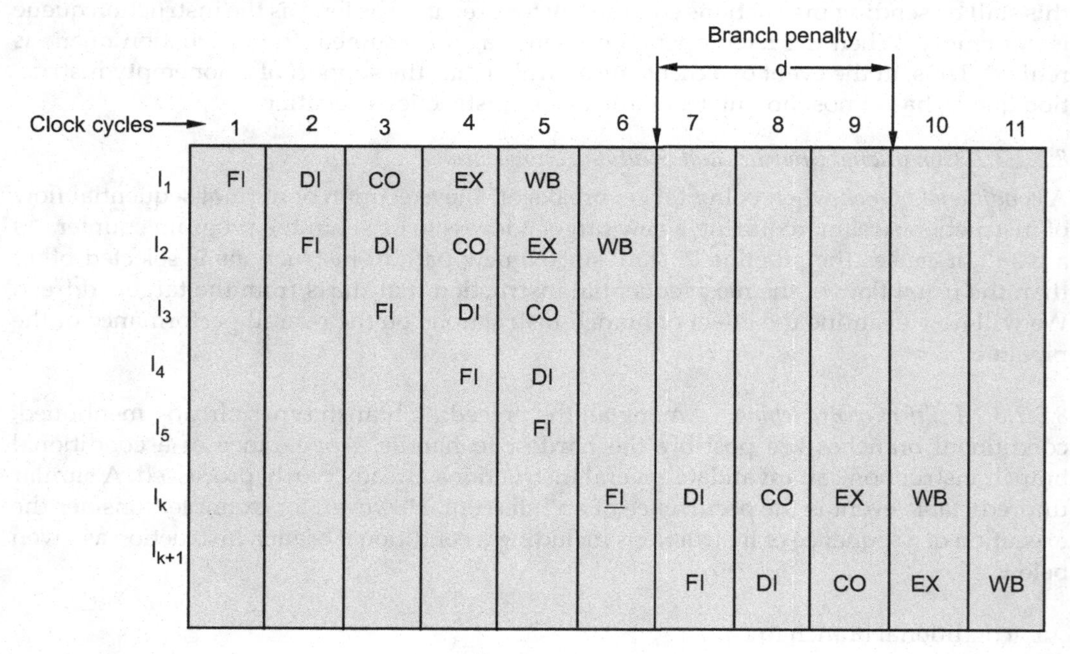

FIGURE 8.14
Branch penalty for a conditional branch.

i. Prefetch branch target
ii. Multiple streams
iii. Loop buffer
iv. Branch prediction
v. Delayed branching.

Here the instruction pipelining is implemented using certain mechanisms. This includes a *brute-force* approach to replicate the initial portions of the pipeline by introducing *instruction buffers* at the instruction fetching stage. While the second stage in the pipeline is executing the instruction, the fetching stage takes advantage of any unused memory cycles to fetch and buffer the instruction. This is called ***instruction prefetch*** or ***fetch overlap***. In one memory-access time, a block of consecutive instructions are then fetched into a prefetched buffer. This block access can be achieved using a *cache memory* or even using *interleaved memory* modules to shorten the effective memory-access time, and hence speedup the pipeline operation. RISC processor MIPS 4000 uses this approach.

8.5.2.3.2.2.1 Prefetch Branch Target When a conditional branch is recognised at the *fetch stage*, the target instruction of the branch is prefetched, in addition to the instruction following the branch instruction as usual. This target is then saved until the branch instruction is executed. If the branch is now actually taken after checking the given condition, the target has already been prefetched and is available. The large IBM system 360/91 model has used this particular approach.

8.5.2.3.2.2.2 Multiple Streams In this type of design, the processor fetches both possible paths, and hence, *two sets of buffer pairs*, namely, *sequential buffer* and *target buffer*, are employed to service the pipeline, making use of two different streams. In normal execution, a pair of *sequential buffers* is used to load sequential instructions from the next sequential address of the branch instruction for in-sequence pipelining. At the time of detecting a conditional branch instruction, the sequential buffers are filled with next block of sequential instructions after the branch instruction as usual, and the target buffers are filled with the instructions from the branch target address of the branch instructions for out-of-sequence pipelining. This is illustrated in Figure 8.15. Once the branch condition is executed and the branch decision is made, appropriate instructions are taken from one of the two respective buffer pairs, and the instructions in the other buffer pairs, hence, will be simply discarded. In each pair, one buffer is used to load instructions from memory,

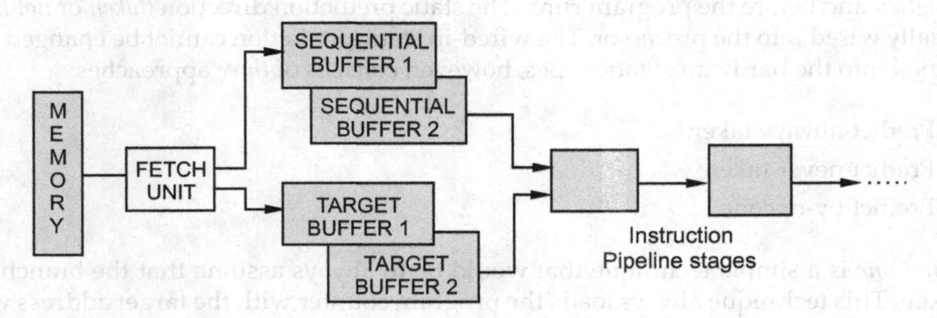

FIGURE 8.15
The use of sequential and target buffers.

and the other buffer in the same pair is used to feed instructions in the pipeline. In this way, the two buffers in a set alternate to prevent a collision between instructions *flowing into* and *flowing out* of the pipeline.

This strategy of double buffering to have two or more pipeline streams as described ensures a constant flow of instructions and data to the pipeline and reduces the time delays caused by the draining and refilling of the pipeline and thereby improves the performance. Some amount of performance degradation, however, is still unavoidable whenever the pipeline is drained at any point of time. This approach has been successfully implemented in IBM 370/168 and IBM 3033 large machines.

8.5.2.3.2.2.3 Loop Buffer In case of sequential instructions contained in a small loop, *a loop buffer* is used as a prefetched buffer. A loop buffer basically is a very-high-speed small memory attached with instruction *fetch stage* of the pipeline and containing ***n*** most recently fetched instructions in sequence. Instructions in the loop body are prefetched as usual and will be executed in sequence if there is no branching out of the loop, until all iterations complete execution. If a branch is detected which falls within the loop boundary, and if the target instruction is already in the loop buffer, unnecessary memory access can be avoided; otherwise standard procedures are followed to handle branch conditions. In essence, the loop buffer works in a similar manner in principle to an *instruction cache* dedicated for instruction execution. The differences are that the loop buffer only retains instructions in sequence, while the cache contains instructions in the most recentlyused order. The loop buffer is much smaller in size, and hence the cost is lower, but the speed is higher. Both CDC 6600/CDC 7600/star-100 and CRAY1 supercomputer have exploited the loop buffer technique. Motorola 68010 implemented a loop buffer in a specified form for executing a three-instruction loop using the DBcc (*d*ecrement and *b*ranch on *c*ondition *c*ode) instruction. A three-word buffer is maintained, and the processor executes these instructions repeatedly until the loop condition is satisfied.

8.5.2.3.2.2.4 Branch Prediction Another increasingly popular and more powerful technique is to *predict* the outcome of a branch decision before the branch is actually executed. Therefore, based on a particular prediction, either the sequential path or the target path is chosen beforehand for execution. Although the chosen path often reduces the branch penalty, it may also increase the penalty in case of incorrect prediction. Various techniques can be used to predict whether a branch will be taken. There are two types of predictions: ***static*** and ***dynamic***.

In ***static prediction***, a fixed decision for prefetching one of the two paths is made during *compilation* and before the program runs. The static prediction direction (*taken* or *not taken*) is usually wired into the processor. The wired-in static prediction cannot be changed once designed into the hardware. Static types, however, consists of *three* approaches:

1. Predict-always-taken
2. Predict-never-taken
3. Predict-by-opcode.

The *first one* is a simple technique that would be to always assume that the branch will be taken. This technique always loads the program counter with the target address when a branch instruction is encountered. The *second one* assumes that the branch will not be taken at all and continues to fetch instructions in sequence as usual. The *third approach* makes the decision based on opcode of the branch instruction. It is assumed that the branch will be taken for some branch types based on opcodes, and automatically choose the target path and sequential path for the rest of branch types. It has been observed that the success rate with this strategy is greater than 75%. However, if the chosen path is wrong, the pipeline is drained, and the instructions corresponding to the correct path are fetched; the penalty is paid. Each of these three approaches takes its own individual specific decision before the arrival of any conditional branch instruction. Such a decision can even be modified allowing the programmer or compiler to select the direction of each branch on a *semi-static* prediction basis. The complexity of the compiler in that situation will be naturally increased. Static prediction methods usually require little extra hardware. They do not consider the runtime history generated from previously executed branch instructions during the program execution upto the time that the next conditional branch instruction appears. The **Motorola 68020** and the **VAX 11/780** use the *predict-never-taken* approach.

A good example in the use of a static prediction mechanism possibly lies in the implementation of **Intel Pentium 4** processor. The architecture of Pentium 4 consists of two cells. The outer cell of this processor is a CISC-like cell that fetches the instruction from the memory strictly *in the order* of the program submitted by the user, thereby resembling a CISC-like approach. The inner shell of this processor, however, consisting of a 20-stage pipeline strictly follows the RISC philosophy (for more details, see superscalar architecture) using RISC-like micro-operations (instructions) that are obtained by translating (decoding) each instruction of a user-submitted program by a fetch/decode unit. The outer shell of Pentium 4 exploits *static branch prediction* using the front-end BTB (branch target buffer) that contains the instructions of a user-submitted program fetched already from memory by fetch/decode unit, to supply the needed instructions, based on the outcome of the execution of branch instruction.

Dynamic (branch strategies) ***prediction*** attempts to improve the accuracy of prediction during the execution of the program by making a decision based on the profile information (history) collected from the previously executed branches. For example, a simple technique would be to record the history of the paths taken by each of the last two branch instructions. If the executions of last two branch instructions have chosen the same path, that path will then be chosen for the execution of the current branch instruction. But if the two paths do not match, one of the paths could be then chosen at random for the current one. The amount of history to be recorded at runtime to keep track of the past behaviour of the branch instructions should be limited; otherwise this will, in reality, be infeasible to implement. Therefore, most dynamic predictions are determined with limited recent history accumulated in dedicated additional hardware.

One such approach known as *counter-based branch prediction* mechanism to implement a prediction scheme is to associate an *n-bit counter* with *each* branch instruction. In this method, after executing a branch instruction for the first time, its counter C is set to be a threshold T if the target path is taken or to $T - 1$ if the sequential path is taken. From then on, whenever, the branch instruction is about to be executed, this history is consulted for prediction. If $C \geq T$, then the target path is taken; otherwise the sequential path is followed. The counter value C is updated after the branch is resolved. If the correct path is the target path, the counter is incremented by 1; if not, C is decremented by 1. If C ever reaches the upper bound, i.e. $2^n - 1$, C will no longer be incremented, even if the target path is correctly predicted and chosen. Likewise, C is never decremented if it reaches the lower bound, i.e. 0. Observations and studies over real-time environments dictate that the performance of a 2-bit predictor ($n = 2$) and the value of $T = 2$ is almost the same as that of predictors

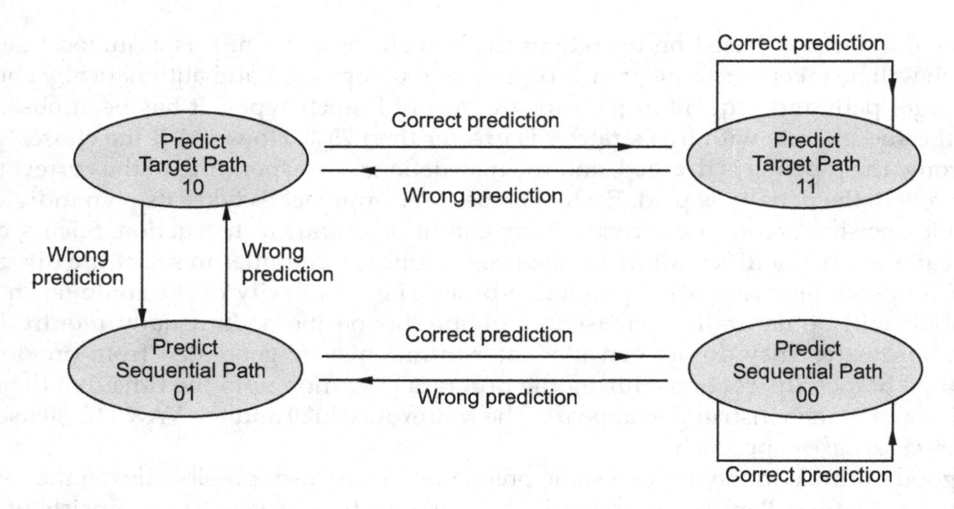

FIGURE 8.16
State transition diagram of a 2-bit predictor.

with a higher number of bits (higher value of *n*). Figure 8.16 represents a state transition diagram of the possible states in a 2-bit predictor.

Now, the obvious question that arises is, strategically *at which stage* in the pipeline will the branch prediction action be taken during runtime? Dynamic branch strategies can be categorised into *three* major classes:

- One class predicts the branch direction based upon information available at the decode (DI) stage.
- Another class uses a cache to store *target address* at the stage the effective address of the branch target is computed (CO).
- The third scheme uses a cache to store *target instructions* at the fetch stage (FI).

In fact, whatever strategy is being followed, it is to be noted that all dynamic predictions are accordingly adjusted dynamically as the execution of the program gradually proceeds.

Dynamic prediction is based on *speculation*, and hence, branch prediction with speculative execution requires extensive ILP — supported by temporary *data registers, BTB, multiple e-units,* and *so forth* that together consequently increase the hardware complexity. But this technique, if implemented adequately and reasonably, would then subsequently require relatively less work at the time of compilation. Unfortunately, the then existing hardware technology was not capable enough to realise this technique properly. This is one of the reasons that branch prediction techniques were not widely used until the 1990s, when ongoing hardware technology further improved and subsequently matured down the years sufficiently to support it. In general, dynamic prediction usually offers better performance than its counterpart static prediction and also provides a greater degree of object code compatibility, since decisions are made after the time of compilation (runtime). This technique has, however, been commercially implemented with utmost success by both Intel in **Pentium Pro** (as also in the original Pentium) and also in **Pentium 4** and by **Motorola** in its **68060** processor introduced sometime in mid-1990s. The ***Pentium*** processor uses a simple form of branch prediction. It chooses the same direction that it chose the last time when the branch was executed. This requires a table of last branch address values to be maintained for each branch instruction. For a program loop, the predicted direction is thus correct for all branches after the first branch, until the loop is exited.

A brief description of the implementation of branch prediction in Intel Pentium 4 is given in the website: http://routledge.com/9780367255732.

8.5.2.3.2.2.5 Delayed Branching Observations on the branch penalty being paid reveal that the branch penalty would be significantly reduced if the *delay slot* could be minimised to attain an almost zero penalty. The purpose of the delayed branching scheme is to eliminate or at least to significantly reduce the adverse effect of the branch penalty. In this type of design, a *certain number of instructions* after the branch instruction are fetched and executed regardless of which path will be chosen for the branch. These instructions should be independent of the outcome of the branch instruction. A delayed branch of k cycles thus allows at most $k - 1$ such useful instructions' execution, irrespective of the fate of the execution of the branch instruction; otherwise, a zero branch penalty cannot be achieved. A delayed branch for two cycles ($k = 2$) is illustrated in Figure 8.17 which shows that one ($k - 1 = 2 - 1 = 1$) delay instruction is required for this scheme.

For example, a processor with a branch delay of k executes a path containing the next k–1 sequential instructions and then either continues on the same path or starts a new path from a new target address. As often as possible, the compiler tries to fill the next k–1 instruction slots after the branch instructions with the instructions from the rest of the program that are independent from the said branch instruction. If no such instruction is available, NOP instructions are placed in any remaining slots. Consider the following example:

```
i1 LD R1, I
i2 LD R2, J
i3 BrZr R2, i7   → branch to i7 if R2=0
i4 LD R3, K
i5 ADD R4, R2, R3   → R4 = R2 + R3
i6 MUL R5, R1, R4   → R5 = R1 × R4
i7 ADD R4, R1, R2   → R4 = R1 + R2
```

Assuming a delayed branch for three cycles ($k = 3$) when the branch condition is resolved at the operand computation (CO) stage, the compiler modifies this code by moving the instruction i_1 and inserting an NOP instruction after the branch instruction i_3. The modified code is as follows:

```
i2 LD R2, J
i3 BrZr R2, i7
```

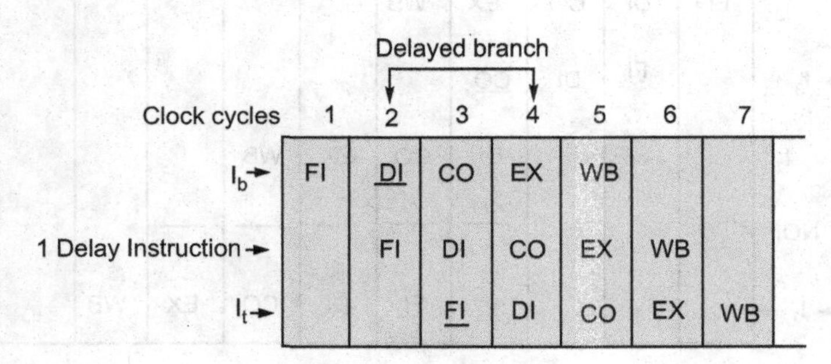

FIGURE 8.17
A delayed branch for two cycles when the branch condition is resolved at the DI stage.

```
i1 LD R1, I
NOP
i4 LD R3, K
i5 ADD R4, R2, R3
i6 MUL R5, R1, R4
i7 ADD R4, R1, R2
```

It can be seen in the modified code that the instruction i_1 is executed regardless of the outcome of the branch instruction. The instruction i_4 could be taken in place of NOP. The technique as used here is similar to that used for *software interlocking* (discussed earlier). NOP behaves here as a filler. The corresponding pipeline stages for this modified code are illustrated in Figure 8.18.

The delayed branch strategy was, however, explored with the advent of RISC machines and is often employed in these processors now. Here, the processor computes the result of conditional branch instructions before any unusable instructions have been prefetched. With this approach, the processor always executes the single instruction that immediately follows the branch. This keeps the pipeline full to maximise the utilisation of the instruction pipeline while the processor fetches a new instruction stream. The delayed branch strategy was, however, subsequently perceived to be less appealing for many reasons as technological developments progressed, when more powerful and productive superscalar machines (described later in this chapter) were introduced. This strategy was then found to be unconducive and perhaps less appropriate to the basic design philosophy based on which the relatively advanced superscalar machine was built.

The branch prediction technique, on the other hand, is however, considered to fit mainly in the domain of pre-RISC culture. It regained importance when the more advanced superscalar machine, under certain compulsions due to its design philosophy, started to exploit this technique for its own convenience. Some processors, like ***PowerPC 601***, use a simple static branch prediction technique. More sophisticated superscalar processors such as the ***PowerPC 620***, the ***Pentium 4***, and most **other superscalar machines** use the traditional dynamic branch prediction technique based on accumulated branch history analysis to improve their efficiencies.

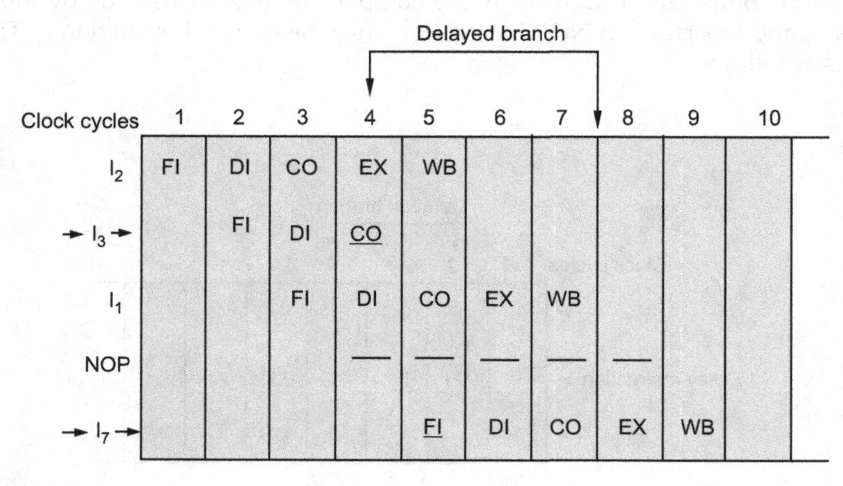

FIGURE 8.18
Code modification for a delayed branch of three cycles when the branch condition is resolved at the CO stage.

A solved real-life example showing the usefulness of branch prediction method is given in the website: http://routledge.com/9780367255732.

8.5.3 Arithmetic Pipeline

The arithmetic/logic unit (ALU), a part of the CPU, carries out all types of both fixed-point and floating-point arithmetic and logical operations with finite precision by an ***integer unit*** (IU) and an in-built ***floating-point unit*** (FU) or by a separate coprocessor to speed up the floating-point operation. However, the advanced RISC microprocessors usually provide the hardware for all types of arithmetic operations in both fixed-point as well as floating-point formats on the same processor chip. Such arithmetic units can be pipelined to maximise performance. The arithmetic units that perform *scalar* operations accept one pair of operands at a time and are called *scalar arithmetic units*. These units, when pipelined, are usually controlled by software loops. *Vector arithmetic units*, on the other hand, accept a set of vector operands at a time, and these units, when pipelined, are designed with pipelined hardware under direct hardware or firmware control. Vector hardware pipelines are often implemented as an additional option on an existing scalar processor unit or as a separate additional stand-alone processor attached to the main unit driven by a control processor. Both scalar and vector pipelined processors are found in extensive use in large mainframes and also in supercomputers.

An arithmetic pipeline is commonly used for implementing complex arithmetic functions like floating-point addition, multiplication, and division. These operations can be implemented with some form of hardware to carry out the basic add and/or shift operations. A pipelined multiply unit is essentially an array multiplier with *special adders* designed to generate the partial products in a way to reduce the carry propagation time. Floating-point operations can be decomposed into consecutive subfunctions and corresponding suboperations.

8.5.3.1 Adder Pipeline Design

The floating-point adder pipeline is constructed with two normalised floating-point binary numbers as input, which are:

$$A = M_1 \times E_1$$

$$B = M_2 \times E_2$$

Where M_1 and M_2 represent the mantissas and E_1 and E_2 are the exponents. The floating-point addition and subtraction can be performed by the pipeline consisting of mainly four stages (segments), with one suboperation for each stage, namely, *comparison of the exponents, mantissa alignment accordingly, mantissa addition/subtraction* and *result normalisation*. Latches are, however, placed between these stages in order to store the intermediate results.

A brief description of the function being executed by each such stage is provided with an appropriate figure in the website: http://routledge.com/9780367255732.

8.5.3.2 Multiplication Pipeline Design

To speed up the multiplication operation, the pipelined architecture requires the use of a *carry propagation adder/carry lookahead adder* (CPA/CLA) which adds partial products to generate the result (see Chapter 7). The pipelined architecture can alternatively use a *carry save adder* (CSA) to add two or more *n-bit* numbers such as *X*, *Y*, *Z*, etc., expressed as $X = (x_{n-1}, x_{n-2}, \ldots, x_1, x_0)$, and as usual produce one *bitwise* sum output number denoted as

$$S' = (0, S_{n-1}, S_{n-2}, \ldots, S_1, S_0)$$

where $S_i = x_i \oplus y_i \oplus z_i$ and a carry output $C = (C_n, C_{n-1}, \ldots\ldots, C_1, 0)$.

The leading bit of the bitwise S′ is always a 0, and the tail bit of the carry vector C is always a 0. The CSA performs bitwise operations (⊕) simultaneously on all columns of digits to produce the output numbers S′; the carries are not allowed to propagate; instead they are saved in a carry vector C. The result is finally obtained by adding the output number (S′) and the carry vector (C) using a CPA/CLA.

Carry–save multiplication is well suited to pipelined implementation. The CPA and CSAs can be used together to implement the pipeline stages of a fixed-point multiplication unit. Figure 8.19 depicts a pipelined architecture for multiplying two unsigned 4-bit numbers using CSA and CPA/CLA. The first stage generates the partial products P_1, P_2, P_3, and P_4. The right-hand side of Figure 8.19 shows how P_1 is generated, and the other partial products can be generated in the same way. The second and the third stages add the partial products through a *Wallace tree* of CSAs, and the final stage is a CPA which adds up the last two numbers to produce the final product P.

For an 8-bit number, four such stages of CSA are required. Each level of the CSA can be realised with a two-gate-level logic. The synchronisation of CSA and CPA in operation is one of the dominant factors in determining the number of pipeline stages as well as the clock period to be used. ***Motorola 68040***, a member of 68000 series of one-chip 32-bit microprocessors introduced sometime in the 1990s, implements the carry–save multiplication method as discussed above.

8.6 Pipeline Control and Collision-Free Scheduling

In order to achieve the maximum pipeline utilisation, it is extremely important to control the sequence of events presented to the pipeline by way of appropriately scheduling the events. Successful scheduling of events requires *two major issues* to be resolved:

i. The initiation between two successive tasks should have *minimum average latency* (MAL) (average latency has already described).
ii. These initiations should not cause any collision at any stage of the pipeline that eventually lead to temporary disruption of the normal flow of execution.

In other words, the main objective of the pipeline control is to realizes *collision-free scheduling* with *MAL*. This pipeline design theory was originally developed by Davidson (1971) and his students.

8.6.1 Control Scheme: Collision Vectors

Inspection of the reservation table always provides the set of forbidden latencies as well as the set of permissible (nonforbidden) latencies. A *forbidden list* F is simply a list of integers consisting of these forbidden latencies. For a reservation table with *n* columns (clock

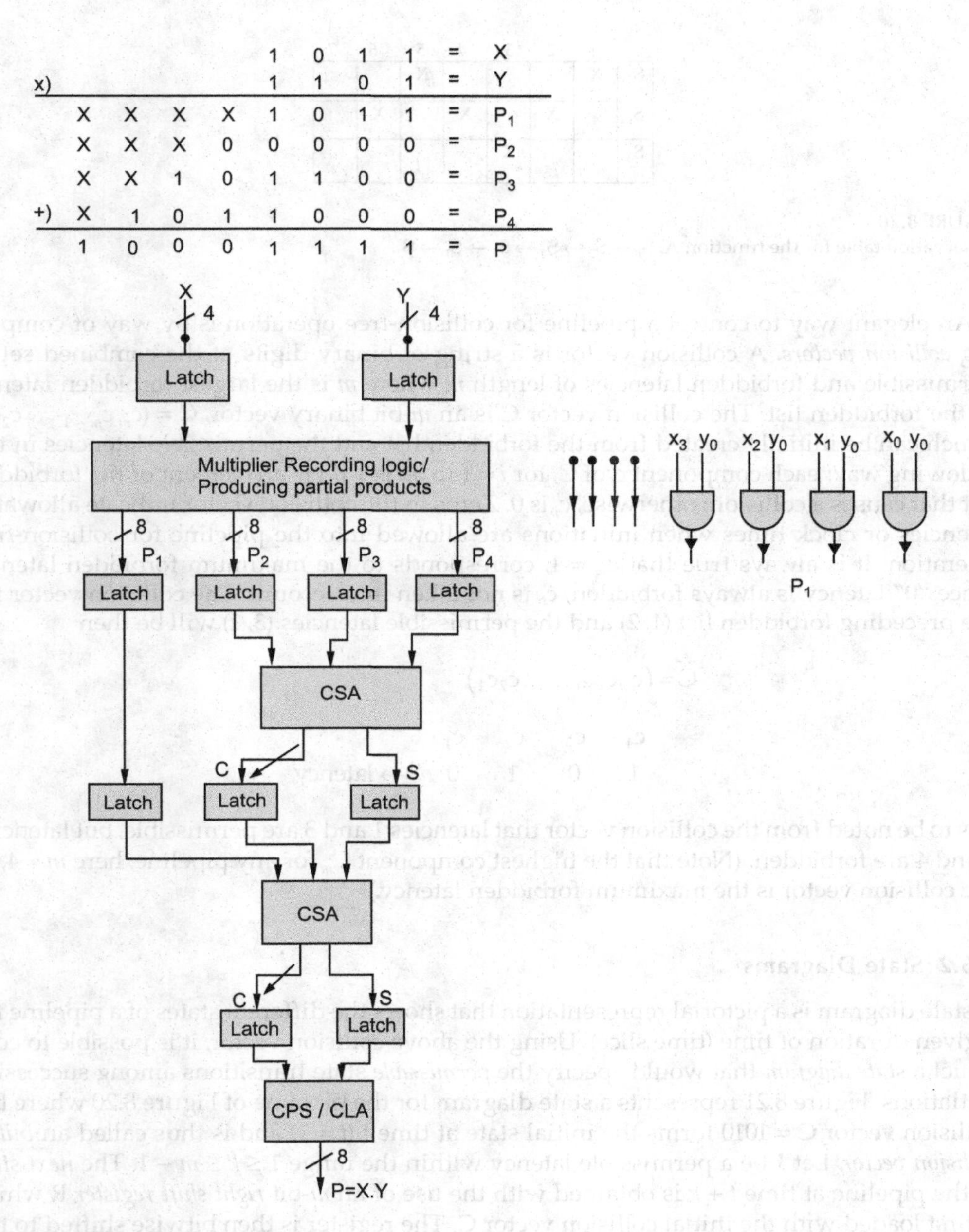

FIGURE 8.19
A pipeline unit for fixed-point multiplication of 4-bit integers.

cycles), the *maximum forbidden latency* $m \le n - 1$, as illustrated in Figure 8.20; the permissible latency p is always preferred to be as small as possible in the range $1 \le p \le m - 1$ to realizes high performance. Figure 8.20 has the forbidden list (4, 2) and the permissible latencies (3, 1). With static pipelines, zero is always considered a forbidden latency, since it is impossible to initiate two jobs at the same time (0 latency) in the same pipeline. But a permissible latency of 1 can always be achieved.

	1	2	3	4	5	6	7
S_1	X				X		
S_2		X		X		X	
S_3			X				

FIGURE 8.20
Reservation table for the function A: $S_1 \rightarrow S_2 \rightarrow S_3 \rightarrow S_2 \rightarrow S_1 \rightarrow S_2$.

An elegant way to control a pipeline for collision-free operation is by way of computing *collision vectors*. A collision vector is a string of binary digits of the combined set of permissible and forbidden latencies of length *m* where *m* is the largest forbidden latency in the forbidden list. The collision vector C is an *m*-bit binary vector, $C = (c_m c_{m-1} \ldots c_2 c_1)$, which can be initially created from the forbidden list and the permissible latencies in the following way: each component c_i of C, for $i = 1$ to m, is 1 if i is an element of the forbidden list that causes a collision; otherwise, c_i is 0. Zeros in the collision vector indicate allowable latencies or clock times when initiations are allowed into the pipeline for collision-free operation. It is always true that $c_m = 1$, corresponds to the maximum forbidden latency. Since "0" latency is always forbidden, c_0 is not taken into account. The collision vector for the preceding forbidden list (4, 2) and the permissible latencies (3, 1) will be then

$$C = (c_m c_{m-1} \ldots\ldots c_2 c_1)$$

$$\begin{array}{cccc} c_4 & c_3 & c_2 & c_1 \\ 1 & 0 & 1 & 0 \end{array} \rightarrow \text{latency}$$

It is to be noted from the collision vector that latencies 1 and 3 are permissible, but latencies 2 and 4 are forbidden. (Note that the highest component C_m for any pipeline, here $m = 4$, in the collision vector is the maximum forbidden latency.)

8.6.2 State Diagrams

A state diagram is a pictorial representation that shows the different states of a pipeline for a given duration of time (time slice). Using the above collision vector, it is possible to construct a *state diagram* that would specify the *permissible* state transitions among successive initiations. Figure 8.21 represents a state diagram for the pipeline of Figure 8.20 where the collision vector C = 1010 forms the initial state at time t ($t = 1$) and is thus called an *initial collision vector.* Let k be a permissible latency within the range $1 \leq k \leq m - 1$. The *next state* of the pipeline at time $t + k$ is obtained with the use of an *m*-bit *right shift register* R which is first loaded with the initial collision vector C. The register is then bitwise shifted to the right. Each 1-bit shift corresponds to an increase in the latency by 1. When a 0 bit emerges from the right end after k shifts (i.e. $c_k = 0$), it indicates that k is a permissible latency. Similarly, when a 1 bit emerges from the right end after k shifts (i.e. $c_k = 1$), it indicates that k is a forbidden latency. Each time a right shift is carried out,a logical 0 enters the left end of the shift register.

Note that the initial state has 0 in positions 1 (c_1) and 3 (c_3) which means that they are permissible latencies. Therefore, a new datum can be initiated to the pipeline after 1 or 3 clock cycles. The *next state* after *i* shifts is thus obtained by bitwise-ORing the initial collision vector with the *i*-shifted register contents. This is explained in Figure 8.21.

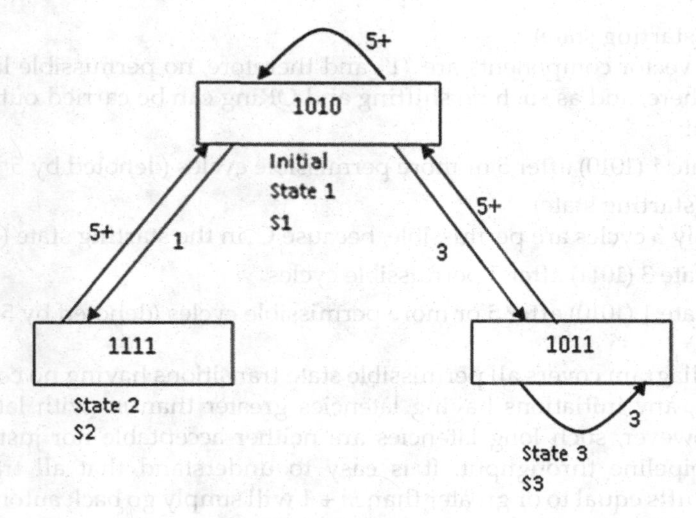

FIGURE 8.21
State diagram obtained from the reservation table in Figure 8.20.

ORing is necessary because the new initiation enforces a new constraint on the current status of the pipeline. The actions *shifting and ORing* ensure that the current content of register R defines all the collision possibilities due to either ongoing pipeline operations or the newly initiated one.

Whenever a new collision vector is generated from an existing collision vector in the state diagram, an arc is usually drawn between them as shown in Figure 8.21. The arc is labeled by latency *i*. The process of generating new collision vectors continues until no more can be formed. The state diagram as obtained with the initial state (1010) shows that only two outgoing transitions are possible, corresponding to the two permissible latencies 1 and 3 in the initial collision vector (1010). In Figure 8.21, from the initial state (1010), one next state (1111) is reached after one right shift of the register R and after ORing it with the initial state (1010). Similarly, from the initial state (1010), the next state (1011) is reached after three right shifts of the register R and after ORing it with the initial state (1010). A stepwise computation in this regard is, however, shown below to describe how Figure 8.21 is finally obtained.

State 1: 1010 (initial state S_1) with permissible latency cycles are 1 and 3.

Using the argument given above, it then follows that the initial state (the first starting state) will then

Reach state 2 (1111) after 1 permissible cycle

Reach state 3 (1011) after 3 permissible cycle

Reach state 1 (1010) after 5 or more permissible cycle (denoted by 5^+).

[When the number of shifts is m (here $m = 4$, the maximum forbidden latency), it is clearly the forbidden latency as $C_m = 1$, and hence, it cannot be used in right shift according to the explanation given above. But when the number of shifts is $m + 1$ (here, $m = 4$, the maximum forbidden latency) or more, all transitions are redirected always back to the initial state regardless of their respective starting state, i.e. irrespective of which state the transition actually starts from.]

State 2: 1111 (starting state)

Here all vector components are "1", and therefore, no permissible latency cycle is present here, and as such no shifting and ORing can be carried out henceforth. However, it;

reaches state 1 (1010) after 5 or more permissible cycles (denoted by 5^+).

State 3: 1011 (starting state)

Here, only 3 cycles are permissible, because C_3 in the starting state (1011) is "0".

Reaches State 3 (1011) after 3 permissible cycles.

Reaches State 1 (1010) after 5 or more permissible cycles (denoted by 5^+).

Thus, the state diagram covers all permissible state transitions having no collision. Within a state diagram, any initiations having latencies greater than m (with latencies m^+) are permissible. However, such long latencies are neither acceptable nor justified from the standpoint of pipeline throughput. It is easy to understand that all transitions with the number of shifts equal to or greater than $m + 1$ will simply go back automatically to the initial collision vector (state).

8.6.3 Greedy Cycles and Minimum Average Latency (MAL)

From the state diagram, we can define a latency cycle (circuit) which is an alternating sequence of collision vectors and arcs, C_0, a_1, C_1, ..., a_n, C_n in which arc a_i connects collision vector C_{i-1} to C_i, and all the collision vectors in the cycle are distinct except the first and last. (A similar concept is found in graph theory where circuits are formed with nodes and edges.) For convenience, we are at present denoting the edges by latencies. There are many such latency cycles that can be traced from the state diagram. For example, a few legitimate cycles from Figure 8.21 are (3), (5), (1,5), (3,5), (1,5,5), (3,3,5), ... and many more. Among these cycles, there are certain cycles in which each state appears only once. Such cycles are called *simple cycles*. From the set of cycles already mentioned above, the cycles (3), (5), (1,5), and (3,5) are simple cycles. The cycle (1,5,5) is not simple because it travels through the state (1010) twice. Similarly, the cycle (3,3,5) is not simple because it repeats the state (1011) twice.

Among these simple cycles, those cycles whose edges are all made with minimum latencies from their respective starting states are called ***greedy cycles*** (optimal latency cycles). The cycles (3) and (1,5) are *greedy cycles*. Such cycles must be simple, and their average latencies must be lower than those of other simple cycles. The average latency of the cycle (1,5) is (1+5)/2 = 3, which is lower than that of the other two simple cycles (5) and (3,5) [(3,5) is (3+5)/2 = 4,] and hence, these two simple cycles are not considered as greedy cycles.

At least, one of the greedy cycles (sometimes it may also be more than one) derived from the set of simple cycles will ultimately yield MAL that could be used in collision-free pipeline scheduling. Here, the MAL in Figure 8.21 is 3, corresponding to either of the two greedy cycles (3) and (1,5). Greedy cycle (3) has a constant latency 3 which equals the MAL for executing the function A without causing collision. The successive initiation of tasks in the pipeline could then be carried out using this MAL to maximise the throughput of the pipeline. The greedy cycle yielding the MAL is thus the ultimate choice for collision-free scheduling of pipeline events. Sometimes, a less efficient cycle (here it is 5) may also be chosen in order to reduce the implementation complexity of the pipeline's control circuit.

8.6.4 Dynamic Pipeline Scheduling

So far we have discussed the scheduling of a static pipeline, where a particular function has to be evaluated avoiding collisions, while different input data are entering the pipeline. With a dynamic pipeline, the situation is even more complicated. Because it is possible that different functions with different input data may be present in the pipeline at the same time. Therefore, the collisions that may arise between these data must be considered as well. However, as with static pipelines, dynamic pipeline scheduling also follows the same course of actions. Here, the procedure begins with the compilation of a forbidden list from the function reservation tables for all the functions present. Next the collision vectors are to be obtained, and from there, the state diagram is developed, and finally MAL is computed and then chosen to schedule the pipeline.

Delay Insertion: A lot of strategies can then be implemented for further decreasing the MAL thus obtained. One of such approaches is to insert *noncomputational delay instruction* to optimise pipeline scheduling. The purpose of a delay insertion is to modify the existing reservation table, thereby yielding a new collision vector. This leads to a modified state diagram, which may produce greedy cycles meeting the lower bound on the MAL. Scheduling on the pipeline can then be carried out using this MAL thus obtained. At this point, we will not get into further details of this area which is considered to be outside the scope of the present context. However, interested readers may refer to Kogge (1981) and Smith (1989) for a better understanding of this topic.

8.6.4.1 Implementation

The pipeline architecture has been best extracted and fully exploited in various ways, particularly in RISC machines, because of certain inherent principles that are strictly obeyed in their architecture and because the design of this class of machines is very conducive for excellent implementation of pipelines. However, CISC machines also are not far behind in this regard. Pipelining has been efficiently implemented in CISC-like microprocessors, such as **Intel** 80486 and its successive versions; higher versions of **Motorola** 68000 series; minis and superminis like **DEC** PDP 11/34, PDP 11/780, VAX 8600, and **IBM** AS 400 series; mainframes like **IBM** 370 series, **IBM** 303X series, **IBM** 308X series, and **IBM** 390 series of machines; and many other machines of different classes and categories launched by different manufacturers.

8.6.4.1.1 IBM 3033/308X/3090 Series Pipelining

The core architecture of versatile IBM 370 system has been followed in the architectural design of these advanced machines with additional hardware support that has been mainly provided to implement improved instruction pipelining scheme to enhance their performance.

A brief description with a block diagram of only the portion of the entire scheme that has been implemented to realise the desired instruction pipelining is given in the website: http://routledge.com/9780367255732.

8.7 Superpipeline Architecture

The performance and throughput of this scalar-based pipeline are influenced and determined by a number of parameters, the most *fundamental parameter being* the *pipeline cycle* (*base cycle for the completion of each stage*) which was assumed to be 1 time unit for this type of pipeline design. Although the other parameters, namely, *latency in instruction issue, rate of instruction issue,* etc., contribute equally to pipeline performance, they are again dependent on this fundamental parameter. Incidentally, it has been observed that many different pipeline stages complete their scheduled tasks much earlier than the prescribed times, say less than half a base cycle (one clock cycle time). Thus, a doubled internal clock speed in this case could allow the execution of two tasks to be completed in one external base cycle. This observation along with other similar aspects has ultimately compelled further modification, essentially by way of incorporating higher resolution in pipeline stages (more number of pipeline stages), which also eventually fulfills the primary requirement of fundamental parameter's refinement. Consequently, the ILP is greatly increased which, in turn, enhances the capability of feeding more number of instructions than usual into the base pipeline for execution within the same prescribed duration of time (Jouppi, N.).

Superpipeline is such an approach which makes use of more and more fine-grained pipeline stages (more than *k*) so that more instructions can be in the pipeline at the same time, thereby increasing ILP. The term *superpipelining* was first coined sometime in 1988 (Jouppi, 88). This approach not only leads to increase in parallelism due to pipelining, but also offers parallelism within the execution of a single instruction. Existing available hardware of scalar-based machine is simply utilised here several times per cycle, and pipeline registers are inserted to split up and separate each such pipe stage.

A superpipeline processor thus has an instruction pipeline with more nonoverlapping finer stages (deeper pipeline) than a typical instruction pipeline design; each stage is then automatically entrusted with lesser amount of work to be done. Consequently, the pipeline cycle time (fundamental parameter) is reduced, and the amount of such reduction, of course, depends on the number of stages being increased. If each stage of a superpipelined processor is obtained by decomposing each stage of our existing scalar-based pipeline into *n* number of nonoverlapping parts, the superpipelined processor thus obtained is said to be of ***degree n***. As a result, the pipeline cycle time is dropped down to *1/n* against the existing cycle time 1. The latency becomes obviously *1/n* that indicates *n* number of instructions can now be initiated successively in the pipeline against the existing initiation of only one instruction. This, of course, requires the service of a more sophisticated control circuit, and a higher-degree superpipeline is after all possible provided a high-speed clocking mechanism is available. However, superpipelined designs emphasise ***temporal parallelism*** by way of *overlapping multiple operations* on a single common piece of hardware, using fine-grained pipelined functional units built with faster components to obtain faster clock cycles in order to offer improved performance.

Superpipelined architectural design has been enormously exploited mostly in RISC machines, but it has been found to be well implemented also in CISC machines for a long time, namely, in **CDC 7600**, **Cray 1** of Cray systems, and **NEC** and **Fujitsu supercomputers**. **MIPS R4000**, a superpipelined single-issue 64-bit RISC machine used an eight-stage pipeline instead of one with five stages. Intel **Pentium 4** (the inner shell of this processor) uses a 20-stage pipeline.

A superpipeline approach is found to have certain ***benefits***. The single functional unit and its associated circuit occupies less space and requires less logic on the chip than its counterpart design based on the superscalar approach (to be discussed next). This extra space available on the chip offers room to include specialised circuitry to realise higher speeds, allow large caches (I-cache and D-cache), and provide wide datapaths for further performance improvement.

A brief description of superpipeline architecture with the respective figure is given in the website: http://routledge.com/9780367255732.

8.7.1 Superpipeline Performance

With an ordinary scalar pipeline machine having k stages, the minimum time required to execute N independent instructions is (assuming pipeline cycle time for the completion of each stage is one clock cycle)

$$t(1,\ 1) = k + (N - 1) \text{ clock cycles}$$

With a superpipelined machine of degree n having the same k stages in the pipeline (assuming base pipeline cycle time for each stage to complete is one clock cycle), the minimum time required to execute the same N instructions is

$$t(1, n) = k + (N - 1)/n \text{ clock cycles}$$

Thus, the speed-up gained by a superpipelined machine over the *base machine* is

$$S(1, n) = \frac{t(1,1)}{t(1,n)} = \frac{k + (N - 1)}{k + (N - 1)/n} = \frac{n(k + N - 1)}{nk + N - 1}$$

It is obvious that the speed-up $S(1, n) \rightarrow n$ as $N \rightarrow \infty$.

8.8 Superscalar Architecture

The instruction pipeline mechanism has been further enhanced to obtain even higher throughput than superpipelined design exploiting more ILP. One such popular approach to accomplish this parallelism, which is implicit in sequential (ordinary) computer programs, is called *superscalar* approach. The term *superscalar* first appeared sometime in 1987 and simply refers to a machine being designed to increase the performance of the execution of scalar instructions. Incidentally, this design arrived on the heels of the RISC architecture. Although the simplified instruction set architecture of a RISC machine readily lends itself to superscalar techniques, the superscalar approach can, however, be used on either a RISC or CISC architecture. A superscalar architecture essentially is a more aggressive approach to equip the processor with, and in essence, replicating the scalar-based pipeline so that two or more instructions at the same stage of different pipelines can be processed simultaneously.

The superscalar approach essentially exploits ***spatial parallelism,*** which means the concurrent execution of multiple operations on separate pieces of hardware (i.e. multiple execution units, where each unit again is usually pipelined with a number of different functional stages, giving rise to a scalar-based pipeline). A superscalar processor thus contains one or more instruction pipelines, each consisting of a set of functional units such as an integer addition unit, floating-point addition unit, multiplication unit, division unit, and graphic unit in a single CPU. These different functional units execute multiple instructions per clock cycle (ILP) while several instructions are simultaneously issued or dispatched to these different functional units (*machine parallelism,* as there is a number of parallel pipelines) (Jouppi, N.).

When m instructions ($m = 2$) as shown in Figure 8.22 are issued or dispatched per clock cycle and the ILP is m so as to utilise the pipeline for fully parallel pipelining operations,

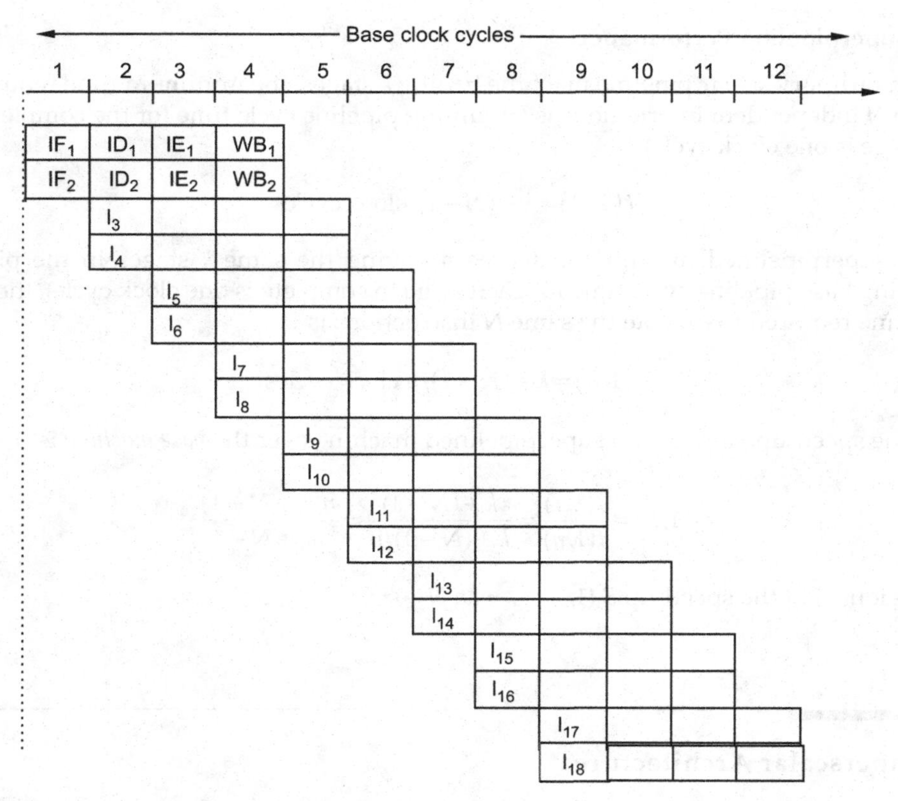

FIGURE 8.22
A superscalar CPU with two four-stage instruction pipelines ($m = 2$).

the superscalar machineis said to have degree m. In fact, Figure 8.22 actually shows a superscalar implementation capable of executing two instances of each stage in parallel. However, an even higherdegree of superscalar implementation is, of course, possible. But when $m = 1$, the superscalar processor eventually degenerates to our ordinary scalar-based pipeline machine. To maximise this ILP, compiler-based optimisation coupled with an appropriate hardware architectural technique can be envisaged.

To handle these simultaneous multiple instructions to be issued, a well-equipped control mechanism is required that can preserve the execution order of dependent instructions to ensure valid result and, on other hand, be able to avoid such factors that may cause pipeline stalling. Such a mechanism can be implemented either by *scoreboarding method* or by *Tomasulo's method* which we have already discussed. In fact, most of the RISC or near-RISC (advanced CISC) processors are based on the superscalar approach and employ the scoreboarding method.

A superscalar processor of degree m has a larger number of different functional units so that m *pipelines* can operate concurrently to achieve a speed-up factor approaching m compared to a scalar-based pipeline CPU. Of course, in some stages in the pipeline, these functional units may be shared by multiple pipelines on a dynamic basis. As usual, several factors, arising out of either being inherent in user's program or being in the machine design, attempt to create such hindrances that eventually limit the parallelism from attaining its optimal level. These factors are mainly *true data dependency, procedural dependency, resource conflicts, output dependency, and similar others.* Figure 8.22 shows the speed-up of a superscalar CPU taking the same base clock cycle used in superpipelined and sequential CPU. It is evident that the superpipelined and superscalar implementations have the same number of instructions being executed at the same time in the steady state. However, the superpipelined processor starts to fall behind the superscalar one at the beginning of the program execution and at the time of handling branch instruction.

A brief descriptionof superscalar architecture with the respective figure is given in the website: http://routledge.com/9780367255732.

8.8.1 Requirements and Essential Components

To keep *m* pipelines busy, it is required that the CPU has to fetch at least *m* instructions per clock cycle to feed *m* parallel pipelines in such a way that fetching, decoding, and execution of *instructions* can smoothly run in parallel in the processor's functional unit. This requires certain rules (protocols) collectively known as the ***instruction-issue policy*** to be followed at the time of issuing instructions. Thus, the critical issue that entwined with superscalar design imposes more stress and places heavy demands on *instruction-fetch logic.* To assist the *instruction-fetch logic* in handling a high volume of instructions, the system provides a large and fast *instruction-only cache **(I-cache)*** for program storage as well as a supporting *data-only cache **(D-cache)*** for operand storage. To strengthen instruction fetching further, an additional storage unit, ***instruction buffer,*** is used to queue the prefetched and (partially) decoded instructions, to be issued later at the right time to the various units in the pipeline for needed execution. Besides, a ***lookahead window*** with its own fetching and decoding logic is used for instruction lookahead in certain situations (out-of-order instruction issue, discussed later) to realize a more enhanced pipeline throughput.

The instruction-issue policy is determined on the current status of the processor which looks ahead of the current point of execution, thereby deciding to locate instructions in the buffer that can be sent to the pipeline for execution. The decision being taken considers a lot of factors in this regard. In fact, the issue of instructions from a single source stream (I-cache) is subject to the detection of the *type of the instruction*, a *data dependence relationship among the successive instructions, procedural* (*control*) *dependencies in case of branch instruction*, and *resource constraints.* Taking all these factors into account, it is really difficult to *schedule multiple pipelines* simultaneously avoiding pipeline stalling and minimizing pipeline idle time, particularly in a situation of having a single source stream. However, *the type of the instructions* to be issued mostly depends on the type of the resources available in the pipeline at that particular time. *Data dependencies* between successive instructions are normally detected by a compiler, could be resolved at the compilation stage (compiler-based optimisation), and finally be made available at the time of scheduling the pipeline. *Procedural dependencies* (control hazard) are handled here in a way similar to that for a scalar-based pipeline. However, in a superscalar machine, fixed-length instructions that are found reasonably suitable and more readily applicable to handle procedural dependencies are preferred, as observed in a RISC or RISC-like (advanced CISC) architecture. *Resource conflicts* can be resolved relatively easily by simply increasing (duplicating) resources, or they can also be minimised by subpipelining the respective affected functional unit, as is found in many RISC chips, especially in Motorola 88000 series, where almost all the functional units themselves are pipelined.

A brief description of this topic with the respective figure is given in the website: http://routledge.com/9780367255732.

8.8.2 Multipipeline Scheduling

Instruction-issue policy schedules the instruction to be fetched and subsequently completed in order to maximise the utilisation of various pipeline elements, thereby largely enhancing the performance of a superscalar processor. Although these policies always attempt to preserve the order of the instructions in the user's program as far as possible, but sometimes they may need to alter the orderings (fetching ordering, execution ordering, etc.) with respect to the ordering found in a strict sequential execution to keep most of the pipeline functional elements busy. In doing so, they also ensure that the result is correct and at the same time nullifies the bad effect of various dependencies and conflicts (as already discussed) that may also arise (Sima, D.).

When instructions are issued as per the program order, it is called an *in-order issue*; otherwise it is called an *out-of-order issue*. Likewise, if the instructions are completed as per the program order, it is called an *in-order completion*; otherwise, an *out-of-order completion* results. An in-order issue is comparatively easier to implement but may not result in optimal performance. Moreover, an in-order issue may sometimes end in an in-order or out-of-order completion. An out-of-order issue, however, usually results in an out-of-order completion. In general, instruction-issue policies can be summarily grouped into the following categories:

i. In-order issue with in-order completion
ii. In-order issue with out-of-order completion
iii. Out-of-order issue with out-of-order completion.

In-order issue with in-order completion is, however, the simplest one to implement, but it is seldom used as it is difficult to maintain both the order of issue and order of completion even in a conventional scalar processor. *In-order issue with out-of-order completion* policy requires more complex instruction-issue logic to implement *out-of-order* completion, but it has the advantage that *out-of-order* completion allows any number of instructions in the execution stages at any point of time up to of course the *maximum degree* of machine parallelism. This approach is found in use in both scalar and superscalar processors. In-order issue of instructions, as usual, causes the execution to stall when there is a resource conflict or a data dependency or a procedural dependency which is particularly more difficult to deal with, especially at the time of *interrupt servicing* and *exception handling*. *Out-of-order issue with out-of-order* completion has several distinct advantages. But it also suffers from the same constraints already described including *output dependence* and *antidependence* which are mainly due to storage conflicts. To maximise the use of registers, ***register renaming*** (duplication of resources) is used here, and more functional units are also added. Although this policy gives the processor more freedom to exploit parallelism, thereby offering enhanced performance, still it is very expensive to realise optimal scheduling. ***Intel Pentium Pro***, and also the other upward versions from Intel, however, implemented this technique in their architecture.

Control parallelism, as obtained from pipelining or the use of multiple functional units, is once again limited by the pipeline length and by the extent of multiplicity of functional units. However, both pipelining and functional parallelism are handled by the hardware automatically, requiring no special software action to activate them.

Superscalar designs exploit ***spatial parallelism*** by way of duplicating hardware resources and hence are better served by CMOS technology (providing storage space) requiring more transistors for its circuit design, obviously making a compromise with the clock cycle rates (speed parameter). An ideal superscalar processor must have simple data-dependence checking, a small lookahead window, and an optimizing compiler with the provision to implement register renaming techniques, along with *scoreboarding* mechanism to extract maximum ILP using the available hardwired parallel pipelines.

A brief description of this topic with each of these three cases ((i), (ii), and (iii)) has been provided separately with appropriate figures in the website: http://routledge.com/9780367255732.

8.8.3 Superscalar Performance

With an ordinary scalar pipeline machine having k stages, the minimum time required to execute N independent instructions is (assuming pipeline cycle time for the completion of each stage is one clock cycle)

$$t(1,\ 1) = k + (N - 1) \text{ clock cycles}$$

With a superscalar machine of m issues having the same k stages in the pipeline (assuming base pipeline cycle time for the completion of each stage is one clock cycle), the minimum time required to execute the same N instructions is

$$t(m, 1) = k + \frac{N - m}{m} \text{ clock cycles}$$

The second term corresponds to the time required to execute the remaining $N - m$ instructions through m pipelines at the rate of m instructions per cycle. Thus, the ideal speed-up gained by a superscalar machine over the base machine is

$$S(m,1) = \frac{t(1,1)}{t(\mathrm{m},1)} + \frac{k + (N - 1)}{k + (N - m)/m} = \frac{m(N + k - 1)}{N + m(k - 1)}$$

It is obvious that the speed-up $S(m, 1) \rightarrow m$ as $N \rightarrow \infty$.

8.8.4 Superscalar Processors: Key Factors

Based on the discussions presented, it can be inferred that the superscalar operation requires some specific hardware support along with a few predefined policies and mechanisms to exploit ILP hidden in the program it executes. Some of the ***key factors*** that must be taken into consideration while implementing this approach are as follows:

- Appropriate mechanisms for initiating, or issuing, multiple instructions in parallel.
- Instruction fetching strategies that simultaneously fetch multiple instructions require the presence of specific pipeline fetch and decode stages. These stages must be available to implement the already chalked-out strategies.
- Suitable befitting techniques are needed to implement branch prediction logic (*control dependencies*) to reduce the negative impact of branch instructions on pipeline efficiency.

- Appropriate logic for determining true *data dependencies* involving register values among the operands of the active executing instructions must be implemented to avoid conflicting use of registers.
- Availability of sufficient hardware resources to carry out parallel execution of multiple instructions thus fetched, including different multiple pipelined functional units, and befitting memory hierarchies capable of simultaneously servicing concurrent multiple memory references.
- Appropriate ordering of the instructions of the program(submitted for execution) in an order other than that specified by the program (being executed),without damaging the semantics of the executing program, for the sake of smooth execution of the instructions to improve CPU's performance.
- Suitable mechanisms to keep track of process states and appropriate measures to keep the process states in correct order.

8.8.5 Implementation: Superscalar Processors

Although the superscalar architecture implementation was originally tied-up to and fully exploited in various ways befitting RISC philosophy, but its principles were later used in CISC machines also. The first commercially available superscalar processor introduced sometime in 1989 was **Intel 80960 CA** (RISC) operate at 25 MHz clock rate, and later came its enhanced version **80960 MM**. But none of these was actually a full-fledged stand-alone processor, and they were used as an integrated part of a master equipment, mostly intended for real-time embedded system control and multiprocessor applications. Intel ***Pentium OverDrive*** (P24T) processor was probably the first one in the CISC line of processors with its in-built *two-issue* superscalar technology providing dual-integer-processing units using branch prediction mechanism for improved performance. From then, the superscalar technology was continuously nurtured by Intel, and at last, sometime in 1995, Intel launched ***Pentium Pro*** (also named as P6 processor) with a full-fledged true superscalar design. After that, each of the subsequent versions of Pentium till the arrival of the current one, the ***Pentium 4***, has gone through constant enhancement and more and more fine-tuning to further improve the underlying superscalar design. **Motorola** implemented this superscalar architecture in the design of its top-of-the-line CISC processor, namely **68040**, and thereafter in the latest member of its 68000 CISC family, the **68060** processor, with clock rates ranging from 50 to 75 MHz. The Motorola RISC **88110** was an early superscalar processor designed with 1.3 million transistors in a 299-pin package and driven by a 50-MHz clock. It actually used a combination of three chip set: a CPU chip (88100) and two cache chips (88200) in a single-chip implementation. The **MIPS R10000** introduced in 1996 uses a 64-bit MIPS IV architecture, fully compatible with its older 32-bit versions R2000/3000 series, and is one of the most powerful single-chip superscalar microprocessors. This design uses around 6.7/6.8 million of transistors operating at a clock frequency of 200 MHz with the issue of *four instructions* per clock cycle using out-of-order issue and out-of-order completion policy that ultimately offers a net throughput of 800 million instructions per second. **PA-RISC 8500** introduced by Hewlett Packard (HP) is one of the most powerful processors and is comparatively superior to most of the contemporary superscalar processors released by different vendors. All the useful and prominent features of a superscalar processor have been realised by a 0.25 μm fabrication process using a total transistor count of over 120 million that leads to an ultimate increase in the system clock frequency and created enough room for on-chip caches. The beauty of this architecture is that it is achieved in a single and simple hardware structure. The **UltraSPARC** is a superscalar processor (which is also superpipelined and is discussed in the next section) enriched with many salient features. The **UltraSPARC III** has been fabricated with 0.18 μm technology having a clock speed in the range of 750–900 MHz and is targeted to attain around 1.5 GHz in the forthcoming days. This chip is the most suitable for use in multiprocessor configuration, with a provision for attaching hundreds of such processors. **IBM RISC System/6000** is a RISC-like superscalar system launched in 1990 built on the idea of IBM 801 system and the PC/RT architecture. It imposes more stress and places heavy demands on instructions and data flow between memory, registers, and the other functional units.

A brief description of each of these processors mentioned above has been provided separately with appropriate figures given in 8.8.5.1–8.8.5.7 in the website: http://routledge.com/9780367255732.

8.9 Superpipelined Superscalar Processors

A superscalar processor can be improved further to obtain a more enhanced pipeline performance using the superpipelining approach in its functional stages. When a machine executes m instructions every cycle (superscalar of degree m) with a pipeline cycle $1/n$th (n pipeline stages in one base cycle) of the base cycle, it is called a superpipelined superscalar machine of degree (m, n).

Figure 8.23 shows such a machine of degree $(m, n) = (2, 2)$. The maximum level of parallelism (considering machine parallelism and ILP) that can be achieved while all the stages are full is mn as compared to scalar machines.

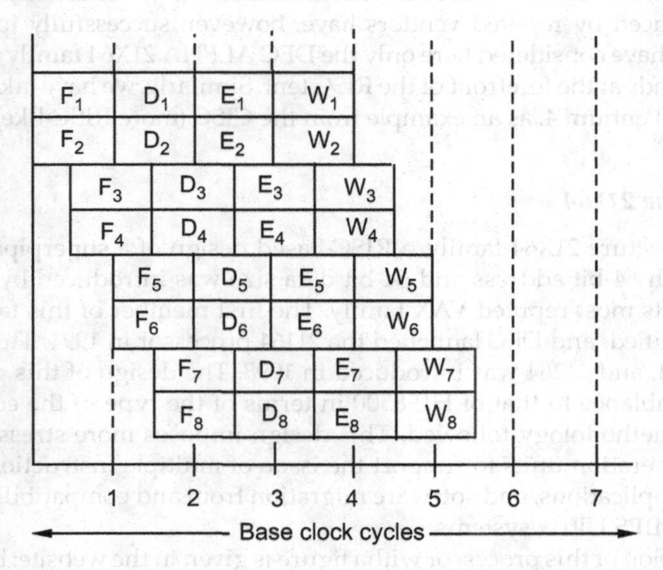

FIGURE 8.23
Superpipelined superscalar execution with degree $m = n = 2$.

8.9.1 Superpipelined Superscalar Performance

With an ordinary scalar pipeline machine having k stages, the minimum time required to execute N independent instructions is (assuming pipeline cycle time for each stage to complete is one clock cycle):

$$t(1, 1) = k + (N - 1) \text{ clock cycles}$$

With a superpipelined superscalar machine of degree (m,n), the minimum time required to execute the same N independent instructions is

$$t(m,n) = k + \frac{N - m}{mn} \text{ clock cycles}$$

Thus, the ideal speed-up gained by a superpipelined superscalar machine over the base machine is

$$S(m,n) = \frac{t(1,1)}{t(m,n)} + \frac{k + (N - 1)}{k + (N - m)/(mn)} = \frac{mn(N + k - 1)}{N + mn\,k - m}$$

It is obvious that the speed-up $S(m, n) \rightarrow mn$ as $N \rightarrow \infty$.

8.9.2 Implementation: Superpipelined Superscalar Processors

Although the superpipelined superscalar architecture was originally implemented in various ways mainly in RISC machines, its principles were equally used later in RISC-like CISC systems also, especially in Intel Pentium machines. Many commercially popular RISC machines introduced by reputed vendors have, however, successfully implemented this principle, but we have considered here only the DEC ALPHA 21X64 family as an example of this class that stands at the forefront of the RISC tent. Similarly, we have taken Intel Pentium series, especially Pentium 4, as an example from the CISC (more RISC-like) tent.

8.9.2.1 DEC Alpha 21X64

The Alpha architecture 21X64 family, a RISC-based design of a superpipelined superscalar processor with 64-bit address and 64-bit data size was introduced by DEC in 1992 as the successor of its most reputed VAX family. The first member of this family was 21064. Later, it was modified, and DEC launched the 21164 processor in 1994. This processor was further upgraded, and 21264 was introduced in 1998. The design of this processor family has a close resemblance to that of HP 8500 in terms of the type of the components used and the design methodology followed. This design imposes more stress on speed, multiple pipelined operation units to support the issue of multiple instructions (superscalar), multiprocessor applications, and software migration from and compatibility with the then VAX/VMS and MIPS Ultrix systems.

A brief description of this processor witha figure is given in the website: http://routledge.com/9780367255732.

8.9.2.2 Intel Pentium 4

Intel Pentium 4 arrived in June 2000 is a superpipelined superscalar processor using 42 million transistors, fabricated with 0.18 µm CMOS process having a clock speed that varies from 1.4 to 1.7GHz. At 1.5GHz, the processor delivers 535 SPEC int2000 and 558 SPEC fp2000 of performance. Unlike that of all the earlier members of Pentium family (P6 architecture) including Pentium 3, its NetBurst architecture is a unique one. The outer shell (frontend) of this processor is like that of CISC, while the inner shell strictly follows the RISC philosophy. The frontend of the processor fetches the instruction from the memory strictly *in the order* of the program submitted by the user, thereby religiously following the CISC–like approach, and nothing else. The remaining steps required for the execution of each instruction are similar to those of a RISC processor. The fetched instruction (executable code of the source program) is then translated (decoded) into one or more fixed-length RISC-like instructions known as *micro-operations*. These micro-operations are then executed by the superpipelined superscalar processor, and at that time, these may be rearranged out-of-order as demanded by the pipeline structure for smooth operation through the pipeline to realise improved performance. Similar to a RISC processor, here also the result of the execution of each micro-operation is stored on the processor's register set following the order of the original program flow.

The Pentium 4 architecture may be viewed as consisting of four basic modules: (i) front-end module, (ii) out-of-order execution engine, (iii) execution module, and (iv) memory subsystem module.

The front-end module of Pentium 4 contains: (i) IA-32 instruction decoder, (ii) trace cache, (iii) microcode ROM, and (iv) front-end branch predictor.

Pentium 4 using a *six-way superscalar design* can dispatch six micro-operations in one cycle that pass through a pipeline built with *hyper-pipelined technology* with pipeline depth extending to 20 stages (*superpipeline*). The machine instructions (executable codes of the source program) are brought from main memory/L3 cache into an on-chip 256 KB non-blocking and eight-way set-associative L2 cache located at the front-end of the processor. The front-end BTB and I-TLB (instruction translation lookaside buffer), however, assist the unit during its fetching operation. The decoder then translates each machine instruction into one to four micro-operations (instructions), each of which is a 118-bit RISC instruction. All these micro-operations generated are then stored in L1 *instruction cache* (which stores up to 12K decoded micro-operations in the order of program execution), also called execution ***trace cache***. In addition, L1 contains an 8 KB data cache. The actual pipeline begins from this L1 cache onwards. The front-end (outer shell) of Pentium 4 uses ***static branch prediction*** using the front-end BTB to determine which instructions are to be fetched next.

The inner shell of Pentium 4 starting from L1 cache onwards actually includes the pipeline and employs a ***dynamic branch prediction strategy*** based on the history of recently encountered branch instructions stored in a four-way set-associative BTB cache having 512 lines. Each entry uses the address of the branch instruction as a tag and also includes the branch destination address that this branch took the previous time and a 4-bit history field. The algorithm being used is referred to as **Yeh's algorithm** which provides a significant reduction in misprediction compared to algorithms that use 2 bits to maintain history. Pentium 4 includes ***out-of-order execution logic***, and this part of the processor reorders micro-operations that are delivered from L1 cache (*trace cache*) in order to enable them to be executed promptly as soon as their input operands are available. Pentium 4 performs ***register renaming*** in its renaming stage within the pipeline that remaps references to the 16 architectural registers (8 floating-point registers plus EAX, EBX, ECX, EDX, EDI, ESI, EBP, and ESP) into a set of 128 physical registers. This stage eliminates false dependencies caused by a limited number of architectural registers while preserving the true data dependencies (reads after writes, RAW). Pentium 4 also employs ***schedulers*** that are responsible for retrieving micro-operations from the micro-operation queues and dispatching these for execution following first-in-first-out (FIFO) ordering whenever there is a competition between different concurrent micro-operations for a specific execution unit, thereby often favouring in-order executions. The ***execution units*** of Pentium 4 retrieve operand values from L1 data cache as well as from integer and floating-point register files which are considered to be the source of pending operations to be executed by the execution units. Two ALUs on this processor are clocked at twice the core processor frequency that allows the basic integer instructions such as Add, Subtract, logical AND/OR, etc. to be executed in half a clock cycle.

A brief description of this processor with its figure is given in the website: http://routledge.com/9780367255732.

8.10 VLIW and EPIC Architectures

The traditional machine instructions specify mostly one operation each, e.g. *load, store, add, multiply, etc.* As opposed to this norm, an instruction may support multiple operations, which would then necessarily require a larger number of bits to encode. Therefore, processors with this type of instruction word are said to have a ***very long instruction word*** (VLIW). When the processor architecture is designed and developed based on this principle, it is known as a VLIW processor. As this processor uses more functional units than a usual superscalar processor, the cycles per instruction (CPI) of a VLIW processor can be even lower. This architecture could, however, be thought of as a *hybrid* form of both *horizontal microcoding* (see Chapter 6) and *superscalar processing.* The concept of horizontal microcoding has been implemented using long instruction words, typically hundreds of bits (256–1024 bits per instruction) in length. Superscalar processing is realised by providing multiple functional units that are used in parallel within this processor. In VLIW, each instruction specifies multiple operations. Different fields of the long instruction word carry different opcodes (see horizontal microprogramming, Chapter 6) that are dispatched to the different functional units simultaneously and can be executed in parallel with independent data operands. Normal programs written usually in conventional short instruction words (i.e. 32 bits) must be clubbed together and then compacted in a legible way to form the VLIW instructions. This code compaction must be done by a compiler which will take all possible measures to optimise the generated object code including branch prediction for smooth operation of these long-word instructions (Colwell, R. P., et al.).

Further refinement of this concept ultimately leads to what is called ***explicitly parallel instruction computer*** (EPIC) (Kathail, B.). The EPIC instruction format, however, is more flexible than the fixed format of multi–operation VLIW instructions; for example, it may allow the compiler to explicitly encode dependences between operations.

The ultimate objective behind both VLIW and EPIC processor architectures is to assign to the compiler the primary responsibility to exploit the plentiful hardware resources available in the processors in parallel. Theoretically, this would not only reduce the complexity of the processor hardware, but also provide overall increased processor throughput. Thus, this approach could be considered, at least in theory, as an emergence of a ***third alternative*** being offered, apart from the existing RISC and CISC styles of processor architecture.

It is not unfair to say that VLIW and EPIC concepts, in general, have not kept their original commitment. ***Intel Itanium 64-bit processors*** (McNairy, C.) make up the most well-known processor family of this class. Experiences with this processor have revealed, as was also argued in many other areas, that processor hardware does not really become simpler, even when the compiler usually bears primary responsibility for the detection and exploitation of ILP . Events such as interrupts and cache misses still remain unpredictable, and therefore, execution of operations at runtime cannot religiously follow the static scheduling specified by the compiler in the VLIW/EPIC instructions; dynamic scheduling is thus urgently required.

These processors have been mostly implemented with microprogrammed control. The clock rate is thus slow due to use of ROM. The execution becomes even slower since some instructions may require a large number of microcode-access cycles to complete their operation. Although VLIW machines are found to behave in a manner very similar to that of superscalar machines, they essentially have some notable differences:

- The code density of the superscalar machine is better than that of VLIW.
- The available ILP in a superscalar machine is comparatively lesser than that which can be realised in a VLIW machine. This is mostly due to the fact that the VLIW instruction often includes bits of non executable operations in order to keep its format religiously fixed. The superscalar processor in this regard issues only instructions of executable operations.
- The ILP in a VLIW machine is totally accomplished at the time of compilation, and the performance of a VLIW processor depends heavily on the efficiency of the code compaction. Superscalar machines possess different architectural characteristics and provide ILP differently in this regard.
- The CPI of a VLIW processor can be lower than that of a superscalar processor.
- The decoding of VLIW instructions is relatively easier than that of superscalar instructions.
- The object code used in superscalar machines is mostly compatible with a large family of nonparallel machines. VLIW machines, on the other hand, exploit different amounts of parallelism that usually would require wide varieties of instruction sets.

However, in summary, it can be concluded that a VLIW processor still seems to be an extreme of a superscalar processor in which all independent or unrelated operations are already synchronously compacted in advance before a run.

The ***distinct advantage*** of VLIW and EPIC architectures lies in their relative simplicity in hardware structure and the underlying instruction set. Parallelism is explicitly encoded in the long instruction word that consequently eliminates the need for additional appropriate hardware and software required to detect parallelism.

One of the ***major shortcomings*** of VLIW architecture is its lack in compatibility with conventional hardware and software that consequently puts this architecture in a position of not being able to provide good performance. As the working of VLIW/EPIC architecture greatly depends on compiler-generated ILP, the practical difficulty of such an approach is that the source program often may have to be *recompiled* even for a different processor model of the same processor family. The reason is simple: such a compiler depends not only on the instruction set architecture (ISA) of the processor family, but also on the hardware resources provided in the specific processor model for which it generates codes.

But for highly computation-intensive applications (operations involving matrices) which usually intend to run on specified hardware platforms, this strategy may then well be feasible and even work nicely, and consequently, it may yield significant performance benefits. Such special-purpose applications can even be fine-tuned for a given hardware platform and then could be run profitably on a regular basis for long periods on the same dedicated platforms.

In case of commonlyused programs, such as word processors, spreadsheets, web browsers, etc., they must run without any such recompilation on all processors of a specific family. Most users of software actually do not have source programs to recompile, and all the processors of a specific family are expected to be compatible with one another in terms of instruction sets. Therefore, the role of compiler-generated ILP is limited in the case of widely-used general-purpose application programs of the types mentioned.

Furthermore, the ILP that is explicitly implanted in the compacted code may require different amounts of latency by different functional units, even though the related instructions are issued at the same time. This leads to a situation that the same VLIW architecture when implemented differently in different machines becomes binary-incompatible with one another. That is why the VLIW architecture has never entered the mainstream of computers. Although the idea seems sound in theory, its extreme dependence on code compaction and trace-scheduling compilation for improved performance has eventually prevented it from being widely accepted and hence dropped from favour, especially in the arena of commercial applications.

8.10.1 Instruction Bundles: The Intel IA-64 Family

Similar to VLIW/EPIC concept, one notable feature out of many distinctive aspects of IA-64 is that *three 41-bit* instructions are grouped into a *128-bit bundle*, along with a 5-bit field called the *template*, which specifies compiler-derived information about how instructions can be executed in parallel. For example, one of the template codes indicates the location of a *stop*, which marks the end of a group of instructions that can be executed in parallel. Such a group may extend over a number of bundles. Information in the templates is used by the processor to schedule the parallel execution of such grouped instructions on multiple functional units to achieve a *superscalar* operation, thereby exhibiting a close resemblance with the characteristics of EPIC which itself is considered as an extension of the concept of VLIW instruction set design (Colwell 1988).

8.11 Thread-Level Parallelism: Multithreading

Instruction–level parallelism (ILP) can be effectively realised by means of pipelining, superpipelining, and then more aggressively by the use of superscalar architecture employing more hardware resources within the processor. To further improve the throughput, designers have created even more complex mechanisms, such as the execution of some instructions in an order different from the way they occur in the instruction stream (out-of-order execution). All these architectures, however, extract ILP by means of a single stream of executing instructions, i.e. a single thread of execution. These processors are often found to have suffered from many different types of dependencies among machine instructions that often limit the amount of parallelism thus obtained. The dependences may be resource dependencies, control dependencies caused mostly by the presence of conditional branch instructions, or true data dependencies (RAW). Here, we are not considering WAR and WAW, because they can be efficiently handled using some form of register renaming.

Out of many possible alternatives, one efficient way to minimise the burden of dependencies and thereby yield greater ILP without much increase in circuit complexity or power consumption is called ***multithreading***. In essence, the instruction stream is divided into several smaller streams of execution, known as *threads*, such that the threads can be executed in parallel. With appropriate hardware support within the processor, it is possible to combine instructions from *multiple independent threads of execution*. Such hardware that supports in realising multithreading would provide the processor with a pool of instructions, in various stages of execution, which have a relatively lesser total number of data dependencies amongst them, since the threads are mostly independent of one another. Moreover, as critical control dependencies are also being separated into a number of threads, less aggressive branch prediction (a method to resolve control dependencies) is really needed.

For example, consider a processor with instruction pipeline consisting of eight stages and with targeted superscalar performance of four instructions completed in every clock cycle. Now, assume that these instructions come from *four independent threads* of execution. Then, on average, the number of instructions within the processor at any point of time *from one thread* would be $(4 \times 8) \div 4 = 8$.

One *distinct advantage* of such hardware-supported multithreading is that the *pipeline stalls* can be here effectively utilised. If one thread, say, for access to main memory,while running into a pipeline comes to a stall, then another thread that is ready can make use of the corresponding processor clock cycles, which would otherwise be simply wasted. Thus, a hardware-supported multithreading approach eventually emerges as an important *latency-hiding technique*.

To provide required support for multithreading, the processor must offer a separate *program counter* for each thread of execution to be executed concurrently. In fact, here the processor designs essentially differ in the amount and the type of additional hardware used to support concurrent thread execution. In general, instruction fetching here takes place in terms of a thread. The processor treats each thread separately and may use a number of techniques for optimizing single-thread execution including branch prediction, out-of-order execution, register renaming, and superscalar approaches. In fact, the thread–level parallelism (TLP), when appropriately combined with ILP is expected to yield an even more improved performance.

The processor design must also include the mechanisms to *switch between threads*, mostly either on the occurrence of a pipeline stall or while scheduling in a round robin manner. Similar to the operating system switching between running processes, in this case, the hardware context of a thread within the processor must be preserved. This basically includes the full set of registers (programmable registers and those used in register renaming), program counter, stack pointer, relevant memory map information, interrupt control bits, protection bits, and others. For *N*-way multithreading support, the processor must store at a time the thread contexts of *N* executing threads. When the processor switches, say, from thread *X* to thread *Y*, the control logic must ensure that the execution of subsequent instruction(s) occurs with reference to the context of thread *Y*. It is to be noted that thread contexts need not be always saved and later restored. As long as the processor *preserves* multiple thread contexts within itself, all that is required is that the processor be able to switch between thread contexts from one cycle to the next.

The enormous varieties of multithreading designs have been, however, realised in both commercial systems as well as experimental systems. Depending mainly on the specific strategy that is adopted for *switching* between threads, hardware support for multithreading may be classified into the following *four principal approaches*:

- **Coarse-grained multithreading**: This is also sometimes referred to as *block multithreading*. Here, the instructions of a thread are executed successively until any event that may cause delay occurs; one such event is a cache miss. As a result, this event then induces a switch to another thread. This approach is, however, found effective mostly on an *in-order issue* processor.
- **Fine-grained multithreading**: This is also known as *interleaved multithreading*. Here, the processor deals with two or more thread contexts at a time, switching from one thread to another at each clock cycle. If a thread is blocked at any time for some reason, say, due to data dependencies or memory latencies, that thread is skipped and set aside, and a thread that is ready is then executed at once.
- **Simultaneous hardware multithreading (SHMT)**: Here the instructions are simultaneously issued from multiple threads to the execution units of a superscalar processor. This, however, combines the capability of superscalar to issue higher instructions with the use of multiple thread contexts.
- **Chip multiprocessing**: Here,a number of processors called ***cores*** are replicated into a single piece of CPU chip (a single piece of silicon, called a die). Typically, each processor core here consists of all of the components that an independent processor does have, and as such, each processor core is able to independently handle separate threads. The advantage of this design approach is that the available logic area and the on-hand transistor count are effectively used here without entering too far into an ever-increasing complexity in pipeline design. This design approach is historically referred to as ***multicore design***, which has been described in detail in the next section.

It is important to note that in the case of the first two approaches mentioned above,i.e.in *coarse-grained multithreading* and *fine-grained multithreading*, instructions from different threads are not executed simultaneously. Instead, each processor is able to quickly switch from one thread to another, using a different set of registers and other context information. As a result, this somehow provides better utilisation of processor resources and helps to avoid significant penalty due to cache misses and other latency aspects. The third one, the SHMT approach, however, carries out true simultaneous execution of instructions from different threads, using available replicated resources needed for execution. Chip multiprocessing, however, also provides simultaneous execution of instructions from different threads.

In this connection, the following observations are of immense interest. Figures 8.24 to 8.28 illustrate some of the possible alternatives in pipeline architecture in different types of processors that involve multithreading and contrast these approaches with the corresponding ones that do not use multithreading. In all the figures, each horizontal row represents the potential issue slot or slots for a *single execution cycle*; that is the width of each row corresponds to the maximum number of instructions that can be issued in a single clock cycle. (Issue slots are defined as the positions from which instructions can be issued in a given clock cycle. Instruction issue is the process of initiating instructions for execution in the processor's functional units. But this occurs only when an instruction moves from the decoding stage of the pipeline to the first execution stage of the pipeline.) The vertical dimension, however, represents the time sequence of clock cycles. An empty (shaded) slot represents an unused execution slot, and an N indicates a NOP in one pipeline.

8.11.1 Scalar Processor

Figure 8.24 shows different approaches with a scalar (single-issue) processor:

- **Single-threaded scalar**: This is the simple pipeline found in traditional CISC and RISC machines as depicted in Figure 8.24a with a single thread of execution, i.e. with no multithreading.
- **Coarse-grained multithreading**: In this case, a single thread is executed until a latency event occurs that would stop the operation of the executing thread and the pipeline, at which time the processor switches to another thread that is ready for execution. This is shown in Figure 8.24b.
- **Fine-grained multithreading**: This is possibly the easiest approach to implement multithreading. Here, the processor switches from one thread to another thread at *each clock cycle*, thereby attempting to keep the pipeline stages fully occupied or close to fully occupied. The processor hardware here is adequately equipped to bring about such switching from one thread context to another between cycles. This is illustrated in Figure 8.24c.

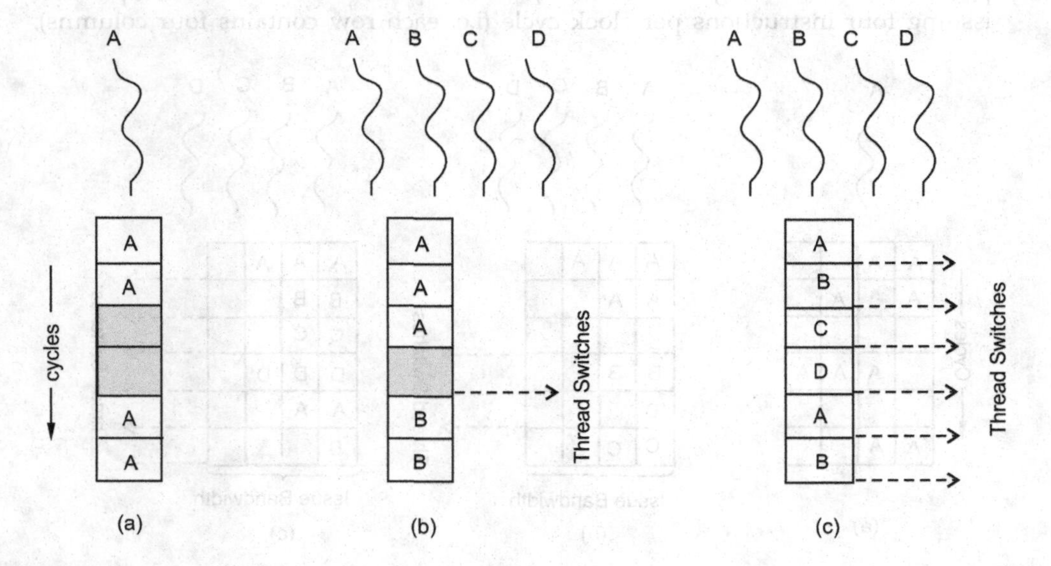

FIGURE 8.24
Executing multiple threads in scalar pipelined processor. (a) Single-threaded scalar, (b) coarse-grained multithreading scalar, and (c) fine-grained multithreading scalar.

It is interesting to observe a difference between two different forms of multithreading as depicted in Figure 8.24b and 8.24c. Figure 8.24b shows a situation in which the time to perform a thread switch is one cycle (shown as shaded portion), whereas Figure 8.24c indicates that thread switching occurs at zero cycle. This is due to the fact that in fine-grained multithreading, it is assumed that there are no control or data dependencies between threads, which not only simplifies the pipeline design but at the same time also allows a thread switch to occur without any delay. On the other hand, coarse-grained multithreading, depending on the specific design and implementation, may require a clock cycle to perform a thread switch as shown in Figure 8.24b. This happens, if a fetched instruction triggers the thread switch and must be then thrown out of the pipeline.

While fine-grained multithreading appears to offer better processor utilisation than its counterpart course-grained multithreading, it does so at the time of single-thread execution. In case of multiple-thread execution, there may be a situation of contention of cache resources amongst the threads being executed which may eventually culminate into an unpleasant cache miss for a given thread(s).

For parallel execution, more options and opportunities can be obtained if the processor is able to issue multiple instructions per cycle. A number of principal alternative approaches in processor design are illustrated in the following section. In each case, it is presumed that the processor hardware is capable of issuing four instructions per clock cycle. However, in all these cases, only instructions from a single thread are issued in a single cycle. The popular alternatives are as follows.

8.11.2 Superscalar Processor

- **Traditional superscalar**: This approach is the primary form of superscalar processor with a single thread of execution that we have already discussed in previous sections, i.e. without multithreading. This is depicted in Figure 8.25a. Here, it is presumed that the single-threaded superscalar processor hardware is capable of issuing four instructions per clock cycle (i.e. each row contains four columns).

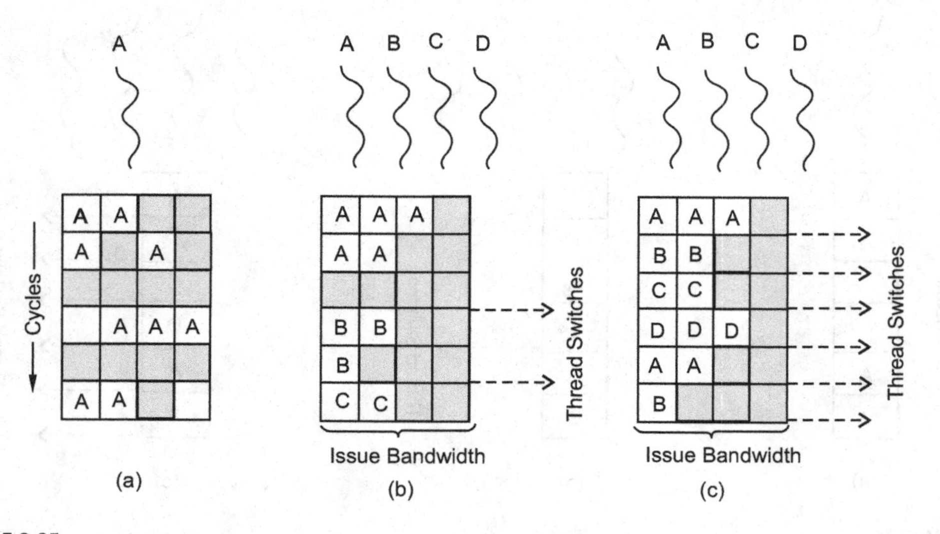

FIGURE 8.25
Approaches to executing multiple threads in superscalar pipelined processors. (a) Single-threaded superscalar, (b) coarse-grained multithreading superscalar, and (c) fine-grained multithreading superscalar.

However, in all these cases, only instructions from a single thread are issued in a single cycle. This architecture was probably the most powerful approach in providing parallelism within the processor in the relatively recent past. It is to be noted that during some cycles, not all of the available issue slots are occupied (e.g., first row of Figure 8.25a). In fact, during these cycles, the number of instructions issued is less than the maximum number for some reasons. This situation is sometimes referred to as *horizontal loss*. In some other instruction cycles, no issue slots are ever used (third and fifth row of Figure 8.25a). These are the cycles when no instruction at all can be issued; this is often referred to as *vertical loss*.

- **Coarse-grained multithreaded superscalar**: In this approach, during each cycle, as many instructions as possible are issued from only one thread, and coarse-grained multithreading (as already discussed) is used. This is shown in Figure 8.25b.
- **Fine-grained multithreaded superscalar**: Here too, during each cycle, as many instructions as possible are issued from a single thread. In this approach, potential delays due to thread switches are eliminated as already mentioned. However, the number of instructions to be issued in any given cycle is again limited by dependencies that exist within any given thread. This is depicted in Figure 8.25c.

8.11.3 VLIW Processor

- **Traditional VLIW**: A VLIW architecture, namely, ***Intel Itanium 64-bit processors*** (IA-64), already described in detail in Section 8.10, places multiple instructions in a single word. Typically, a VLIW is realised essentially by the compiler, which places operations that may be executed in parallel in the same word. In a simple VLIW processor, as shown in Figure 8.26a, if it is not possible to completely fill the word with instructions to be issued in parallel, then NOP is used in the remaining space of the VLW word.
- **Coarse-grained VLIW**: This approach should necessarily provide efficiencies similar to those provided by coarse-grained multithreading on a superscalar architecture. Figure 8.26b illustrates a representative pattern of coarse-grained multithreading operation on this type of processor.

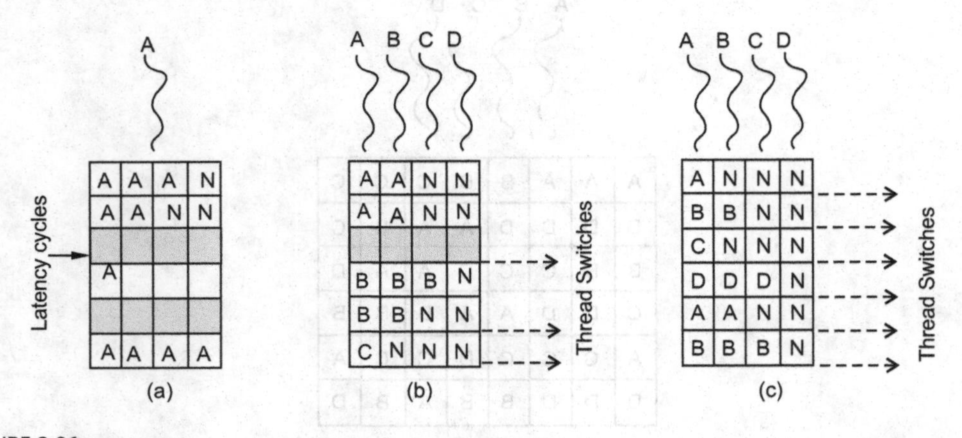

FIGURE 8.26
Schematic diagram of executing multiple threads in a VLIW processor. (a) Traditional VLIW, (b) coarse-grained multithreading VLIW, and (c) fine-grained multithreading VLIW.

- **Fine-grained VLIW**: This approach should necessarily provide efficiencies similar to those provided by fine-grained multithreading on a superscalar architecture. Figure 8.26c depicts a representative example of fine-grained multithreading operation on this type of processor.

The most recent trend towards processor design by way of effective utilisation of chip area as well as efficient use of enormous number of transistors (gates) available within the chip after balancing all the critical design and operational factors, is to use two notable approaches, namely, ***SHMT*** and ***chip multiprocessors*** (also known as *multicore processors*). Both these approaches, however, enable the parallel, simultaneous execution of multiple threads. The details of these two approaches are discussed in detail in the following sections.

8.11.4 Simultaneous Hardware Multithreaded Processor (SHMT)

Figure 8.27 exhibits an SHMT processor capable of issuing eight instructions at a time (i.e.eight columns in a row). If one thread has a high degree of ILP, it may be able to consume all of the horizontal slots on some cycles. However, on other cycles, instructions from two or more threads may be issued. If adequate threads are active, then it would usually be possible to issue the maximum number of instructions in any cycle and thereby yield a high level of processor efficiency.

8.11.5 Chip Multiprocessors (Multicore Processors)

Figure 8.28 illustrates a CPU chip consisting of four *core processors*; each of these here is assumed to have a two-issue *superscalar processor* for the sake of convenience (two columns for each of these four cores). Each processor is assigned a thread, from which it can issue up to two instructions per cycle. However, each core processor could have been alternatively built with the *SHMT approach* also in place of a superscalar design. But determining which design strategy is to be exploited in the design of the processor core at the time of CPU fabrication is entirely a critical design trade-off after considering many other related important issues apart from the primary objectives that the

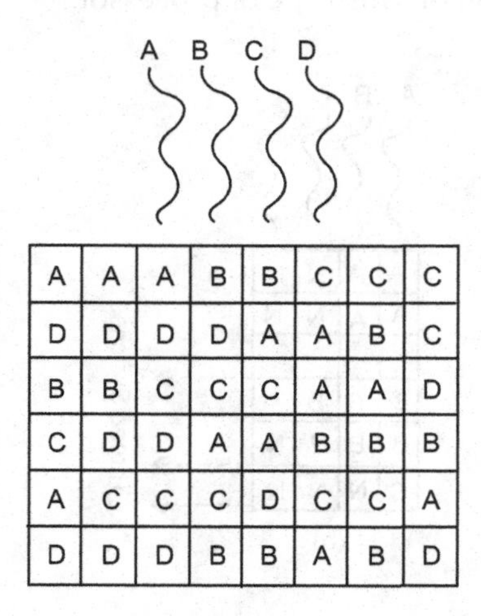

FIGURE 8.27
Simultaneous hardware multithreading (SHMT).

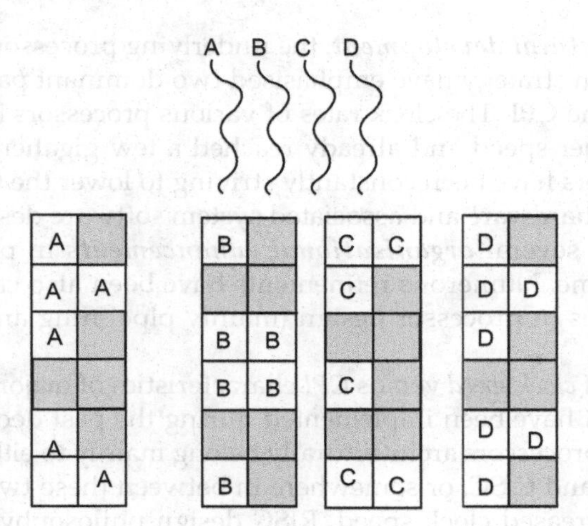

FIGURE 8.28
Chip multiprocessor (multicore processor).

processor would intend to meet. Various types of multicore processors have been discussed in detail in Section 8.12.

From the above discussions, it can be summarily concluded that the multithreading approach when exploited in any form of processor design would always provide better processor efficiency when compared to the corresponding design approach with no multithreading. It is found that a chip multiprocessor is always superior to any form of superscalar processor with the same instruction issue capability, because the horizontal losses will be greater for the superscalar processor. In addition, it is also observed by comparing Figures 8.27 and 8.28 that a chip multiprocessor with the same instruction issue capability as an SHMT processor cannot achieve the same degree of ILP (as there exist empty slots shown shaded in Figure 8.28). This is due to the fact that the chip multiprocessor is not able to hide latencies by issuing instructions from other threads. That is why the multithreading design approach within each of the processor cores of a chip multiprocessor (multicore processor) nowadays is ultimately a strategic design consideration. Some of the contemporary processors found in use implement this design approach.

8.12 Multicore Architecture

8.12.1 Background

Over the last couple of decades, enormous advancements and colossal achievements have been made in various areas of computer technology. These advancements have had an immense impact on processor designs and system developments. The processor implementation technology has also rapidly evolved over decades due to natural consequences of the steady advances in VLSI technology. As a result, various processor families have been introduced, with continuous exponential increase in execution performance. To achieve such performance improvement, major *architectural upgradation* as well as important *organisational enhancements* in processor designs is continuously progressing with no indication of any slowdown.

In regard to ***architectural development***, the underlying processor design principle and related implementation strategy have emphasised two dominant parameters, namely, the *clock rate* (*speed*) and the CPI. The clock rates of various processors have moved from low to a considerably higher speed and already reached a few gigahertz. The other trend is that processor designers have been constantly striving to lower the CPI rate further using numerous innovative hardware and associated system software design approaches.

On the other hand, several ***organisational enhancements*** in processor design have taken place by this time. Numerous refinements have been also carried out in terms of organisational changes in processor design (mainly pipelining and related aspects) on the chip.

A broad spectrum of *clock speed* versus *CPI* characteristics of major categories of contemporary processors that have been implemented during the past decade or so is shown in Figure 8.29. All these processors architecturally belong mainly to either of the two distinct classes, namely RISC and CISC, or somewhere in between these two. While CISC design principle relies on increased clock speed, RISC design philosophy, however, religiously attempts to attain a lower CPI.

In the CISC tent, at present there is the dominant presence of top-of-the-line conventional processors like the *Intel Pentium, Motorola M68060, VAX/8600, IBM 390,* etc. With advanced implementation techniques, the clock rate of today's CISC processors has significantly increased to the tune of a few GHz. The CPI of different CISC processors, however, is different and may be even 20 on the higher side. That is why the CISC processors are located at the upper part of the design space (Figure 8.29).

In the RISC class, there are several examples of amazingly fast processors, such as, *ARM, SPARC, Alpha, PA-RISC, MIPS, Power series, etc.* With the use of efficient advanced pipeline approaches, the effective CPIs of RISC processors, on an average, have been reduced to one or two cycles.

The ***organisational changes*** that have taken place until now in processor design in both RISC and CISC processor architectures have primarily targeted in an ultimate increase in ILP to accomplish more work in each clock cycle. Several such changes that have been implemented most successfully include the following in chronological order:

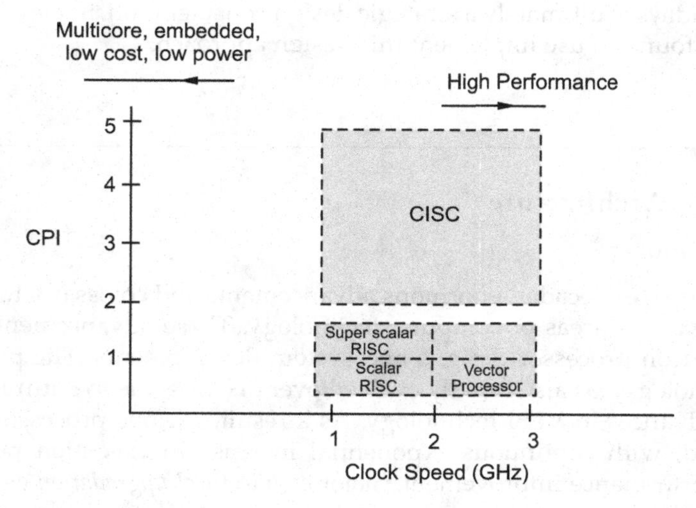

FIGURE 8.29
CPI versus clock speed of major categories of processors.

Pipelining: This approach in processor design (already described in detail), in fact, gives birth to a concept known as *ILP* that eventually enhances the processor performance and thereby improves the throughput significantly without using any faster hardware components.

Superscalar: In order to further increase the ILP — obtained from a traditional pipeline design approach, designers have constructed *multiple pipelines* within a processor by way of replicating functional resources (multiple functional units). This has been made possible mostly due to the availability of the more superior contemporary VLSI technology. As a result, this facilitates parallel execution of more instructions per clock cycle using the available parallel pipelines (presence of more than one pipeline) as long as pipeline hazards are avoided. Superscalar processors of both RISC and CISC classes allow multiple instructions to be issued simultaneously during each clock cycle. Thus, the effective CPI of a superscalar processor should be lower than that of a scalar RISC processor. The clock rate of a superscalar processor, however, matches that of a scalar RISC processor.

The processors in *vector supercomputers* use multiple functional units for concurrent scalar and vector processing. The effective CPI of a processor used in a supercomputer should be very low, and hence the processor is positioned at the lower right corner of the design space (Figure 8.29).

SHMT: Introduction of the thread concept and subsequently its successful implementation in processor design once again lower the effective CPI. Use of replicated register banks in the pipeline organisation enables multiple threads to be concurrently executed by sharing pipeline resources during each clock cycle.

All these attempts that are made in the advancement of processor design, both architecturally and organisation-wise, however, have been ultimately targeted at somehow increasing the performance of the system but of course at the cost of increasing complexity. Implementation of pipelining in the organisation of processor design to obtain improved performance, however, has invited more complexity than that is in the design of a conventional nonpipelined processor. Inclusion of more stages in the pipelined design (superpipeline) while offers better performance but increases the underlying design complexity which, in turn, challenges its viability. However, there always exists a practical limit as to how far this trend of including even more stages in the pipeline can be continued.

In case of superscalar organisation, performance enhancement has been achieved primarily by means of increasing the number of parallel pipelines. This necessarily requires additional logic for managing the regular multipipeline operations and avoiding the hazards to extract as much output as possible. Still, the full use of multiple pipelines cannot be realized due to the presence of numerous types of hazards as well as prevailing resource dependencies. Consequently, there are diminishing returns as the number of parallel pipelines being provided gradually increases.

With SHMT organisation, managing multiple threads and appropriately scheduling them over a set of available pipelines using additional logic often limits the number of threads as well as the number of pipelines that can be effectively utilised. This also starts to pay diminishing returns as the number of concurrently executing threads and the number of parallel pipelines is gradually increased.

After the successful and effective introduction of ILP — in the processor design sometime in the late 1980s by means of exploiting pipelining, and then superscalar techniques, the SHMT approach and subsequently its far more fine-tuned implementation ultimately resulted in a steep rise in the performance improvement of processors of all kinds.

This was observed till the beginning of 2000 AD. After that, no appreciable improvement in the processor performance has ever been attained. This is due to the fact that the increasing difficulty in designing, fabricating, and debugging of the chips with contemporary technology to deal with the increasing complexities of all logical issues has by far reached its limit. Most of the chip area is now occupied with coordination and signal transfer logic. As a result, the effective implementation of ILP and machine-level parallelism appears to have reached an extent from which it would not be practically possible to extend it any further profitably.

Power Consumption Considerations: Another important aspect related to processor design following ILP philosophy is the ***power density*** consideration. While constantly attempting to maintain improved performance, the designers were bound to use more transistors on more densely packed chips to realise higher intelligence and higher clock frequencies. This consequently resulted in shortening the electrical path length which has considerably increased the operating speed. The negative impact is that the power requirements have increased exponentially and chip density and clock frequency have gradually gone up, and beyond a certain point, the power consumption of the chip rises disproportionately fast with clock speeds. Such increases in *power density* (W/cm^2) will eventually result in generation of enormous heat. The teething problems of dissipating this huge amount of heat on such high-density, high-speed chips pose a serious design issue that firmly limits chip density from going further ahead beyond a certain point.

One possible way to regulate the high power density on a chip is to use more of the available chip area for cache memory. The reason is that memory transistors are relatively smaller and have a power density an order of magnitude lower than that of logic. This is illustrated in Figure 8.30a. Figure 8.30b shows that the percentage of chip area devoted to memory has grown to exceed 50% as the chip transistor density has increased (Borkar 2003).

With increasing chip density, the power consumption trend is constantly rising. As shown in Figure 8.31 (Borkar 2007), it is expected that in a couple of years, the microprocessor chip density would be 100 billion transistors on a 300 mm^2 die. With normal assumption that about 50%–60% of the chip area would be devoted to memory, the chip will then support on-chip cache memory of about 100 MB and leave over 1 billion transistors available for logic use.

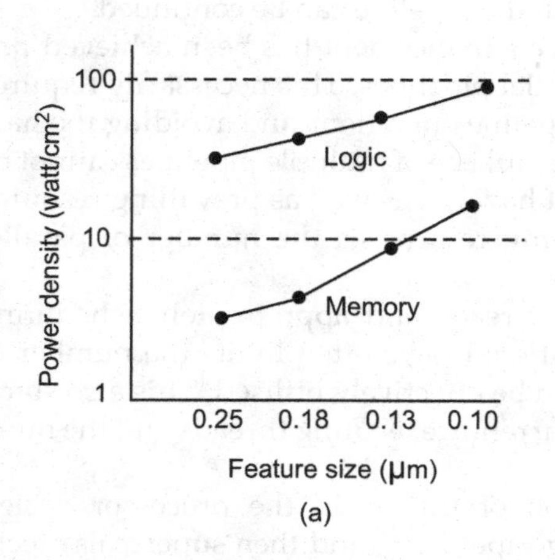

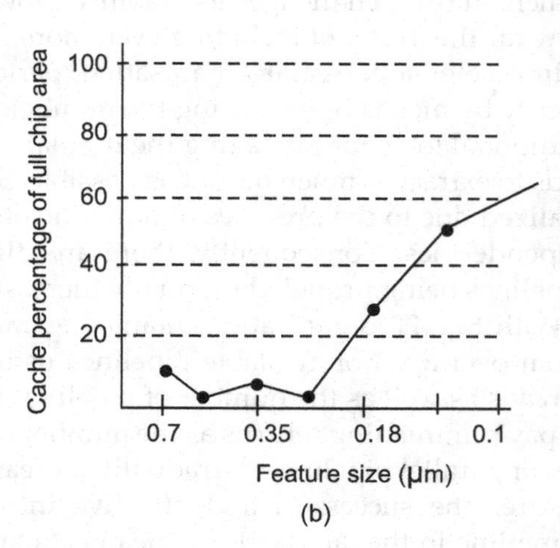

FIGURE 8.30
Power and memory considerations on processor chip. (a) Power density and (b) chip area.

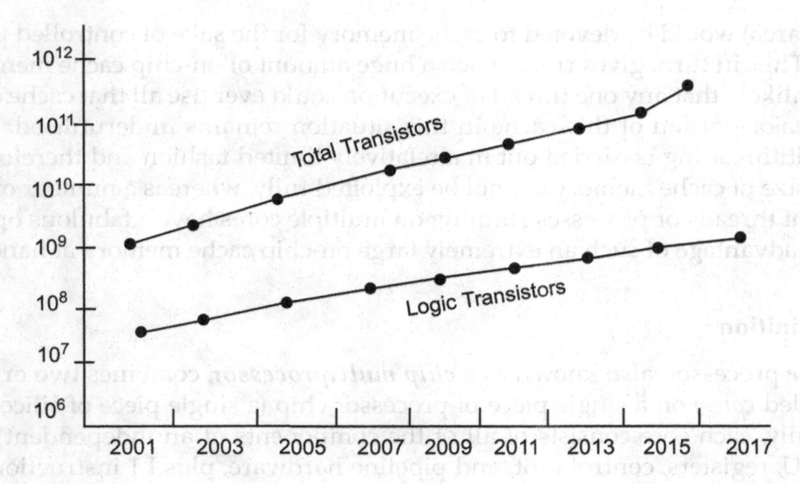

FIGURE 8.31
Utilisation of transistors in a processor chip (Borkar, 2007).

In spite of the presence of all these conflicting aspects, processor clock speeds have, however, reached as high as 4GHz in recent years. But it has also been observed that processor performance does not scale with clock speeds. Out of many prevailing reasons, one of the main ones for this is that the relative cost of a cache miss is greater at higher processor speeds.

In view of all these factors, there have been several attempts in relative leveling off of processor clock speeds in recent years, while greater attention is being paid to how best to design the chip to most effectively utilise such an enormous number of transistors available on it. As already mentioned, there are critical limits to the effective use of techniques such as superscalar processors and SHMT. Given that a large number of transistors can be fabricated on a chip, it follows that huge performance benefits can be derived in one way by integrating system functions *on a chip,* even if it is not possible to continue to push clock speeds higher. Another outcome of these technological factors is that system performance can be more easily enhanced by employing *multicore processors, systems-on-a-chip* (SoCs, e.g. Tilera's TILE64 system, a 64-core processor for embedded applications, in which each chip consists of a regular 8 × 8 grid of tiles. Each tile on this chip has its own general-purpose processor core, L2 cache, and a nonblocking mesh router to provide for communication with other tiles on the chip and for off-chip data traffic with main memory, I/O devices, and networks), *stream processors,* and *larger two-level on-chip cache* memories than by pushing a single processor to its technological performance limits. These are only a few of many other possibilities that can be cited as an outcome of ongoing architectural developments.

In general terms, all these experiences in recent decades have culminated in what has been encapsulated in a rule of thumb known as **Pollack's rule** (99), which states that *performance increase is roughly proportional to square root of increase in complexity.* In other words, if the logic is made double in a processor core, then it delivers only 40% more performance. In principle, the use of multiple cores has the potential to provide near-linear performance improvement with the increase in the number of cores. That is why the introduction of the multicore architecture in the evolution process of the processor design, in fact, became inevitable.

Multicore organisation is also preferred when power consumption of the chip is taken into active consideration. As already mentioned, that most of the chip area (around 50%–60%

of the chip area) would be devoted to cache memory for the sake of controlled power consumption. This, in turn, gives rise to such a huge amount of on-chip cache memory that it becomes unlikely that any one thread of execution could ever use all that cache effectively. In fact, a major portion of this cache in this situation remains underutilised. Even with SHMT, multithreading is carried out in a relatively limited fashion and therefore such an enormous size of cache memory cannot be exploited fully, whereas a number of relatively independent threads or processes running on multiple cores have a fabulous opportunity to take full advantage of such an extremely large on-chip cache memory at hand.

8.12.2 Definition

A ***multicore*** processor, also known as a ***chip multiprocessor***, combines two or more processors (called *cores*) on a single piece of processor chip (a single piece of silicon, called a die). Typically, each core consists of all of the components of an independent processor, such as ALU, registers, control unit, and pipeline hardware, plus L1 instruction and data caches. In addition to the multiple cores, the contemporary multicore chips also includes L2 on-chip cache and, in some cases, even L3 on-chip cache. A multicore processor however provides, in essence, some kind of a ***structural parallelism*** within a processor.

8.12.3 Design Issues

The primary design choice for the multicore organisation of a single processor involves certain critical factors, such as the number of processor cores to be used in the chip, the number of levels of cache memory to be included, and the extent to which the cache memory is to be shared (Olukotun, K., et. al.). The fundamental parameters that are embraced with these factors can be considered as a trade-off involved in this context that can be expressed in a more simplified way as

Total chip area = number of cores × chip area per core

or

Total transistor count = number of cores × transistor count per core.

For the sake of simplicity, it has been assumed that the cache and interconnect area (and transistor count) can be considered proportionately on a per core basis.

At a given time, VLSI technology limits the left-hand side in the above equations while the designers must decide as to what weight needs to be given individually to the two factors appearing on the right-hand side of the said equation. Aggressive exploitation of ILP, with multiple functional units and more complex control logic, increases the chip area (and transistor count) per processor core. Consequently, the number of cores that could be realised within the processor would be fewer. Alternatively, in a design approach catering to a different category of target applications, the designers may select simpler cores consuming lesser chip area per core, thus using lesser number of transistor counts and consequently placing a larger number of such simpler cores on a single chip. Of course, practical system design would actually involve such issues which are much more complex and even more critical than these; still a basic design issue is observed here: For the targeted application and desired performance, how should the designers divide available chip resources among processor cores and, within a single core, among its various functional components? (Jerraya, A., et al.)

8.12.4 Multicore Organisation

Organisation of multicore systems, as being viewed, essentially includes *three* main decisive parameters mentioned below:

- The number of processor cores on the chip
- The number of levels of cache memory to be employed
- The amount of cache memory that is to be shared.

Each parameter is again specified by its number, size, capacity, capability, its placement, and also the way in which it would interact with the others and will eventually determine the type of the organisation that the multicore system would have.

Four general organisations of the contemporary multicore systems are depicted in Figure 8.32. Figure 8.32a illustrates an organisation in which the only on-chip cache is L1 cache which is again divided into instruction and data caches, with each core having its own dedicated L1 cache. This type of organisation is often found in some of the earlier multicore computer chips and is still seen in use in some embedded chips. An example of this organisation is the **ARM11 MPCore**. The organisation as shown in Figure 8.32b indicates that there is enough room available on the chip that enables it to build a separate

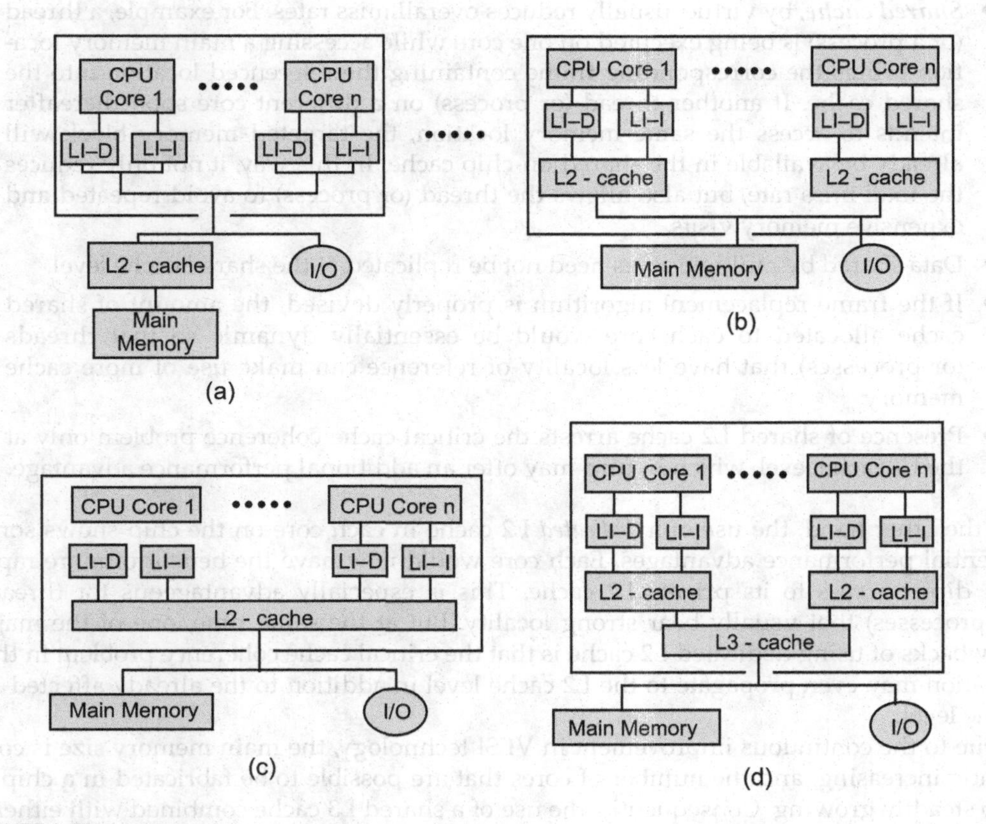

FIGURE 8.32
Different possibilities of multicore organisation. (a) Dedicated L1 cache, (b) dedicated L2 cache, (c) shared L2 cache, and (d) shared L3 cache.

on-chip dedicated unified L2 cache, apart from the existing on-chip L1 split cache (I-cache and D-cache) in each core. This type of multicore organisation is found in **AMD Opteron**. Figure 8.32c shows an arrangement almost the same as that shown in Figure 8.32b but with a little difference that the on-chip L2 cache in Figure 8.32c is a larger one and is used here in a manner shared by all cores. The **Intel Core Duo processor UltraSPARC T2** has this organisation. With the increasing potential of VLSI technology that provides the system designer with an abundance of hardware capabilities, the amount of space now available in the chip as well as the total transistor count obtainable on the chip continues to grow. This provokes the designers to include, for the sake of improving the performance further, a separate *shared* unified L3 cache, apart from usual dedicated L1 and L2 caches for each core processor. This organisation is illustrated in Figure 8.32d. The implementation of **Intel Core i7** is an example of this organisational approach.

Now the burning question arises as to how these on-chip caches will ultimately be deployed. Whether the on-chip cache (especially L2 cache) would be used in a *shared* or *dedicated* manner to realise a higher net throughput; this is essentially a typical design trade-off. Both the shared and the dedicated approaches, however, have their own merits and drawbacks. But the use of *shared on-chip* L2 cache usually shows several distinct advantages over exclusive dependence on corresponding dedicated caches. Some of the notable advantages are as follows:

- ***Shared cache***, by virtue, usually reduces overall miss rates. For example, a thread (or a process) is being executed on one core while accessing a main memory location brings the corresponding frame containing the referenced location into the shared cache. If another thread (or process) on a different core soon thereafter intends to access the same memory location, the targeted memory block will already be available in the shared on-chip cache. In this way, it not only reduces the total miss rate, but also allows the thread (or process) to avoid repeated and expensive memory visits.
- Data shared by multiple cores need not be replicated at the shared cache level.
- If the frame replacement algorithm is properly devised, the amount of shared cache allocated to each core would be essentially dynamic so that threads (or processes) that have less locality of reference can make use of more cache memory.
- Presence of shared L2 cache arrests the critical cache coherence problem only at the L1 cache level, which in turn may offer an additional performance advantage.

On the other hand, the use of a *dedicated* L2 cache in each core on the chip shows some potential performance advantages. Each core would now have the benefit of more rapid and direct access to its private L2 cache. This is especially advantageous for threads (or processes) that usually bear strong locality. But at the same time, one of the major drawbacks of using dedicated L2 cache is that the critical cache coherence problem in this situation may even propagate to the L2 cache level in addition to the already affected L1 cache level.

Due to the continuous improvement in VLSI technology, the main memory size is constantly increasing, and the number of cores that are possible to be fabricated in a chip is also steadily growing. Consequently, the use of a shared L3 cache combined with either a shared or a dedicated L2 cache per core is likely to provide a higher net throughput than simply straightaway employing an extremely large size of shared L2 cache in place.

Last but not least, another important design decision in the multicore system organisation is whether the individual cores will be *superscalar* or will implement SHMT. In recent years, there is a clean shift in system design away from structural parallelism (superscalar) and towards support for fine-grained structural parallelism (hardware threads). The basic driver behind such a shift is simple: *achieving maximum performance for a given system cost*. Development of multicore SHMT architecture is a clear consequence of such a shift. For example, the ***Intel Core Duo*** uses superscalar cores, whereas the more advanced ***Intel Core i7*** uses SHMT cores. One of the main reasons behind this approach is that the SHMT has the outcome of scaling up the number of hardware-level threads that the multicore system provides. Thus, a multicore system with four cores and the SHMT that supports three simultaneous threads in each core appears to be the same as a multicore system with 12 cores at the application level. As the software development for a given range of applications on a platform having steady growth in VLSI capabilities proceeds to meet the grand challenges of fully exploiting parallel resources, an SHMT approach in this scenario appears to be more conducive and supportive than its counterpart, the superscalar approach.

8.12.5 Basic Multicore Implementation: Intel Core Duo

Intel has continuously introduced a number of more modern multicore products using the constantly evolving more advanced VLSI technology to achieve an even higher performance while preserving the backward compatibility to keep up with its family concept. These processors with their standard CISC instruction set have combined RISC design standard techniques, such as *micro-operation pipeline*, *multiple functional units*, and *out-of-order sequencing*, in their internal structure. The original **Core** brand refers to Intel's 32-bit mobile dual-core X86 CPUs that were derived from a more enhanced version of the Intel P6 microarchitecture, the ***Pentium M*** branded processors. The Core brand comprised *two* branches: the **Duo** (dual-core) and **Solo** (Duo with one disabled core, which replaced the Pentium M brand of single-core mobile processors). It emerged in parallel with the ***NetBurst*** microarchitecture (Intel P68) of the Pentium 4 brand and was a precursor to the 64-bit Core microarchitecture of **Core 2** branded CPUs.

Intel's first dual-core mobile (low-power) processor was launched on January 6, 2006, by the release of the 32-bit **Yonah** CPU using a fabrication technology of 265 nm, and it was targeted mainly for laptop use. Its dual-core layout, contrary to its name, had more in common with two interconnected Pentium M branded CPUs packaged as a single die (piece) silicon chip (IC) than with the subsequent 64-bit Core microarchitecture of Core 2 branded CPUs. Within a short period, Intel however launched many other similar processors having almost identical architectures but with improved fabrication technology consisting of two X-86 superscalar core processors (that is why it is called **Core Duo**) with no provision for Intel's own hyper-threading technology. The Core Duo processor was operated at a speed of 1.66 GHz, and as usual, it was equipped with a dedicated L1 cache offered to each core and a large shared L2 cache. The width of the instruction set was 32 bits.

In this section, we will be looking at only the basic multicore architectures, the notable ones, the Intel Core Duo with a brief description of all its key elements and their salient features with respective activities to give an overall understanding in relation to fundamental multicore aspects. The general structure of Intel Core Duo with its major functional elements is shown in Figure 8.33.

Each core here is equipped with a separate independent **thermal control unit** to manage power consumption and related dissipation of heat generated in this kind of high-density chip to yield maximum processor performance within the confines of existing

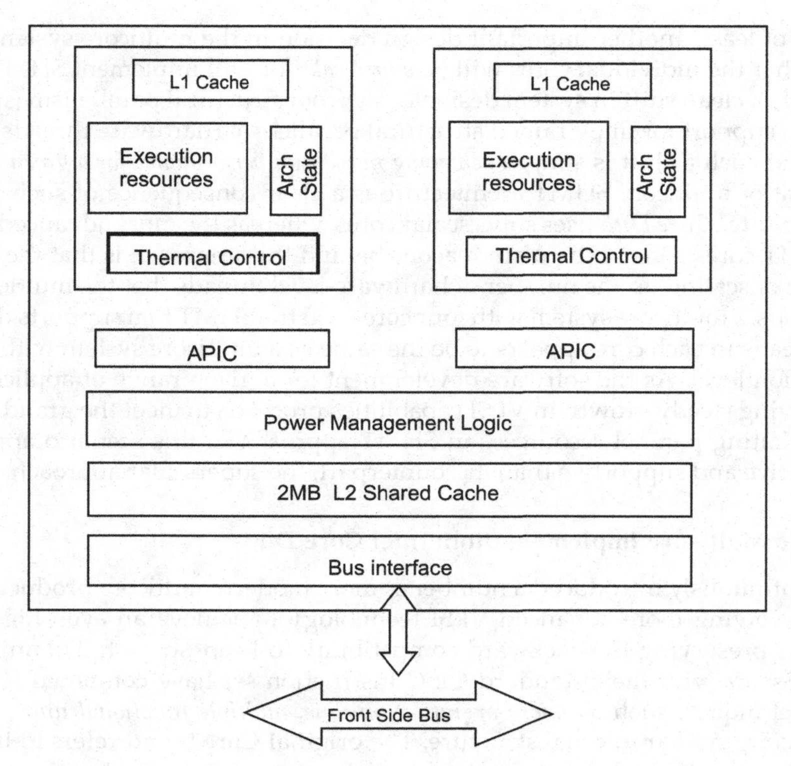

FIGURE 8.33
Schematic block diagram of Intel Core Duo.

thermal constraints. This unit also improves ergonomics with a cooler system and lower fan acoustic noise and monitors digital sensors for high-accuracy die (chip) temperature measurements. The maximum temperature of each core (an independent thermal zone) is reported separately via dedicated registers that can be polled by software. If the temperature in a core at any point of time exceeds a threshold, the thermal control unit at once reduces the clock rate for that core to reduce power consumption and thereby trim down heat generation.

The **Advanced Programmable Interrupt Controller** (APIC) performs many important functions including the following:

- The APIC provides interprocessor interrupts, which allow any process to interrupt any other processor or set of processors. When a process (or a thread) in one core generates an interrupt, it is received by the local APIC and routed to the APIC of another core as an interrupt to that core.
- The APIC accepts I/O interrupts and routes these interrupts to the appropriate cores.
- Each APIC is equipped with a timer which can be set by the operating system to generate an interrupt to the respective local core.

The **power management logic** unit takes care of the power consumption and monitors thermal conditions and CPU activity, and adjusts voltage levels, thereby increases battery life for mobile platforms, such as laptops mobile phones, etc. The VID voltage (Voltage Identification used by Intel; which is the set/stock voltage for a given clock speed) range within which the processor chip operates is, however, 1.0–1.212 V. This unit includes an advanced *power-gating capability* that allows an ultra-fine-grained logic control to turn on individual processor logic subsystems if and only if they are needed. In addition, many buses and arrays are split so that data required in some modes of operation can be put in a low-power state when not needed. However, the power consumption still remains around 15 watt.

As usual with all multicore systems, each core here also has its own dedicated **L1 split cache** consisting of a 32-KB instruction cache and a 32-KB data cache. The processor includes a *shared* 2 MB **L2 cache**. The cache logic permits a dynamic allocation of shared cache space based on the needs of the currently executing core, so that one core can be assigned even up to 100% of the L2 cache space. The L2 cache includes logic to support the MESI protocol (to resolve cache coherence, see Chapter 4) for the attached L1 caches. When a cache-write is carried out at the L1 level, the cache line gets the M (modified) state when a processor writes to it; if the line is not in E (exclusive) or M state prior to writing to it, the cache sends a *Read-For-Ownership* (RFO) request that ensures that the line is present in the L1 cache and is in the 1 state in the other L1 cache. The Intel Core Duo, however, extends this protocol to accommodate the particular situation when there are multiple Core Duo chips organised as a symmetric multiprocessor (SMP) system (discussed in Chapter 10).

Intel Core Duo is equipped with an **arbiter bus**; the *bus interface* controls L2 cache and provides the connection to the external bus, known as the *front-side bus*, which operates at a speed of 667 MHz with no provision of parity. This bus in turn connects to the main memory, I/O controllers, and other processor chips.

More details about Intel Core Duo processor are given in the website: http://routledge.com/9780367255732.

8.12.6 Intel Core 2 DUO

The majority of the desktop and mobile Core 2 processor variants were Core 2 Duo with two processor cores on a single Merom, Conroe, Allendale, Penryn, or Wolfdale chip. The Allendale was launched sometime in January, 2007, using a 265 nm fabrication technology, and Wolfdale was launched sometime in January, 2008, using a 265 nm fabrication technology. All these came with a wide range of performance and power consumption, starting with the relatively slow ultra-low-power Uxxxx (10 W) and low-power Lxxxx (17 W) versions to the more performance-oriented Pxxxx (25 W) and Txxxx (35 W) mobile versions and the Exxxx (65 W) desktop models. The mobile Core 2 Duo processors with an "S" prefix in the name are produced in a smaller μFC-BGA 956 package, which allows building more compact laptops. Each version comes with a number of products (chips), each product uses a chronologically higher product number than its predecessor. Products with higher numbers given to a specific version as mentioned above usually refers to a better performance which depends largely on the core, the clock frequency of the front-side bus, and the size of the second-level cache which are again modelspecific. Core 2 Duo processors typically use the full L2 cache of 2, 3, 4, or 6 MB available in the specific stepping of the chip, while versions with the reduced cache size are sold for the low-end consumer market as Celeron or Pentium Dual-Core processors. Like these processors, some low-end Core 2 Duo models also disable features such as **Intel Virtualization Technology**.

8.12.7 Intel Core 2 Quad

Core 2 Quad processors are multichip modules consisting of two dies similar to those used in Core 2 Duo, forming a quad-core processor. This allows nearly twice the performance of a dual-core processor at the same clock frequency in ideal conditions. Initially, all Core 2 Quad models were versions of Core 2 Duo desktop processors, Kentsfield derived from Conroe launched in January 2007 with 465 nm fabrication technology and Yorkfield derived from Wolfdale launched in March 2008 using a 445 nm fabrication technology, but later Penryn-QC was added as a high-end version of the mobile dual-core Penryn. The ***Xeon 32xx*** and ***33xx*** processors are mostly identical versions of the desktop Core 2 Quad processors and can be used interchangeably.

A brief description of Intel Core 2 Quad with a relevant table is given in the website: http://routledge.com/9780367255732.

8.13 Multicore with Hardware Multithreading

With continuous advancement in VLSI technology, the aggregate functionality that can be built into a single chip has been observed to be growing steadily with time. Consequently, multiple processor cores as well as provision for hardware support for multithreading on a single chip can both be seen as natural consequences of the steady advances in VLSI technology. Both these developments, however, eventually address the needs of important segments of modern computer applications and workloads.

Depending on the specific strategy that is adopted for *switching* between threads, hardware support for multithreading may be classified under any one of the following:

- ***Coarse-grain multithreading*** refers to switching between threads only on the occurrence of a major pipeline stall, which may be caused by, say, access to main memory, with latencies of the order of a hundred processor clock cycles.
- ***Fine-grain multithreading,*** also known as *interleaved multithreading*, refers to switching between threads on the occurrence of any pipeline stall, which may be caused by,say,L1 cache miss. But this term would also apply to designs in which processor clock cycles are regularly being shared (i.e. switched) amongst executing threads, even in the absence of a pipeline stall.
- ***Simultaneous multithreading*** refers to machine instructions that *issue two (or more) threads in parallel in each processor clock cycle*. This would correspond to a multiple-issue processor where the multiple instructions being issued in a clock cycle come from an equal number of independent execution threads.

The recent scenario in the processor design approach is thus observed to shift towards multicore chips with hardware multithreading. This organisational design approach thus results in mainly two types of important performance benefits:

- Multicore chips with hardware multithreading can exploit a broader range of *structural parallelism* inherent in applications. The processor core in a multicore chip operates mainly in a shared memory mode. However, message passing, which works independently of physical locations of processes or threads, also provides a natural software model to exploit the structural parallelism present in an application.

- A multicore system with hardware multithreading also supports the *natural parallelism* which is always present between two or more *independent* programs running on a system. Even two or more operating systems can share a common hardware, in effect, providing multiple ***virtual computing environments*** to users. Such virtualisation makes it possible for the system to support more complex and composite workloads, thereby resulting in better system utilisation and an appreciable return on investment.

With the continuously increasing power of VLSI technology in abundance, the development of *multicore* SoCs, even in the presence of ample multithreading, also became inevitable, since there are practical limits to the number of threads (multithreading) a single processor core can support. Each core on the advanced Sun UltraSPARC T2, a *multicore* SoC processor, for example, supports eight-way fine-grained multithreading, and the chip has eight such cores. Still, much of the available aggregate power of a multicore processor remained unutilised. Multicore chips, however, till now keep their promise of providing higher net processing performance per watt of power consumption.

SoCs are one of the examples of fascinating design trade-offs. However, for any actual task of practical processor design, it is necessary to make many design choices and trade-offs, then validate the chosen design thus arrived at using simulations, and finally complete the design in detail to the level of logic circuits.

8.13.1 IBM Power 5

One of the recent past releases that came from IBM following the line of its Power architecture is the introduction of IBM Power5 processor chip, launched in 2004 in a system introduced by IBM. It was primarily intended to compete mostly against the ***Intel Itanium 2*** and to a lesser extent the ***Sun Microsystems UltraSPARC IV*** and the *Fujitsu SPARC64* in the high-end enterprise server market. It was eventually superseded in 2005 by an improved iteration, the Power5+.

Power5 is a second generation dual-core (multicore) processor which uses the SHMT approach on both the two separate processor cores. It is interesting to note that at the time of its design, with certain objectives to be attained, the designers worked with various possible alternative design approaches individually and simulated them accordingly. They observed that having two two-way SHMT processor cores on a single chip yielded a performance that was superior to that of a single four-way SHMT processor. They also examined the result of the numerous simulations thus obtained, which revealed that additional multithreading, if carried out even more, with available hardware resources that support two threads, might lead to the degradation of the processor's performance, mainly because of critical *cache thrashing*, as data from one thread always attempts to displace data needed by another thread. That is why the Power5 is built using SHMT with only two concurrent threads on each of two separate processor cores. The Power5 is equipped with *power management facility* that can reduce power consumption below the standard single-thread level, yet this facility has no performance impact as such. For low-priority threads, it can switch to low-power mode. Moreover, the SHMT can even be made disabled to optimise the current workload. Several Power5 processors in high-end systems, however, can be again coupled together to act as a single *vector processor* by a technology called **ViVA (Virtual Vector Architecture)**.

Power5+: This processor is an improved upgrade of the existing Power5 with a newer 90nm fabrication process. It was initially made for only lowering the power consumption, so its die size was decreased from 389 mm^2 to 243 mm^2. Clock frequency was maintained between 1.5 and 1.9 GHz. Subsequent versions of Power5+, however, raised the clock frequency to 2.2 GHz in 2006 and further to 2.3 GHz later in 2006. This processor was packaged in the same way as previous Power5 microprocessor was packaged, but it was also available in a quad chip module (QCM) containing two Power5+ dies and two L3 cache dies, one for each Power5+ die. These QCM chips, however, ran at a clock frequency between 1.5 and 1.8 GHz.

A brief description of Power5 architecture with relevant figures is given in the website: http://routledge.com/9780367255732.

8.13.2 Intel Core i7

The Intel Core i7 processor named *Bloomfield* was introduced in November 2008 using a fabrication technology of 445 nm. Soon after in July 2010, the *Gulftown* with a fabrication technology of 632 nm, then *Scandy Bridge* in January 2011 with a fabrication technology of 432 nm, and finally *Ivy Bridge* with a 422 nm fabrication technology in April 2012 were released. Each product in chronological order of its release came with some new marvelous features assisted largely by an enhanced core, increased bus system clock frequency and larger size of caches provided at different levels to support the continuously growing demands of current applications. However, all these chips have *four* X-86 SHMT processor cores, each with a dedicated split L1 cache (32KB I-cache and a 32KB D-cache) and a dedicated unified 256KB L2 cache and with a large 8MB L3 cache shared by all four cores.

One elegant mechanism that Intel exploits to make its cache usage more effective is ***prefetching***, to fill the caches speculatively with data which is likely to be required soon. It is observed that the Core i7, however, improves on L2 cache performance, better than that of Core 2 Quad shared L2 cache, with the use of the dedicated L2 cache supported by a shared L3 cache with a relatively high speed. The Core i7 chip supports **two forms** of external communication to other chips. One is the **DDR3 memory controller** that connects the DDR main memory onto the chip. The interface supports three channels; each one is 8 bytes (64 bit) wide for a total bus width of 192 bits, for an aggregate data rate of up to 32 GB/s. With the memory controller on the chip, the front-side bus is not needed and hence is eliminated. The other one is the Quick Path Interconnect (QPI) which is a cache-coherent, point-to-point link-based electrical interconnect specification for Intel processors and chipsets. It provides high-speed communication between connected processor chips. The QPI links operate at 6.4 GT/s (giga transfers per second). At 16 bits per transfer, that adds up to 12.8 GB/s, and since QPI links involve dedicated bidirectional pairs, the total bandwidth is 25.6 GB/s.

8.13.2.1 Distinctive Features of Intel Core i7 900-Series Processors

- Intel, however, launched three Core i7 900-series processors in succession differing mostly in processor frequency and having model numbers Core i7 920 (2.66 GHz), Core i7 950 (3.06 GHz), and Core i7 975 (3.33 GHz). All the major features of these three processors are otherwise more or less the same. Each of these processors provides ***four complete execution cores*** in a single processor package.

- Each processor using ***Intel Hyper-Threading (HT) technology*** delivers two processing threads per physical core for a total of eight simultaneously executing threads, thereby offering a massive computational throughput.
- ***Intel Turbo Boost technology*** used in each processor facilitates the dynamic increase of the processor frequency as needed, by taking advantage of thermal and power headroom when operating below specified limits. This results in automatic yielding of more improved performance when it is needed the most. This Turbo Boost technology provides *single-core performance* up to 2.93 GHz for Core-i7 920 processor, 3.33 GHz for Core-i7 950 processor, and 3.6 GHz for Core i7 975 processor.
- The large last-level L3 (*8MB*) ***Intel Smart cache*** enables dynamic and efficient allocation of shared cache space to all four cores to almost match their every needs.
- An ***on-chip integrated memory controller*** supporting 3 DDR3 channels operated at 1066 MHz offers dramatic memory read/write performance through an efficient prefetching algorithm, lower latency, and higher memory bandwidth, thereby making the Intel Core i7 processor family ideal for data-intensive applications.
- All these processors include the ***full SSE4*** (128-bit) ***instruction set***; each instruction is issued at a throughput rate of *one per clock cycle*, thereby improving the performance of a broad range of multimedia and computation-intensive applications.
- ***Intel Quick Path interconnect*** *(Intel QPI)* attached with the Intel Core i7 900 processor series increases bandwidth and lowers latency, achieving data transfer speeds as high as 25.6 GB/s.
- The type of socket being used with each such processor is LGA 1366.

A brief description of Intel Core i7 architecture with a relevant figure is given in the website: http://routledge.com/9780367255732.

8.13.3 Sun UltraSPARC T2 Processor

Since mid-1980s, many powerful RISC processors including SPARC family of processors that came from Sun Microsystems have been introduced with many innovative ideas using a relatively much simpler design with effective instruction pipelines and efficient usage of on-chip cache memory. The original SPARC processor was a 32-bit RISC processor with load-store architecture, relatively simple addressing modes, and register-to-register arithmetic logic machine instructions in three-address format (see Chapter 9, Sun SPARC, for more SPARC processors details).

The **UltraSPARC T2**, a 64-bit enhanced version of its predecessor UltraSPARC T1, is the newest *multicore* processor with SHMT and SoC version of UltraSPARC with extensive on-chip support for *multithreading, networking, I/O,* and other key functions. The T2 chip has an area of just under *3.5 cm*^{2}, fabricated with *65 mm* line width using about *500* million transistors, and can operate at *1.4 GHz* with a *1.1 V* supply. The normal power consumed by the chip is about 95 W; still, on a per thread basis, this power consumption works out to be quite low.

The architecture of T2 processor consists of an *eight-processor core* in which each core supports *eight-way fine-grained multithreading*. Each such core has its own data paths, register sets to support execution of multiple threads, two integer operation units, and a floating-point unit. In addition, each core has hardware provided for ***cryptography*** and ***graphics***. Over all, the chip supports 64 parallel threads and exhibits comparatively a high level of thread – level parallelism (TLP) but not necessarily much ILP. Besides having as usual a dedicated L1 instruction cache and data cache with each core shared by all eight threads executing on each core, the processor also contains a crossbar switch, a 4 MB shared L2 cache organised in the form of eight parallel banks for faster access, and extensive support for I/O and networking. Therefore, it is in fact an SoC. Since the threads run independently of each other (single thread of execution) and share all hardware resources, each thread behaves here as a *virtual processor* in its own right. Thus the single chip is considered to be able to support 64 *virtual systems*.

The design of the UltraSPARC T2 is primarily targeted towards computation-intensive applications with a high degree of multithreading. Apart from back-end servers, its capabilities include network devices such as *packet routers, switches* for local area networks (LANs), *graphics* and *imaging applications*, and other similar facilities.

Last but not least, one unique feature is that the complete design of UltraSPARC T2 chip has been made available on the web to researchers and developers under an *open source* agreement. The stated objective behind this decision made by Sun Microsystems is to encourage further innovations in processor design and its applications around the world.

A brief architectural description of Ultra SPARC T2 with a relevant figure is given in the website: http://routledge.com/9780367255732.

8.14 Summary

The performance of a CPU has been greatly improved by changing its architectural design, decomposing the underlying operations of its program-control and execution logic into a number of consecutive stages in sequence organised in the form of a *pipeline*. A typical *m*-stage instruction pipeline overlaps the execution of up to *m* separate instructions, thereby allowing the performance level to attain nearly one CPI. Smooth operation of a pipeline without collision is obtained using necessary pipeline control and effective scheduling with no hazards at the time of issuing instructions. Numerous methods have been devised, sometimes by means of mechanisms with the use of additional hardware and software support, to keep the pipeline free from any hazards and their ill effects. This base (scalar) pipeline architectural design has been further modified to achieve an even better performance by using more stages in the pipeline; this offers a reduced pipeline cycle and thereby increases the rate of instruction issue. This approach when implemented over base (scalar) pipeline results in a *superpipeline* (or deeper pipeline). Due to the availability of faster hardware in abundance with reduced cost, the architectural design of base (scalar) pipeline has been further modified by way of using multiple-instruction pipelines within the CPU resulting in a superscalar design that ultimately realizes an even higher throughput, since a CPI level less than 1 has been achieved. The superscalar processor has been modified once again to yield an even higher throughput by making each of the multiple-instruction pipeline superpipelined. ***Intel Pentium 4*** processor implemented this architectural design approach and is recognised as a *superpipelined superscalar processor*.

An alternative in the architectural design of the processor has been considered to yield an even higher throughput with the use of multiple operations in each machine instruction, instead of mostly one operation as is generally specified in traditional machine instructions. The processor developed following this approach and having a long instruction word is known as the *VLIW processor.* A little further refinement of this concept injecting more flexibility than the fixed format of multi–operation VLIW instruction ultimately leads to what is called *explicitly parallel instruction computer* (EPIC). ***Intel Itanium 64-bit processors*** make up the most-well-known processor family of this class.

Instruction-level parallelism (ILP), by far, has been realised to the highest possible extent, but it often experiences critical dependencies between machine instructions that summarily limit the expected amount of parallelism. One elegant way to reduce the aggregate of dependencies and thereby yield greater ILP without much increase in circuit complexity or power consumption is called *multithreading* or *thread-level parallelism* (*TLP*). Many modern high-performance processors, irrespective of whether they are scalar, superscalar, VLIW, or multicore processors of today, employ this marvelous approach in their architectural design.

With the radical and rapid advancement in the development of VLSI technology, enough room became available within the processor chip, with a very large number of available unutilised transistors (gates) on it. All the performance improvement approaches, like superscalar and VLIW processors, by this time have already reached their extreme saturation point in terms of their effective implementation. This has given rise to the appearance of inevitable *multicore architecture* in the evolving process of the processor design. A multicore processor, also known as a *chip multiprocessor,* combines two or more independent processors (called cores) on a single piece of processor chip (a single piece of silicon called a die). Typically, each core consists of all of the components of an independent processor plus L1 instruction and data caches and may also have L2 on-chip cache and in some cases even L3 on-chip cache. A multicore processor, in essence, provides some kind of a *structural parallelism* within a processor. Multicore architecture, however, has been further enhanced using hardware multithreading in its architectural design to offer an even higher throughput. **IBM Power5 series, Intel Core i7,** and **Sun UltraSPARC T2** processors are only a few of such high-performance processors that have implemented this approach in their processor design.

Exercises

8.1 What is pipelining in the processor design? Draw a space-time diagram for a five-segment linear pipeline showing the time it takes to process seven tasks. What would be the speed-up ratio? Find out the theoretical maximum speed-up that can be obtained from this pipeline. What are the reasons for which theoretical maximum speedup cannot be obtained in reality?

8.2 What are the major areas where the pipeline organisation can be suitably applied? Derive an expression for the bandwidth of a k-stage pipeline executing n number of instructions. What do you mean by latency of a pipeline? Why it is considered an important parameter?

8.3 What is the difference between ILP and machine-level parallelism?

8.4 Consider the execution of a program of 100 instructions by a linear pipelined processor with a clock rate of 20 MHz. Assume that the instruction pipeline has five stages and that one instruction is issued per clock cycle. There are no hazards, and all the penalties due to any branch instructions and out-of-sequence executions are ignored.

a. Calculate the speed-up factor in using this pipeline to execute the program and compare it with that obtained using an equivalent nonpipelined processor with an equal amount of flow-through delay.

b. What are the efficiency and throughput of this pipelined processor?

8.5 A pipelined processor has five-segment linear pipeline with a clock cycle of 10 ns, and one instruction is issued per clock cycle. A task of 100 instructions is processed through it. A nonpipelined system takes 50 ns to process one instruction of the same task. Ignoring all the penalties,

a. Determine the speedup ratio of the pipeline.

b. What are efficiency and throughput of this pipelined processor?

8.6 Define a reservation table. What is the basic difference between such a table and a space-time diagram? Consider the following pipeline reservation table.

1	2	3	4
×			×
	×		
		×	

a. What are the forbidden latencies?

b. Draw the state transition diagram.

c. List all the simple cycles and greedy cycles.

d. Determine the optimal constant latency cycle and the minimal average latency.

e. Determine the throughput of this pipeline if the pipeline clock period be taken as $\tau = 15$ ns.

8.7 Consider the following reservation table for an arbitrary four-stage pipeline with a clock cycle of $\tau = 25\,\text{ns}$.

1	2	3	4	5	6
×					×
	×		×		
		×			
			×	×	

a. List the forbidden latencies and the initial collision vector.
b. Draw the state transition diagram for scheduling the pipeline.
c. List all the simple cycles from the state diagram.
d. Identify the greedy cycles among the simple cycles.
e. Determine the MAL associated with the shortest greedy cycle.
f. Determine the pipeline throughput corresponding to the MAL and given τ.
g. Determine the lower bound on the MAL for this pipeline.

8.8 Discuss briefly with examples the factors that limit the performance enhancement of a pipelined processor.

8.9 What do you mean by pipeline hazards? State and explain the different type of hazards being observed.

8.10 Define the following terms in brief:

Resource conflicts

Procedural dependency

True data dependency

Instruction window.

8.11 "Data dependency in the program prevents the smooth operation of the pipeline". Discuss this issue with examples and give any two methods to solve this problem.

8.12 "Data hazards may be classified into a number of types". How many types are there? Briefly describe each type with examples. What do you mean by antidependency and output dependency? How do they affect the smooth operation of the pipeline?

8.13 What do you mean by dynamic scheduling? Discuss any one method to implement dynamic scheduling to resolve data dependency.

8.14 "The performance of the pipeline is adversely affected due to the presence of control hazards". Define *control hazard.* What are the factors that create such a hazard? What is the delay penalty associated with a branch instruction? By how much can the use of forwarding paths reduce this penalty?

8.15 "The existence of conditional branches in the program is one of the factors creating a control hazard. State and explain with example the effect of a conditional branch on the pipeline performance. What are the different techniques that have been proposed to reduce its effect? Discuss one such technique in detail.

8.16 "Branch prediction is another increasingly popular and more powerful technique to resolve the control hazard arising due to the presence of conditional branches". Discuss this technique in detail with appropriate example.

8.17 An assembly language program is given below.

```
I1: Move R4,R5       /Move the content of R5 to R4 /.
I2: Move R7,(R4)     /Move the content of memory addressed
                        by R4 to R7/.
I3: Sub R4, R4, 5    /Subtract 5 from contents of R4 and
                        store the result in R4/.
I4: Load R8,(R4)     /Load the content of memory addressed
                        by R4 to R8/.
I5: BLE R7, R8, L5   /Branch to L5 if content of R8 is
                        greater than the contents of R7
I6: MUL R8, R8, 7    /Multiply the content of R8 by 7 and
                        store the result in R8/.
```

Show the different types of data dependencies, such as WAW, RAW, and WAR, that exist in the above program.

8.18 Consider the following assembly language program that is processed by a four-stage instruction pipeline having the stages S_1 (instruction fetching and decoding), S_2 (operand loading), S_3 (execution), and S_4 (operand storing):

```
       LD R3,#X            / Load constant X into general register R3/
       LD R4,#Y            / Load constant Y into general register R4/
       LD R5,#Z            / Load constant Z into general register R5/
       LD R8,#0            / Clear general register R8/
       BEQ R3, R4, adr1    / If R3 = R4 then go to adr1/
       ADD R8,R3,R4        / Add R3, R4 and R8 and store the result to R8/
       MUL R8,R8,R8        / Square the contents of R8
       MUL R8,R8,#1        / Increment R8 by one
adr1:  ST m(R2),R8         / Store R8 in the memory location addressed by R2/
```

Assume that stage S_2 can only support data reads (from registers and/or memory) and that data reads and writes can occur only in stage S_4.

Identify all RAW, WAR, and WAW hazards that are present in this case.

8.19 A pipelined processor has two branch delay slots. An optimizing compiler can fill one of these slots 85% of the time and can fill the second slot only 20% of the time. What is the percentage improvement in performance achieved by this optimisation, assuming that 20% of the instructions executed are branch instructions?

8.20 Discuss an example that uses delayed load with a four-segment pipelined processor.

8.21 What are the salient characteristics of the superscalar approach in the design of a processor?

8.22 Distinguish between a scalar processor and superscalar processor in terms of *instruction issue, pipeline architecture,* and *processor performance.* What is parallel instruction issue under ideal circumstances?

8.23 Explain with an arbitrary diagram the major difference between a pipelined and a superpipelined processor. Explain why a superpipelined processor effectively offers a lower performance than a superscalar processor of small degree.

8.24 What will be the speed-up ratio of a superpipelined superscalar processor of degree (*m,n*) over a base scalar processor of degree (1, 1)? What are the practical limitations that prevent the growth of the superscalar processor of degree *m*? What are the practical limitations preventing the growth of a superpipelined processor of degree *n*?

8.25 What are design trade-offs between a large register file and a large D-cache? Why are reservation stations or reordering buffers needed in a superscalar processor?

8.26 What are the key characteristics of a superscalar processor organisation? What are the factors that limit the degree of superscalar design?

8.27 Compare superpipelined, superscalar, and VLIW processors with relevant examples.

8.28 What are the forces that drive designers to move to a multicore organisation rather than increasing parallelism within a single processor?

8.29 Why is there an increasing trend to offer more chip area to implement cache memory?

8.30 Discuss the primary design issues in a generic multicore organisation.

8.31 What are the advantages of using a shared L2 cache for cores of the processor chip compared to using a separate dedicated L2 cache for each core.

8.32 Why is SHMT approach considered superior than superscalar design when implemented within each core of a multicore organisation?

8.33 Enunciate the differences observed between simple instruction pipelining, superscalar design, and SHMT.

Suggested References and Websites

Borkar, S. "Thousand core chips – A technology perspective." *Proceedings of ACM / IEEE Design Automation Conference*, 2007.

Circello, J., et al. "The superscalar architecture of the MC68060." *IEEE Micro*, vol. 15, pp. 10–21, April 1995.

Dulong, C., "The IA–64 Architecture at work." *Computer*, vol. 31, no. 7, pp. 24–32, July 1998.

Hammond, L. and Nayfay, B. "A single–chip multiprocessor. "*Computer*, vol. 30, no. 9, pp. 79–85, 1977.

Hennessy, J. L. and Patterson, D. A. *Computer Architecture: A Quantitative Approach*. San Mateo, CA: Morgan Kaufmann, 1990.

Hwang, K. and Xu, Z. *Scalable Parallel Computing: Technology, Architecture, Programming*. New York: WCB / McGraw-Hill, 1998.

IEEE Micro, vol. 20, no. 5, six articles on the IA–64 Architecture and the Itanium Processor, September / October 2000.

Jouppi, N. and Wall, D. "Available instruction–level parallelism for superscalar and superpipelined machines." *Proceedings Third International Conference on Architectural Support for Programming Languages and Operating Systems*, April 1989.

Kathail, B., Schlansker, M., and Rau, B. "*Computing for EPIC Architectures*. "*Proceedings of the IEEE*, November, 2001.

Kogge, P. M. *The Architecture of Pipelined Computers*. New York: McGraw-Hill, 1981.

Omondi, A. *The Microarchitecture of Pipelined and Superscalar Computers*. Boston, MA: Kluwer, 1999.

Olukotun, K., Hammond, L., and Laudon, J. *Chip Multiprocessor Architecture: Techniques to Improve Throughput and Latency*. San Rafael, CA: Morgan & Claypool, 2007.

Ramamoorthy, C. V. and Li, H. F. "Pipeline architecture. "*ACM Computing Survey*, pp. 61–102, March 1977.

Smith, M.; Johnson, M.; et.al. " Limits on multiple Instruction Issue." *Proceedings, Third International Conference on Architectural Support for Programming Languages and Operating Systems*, April, 1989.

Stallings, W. *Computer Organisation and Architecture*, Indian ed. Dorling Kindersley India Pvt. Ltd., 2010.

Intel website: intel.com

Intel Developer's Page: Provides the *Intel Technology Journal* and includes a starting point for obtaining Pentium information.

Intel Corp. "*Pentium Pro and Pentium II Processors and Related Products.*" Aurora, CO, 1998.

9

RISC Architecture

LEARNING OBJECTIVES

- To provide an overview of the characteristics of CISC (complex instruction set computer) architecture and its drawbacks.
- To get an understanding of RISC (reduced instruction set computer) architectures: its definition and features.
- To enunciate the different aspects that are relevant to the RISC versus CISC debate.
- To articulate the design issues and instruction set of RISC processors, including the instruction format and addressing scheme.
- To compare a RISC processor with a contemporary CISC processor in terms of important architectural design parameters.
- To learn about different types of leading representative RISC processors.

With the advent of the revolutionary inventions in the field of electronics, such as integrated circuits (ICs), small-scale integration (SSI), large-scale integration (LSI), and very-large-scale integration (VLSI), the digital hardware became extremely powerful and equally cheaper that consequently accelerated the moves towards gradual emergence of more modern computers with versatile architectures when compared to those of early days. Successful implementation of pioneering concept of *microprogramming* in the computer architecture has abruptly changed the entire scenario. The microarchitectures of the system then became extremely straightforward and simple enough that made the machines easier to program. More number of new instructions, not all are used regularly, was then continuously added in the instruction set to make the system more upgraded. As a result, the instruction set gradually went on highly complex, and so the microprogram which ran on them became lengthy and equally complicated that started to constantly create hindrances in the normal execution of the regular frequently used operations. The net effect is that the performance was observed to have notably degraded from its normal expected level. This newly introduced instruction set was called an *orthogonal instruction set*, and this basic set of features is now called a *complex instruction set computers*, or in short, CISC (pronounced "sisk").

At this juncture, a number of designers started to argue that computers had become much too complicated over the years and that all those designs should be thrown out, and new designs should be started afresh. They recommended that computers must use a set of fewer, simpler, and less orthogonal instructions that would be looked almost like a small subset of the machine instructions of the existing *CISC*. As these simpler instructions do not require complex decoding, they could be executed much faster within the CPU (central processing unit), avoiding as much slower memory access as possible. All these thoughts together with other relevant aspects when religiously followed, eventually gave birth to a different form of architecture, historically known as *RISC* (*reduced instruction set computer*). RISC machines are really among the most interesting and potentially have one of the most innovative new design concepts that can be considered a renaissance in the revolutionary process of architectural evolution of computers. Its introduction has ultimately challenged the traditional approaches followed in the architectural design of CPU by the computer architects of those days. In this chapter, we introduce the major characteristics of CISC and RISC architectures, most of the salient features of CISC and RISC machines, and finally, a comparison between RISC and CISC. We then present the instruction set and instruction format of a few popular representative RISC processors introduced by reputed manufacturers.

9.1 Background: Evolution of Computer Architecture

The earliest digital computers starting from ENIAC to CDC 6600 (1962) were extremely simple having a few machine instructions with only one or two addressing modes, directly executed in hardware, and have used very simple high-level languages. The introduction of IBM 360 ***mainframe*** series in 1964 with a marvellous new idea of injecting *microprogramming* in the computer architecture has radically changed the microarchitectures of the system, making it straightforward and simple enough to program it at ease. It consequently ventilated a sparkling idea of a family concept with all its merits to evolve. More number of new instructions was then continuously added in the instruction set to exploit the constantly evolving more superior hardware technology made in-built to make the system more advanced providing many salient features. This newly introduced instruction set was called an *orthogonal instruction set*, and this basic set of features is now called a *complex instruction set computers*, or in short, *CISC*. These advanced machines, however, became capable of handling more modern high-level languages enriched with many powerful features to support a diverse spectrum of evolving application areas for users of different disciplines. In addition, migration of many functions from software implementation into its equivalent hardware realization was also made possible. While all these ongoing activities were appeared to be amazing, but in reality, it invited a lot of problems as this provoked the designers to continuously append many additional features on the existing architecture that were really seldom used. As a result, the internal complexity of the computer and its instruction set gradually went on increasing and is also highly complex. Consequently, the microprograms that ran on them became lengthy and equally complicated that started to constantly create hindrances in the normal execution of the regular frequently used operations. The net effect is that the performance was observed to have notably degraded from its normal expected level. Nevertheless, most of the contemporary computers still follow this concept in one form or the other.

This approach quickly became very popular, and almost immediately, worldwide contemporary CISC computers, both mainframes and minicomputers, were introduced by giant manufacturers, including Burroughs, Univac, NCR, CDC, Honeywell, and DEC.

Even microprocessors from Intel (X-86 family) and Motorola (68000 family) have exploited this approach in abundance with minimal architectures that rapidly progressed and even exceeded the complexity of minicomputers and mainframes of those days. The domain of microprogramming usage, however, got continuously expanded. Most of the frequently used library procedures were then injected into the microprograms to negotiate their calls that summarily reduced the otherwise required frequent visit to slower main memory, thereby substantially increasing the speed of the operations. But the complexity of the underlying microprograms got gradually increased. In addition, arrays and records are dealt with using special *addressing modes*. Large parts of *procedure calls*, including parameter passing, stack handling, register saving, etc. are now transferred into the microcode (microroutines). Sophisticated high-level languages with powerful constructs, such as **if**, **while**, **case**, etc. required additional instructions in the instruction set for their effective translation by a compiler. As the ultimate goal of the CISC architecture was targeted to provide a single machine instruction for each statement written in a high-level language, the machine instruction set then became more and more complex, and the burden on the microprograms gradually became heavy. Almost everyone, however, started to believe this trend as positive and an ultimate one, and was sure enough for it to be continued for many years to come.

In the late 1970s, the existing technology began to radically improve, thereby offering much faster processor and relatively speedy semiconductor RAM memory. At the same time, it has been gradually felt by the manufacturers to writing/upgrading, debugging, and maintaining all those intricate microcodes used as a sole backbone to be immensely difficult to continue. A stage thus had been set for someone to realize that computers could be made as simple as possible so as to run a lot faster. This demanded a total elimination of the existing interpreter (microprograms) from the architecture that stood as a stumbling block. A different approach thus started to evolve instead that deliberated each program to be compiled straightaway, close to machine codes, and to be executed out of the fast semiconductor memory by the hardware directly.

More details about this topic are given in the website: http://routledge.com/9780367255732.

9.2 Characteristics of CICS and Its Drawbacks

In summary, the ***major characteristics*** of *CISC architecture*, worthy of mention, are as follows:

- A small set of 8–24 general-purpose registers (GPRs);
- A large number of instructions, typically ranging from 100 to 350;
- Some instructions that perform specialized tasks and are seldom used;
- Instructions that handle and manipulate operands in memory;
- Different types of instruction/data formats of variable length;
- A wide variety of available addressing modes, typically ranging from 5 to 20 different types;
- Many high-level language constructs/statements are directly implemented in hardware/firmware (microprogram).

9.2.1 Drawbacks

- Machine language gets larger and more complicated;
- The microprogram (the interpreter) gets bigger and complex, and hence runs slower;
- More instructions in the instruction set mean more time spent decoding the opcodes;
- Large number of addressing modes cause address analysis to use *microprocedures* (a set of microcode) repeated number of times;
- Many of all kinds of instructions and modes included are seldom in use;
- Maintenance of gradually large and complex microcode is extremely difficult.

9.3 RISC: Definition and Features

The term *RISC* was, however, first used by David Patterson, University of California, Berkley, in the early 1980s, while investigating with a group of his colleagues the design of their RISC-1 architecture. A RISC machine is essentially just a computer with a small number of vertical microinstructions (not microprograms). The philosophy behind it is that user programs after compilation will be translated into sequences of these microinstructions, which are then to be executed directly by the underlying hardware, with no intervening interpreter (microprogram). Eliminating this level of interpretation is ultimately the secret of RISC machine's enhanced speed. The result is that the simple things that programs actually do, like adding two registers, can now be completed in a single microinstruction, in contrast to 8 *horizontal* microprogramming or 15 *vertical* microprogramming instructions of the fastest CISC machine, a straightway winning factor of 10. In fact, before the invention of microprogramming, all computers were essentially RISC-like machines with simple instructions directly executed by the underlying hardware.

A RISC instruction set typically contains less than 100 instructions with only 3–5 simple addressing modes, and has a fixed instruction format (32 bits). Most instructions are register based and are executed mostly in one cycle under hardwired control. Memory access is done only by *load/store* instructions. Since the complexity of the instruction set has been greatly reduced, a higher clock rate and a lower CPI (cycles per instruction) have been realized that ultimately give rise to higher MIPS (million instructions per second) ratings. A large register file consisting of at least 32 registers (nowadays, it is around 100 or even more) is used for general purpose as well as for faster context switching among multiple users.

RISC machines are fast due to the advancement in software, but not in hardware. The improvement in ***optimizing compiler*** *technology* now makes it possible for RISC to generate microcode directly, skipping the interpreter. RISC machines are somewhat simpler than even vertical microarchitectures (CISC). Recently, engineers have found ways to further compress the reduced instruction sets to fit it even in smaller memory systems (e.g. the ARM's (**Advanced RISC Machines'**) *Thumb* instruction set encoded into a 16-bit halfword format). In applications that do not need to run older binary software, compressed RISCs are expected to be coming out to dominate the future market. The RISC design has further gained power by way of pushing some of the less frequently used operations into its software. A RISC design, for any given level of performance, always has a much smaller "gate count" (number of transistors), the main driver in overall cost/performance considerations. In other words, a fast RISC chip is much cheaper than a corresponding compatible fast CISC chip. Initially, RISC chips prevailed over the market for 32-bit ***embedded systems.*** Smaller RISC chips, however, are becoming even more common in the cost-sensitive 8-bit embedded system market. The main market for RISC CPUs has been for systems that require low power or small size and low cost.

9.4 Representative RISC Processors

Generic RISC processors are called the *scalar RISC* processors since they are designed to issue one instruction per cycle. The first RISC machine in *1974* was an experimental project ***IBM/801*** ECL, which was further enhanced by IBM in *1986* as ***PC/RT*** processor, and later, IBM released the RISC System ***RS/6000*** line of computers in *1990* using ***POWER*** architecture. A group of Berkeley scientists led by David Patterson in *1980* designed ***RISC I*** and then ***RISC II*** within a short period. In *1981,* John Hennessy at Stanford designed and fabricated a somewhat different RISC chip named the *MIPS* (***Mi***croprocessor without ***I***nterlocking ***P***ipe ***S***tages), and formed a company that introduced MIPS RX000 series with the first member *MIPS R2000* and subsequently, ***MIPS R3000, MIPS R4000 (1991),*** and finally ***MIPS R10000 (1994).*** In *1987,* ***SUN Microsystems*** developed an open microprocessor design ***SPARC*** (***S***calable ***P***rocessor ***ARC***hitecture), architecture, and not a chip. About half a dozen semiconductor vendors, such as ***Fujitsu, Cypress, LSI Logic Inc.*** and ***Texas Instruments,*** have been given licence to produce SPARC chips using different technologies (CMOS, ECL, GaAs (gallium arsenide), gate array, VLSI, etc.). A GaAs SPARC was reported to yield a 200-MIPS peak at 200-MHz clock rate. In *1988,* ***Motorola*** launched RISC ***88000,*** and in **1990,** Motorola launched ***M88100.*** From **1993** onwards, Motorola implemented their ***7XX*** and ***7XXX*** lines of processors in the line of ***PowerPC*** processors (6XX lines, such as 601, 603, 604, etc.), with *MPC* prefix on their labelling. They introduced one of the latest versions of the processor in the ***MPC7XXX*** line, the ***MPC7450.*** In *1989,* ***Hewlett–Packard*** (HP) announced their ***PA-RISC*** with a version ***PA-RISC 1.1 (1990),*** and then changed the design significantly to release ***PA-RISC 2.0 (1996)*** with 64-bit extensions. In *1989,* ***Intel*** released its ***i860*** using *VLIW* architecture, and later ***Intel 80960*** microprocessor, a member of the ***i960*** family of pipelined 32-bit RISCs. In ***1993, Apple, IBM,*** and ***Motorola*** jointly developed single-chip microprocessor: the ***PowerPC;*** the first member was the *PowerPC 601* processor. This family, however, included the ***603, 604, 620,*** and the other models of the ***6XX*** line, which shared a common architecture. In *1992,* ***DEC*** introduced the ***Alpha*** architecture and later was acquired by ***Compaq,*** and Alpha processors have since then been labelled with the numbering sequence ***21X64,*** having X = 0, 1, and 2. ***Advanced micro device (AMD)*** has imbibed RISC culture and launched ***AMD 29000 (1990)*** using floating-point unit (FPU) outside the CPU chip (off-chip) like SPARC.

In the early **1980s, ARM Limited,** evolved out of *Acorn Computers, Inc.* has designed a family of microprocessors which were mainly low-powered and low-cost ones, and were mostly targeted for applications in the domain of embedded systems, such as mobile telephones, communication modems, portable digital fault-detection accessories, and for similar other pieces of equipment. The ARM designs are usually integrated with application-specific hardware on the same chip. That is why, the companies engaged in manufacturing embedded systems and customized application-specific products can incorporate the designs of ARM processor into their products under licence from ARM Limited. This company later collaborated with **DEC** (later, in the folds of the **Compaq Company**) to produce **Strong ARMCPU**. The **ARM SA-110 CPU**, however, is now manufactured by **Intel**.

9.5 RISC Characteristics

The ***major attributes*** of RISC that have been defined most recently are as follows:

- Simple instruction set consisting of relatively few *instruction types* and only some *addressing modes*;
- Fixed and easily decoded instruction formats;
- Fast single-cycle instruction execution;
- *Hardwired* rather than microprogrammed control;
- Limited memory access, mainly, confined to load and store instructions;
- Large number of registers, to minimize very often interactions with memory;
- Massive pipelining, a key to speed up RISC machines;
- Use of optimizing compilers to yield enhanced performance of the resulting object code.

In fact, several of these RISC attributes are closely interrelated. The major thrust to achieve desired level of performance is, however, entrusted on compilation process that requires the machine architects and compiler writers to cooperate closely in the design process.

The details of this discussion are given in the website: http://routledge.com/9780367255732.

9.6 The RISC Impacts and Drawbacks

RISC processor has several advantages due to some of its inherent characteristics that provide attractive outcomes than its CISC counterpart. Incidentally, some of these advantages eventually culminate to also create its several drawbacks as experienced by the designers as well as by the users. A few of all these are as follows:

- Simple instruction set with reduced complexity provides higher clock rate and a lower CPI, providing higher MIPS ratings;
- Use fewer transistors, even considering large register files makes the design easier and simpler with perhaps lesser bugs than usually found in microcode in CISC;
- Use of relatively few transistors occupies very little space than conventional CISC using silicon chips and is thus well suited to make use of very-high-speed chips based on GaAs instead of silicon;

- Availability of enough space within the die (chip) facilitates to include several extra functionalities, such as memory management unit (MMU) and floating-point arithmetic units, on the same chip;
- High-level language compiler produces more efficient codes (microinstructions) which are then directly executed by the underlying hardware, with no intervening interpreter (microprogram);
- Loading of instructions in a RISC processor is simple and decoding is simplified as opcodes and address fields are located in the same positions for all instructions;
- As RISC is simpler than corresponding CISC, a new RISC processor can be designed, developed, and tested more quickly that summarily reduces the time between design and shipment – a crucial issue in a rapidly moving and competitive industry.

9.6.1 Drawbacks

- The CISC architecture is so compatible that it is identical from micro (PC) to mainframe, thereby offering a complete range of the family of machines. The smaller models can even be microprogrammed to achieve their functionality in software (microcode), while the larger models can be hardwired for gaining speed. This marvellous concept is not found in RISC architecture;
- A RISC processor suffers from the absence of some useful sophisticated instructions that are mostly available in contemporary CISC;
- The presence of hardwired control appears to be a *two-edged sword*. On the one hand, it expedites the speed of the execution, and on the other hand, it is considered as another shortcoming of RISC since it is less flexible and more error-prone;
- Increase in RISC program length results in more instruction traffic and larger size of memory demand that might create reasons to drop RISC from favour;
- Use of a large register file is another concern causing problems in RISC. Although a larger register set can hold more intermediate results and reduce the *to-and-fro* data traffic between the faster CPU and slower memory, the register decoding system itself becomes more complicated.
- In fact, it is really difficult to determine the optimal sizes of register set, I-cache, and D-cache, which are directly related to the performance of the RISC.

9.7 RISC versus CISC Debate

The debate between RISC and CISC designers has lasted for more than a decade. The RISC camp began to claim that CISC had reached a dead end and should be abandoned. The designers and associates of CISC have taken a close and critical look at RISC technology, and found problems that the RISC proponents have tried to sweep under the rug. To enter the heart of the RISC–CISC controversy, we present here the arguments offered by both sides, to let the readers form their own views, draw their own conclusions, and make an informed judgement of their own.

9.7.1 Running Programs in High-Level Languages

- The benchmarking between RISC and CISC with regard to the performance of high-level language program execution reveals that *procedure calls* are better handled by RISC than by CISC, but *jumps* in a program (GO TO) are handled better by CISC;
- RISC outperforms CISC in case of *recursive programs,* but is rudimentary (primitive) in *I/O handling* and hence is far behind CISC;
- CISC outclasses RISC with regard to *floating-point* calculation and even in *integer multiplication* and *division* with all three varieties: single (32-bit), double (64-bit), and extended (128-bit) precision;
- Better *design of compiler* in RISC does a great job with better use of registers and factored out CISC in executing users' job;
- CISC machines while running on versatile operating system (like VMS in VAX) clearly elbowed out RISC. But how this superior performance of CISC could be then considered, being the credit of only of its underlying architecture.

Hence, the set of programs being run for benchmarking plays an important role in declaring which one is faster. Thus, an artificially designed synthetic benchmark program could be used for this purpose, which typically does *no* I/O at all, and really tests the compiler and CPU performance. Such a synthetic benchmark that measures *floating-point performance in* ***whetstones*** (a statistically *average floating-point* instructions) *per second,* and later in ***dhrystones*** (measured only *integer computations*), can give at least a rough measure to compare RISC and CISC machine performances, accepting the factors such as compiler quality, register allocation, cache performance, etc. which can affect the end result.

A pilot test with four RISC chips from four vendors when executed based on this idea has surprisingly outperformed the IBM's top-of-the-line 3090/200 mainframe (CISC). Hence, it becomes hard to deny that RISC machines are really very fast.

9.7.2 Technology of the Components

The *inherent technology* on which these two categories of machines have been built is also an important factor in this comparison. RISC used *ECL transistors* that switch much faster than *CMOS transistors* used in CISC. These two machines used different clock rates; different buses; different types of memory chips, including caches; different MMU; different pipelining techniques etc.

Technology-independent measure is also not a full proof as it first appears. Pre-fetching of instructions often adds extra bus load that may have an impact on the overall performance.

9.7.3 Role of Large Register File

The execution of simulated programs on different types of register model, such as

- Overlapping register windows (large VAX: CISC and SPARC: RISC),
- Non-overlapping register windows, and
- A single set of registers,

reveals that overlapping register windows offer much faster operation, but for recursive programs, they actually slow down. The conclusion is that overlapping register windows are a good idea, but this tells very little about RISC architecture versus CISC architecture.

The comparison of an overlapping register as in CISC (VAX) with an overlapping register used in RISC is not fair, because CISC would take up much more *chip area* than the RISC. Put in other words, if a RISC designer were given the same amount of chiparea that CISC machine enjoys, he/she could either include a much larger register file, an on-chip memory cache, or other features that could yield even better performance.

However, the boundary between RISC and CISC architectures has gradually faded away, because both are now implemented with the same hardware technology. For example, the ***Motorola 88100, Intel Pentium,*** and ***VAX 9000*** are built with a mix of some useful features taken from both the RISC and CISC camps.

It can be judiciously concluded that it is the applications that will ultimately decide and determine the best choice of a processor architecture. It can be convincingly predicted that the evolving architecture of tomorrow's processor might even be a hybrid, taking some of the positive features from both the tenets.

9.8 RISC Design Issues

The ***major issues*** that are encompassed with RISC design can be roughly summarized as follows:

- Analyze the applications to identify the key operations;
- To execute these key operations design an optimal data path;
- Using the devised optimal data path, design appropriate instructions;
- Add new instructions only if they do not slow down the machine;
- Repeat this process for other resources.

The optimal data path needed to execute the key operations involved mainly in commercial and scientific applications, including CAD and robotics, mostly includes the following:

a. The registers;

b. The ALU (arithmetic and logical unit);

c. The buses connecting them.

The time required to fetch the operands from the registers, run them through the ALU, and store the result back into a register is called the ***data path cycle time.*** This time should be made as short as possible. In fact, the rule says; *sacrifice everything to reduce the data path cycle time.* Whenever an attractive new feature is required to be added, if it increases this cycle time, it is probably not worth having.

Based on the data path as designed, the appropriate machine instructions are then to be developed. Only a few instructions and equally few addressing modes are typically considered. Additional instructions, if required, should only be added if they are found to be frequently used, and do not degrade the performance of the most important ones.

Finally, the same process should be repeated for other resources within the CPU, such as cache memory, MMU, FPU, co-processors, and similar other ones.

9.9 RISC Instruction Set

RISC processors introduced by different manufacturers essentially differ in their underlying instruction set. Each processor is distinguished from the others in respect of having some distinctive features in this set, exclusively of its own, that contribute to the beauty of the respective processor. However, a generic RISC processor is observed to include mostly the following types of instructions (a representative list, and not an exhaustive one) in its instruction set.

- **Register–register instruction**, in which all the source operands as well as the destination operands are located in specified CPU registers.
- **Register–immediate instruction**, in which one of the source operands is a signed *constant* for all types of data transfer operations as well as for all types of arithmetic operations, and the other source operand is in a specified CPU register. The destination operand, as usual, resides in the CPU register also.
- **Arithmetic and logical instructions** are found in many different forms as well as in many different versions in different RISC processors.
- **Floating-point instructions** are usually executed, but not necessarily, by a coprocessor chip. This instruction is executed mostly by a separate FPU equipped with many CPU registers (usually 32 registers, each of 32 bits). While FPU is in execution, CPU can continue its execution with other things in parallel.
- **Load and store instructions** can only access main memory; all other instructions must have their operands in CPU registers to avoid repeated visits to comparatively slower memory. Load/store instructions, however, may be of several types that appear in various forms in different RISC processors introduced by different manufacturers.
- **Jump instruction** has its operand with a constant value and the address of the target location.
- **Call instruction**, in essence, the procedure call, is probably the most time-consuming operation in a compiled high-level language program. Different RISC processors, however, have efficiently implemented these operations in many different ways within the confinement of their underlying architectures. Two significant aspects are attached with this instruction that are found to be different, and thus, differently implemented in different RISC processors, such as the *number of parameters and variables* (both local and global) that a procedure deals with, and the *depth of nesting*.
- **Branch instructions** are available in the instruction set of all RISC processors, but these instructions are usually found to differ in different RISC processors with respect to various features that they normally contain. However, they are always designed in such a way as to ultimately speed up the performance using the underlying pipeline organisation, and also take the assistance of the prescribed compiler that optimizes the generated object code at the time of program compilation.

- **Read and write instructions** are normally used to read and write the special registers.
- **Save and restore instructions** sometimes are also used as separate instructions in some RISC processors. They usually manipulate only the register window and stack pointer.

Apart from these common ones, there exist several other instructions and also some special instructions that are essentially different and are differently implemented in different RISC processors.

9.10 RISC Instruction Format

The basic instruction format used in a generic RISC machine as shown in Figure 9.1 is:

1. 7–bit opcode
2. Two 5–bit registers (DEST and SOURCE)
3. A mode bit (I)
   I = 0, not immediate
   I = 1, Immediate
4. A Condition code bit (C)
   C = 0 Don't set condition code
   C = 1 Set condition code.

For ordinary instruction, like ADD, the operands depend on the *I*-bit. If $I = 0$, this is called *register addressing*. One operand is taken from the source register, the second operand is taken from the register specified by the low-order 5bits of the OFFSET field, and the result is stored in the destination register. If $I = 1$, the second operand is a 13-bit constant, giving *immediate addressing*.

9.11 RISC Addressing Mode

The low-order 5 bits of the OFFSET field specify the register ($2^5 = 32$ registers), and this fact that Register 0 is hardwired to the constant 0.

- **Indexed addressing**: The OFFSET is added to the source register to form the effective memory address.
- **Register indirect**: If OFFSET is 0, then *indexed addressing* reduces to *register indirect* addressing.

Bits	7	1	5	5	1	13
	OPCODE	C	DEST	SOURCE	I	OFFSET

FIGURE 9.1
Generalized RISC instruction format.

- **Register direct**: If the register 0 is specified, i.e. low-order 5-bits is not 0, then 13-bit offset field gives ($2^{13} = 8K$) direct addressing of the bottom 8K of memory, which is useful for accessing global variables.
- **Other modes**: They can be constructed at runtime by building an address in a register and then using register indirect or indexed addressing.
- **PC-relative conditional JUMP**: This is realized by concatenating or adding the low-order 3 fields ($13 + 1 + 5 = 19$) to form a 19-bit signed offset; the DEST field then specifies the condition.

9.12 Register Windows: The Large Register File

RISC architecture, by virtue of its guiding philosophy, always provides a large number of physically small registers that form register files. One of the main objectives of using such a large set of registers is to hold the most frequently accessed operands in the registers, thereby aiming to reduce the required number of frequent visits to slower main memory as far as possible, for the sake of performance improvement.

Registers are mostly used to hold the operands of executing programs (procedures). When switching of procedure (process switch) occurs due to a procedure call, all the registers used by the *calling procedure* need to be saved in order to free them, to be reused for other purposes. On return from the *called procedure*, all the registers which were saved once again need to be restored (loading back to the corresponding registers) to allow the calling procedure to again continue from the point where it left. Saving and restoring of information stored in the registers is somehow a problematic and time-consuming one that often limits the overall performance to attain the desired level. To get rid of this additional overhead, large number of registers available in the processor are logically divided into multiple small sets of registers, and each such set is called ***register window*** that can be assigned to each individual procedure. At any given time, an application program sees only a particular register window (a set of specific registers) allocated to them. Incidentally, a typical calling procedure usually employs a few parameters to be passed that are adequately supported by this small number of fixed registers present in a window. The net effect is that in the event of a procedure call, the processor is then simply switched from one register window to a different appropriate one, rather than saving/restoring all registers individually in memory, thereby avoiding the critical huge overhead associated with this activity.

This concept is explained in Figure 9.2. Each register window here is formed with a set of a certain number of registers and is divided into three predefined fixed-size sections. Consider the procedure at level L. Its in-registers holds parameters which are passed down from the procedure (level $L - 1$) that called the *current procedure*, and holds the results that are to be passed back up from the current one to its caller (level $L - 1$). Local registers are used by local variables of the respective procedure. Out-registers are used to exchange parameters and results with the next lower level (level $L + 1$). It is to be noted that the out-registers at one level are physically the same as the in-registers at the next lower level. This notion of using overlapped register windows is perhaps one of the most sparkling features introduced first by Berkeley RISC architecture that permits parameters to be simply passed without the actual movement of data. It is once again reiterated that except for the overlap, the registers used at two different consecutive levels are always physically distinct.

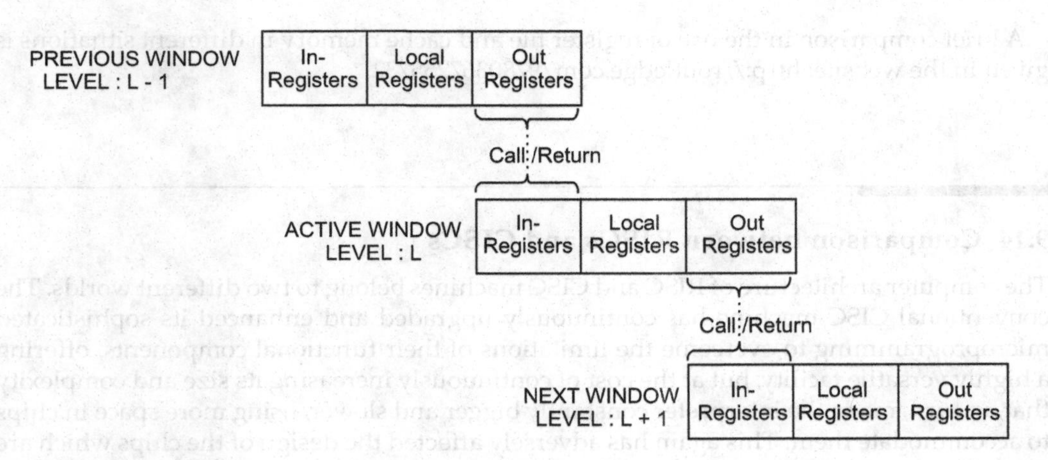

FIGURE 9.2
Three overlapping register windows.

Since the number of nested procedure (depth of levels) activations hardly changes beyond a small range, and is observed to mostly remain bounded over a considerable duration of time, the number of register windows needed for this purpose is also thus kept limited, in the processor design. These register windows can then be employed to hold only a limited number of most recent procedure activations. Other prevailing age-old activations can be saved in memory and later could be restored at an appropriate time. In actual implementation, the register windows as well as the ways they are operated as described, however, can be realized by arranging the register windows in the form of a circular organisation (depicted in SPARC processors, Figure 9.5 available on the website). Two notable examples can be cited where this approach has been successfully implemented in practice, namely ***Sun SPARC architecture*** and the IA-64 architecture used in ***Intel's Itanium processor***.

9.13 Register File and Cache Memory

While the register files are used as temporary memory storage to hold the most frequently used operands, seemingly appearing almost cache-like, they provide relatively much *faster operations* when operands are needed. One of the main reasons is that registers are *tightly* connected with the ALU, whereas the on-chip caches are comparatively *loosely* coupled. Although optimization in cache organisation and availability of today's multilevel on-chip caches provide enough support to the CPU to yield rather fast operation, the ultimate objectives in the use of these two important hardware resources are somewhat different, which again mostly depends on the situations where they can be exploited most profitably. Usually, these two vital resources address different domains, conducive individually to them, based on their inherent characteristics. That is why, a proper choice between a large window-based register file and a cache is not very straight off. However, a moderate comparison in respect of their effective usage can be summarized based on their capabilities in the context of different situations that are often encountered during the execution of applications, in general.

A brief comparison in the use of register file and cache memory in different situations is given in the website: http://routledge.com/9780367255732.

9.14 Comparison between RISCs and CISCs

The computer architecture of RISC and CISC machines belong to two different worlds. The conventional CISC machine has continuously upgraded and enhanced its sophisticated microprogramming to overcome the limitations of their functional components, offering a highly versatile facility, but at the cost of continuously increasing its size and complexity that, in turn, made the interpreter constantly bigger and slower, using more space in chips to accommodate them. This again has adversely affected the design of the chips which are basically made of silicon transistors having relatively slower switching times. In fact, CISC complexity was exposed and came to the notice of the public when a design flaw affecting the floating-point division instruction of the Pentium was discovered in 1994. The cost to Intel for this bug, including the replacement cost of Pentium chips already installed in PCs, was about $ 475 million. RISC machines, on the other hand, started their design approach from a different perspective. This machine is essentially a computer with a *small number* of *vertical microinstructions* which can be directly executed by the hardware with no intervening interpreter. The optimization in compiler technology used in RISC machines, however, generates microcode directly at an acceptable level taking the necessary help from its hardware designers. In essence, the fundamental differences that exist between these two classes of machines are summarized in Table 9.1.

The major architectural distinction between a typical RISC processor and a typical CISC processor is given in the website: http://routledge.com/9780367255732.

TABLE 9.1

Salient Characteristics of CISC and RISC Architectures

Architectural Characteristics	Complex Instruction Set Computers (CISC)	Reduced Instruction Set Computers (RISC)
Instruction set size	Large set of instructions.	Small set of instructions, mostly register-based.
Instruction formats	Instructions with variable formats, 16–64 bits per instruction.	Instructions with fixed format.
Addressing modes	Normally 12 – 24.	Limited to 3 – 5.
GPRS	Mostly ranges from 8 to 24 GPRs.	Large number of GPRs ranging from 32 to 192.
Cache design and usage	Use of unified cache. Recent trend to use split caches for instructions and data.	Mostly use of split caches. Separate caches for instructions and data.
Processor speed (clock rate and CPI)	33–66 MHz initially and may be a few GHz or more in recent release, and CPI mostly between 2 and 15.	33–66 MHz initially and may be a few GHz or more in recent release and CPI mostly between 1.5 and 2.
Memory references	Many instructions may visit slower memory multiple times per instruction.	Memory visits require only in *load/store* instructions.
Software attributes	Complexity lies in the *microprogram development*.	Complexity lies in optimization of the compiler.
CPU control design	Most are microcoded, but recent CISC also uses hardwired control.	Almost all hardwired control without using control memory

9.15 RISC Pipelining

The instruction pipelining principles (see pipeline architecture, Chapter 8) are found quite conducive to the inherent characteristics of RISC architecture, and hence, massively used since the first part of the 1980s in RISC machines, including RISC scalar and superscalar processors with adequate hardware support, as one of the major contributors to achieve RISC's high performance. The pipeline design thus implemented in the instruction cycle of RISC processor may be composed of some basic stages, where each stage, in turn, may be decomposed again into a number of substages in actual implementation for its smooth operation to yield an even better outcome. As the RISC architecture always has a smaller, simpler, and a regular instruction set, the design of phasing an instruction cycle into four or five stages, and even more nowadays to execute an instruction, is found quite appropriate that eventually offers an average potential speed-up of a factor of almost 4 or 5, and sometimes even more. These basic stages, thus, are as follows:

- Instruction fetch unit;
- Instruction decode unit;
- Operands address calculation and operand fetching;
- Execution of the instruction;
- Write back of the result at the destined location.

The responsibilities carried out individually by each such stage are described in the website (http://routledge.com/9780367255732) and more elaborately explained in Chapter 8.

9.16 RISC and CISC Union: Hybrid Architecture

Till the mid-1990s, processor architects were divided into two opposite tents. While CISC designs were mostly preferred mainly due to immense success of microprograming and a wide availability of existing versatile software, the RISC design was favoured chiefly with its simplicity and efficiency. However, the demarcation between them is gradually becoming faded away with passing days, and a trend eventually started to include one's useful features into the architecture of other. Many modern CISC processors are observed to include more number of GPRs (essentially a RISC ideal) in order to improve their efficiencies. Intel, from its Pentium III, included an additional set of eight 128-bit vector registers to implement SIMD-based multimedia (SSE technology, see chapter 3) applications. AMD's new X-86 chips also added further eight GPRs and eight extra SSE register for multimedia. This trend is now found to be continued, and in fact, the successor of Intel Pentium and Itanium IA-64 series will move further by adding 128 GPRs.

Moreover, the Pentium and Athlon family of processors now exploit a CISC–RISC hybrid architecture that uses a type of decoder to convert the CISC instructions into corresponding simpler RISC instructions before execution. These are then executed very fast by an embedded massively pipelined RISC core, equipped with many performance-enhancing hardware and software facilities. Traditionally, these have been possible only in a true RISC design. These hybrid processors are fully compatible with the software developed for their CISC predecessors, yet they can equally compete against processors based on true RISC designs. This concept expects to be enhanced more in the days to come, although these processors are not at all suited for popular mobile and embedded applications. On the other hand, contemporary RISC processors also indulged themselves to move more towards CISC-like designs by including more instructions and added functions (against RISC culture) than old CISC designs. For example, Motorola G4 processor used in power Macs and eMacs adds 162 new instructions to the existing RISC architecture for its Alti Vac unit to more efficiently handle *multimedia* and *digital signal processing* applications.

9.17 Types of RISC Processors

RISC machines available from different manufacturers exploit many design choices, of course, within the constraints of RISC; each one is different with many lucrative features absolutely of their own; each is equally competent, but experts still can only arrive at dissimilar conclusions. Some of these machines also made a compromise and included several useful features of CISC in their design to make them more versatile. In fact, the field of RISC design is full of pitfalls and choices, and no doubt, no common conclusions can yet be reached. The subject is still open and invites much activity for the exploration in the years to come. We cover here a brief architectural detail of only a few of them worthy of mention, for a clear understanding about the strength and capability of RISC architectures.

9.17.1 PowerPC Processors

In the early 1990s (1993), Apple, IBM, and Motorola jointly developed a family of single-chip microprocessors, the PowerPC, following the line of existing IBM RISC System RS/6000 series of computers. This family includes the members 601, 603, 604,620, and also other models that mostly exhibit certain common features, but few models have even something more. Interestingly, although PowerPC follows the typical RISC designs having a fixed instruction length of 32-bit word, it includes a variety of formats and addressing modes, and has a substantially large number of instructions – more than 200 distinct types that goes against the RISC culture and philosophy.

A brief detail of all PowerPCs with associated figures is given in the website: http://routledge.com/9780367255732.

9.17.2 SPARC Family of Processors

The *SUN Microsystems Corporation* in 1987 has introduced its first open microprocessor architecture, and not an implementation of a chip, popularly known as ***SPARC*** (*S*calable *P*rocessor ***AR****C*hitecture). Different technologies (CMOS, ECL, GaAs, gate array, VLSI, etc.) and different specifications were used by different licensed manufacturers, such as ***Fujitsu, Cypress, LSI Logic, Inc.*** and ***Texas Instruments***, to fabricate the chip, but the basic instruction set architecture has been remained the same.

SPARC is basically a 32-bit design consisting of integer unit (IU), FPU, an optional user-supplied co-processor, a MMU, various sizes of off-chip cache, and different types of memory organisation. Use of an ***overlapping register window*** scheme using 32 visible and accessible registers, called *R0 to R31* is one of its marvellous features. It works on boundary-aligned 32-bit words and supports a paged linear address space of 2^{32} individually addressable 8-bit bytes. Memory is a *big-endian*, like *Motorola 68000* family. Although SPARC design is truly a uniprocessor architecture, it has kept the provision to connect multiple SPARC chips to build a symmetric multiprocessor (SMP) system with common shared main memory. Special instructions have been thus included to handle multiprocessor synchronization and other related similar issues.

9.17.2.1 UltraSPARC Processors

The UltraSPARC is a ***super pipelined superscalar processor*** enriched with some salient features, not available with ordinary SPARC. Superscalar processor consists of two independent pipelines: each pipeline consists of *nine stages*, and each stage is completed strictly in one processor clock cycle. Each pipeline consists of *two execution units*: one for integer operations with its own register set and one for floating-point operations having its own different register set, and they can be operated simultaneously in parallel through its own pipeline. As a result, a total of four new instructions can enter the execution phase every clock cycle. The processor uses two levels (multilevel) of cache: an *external cache* (E-cache) and two internal caches, one for instruction (I-cache) and one for data (D-cache). The MMU has *two translation lookaside buffers* (page table storage): one for instructions (iTLB) and one for data (dTLB). The UltraSPARC series of processors handles both addresses and data as 64-bit values, but maintains downward compatibility to accommodate everything of earlier 32-bit versions. A recent release, the **UltraSPARC III** fabricated with 0.18-μm technology having a clock speed in the range of 750–900 MHz, is enriched with many useful advanced features. It is targeted to attain around 1.5 GHz in the forthcoming days.

A brief detail of all SPARC and UltraSPARC processors with related figures is given in the website: http://routledge.com/9780367255732.

9.17.3 MIPS Processors

MIPS, the Stanford chip, slightly modified from Berkeley RISC chip, was initiated by Hennessy who subsequently formed MIPS Computer Systems (later a division of Silicon Graphics) in 1984 to introduce **MIPS I** as the first member of the MIPS RX000 series of microprocessors. Later, the 32-bit MIPS R2000 in 1985, followed by architecturally identical R3000 in 1988 differing only in speed and price, and later 64-bit R4000 in 1991, was launched. Later members of the same series, like R10000 announced in 1994, add numerous architectural extensions and far more complex instruction pipelines. Subsequently in 1999, 32-bit MIPS 2 and later 64-bit MIPS 64 were released.

The major components of this single-IC microprocessor family include a register file of 32 GPRs, each having 32-bits, and the processing logic to perform the basic fixed-point arithmetic/logic functions using 32-bit operands. Floating-point operations are executed by an on-chip or off-chip FPU supported also by an optional floating-point co-processor obeying IEEE 754 standards. MIPS never impose any *earmarking* on any of the registers in the chip for any special purposes. As a result, the local/global variables and the input/output parameters can all be put in any of the registers for even faster execution. A unit called *system control co-processor* provides communications with external memory (both cache and main memory), and also an automatic address translation logic supported by special-purpose arithmetic circuits to perform address computations required to handle virtual memory system. MIPS uses a *deeper instruction pipeline* of *five stages*, compared to *four stages* used in its contemporary counterpart ordinary SPARC. Each instruction in the instruction set is of 32 bits and word-aligned. The address space is 2^{32} bytes (4 gigabytes) and byte addressable, while the upper 2-gigabyte address space is reserved for the operating system. The memory can be configured either as *big-endian* or as *little-endian* by selecting a pin on the chip, thereby satisfying the users of both tents. The machine supports paged virtual memory management. In general, MIPS architecture has been built up using different trade-offs where more thrusts have been put on software that ultimately resulted in creating problems for the software designers, particularly in the optimization of compiler design. Still, it has been done only for the sake of making the hardware simpler with faster operation to realize enhanced performance.

A brief detail of MIPS processors with related figures is given in the website: http://routledge.com/9780367255732.

9.17.4 PA-RISC Processors

The family of RISC processors developed by HP, commonly known as *Precision Architecture RISC processors* (*PA-RISC*), was first launched in 1986 using the original design principles with almost no change of its existing Precision Architecture (CISC). The initial PA-RISC perhaps has the most *unusual features*, and it was then slightly expanded in its version 1.1 with improved capabilities launched in 1990, and subsequent enhancement with significant upgrades on it was the PA-RISC version 2.0, shipped in 1996. It was this version on which the successive releases of the family of processors PA-RISC 8000, 8200, and 8500 were based on. In fact, the HP-PA 8000 implementation was somewhat more ambitious in organisation than other contemporary multiple-issue processors, such as Alpha 21164 and Ultra SPARC.

The PA-RISC 8500 processor, a member of the 8000 family, fabricated with 0.25-μm technology is a ***four-way superscalar implementation*** equipped with on-chip split cache into one quarter of chip die. This processor is fully scalable and supports *multiprocessor configuration*, thereby allowing several CPUs to be used in the system. An advanced processor enhancing 8500 model, such as PA-RISC 8700 having a clock rate of 750MHz, and soon after, 8700+ with a higher clock rate of 875MHz, and later PA-RISC 8800, has been launched raising the clock rate to 1GHz with ***two CPUs*** on a single chip that essentially gave rise to a concept of **multicore** which Intel started to introduce only recently in its latest Pentium model. The bandwidth also has been considerably increased from 1.6 GB/s to 6.4 GB/s per chip. Many other attractive features have been incorporated in the design process that presently lies outside the scope of this discussion. The beauty of PA-RISC is that it is realized in a single simple hardware structure with almost no complexity in the underlying design, and sufficient provisions have been kept to make use of more unused available chip space for incorporating many additional attractive features in the years to come.

A brief detail about the generic architectural design of PA-RISC processor with figures is given in the website: http://routledge.com/9780367255732.

9.17.5 ARM (Advanced RISC Machine) Processors

Today's ARM technology has its origin in the Acorn RISC Machines (ARM), a microprocessor developed by Acorn Computers Company in the United Kingdom sometimes in early 1980s. Later, the family name has been changed to *Advanced RISC Machines*, a new company known as *ARM Ltd.* promoted by Acorn, VLSI, and Apple Computer as founding partners, of course, without changing its original acronym, ARM. This processor series started with ARM 1 and has till now reached ARM 11.The primary target of ARM family was aimed for use in low-cost, small-size, high-performance, and low-power systems and applications such as portable computers and games. Now, it is mainly intended for mobile telephones, data communication (protocol converter), portable instruments and computers, smart cards, and embedded system applications; the target device is battery-powered requiring voltages in the range of only 1–3 V. Apple Corporation in 1993 employs the ARM6 microprocessor for their small personal computer system, and nowadays, ARM chips are the processors in Apple's popular **iPod** and **iPhone** devices. Due to the inherent simplicity of the design and low gate count (only a fraction of total transistor count that Pentium chips generally use), this processor is now the industry leader in low-power processing on a watt per MIP basis.

The ARM designs are sometimes integrated with application-specific hardware (ASIC) on the same chip called the ***cores***. Two different forms are available in ARM design: ***hard macrocell*** provides only a detailed physical layout of a particular chip fabrication process, and ***synthesizable*** form indicates the high-level language software module that can be synthesized in the realization of the target technology. The ARM7TDMI processor is a hard macrocell, and its counterpart, the ARM7TDMI-S, is a synthesizable processor.

ARM is a *32-bit processor* with both its data words and address words of 32 bits long. This processor has a total of 37 registers, out of which 15 registers labelled R0 through R14 are *GPRs* and R15 is the *program counter*. Of these registers, R0 through R7 and R15 are visible, and are shared by all modes. The other *15 GPRs* called *banked registers* are used at the time of process switching or context switching. The remaining six are the *program status register*. The contents of these registers are, however, available to the user, even in the system modes. The *current program status register* (CPSR) holds the *condition code flags*. Apart from user mode, ARM provides several other *privileged modes*, such as *supervisor mode, abort mode, undefined mode, fast interrupt mode*, and *interrupt mode*, for various exception handling. The ALU employs a combinational logic for *addition* and *subtraction*, and a combinational shift circuit using sequential shift-and-add method for *multiplication* and other operations. A separate *address–incrementer circuit* realizes address–manipulation operations such as PC: = PC + 1, independently of the ALU. A *barrel shifter* is provided for fast shift operation with a large number of shifts using a single instruction.

The architectures of ARM processor are classified as either ***processor cores*** or ***CPU cores***. A *processor core* contains only a processor (both PCU and DPU) and its associated addresses, and the required data bus connections. ARM7TDMI, ARM9TDMI, and ARM10TDMI processors belong to this class. A *CPU core* contains the processor along with cache (on-chip) and memory management components. ARM720T, ARM920T, ARM1020E, and **StrongARM** (developed in collaboration with DEC) processors fall into this category. Processors belong to these two different classes having different architectural designs, different clock rates, different bus widths, and different sizes of both on-chip and off-chip caches, and above all, they implement various number of stages in instruction pipelines ranging from *three* to *six* stages. The cache organisation is constantly upgraded with continuous evolution of ARM family that reflects the relentless pursuit of further improvement in performance. While ARM 7 models use a unified L1 cache, all other subsequent models, however, use a split cache (I-cache and D-cache) with set-associative cache design, but varying in the degree of associativity and also in the block size. Most of its members use a logical cache, while a few others use a physical cache. The processor supports both **little-endian** and **big-endian** memory organisation, and is *byte addressable,* using 32-bit addresses for both data words and address words. I/O devices are *memory-mapped,* sharing the same address space with memory. As a result, all the instructions used for CPU–memory transfers are also used for I/O operations. Several internal buses are provided through which data are efficiently transferred between the data processing circuits and the DPU's registers.

A brief detail about the generic architectural design of ARM processor with figures is given in the website: http://routledge.com/9780367255732.

9.17.6 Motorola Processors (MC 88000)

In **1988, Motorola** launched RISC **M88000** followed by the **M88100** sometimes in *1990,* built on 1-μm HCMOS technology using 1.2 million of transistors with an operative clock rate of 20 MHz (actually 50 MHz) having mixed features taken from both the RISC and CISC tents. It combined the three-chip set, *one CPU* (88100) chip and ***two cache*** (88200) chips, in a single-chip implementation. The 88100 is a symmetrical *two instruction-issue* **superscalar processor** with a separate 32-bit IU and a FPU having their own sets of registers, and the operations in these units can even be carried out in parallel, if there is no dependency in the logical flow of the operation. In addition, there are two other separate units, namely the ***instruction unit*** and ***data memory unit*** which *loads* and *stores* operands between processor and external memory. All these functional units share the same register file with no conflict by using an innovative mechanism known as register ***scoreboarding*** (see Chapter 8). Memory organisation is **big-endian** and the memory systems support *segmented paged virtual memory* management. Two cache MMUs (CMMUs), one for data and one for instruction, are implemented as off-chip caches (M88200), and MESI protocol is used here to resolve cache coherence problem in its multiprocessor organisation. The processor provides demanding ***3-D colour graphics*** image rendering and **digital signal processing** applications.

Motorola MPC7450: The PowerPC 6XX line culture of Motorola in collaboration with Apple and IBM has been continued by Motorola leading to an implementation of the new MPC7XX and 7XXX lines of processors with a new prefix MPC (M stands for Motorola and PC stands for PowerPC) on the labelling. The notable one in the line of MPC7XXX is MPC7450, released in early 2001 with clock rates up to 733MHz is a *four instruction-issue* **superscalar processor** with a relatively deeper pipeline up to *seven stages*. This processor contains *eleven functional units* such as *four* IUs, *one* FPU, *four* special type of units (named as *Alti Vec*) that can perform parallel arithmetic operations on a particular type of vector data operands (packed form), *one* load/store unit, and *one* branch unit. The cache design here is really peculiar with both multilevel on-chip and off-chip caches. The Level 1 (L1) on-chip *split cache* consists of separate 32 KB for instruction and data and uses an eight-way set-associative technique. The Level 2 (L2) cache is also on-chip, but a *unified* one having a size 256 KB and is also eight-way set-associative. The off-chip Level 3 (L3) cache can be of 1MB or 2MB and is accessed over a 64-bit bus. Since both L1 and L2 are on-chip, L1–L2 handshaking performs at the rate of processor speed using a data path of 256-bit width.

A brief detail of the Motorola processor with figure is given in the website: http://routledge.com/9780367255732.

9.18 Comparison of Four Representative RISC Machines

We have discussed the architecture, organisation, and related hardware support available in a few popular RISC machines that stand at the forefront of the RISC culture. The addressing modes, instruction formats, a few salient features, and some key instructions available to each individual one have been briefly explained. We will now concentrate on enumerating the differences in scalar versions of four representative RISC machines, namely PowerPC, MIPS, SPARC, and PA-RISC. Fortunately, there exist so many similarities that make it possible to roughly cover the differences in these four architectures within these few pages. We are deliberately keeping the ARM processor outside this comparison, because this processor has many different forms and architectures that are mostly dissimilar with those under discussion. However, we are trying to exhibit a presentable comparison between these four representative systems on their salient features in the areas of both hardware and software. Superscalar versions of these machines have a diverse spectrum of features and hence have been kept outside the purview (domain) of this comparison.

A brief detail of the said comparison presented in a tabular form is given in the website: http://routledge.com/9780367255732.

9.19 Summary

The emergence of RISC machine is considered as a renaissance in the area of computer architecture although many truly remarkable innovations have already been made in this area by this time. The RISC concept actually starts afresh following a completely different philosophy from the existing CISC culture, and attempts to remove most of the drawbacks and the retarding forces that a CISC processor often experiences during its operation. The fundamental design issues of a generic RISC processor with its typical characteristics and salient features along with the instruction set, including the distinctive addressing modes and their peculiarities, have been discussed. The presence of registers in abundance and their specific organisation, along with the availability of multiple-level on-chip caches supported by powerful matching pipeline techniques, are considered as one of the primary reasons behind its enormous strength and potential capabilities, without using much faster components as such. In fact, contribution of optimizing compiler and other innovative software is considered as one of the secrets of RISC's ultimate success, although fast well-organised strongly regimented hardware is present. The comparison between a RISC processor and a contemporary compatible CISC processor has been underlined in terms of basic architectural components they employ, and the overall performance they offer. A few leading representative RISC processors introduced by different vendors that followed different implementation approaches have been illustrated along with their respective distinct characteristics in their specific architectural designs, and the instruction set they provide. Lastly, a comparison between four representative RISC processors introduced by different vendors which are considered to be at the forefront of this class, excepting ARM processors for some reasons, is tabulated in terms of some fundamental architectural parameters that often influence the performance of a processor of this kind.

Exercises

9.1 Discuss the various factors that led to the development of RISC processor from the conventional mostly used CISC processor. "Too many addressing modes available in the instruction set of a computer is not a bliss but a curse": give your comments.

9.2 What are the major characteristics of CISC architecture? Why would you consider some of these characteristics as the drawbacks of CISC?

9.3 Compare and contrast the RISC and CISC processors in terms of their major characteristics.

9.4 Compare and contrast the RISC and CISC processors from the architectural and technological points of view.

9.5 Compare the instruction set architecture in RISC and CISC processors in terms of instruction formats, addressing modes, and CPI.

9.6 Explain with a diagram the generalized instruction format of a RISC processor. State and describe the most frequently used addressing modes which are quite common to almost all such processors.

9.7 Distinguish between scalar RISC and superscalar RISC in terms of instruction issue, pipeline architecture, and processor performance.

9.8 Why 32 GPRs do mostly used in RISC IUs? Explain the relationship between the IU and the FPU in most RISC processors with scalar and superscalar organisations.

9.9 What do you mean by overlapped windows? Explain how to use the overlapped windows for parameter passing between the calling procedure and the called procedure? Write down the advantages that can be derived from the use of register windows.

9.10 Suppose a RISC machine has a large on-chip cache, with a hit ratio of over 90%. Further assume that LOAD instructions can complete in 1 cycle on a cache hit and 2 on a cache miss. Does it matter if this machine has hardware interlocking? Explain.

9.11 What are the salient architectural features of PowerPC, which are common to almost all its different models? "PowerPC instruction set is a clear departure from the conservative RISC culture". Discuss the instruction set of PowerPC in the light of the above statement.

9.12 Explain the concept of register windows implemented in the SPARC architecture. What are the design trade-offs between a large register file and a large D-cache?

9.13 Write down the features that sense you to consider SPARC as a scalable architecture.

9.14 Why are reservation stations or reorder buffers needed in a superscalar processor?

9.15 The MIPS processors have no condition codes. Does it mean that while generating code at the time of compilation for

if x < y then ....

takes more instruction on MIPS than on the SPARC? Explain.

9.16 "SPARC is not a chip but an architecture". What are the different technologies used to realize this architecture? Give an overview of the SPARC instruction set with reference to its strength and drawbacks.

9.17 The most common instruction in any assembly language program is moving a constant to a register. Can the SPARC do this in one instruction? If so, how? If not, how it achieves and how many does it take?

9.18 "UltraSPARC processors are enriched with some salient features". Briefly explain those features.

9.19 What are the distinct instruction features found in PA-RISC that are not common to other contemporary RISC processors? Explain.

9.20 "Motorola RISC processors possess some nice innovative features in their architectural design which are not very common to its other contemporary competitors". State and briefly explain those features.

9.21 ARM processors nowadays are in extensive use in some specific areas. Briefly state those areas and explain the reasons in short for such high acceptance in those areas.

Suggested References and Websites

Communications of the ACM, vol.37, no. 6, eight articles on the PowerPC, June 1994.

Furber, S. *ARM Systems–on–Chip Architecture.* Harlow, England: Addison-Wesley, 2000.

Patterson, D. A. "Reduced instruction set computers." *Communications of the ACM,* vol. 28, pp. 8–21, January 1985.

ARM website: arm.com

IBM website: ibm.com

10

Parallel Architectures

LEARNING OBJECTIVES

- To discuss, in detail, Flynn's classification of computer architectures: SISD, SIMD, MISD and MIMD.
- To introduce parallel computers and their environment.
- To describe interconnection networks and their different types.
- To learn about classification of MIMD architecture
  - tightly-coupled multiprocessor (Shared-memory, Symmetric multiprocessor (SMP),
  - UMA model as parallel processing systems
  - loosely-coupled multiprocessor (distributed-shared memory, NUMA model) and
  - its commercially popular variants (CC-NUMA model)
- To know about the features of multicomputers, a network of computers; its different architectural models including computer networks and distributed systems.
- To explain cluster architecture of computers, its classification, and different methods of clustering, including the role of the operating system that drives it.
- To describe blade servers in brief, and its enormous impact in large-scale distributed systems built with cluster architecture.
- To introduce SIMD computer organisations including SIMD vector processors and SIMD array processors.
- To gain knowledge about massively parallel processing (MPP) systems, and also scalable parallel computer architectures.
- To introduce the different types of general-purpose supercomputers and the software environment they require.

10.1 Introduction

The performance of a traditional computer, as proposed by Von Neumann, with a single CPU that issues sequential requests to memory, one at a time, executing scalar data has steadily progressed in steps with continuous implementation of faster hardware technologies and techniques, innovative designs in processors, and through numerous novel ideas used in system software design and its optimization. By this time, , due to the availability of more modern sophisticated hardware resources, computer architecture has progressed through mostly many evolutionary rather than revolutionary changes over the last few decades. The notable ones are ***pipeline architecture*** (instruction—level parallelism), and then the ***superscalar architecture*** (machine parallelism) using several pipelines with multiple functional units within one chip. Continuous advancements of these two approaches and their exhaustive use in gradual steps have been rigorously implemented up to their extreme limits in the on-going evolutionary process of computer architecture. The ever-increasing speed in processor and memory technologies coupled with increased capability and progressive reduction in size was then gradually approaching their physical limits. Revolution in VLSI technology, however, continued, which ultimately further made it feasible to develop multiple processor cores within a one-chip powerful microprocessor, historically known as ***multicore***, as well as larger-capacity RAM at a reasonable cost. In addition, many other significant improvements have been also made in the areas of other resources involved in computer architecture. As a result, many different forms of advanced computer architecture, leaving the line of traditional approach have been experimented with; some of them have already been implemented, while others are still in the process of further befitting developments that summarily make abrupt changes in the conventional concept of computer architecture.

In spite of these achievements, computing demand was observed to be still remaining ahead of the available computer facilities. Moreover, some real-life applications gradually evolve by this time, which require specific system supports that were absolutely beyond the existing capabilities of the fastest contemporary machines. Real-life processes are those, such as, in disciplines including aerodynamics (Aircraft control, Ballistic missile control), seismology (Seismic data processing), meteorology (Weather forecasting), fluid-flow analysis, computer-aided design (CAD), nuclear / atomic physics problems, and many more of similar types in different disciplines. With relentless progress in more advanced VLSI technologies, the cost of the computer hardware has sharply come down. Consequently, all-out attempts were then taken by computer designers aiming to ultimately assume that a computer should consists of *some number of* control units, ALUs, and memory modules that can operate *in parallel* without substantially increasing the cost, and the results thus obtained have been eventually found to be spectacular. This idea, when implemented, resolved many of the pending issues by exploiting a new concept known as ***processor-level parallelism***. The use of multiple CPUs to achieve very high throughput, and fault tolerance / or reliability, opens up a new horizon, popularly known as ***parallel organisation***. Moreover, while the failure of the CPU is almost always fatal to a single-CPU computer, a parallel system can be so designed that it can continue functioning, even in the event of failure of one of its CPUs, perhaps with a reduced performance level.

Numerous questions now start to arise as to (*parameters*):

i. How many complete CPUs or processing elements would be present? What will be their characteristics and size?
ii. Whether the number of control units would be one or more than one?
iii. How the memory module would be constructed; what will their characteristics and size be? Above all, how they will be organised?
iv. Last but not the least, how all these resources would be interconnected?

Some designs envisaged an idea to build a machine with a relatively small number of powerful CPUs, to be interconnected with an efficiently highbandwidth for communications. Others used a large number of ALUs that were strongly connected and worked in unison, monitored by a single control unit. Another possibility that has been explored was to build a machine using many small stand-alone systems / or workstations, connected by a local area communication network. Numerous intermediate architectures have been also experimented with. Above all, each design was found to have its own strengths and drawbacks if the implementation complexity and the cost factor were taken into account, apart from the considerations of the environment with regard to their usage. Last but not the least, is that when a system is built with many processors for various computations to proceed in parallel, it is tedious to break an application down into small tasks, and then assign each one to individual processors for in-parallel execution. Scheduling of tasks in multiple processors, control of the functional line, as well as coordination over the entire execution demand the services of additional sophisticated hardware and related software techniques.

10.2 Classification of Computer Architectures: Flynn's Proposal

Computer architecture, however, has progressed through many different forms by this time, and finally has been able to include multiple CPUs in its architectural design, which consists of a set of $n > 1$ processors (CPUs) $P_1, P_2, P_3, \ldots P_n$, and $m > 0$ shared / distributed, or shared as well as distributed main memory units $M_1, M_2, \ldots, M_m$ that are interconnected using various forms of design approaches. In addition, different types of emerging I/O processors associated with numerous categories of I/O devices are also involved in the core architecture. The net result was that a new form of computer architecture known as ***parallel computers*** eventually evolved. Using this architecture, a different type of more sophisticated application processing known as ***parallel processing*** came into use, which is usually defined in terms of both multiple processors (processing elements) and multiple processes of the instructions.

Flynn's Classification: The world of computer architecture was then found flooded with many different proposals having numerous forms of architectural designs with multiple CPUs (or processing elements) and multiple memory modules. It started from relatively very simple design approaches to the most sophisticated and versatile ones that indulged in comparatively complex design methodologies involving various amounts of basic hardware processing resources (like CPU, main memory and I/O). But whatever be the form of the architecture, the fundamental approach in any of its form is that a processor, after fetching instructions and operands from memory M (main memory or cache), executes the instructions on the related operands (data), and finally place the results in M. The instructions being executed form an *instruction stream* that flows from M to the processor. The data which is involved in the execution forms another stream, called *data stream* that flows to and from the processor, before and after the execution. Numerous schemes have been, however, tried to classify various forms of existing computer architectures, and the only scheme that is well accepted is that of Michael J. Flynn who proposed a broad classification (rather a crude approximation) based on the number of simultaneous instructions and data streams handled by the processor during the execution of the program.

Considering all the constraints and the bottlenecks that the system architecture usually encounters while executing parallel processing, let I_m and D_m be the *minimum* number of active instruction and data streams that are being operated by the processor at its *maximum* capacity, exercising its full degree of parallelism. These two streams I_m and D_m, to some extent, are independent, and thus *four combinations* of these two can exist, which classify the world of computers into *four broad categories* as proposed by Flynn based on the values of I_m and D_m associated with their CPUs.

i. Single Instruction stream Single Data stream (SISD): $I_m = D_m = 1$, The classical, sequential Von Neumann computer with a *single processor* capable of executing only scalar arithmetic operation using a single Instruction stream and one Data stream at a time, fall into this category. All types of machines – whether pipelined, superscalar, or any combination of these two architectures – however, also belong to this category.

ii. Single Instruction stream Multiple Data stream (SIMD): $I_m = 1, D_m > 1$, Here, a single instructionis executed at a time, and hence, requires a single control unit. Multiple data sets are concurrently executed by this single instruction, and hence, multiple ALUs are needed. This category is again divided into *two subgroups*: The first one is for *numeric supercomputers*, and the second one is for *parallel-type machines* that operate on vectors performing the same operation (single instruction) on each vector element (multiple data). Early parallel computers,such as the Array processors ILLIAC IV, ICL DAP (Distributed Array Processor) that have a single master control unit used to broadcast each instruction to many independent ALUs, each containing their own data for simultaneous execution.

iii. Multiple Instruction stream Single Data stream (MISD): $I_m > 1, D_m = 1$, Multiple instructions operate on same single piece of data. A single data stream passes through a pipeline and is processed by different (micro) instruction streams in different segments of the pipeline. CRAY-1, CYBER205, etc. are such types of pipeline processing computers. *Fault-tolerance computers* also fall into this category where several CPUs process the same data using different programs to detect and eliminate faulty results.

iv. Multiple Instruction stream Multiple Data stream (MIMD): $I_m > 1, D_m > 1$, Multiple instruction streams executed simultaneously on multiple data streams. This demands that multiple independent CPUs operating as part of a larger system execute several programs simultaneously, or different parts of a single program concurrently. All multiprocessors, including most parallel processors, belong to this category. MIMD is again subdivided into *two categories*: Those MIMDs that use *shared-primary memory* are called *Multiprocessors*. Those that do not use this are known as *Multicomputers*, Private Memory Computers, or Disjoint Memory Computers.

Figure 10.1 illustrates a framework of different categories of machines. Flynn's taxonomy ends here, projecting a somewhat subjective view point, which is essentially a behavioural one. This classification, although schematically distinguishing computers by category in the way of making a distinction between instruction (control) and data, is still considered a course model (rough estimate) as some machines nowadays are often found to be hybrids of some of these categories. In a CM-5 machine for example, we have observed this hybrid approach where a universal architecture has been realized, combining the useful features

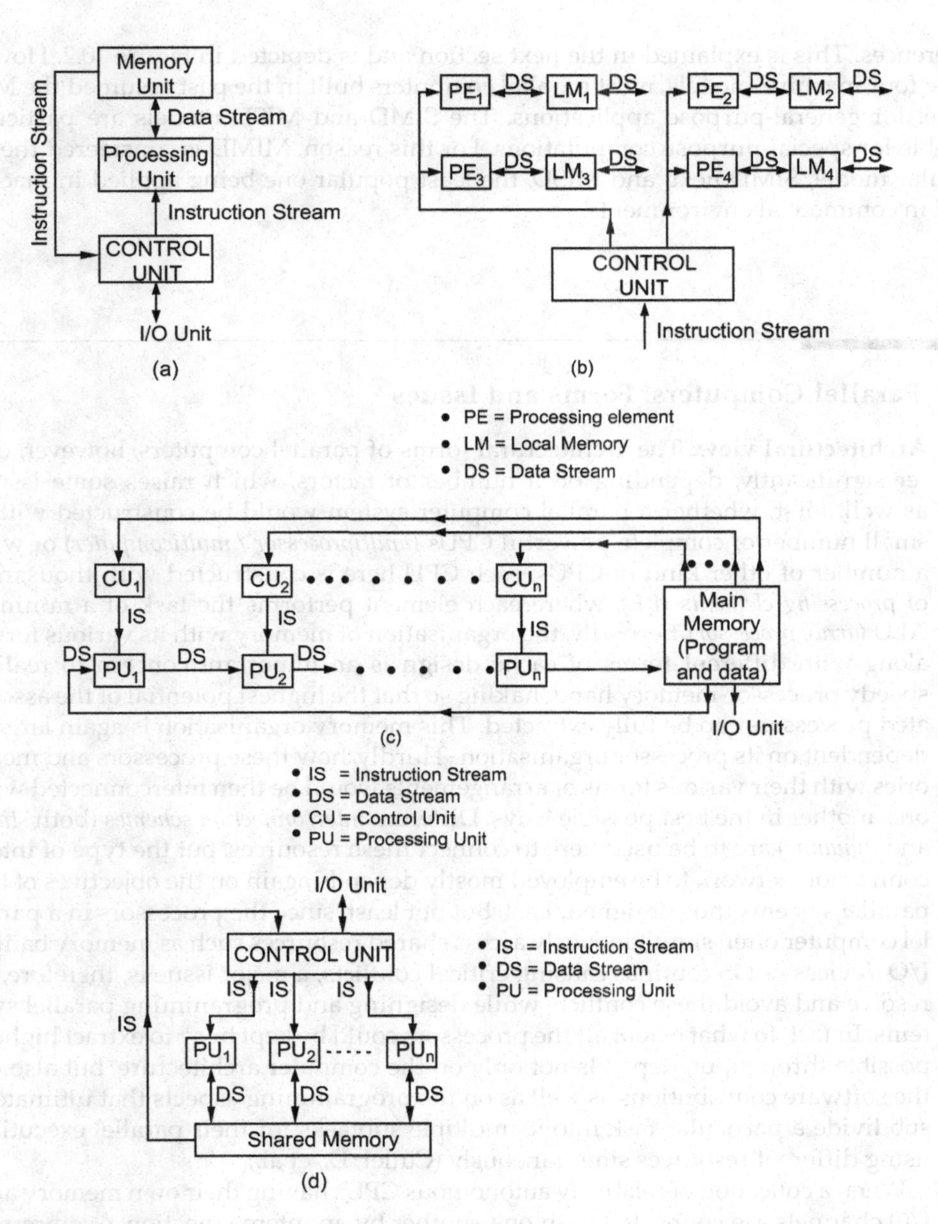

FIGURE 10.1
Flynn's classification of computers. (a) SISD uniprocessor framework, (b) SIMD framework (multiple processing element), (c) MISD framework (multiple control unit), and (d) MIMD framework (with shared memory).

and advantages of both SIMD and MIMD machines. Nevertheless, this classification is useful in that it provides a framework for the design space. However, it fails to provide anything that could indicate a computer's structure.

Flynn's classification can be once again extended to further split each category into subgroups. In the following section, we further divide each category in the domain of parallel computers (SIMD and MIMD) into even more sub-classes, based on their architectural differences. This is explained in the next section and is depicted in Figure 10.2. However, of the four machine models, most parallel computers built in the past assumed the MIMD model for general-purpose applications. The SIMD and MISD models are particularly suitable for special-purpose computations. For this reason, MIMD is considered the most popular model, SIMD next, and MISD, the least popular one being applied in machines used in commercial environments.

10.3 Parallel Computers: Forms and Issues

i. **Architectural view**: The architectural forms of parallel computers, however, differ significantly, depending on a number of factors, which raises some issues as well. First, whether a parallel computer system would be constructed with a small number of complete powerful CPUs (*multiprocessor / multicomputer*) or with a number of other kind of CPUs; each CPU here is constructed with thousands of *processing elements* (PE), where each element performs the task of a minimal ALU (*array processor*). Secondly, the organisation of memory with its various forms along with different forms of cache design is an important concern to realize speedy processor-memory handshaking so that the highest potential of the associated processors can be fully extracted. This memory organisation is again largely dependent on its processor organisation. Thirdly, how these processors and memories with their various forms of arrangements would be then interconnected with one another in the best possible ways. Different *interconnection schemes* (both *static* and *dynamic*) are to be used here to connect these resources, but the type of interconnection network to be employed mostly depends again on the objectives of the parallel systems thus designed. Last, but not least, since the processors in a parallel computer often simultaneously access shared resources such as memory banks, I/O devices or OS routines causing critical conflicts, a major issue is, therefore, to resolve and avoid these conflicts while designing and programming parallel systems. In fact, to what extent all the processors could be kept busy to extract highest possible throughput, depends not only on the computer architecture, but also on the software contributions as well as on the programming aspects that ultimately subdivide a particular task into its multiple subtasks for their parallel execution using different resources simultaneously (Culler, D., et al.).

   When a collection of relatively autonomous CPUs having their own memory and I/O channels are connected with one another by an interconnection / communication network (e.g., LAN), they are called ***loosely coupled systems***. Such systems are often called ***distributed computer systems*** or even are called ***multicomputer***. Its exact opposite is the system in which multiple CPUs with private/shared memory modules and common I/O devices being connected with one another over high-bandwidth interconnection networks are called ***tightly coupled systems***, and usually fall into the category, historically referred as ***multiprocessors***.

ii. **Application view**: Parallel computers while being used in application areas raise a simple question as to what is to be run in parallel. A variety of possibilities can be observed here. Some parallel computers are so designed that they allow multiple independent jobs to run simultaneously (in parallel) with very little or almost no communication between one another. *On-line transaction processing* (OLTP) run on a computer system having multiple complete CPUs that can handle a few hundred users at remote terminals deployed in the area of banking operation, railway reservation, etc. fall into this category. A different type of application environment has also been observed where a single job consisting of many parallel processes is executed by a parallel computer (Lewis, T. G.). Consider for example, a chess program that analyzes a given board pattern by generating a list of legal moves that can be made from it, and then forking off parallel processes to repeatedly analyze each new board in parallel. Here, the application is a single user-oriented one, but many processes are to be run in parallel and not in a sequential manner one after another. Another type of real-time application is *weather forecasting* in which a single instruction of a program is executed on multiple different data simultaneously in the form of vector computation.

These three representative examples express only a few of a wide variety of parallel processing. Although it is really hard to differentiate precisely between the various forms of parallel processing, these three examples, at least, differ in what is sometimes called ***grain size***. Grain size usually refers to the type of the task, the design of its algorithms, and the related software. When large pieces of software are run in parallel with little or almost no communication between the pieces, it is called ***coarse-grained parallelism***. Large banking operations, reservation systems belong to this category. The opposite extreme of this approach, as found in vector processing is called ***fine-grained parallelism***. Problems in the areas of artificial intelligence, weather forecasting, computer-aided design, ballistic missile control, etc. fall into this category. It has also been observed that in most cases, while problems with coarse-grained parallelism are well–suited for *loosely-coupled systems*, problems with fine-grained parallelism work excellent on *tightly-coupled systems*. However, there is as such no general rule in this regard, and it can be considered only as a guide at best.

10.4 Parallel Computers: Its Classification

Multiple-CPU computer systems in the MIMD category consists of a set of $n > 1$ processors (CPUs) P_1, P_2, P_3, … P_n, and $m > 0$ shared / distributed / or shared as well as distributed (main) memory units M_1, M_2, …, M_m that are interconnected using various forms of design approaches. Different types of emerging I/O processors attached with varieties of I/O devices are also regularly involved in the design. Nowadays, they eventually play a dominant role due to the constantly increasing user density to cope with, and numerous types of evolving applications to handle; still for the sake of simplicity and convenience, they are kept outside the purview of this discussion. The classification is simply made here on a broad view, deliberately avoiding a spectrum of possibilities in details. Figure 10.2 exhibits a broad view of the different classes of computer architectures as proposed by Flynn as well as the sub-classifications that exist in each of them.

The conventional **SISD** computer architecture with all its variants, such as, pipelined, superpipelined, and superscalar machines is all *uniprocessor*, where $n = m = 1$, and the processor–memory interconnections are mostly made over the system bus. The **SIMD** architectural design for computer systems appeals more directly to a class of special-purpose

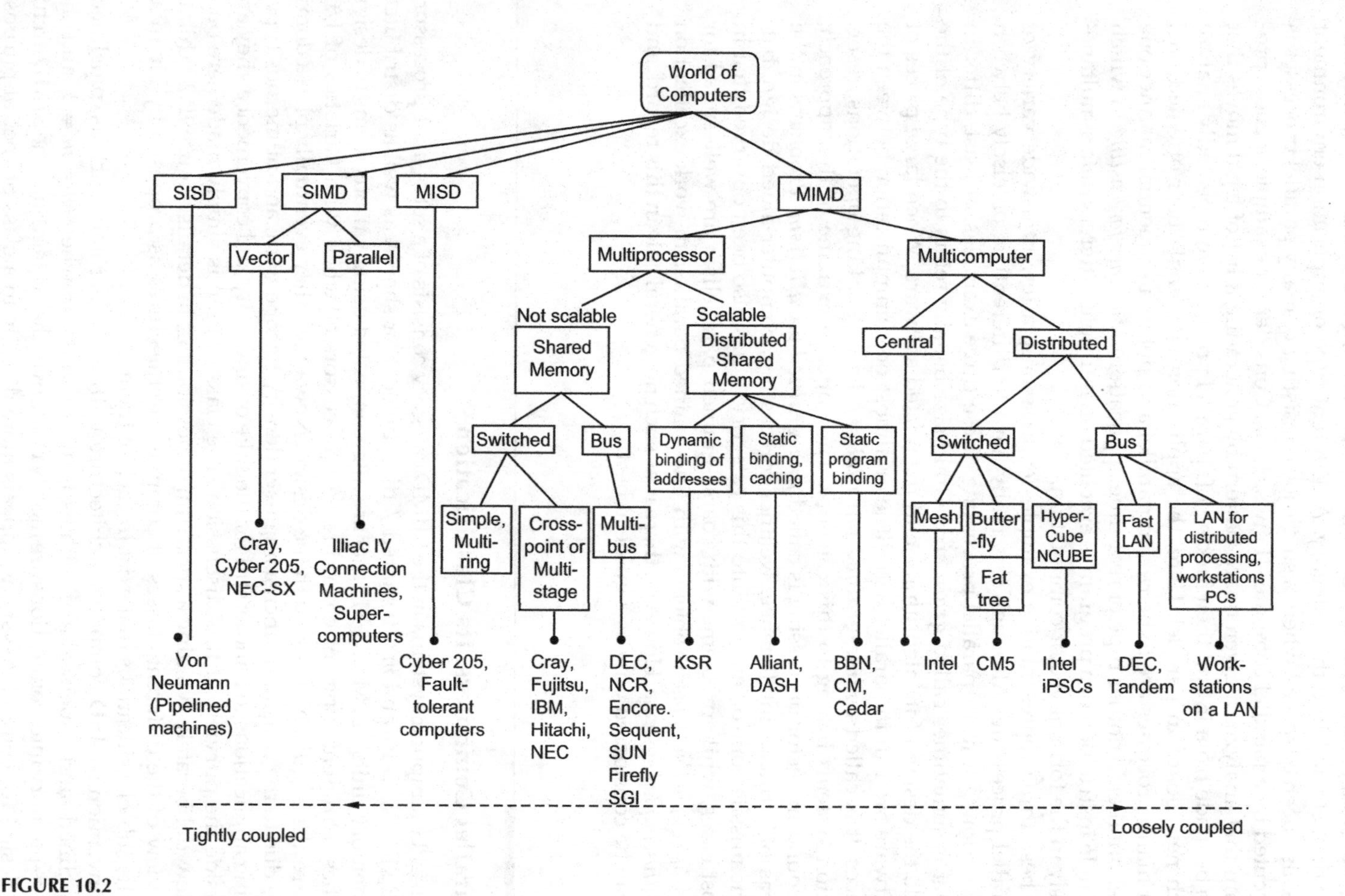

FIGURE 10.2
A taxonomy of parallel computers with some representative systems.

applications, mostly used for synchronized vector and array processing that exhibit an obvious form of ***data parallelism***. An SIMD computer thus exploits ***spatial parallelism*** rather than *temporal parallelism* as in a pipelined computer. SIMD computing is realized through the use of an array of *processing elements* (PEs) synchronized by the same control unit. This design concept has been used in the architecture of pipelined multivector supercomputers, and in SIMD ***array processors***. Associative memory can also be used to build SIMD associative processors. Only in the **MIMD** category, various forms of processor–processor interconnections and processor–memory interconnections have been rigorously exploited that give rise to two distinct categories: those MIMDs use shared primary memory, usually called ***multiprocessors***, and those that do not, sometimes called ***multicomputers*** or *Private Memory Computers*, or even *Disjoint Memory Computers*. The primary difference between them is essentially, that in a *multiprocessor*, there is usually a single virtual address space that is shared by all CPUs. All the machines share the same memory.

A ***Multiprocessor*** with *n* number of CPUs can be so designed to achieve as close to *n* times speed up as possible, with a reduced overall system costs. Due to the advent of more powerful VLSI technology from sometime in the mid-1980s onwards, the conventional definition and the traditional architecture of the multiprocessor system have been radically changed. It then became feasible to develop one-chip powerful multiprocessors (**multicore**, also known as *chip multiprocessor*) and larger-capacity RAM at reasonable cost. Large-scale multiprocessor architectures then started to evolve with multiple memories that are now *distributed* with the processors. Here, each CPU, while can access its own *local memory* quickly, accessing the other memories (*global*) connected with other CPUs is also possible but is relatively slower. In addition, there is also a larger shared *remote memory* accessible to all processors. That is, all these physically-separated memories can now be addressed as one logically shared address space, meaning that any memory location can be addressed by any processor, assuming that it has the correct access rights. This, however, does not discard the fundamental *shared memory* concept of multiprocessor but rather supports it in a broader sense, ventilating the concept of ***distributed shared memory***.

In fact, multiprocessors are classified by the organisation of their memory systems (shared memory and distributed shared memory) as well as by the interconnection networks (dynamic or static) being used. *Centrally-shared memory* multiprocessors, also known as **UMA** (uniform memory access), use a limited number of processors located relatively closely, and has a single address space. They are often called ***tightly-coupled multiprocessor*** or sometimes referred as ***parallel processing system***. They are mostly not scalable, or may be scalable to a very limited extent, constrained by the bandwidth of the shared memory. If all the CPUs in this system are made identical and each CPU is allowed to execute either OS code or user program, then this system is called **Symmetric Multiprocessor** (SMP). Any communication between the processors in this system usually takes place through the shared memory. When a message is sent from one processor to another, the delay experienced is short, and the data rate is high, since the CPU chips are likely to be placed on the same printed circuit board and connected by wires etched in the board. Although these systems can be employed for general-purpose multiuser applications, they tend to be used to work more on a single program (or problem), which is already subdivided into a series of multiple subtasks for their parallel execution, using different resources simultaneously to achieve maximum speed up. The other type is *distributed-shared memory* (DSM) multiprocessors, which are often called ***loosely-coupled multiprocessor***, or sometimes also known as *Scalable shared memory architecture*, and may also be referred to as *distributed computing systems*. These multiprocessors, on other hand, use **NUMA** (Non Uniform Memory Access) mechanism to access physically-separated memories (local, global, and remote) that can be addressed as one logically shared address space. However, multiprocessor systems are best suited for general-purpose multiuser applications where the major thrust is on programmability. Shared-memory multiprocessors can form a very cost-effective approach, but latency tolerance while accessing remote memory is considered a major shortcoming. Lack of scalability is also a serious limitation of such a system.

In a ***Multicomputer,*** in contrast, each individual machine consisting of processor–memory–I/O module, forming a **node,** is essentially a separate stand-alone autonomous computer. Therefore, this arrangement consisting of a collection of ***loosely-coupled*** interconnected independent machines (nodes) is therefore rightly called *multicomputers* or ***distributed computing systems.*** Typically, memory as well as I/O is distributed here among the processors. The nodes in the system are, however, additionally equipped with communication interfaces so that they can be interconnected with one another by a communication network in the proposed arrangement. *Multicomputers* having multiple disjoint address spaces are *scalable* with distributed memory, and physical communication between these fundamentally independent machines (processors) is done by explicitly *passing messages* across the communication network that interconnects these individual machines. Consequently, the inter-machine message delay is relatively large and the data rate is comparatively low. However, multicomputers, also called *message passing machines,* usually yield cost-effective higher bandwidth, since most of the accesses made by each processor are mainly to its local memory, thereby reducing the latency on average that eventually results in increased processor performance.

Multicomputers (Distributed computing systems) can be further classified into two different categories: **Homogeneous** and **Heterogeneous**. In a *homogeneous multicomputer,* all processors are the same, and generally have access to the same amount of private memory. There is essentially only a single interconnection network that uses the same technology with each computer present in this arrangement. Homogeneous multicomputers are thus often found in use more as *parallel systems* (working on a single problem), just like *multiprocessors.* A *heterogeneous multicomputer,* in contrast, may contain a variety of different independent computers, which, in turn, are connected through different networks. For example, a large distributed computing system may be built up from a collection of different small local-area computer networks, and these networks are then interconnected through a different communication network such as an FDDI (*FibreDistributed Data Interface*) or ATM-switched backbone.

Another most-frequently used taxonomy is in regard to the **interconnections** made between processor–processor, and processor–memory in multiple-processor computer architecture as cited in Figure 10.2. Both multiprocessors as well as multicomputers individually can again be divided into two categories based on the types of interconnection architecture being used as shown in Figure 10.2. These two types are defined as ***bus*** and ***switched.*** By bus, it is meant that there is a single network, backplane, bus, cable or other medium that connects all the machines. For example, commercial cable company usually runs a wire down the street, and all the subscribers have their own taps from it to run their own televisions. **Switched systems** do not have such a single backbone (like cable television). Instead, there are individual wires from machine to machine, and a module being used in between to provide them such interconnections has different patterns of arrangement of its switching elements internally of their own that provides and determines what types of interconnections are to be made between processor–processor, and processor–memory, and also with the external I/O devices. Usually, messages move along the wires, with an explicit switching decision made internally at each step inside the module to route the messages along one of the outgoing wires. An example of this type of arrangement is the worldwide public telephone system organisation. The numerous types of internal arrangement of the switching components made inside the module give rise to different types of its internal topology, which are popularly known as ***interconnection networks***. Some of the commonly-used interconnection networks will be discussed in the next section describing their topology, usage, as well as a few of their merits and drawbacks.

Multicomputers (distributed computing system), on the other hand, due to their loosely-coupled architecture facilitate its member processors to be located even far from each other in order to cover a wider geographical area. In addition, this system is more freely expandable (scalable), and theoretically can contain any number of interconnected processors with no limits as such. They are designed primarily to allow many users to work together on many unrelated problems, but occasionally in a cooperative manner that involves sharing of resources. Incidentally some *fibre-optic* based multicomputers (loosely-coupled) nowadays have been found to also work very closely at memory speeds (tightly-coupled). Therefore, the terms "tightly-coupled" and "loosely-coupled" although bear some useful concepts, any distinct demarcation between them is difficult to maintain because the design spectrum is really a continuum.

Last but not the least, with rapidly increasing power of VLSI technology that provided the system designer with an abundance of hardware capabilities and numerous options within affordable cost, the architectural evolution in recent years is being observed to progress continuously. It sets a trend that facilitates the evolution of some new forms of modern parallel architectures (e.g. **cluster architecture**), which actually lie in between the conventional multiprocessors and traditional multicomputers. Consequently, it is viewed that the boundary and the demarcation line which yet exists in between multiprocessors and multicomputers is gradually fading away and eventually this distinction may be destined to vanish completely.

10.5 Parallel Computers: Its Environment

Development of parallel computers (both multiprocessors and multicomputers) alongwith the required supporting environment for their working can be illustrated by a layered representation as depicted in Figure 10.3, based on a classification as proposed by Ni Lionel (1991). Hardware configuration of these computers usually differs from machine to machine, even those of the same model. The address space of a processor in a computer system varies widely among different architectures that depend mostly on the memory organisation, which is again machine-dependent. All these features along with other similar ones are to be decided by the architects at the time of design, to meet the objectives and match the targeted application domains.

A user usually has an architecture-transparent view, and always intends to develop application programs and programming environments, essentially to be machine-independent so that those can be ported to many different computing environments straightaway, or may be with minimum conversion costs. The system programmers, on other hand, are more inclined to explore the different attributes provided by the hardware, and try to make use of them effectively. The operating system (OS) itself and its various functionalities, including communication mechanisms used by different processing resources, are different for different class of machines, and are mostly machine-dependent. *High-level languages* and associated *compilers* or *communication models* mostly depend on the

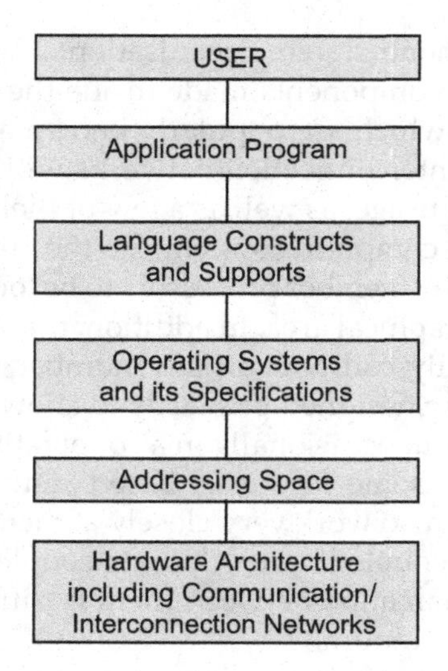

FIGURE 10.3
A schematic layered view of a parallel system environment.

architectural aspects, and options are utilized in their implementation. The use of *optimizing compilers* can facilitate the extraction of the best of the system facilities in favour of the languages being used, and thereby improve the system performance as well. However, the OS support and compilers should be designed in a befitting manner so that as many architectural constraints as possible faced by the users can be removed or, at least, can be reduced. Programming languages such as Linda, Ada, Lisp, Pascal, C, C++, Fortran, and others mostly offer an architecture-transparent communication model for parallel computers, and usually provide numerous features for ease and convenience of application program development. All these and similar other aspects are essentially involved in developing the computing environment, and a proficient architect, after all, should view the entire scenario as a whole, and not by individual parts (Zorpetta, G.,).

10.6 Interconnection Networks

An Interconnection Network is essentially a pathway using which the resources can establish communications between themselves. It allows information to flow either between any pair of modules in the system or by broadcasting information from one module to many other modules. Various types of such networks exist, and can also be designed in numerous ways. The selection of a particular network from a spectrum of choices is determined on the basis of its various *parameters*; such as, *cost, topology, bandwidth, latency, switching method, timing protocol, control strategy, effective throughput, scalability, hardware complexity,* and above all, *ease of implementation.*

- Interconnection networks employed for local connectivity usually have a *symmetric topology* while connectivity required for remote access has *irregular topologies.*

- The *bandwidth* refers to the data transfer capacity (or rate) of the transmission link, and is expressed in terms of bits or bytes per second.
- *Latency* refers to the time delay faced by a unit of information while being transmitted through the network.
- *Switching method* expresses the manner of transferring data in a network which is either *circuit switching* or *packet switching. Circuit switching* is the traditional way to offer a connection-based service. Once a connection is established, the resource is granted a path in the network, it then takes control of the path, reserving the total bandwidth for the entire duration of the data transfer. The alternative is *packet switching* in which the information is broken into small packets, individually competing for a chosen path in the network. In general, packet-switched approach allows more use of the bandwidth, and is the traditional means to support *connectionless* communication.
- Network operation is monitored by a certain *timing control,* which can be either *synchronous* or *asynchronous.* Synchronous networks are controlled by a global clock that synchronizes all network activities. Asynchronous networks use handshaking or interlocking mechanisms to coordinate fast and slow devices that issue requests over the same network.
- The *control strategy* used to operate a network can be classified as *centralized* or *distributed.* Centralized control uses a global controller that receives requests from all the resources attached to the network and grant access to the network to one or more requesters. A distributed control system, on the other hand, allows the local devices to independently handle all the requests.
- *Effective throughput* is the *actual rate* of data transfer which is usually less than the available bandwidth, since the network does not carry data all the time.
- *Scalability* allows an ideal network to have the provision for further modular expansion (scalable) to meet on-going increasing demands, of course, with a proportional increase in cost for increased performance with use of additional machine resources.
- *Hardware complexity* refers to the design and subsequent implementation costs such as those for wires, connectors, switches, arbitration mechanisms, and interface logic.

10.6.1 Interconnection Network: Different Types

The *topology* of an interconnection network can be of two types: ***Static*** or ***Dynamic.***

Static networks are direct networks that are formed of point-to-point direct connections which are fixed, once it is built, and cannot be changed during program execution. These types of networks are used for fixed connections among subsystems of a centralized system, or are found useful for building distributed systems (multiprocessors, multicomputers), and are also used in SIMD machines where the communication patterns are predefined or predictable or realizable with static connections. Static networks give rise to different popular topologies that mainly include *linear array, ring, tree, star, mesh, hypercubes,* etc.

Dynamic networks give rise to connections, dynamically changeable, providing all possible communication patterns based on program demands. These networks can be implemented in many different ways using switching elements which can be dynamically configured to satisfy the communication demand during the execution of user programs.

The realization of numerous networks is, however, primarily attributed to the cost, and its related performance. The performance is measured in terms of its *bandwidth, data transfer rate, latency,* and above all, the patterns of communication. In increasing order of cost and performance, these networks include *bus systems, multistage interconnection networks* (MIN), and *crossbar switch networks,* which are often used in shared-memory multiprocessors, and in SIMD computers for inter-PE (Processing Elements) data routing. In fact, the organisation of a multiprocessor system ultimately may be of various types that are mostly based on the different kinds of dynamic interconnection networks being used.

Here, we will describe in brief some of the commonly-used interconnection networks of both types that may be implemented in multiple-processor system organisations.

10.6.1.1 Hierarchical Common (Shared) Bus Systems

The most obvious, simplest, and comparatively economical means for interconnecting a number of modules to construct a multiprocessor machine is to use a digital bus system that consists of a hierarchy of different common buses. The structures and interfaces of these buses are primarily the same as for a single-processor system that uses a bus interconnection. The CPUs, memory modules, and various I/O units are directly attached to this bus system, and time-share its communication facilities. Each bus is formed with a number of signal, control, and power lines. Different buses of different types and characteristics are employed to perform different interconnection functions.

The hierarchy of bus systems is built up as depicted in the Figure 10.4. The bus systems are packaged at different levels; at each level the bus is dedicated to a particular source–destination pair for the full duration of the requested transfer. Since several modules are connected at each level of the common bus, and every module competes to gain control over bus access, it is thus essential to have an efficient *bus arbitration mechanism* (see Chapter 5) to resolve the contention (conflicts) and delays. However, the bus system generally consists of local buses, backplane buses, and I/O buses.

Local buses are implemented on *printed circuit* boards. Normally, each CPU is provided with a local bus attached with a local memory unit containing part of the shared address space and can also support local I/O subsystems. This system configuration, while providing a common communication path among the major components, removes most of the routine memory traffic from the main system bus that can now be devoted more for interprocessor communications. The global memory system is implemented on the memory board that uses a dedicated memory bus to connect memory modules with the interface logic. I/O board implements the I/O processor (IOP) that uses its own local data bus to communicate with various peripheral devices using different lines that may be at different layers in the printed circuit board. The communication board also has its own local data bus to interface with the network for remote access. ***Intel Pentium*** while used in shared-bus multiprocessor organisation, the local bus is then generally designed with standard buses like the PCI bus, VESA Local Bus (VLB).

A ***Backplane*** is a printed circuit on which many connectors are attached. These connecters are used to mount various functional plug-in boards. Motherboard, just being a passive receptacle for other boards (CPU board, memory board, etc.), can be called a *backplane.* Since the system bus is the common pathway to communicate between the resources, and also among all plug-in boards, it is constructed on the backplane with shared signal paths and utility lines. Several popular backplane buses, such as Multibus II, VME bus, Future bus +, PCI, etc., have been developed. Their standards, specification and other details have already been explained in Chapter 5.

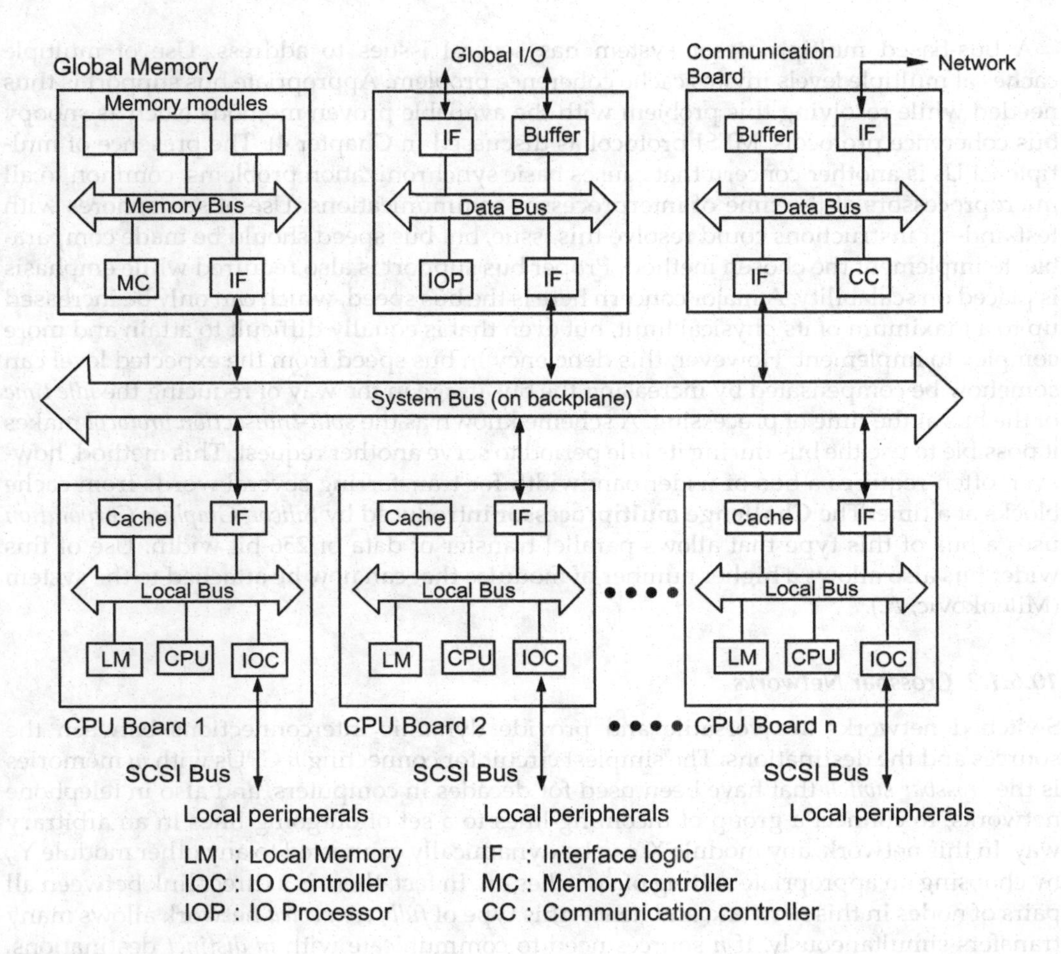

FIGURE 10.4
Shared bus multiprocessor at board level, backplane level and I/O level with global and local resources.

The **I/O bus** is used by the I/O subsystems to connect input/output devices. Local I/Os generally use Small Computer System Interface (SCSI) bus, and nowadays, mostly USB or some other type to connect components like Disk, tape units, printers, etc., to an individual processor through an *I/O controller.* The global I/O is provided with the use of a separate I/O board that allows the computer to connect more multiple devices through an independent *I/O processor* (IOP). The IOP is interfaced with the system bus via a fast communication bus like VME, FDDI optical bus, Ethernet, etc.

A hierarchical bus system, however, allows a maximum of about 100 processors to build a medium-sized multiprocessor system. Digital buses are used in commercial systems ranging from small workstations to mainframe and multiprocessor systems. Speed limitations of this system have been overcome to a large extent by providing each CPU with a cache memory to drastically reduce the number of its system bus visits and accesses. Although the bus approach has been expanded with the use of PCI SIG (PCI Special Interest Group, *see* Chapter 5 for details), it is still limited mainly due to the packaging technology being employed.

A bus-based multiprocessor system has several issues to address. Use of multiple caches at multiple levels invites cache coherence problem. Appropriate bus support is thus needed while resolving this problem with the available proven methods (such as snoopy bus coherence protocols, MESI protocol as discussed in Chapter 4). The presence of multiple CPUs is another concern that causes basic synchronization problems, common to all microprocessors at the time of interprocessor communications. Use of semaphores with test-and-set instructions could resolve this issue, but bus speed should be made comparable, to implement the chosen method. Proper bus support is also required while emphasis is placed on scalability. A major concern here is the bus speed, which can only be increased up to a maximum of its physical limit, but even that is equally difficult to attain and more complex to implement. However, this deficiency in bus speed from the expected level can somehow be compensated by increasing the bus usage in the way of reducing the *idle time* of the bus at the time of processing. A scheme known as the *split-transaction protocol* makes it possible to use the bus during its idle period to serve another request. This method, however, often requires a bus of wider bandwidth for transferring several words from cache blocks at a time. The **Challenge multiprocessor** introduced by *Silicon Graphics Corporation* uses a bus of this type that allows parallel transfer of data of 256-bit width. Use of this wider bus also allows a higher number of modules that can now be attached to the system (Milenkovic, A.).

10.6.1.2 Crossbar Networks

Switched networks are versatile and provide dynamic interconnections between the sources and the destinations. The simplest circuit for connecting n CPUs with m memories is the *crossbar switch* that have been used for decades in computers, and also in telephone networks, to connect a group of incoming lines to a set of outgoing lines in an arbitrary way. In this network, any module X_i can be dynamically connected to any other module Y_k, by choosing an appropriate setting of switches on. In fact, there is a direct link between all pairs of nodes in this network, and hence, this type of *fully connected* network allows many transfers simultaneously. If n sources need to communicate with m *distinct* destinations, then all of these transfers are possible independently and asynchronously due to having all possible (all permutations) paths and it also permits all the transfers to take place concurrently. Since, any transfer can be made, and is not prevented at any point of time; the crossbar network is sometimes called a ***nonblocking network***.

Figure 10.5 shows just a simple single switch that is placed at each intersection of a horizontal and vertical line called a *crosspoint*. This switch can be electrically opened or closed, depending on whether the horizontal and vertical lines are to be connected or not. In an actual multiprocessor implementation, the paths through the crossbar network are much wider, which implies that multiple switches are needed at *each crosspoint*. Moreover, while a total of n modules are interconnected in such a network, the number of *crosspoints* becomes n^2 and this makes the total required number of switches very high as n gradually increases, resulting in a less cost-effective network and also requiring extensive hardware in implementation. This awful property thus causes to restrict the widespread use of crossbar networks in general, only being found profitable in use in mostly small or medium-sized systems. The advantage is its *nonblocking* nature that requires no advance planning. Moreover, processor interfaces and memory port logic becomes simpler and cheaper, and crosspoint switches being used are also equipped with arbitration logic as well as conflict resolution mechanism while many other logics ultimately bear all such responsibilities required for hassle-free communications, in this approach.

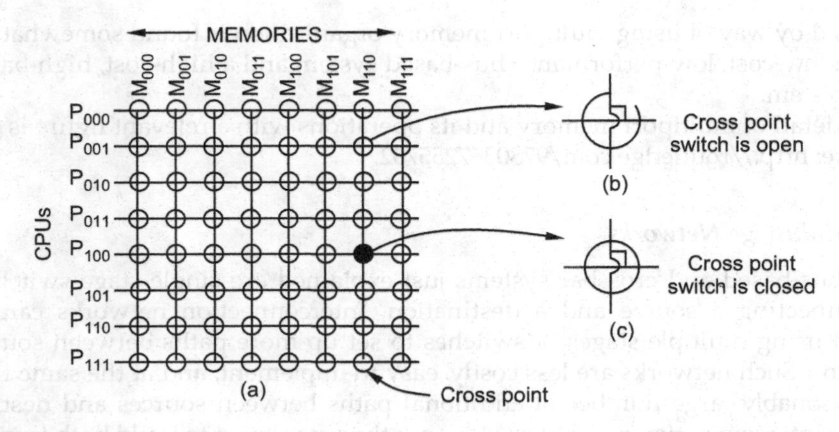

FIGURE 10.5
(a) An 8 × 8 Crossbar Switch, (b) An open crosspoint, and (c) A close crosspoint.

In an $n \times n$ crossbar networks, n processor can send at most n memory requests independently and asynchronously. Two situations may arise. If all the requests come from n processors to n distinct memory words, then those can be delivered simultaneously in each memory cycle. This is possible when the memory modules are made *n-way interleaved* to allow overlapped access. If all, or some of the requests are destined for the same memory module at the same time, then only one of the requests will be serviced at a time.

With the use of unary (can be set on or off) switches, crossbar networks are single-stage nonblocking. A single-stage crossbar network is not expandable once it has been built. One such implementation is found with larger crossbar switches in ***Sun's E10000*** system in which 16 four-processor nodes are connected by a 16 × 16 crossbar switch. Crossbar switch mechanism has been implemented even *on-chip* in a recent release from Sun in its ***UltraSparc T2*** multicore multithreaded high-end processor (see Chapter 8).

However, a *multilevel* crossbar switch can also be used for more modules (larger configuration) to connect where a crossbar switch at level 1 connects a crossbar switch at level 2 and so on. Such systems have been found in use in large computers, like ***Fujitsu VPP500*** *(a vector parallel processor)*, ***Hitachi SR 8000*** and ***NEC SX-5*** systems. In these systems, a different type of large crossbar (224 × 224) network is used for processor–processor (interprocessor) communication. Here, the nodes are the processors (or processing element PEs with attached memory) and the CPs (control processors) which are used to monitor the entire system operations as well as the crossbar network operation. In this network, only one crosspoint switch can be set as on in each row and in each column. In all *Cray multiprocessors*, crossbar networks are used between the processors and memory banks. The banks use 64-, 128-, or 256-way of interleaving (in Cray Y-MP) that can be accessed via four ports.

10.6.1.3 Multiport Memory

One of the major drawbacks of a crossbar networks is its high cost and cumbersome wiring, when it is built in a multiprocessor. Many mainframe multiprocessors thus use a multiport memory organisation to reduce the load, and thereby to lower the cost and complexities of the crosspoint switch being used. This approach summarily relieves the crosspoint switches from its additional burden of handling all arbitrations and conflict resolutions, and entrusts it to the memory controller attached with the memory modules. The solution thus offered by way of using multiport memory organisation is found somewhat midway between a low-cost, low-performance bus-based system and a high-cost, high-bandwidth crossbar system.

A brief detail of multiport memory and its operations with a relevant figure is given on the website: http://routledge.com/9780367255732.

10.6.1.4 Multistage Networks

Both the bus-based and crossbar systems just explained are single-stage switching networks connecting a source and a destination. Interconnection networks can also be developed using multiple stages of switches to set up more paths between sources and destinations. Such networks are less costly, easy to implement, and at the same time provide a reasonably large number of additional paths between sources and destinations. Multistage interconnection networks (MIN) are thus employed to build both larger multiprocessor systems and SIMD computers. A generalized multistage network is illustrated in Figure 10.6. Various classes of MINs can, however, be constructed that differ in the switch modules being used, and in the manner of wiring the inter-stage connection (ISC) patterns being built up (Adams, G. B. III).

The simplest design is based on the humble 2 × 2 switching elements as shown in Figure 10.7a. The switches can be dynamically set to establish the desired connections

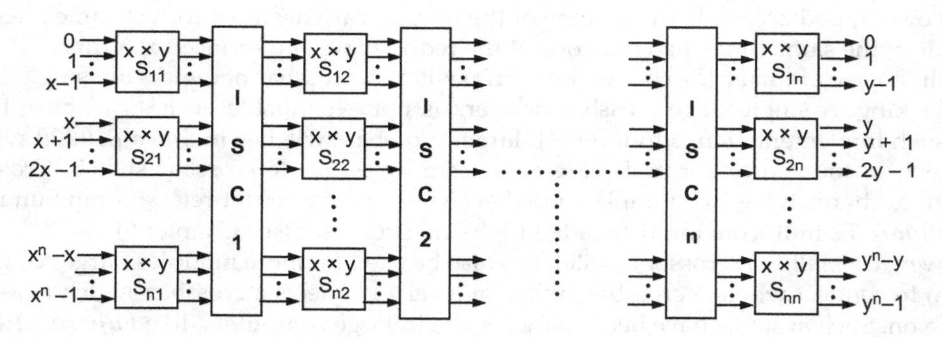

FIGURE 10.6
A generalized structure of a multistage interconnection network (MIN) with $x \times y$ switching elements and interstage connection patterns (ISC1, ISC 2, ... ISC n).

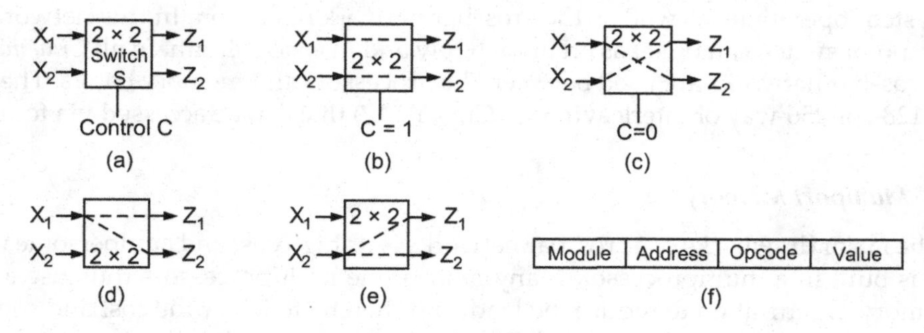

FIGURE 10.7
Switching element with different states and message format. (a) Control C, (b) Direct, (c), Crossover, (d) Upper broadcast, (e) Lower broadcast, and (f) Message format.

between the inputs and outputs to construct dynamic interconnection networks. Each switch S has a pair of input data buses X_1 and X_2, a pair of output data buses Z_1 and Z_2, and some control logic (not shown). All four buses are identical and can function as processor–processor or processor–memory links. Messages arriving on either input line can be switched to either output line. Various combinations of the *switch states* implement different *permutations, broadcast* or *other connections* from the inputs to the outputs.

When $Z_1 = X_1$ (Z_1 is connected to X_1) and $Z_2 = X_2$, it is called a *direct* or *through state* T of S. This is shown in Figure 10.7b. When $Z_1 = X_2$ and $Z_2 = X_1$, itis called a *crossover state* X of S as shown in Figure 10.7c. The broadcast state is shown in Figure 10.7(d) and (e). While switch states may be of various types, messages for our purposes will contain upto four parts as shown in Figure 10.7f. In message format as shown, the *Module* field tells which memory to use. The *Address* specifies an address within a module. The *Opcodes* gives the operations to be performed, such as, READ or WRITE. The optional *Value* field contains an operand, such as a 32-bit / 64-bit word to be written. The switching element examines the *Module* field and uses it to determine the direction of transfer, either to Z_1 or to Z_2.

By using this 2 × 2 switch as a building block, it can be arranged in many different ways to build larger *multistage networks* (Wu, C. L., et al.) that can be used in *massively parallel computing systems.* The ISC (inter-stage connection) patterns also are of various types that are often used to include *Multistage Crossbar, Perfect shuffle, Multiway shuffle, Butterfly, Cube connection,* etc. Some of these ISC patterns are illustrated here in brief.

10.6.1.5 Omega Network (Perfect Shuffle)

The economy class Omega network can be constructed by any of four possible connections of 2 × 2 switches as shown in Figure 10.7 by setting the control signals of the switching elements in various ways. The processor–processor or processor–memory connections can be established with this network, which depends on the number of stages, the fixed connections linking the stages, and the setting of the switching elements. Here, we have connected eight CPUs to eight memories using 12 switching elements that are arranged into three stages and intended to provide dynamic connections among eight processors and eight memories as shown in Figure 10.8. More generally, to connect *n* CPUs with *n*

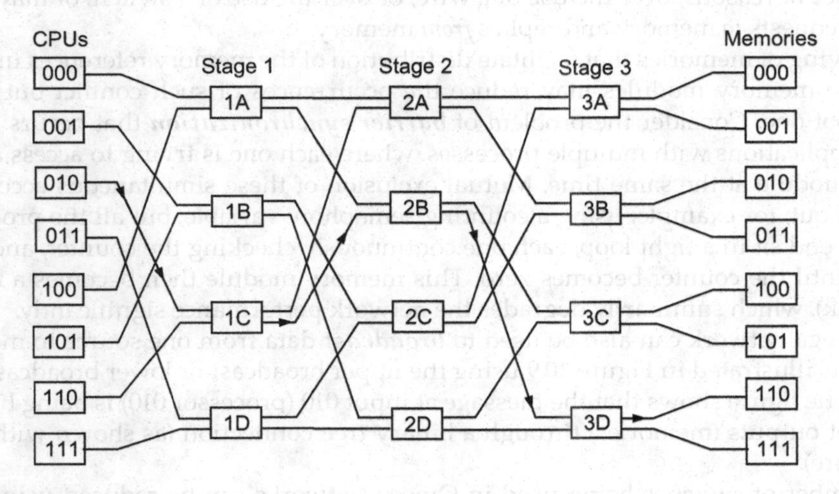

FIGURE 10.8
An 8 × 8 Omega network built in three-stage with 2 × 2 switches.

memories, the number of stages needed is $\log_2 n$ with $n/2$ switching elements per stage; each element in each stage is individually controlled, and a total of $[(n/2) \times \log_2 n]$ simpler 2×2 switching elements are thus here required. This is, however, far better than n^2 crosspoints (crossbar network), especially for high values of n.

The wiring pattern of omega network is often called *perfect shuffle*, since the mixing of signals at each stage resembles a deck of cards that is being divided in half, and then mixed card-for-card. Control logic associated with the MIN sets the switch states dynamically to service the interconnection requests issued from processors / memories. This can be viewed as *self-routing*. A particular MIN state after setting is retained for a long duration to allow at least one package to be transferred through the network. The state then changes to match the source–destination requirements of the next set of packages, and so on. It can be assumed that the processor can use a buffer to temporarily store or queue up its outgoing packages until the MIN is ready to transfer them. The processor, on the other hand, accepts incoming packages as soon as they arrive.

To see how the omega network works, suppose that CPU010 wants to read a word from memory module 110. The CPU containing 110 in the *Module* field sends a READ message to switch 1C. Data routing is controlled by inspecting the destination code (here 110) in binary. When the ith high-order bit of the destination code is a 0, a 2×2 switch at stage i connects the input to the upper output. Otherwise, the input is directed to the lower output. The CPU (010) connected to switch 1C takes the first bit (i.e. left most) of 110 and uses it for routing. The message is here routed via the lower output to 2B. The route of this communication is indicated by an arrow in Figure 10.8.

All the switches in second-stage including 2B, use the second left most bit of the destination for routing. This second leftmost bit here is also a 1 (in 110); the message is now forwarded via the lower output of 2B to 3D. Now, the third bit is tested, and is detected to be a 0. Consequently, the message goes out on the upper output of 3D and reaches the targeted memory 110.

If two requests from two CPU simultaneously want to access the same memory module, conflicts will occur, and one of them would have to wait till the other finishes, causing communication delays, and consequently, lowers the effective bandwidth. Unlike the crossbar switch, the omega network thus is a ***blocking network***. Conflicts can arise due to a number of reasons; over the use of a wire, or over the use of a switch, or may even be between requests *to* memory and replies *from* memory.

Interleaving of memories that facilitate distribution of the memory references uniformly among the memory modules may reduce the occurrences of such conflict but it is not a full–proof one. Consider the problem of ***barrier synchronization*** that occurs in many parallel applications with multiple processes, where each one is trying to access a certain memory module at the same time. Mutual exclusion of these simultaneous accesses can be carried out, for example, using a counting semaphore variable, but all the processes at their own end sit in a tight loop, each one continuously checking the counter, and cannot proceed until the counter becomes zero. This memory module then becomes a ***hot spot*** (bottleneck), which summarily degrades the network performance significantly.

The Omega network can also be used to ***broadcast*** data from one source to many destinations as illustrated in Figure 10.9 using the upper broadcast or lower broadcast switch settings. The figure shows that the message at input 010 (processor 010) is being broadcast to all eight outputs (memories) through a binary tree connection (as shown with arrows in the figure).

The number of switches being used in Omega networks can be reduced using a trick that is employed in some multiprocessors. Instead of having $\log_2 n$ stages, there is only

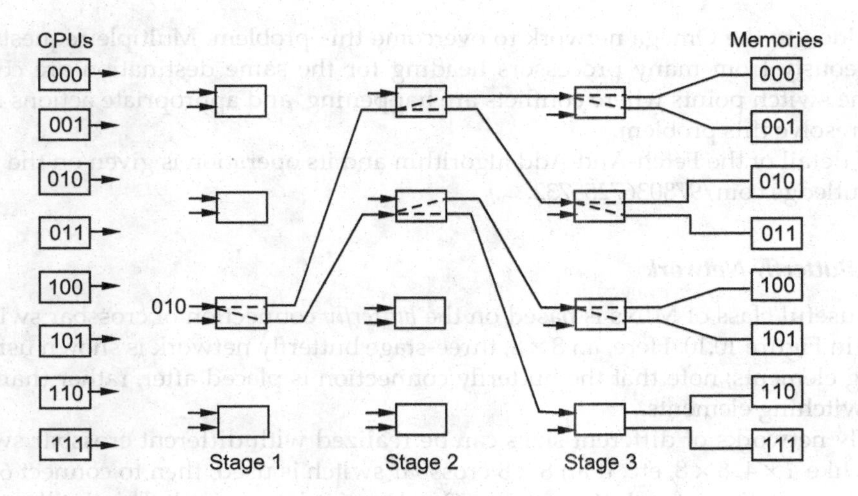

FIGURE 10.9
Broadcast capability of Omega network (8 × 8) built in three-stages with 2 × 2 switches.

one stage in the network. The output thus obtained from this stage is once again fed back into the input. Thus, each message traverses $\log_2 n$ passes (not stages) over the network. Such a design is called a ***recirculating Omega network***. The major disadvantage is that pipelining cannot be implemented (single stage). Hence, in a multistage Omega network, while upto $(n/2) \log_2 n$ messages can be simultaneously switched, in a recirculating network, it is certainly less, and the maximum is only $n/2$.

The Omega network thus has two major ***drawbacks***: namely, the *blocking* problem, and the *hot spot* issue that need to be addressed. The blocking problem (already discussed) in Omega networks has been resolved by Benes Networks. The solution of *hot-spot* problem has been, however, dealt with the Fetch-And-Add Approach.

10.6.1.6 Benes Network

The blocking problem in Omega networks can be solved by Benes network, adding more stages using here a five-stage network (in place of three stages) with additional hardware that makes it nonblocking. In an Omega network, there is exactly *one path* from any CPU to any memory, whereas in a Benes network, there is *more than one path* and many other alternatives.

A brief detail of Benes network and its operations with relevant figure is given on the website: http://routledge.com/9780367255732.

10.6.1.7 The Hot-Spot Problem

10.6.1.7.1 Fetch-And-Add Approach

If multiple requests are issued simultaneously from many processors to access a certain memory module, the module may appear as a *hotspot*. Use of a lock or semaphore can make these simultaneous accesses mutually exclusive, but at the cost of wastage in *busy waiting* of the processors, due to continuous checking of the semaphore to gain access to the module. Moreover, this semaphore, as a result, may itself become a *hot spot*, since it is shared by many processors. As an alternative, a combining mechanism thus has been proposed that can be added to the Omega network to overcome this problem. Multiple requests issued simultaneously from many processors heading for the same destination are combined here at the switch points where conflicts are happening, and appropriate actions are then taken to resolve this problem.

A brief detail of the Fetch-And-Add algorithm and its operation is given on the website: http://routledge.com/9780367255732.

10.6.1.8 Butterfly Network

Another useful class of MINs is based on the *butterfly* connection of crossbar switches as depicted in Figure 10.10. Here, an 8 × 8, three-stage butterfly network is shown using 2 × 2 switching elements; note that the butterfly connection is placed after, rather than before, the *n*/2 switching elements.

Butterfly networks of different sizes can be realized with different crossbar switching elements like 4 × 4, 8 × 8, etc. If an 8 × 8 crossbar switch is used, then to connect 64-input, a network consisting of only two stages ($2 = \log_8 64$) is required. This is illustrated in Figure 10.11. For more inputs to connect, either more number of stages or more-way switching elements are needed to build the network.

Drawbacks of Multistage Networks: The ***two major drawbacks*** of multistage networks are: the cost of the switching elements being used (either 2 × 2 or 4 × 4 or 8 × 8) associated with the additional expenditure due to increased wiring, and the penalties being paid to implement the required atomic operation while providing concurrent accesses attempted by different sources to the same destination. These two drawbacks can, however, be overcome only with the use of cheaper and faster switching technology, if available, and in that situation, this network with all its merits can be employed to build up large-scale *UMA* (Uniform memory access) multiprocessor systems since the delay for paths connecting any two modules is always same. Interest in these networks gradually increased, and attained a peak in the 1980s, and diminished significantly in the past few years. Other schemes which we will be discussing next, gradually came in favour for having many distinct advantages, and hence, have become more attractive.

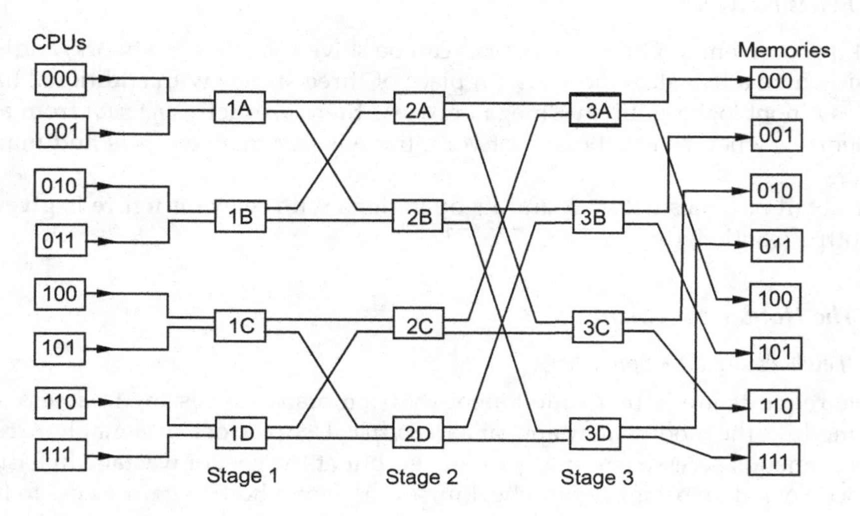

FIGURE 10.10
An 8 × 8 Butterfly switch network with 2 × 2 Crossbar switches.

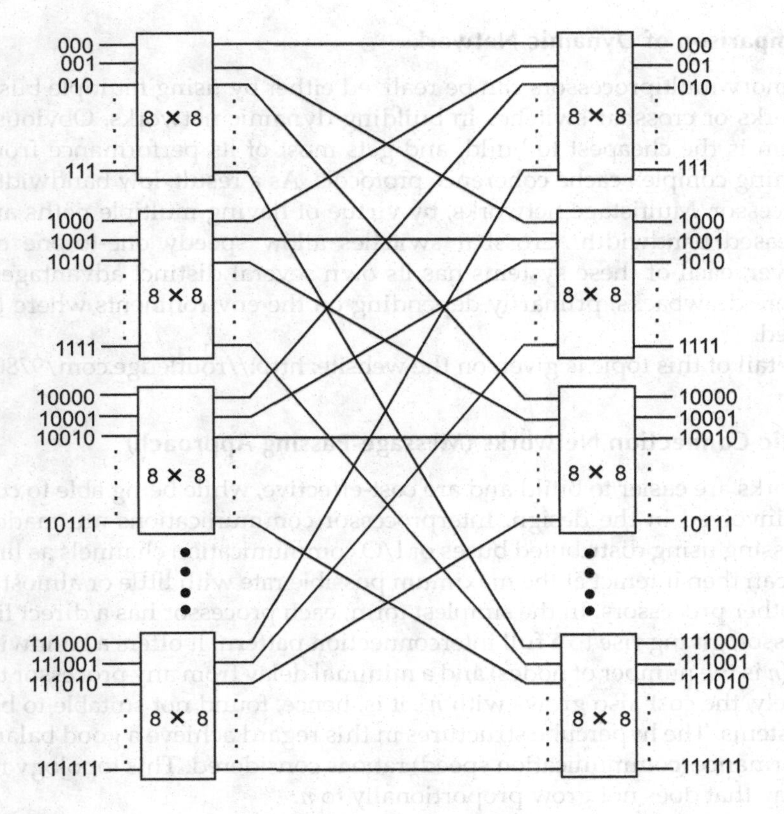

FIGURE 10.11
A 64 × 64 Butterfly switch network with 8 × 8 Crossbar switches using two-stage and eight-way shuffle interstage connections.

10.6.2 Implementation of Multistage Networks

A high-speed 128-port *Omega network* have been used by IBM in its **RP3 multiprocessor** having 512 processors (nodes) that offers a bandwidth of 13 Gbytes/sec at 50 MHz clock rate. ***Cedar multiprocessor*** also used *multistage Omega networks* in their systems to build up the interconnection networks. The ***Alliant FX/2800*** uses *crossbar interconnects* to communicate between seven 4-processor (Intel 860) boards plus one I/O board, and eight interleaved shared cache boards. *Multilevel crossbar switch* have been used to connect larger configuration in ***Fujitsu VPP500*** *(vector parallel processor)*, ***Hitachi SR 8000*** and ***NEC SX-5*** large systems. In all ***Cray multiprocessors,*** crossbar networks are used between the processors and memory banks. The banks use 64-, 128-, or 256-way of (in Cray Y-MP) interleaving that can be accessed via four ports. The ***BBN Butterfly processors (TC 2000 series)*** from BBN Advanced Computers use 8 × 8 crossbar switch modules to build a three-stage 512 × 512 butterfly network for a 512-processor model that offers a maximum inter-processor bandwidth of 2.4 Gbytes/sec at 38 MHz clock rate with a 1-byte data path. Using Motorola RISC processor 88100, the crossbar switch modules have been used to design a NUMA machine for real-time applications. ***IBM RS/6000 SP*** multiprocessor also encouraged the use of multistage network as one of its several options for interconnecting the groups of processors. ***IBM ES/9000*** large system uses crossbar network to connect I/O channels and shared memory.

10.6.3 Comparison of Dynamic Networks

Shared-memory multiprocessors can be realized either by using multiple buses or multistage networks or crossbar switches in building dynamic networks. Obviously, the bus-based system is the cheapest to build, and gets most of its performance from snooping caches running complex cache coherence protocols. As a result, low bandwidth is offered to each processor. Multistage networks, by virtue of having multiple paths and switches offers increased bandwidth. Crossbar switches allow speedy one-to-one communication. However, each of these systems has its own several distinct advantages as well as certain severe drawbacks, primarily depending on the environments where they will be actually used.

A brief detail of this topic is given on the website: http://routledge.com/9780367255732.

10.6.4 Static Connection Networks (Message-Passing Approach)

Static networks are easier to build and are cost-effective, while being able to control many processors involved in the design. Interprocessor communications are made by way of message passing using distributed buses or I/O communication channels as links. Nearby processors can then interact at the maximum possible rate with little or almost no interference from other processors. In the simplest form, each processor has a direct link to every other processor, giving rise to a full interconnection pattern. It offers a bandwidth proportional to n^2 (n is the number of nodes) and a minimal delay from any processor to any other. Unfortunately, the cost also grows with n^2, it is, hence, found not suitable to be employed for large systems. The hypercube structures in this regard achieve a good balance as far as cost / performance (communication speed) ratio is considered. This topology results in an average delay that does not grow proportionally to n.

10.6.4.1 Hypercubes

Hypercube is a binary n-cube architecture in which an n-cube consists of $N = 2^n$ nodes that are connected in an n-dimensional cube with two nodes per dimension. Each node is represented by small circle that consists of a processor, local memory and communication circuits. Each processor P_i has bi-directional links to n other neighbouring processors; these links actually form the edges of the hypercube. (Hayes, J. P., et al.). In an n-dimensional hypercube, each node is directly connected to immediately adjacent n neighbours. The cube's side is of length 1, so each co-ordinate is either a 0 or a 1. Moving from any arbitrary vertex to any of its adjacent vertices changes exactly one co-ordinate, so adjacent vertices differ in exactly one position. This is illustrated in Figure 10.12a. A set of 2^n distinct n-bit binary addresses can be assigned to the processors in such a way that the address of P_i differs from that of each of its neighbours in exactly 1 bit. Figure 10.12b illustrates a 3-cube hypercube with 8 nodes. A 4-cube can be formed by interconnecting the corresponding nodes of two 3-cubes as shown in Figure 10.12c. The ***node degree*** of an n-cube equals n.

The ***diameter*** of the n-cube hypercube is the distance between the two furthest points which is also equal to n. The diameter of an n-node hypercube is $\log_2 n$ compared to $2n$ for grid or mesh (discussed later) network. For example, when $n = 1024$, the diameter drops from 2048 ($2n = 2 \times 1024 = 2048$) for grid or mesh to 10 ($\log_2 1024 = 10$) for hypercubes, thereby resulting a sharp decrease in the mean delay that appears to be significant in favour of hypercube topology.

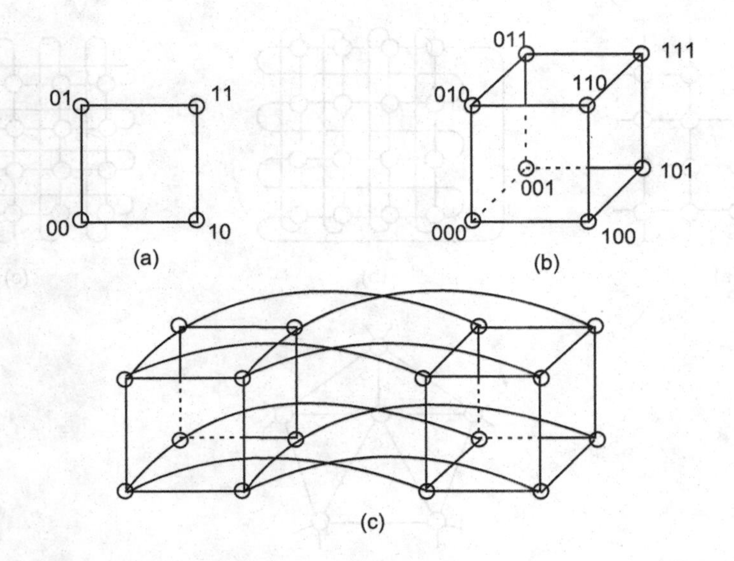

FIGURE 10.12
(a) 2-cube two-dimensional; (b) 3-cube three-dimensional; (c) 4-cube formed by interconnecting two 3-cubes (four-dimensional).

Since alternate paths are available at each node, routing messages through the hypercube is particularly easy. If the node N_x wishes to communicate with node N_y, it can proceeds in many ways. In general, an n-dimensional hypercube has n routing functions, defined by each bit of the n-bit addresses. One of them is as follows: the binary addresses of the source, x, and the destination y, are compared from *least to most* significant bits. Suppose that the nodes differ first in position p. Node N_x then sends the message to its neighbour whose address k, differs from x in bit position p. Node N_k then forwards the message to the appropriate neighbour using the same address comparison scheme. The maximum distance that any message needs to travel in an n-dimensional hypercube is n hops.

Hypercube has then constantly enhanced making use of contemporary technological developments, incorporating many useful features to make it more attractive, and thus passed through different generations with passing days. It has been used in ***Intel iPSCs*** with a seven-dimensional cube to connect up to 128 (= 2^7) nodes. NCUBE machines in ***NCUBE/10*** had up to 1024 (= 2^{10}) nodes in ten-dimensional cube. Thinking Machine Corporation implemented hypercube networks in their popular ***CM-2 machine***. But, due to poor scalability and difficulty in packaging higher-dimensional hypercubes, and above all, with the emergence of more attractive alternatives, the hypercube networks gradually lost much of their popularity in the early 1990s, and were finally replaced by other emerging better architectures.

A brief detail of the advantages of hypercube is given on the website: http://routledge.com/9780367255732.

10.6.4.2 Mesh and Torus

One of the most natural elegant ways of interconnecting a large number of nodes is by means of a *mesh*. A 4 × 4 mesh network with 16 nodes is shown in Figure 10.13a. The links between the nodes here also are bi-directional. Routing of a message in a mesh network can

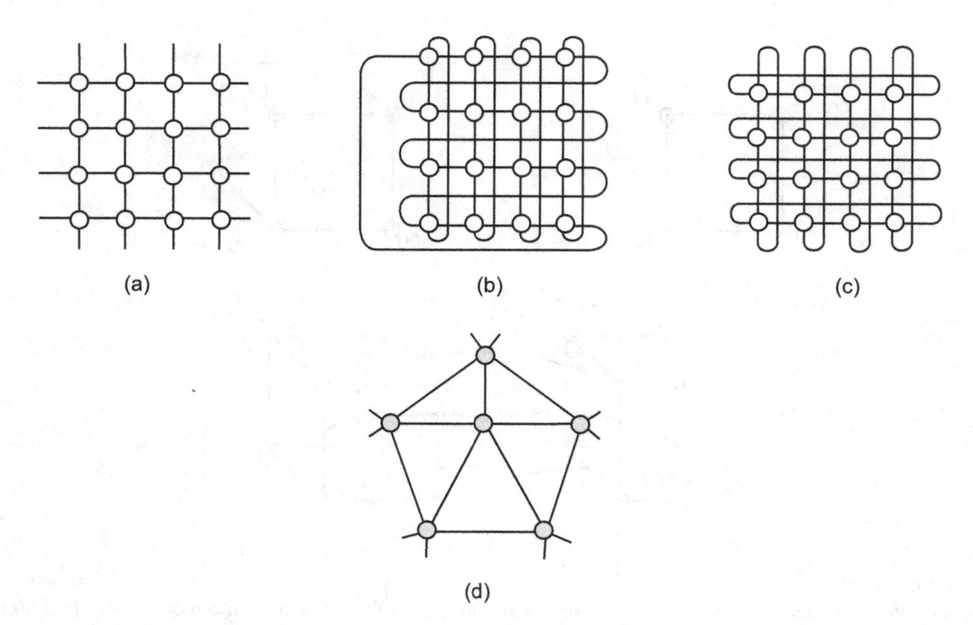

FIGURE 10.13
(a) Mesh, (b) Illiac 4 × 4 mesh, (c) Torus, and (d) Systolic array.

be carried out in several different ways. One of the simplest and most effective approaches is to choose the path between any source node N_x and a destination node N_y such that the transfer first takes place in the horizontal direction from N_x towards N_y. When the column in which N_y resides is arrived at, the transfer then starts to proceed vertically along this column. Examples of mesh-based multiprocessors are ***Intel Paragon*** (2-dimensional) replacing its hypercube predecessors, ***MPP, CM-2, DAP, Tandem Himalaya Series*** (1993) etc., and many other experimental machines.

Mesh-based architectures have a lot of variations by way of allowing wraparound connections. The Illiac IV assumed a 4 × 4 Illiac mesh with a constant node degree of 4, and a diameter of 7 as shown in Figure 10.13b. The Illiac mesh is topologically equivalent to a ***chordal ring*** of degree 4.

If the wraparound connection is made between the nodes at the opposite ends as shown in Figure 10.15c, the result is a network that consists of a set of bi-directional rings in the X-direction, connected by a similar set of rings in the Y-direction. This is called a ***Torus.*** This topology combines the ring and mesh, and can be extended to higher dimensions. In general, an $n \times n$ binary Torus has a node degree of 4, and a diameter of $2 \lfloor n/2 \rfloor$ which is one half of that of the mesh. This results in a considerable reduction in the average latency while information is transferred, but at the cost of greater complexity. Such a network has been found in use in ***Fujitsu's AP 3000*** machines. Torus of higher dimension such as a three-dimensional Torus is observed in use in ***Cray MPP T3E*** multiprocessor.

10.6.4.3 Systolic Arrays

This is especially a class of multidimensional pipelined array architectures designed for implementing fixed algorithms. In Figure 10.13d what is shown is a systolic array specially designed for performing matrix multiplication. In this example, the interior node degree is 5. Static systolic arrays, in general, are pipelined with multidirectional flow of data streams. With fixed interconnections and synchronous operation, a systolic array matches the communication of the underlined algorithm. For special applications like signal/image processing, the systolic arrays may offer a better performance / cost ratio. However, its structure has limited applicability, and can be very difficult to program. Still, systolic array topology may be used in building VLSI array processors, and also have been found in use in commercial machine ***Intel iWarp*** system.

10.6.4.4 Ring

One of the simplest topologies of interconnection networks is a *ring* which is obtained by connecting the two terminal nodes of a linear array with one extra link. This is illustrated in Figure 10.14a. The links used in a ring of N nodes may be unidirectional, and the diameter would be then N. When the link being used is bi-directional, the diameter is $\lfloor N/2 \rfloor$. Ring is symmetric in topology with a constant node degree of 2. Ring is easy to construct and implement. The link of a node is kept sufficiently wide to accommodate a complete packet in parallel for both of its neighbours.

The **token ring** of IBM has a topology in which a message circulates along the ring through the nodes until it reaches the destination with a matching token. Whether pipelined or packet-switched, rings have been used as an interconnect in ***CDC multiprocessors*** and in the ***KSR-1*** and ***KSR-2*** computer systems from Kendal Square Research. ***Exemplar V2600*** from Hewlett-Packard also uses ring networks.

It is not desirable to construct a long ring to connect many nodes because the latency in transfer would be unacceptably large. Rings can also be used as building blocks in other topologies such as hypercubes, meshes, trees and fat trees. When rings are used at any level of these topologies, the result is a hierarchy of rings that looks lucrative for communication but there may be a severe bottleneck at the highest-level ring.

Various forms of rings can be obtained by increasing the node degree. If the node degree can be increased from 2 to 3, 4, and even more, we obtain different ***chordal rings***. One such chordal ring with node degree 3 is shown in Figure 10.14b. If the extra links of a node in a ring go to those nodes having a distance equal to an integer power of 2, we obtain a ring known as ***barrel shifter***. This implies that a node i is connected to j, if $|j - i| = 2^k$ for $k = 0, 1, 2, \ldots, n - 1$ in a network having nodes $N = 2^n$. The node degree of a barrel shifter moderately increases with a notable decrease in diameter when compared to a *chordal ring*, and

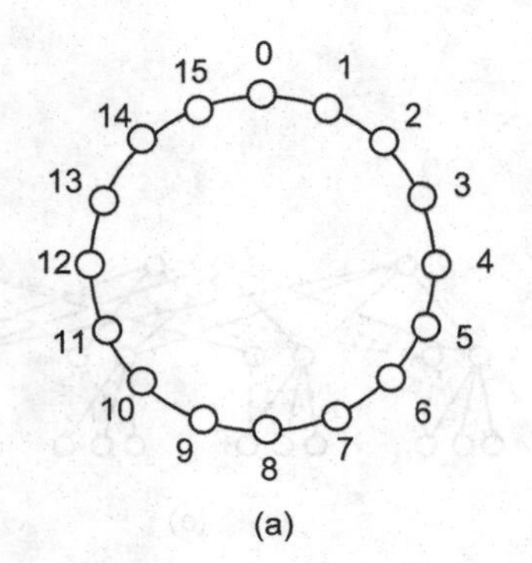

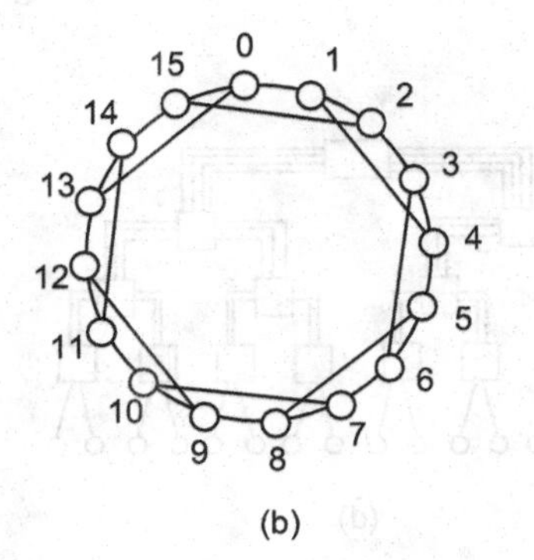

FIGURE 10.14
Ring-based networks. (a) Ring, and (b) Chordal ring.

is not too complex to implement. When more links are added, the complexity increases, but higher the node degree, shorter the network diameter and that ultimately results in reduction of latency. In the extreme, in the ring having a completely connected network (each node is directly connected to all other nodes in the ring) consisting of N nodes, the node degree is the maximum, that is $N - 1$, with the shortest possible diameter of 1 but it would be most complex and equally costly to implement.

10.6.4.5 Tree and Star

Tree topology is another form of interconnection networks that is hierarchically structured. Tree may be of various types depending on the degree of its intermediate nodes but only one path at a time can be established through a given node in the tree. Binary tree in this regard has special importance when the communication aspect is concerned. A complete binary balance tree of k-level can accommodate a total number of nodes $N = 2^k - 1$. The maximum node degree is 3 and the diameter is $2(k - 1)$. Figure 10.15a illustrates a specimen of a ***binary tree***. Figure 10.15b, however, shows a ***multi-way tree***. Networks realized in the form of a tree work satisfactorily if most of the communications occur with nodes having close proximity, otherwise the root node may become a bottleneck. A tree structure having constant node degree can be attributed as scalable architecture; binary tree is such a tree, although its diameter is comparatively long, but at the same time easier to implement.

The ***star*** is basically a tree of *two-level* as shown in Figure 10.15c. The node degree is very high which is $d = N - 1$ for a star having N nodes. The star structure is best suited in systems having centralized architecture with a supervisor (master) node.

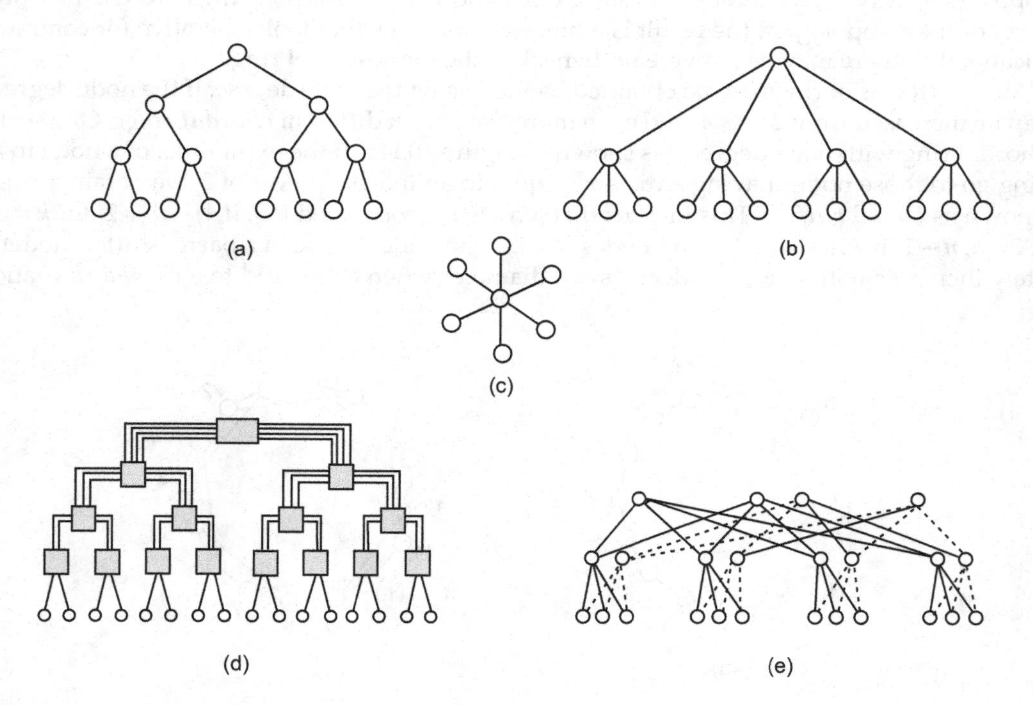

FIGURE 10.15
Tree-based networks. (a) Binary tree, (b) Multiway tree, (c) Star, (d) Binary fat tree, and (e) Multiway fat tree.

10.6.4.6 Fat Tree

The drawbacks of a (binary) tree having the possibility of a bottleneck at the upper levels have been reduced by way of increasing the number of links in the upper levels of a tree. The modified tree is known as *fat tree*, as introduced by Leiserson in 1985 in which each node in the tree (except at the top level) has more than one parent. Two types of fat tree are shown in Figure 10.15d and e that indicates the channel width of a fat tree increases as one traverse from leaves to the root. The fat tree structure-wise resembles a natural tree in which branches get thicker towards the root. A fat tree structure has been applied in ***Connection Machine*** *CM-5* from Thinking Machine Corporation replacing the previous hypercube implementation which was used in its earlier version ***CM-2***.

10.6.4.7 Transputer

Transputer is a tiny RISC-type microprocessor chip used as a building block to realize interconnection networks for parallel computers. It consists of a small processor having an extremely reduced instruction set consisting of only 16 instructions, like LOAD, STORE, ADD etc., and all instructions are 1 byte long having 4 bits opcode for each of 16 ($2^4 = 16$) instructions. It contains a total of six 32-bit registers, namely, the program counter, a stack pointer, an operand register, and a register file of 3 registers, an on-chip RAM of 4K bytes (depending on model), and four pairs of simplex (i.e. unidirectional) I/O channel interfaces. External memory can also be added to increase its capacity. Simple message-passing hardware protocol is used for interprocessor communications that provides a very small delay of about 6 µs for short messages of 32-bits, and a fairly high bandwidth (megabytes / sec) for long messages. The ***INMOS*** Transputer chip is such a compute-communication microprocessor. This design is much attractive from a packaging point of view due to its requirement of only a small number of pins (8 pins for 8 links) that can be packed tightly on a printed circuit board. Since most networks have a node degree less than 4, networks which have a constant node degree of 4 or less can fairly select a transputer (such as ***T800***) as its building block.

A brief detail of the transputer with the figure is given on the website: http://routledge.com/9780367255732.

10.6.5 Comparison of Static Networks

Although most of the merits and drawbacks of each topology have been mentioned in their respective discussions, a few points still have been summarized here for a better understanding of static networks.

Low-dimensional networks such as ***mesh*** and ***ring*** being characterized by point-to-point links connecting adjacent nodes can operate better under non-uniform loads because they can have more resource (adjacent wires) sharing. These nodes can be driven at high clock rates due to having lower average block latency yielding a higher maximum throughput. Packet switching mechanism is employed using store-and-forward method. **Ring** network, by virtue of its topology, possesses broadcast capability, while broadcasting in a mesh network is more difficult due to its inherent topological limitations. Both work well in multiprocessors of smaller configuration, and have almost uniform scalability without difficulty. However, incremental expansion is comparatively limited in a ring than in a mesh network, but a ring is simpler to implement. Mesh systems are, however, found to be suitable for use in both small and in very large systems, while ring can support at best a moderate size.

In a high-dimensional network such as in ***binary n-cube*** (hypercube), all connecting wires are dedicated to a particular dimension, and cannot be shared between dimensions. This may give rise to uneven traffic load on certain wires when the others lying in different dimensions are relatively idle. Circuit switching mechanism is employed using a *worm-hole* method (a hardware routing) in order to minimize communication delay caused by the increase in its diameter. The presence of a higher number of links affects the network cost, and creates difficulty in packaging. However, for larger multiprocessors, this network can be used when each of its nodes can be connected to a smaller multiprocessor having dedicated I/Os and / or using shared I/Os that can operate in parallel to satisfy massive computation demand. This gives rise to an architecture known as ***multi-multiprocessor***.

10.6.6 Hybrid (Mixed Topology) Networks

Several different interconnection network topologies have been already discussed along with their merits and drawbacks. In quest of more enhanced performance than what a particular topology could offer, and that too at a competitive price, the designers of multi-processors and of MPP (massively parallel processing) machines were inclined to develop a mixed topology to obtain a relatively superior performance. This mixed topology would be realized taking useful features and characteristics of different topologies with an appro-priate blending in implementation. For low-end multiprocessors, bus and crossbar could be individually chosen as one member in the mixed topology to build an interconnection network. A cluster of processors (2–8) could be connected with each other using a bus or a crossbar to form a node; these nodes can then be interconnected using another suitable topology to form a larger system.

Hewlett-Packard's ***Exemplar V2600 system*** uses nodes where each node is made up of multiple processors being connected by a *bus*. These nodes are then interconnected using a *ring network. Data General's* ***AV2500 system*** follows the same approach using the same type of topology. ***Alliant FX/2800 system*** follows the same approach using a different type of topology. ***Compac's AlphaServer SC*** uses nodes of multiple processors connected by a *crossbar switch* and these nodes are then interconnected using a *fat tree* network.

Various sizes of multiprocessors ranging from relatively smaller to comparatively larger are found in market, demanded by users engaged in processing diverse spectrum of applications. Many machines in the market have about 4–128 processors. For very small systems, say, up to 16 processors, the most effective and reasonable choices are a shared bus or a crossbar switch (dynamic interconnection networks). For moderate size of multi-processors, the obvious choice is mesh networks. For handling massively parallel process-ing (MPP) systems, some of the static topologies (n-cube) are more scalable in some specific applications. For general-purpose computing, large-scale MINs or crossbar networks are suitable, and may become more viable depending on the advancement in electronics tech-nology (Siegel, H. J.).

10.6.7 Important Characteristics of a Network

Several different network topologies with their own merits and drawbacks are involved in building various types of interconnection networks used in a multiple-processor system. The selection of a particular interconnection network is very critical to decide, mostly depending on the environment where it will be used, along with fulfilment of certain fun-damental objectives and requirements. The other salient features present in a particular topology determine its superiority over others in practical situation. Different topologies, however, attain different levels of certain underlying characteristics (Feng, T. Y., et al.). Some of the important characteristics are: (i) network speed and its throughput; (ii) the routing mechanism including broadcast and multicast facilities; (iii) scalability; (iv) fault tolerance; (v) reliability; (vi) ease of implementation and related costs.

However, topological equivalence has been established among a number of network architectures. The most important for an architecture to survive in future systems is mostly its *packaging, efficiency*, and *scalability* to allow modular growth.

A brief explanation of each of these characteristics is given on the website: http://routledge.com/9780367255732.

10.7 Multiprocessor Architectures

Multiprocessor are systems with multiple CPUs capable of independently executing different tasks in parallel. Apart from having *shared common memory* or *unshared distributed* memories, these processors also share resources such as, communication facilities, I/O devices, system utilities, program libraries, databases, and similar others. They are operated under the control of an integrated operating system that provides interaction between processors and their programs at the job, task, file, and even in data element level. Multiprocessors can be *classified* in a number of ways: (i) The number of CPUs present in a system. *Modestly parallel* systems contain 2 to about 30 processors, while *massively parallel* systems can contain even thousands of such processors. (ii) The patterns of interconnections that are created between the CPUs and the memory modules and that too, whether the memory modules would be *centrally shared* or *distributed shared*. (iii) The way the multiple CPUs themselves will be interconnected with one another. (iv) The form of interconnection networks (i.e. whether static or dynamic) to be used. Still, many other aspects remain that are considered at the time of *multiprocessor implementation*.

Multiprocessor systems are best suited for general purpose multi-user applications where major thrust is on programmability. Shared-memory multiprocessors can form a very cost-effective approach, but latency tolerance while accessing remote memory is considered a major shortcoming. Lack of scalability is also a key limitation of such a system. Distributed shared memory (DSM) multiprocessors, however, address all these issues, and resolve most of all these drawbacks to a considerable extent by way of providing an extended form of stringent shared-memory multiprocessor architecture.

10.7.1 Shared-Memory Multiprocessor

The multiprocessor architecture in which the primary memory is shared is usually called ***shared-memory multiprocessor***. In this multiprocessor, there is only a *centrally shared global memory* having a single virtual address space that is accessed and shared by all processors. In addition, each processor may also have an extra private cache (local memory) to further speed up the operation. Peripherals can be attached in some other form of sharing. These multiprocessors are mostly not fit to be scalable; they are sometimes referred to as ***tightly-coupled*** since high-bandwidth communication networks are used to extract a high degree of all types of resource sharing (Catanzaro, B.).

Shared memory does not always mean that there is only a single centralized memory that is to be shared. In fact, the advent of constantly emerging more powerful VLSI technologies since mid-1980s onwards, offered an abundance of hardware capabilities and numerous options within affordable cost. The conventional definition and the traditional architecture of the multiprocessor have been then radically changed. It became feasible to develop even one-chip powerful multiprocessor (***multicore***, also known as ***chip multiprocessor***, see Section 8.12.2) and larger capacity RAM at reasonable cost. Large-scale multiprocessor architectures then started to evolve with multiple memories that are now distributed with the processors. Here, each CPU can quickly access its local memory, and accesses to the other memories connecting with other CPUs are also possible, but are relatively slower in operation. That is, these physically-separated memories can now be addressed as one logically shared address space, meaning that any memory location can be addressed by any processor, assuming that it has the approved access rights. This, however, does not discard the basic shared memory concept of multiprocessors; rather this extends it in a broader sense ventilating the concept that is known as *distributed shared memory*. Multiprocessors having **distributed shared memory** (DSM) are, however, relatively scalable and sometimes referred to as ***loosely-coupled multiprocessors***. Here also, peripherals can be attached in some other suitable form of sharing (Vranesic Z.G., et al.).

These two types of multiprocessors, due to their inherent architectural differences, can also be differentiated primarily in terms of their speed and ease with which they can interact on common tasks. Irrespective of the organisation of the multiprocessor, there actually exist two primary points of *contention*, namely the shared memory, and the shared communication network through which all sorts of interactions are made. Common cache memory (apart from using extra private cache (local memory) to each processor to further speed up its own activities) is often employed to reduce such contentions. These two types of multiprocessors also differ in the types of interconnection network being used that eventually puts a significant impact on the bandwidth and saturation of system communications, thereby directly influencing the system performance as a whole (Mak, P., et al.). In addition, the other associated important issues, such as, cost, complexity, interprocessor communications, and above all, the scalability of the presented architecture need to be considered. However, the kind of interconnection network being used in both of these two types of multiprocessors may be in the form of:

- Common (hierarchical) bus
- A crossbar switch
- Hypercube
- Multistage switches (network) or in some other form

Shared-bus systems are relatively simple and popular, but their scalability is limited by bus and memory contention. Crossbar systems allow fully parallel connections between processors and different memory modules, but their cost and complexity grow quadratically with the increase in the number of nodes (see Crossbar switch]. Hypercubes and multilevel switches are scalable and their complexities grow only logarithmically with the increase in number of nodes.

Multiprocessors while using *shared memory* or *distributed shared memory* give rise to *three* different models that differ mainly in how the memory and peripheral resources are to be connected; *shared* or *distributed*. Three such common models are found, namely:

(i) Uniform Memory Access (**UMA**), (ii) Non-Uniform Memory Access (**NUMA**), (iii) No-Remote Memory Access (**NORMA**) (Yew P. C., et al.).

10.7.2 Symmetric Multiprocessors (SMP): UMA Model

In a centralized shared-memory multiprocessor known as UMA (Uniform Memory Access), each of *n* processors can uniformly access any of the common *m* memory modules at any point of time. In Figure 10.16 a, the CPUs, the memory, and the I/O subsystems are connected using an interconnection network in the form of a common (hierarchical) bus. Here, each CPU chip contains an on-chip level 1 (L1) private cache, and in addition, a CPU may also have an on-chip dedicated or shared L2 cache apart from using an off-chip L3 cache. Whatever be the architectural improvement be carried out, of course, within the confines of the basic concept to speed up the execution, the shared bus can service only one request out of many already arrived at. Hence, the bus itself eventually becomes a hot spot creating a severe bottleneck that summarily limits the performance of the entire system. The CPU would then have to face an unpredictable delay while accessing the shared memory.

In order to minimize this hot-spot problem, Figure 10.16 b shows an alternative design that uses a crossbar network to connect the CPUs and the I/O subsystems with the memory units. Here, the CPUs and the I/O subsystems face relatively less delays in accessing memory, since the crossbar network as usual provides several alternative connections that can be used in parallel as long as they do not conflict in their source or destination entities. The delays caused in the crossbar switch would also be more predictable than those of the bus. System performance, however, would be appreciably better than that using a bus as an interconnection network.

The UMA model of multiprocessors can be again divided into two categories: *Symmetric* and *Asymmetric*. When all the processors in the system share equal access to *m* shared memory modules as well as to all shared I/O devices through the same channels or

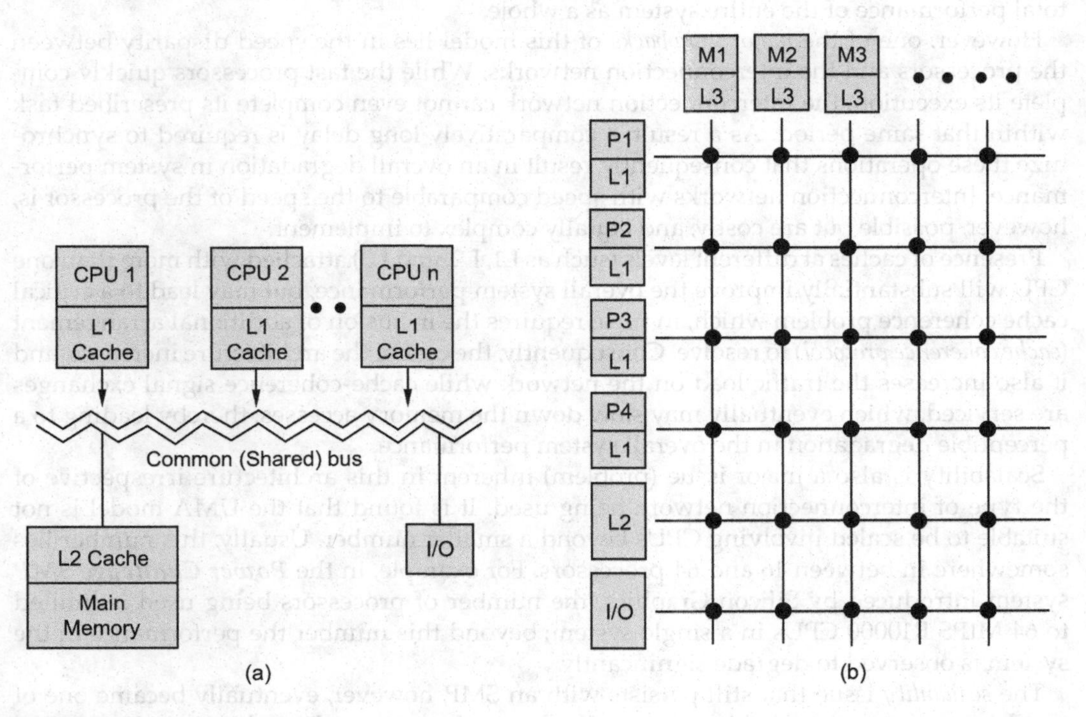

FIGURE 10.16
A scheme of UMA model (shared-memory multiprocessor) (a) System bus and (b) Crossbar Switch.

through different channels that provide paths to the same devices, the multiprocessor is called a ***symmetric multiprocessor*** (**SMP**). This is illustrated in Figure 10.16a and b using interconnection network in the form of a common (hierarchical) bus and crossbar network, respectively. However, other forms (types) of interconnection network can also be employed in place of crossbar network. In this category, all the processors are allowed to run all sorts of interrupt service routines and other supervisor-related (kernel) programs. In the ***asymmetric category,*** not all but only one or a selective number of processors in the multiprocessor system are permitted to additionally handle all I/O and supervisor-related (kernel) activities. Those are treated as *master processor(s)* that supervises the execution activities of the other remaining processors, known as attached processors. However, all the processors here also have uniform access to any of *m* shared memory modules as usual.

UMA model is easy to implement and is found to be suitable in the general-purpose multi-user applications under time-sharing environment, as well as also in parallel processing applications. Parallel processes here must communicate by software using some form of message passing by putting messages into a buffer in the shared memory, or by using lock variables in the shared memory. Interprocessor communication and synchronization are normally carried out using shared variables to be located in the common memory.

One *distinct advantage* of SMP is that it continues to operate even in the event of certain failures of some CPUs, but, of course, affecting only with a graceful degradation in the performance of the entire system. Failure of a processor in most of the situations is not so severe to the operation of other processors present in the system if it is not executing the kernel code at the time of failure. At best, only the process(es) availing the service of the failed processor would be affected, and the other processes henceforth will be barred from getting the service of the failed processor, and that may only affect, to some extent, the total performance of the entire system as a whole.

However, one of the *major drawbacks* of this model lies in the speed disparity between the processors and the interconnection networks. While the fast processors quickly complete its execution, the interconnection network cannot even complete its prescribed task within that same period. As a result, a comparatively long delay is required to synchronize these operations that consequently result in an overall degradation in system performance. Interconnection networks with speed comparable to the speed of the processor is, however, possible but are costly, and equally complex to implement.

Presence of caches at different levels (such as L1, L2, and L3) attached with more than one CPU will substantially improve the overall system performance, but may lead to a critical cache coherence problem which, in turn, requires the inclusion of additional arrangement (*cache coherence protocol*) to resolve. Consequently, the cost of the architecture increases, and it also increases the traffic load on the network while cache-coherence signal exchanges are serviced, which eventually may slow down the memory accesses, thereby leading to a perceptible degradation in the overall system performance.

Scalability is also a major issue (problem) inherent in this architecture irrespective of the type of interconnection network being used. It is found that the UMA model is not suitable to be scaled involving CPUs beyond a smaller number. Usually, this number lies somewhere in between 16 and 64 processors. For example, in the ***Power Challenge SMP*** system introduced by Silicon Graphics, the number of processors being used is limited to 64 MIPS R10000 CPUs in a single system; beyond this number the performance of the system is observed to degrade significantly.

The *scalability* issue that still persists with an SMP, however, eventually became one of the driving motivations behind the development of a large-scale multiprocessing system while retaining most of the flavour of SMP intact. The ultimate outcome is the emergence of a new architecture, what is known as the NUMA architecture. The basic objective of NUMA is to maintain a transparent system-wide memory while permitting even multiple multiprocessor nodes in the architecture; each such node however, is itself a multiprocessor consisting of a collection of multiple processors (CPUs) equipped with its own resources and local interconnect system.

10.7.3 Distributed Shared Memory Multiprocessors (DSM): NUMA Model

The two popular approaches that eventually converted to numerous commercial products for providing a multiple-processor system to support applications are SMPs and Clusters (to be discussed in Section 10.9). Another architectural approach that has drawn considerable interest in this area is the non-uniform memory access (NUMA) architecture.

NUMA is a comparatively more attractive form of a shared-memory multiprocessor system where the shared memory is physically distributed (attached) directly as the local memory to all processors so that each processor can sustain a high computation rate due to faster access to its *local memory.* A memory unit local to a processor can be globally accessed by the other processors with an access time that varies with the location of memory word. In this way, the collection of all local memories attached to individual processors forms a global address space shared by all processors. NUMA machines are thus legitimately called ***distributed shared-memory*** **(DSM)** or *scalable shared-memory* architectures. A slightly different implementation of NUMA multiprocessor is even supplemented by a physical *remote* common memory, in addition to its existing usual distributed memory that is local to a processor but global to other processors. As a result, this scheme forms a memory hierarchy where each processor has the fastest access to its local memory. The next is its access to global memories that are individually local to other processors. The slowest is the access to common remote memory. Figure 10.17 exhibits a representative architectural scheme of such a NUMA system in which the large rectangle encloses a node of the system. A few of the many commercial products introduced on the basis of NUMA architecture are, namely, the ***BBN TC-2000***, which uses 512 *Motorola 88100 RISC processors* and a *butterfly network* for interconnections. The ***Silicon Graphics Origin NUMA*** *system* is designed to support up to 1024 *MIPS R10000* processors, and the ***Sequent NUMA-Q*** is built, with up to 252 *Pentium II* processors.

Similar to SMP, the architecture used in the NUMA model must ensure coherence between caches attached with CPUs of a node, as well as the coherence between existing non-local caches. Consequently, this requirement, as usual, consumes part of the bandwidth of interconnection networks that eventually may cause memory accesses to be slowed down (Stenstrom, P., et al.).

Usually, the nodes in the NUMA (DSM) architecture are typically high-performance SMPs. Each such SMP contains around 4 or 8 CPUs, and all SMPs are then connected by a high-speed global interconnection network. Due to the availability of this non-local communication network, the NUMA architecture becomes fairly scalable, and it is now possible to add more nodes, whenever needed, to obtain a more improved performance (Grindlay, R. et al.). The actual performance of a NUMA system, however, depends mostly on the non-local memory accesses made by the processes following the memory hierarchy during their execution. This issue, however, lies within the domain of OS that would be addressed in the next section.

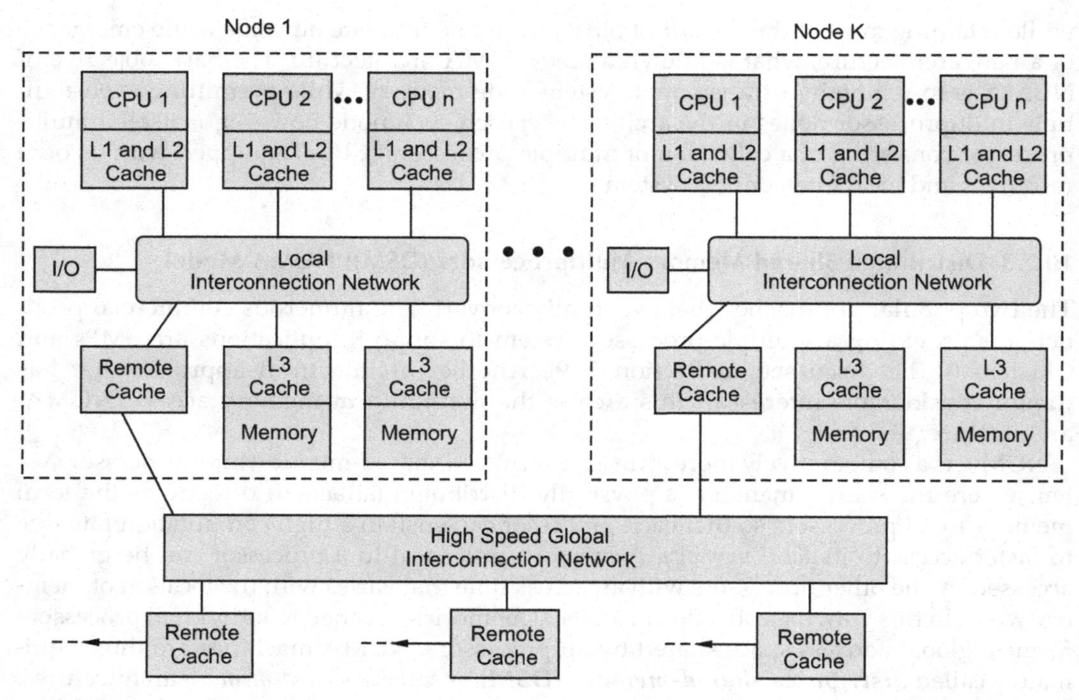

FIGURE 10.17
A scheme of NUMA model (Distributed shared memory architecture).

10.7.4 Cache-Coherent NUMA: CC-NUMA Model

The wide acceptance **of NUMA** architecture-based machines in the commercial market encouraged designers to redefine and reconstruct the existing NUMA architecture that would be mostly free from its inherent drawbacks. One of the critical shortcomings in the traditional NUMA model is the cache coherence problem between caches attached with CPUs within a node, as well as inconsistency between existing non-local caches (caches attached with other nodes). As a result, the popular commercial products that were subsequently launched without having a cache-coherence problem are **CC-NUMA systems** (Stenstrom, P., et al.), in which cache consistency is rigorously maintained among the various caches attached with their respective processors (Scheurich, C., et al.). While CC-NUMA systems are quite distinct from both **SMPs** and **Clusters** (to be discussed in Section 10.9), they are still considered to be more or less equivalent to Clusters, although they are often referred to in the commercial literature as CC-NUMA systems.

As usual, multiple independent nodes, each of which is essentially an SMP organisation, is the basic building block of the overall CC-NUMA organisation. Figure 10.18 depicts such a representative CC-NUMA organisation, in which each node contains multiple processors, each of which with its own L1 and L2 caches, plus main memory (Lee, R. L., et al.). While each node consisting of multiple processors is connected with a private local main memory of their own, from the point of view of all the processors present in the entire system, there is only a single addressable transparent system-wide memory in which each location has a unique system-wide address. In fact, when a processor initiates a memory access, if the requested memory word is not in the processor's cache (considering both L1 and L2), the L2 cache initiates a fetch action across the local bus to get the desired

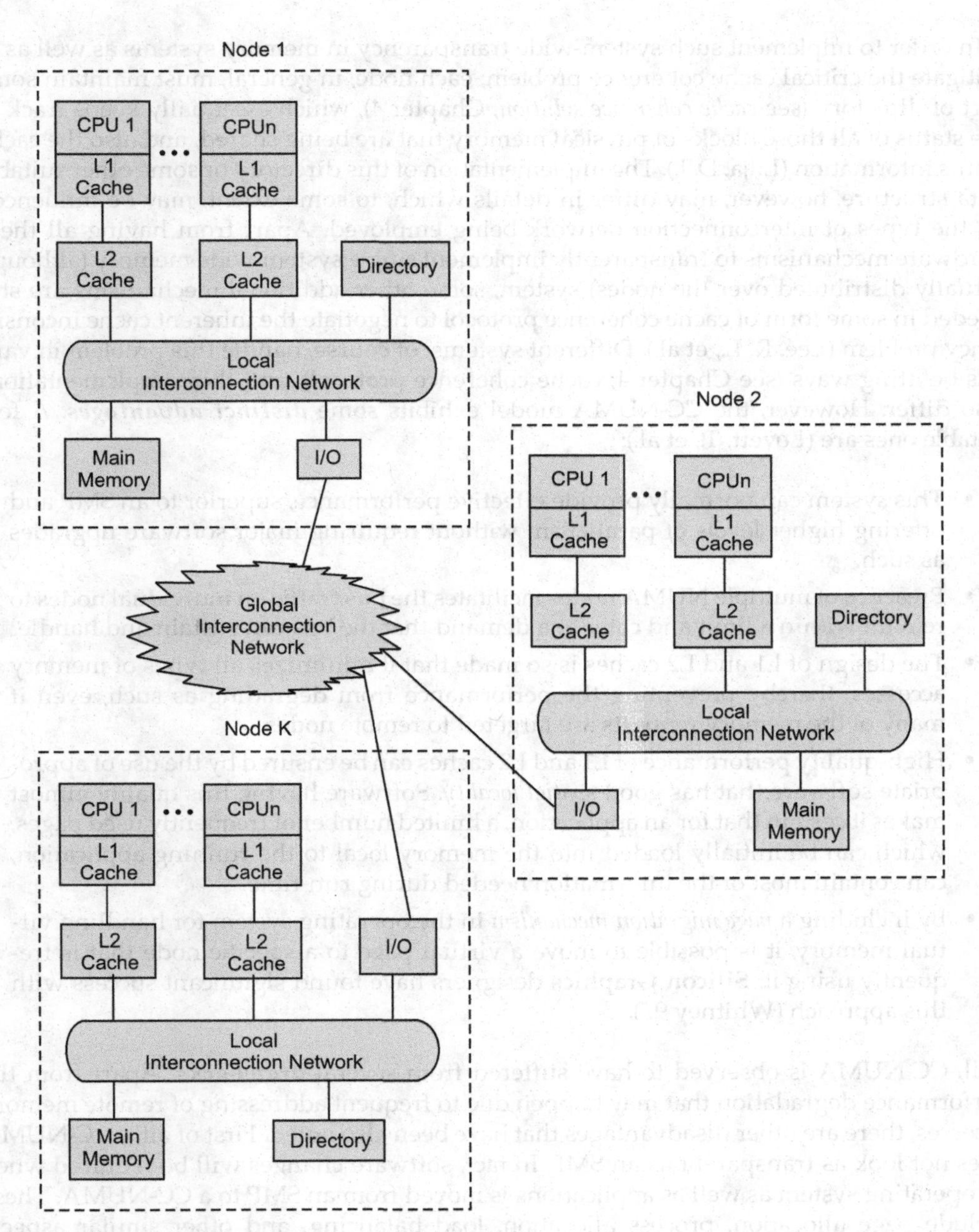

FIGURE 10.18
A scheme of a CC-NUMA organisation.

word from the local portion of the main memory. If the desired word is not found, then the global portion of the main memory is consulted. An automatic request is then issued across the interconnection network to fetch the word; it is then delivered to the local bus of the requesting processor, and then is finally delivered to the requesting cache on that bus. The whole proceedings are carried out most transparently in an automatic manner (Scheurich, C., et al.).

In order to implement such system-wide transparency in memory systems as well as to mitigate the critical cache coherence problem, each node, in general, must maintain some sort of directory (see *cache coherence solution*, Chapter 4), which essentially keeps track of the status of all those blocks of physical memory that are being shared, and also the cache status information (Lilja, D. J.). The implementation of this directory or some other suitable data structure, however, may differ in details which, to some extent, may be influenced by the types of interconnection network being employed. Apart from having all these hardware mechanisms to transparently implement single system-wide memory (although actually distributed over the nodes) system, some other additional mechanisms are still needed in some form of cache coherence protocol to negotiate the inherent cache inconsistency problem (Lee, R. L., et al.). Different systems, of course, handle this problem in various befitting ways (see Chapter 4; cache coherence protocol), and their implementations also differ. However, the CC-NUMA model exhibits some ***distinct advantages***. A few notable ones are (Lovett, T., et al.):

- This system can normally provide effective performance, superior to an SMP and offering higher levels of parallelism without requiring major software upgrades as such.
- Presence of multiple NUMA nodes facilitates the bus traffic on individual nodes to remain within a limit and cater to a demand that the bus can sustain and handle.
- The design of L1 and L2 caches is so made that it minimizes all types of memory accesses, thereby preventing the performance from degrading as such, even if many of the memory requests are targeted to remote nodes.
- High-quality performance of L1 and L2 caches can be ensured by the use of appropriate software that has good *spatial locality*. Software having this quality almost makes it certain that for an application, a limited number of frequently used pages, which can be initially loaded into the memory local to the running application, can contain most of the information needed during run-time.
- By including a *page migration mechanism* in the operating system for handling virtual memory, it is possible to move a virtual page to a specific node that is frequently using it. **Silicon Graphics** designers have found significant success with this approach (Whitney 97).

Still, CC-NUMA is observed to have suffered from several ***drawbacks***. Apart from the performance degradation that may happen due to frequent addressing of remote memory accesses, there are other disadvantages that have been also noted. First of all, a CC-NUMA does not look as transparent as an SMP. In fact, software changes will be required when an operating system as well as applications is moved from an SMP to a CC-NUMA. These include page allocation, process allocation, load balancing, and other similar aspects that are normally carried out by these operating systems. The second one is essentially that of its *availability*. Actually, this is rather a complex issue and depends mostly on the exact implementation of the CC-NUMA system within the periphery of present hardware improvement (Stenstrom, P., et. al).

10.7.5 No Remote Memory Access (NORMA)

Multiprocessors having a *global memory* system allow any processor to access any memory module without any hindrance created by another processor. A different way of organizing

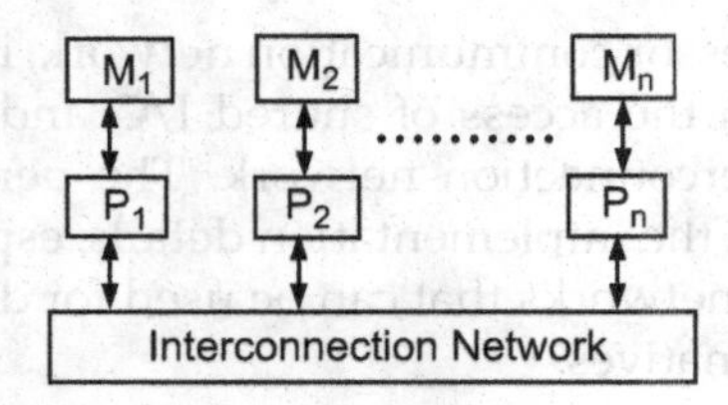

FIGURE 10.19
A distributed memory system multiprocessor.

this system can also be possible, as shown in Figure 10.19. All memory modules here serve as private memories for the processors that are directly connected to them (in contrast to single system-wide memory forming with all the memories connected to other processors or distributed over the nodes). No processor can access a global memory without the cooperation of the concerned remote processor. This cooperation can normally take place in the *form of messages* exchanged by the processors. Such systems are often called NORMA or ***distributed memory systems*** with a ***message-passing protocol***.

10.7.6 General-Purpose Multiprocessors

A generalized multiprocessor system can be envisaged by combining the salient features of the UMA, NUMA and CC-NUMA (also COMA) models, as already discussed, that mainly use processors and memory modules as major functional units. But, any multiprocessor must also provide extensive I/O capability to manage the growing user density in its working environment, and hence, it should not be kept set aside in the design concept of the generalized multiprocessor system organisation. Above all, the system must include efficient interconnection networks for fast communication among multiple processors and shared memory, I/O, and peripheral devices. A high-level view of such a possible multiprocessor organisation is depicted in Figure 10.20.

Each processor P_i is provided with its own local memory and private cache. Multiple processors are connected to shared-memory modules through an interprocessor-memory interconnection network. Processors are themselves directly interconnected with one

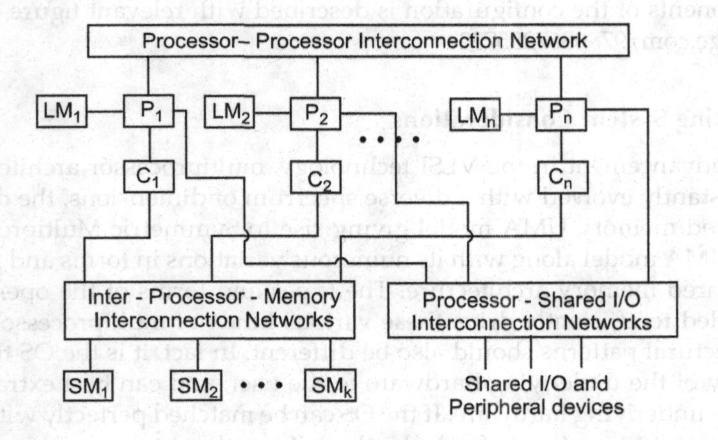

FIGURE 10.20
Generalized multiprocessor system with interconnection structures using local memory, cache memory, shared memory and shared peripheral devices.

another through an interprocessor communication network, instead of through the shared memory. The processors gain the access of shared I/O and peripheral devices through a separate processor-I/O interconnection network. The performance and cost of these machines critically depend on the implementation details, especially, on the efficiency and capability of various types of networks that can be used for different purposes of communication, with numerous alternatives.

10.7.6.1 Implementation: A Mainframe SMP (IBM z990 Series)

An SMP realized with the use of a single shared bus is probably one of the most common arrangements for PCs and workstations. But with this arrangement, the single bus becomes a bottleneck while managing growing traffic load, and also critically affects the scalability of the design. An alternative approach has however been used in a recent implementation introduced by IBM in its z-series mainframe family called the z990 series (Siegel, 04, Mak, 04). This family of systems is absolutely scalable, spanning a range from a uniprocessor with one main memory card to a high-end system consisting of 48 processors and 8 memory cards. The system contains a particular type of unit, called **book,** which is a pluggable unit that can contain up to 12 processors with up to 64 GB of memory, I/O adapters, and a 32 MB L2 cache in a *system control element* (SCE) which connects other elements. The z990 system comprises of *one* to a maximum of *four* such books. Each book contains two self-controlled memory cards (for a total of eight such cards in four books across the maximum configuration), each card can handle its own memory accesses at relatively high speeds. Each processor chip is dual-core consisting of two identical CISC superscalar microprocessors used as a central processor (CP). Each processor is connected to a single memory card (actually to L2 cache) by a *point-to-point* links (switched interconnections) and each L2 cache, in turn, has a link to each of the two memory cards on the same book via the *memory store control* (MSC). Instead of using bus, point-to-point links (switched network) are also employed here to connect I/O channels. Besides, many other additional nice features, however, have been also included in the design that altogether contributes to yield significant performance improvement as well as an appreciable reduction in average bus traffic.

A brief detail of a representative scheme of the IBM z990 multiprocessor structure with the key components of the configuration is described with relevant figure onthe website: http://routledge.com/9780367255732.

10.7.7 Operating System Considerations

Due to rapid advancement in the VLSI technology, multiprocessor architectures by this time have constantly evolved with a diverse spectrum of dimensions, the dominant ones being the shared memory UMA model giving rise to Symmetric Multiprocessor system (SMP), and NUMA model along with its numerous variations in forms and patterns using distributed shared memory architecture. The types and forms of the operating systems (OS) thus needed to efficiently drive these various kinds of multiprocessors having dissimilar architectural patterns should also be different. In fact, it is the OS that essentially projects a view of the underlying hardware to the user, and can best extract the highest potential of the underlying hardware. If the OS can be matched perfectly with the underlying hardware architecture (made for each other), it can then stage a magical performance going even beyond the highest level of expectations, and be able to extract the best out of the hardware capabilities provided. However, the details of the various types of operating systems to be employed to drive these different kinds of multiprocessor models is outside the scope of this book at present; interested readers are hereby referred to consult any standard text book in this area, including the one by the same author, (Operating Systems; Chakraborty, P, Jaico Publications, 2011).

The need of an appropriate operating system to efficiently drive each of the various kinds of multiprocessors having dissimilar architectural forms is described on the website: http://routledge.com/9780367255732.

10.8 Multicomputer Architectures

The first group of MIMD is a multiprocessor with all its variants, and the second group of MIMD is the evolution of machines in which physical memory is distributed among the processors to support large processor counts. For this reason, this form of MIMDs of those days have also been called *no-remote-memory-access* (NORMA) machines. Typically, memory as well as I/O is distributed among the processors to reduce memory-access latency to yield a cost-effective higher bandwidth. Therefore, each individual processor–memory–I/O module thus here forms a unit, called a ***node***, which is essentially a separate standalone autonomous computer. A machine (arrangement), when built with these nodes (multiple computers) together, is thus rightly called a ***multicomputer***, or a loosely-coupled **distributed computing system**. The nodes in this arrangement (system) have network interfaces to connect with one another through an interconnection network. They communicate with one another, and also with the non-local (local to some other node) memories by explicitly *passing messages* across this interconnection network. As a result, the inter-node message delay is relatively large, and the data rate is comparatively low. Peripherals can also be attached using some other form of sharing. The basic structure of a distributed-memory multicomputer is shown in Figure 10.21.

Multicomputers when built with such interconnected self-sufficient nodes offer a very pleasant environment with a cost-effective high throughput for applications in which multiple independent jobs are simultaneously executed requiring little or almost no communication with one another. This type of multicomputers, however, provide a *coarse-grained parallelism* that proposes a ***system-level parallelism***. An extension of this approach can even be to have each node to further contain a small number of processors (2–8) instead of a single processor. These processors may be interconnected locally by a small bus or by a suitable different interconnection technology, which is often less scalable than the global interconnection networks being used. While this arrangement of multiple processors in a node together with a private memory and a local network interface may be a more useful approach from a cost-efficiency point of view, it is not fundamental to how these machines work.

Many variations of this basic scheme are also possible, including private cache memory, which is offered to the processor on each individual computer (node) to negotiate *two* primary points of *contention*, such as on the distributed memory, and on the shared communication network itself. However, the major *architectural differences* between the ***distributed-memory multiprocessor*** (such as ***NUMA model***), and the various types of ***multicomputers*** (distributed memory) are essentially how global communication is made, and also the logical architecture of the distributed memory itself. Usually, high-bandwidth communication networks are used in multicomputers to extract a high degree of resource

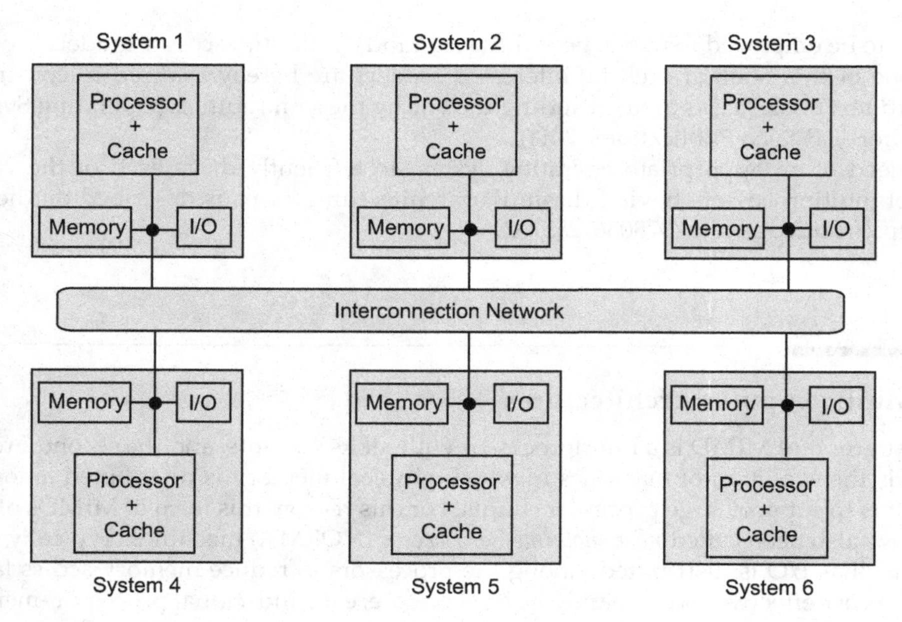

FIGURE 10.21
Basic architecture of a distributed-memory multicomputer.

sharing, of course, within the confines of its cost, complexity, inter-system communications, and similar other aspects. Summarily, the parallelism as obtained in a multicomputer is certainly more scalable (scalability) than that in a distributed shared memory multiprocessor (NUMA).

A brief detail of this topic is described on the website: http://routledge.com/9780367255732.

10.8.1 Design Considerations

Many different issues are entwined with multicomputer design. While solving these issues, different approaches may be taken,which again give rise to different categories of multicomputer, when a particular category is found most suitable and viable to handle a particular application environment. However, some of the most important fundamental issues are:

- A *Message passing mechanism* is required for processor–processor communication in multicomputer organisation that may be again of two forms; **synchronous mode,** which can be thought of as a *remote procedure call* (RPC), and the other form is **asynchronous mode**. However, the message passing mechanism used in different machines are also of fairly diverse types. The RPC is a ***software system*** that actually encapsulates the entire message-passing mechanism as well as the details of sending and receiving messages. Another way for message passing is to use ***hardware routers*** as are found in many modern multicomputers. A series of routers and channels are involved here in the message passing between any two participating nodes. Nodes of different types can also be mixed in this arrangement, which gives rise to a different category of multicomputers known as ***heterogeneous multicomputers***. The internode communications in a heterogeneous multicomputer are realized in a different way by introducing *compatible data representations* and *message-passing protocols.*

- *The types of network* and the *nature of interconnections and communications* are to be used between the nodes of a multicomputer. Various types of static network (already discussed) having different topologies, namely Ring, Tree, Mesh, Torus, Hypercube (nCube) etc. can be used to construct different types of multicomputers. Diverse patterns of communication such as point-to-point, broadcasting, multicasting, various permutations, etc. can be exploited in these multicomputers to satisfy certain performance targets. Other similar existing issues are also to be settled, and different approaches may however, be then taken to solve these issues that in turn, give rise to different categories of multicomputer to handle different types of application environments.
- Last but not the least is targeted over an issue, as how to build the fastest multicomputers (multiprocessors); whether using many small processors or a smaller number of faster processors to attain at least an expected level of parallel processing. After a prolonged debate, the road to reach the target is, however, assumed mainly relying on expected advancement in our ability to program parallel machines, and at the same time anticipating continued progress in microprocessor (both RISC and CISC) performance, and also looking forward to appreciable advancements in the design of parallel architecture. The goal with regard to processor performance was set for a computer capable of sustaining at least a teraFLOPS (one million MFLOPS), and that was expected to be availed by 1995. It was observed that a machine in 1994, however, already achieved 140,000 MFLOPS (0.14 TFLOPS) using a 1904-node Paragon machine containing 3808 small processors.

More details of this topic are described on the website: http://routledge.com/9780367255732.

10.8.2 Multicomputer Generations

Multicomputers have gone through a series of innovative developmental stages that can be narrated in terms of its several generations, targeting ultimately towards massive parallel processing (MPP). The *first generation* (1983–1987) was built up with multiple microprocessors using popular static interconnection networks (hypercube-based) and software-controlled message passing mechanisms, thereby offered an attractive cost / performance that eventually put a strong challenge against the most popular versatile mainframe machines of those days from multinational giants, like IBM, NCR, HP, DEC, and others.

The *second generation* (1988–1992) of multicomputers used faster microprocessors including RISC processors with advanced interconnection network, *mesh topology*, and *hardware-supported routing* schemes (*wormhole routing*) replacing the slower software-controlled *message-passing technique*. As a result, the communication latency came down to less than 5 μs from 6000 μs for global communication and that for local communication to 5 μs from 2000 μs, and the speed up was around ten times on average than its predecessors as a whole. This generation, however, offered *medium-grain parallelism* implemented at the processing level of tasks, procedures, and subroutines to balance this computational granularity with the existing communication latency for the sake of realizing proper synchronization.

The *third generation* (1993–1997) is a *fine-grained multicomputer* with a greater number of more powerful processors, faster routing channels, and moderate size of private memory to each processor, targeting ultimately towards *massive parallelism* by exploring *data parallelism* as well as *instruction-level parallelism*. Adequate supports from the language as well as the run-time software environment created by OS have been provided along with the use of optimizing compiler, which can automatically detect the parallelism and translate source code to an acceptable parallel form to be recognized by the run-time system during execution. Here, the trade-off is now simply between the level of parallelism required and the involvement of allied cost to realize that.

Today's multicomputers, while offering massive parallelism (MPP machines), have been further enhanced by way of combining the private virtual address spaces already distributed over the nodes into a globally shared virtual memory. To extract even more fine-grained parallelism with minimum possible latency in processor–memory handshaking, various levels of caches have been included, thereby replacing page-oriented message passing mostly by cache block-level mechanisms.

A brief detail of this topic is described on the website: http://routledge.com/9780367255732.

10.8.3 Different Models of Multicomputer Systems

Many variations of the basic scheme to construct multicomputers are used to build various models of this system. These models, however, can be broadly classified into many different categories, of which some common ones are:

1. ***Minicomputer model*** **(Peer-to-peer)**: This organisation usually consists of a few stand-alone minicomputers (may even be large supercomputers as well) interconnected by a suitable communication network, each may have several interactive terminals to accommodate multiple users and allow them to individually work concurrently on many *unrelated problems*, accessing remote data and resources, if required, that are collectively stored and managed on some other machines. The need to place individual objects and retrieve them, and to maintain replicas amongst many computers in order to distribute the load and to provide resilience in the event of individual machine faults or communication link failure, summarily renders this architecture relatively more complex than its counterparts, other popular forms of relatively simple architecture.
2. ***Workstation model***: A workstation is simply a stand-alone autonomous small computer (usually PC) equipped with its own disk (diskful workstation) and other peripherals, providing a highly user-friendly interface to a single-user. Several such workstations scattered over a wide area when interconnected by a communication network, give rise to multicomputers, normally called ***Workstation model***. Here, each machine, apart from its own use, might provide services to the others on request, and processes that run on separate machines can exchange data with one another through the network.
3. ***Workstation-server model***: This model, also commonly known as the ***client–server model***, comprises of mostly *diskless* (but a few may be *diskful*) workstations and a few minicomputers essentially used as servers. With the availability of relatively low-cost high-speed networks, diskless workstations have been found to be more convenient in the network environment than their counterpart diskful workstations, since it is easier to maintain and enhance both hardware and system software on only a few large disks at the minicomputers' (servers') end, than for many small disks attached with many diskful workstations geographically scattered over a large area as was the case with the ordinary workstation model

already discussed. Here, normal computation activities required by the user's processes are performed at the user's home workstation. But, a client (a user located on any workstation) process can issue a request to a server process (which is a minicomputer in this case, or a specialized machine) for getting some service, and the server complies with the request by executing it, and finally sends back a corresponding reply to the client process after processing the request. Therefore, this model does not require any migration of the user's processes to the target server machine for getting the work executed by those machines. Several variations can, however, be derived on the above models from the consideration of the following factors:

- deployment of multiple servers and caches to increase performance and resilience;
- use of low–cost computers with limited hardware resources to fulfill users' needs that are equally simple to manage;
- the use of mobile code and mobile agents;
- to add or remove mobile devices in a convenient manner, as and when is required.

The client–server model, however, can be realized in a variety of hardware and software environments with typical characteristics, distinct from other types of distributed computing systems, and thus have become increasingly important, mainly for providing an effective general–purpose means for the sharing of information and resources. Moreover, there is as such no clear difference between a client and a server process, and that both the client and server processes sometimes can even run on the same computer. Some processes are often found both client and server processes. That is, a server process may sometimes use the services of another server (as in the case of three–tier architecture), thereby appearing as a client to the latter.

4. ***Processor–Pool model***: The workstation–server model nicely fits with the most common environment in which many almost–evenly distributed users having their own relatively small work load requiring mostly uniform computing power and shared resources can be simultaneously accommodated by way of allocating a processor to each such user. But, there are certain applications that have been observed to place demands once in a while for a massively–large amount of computing power over a relatively short duration of time. Processor–pool model is the right choice in such situation that befits to handle an environment in which all the processors are then clubbed together to be shared by the users as and when needed. The pool of processors can, however, be built with a large number of microcomputers and minicomputers (small mainframes) interconnected with each other over a communication network. Each such processor in the pool is, however, equipped with sufficient amount of memory of its own, and is capable of independently executing any program of the distributed computing system.

   Usually, in the processor–pool model, no terminal is directly attached to any processor; rather, all the terminals are attached to the network via an interface, a special device (a terminal controller, may also be in–built within the communication network). As a result, the user can access the entire system, and never logs onto a particular machine as a home machine, from any of the terminals which are usually small diskless–workstation, or may be a graphic terminal, such as X–terminals. In fact, when a job is submitted, it will be received by a special server (commonly called *run server,* a part of the related operating system) that manages, schedules, and allocates one or an appropriate number of processors from the pool to different users depending on the prevailing environment or on a demand basis. In some situations, more than one processor can be allocated to a single job that can run in parallel, if the nature of the job supports it (i.e. if the job can be decomposed into many mutually exclusive segments, one processor can then be assigned to each such segment, and all these segments can then eventually run in parallel on different processors). As usual, the processors are deallocated from the job when it is completed, and returned to the pool for other users to utilize them.

Processor–pool model, by virtue, provides much better utilization of *total* available processing power of a distributed computing system than any other model that supports the home–machine concept, in general. This is due to the fact that the entire processing power of this system in this model is available for any logged–on users, if needed, whereas this does not hold good for any other model including the most important one, the workstation–server model in which several workstations at times may be lying idle, but still they cannot be utilized for other jobs running at that time. Despite having many other advantages, one of the *major shortcomings* of this model is the presence of a relatively–slow speed interconnection network that communicates between the processors where the jobs are to be executed, and the terminals via which the users interact with the system. That is why this model is usually considered to be not suitable for the general environment in which typically high–performance interactive applications run. A distributed computing system based on processor–pool model has been implemented in one of the reputed system, ***Amoeba*** [Mullender, et al 1990].

Multicomputers are best suited primarily to cater general–purpose multi–user applications in which many users are allowed to work together on many unrelated problems, but occasionally in a cooperative manner that involves sharing of resources. Such machines usually yield cost–effective higher bandwidths, since most of the accesses made by each processor in individual machine are often to its local memory, thereby resulting in reduced latency that eventually yields increased system performance. The nodes in the machine are equipped with required interfaces so that they can be always connected with one another through a communication network.

In contrast to the tightly–coupled multiprocessor system, the individual computers forming the multicomputer system can be located far from one another, thereby covering a wider geographical area. Moreover, in tightly–coupled systems, the numbers of processors that can be effectively and efficiently employed is usually limited, and are constrained by the bandwidth of the shared memory, resulting in restricted scalability. Multicomputer systems, on the other hand, having loosely–coupled architecture are more freely expandable in this regard, and theoretically can contain any number of interconnected computers with no limits as such. On the whole, multiprocessors tend to be more tightly–coupled than multicomputers, because they can exchange data almost at memory speeds, but some fiber–optic based multicomputers of today have also been found to work closely at memory speeds.

A more detail of each of these models in this topic with their respective figures is described in the web site.

10.8.3.1 Multitiered Architecture: Three–tier Client–Server Architecture

The commonly–used conventional client–server architecture comprises of two–level or tiers: a *client tier* and a *server tier*. The concept of a three–tier architecture is also increasingly popular and common. In this architecture, the application software is usually distributed among three types of machines, namely, a *user machine*, a *middle–tier server*, and a *backend server*. The user machine in this model is typically a ***thin client*** (a PC or workstation). The middle–tier machines are separate servers that contain programs which form part of the processing level, and are supposed to be essentially *gateways* between the thin user clients and a variety of backend servers. Here, although this middle–tier machine is a server to the thin client, but sometimes it needs to act as a client to the backend servers. The middle–tier machines can also be employed to convert protocols and can map from one type of database query to another. In addition, the middle–tier machine can even integrate/merge results from different data sources. In effect, the middle–tier machines works as an interface between the desktop applications and the background legacy applications (or evolving corporate–wide applications) by mediating between these two different worlds. The number of tiers, can be even extended beyond three depending on design principles in relation to what type of distributions are to be provided to service the execution requirements. In effect, multitier architecture is used to process an application by dividing it into a user–interface, processing components, and data level in a distributed manner, equivalent to organizing it as a client–server model. This type of distribution is often referred to as ***vertical distribution***.

A typical usage of a three–tier architecture is observed in *transaction processing* in which a separate process, called the *transaction monitor* located on middle–tier server coordinates all transactions across possibly different data servers. Here, the thin client (workstation) is, however, equipped with a local disk of its own that contains part of the data, and runs most of the applications. But all operations on files or database entries go to the respective servers. Conclusively, it can, however, be inferred that the interaction between the client and the middle–tier server as well as that between the middle–tier server and the backend server also follow the client–server model in which the middle–tier server acts both as a server to the client, and also as a client to the backend server. Thus, the middle–tier system acts at times as a client or as a server. Many examples of a distributed computing system based on the workstation–server model can be cited; an earlier one perhaps is the V–System [Cheriton, 1988].

10.8.4 Computer Networks

Many variations of multicomputer design methodology exist in order to achieve different target objectives. When a collection of low-cost stand-alone autonomous computer systems having same or different environments, each is capable of making communication and cooperation, are interconnected by means of using *general-purpose communication links* and common software protocols (replacing the specialized interconnection network) to exchange information between the processes (programs) running on different machines, it is then popularly known as ***computer networks***. Figure 10.22 illustrates a representative scheme of such a local-area computer network. Each autonomous machine in this arrangement, running under its own *operating system*, is able to execute its own application programs to support its own local users, and constitutes as well the computational resources to the networks, is often referred to a *nodes, sites, host, computer, machines,* and so on. The nodes may be *homogeneous* or *heterogeneous* spanned over a geographical area, and may vary in size and function. Size-wise, a node may be a *small microprocessor, a workstation*, a *minicomputer*, a *mainframe*, or a *large supercomputer*. Function-wise, a node may be a single-user personal computer, a general-purpose time-sharing system, or a dedicated system (such as, a file server or a print server) without any provision for interactive users.

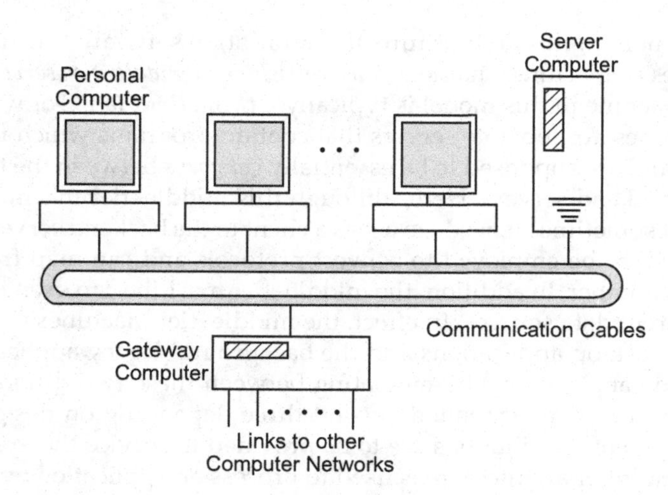

FIGURE 10.22
Block diagram of a local-area computer network.

There is essentially no coordination at all among these nodes, but they are all connected to an interconnection network so as to fulfil a few system-wide requirements. Precisely, a computer network is a typical form of network of computers, and that is why, all networks of computers are not always computer networks.

While each machine in the computer network has its own private operating system, but the entire network arrangement is driven by a separate *loosely-coupled operating system*, known as **Network Operating System** (NOS) that also preserves the autonomy and other aspects of each individual machine intact. This NOS is simply an adjunct to the local operating systems and never encroaches into the internal working of any individual machines to manage them. Its ultimate target is to provide resource sharing by enabling the users to make use of the facilities and the services available on a specific machine apart from the one of its own, i.e., it provides an additional support that local services are made available to remote users. Here, the user is quite aware that there are multiple independent similar or dissimilar computers, and must accordingly deal with them explicitly. Typically, common communication architecture (protocol) is used to support these network applications.

Computer networks have ***several advantages*** over the large mainframe (centralized) computers in certain application areas. It provides large pieces of software to run in parallel under same or different types of operating systems with little or almost no communication between the pieces, thereby offering ***coarse-grained parallelism***. It is cost-wise within affordable limits, and also can be ***incrementally scalable*** providing relatively higher processing facilities. Users can also enjoy shared access to costly computing facilities, such as high-performance supercomputers, specialized I/O equipments via the computer network at reduced cost. Another distinct feature is its ***openness***, meaning that it is easy to configure the arrangement out of many different systems possibly from different vendors. Also, it is easy to add new systems or replace existing ones without affecting the others that stay in place. In other words, a computer network is sufficiently ***extensible***, and also ***flexible***. In addition, a software failure, which is more frequent than hardware breakdown, on a single node cannot prevent the rest of the nodes from communicating, and normal operations in the remaining portion of the network can be continued, but, of course, with graceful performance degradation. Also, the failure of any system, or any fault in interconnection hardware in this arrangement will not affect the other systems to continue, and the ***fault tolerance*** issue here is automatically solved to a great extent.

Computer network also suffers from ***a number of drawbacks***. The communication over the interconnection network (LAN) is obviously *slower*, largely due to the frequent intervention of the operating system at the time of exchanging messages between programs running on different computers. This indicates that a network of workstations does not yield better performance than a self-contained system having a specialized interconnection network of its own. Moreover, large-scale scientific computations often require a task to be partitioned into subtasks, and demand frequent and rapid exchange of results between these subtasks. The time required for such exchanges; as they are essentially slow I/O transfers that summarily limit the usefulness of a computer network. Although special I/O programs and fast LAN are available nowadays, yet it is not still considered as an effective approach in building an MPP system.

10.8.5 Distributed Systems

A multicomputer is essentially a network of computers, by virtue of being a distributed computing system, when managed by a distributed operating system (a specific type of OS), the entire arrangement is then commonly called a *distributed system*, the ultimate of an MIMD architecture. Various definitions of distributed system have been proposed, but none of them found to be absolutely satisfactory. However, a rough characterization in this regard may be: *that a distributed system is one that runs on a collection of interconnected independent computers which do not have shared memories, yet appears to its users as a single coherent system.*

This characteristic is also sometimes referred to as the **single system image**. A slightly different notion of a distributed system is that, it is one that consists of multiple autonomous central processing units (nodes of a network of computers) that work together, yet appears to its users like an ordinary uniprocessor system. The key concept here is *transparency*. In other words, the use of multiple processors should be invisible (transparent) to the user who views the entire system as a **virtual uniprocessor**, and not as a collection of distinct machines (processors). Consequently, any machine anywhere in the domain of the distributed system appears to the user as his/her own home machine, irrespective of its geographical location. A distributed system (Coulouris, G. F., et al.) is thus mainly aimed at making it easy for users to access remote resources (both hardware and software), and to share them with other users in a controlled way. In this way, it facilitates to collaborate and exchange information amongst the users widely dispersed geographically, as is best illustrated by the success of the internet with its underlying hardware and associated software (Colwell, R. P., et al.). Whatsoever it may be, no matter how it is expressed, the leading edge in quest for a distributed system and its development, however, is mainly focused in the area of distributed operating system (DOS). Although some commercial systems have been already introduced, fully functional DOS with most of its required attributes, and hence, full-blown distributed systems are still in the experimental stage in the laboratory.

The *functionality* of a distributed system is ultimately to extract the maximum of the services that are available from the existing multiple resources in the network, and thereby distributes (disperses) the processes of an application across various machines (or CPUs) present in the system to achieve highest possible computation speed-up, efficient utilization of the potentials of available resources (may be of even supercomputers, or high-performance storage system, or any valuable software resources) by way of sharing, whenever possible, effective communication and cooperation between users via the existing communication network, and above all, to provide reliability and fault-tolerance, whenever necessary. Users of this system often also feel the need to use special techniques that facilitate to access resources transparently over the existing communication network.

Distributed systems give rise to the distribution in processing that are basically of two types. *Distributed processing* when organised as equivalent to essentially a client–server application using multitier architecture (see Section 10.8.3, No. 3, Workstation-Server model), this type of distribution is often referred to as ***vertical distribution***. As this type of distribution is merely a way of organizing client–server application, it is not so important in many cases in the distributed system arena. Alternatively, in modern architectures, one of the essential requirements that count is often to distribute both the clients (applications) as well as the servers (resources). This type of distribution is referred to as ***horizontal distribution***. Here, a client or a server or both may be physically split up into logically equivalent part, when each part operates individually on its own share of the complete data set, thereby offering all the functionalities that a distributed system can provide (as already mentioned before) including balancing of load.

A popular example of a horizontal distribution can be exhibited assuming a web server which is replicated across several machines connected in a *peer-to-peer* manner in a local area network. Each machine has the same set of web pages. When a page is modified, a copy of it is immediately placed at each machine. When a request arrives from a different machine connected to this network that handles all such requests, the request is then forwarded to a machine within the network using *round-robin policy*. It is observed that for highly popular websites, this form of horizontal distribution is quite effective, provided enough bandwidth is offered.

Users of distributed system have *user ids* and *passwords* (and not machine identification, like IP address) that are valid throughout the system. This feature makes the communication possible to happen conveniently between users in two ways. First, communication using *userids* automatically invokes the security mechanisms to intervene, and thereby ensures authenticity of communication. Secondly, the users can be remained mobile at will within the arena of the distributed system, and still be able to communicate with other users of the system at ease irrespective of their own present location (i.e., the machine being currently used can be located anywhere within the domain of the distributed system).

However, with the aid and necessary support from the DOS which is a ***tightly-coupled operating system***, the distributed system demystifies some distinctive and unique features that are worthy to be mentioned. Some of them, apart from many other similar ones are:

Transparency and its different aspects: An important characteristic of a distributed system is to hide the fact that its processes and resources are physically distributed across multiple computers. This implies that the existence of a collection of distinct machines (processors) that are connected by a communication network to constitute the system must be made invisible (*transparent*) to its users providing them only a view of virtually a single uniprocessor system image. Complete transparency in this respect is, however, difficult to realize since it includes several different aspects of transparency that must be supported by the respective distributed operating system. However, the eight forms of transparency in this regard as pointed out by the ISO Reference Model for Open Distributed Processing [ISO 1992] are given in Table 10.1.

Each transparency aspect has its own important role that comes into play to provide all the distinct advantages of a distributed system to its users. In fact, all these transparencies

TABLE 10.1

Eight Forms of Transparency as per ISO Reference Model for Open Distributed Processing

Transparency	Significance
Location	Hide where a resource is physically located.
Access	Hide differences in data representation and how a resource is accessed.
Migration	Hide that a resource may be moved to another location, even while in use.
Replication	Hide that a resource is replicated for better performance and reliability.
Concurrency	Hide that a resource may be shared by several competitive users.
Failure	Hide the failure and subsequent recovery of a resource.
Performance	Hide resource distribution and allocation when loads vary dynamically.
Scaling	Hide scaling without affecting the on-going activities of the users

when being put together form the actual beauty of the system that truly enables a user to make use of any resources or processes anywhere in this domain irrespective of its actual geographical location, and in particular, with which system it is physically attached. Aiming to realize distribution transparency while designing and implementing distributed systems is although a nice approach, but it should be taken into account with other aspects; one such is performance. We, however, keep ourselves restricted from any such detailed descriptions as to how these transparencies are actually implemented and subsequent their implications in this regard. Interested readers are, however, suggested to consult any text book on this subject including the book by the same author; Operating Systems, P. Chakraborty, Jaico Publishing House, 2011.

Openness is another important attribute of a distributed system that mainly provides ***interoperability***, *portability*, and *flexibility*. *Interoperability* characterizes the extent by which different implementations of systems or components introduced by different vendors can co-exist and work together by merely relying on one another's services as specified by a common standard implemented through appropriate interfaces. ***Portability*** commonly characterizes to what extent an application developed for a particular distributed system can be executed, almost without any modification, on a different distributed system that implements the same interface as that particular system. ***Flexibility*** implies that it should be easy with minimum effort to configure the system, out of different components possibly introduced by different vendors, and easy to add new components or replace existing ones without affecting the others that stay in place. Flexibility is, however, easier to describe but more difficult to actually attain.

Reliability, a demanding feature of a distributed system is closely associated with the *availability* of its resources that is ensured by way of protecting them against likely faults. But, by virtue of having multiple instances of the resources in distributed system, it also automatically guarantees more reliability in the system, even in the event of certain failures, although these resources otherwise are mainly used to provide required functionalities that a distributed system must have. To obtain higher reliability, this system negotiates faults in ways such as; *to avoid faults, to tolerate faults*, and *to detect faults and subsequent recovery from faults*. Various acceptable popular methods in this regard are available that are in use to deal each of these issues separately. However, to yield increased reliability, it always involves costly extra overhead, and that is why, it becomes a hard task for the designers to decide as to what extent the system would be made reliable so that a good balance of cost versus mechanism can be effectively implemented.

Scalability is one of the key facets of a distributed system that refers to the ability of a given system to expand in size to cover more geographical area, and to span administratively in order to support the increased service load. It is quite common that a distributed system will naturally grow with time by way of adding new machines or even an entire sub-network (may lie geographically apart) to the existing system so that the increased workload can be handled without causing any serious disruption of service or notable degradation in system performance. Obviously, there exist some accepted principles that are to be followed as a guideline while describing and designing scalable distributed systems. Unfortunately, a distributed system that is scalable in all respects often exhibits constant loss in performance as the system gradually scales up.

However, some of the many ***distinct advantages*** of a distributed system that are found worthy to be mentioned are as follows:

- It facilitates to integrate different applications running on different machines (computers) into a single system.
- When properly designed, it scales well with respect to the capacity of the underlying communication network.
- It can also include the hardware of tightly-coupled computer systems including multiprocessors and homogeneous multicomputers.

Some of the major ***disadvantages*** of a distributed system are as follows:

- A true distributed system is more costly and requires much maintenance overhead cost on regular basis, both in terms of hardware and its associated software.
- Its underlying software is extremely complex.
- Total dependence of communication network is a pitfall of such a system.
- Performance is often degraded due to frequent communication between, and migration of the elements involved in the processing.
- Due to increased connectivity and sharing, the system is often exposed to severe threats in security, but provides only little and weaker protection against such threats.

Still, there is a considerable interest worldwide in designing, building, and installing true, versatile distributed systems for various reasons as well as to handle some specific environments (such as, world wide web, web services, applications of large multinational organisation, etc.) that cannot be profitably managed by other types of systems in use.

Modern nearly-distributed systems (often called *cluster architecture*, discussed later in Section 10.9) are generally built by means of using an additional layer of software on top of a *network operating system* (not using a true distributed operating system as should be done on the underlying distributed hardware). This layer, called ***middleware***, provides a means of realizing *horizontal distribution* by which applications and resources (both hardware and software) are made to be physically distributed and replicated across multiple machines (computers) to harness distributed processing. This layer is so designed that it can hide the heterogeneity, thereby ensuring interoperability between different implementations (openness), as well as providing a distributed nature for the underlying collection of machines (computers). Middleware-based distributed systems, however generally adopt a specific model for expressing distribution and communication. Popular models are mainly based on remote procedure calls (RPC), distributed files, objects, and also documents. On the other hand, distributed systems when realized on the basis of only vertical distribution using multitier architecture (Section 10.8.3, No. 3, Workstation–server model) are, however, not at all sufficient to build large-scale systems. That is why the modern systems thus built are generally distributed both in horizontal as well as in vertical sense.

10.9 Clusters: A Distributed Computer System Design

True distributed system is costly to afford, and equally expensive to scale and subsequently to maintain. But its worthy and useful features that essentially provide enormous computational flexibility and versatility cannot be ignored and left unused, particularly in the current scenario of the computing world in which large applications are often distributed and span over a wide geographical area. One of the best examples that can be cited in this regard is the internet with all its underlying hardware and associated software.

Due to constant decrease in hardware prices, low-cost moderately-spanned computer networks gradually became popular, and eventually came into extensive use. In addition, continuous advancements in communication technology further improved the interconnection mechanisms between the systems in a superior way; one of the prime requirements of distributed processing. Consequently, there is another significant development that eventually evolves in the design of distributed computing system, what is called *clustering*. Cluster architecture has evolved relatively recently, and appeared as saviours to mainly manage different types of numerous application areas that otherwise require a true distributed system for their processing. This comparatively low-cost cluster architecture has fully satisfied the user community with a base substitute of a true distributed system which is relatively more expensive. It looks as if the users are made happy with only a glass of grog when the bottle of champagne is beyond their reach!

A cluster can be defined as a group of interconnected, self-sufficient computers ***(multicomputer)*** working together as a ***unified computing system*** that can cast a ***single-system image*** **(SSI)** as being *one single machine* to the outside world. Each individual computer involved in a cluster may be a uniprocessor or a multiprocessor, and is typically referred to as a *node* which can even run on its own without any assistance from the cluster. The use of a multiprocessor as a node in the cluster, although not necessary, however, does improve both performance and availability. Since clusters are composed of independent and effectively redundant computers (nodes), they have a potential for fault-tolerance that makes them suitable even for other classes of problems in which reliability is of paramount interest. The clustering approach, however, can also be considered as an alternative to symmetric multiprocessing that offers effectively unbounded processing power, storage capacity, high performance, and high availability, which could be used to solve much larger problems than what a single machine could do (Pfister, G.). This arrangement, however, has gained much its importance, particularly in the area of server applications, in recent years.

Cluster architecture, while being developed and introduced as a base substitute of a distributed system, is built up essentially on the platform of a computer networks comprising of heterogeneous multicomputer systems operated by loosely-coupled network operating system (NOS). This multicomputer system is, however, composed of interconnected nodes (computers) in which each node individually is a complete, independent, and autonomous system. Our traditional distributed system (as per definition given in the beginning of Section 10.8.5) is run by a distributed operating system (DOS), but a DOS is not supposed to manage a collection of *independent computers*. On the other hand, a network of computers in the premises of computer network being run by network operating system (NOS) never provides the view of a *single coherent system*. So neither DOS nor NOS really qualifies as part of a distributed system in this regard. The obvious question thus arises as to whether it is possible to develop a distributed computing system that could be run in such a way so that it could possess most of the salient features of these two different domains (namely, DOS and NOS environment). Those features are mainly: transparency and related ease of use provided by DOS, and scalability and openness offered by NOS. One viable solution in this regard was thus attempted simply by way of enhancement to the services that a network operating system provides, such that a better support for distribution transparency could be realized. To implement these enhancements, it was then decided to include a layer of software additionally with a network operating system in order to improve its distribution transparency, and also to more or less hide the heterogeneity in the collection of underlying systems. This additional layer being used is known as ***middleware*** which lies at the heart of modern distributed systems of this category being currently built up.

The next obvious question arises as to where this additional layer of software is to be placed. Now a network operating system (NOS) itself offers programming interfaces that are being made use of by many distributed applications. Lots of examples in this regard can be cited; one such example is message communications through sockets that allow processes on different machines to pass messages. Interfaces to the local file system offered by NOS are also used by many applications. All these and many similar others altogether create a situation that makes distribution to be hardly transparent. That is why it is suggested to place this additional layer of software in between the applications and the network operating system, so as to offer a higher level of abstraction. Thus, this layer is physically placed in the middle between applications and NOS as shown later in Figure 10.24, and hence, it is legitimately called ***middleware***.

Cluster architecture comprises complete, independent, and autonomous systems (nodes), each such individual system has its own private main memory; applications do not see a large global memory. In effect, coherency is thus required to be maintained in software rather than in hardware. This memory granularity eventually affects performance, and to achieve maximum performance, befitting software must be tailored to this environment (Baker, M., et al.).

The use of middleware layer of software in the premises of computer networks provides a means of realizing ***horizontal distribution*** by which applications and resources (both software and hardware) are made to be physically distributed and replicated across multiple machines (computers), thereby providing distributed nature to the underlying computer networks. As a result, each individual machine in the networks of computers can now operate on its own share of the complete data set, thereby offering all the functionalities that a distributed system generally provides (as already mentioned before) including balancing of load. This layer is also capable to hide the heterogeneity in the underlying machines (computers), and thus ensuring interoperability between different implementations (*openness*).

In addition, architecture of underlying computer networks in the form of multitier (client–server) organisation inherently offers a ***vertical distribution*** (see multitier architecture, Workstation-server model, Section 10.8.3). Consequently, cluster architecture, when built up with server-based hardware supported by befitting middleware and managed by appropriate operating system (discussed later in Section 10.9.5) can ultimately provide a distribution both in the horizontal as well as in the vertical sense, leading to a feasible implementation of a large-scale distributed system. Middleware-based distributed systems, however generally adopt a specific model for expressing distribution and communication. Popular models are mainly based on *remote procedure calls* (RPC), *distributed files, objects*, and also *documents*.

10.9.1 Distinct Advantage

Clustering method by way of interconnecting independent autonomous computers exhibits many distinct advantages. Clusters offer the following advantageous features that can be realized at a relatively low cost (Weygant, P.):

- High Performance
- Expandability and Scalability
- High Throughput
- High Availability
- Openness

It permits organisation to boost their processing power using standard technology (commodity hardware and software components) that can be acquired / purchased at a relatively low cost. This provides *expandability*, an affordable upgrade path that lets organisations increase their computing power while preserving their existing investment intact by incurring only a little additional expenses. The performance of applications also improves with the aid of a scalable software environment.

A clustering approach offers high ***scalability*** and more ***availability*** of the computing resources present in the cluster. In fact, this approach offers both *absolute scalability* as well as *incremental scalability* which means that a cluster configuration can be easily extended by adding new systems to the cluster in small increments, of course, within the underlined specified limits. Such attempts will not disturb the existing functionalities at all, and no major upgrades as such are required in the environment, still the smaller system can be gradually converted into a larger one that can eventually exceed the capacity and power of even the largest standalone machines to negotiate the constantly on-going increasing demand. In fact, a cluster in this way can then accommodate tens, hundreds, or even thousands of computing systems, each of which, in turn, can even be a multiprocessor (Brewer, E.).

High availability also implies ***high fault-tolerance***. As the cluster consists of a set of interconnected self-sufficient computers (nodes), certain failures of one or more such nodes is not unnatural, but it will not so severely affect the operation of the cluster in question. In fact, it continues to operate even in the event of such failures, but, of course, affecting only with a graceful degradation in the performance of the entire system. In today's many products, fault-tolerance is, however, handled automatically in software. Moreover, clustering also possesses a ***failover*** capability which is realized by the use of a backup computer placed within the cluster to take over the charge of a failed computer to negotiate any such exigency.

A clustering approach, in fact, is also capable of hiding the *heterogeneity* that may exist in the collection of underlying interconnected machines (computers), and thereby ensuring interoperability between different implementations ***(openness)***.

Last, but not the least, clustering approach exhibits a ***superior cost/performance*** ratio. It is possible to build up a cluster with commodity building blocks that can offer an equal or even greater computing power as well as superior performance than a comparable single large machine, and that too at reasonably much lower cost and complexity.

10.9.2 Classification of Clusters

Clusters can be categorized in many different ways. The simplest possible classification is based on whether the computers involved in the cluster share access to the same disks. Figure 10.23a shows a two-node cluster in which there is no common shared disk, but the interconnection is made only by means of a high-speed link that can be used for message exchange to coordinate cluster activity. The link can be a LAN that may be a *dedicated* one only to the participating nodes, or can be *shared* by other computers that lie outside the cluster. Remote client systems must have the provision to link with the LAN or WAN of the server cluster.

Figure 10.23b shows the other alternative of this simple classification which is essentially a shared-disk cluster. In addition to the standard high-speed link between the nodes for message exchange, there is a disk subsystem that is directly linked to the multiple nodes within the cluster. The disk subsystems being used here is a RAID system as shown in Figure 10.24b which is always considered more advantageous than using a single shared-disk. Because the RAID system not only provides high availability to the multiple nodes present in the cluster, but is also free from single-point-of-failure which is always embraced with a single shared-disk. However, clusters can be classified into many ***different categories*** based on various factors, such as:

Application Target

- High performance (HP) Clusters: For Computational Science
- High Availability (HA) Clusters: For Mission-critical Applications

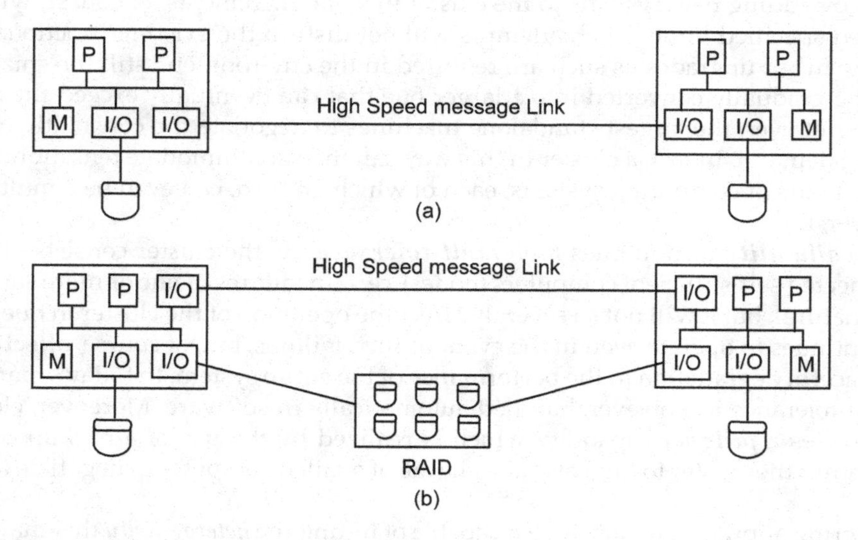

FIGURE 10.23
A schematic block diagram of cluster configuration. (a) Standby server with no shared disk and (b) Shared disk.

Node Ownership

- Dedicated Clusters: Owned by an individual
- Non-dedicated Clusters: Dedicated as a Cluster node

Node Hardware: PC, Workstation, or SMP

- Clusters of PCs (CoPs) or Piles of PCs (PoPs)
- Clusters of Workstations (CoWs)

Node Operating System: Linux, Windows NT, AIX, Solaris, etc.

- Linux Clusters (e.g., Beowulf)
- Solaris Clusters (e.g., Barkley Node of Workstations (NOW))
- NT Clusters (e.g., HPVM)
- AIX Clusters (e.g., IBM SP2)
- Digital VMS Clusters
- HP-UX Clusters
- Microsoft Wolfpack Clusters

Node Configuration: Node Architecture and Type of OS It Is Loaded With

- Homogeneous Clusters: All nodes will have similar architectures and run the same Operating Systems.
- Heterogeneous Clusters: All nodes will have different architectures and run different Operating Systems.

Levels of Clustering: Based on Location of Nodes and Their Count

- Group Clusters (nodes # 2–99): Nodes are connected by SANs (System Area Networks)
- Departmental Clusters (nodes # 10s–100s):
- Organisational Clusters (nodes # many 100s):
- National Metacomputers (WAN / Internet-based): (# nodes: many departmental / organisational systems or clusters)
- International Metacomputers (Internet-based): (# nodes: 1000s to many millions)

Individual clusters may, however, be interconnected once again to form a larger system (cluster of clusters). In fact, the internet itself can be used as a computing cluster. The proliferation of wide-area networks of computer resources for high performance computing has eventually led to the emergence of a new field called ***Metacomputing***. Any discussions over metacomputing, however, at present is outside the scope of this book. Interested readers are hereby referred to consult (Baker, M.).

10.9.3 Different Clustering Methods

Although clustering of computers is a relatively recent significant development in system architectural design methodology, but it quickly gained much importance due to the range of options it can provide in its usage to cater to many different operational environments, particularly, in the area of server applications (Buyya, R.). Based on the functional alternatives that a cluster normally provides, the classification of *clustering method* in respect to its functional line can be described in the following way. This classification in other way also demystifies the objectives of a specific cluster and defines its design requirements as well.

Passive Standby: This method is a classical one in which one computer is used as primary to handle all of the processing loads while the other computer (server) remains inactive and used as a standby that takes charge only in the event of any such failure of the primary one. The primary active machine constantly keeps coordination with the passive one by way of periodically sending it a *"heartbeat"* message. Should the message stop arriving at any point of time, the standby machine assumes that the primary server has failed, and at once takes charge to continue the ongoing operation. This arrangement, however, provides only increased availability, but never improves performance because the standby machine is never involved in normal course of processing activities. Moreover, if the heartbeat message is only used as information exchange between the two systems, and if the two systems do not share common disks, then the standby provides only a functional backup but has no access to the databases operated by the primary.

Active Secondary: The passive standby, however, is at best a standalone connected machine, not involved at all in the working of a cluster, and is therefore, generally not referred to as a *cluster*. Because the term *cluster*, by definition, is reserved for an arrangement in which multiple interconnected standalone computers will be present, all of them actively involved in processing while maintaining to the outside world, the image of a single system. This configuration is, hence often referred to as *active secondary*. However, *three* basic classifications of this clustering can be identified. They are:

i. Separate Servers
ii. Shared nothing
iii. Shared memory

i. **Separate Servers**: In this clustering approach, each computer in the cluster is a self-sufficient separate server with its own disks, and there is no disk sharing between the systems. This arrangement is found to yield high performance as well as high availability. The software to be used in this arrangement must have some form of management as well as scheduling mechanism so that it can attain the incoming client requests and assign them appropriately to its own resources in order to achieve high resource utilization, and at the same time can make the load balanced. It is also desirable for this arrangement to must have a ***failover*** capability that automatically switches an executing application and associated data resources from a failed system to an appropriate alternative system within the cluster so that the ongoing execution of the application can be continued till its completion. Although the performance is slightly degraded in the event of such an undesirable situation, but that also is without loss of any services as such. To accomplish this, data and status must constantly be replicated among systems, or some other suitable arrangement should be made so that each system is aware of, and has access to the most-current data of the other systems. While this arrangement of constant data exchange operations essentially creates high communication traffics as well as increased server load that eventually leads to additional overheads for the purpose of only ensuring high availability, on the other hand, this results in substantial degradation in overall performance at the same time.

ii. **Shared Nothing**: In order to alleviate the additional administrative overhead being incurred in making the arrangement of separate servers as mentioned above, most of the clusters of today consist of servers connected to common disks (Figure 10.23b). In one variation of this approach, known as *shared nothing*, the common disk is partitioned into volumes, and each such volume is owned by a single computer. In the event of failure of any computer system, the cluster must be reconfigured afresh so that some other active computer can gain the ownership of the volume attached to the failed computer. In this way, constantly copying of data among all systems to enable each system to have easy access to the most-current data of the other systems can be simply avoided.

iii. **Shared Disk**: Another approach that many of the clusters of today follow is that the cluster consists of servers connected to common disks shared at the same time by all the servers. This arrangement is known as ***shared disk*** cluster. With this approach, each computer in the cluster has access to all the volumes of the disks. For this to happen consistently, this type of cluster must be equipped with some type of locking mechanisms to implement mutual exclusion to ensure that data can only be accessed by one computer at any point of time.

10.9.4 General Architectures

Formation of a cluster by organizing computers can be accomplished in a variety of ways (Buyya, R.). However, the typical cluster architecture is depicted in Figure 10.24. The individual computers here, which may be homogeneous or heterogeneous, are usually

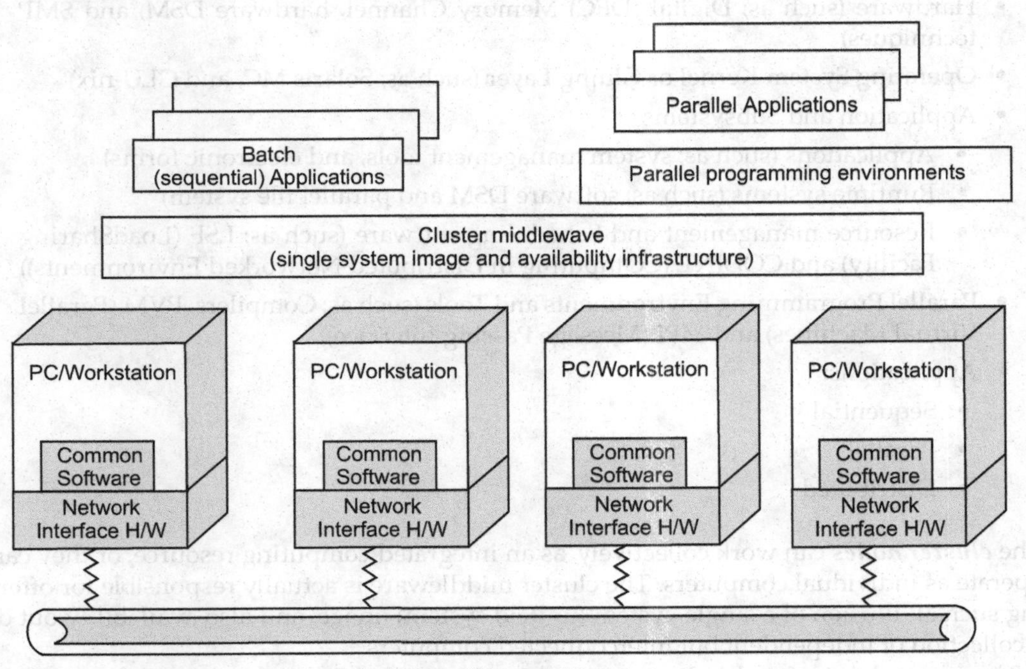

FIGURE 10.24
A schematic block diagram of Cluster Computer Architecture (Buyya, R).

connected by some high-speed LAN or switch hardware to realize faster communication. Each computer in the cluster can run on its own, apart from its operation as a member of the cluster in which it belongs. Although the cluster is built up with a collection of interconnected stand-alone computers, it always projects itself with an illusion as a single system to the outside world. To realize this single-system view, a ***middleware layer*** of software is to be included in each computer, the presence of which enables each computer to operate as a cluster member in unison with the other members, apart from being operated on its own as an independent individual computer. The other functions of the middleware, however, are involved mainly in providing access transparency, distribution transparency, balancing of load, high availability, and many other similar things.

A cluster, in addition, will also be equipped with other software tools to facilitate the cluster to efficiently execute those programs that are capable of realizing *parallel execution*. The following are some *prominent components* that often constitute the infrastructure of a cluster computer:

- Multiple High Performance Computers (PCs, Workstations, or SMPs)
- State-of-the-art Operating Systems (Layered or Microkernel-based)
- High Performance Network / Switches (such as; Gigabit Ethernet or Myrinet)
- Network Interface Cards (NICs)
- Fast Communication Protocols and Services (such as; Active and Fast messages)
- Cluster Middleware (Single System Image (SSI), and System Availability Infrastructure)
- Hardware (such as; Digital (DEC) Memory Channel, hardware DSM, and SMP techniques)
- Operating System Kernel or Gluing Layer (such as; Solaris MC, and GLU-nix)
- Application and Subsystems
    - Applications (such as; system management tools, and electronic forms)
    - Runtime systems (such as; software DSM and parallel file system)
    - Resource management and Scheduling software (such as; LSF (LoadSharing Facility) and CODINE (COmputing in DIstributed Networked Environments))
- Parallel Programming Environments and Tools (such as; Compilers, PVM (Parallel Virtual Machines) and MPI (Message Passing Interface))
- Applications:
    - Sequential
    - Parallel
    - Distributed

The ***cluster nodes*** can work collectively, as an integrated computing resource, or they can operate as individual computers. The cluster middleware is actually responsible for offering such an illusion of a single-system (unified system) image, and also availability out of a collection of independent but interconnected computers.

The ***network interface hardware*** comprising Network / Switches acts as a communication processor, and is responsible for transmitting and receiving packets of data between cluster nodes.

Communication software provides a means of fast and reliable data communications among cluster nodes, and to the outside world. Often, clusters with a special network / switch like Myrinet use communication protocols, such as, active messages for fast communications among its nodes. They potentially bypass the operating system, and thus remove the critical communication overheads, thereby providing direct user-level access to the network interface.

Programming environments can offer portable, efficient, and easy-to-use tools for development of applications. They include message passing libraries, debuggers, and profilers. It should once again be noted that clusters could be used for the execution of sequential as well as parallel applications.

The *middleware* lies at the heart of modern distributed systems and plays a critical role in the realization of numerous required cluster operations. Most middleware is based on some model, or paradigm, for describing distribution and communication. However, some of the major representative functions and services, that are common to and are essentially provided by a middleware used in a cluster, are assumed to be the following (HWANG, K 1999):

- **Single-entry point**: A user logs normally onto the cluster rather than to an individual computer.
- **Single-file hierarchy**: The user views a single hierarchy of file directories under the same root directory. It is provided through a distributed file system, but more advanced middleware have integrated databases into their systems, or otherwise provide facilities for applications to connect to databases.
- **Single control point**: A default node always exists that is used for cluster Management and control.
- **Single memory space**: The presence of distributed shared memory formed by different computers in the cluster facilitates programs to share variables.
- **Single job-management system**: A job scheduler exists in a cluster (not related with any individual computer) that receives jobs from all users submitted to the cluster irrespective of any specification as to on which computer it is submitted to, and to which computer the submitted job will be executed on.
- **Single-user interface**: Irrespective of the workstation through which a user enters the cluster, a common graphic interface would support all the users at the same time.
- **Single virtual networking**: Any node can access any other machine in the cluster, even though the actual physical cluster configuration may consist of several interconnected networks. There always exists a single virtual network operation.
- **Single process space**: A uniform process-identification scheme is used. A process executing on any node can create or communicate with any other process on any local or remote node.
- **Process migration**: Any process running on any node can be migrated transparently to any other node irrespective of its location, which enables the cluster to balance the load existing in the system at any point of time.
- **Single I/O space**: Any node can access any local and remote I/O devices including disks without having any prior knowledge of their actual physical location.

- **Checkpointing**: This function would periodically save the process state, intermediate results, and other related information of the running process that enables it to implement a *rollback recovery* (a failback function) in the event of any fault, and subsequent failure of the system.

Other services similar to these are also essentially required from a cluster-middleware, mainly to cast a single-system image as well as to enhance the *availability* of the cluster.

10.9.5 Operating System Considerations

Whatever be the hardware configuration and arrangement of the cluster, it requires some specific type of software (specifically OS) so that this form of distributed computing system can cast a single-system image to the user. An operating system that can project this view (single-system image), of course, will be of a special type and will be different for different cluster architectures, and is thus essentially required to be fully matched with the underlying respective cluster hardware. *It should be clearly noted that although a cluster software is not a distributed operating system, still it exhibits several useful features that closely resemble those found in true distributed operating systems.* The *cluster software*, while controls the entire operation of all the nodes present in the cluster, spreads the flavour of *parallel processing* by providing a unified system image to the user known as a **single-system image.** It *speeds up* computation rendering parallel processing by exploiting the services of several CPUs (nodes) present within the cluster, and this is accomplished by scheduling and executing independent sub-tasks of an application simultaneously on different nodes within the cluster. While it provides **high availability** through redundancy of available resources such as CPUs and other I/O media, those are also delegated in carrying out effective *load balancing* among the existing computer systems. This software is also adequately equipped to provide enough **fault-tolerance** as well as **failure management**. Apart from being equipped with its usual software, a cluster will also have other software tools supported by the underlying OS, such as, **parallelizing compiler, software interfaces** *and* **programming language interfaces**, *etc.* so that all these together can then create an environment, very similar to that of a true distributed system.

10.9.6 Windows Cluster

The Windows cluster server (also known as Wolfpack) is essentially a *shared-nothing cluster* (as already described), in which each disk volume and other resources are owned by a single system at any point of time. The Windows cluster server (Short, R., et. al) essentially creates an illusion of a single-system image, and is formed based on some fundamental concepts which are mainly the following:

Cluster Service and Management: A collection of software must reside on each node that manages all cluster-specific activity. A cluster as a whole is, however, managed using distributed control algorithms that are implemented through actions performed in all nodes. These algorithms require that all nodes in a cluster must have a consistent view of the cluster, i.e., they must possess identical lists of nodes within the cluster. An application has to use a special Cluster API (Application program interface), DLL (dynamic link library) to access cluster services.

Resource: The concept of resource in Windows is somewhat different. All resources in the cluster server are essentially *objects* that can be actual *physical resources* in the system including hardware devices such as disk drives and network cards, a *logical resource* such as logical disk volumes, TCP / IP addresses, entire applications, and databases, or a resource that can even be a *service.* A resource is implemented as a dynamic link library (DLL), so it is specified by providing a DLL interface. The resources are managed by a *resource monitor,* which interacts with the cluster service via *remote procedure calls* (RPC), and responds to cluster service commands to configure and move a collection of resources. A resource is said to be *online* at a node when it is made connected to that specific node to provide certain services.

Group: A group is a collection of resources managed as a single unit. A resource, however, always belongs to a group. Usually, a group contains all of the elements needed to run a specific application, including the service provided by that application. A group is owned by one node in the cluster at any time; however it can be shifted (moved) to another node in the event of a fault or failure. A resource manager exists in a node that is responsible for starting and stopping a group. If a resource fails, the resource manager intimates the ***failover manager*** and *hands over* the group containing the resource so that it can be restarted at another node.

Fault Tolerance: Windows cluster server provides fault-tolerance support in clusters containing two or more server nodes. Basic fault-tolerance is usually provided through RAIDs of 0, 1, or 5 that are shared by all server nodes. In addition, when a fault or a shutdown occurs in one server, the cluster server moves its functions to another server without causing a disruption in its services.

An illustration of the various important components of the Windows Cluster Server, and their relationships with a single system in the cluster is depicted in Figure 10.25. Individual cluster service is accessed by the respective manager from a lot of existing managers. Each node has a ***node manager*** which is responsible for maintaining this node's membership in the cluster and also the list of other nodes in the cluster. Periodically, it sends messages called *heartbeats* to the node managers on other nodes present in the cluster for the purpose of node fault detection. When one node manager detects a loss of heartbeat messages from another node in the cluster, it broadcasts a message on the private LAN to the entire cluster, causing all members to exchange messages to verify their view of the current cluster membership. If a node manager does not respond or a node fault is otherwise detected, it is removed from the cluster and each node then accordingly corrects its list of nodes. This event is called a ***regroup event.*** The ***resource manager*** concerned about resources now comes into action, and all active groups located in that faulty node are then 'pulled' to other active nodes in the cluster so that resources in them can be accessed. The use of a shared disk, however, facilitates this arrangement. When a node is subsequently restored after a failure, the *failover manager* concerned with nodes decides which groups can be handed over to it. This action is called a ***failback,*** it safeguards as well as ensures resource efficiency in the system. The *handover* and *failback* actions can also be executed manually.

The configuration database used by the cluster is maintained by the ***configuration database manager.*** This database contains all information about resources and groups, and node ownership of groups. The database manager on each of the cluster nodes interacts cooperatively to maintain a consistent picture of configuration information. *Fault-tolerant*

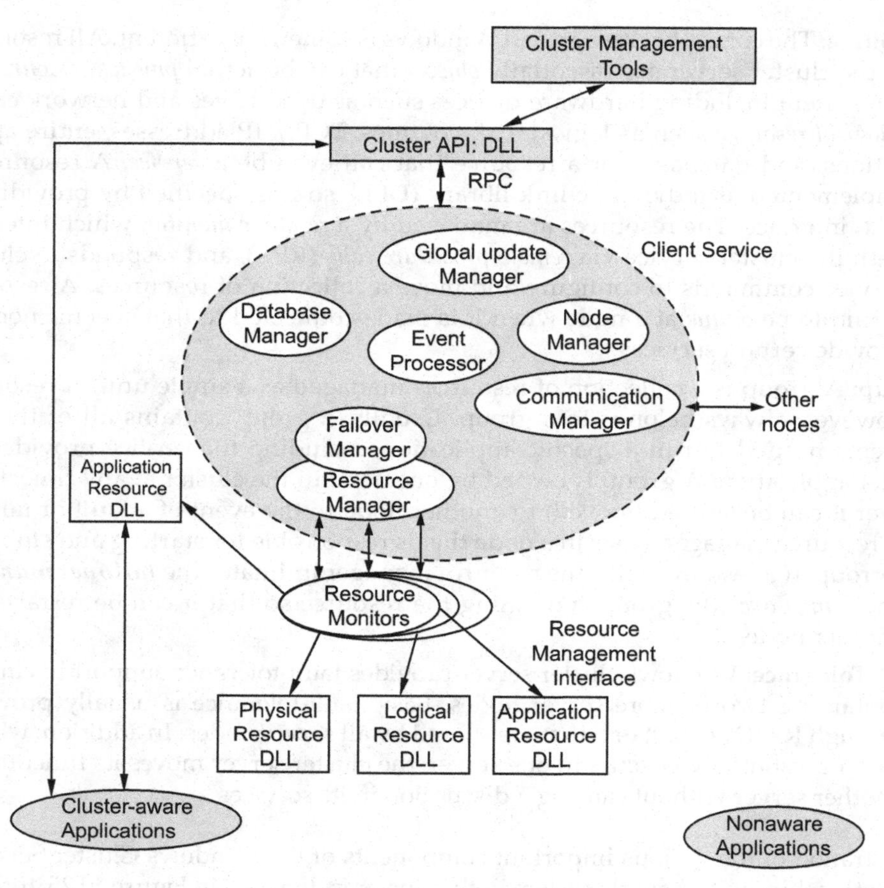

FIGURE 10.25
A block diagram of Windows Cluster Server (SHORT 97).

transaction software is used to assure that changes in the overall cluster configuration before failure, during failure, and after recovery from failure are performed consistently and correctly.

The functions of ***resource manager/failover manager*** have been already described. In effect, it makes all decisions regarding resource groups and takes such appropriate actions as to startup, reset, and failover. In the event of a node failure, the failover managers on the other active nodes cooperate to effect a distribution of resource groups from the failed system to the remaining active systems. When a node is subsequently restored after rectifying its fault, the failover can decide to move some groups taken earlier back to the restored system along with the others. In particular, any group can be configured with a preferred owner. If that owner fails and then restarts, it is desirable that the group in question is moved back to the node using a rollback operation.

There exist many other managers that perform their respective roles in carrying out numerous duties and responsibilities as entrusted. One such processing entity is known as the ***event processor (handler)*** that coordinates and connects all of the components of the cluster service, handles common operations, and controls cluster service initialization. The ***communication manager*** monitors message exchange with all other nodes present in the cluster. The ***global update manager,*** however, provides a service used by other components within the cluster service.

The Windows Cluster Server balances the incoming network traffic load by distributing the traffic among the server nodes in a cluster. This is accomplished in the following manner. The cluster is assigned with a single IP address; however, incoming messages go to all server nodes in the cluster. Based on the current load distribution arrangement, exactly one of the servers accepts the message and responds to it. In the event of a node failure, the load belonging to the failed node is distributed among other active nodes. Similarly, when a new node is included, the load distribution is reconfigured to direct some of the incoming traffic to the new joining node.

10.9.7 Sun Cluster

The *Sun Cluster framework* integrates a cluster of two or more Sun systems. The operating system managing Sun Cluster is essentially a true *distributed operating system* built as a set of extensions to the base Solaris UNIX system. It straightaway casts a single-system image to the user as well as to the applications, and hence, appears as a single computer system on which the Solaris operating system runs. This Solaris operating system obviously provides better availability, and more scalability of services. A schematic representation of an overall Sun Cluster architecture running on existing Solaris kernel is depicted in Figure 10.26. The distinctive components of this system are:

- Object and Communication support
- Global Process Management
- Networking
- Global Distributed File System

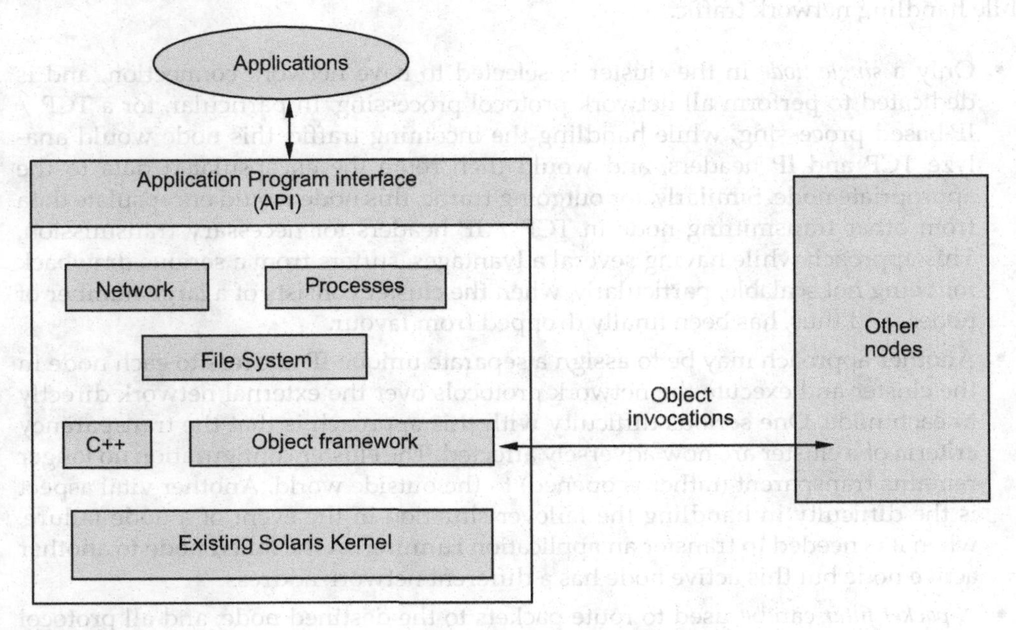

FIGURE 10.26
A block diagram of SUN Cluster Structure.

Object and Communication support: The implementation of Sun Cluster is based on object orientation in which the CORBA (Common Object Request Broker Architecture) object model is used to define objects, and RPC (remote procedure call) mechanism is used to support communication. The CORBA Interface Definition Language (IDL) is used to specify interfaces between MC (cluster) components in different nodes. The elements of MC are implemented in the object-oriented language C++. The use of a uniform object model and exploiting the support of IDL offer a suitable mechanism that facilitates both inter-node as well as intra-node process communication activities. All these things are, however, built on top of a Solaris kernel to work together with virtually no changes required to the kernel, even if any changes are made in any area above the kernel.

Global Process Management: Existing process management is enhanced with the use of the global process management which provides globally unique *process ids* for each process in the cluster so that each node is aware of the location and status of the other processes. This feature is useful in *process migration*, wherein a process during its lifetime can be transferred from one node to another to balance the computational loads, as well as to make ease of computation in different nodes or to achieve computation speed-up. A migrated process in that situation should be able to continue using the same pathnames to access files from a new node. The use of a distributed file system, in particular, facilitates this feature to work. The threads of a single process, however, must be on the same node.

Disk Path Monitoring: All disk paths can be monitored and configured to automatically reboot a node in case of multiple path failure. Faster reaction in case of severe disk path failure provides improved availability.

Configuration Checker: It checks for vulnerable cluster configurations regularly and rapidly, thereby attempts to limit failures due to mis-configuration throughout the lifetime of the cluster.

Networking: A number of *alternative approaches* have been thought of by Sun cluster while handling network traffic.

- Only a *single node* in the cluster is selected to have network connection, and is dedicated to perform all network protocol processing. In particular, for a TCP / IP-based processing, while handling the incoming traffic, this node would analyze TCP and IP headers, and would then route the encapsulated data to the appropriate node. Similarly, for outgoing traffic, this node would encapsulate data from other transmitting node in TCP / IP headers for necessary transmission. This approach while having several advantages, suffers from a serious drawback for being not scalable, particularly, when the cluster consists of a large number of nodes, and thus, has been finally dropped from favour.
- Another approach may be to assign a separate unique IP address to each node in the cluster and execute the network protocols over the external network directly to each node. One serious difficulty with this approach is that the transparency criteria of a cluster are now adversely affected. The cluster configuration no longer remains transparent (rather is opened) to the outside world. Another vital aspect is the difficulty in handling the failover situation in the event of a node failure, when it is needed to transfer an application running on the failed node to another active node but this active node has a different network address.
- A *packet filter* can be used to route packets to the destined node, and all protocol processing is performed on that node. A cluster in such situation appears from

the outside world to be a single server with a single IP address. Incoming traffic is then appropriately distributed among the available nodes of the cluster to balance the load. This approach is found to be an appropriate one for a Sun cluster to adopt.

Incoming packets are first received on the node that has the physical network connection with the outside world. This receiving node filters the packet and delivers it to the right target node over the cluster's own internal connections. Similarly, all outgoing packets are routed over the cluster's own interconnect to the node (or to the one of the multiple alternative nodes) that has an external physical network adapter. However, all protocol processing relating to outgoing packets is always performed by the originating node. In addition, the Sun cluster maintains a global network configuration network database in order to keep track of the network traffic to each node.

Availability and Scalability: Sun Cluster provides ***availability*** which is provided through failover, whereby the services that were running at a failed node are transferred (relocated) to another node. ***Scalability*** is provided by sharing (as well as distributing) the total load across the existing servers.

Multiple storage technologies and storage brands: Solaris Cluster can be used in combination with different storage technologies such as FC, SCSI, iSCSI, NAS storage on Sun or non-Sun storage.

Easy-to-Use Command Line Interface: An object-oriented command line interface (CLI) provides a consistent and familiar structure across the entire command set, making it easy to use and limiting human error. Command logging enables tracking and replay.

Global File System: The beauty and the strength of the Sun cluster is its global file system as depicted in Figure 10.27. This file system is essentially built on the use of *virtual node* (*v-node*) and *virtual file system* (VFS) concepts. The *v-node* structure is necessarily used to provide a powerful, general-purpose interface to all types of file systems. A *v-node* is used to map pages of memory into the address space of a process to permit access to a

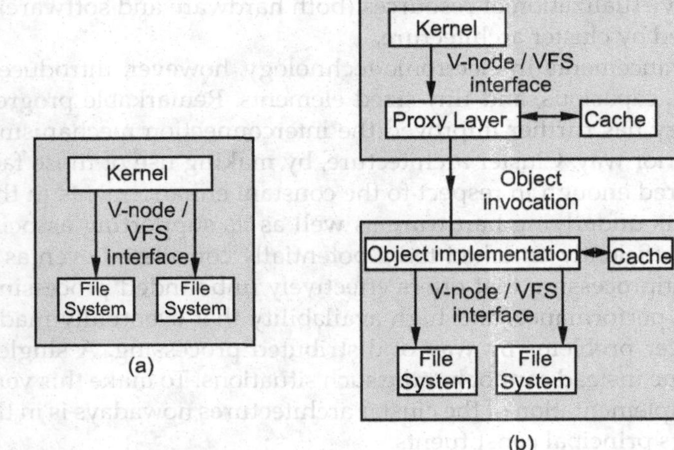

FIGURE 10.27
A scheme of SUN cluster File System Extension. (a) Standard Solaris and (b) Sun Cluster.

file system and to map a process to an object in any type of file system. The VFS interface, however, accepts general-purpose commands that operate on entire files and translates them into actions appropriate for that subject file system. The global file system provides a uniform interface to all files distributed over the cluster.

A process can open a file located anywhere in the cluster, and processes on all nodes use the same pathname to locate a file. In order to implement global file access, MC includes a proxy file system built on top of the existing Solaris file system at the *v-node* interface. The VFS / *v*-node operations are appropriately converted by a proxy layer into object invocations. The invoked object may reside on any node in the system. The invoked object subsequently performs a local *v*-node / VFS operation on the underlying file system. No modification is, however, required either at the kernel level or in the existing file system to support this global file environment. In addition, caching is used once again to facilitate reducing the number of remote object invocations that in turn minimizes the traffic on the cluster interconnect. A few of the Multiple File Systems and Volume Manager that are supported by the Sun Cluster are:

- UFS, VxFS, and ZFS as root file systems.
- HA UFS, HA NFS and HA ZFS, HA QFS and shared QFS (with Oracle RAC), and HA VxFS.
- Global File System on UFS and VxFS,
- SVM, VxVM, and ASM

10.9.8 Blade Servers

The concept of cluster architecture, its development, and subsequently its successful implementation quickly became very popular in application processing that created a huge impact in the computing world across the globe, and gradually became the state-of-the-art in the area of large-scale distributed application processing. The computing culture then started to evolve in a different way giving rise to a marvellous concept what is known as ***cloud computing***. The main theme behind such an innovative computing approach essentially lies in virtualization of resources (both hardware and software) by the way of distribution blessed by cluster architecture.

Continuous advancements in electronic technology, however, introduce more sophisticated, intelligent, capacious, and tiny-sized elements. Remarkable progress in communication technology has further improved the interconnection mechanisms between the systems in a superior way. Cluster architecture, by making use of those facilities, gradually became matured enough in respect to the constant enhancements in the designs and developments of its underlying hardware as well as its supporting associated software. Soon, it became sufficiently enriched to be potentially considered even as an alternative to symmetric multiprocessing that offers effectively unbounded processing power, storage capacity, high performance, and high availability that eventually made it capable of solving much larger problems by way of distributed processing. A single large costlier mainframe machine instead cannot handle such situations. To make this versatility realizable, a common implementation of the cluster architectures nowadays is in the use of blade servers as one of its principal constituents.

A ***blade server*** is literally a server *on a board* containing processors, memory, integrated network controllers, an optional fibre channel host bus adaptor (HBA), and other input/output (IO) ports. Each such server module is usually called "blade" and is self-standing, or multiple blade cards can be housed on a chassis in a rack-mounted fashion to form an even larger server architecture.

Blade server, in essence, is a stripped down computer server with a modular design optimized to reduce the use of physical space, minimize power consumption, and thereby less heat generation, and also fulfils other requirements with lesser components while still offering all the functionalities to be a full-fledged computer by itself. A server blade with all its functional components along with similar other blades, and network connectivity with the required accessories for all these server blades when typically housed in a rack-mountable enclosure share common resources such as cabling, power supplies, cooling fans, etc., thereby reducing both cost and complexities associated with these heads. When compared to a traditional computer server, the blade server offers a superior computational power, better scalability, higher fault-tolerance, and easy portability with an overall reduced cost and less overhead.

Server blade typically comes with 1–4 processors, a minimum random access memory (RAM) of 2GB that can be expanded to at least 8GB supported by an adequate cache memory, and integrated Ethernet. The possible network connections are: 10Base-T Ethernet, 10Base-2 Ethernet, Fast Ethernet, Token Ring, Fibre Channel, and also Fieldbus. Blade chassis normally operate at standard input voltage of 200to 240VAC supporting at least ten such hot-swappable blades being mounted 1 height unit (1U) apart. The chassis provides a CD-ROM drive or an alternate way for installing the operating system and application software. Blade servers typically have a front panel containing a number of informational LEDs as indicators relating to power supply, system activity, and also system failure, which may be general or specific to blade components. These servers are operated by Microsoft windows, Unix, AIX, Linux, Solaris, HP-UX, Mac OS, etc., and also cluster versions of these operating systems.

When substantial banks of several blade servers, located geographically far apart, are connected by a variety of the most popular networking brands and standards including Ethernet and Fibre Channel through the ports available with individual servers, it can then handle massive traffics with less latency, providing enormous storage capacity across this arrangement. Intelligent switches or routers can also be used in between to cover even more geographical area. An example of such a massive blade server arrangement being connected by networks with one another is depicted in Figure 10.28.

The blade servers provide many ***distinct advantages*** that are not found in traditional ordinary servers. A few of these advantages are: relatively lower acquisition cost as well as lower operational cost for deployment and troubleshooting and repair, proportionately lower power and cooling requirements, lesser space requirements, reduced cabling requirements, flexibility and modularity and ease of upgrading (scalability), more efficient out-of-band management, disaster management, and also faster server-to-server communications.

The notable ***disadvantages*** of blade servers are, however, its expensive configuration, i.e., economies of scale, vendor-lock, business case, and also typical requirements for heat dissipations and cooling arrangements.

In the application domain, with the use of blade servers, it is now possible to have the essentials to build and maintain vital solutions to support many ever-demanding areas, namely, cloud computing, email messaging, large databases hosting, ERP hosting, virtualization of resources using supporting environment, file sharing and serving, streaming audio and video content, and many others.

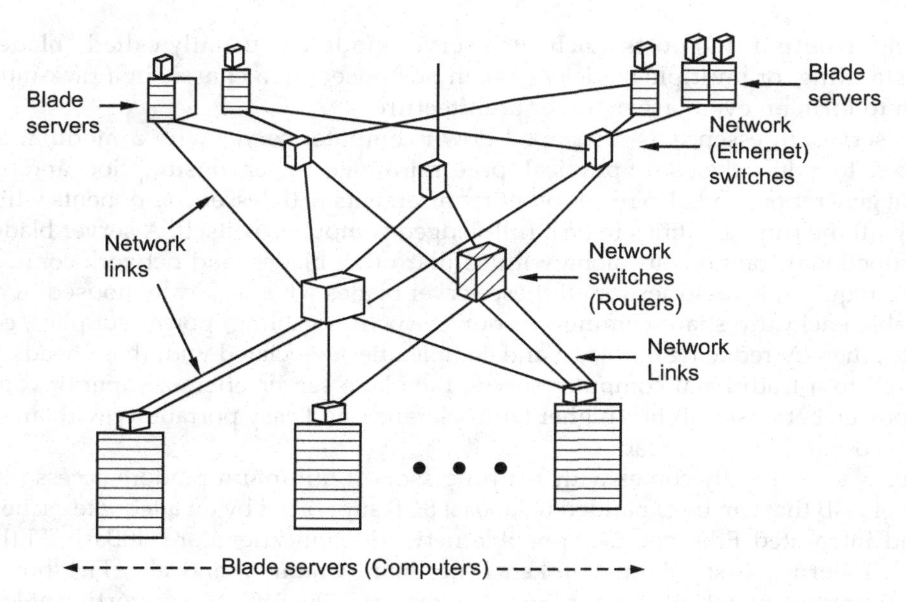

FIGURE 10.28
A massive blade server arrangement using Ethernet configuration.

10.10 SIMD Machines

In an SIMD (Single Instruction Multiple Data) machine, one instruction at any point of time is simultaneously executed by multiple processors (processing elements, PEs) using different data streams of same application. Each processor (or processing element) has its own data memory (local memory, LMs) enabling an SIMD to accommodate multiple data, but a single instruction is fetched from the instruction memory, dispatched by a single control processor. Although this is less general than the MIMD model, since full generality is not desired, it is simpler, cheaper, and typically is a special–purpose for certain classes of problems, and potentially much faster in execution. Representative application areas of an SIMD machines are: daily weather forecasting, ozone modelling for global climate forecast, oceanography, image processing and understanding, air pollution reduction through computational modelling, etc. However, SIMD processors can be classified into two broad categories, namely, *Vector processors* and *Array processors*. Each of these processors will be discussed later in subsequent different sections.

From historical point of view, many of the early *multiprocessors*, like the ***ILLIAC IV***, were essentially SIMD. Those machines were built to execute a single instruction that operates on many data items concurrently, using many functional units. This is, of course, the key idea that is mostly being followed today by more recent SIMDs.

SIMD models have been reviewed and received renewed attention in the 1980s, first by Thinking Machines Corporation (TMC), and then by MasPar (Nickolls, J. R.). It has been observed that this model was not well suited as a general-purpose multiprocessor architecture due to being too inflexible. Moreover, it failed to combat the tremendous performance and cost advantages that the on-going microprocessor technology eventually started to offer. The designers were thus forced to start thinking afresh along different lines that SIMD machines must be built with custom processors using special-purpose designs to handle specific applications that are highly data parallel, and require a limited set of functional units interconnected with one another.

Real SIMD computers need to have a combination of SISD and SIMD instructions. While SISD host computers perform all scalar, non-parallel operations, such as, address calculations, branching, etc., the SIMD instructions are sometimes broadcast to all the execution units (PEs), each of which has its own storage elements (LMs). Individual execution units, for the sake of flexibility however, can be disabled by SIMD instructions. The SIMD programs are executed in lock-step, and this hardware synchronization is enforced by the control unit. Vector operands are loaded simultaneously into the PEs from local memories using a global address with different offsets in local index registers. Storing of vector operands can be performed in a similar manner. Constant data are usually broadcast to all PEs simultaneously.

SIMD machine architecture is thus specified mainly by ***five parameters***; the strength and capability of each parameter and their collective contribution determine the category of SIMD machine. These parameters are:

- The number of ***processing elements*** (PEs) that are available in a machine. ILLIAC IV has 64 PEs, The MasPar MP-1 has 1024 to 16384 PEs in different models, CM-2 from Thinking Machine Corporation uses 65,536 PEs.
- The design of the ***control unit*** (CU) that directly executes a particular set I_1 of instructions which includes the scalar instructions and program flow control instructions. The PEs, however, receive instructions from CU and execute integer / standard floating-point operations over various sizes of data.
- A particular set I_2 of instructions include arithmetic, logic, data routing, masking, and other local operations. CU ***broadcasts*** decoded vector instructions to the PE array for executions, and controls inter-PE communications. Each such instruction is executed by all PEs in parallel over data available in their respective local memories (LM).
- Out of a set of ***masking schemes*** (M), one masking scheme is chosen to partition the set of PEs into enabled and disabled subsets. This scheme is built within each PE, and CU continuously monitors it to set and reset the status of each PE during runtime. This is done to keep some PEs operative (set) for a certain operation while the other PEs are maintained inoperative at the same time.
- Out of a set of available ***Interconnection Network*** (ICN) schemes, one or more than one is chosen to establish inter-PE communications and inter-CU-PE routing. CM-2 machine uses hypercube routers. Each PE is equipped with special hardware to form router node, and all these router nodes on all PEs are wired together to form a Boolean n-cube that are used for routing data among the PEs.

SIMD works best in dealing with vectors (or arrays) in for-loops. Each SIMD instruction is equivalent to an *entire loop* with each iteration computing one of the vector elements (array elements) of the result, updating the indices and branching back to the beginning. This gives rise to massive *data parallelism* that can be further exploited in massive parallel processing (MPP).

10.10.1 SIMD Computer Organisations

Most SIMD computers use a *single array control unit* and *distributed memories,* except for a few that use *associative memories.* The single array control unit decodes the instruction set of an SIMD computer. The processing elements (PEs) in the SIMD array are essentially passive ALUs which execute instructions that are broadcast from the control unit. All PEs must operate in lockstep, synchronized by the same array control unit. The SIMD computer models are usually categorized based on the *memory distribution* and *addressing scheme* being used.

Distributed-memory Model: A distributed-memory SIMD computer consists of an array of PEs in which each PE has its own local memory (distributed memory). This array of PEs is controlled by an array control unit. Instructions and data are loaded into the control memory through the host computer. An instruction is sent from the control memory to the array control unit for decoding. If it is a scalar or program control operation, it will be directly executed by the scalar processor attached to the array control unit. If the decoded operation is a vector operation, it will be broadcasted by the same control unit to all the PEs for parallel execution in the same cycle. The PEs are, however, synchronized in hardware by the control unit. All PEs are interconnected by a data interconnection network as shown in Figure 10.29a, which performs inter-PE data communications, and other routing operations under program control through the control unit.

Most of the SIMD machines have been implemented on the lines of distributed memory model. These machines are also of various types, and found to differ mainly in the use of data interconnection network being chosen for inter-PE communications. In the early days, *mesh architecture* has been the most popular choice, and found in use in ***ILLIAC IV, Goodyear MPP,*** and ***DAP 610.*** Later, superior *hypercube* embedded in a mesh, however, replaced the earlier one as was found in a more advanced machine ***CM 2,*** while a multistage crossbar router, known as **X-Net-plus** has been implemented in ***MasPar MP1.*** Massively parallel SIMD machines of today mostly use *interconnection* or *communication networks* to exchange data between processing elements (PEs).

Shared-memory Model: A variation of this architecture is an SIMD machine that uses *shared memories* among the processing elements. An alignment network managed by the same array control unit is used for the inter-PE memory synchronization (to avoid access conflicts) and communication. ***Burroughs Scientific Processor (BSP)*** follows this approach as shown in Figure 10.29(b) to realize word-parallel SIMD computers, although some SIMD computers, such as, the CM / 200 use bit-slice PEs.

When **associative memory** is used in shared-memory or distributed-memory architecture, they form a special subclass of SIMD computers known as ***SIMD associative processor.*** Depending on the type of associative memory used, the associative processors are again classified into *two* broad classes:

- The fully parallel organisation (PEPE organisation)
- The bit-serial organisation (STARAN organisation)

Both PEPE and STARAN have been discussed in the Associative Memory, Chapter 4.

SIMD Instructions: SIMD computers provide vector and array processing. Vector instructions are mainly for arithmetic, logic, data-routing, and masking operations over vector elements, as already mentioned earlier. In *bit-slice* SIMD machines, the vector operands are nothing but binary vectors, whereas in *word-parallel* SIMD machines, the vector components are 4-, or 8-byte numerical values. All SIMD instructions must use

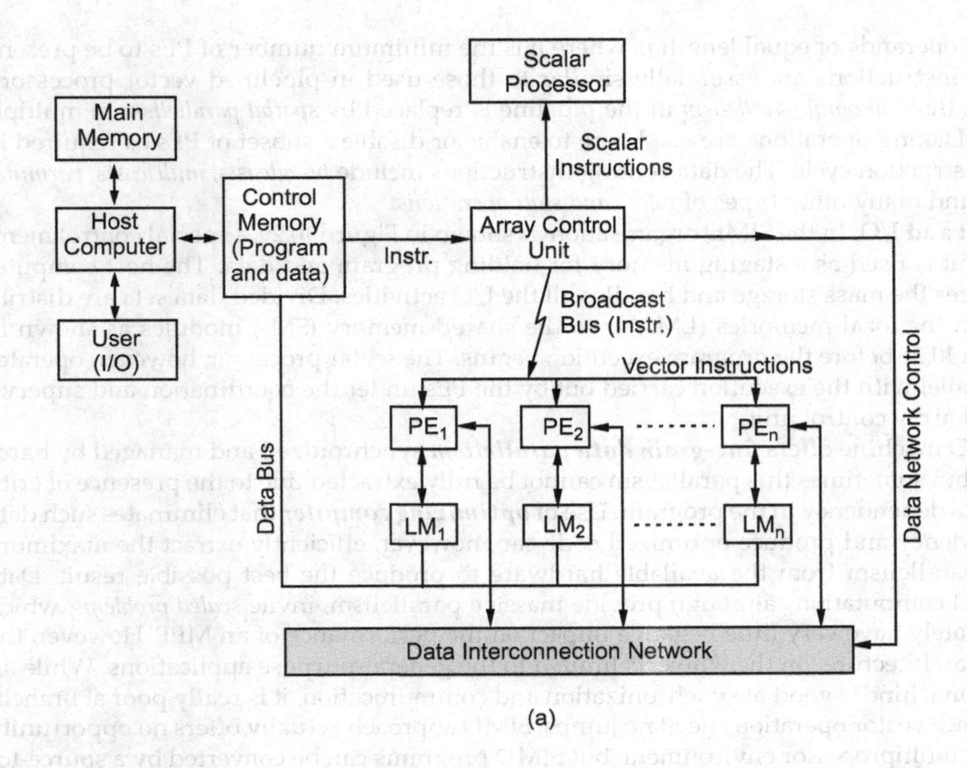

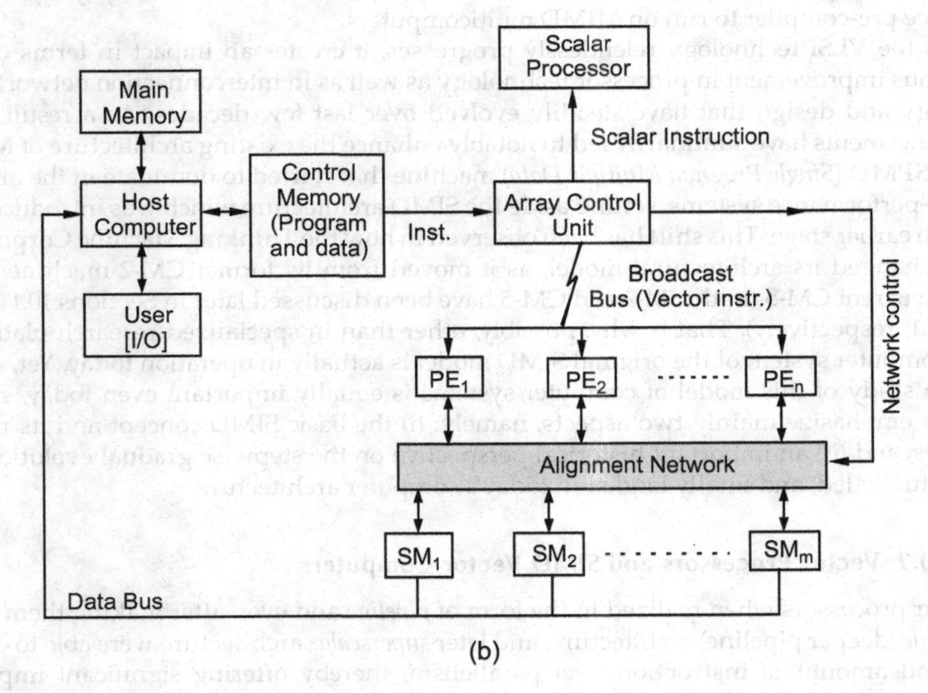

FIGURE 10.29
SIMD Computer with different types of memory arrangement. (a) SIMD machine with distributed local memories (ILLIAC IV type) and (b) SIMD machine using shared-memory modules (BSP types).

vector operands of equal length *n*, where *n* is the minimum number of PEs to be present. These instructions are essentially similar to those used in pipelined vector processors, except that *temporal parallelism* in the pipeline is replaced by *spatial parallelism* in multiple PEs. Masking operations are used here to enable or disable a subset of PEs as required in any instruction cycle. The data-routing instructions include *broadcasts, multicasts, permutations,* and many other types of *rotate and shift operations.*

Host and I/O: In the SIMD organisation as shown in Figure 10.29, a special control memory unit is used as a staging memory for holding program and data. The host computer manages the mass storage and handles all the I/O activities. Divided data sets are distributed in the local memories (LMs) or to the shared memory (SMs) modules as shown in Figure 10.29 before the program execution begins. The scalar processor, however, operates in parallel with the execution carried out by the PEs under the coordination and supervision of array control unit.

SIMD machine offers ***fine-grain data parallelism*** synchronized and managed by hardware, but sometimes this parallelism cannot be fully extracted due to the presence of critical data dependency in the program. Use of ***optimizing compiler*** that eliminates such data dependency and produce optimized code can, however, efficiently extract the maximum data parallelism from the available hardware to produce the best possible result. Data parallel computations although provide massive parallelism, invite *scaled problems,* which fortunately have very little negative impact on the performance of an MPP. However, the SIMD architecture, on the whole, is limited to the general-purpose applications. While an SIMD machine is good at synchronization and communication, it is really poor at branching, since vector operations need no jumps. SIMD approach actually offers no opportunity to be a multiprocessor environment, but SIMD programs can be converted by a source-to-source pre-compiler to run on MIMD multicomputers.

As the VLSI technology relentlessly progresses, it creates an impact in terms of continuous improvement in processor technology as well as in interconnection network technology and design that have steadily evolved over last few decades. As a result, these advancements have summarily led to notably enhance the existing architecture of **MIMD** and **SPMD** (*Single Program Multiple Data*) machine that started to dominate in the arena of high-performance systems, setting aside the SIMD architecture which was introduced at a much earlier stage. This shift has been observed in how the Thinking Machine Corporation has changed its architectural model, as it moved from its former CM-2 machine to the then current CM-5 (both CM-2 and CM-5 have been discussed later in Sections 10.11.2 and 10.12.1, respectively). That is why, possibly, other than in specialized research platforms, no computer system of the original SIMD model is actually in operation today. Yet, a thorough study of this model of computer systems is equally important even today, since it helps emphasize mainly two aspects, namely: (i) the basic SIMD concept and its related issues, and (ii) an important historical perspective on the stepwise gradual evolution that eventually led, and finally landed in today's computer architecture.

10.10.2 Vector Processors and SIMD Vector Computers

Scalar processors when realized in the form of *pipeline* and even after making them *superpipeline* (deeper pipeline) architecture, and later *superscalar* architecture were able to extract a good amount of instruction-level parallelism, thereby offering significant improved performance, but still there are limits on such performance improvement that these approaches could achieve. The two primary factors that cause to limit the performance are: *clock cycle time,* which although can be reduced using superpipelined design, but this design, however, in turn, invites a corresponding increase in CPI due to an increase in pipeline dependencies. The other one is *instruction fetch and decode rate* (also known as ***Flynn bottleneck***) that creates hindrance to fetch and issue many instructions per clock. The net effect is that it has been found difficult to build processors with high clock rates as well as very high issue rates. Memory references are still critical, even with the use of various forms of memory hierarchy and cache architectures that ultimately result in considerable long latency. All these factors together with other related aspects are continuing to be serious with some specific types of applications while in execution.

One form of SIMD architecture that is well-suited for this specific type of applications is the Vector machine having a *Vector processor* that came out with an innovative approach providing high-level operations which work on vectors. A *vector* is basically an ordered set of scalar data items, all of the same type, stored in memory. A vector processor essentially consists of a scalar processor and a vector unit, and this vector unit could be thought of as an independent functional unit capable of executing vector operations efficiently. Scalar instructions are executed on scalar processor, whereas vector instructions are executed on this vector unit. A vector processor actually comprises of special hardware that includes vector registers, functional pipelines, processing elements (PEs), and register counters for performing operations on vectors.

A vector processor executes vector instructions giving rise to vector processing on the ordered set of scalar data items. It is distinguished from scalar processing which operates on one or one pair of data. A typical vector operation might add two 32-element (a_1, a_2, …, a_{32} and b_1, b_2, …, b_{32}) floating-point vectors to obtain a single 32-element (c_1, c_2, …, c_{32}) vector result. The vector instruction is equivalent to an entire loop, in which each of the iterations computes one of the 32 elements of the result and updates the indices and then branches back to the beginning to continue, and all these happen per clock cycle continuously. A schematic structure of a vector ALU is illustrated in Figure 10.30. The conversion from scalar code to vector code is called ***vectorization***.

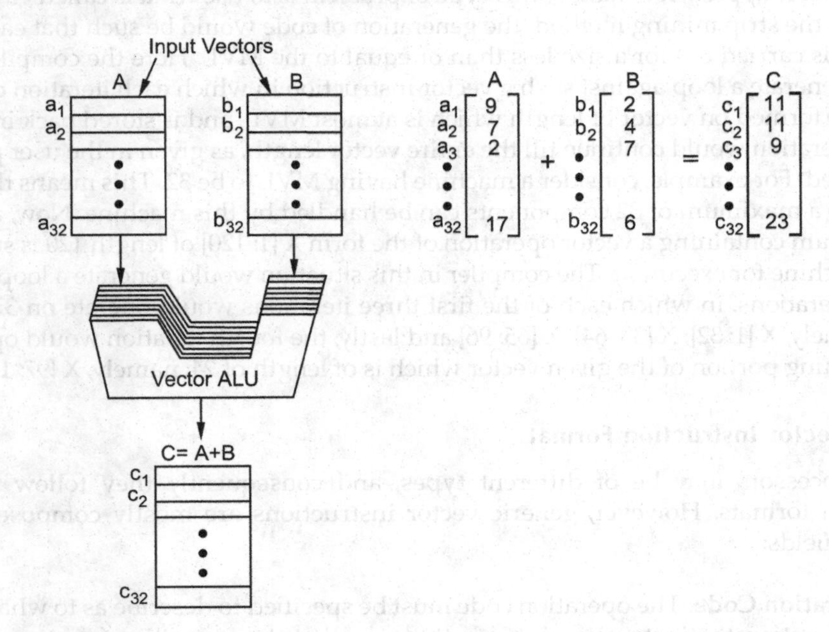

FIGURE 10.30
A vector ALU used in vector addition.

In general, vector processing is much faster and more efficient than its counterpart scalar processing. Both pipelined processors and SIMD computers can perform vector operations. Vector processing, however, reduces software overheads incurred in the maintenance of looping control. In addition, it reduces memory-access conflicts, and above all, it is found to be very conducive to the pipelining and segmentation approaches to generate one result per clock cycle continuously.

10.10.3 Vector Stride

The information being processed by a vector processor may not be in consecutive memory locations. But they are usually ordered to have a *fixed distance* that separates successive elements which are to be merged into a single vector. This distance is called *stride*. The stride for a vector expresses the address increment used to move from one element to the next at the time of vector access. Consider a two-dimensional vector, and assume that a program is written to access rows or columns of this vector. To accomplish this, the stride value is provided. The load unit uses this stride value to find these locations, and load them into the vector registers whenever required. For example, suppose a two-dimensional (2-D) vector of size 8×8 is stored in the main memory in a row-major order. If the stride is set to 1, then we can fetch a row of 8 elements. If the stride is set to 8, then we can fetch a column of 8 elements.

10.10.4 Vector Fitting

In many applications, it is often found that the length of the vectors used in the program could be much larger than the maximum vector length (MVL) available in a vector register machine. Moreover, the length of a vector may not be made out before actual run. For example, a loop of the form "for $i = 1, N$" where the value of N can be ascertained only at run-time. This difficulty of fitting user vector lengths to machine vector lengths is called *vector fitting*.

One possible approach to handle this type of problem is to use what is called ***"Strip mining"***. With the strip mining method, the generation of code would be such that each vector operation is carried out for a size less than or equal to the MVL. Here the compiler would actually generate a loop against such a vector instruction in which each iteration operation is to be performed on vector of length which is at most MVL, and is stored back into memory. The iteration would continue till the entire vector length, as given in the user program, is exhausted. For example, consider a machine having MVL to be 32. This means that a vector having a maximum of 32 components can be handled by this machine. Now, assume a user program containing a vector operation of the form X [1: 120] of length 120 is submitted to this machine for execution. The compiler in this situation would generate a loop containing four iterations, in which each of the first three iterations would operate on 32 components, namely, X [1: 32]; X [33: 64]; X [65: 96] and lastly, the fourth iteration would operate on the remaining portion of the given vector which is of length of 24, namely, X [97: 120].

10.10.5 Vector Instruction Format

Vector processors may be of different types, and consequently they follow different instruction formats. However, generic vector instructions are mostly composed of the following fields:

- **Operation Code**: The operation code must be specified to describe as to what type of operation the instruction intends to do, so that the respective functional unit can be allocated or the multifunctional unit can be reconfigured to perform the specified operation as indicated by this field. Normally, the microcode control is used to assign the needed resources.

- **Base Address**: For a memory-reference instruction, the *base address* is needed to find the beginning of the actual physical memory address for both source operands as well as result vectors. However, if the operands and results of the computation are located in vector register file (collection of registers), then the designated *vector registers* must be specified within the instruction format.
- **Offset (or Displacement)**: The offset field contains the address which is relative to the prescribed base address. Using the given base address and the content of this offset field (positive or negative), the effective memory address of the operand vector is obtained after required computation.
- **Address Increment**: The address increment between the scalar elements of vector operand needs to be specified. Some vector computers, like the supercomputer Star-100 introduced by Control Data Corporation (CDC) in 1973, however, assume this increment to always be 1, i.e., they insist the elements to be consecutively stored in memory all the time. Other computers like TI-ASC allow a variable increment while storing the scalar elements of vector operands. This approach certainly offers a greater flexibility at the time of allocating memory to an application.
- **Vector Length**: The vector length (a positive integer) is always needed, that declares the number of elements present in a vector operand in the form of an ordered set. This is required to assert the end of vector instruction execution.

10.10.6 Vector Instruction: Different Types

Vector computers, apart from executing normal vector operations, also need to do scalar (non-vector) operations, and mixed vector–scalar operations. We use here a function f that performs some operation on each element of a single or two vector operands which may be located in working registers or memory. Recall that a vector operand contains an ordered set of n elements, where n is called the length of the vector. All elements in a vector are scalar quantities of same type, which may be an integer, a floating-point number, a logical value, or even a character. The following *six types* of vector instructions are described here for pipelined vector-register processors:

1. **Vector–vector instructions:** One or two vector operands are fetched from respective vector registers and produce results in another vector register. If A, B, and C are the three vector registers, these instructions can be defined as:

   $A_i := f_1\ (B_i)$ $\quad f_1$ = sine, cosine, square root etc.

   $A_i := f_2\ (B_i, C_i)$ $\quad f_2$ = add, subtract

   Example are: A = Cos (B)

   A = B + C

2. **Vector–scalar instructions**: A scalar operand can be combined with a vector one. A vector–scalar operand can be defined as:

   $A_i := f_3$ (scalar, B_i) $\quad f_3$ = multiply B_i by a constant.

   An example is a scalar product:

   A = $s \times$ B in which each element of B is multiplied by a scalar s to produce vector A of equal length.

3. **Vector reduction instructions**: These instructions take a vector (or two vectors) as input and produce a scalar as output.

   The examples are: scalar : $= f_4$ (A)
   scalar : $= f_5$ (A, B)

   where f_4, for example, includes finding of *maximum, minimum, sum,* and *mean value* of all elements of a vector.

   An example of f_5 is the *dot product,* which performs

   $$s = \sum_{i=1}^{n} a_i \times b_i \text{ from two vectors } A = (a_i) \text{ and } B = (b_i)$$

   Truly speaking, f_5 is not a fundamental one, it can be derived by first multiplying the corresponding elements of two vectors together (f_2), and then summing the result (f_4).

4. **Vector–memory instructions**: This instruction corresponds to vector load or vector store, element by element, between the vector register (A) and the memory M that can be defined as:

   A: $= f_6$ (M) ; where f_6 indicates vector load
   M: $= f_7$ (A) ; where f_7 indicates vector store

5. **Gather and scatter instructions**: These instructions use two vector registers to gather or to scatter vector elements randomly throughout the memory, by the operations:

   $A \times V_0$: $= f_8$ (M) ; f_8 indicates *Gather* operation.
   M: $= f_9$ $(A \times V_0)$; f_9 indicates *Scatter* operation.

   Gather is an operation that fetches from memory the non-zero elements of a sparse vector using indices, which themselves are indexed. *Scatter* does exactly the opposite, storing into memory a vector in a sparse vector whose non-zero entries are indexed.

   The vector register A contains the data and the vector register V_0 is used as an index to gather or scatter data from or to random memory locations. The gather instruction, as illustrated in Figure 10.31, transfers the contents (300, 200, 500, 100, 400) of non-sequential memory locations (203, 201, 206, 200, 208) to five elements of a register A. The base address (200) of the memory is indicated by an address register A_0. The offset from the given base address is retrieved from the vector register V_0. The effective memory addresses are obtained by adding the base address to the indices. The number of elements being transferred is indicated by the contents (5) of a *vector length* register VL. Both A_0 and VL registers are embedded in the instruction.

   The scatter instruction just reverses the operation as being conducted in the gather instruction. This is illustrated in Figure 10.32.

6. **Masking instructions**: This type of instruction uses a *mask vector* to expand or compress a vector to a longer or shorter index vector, respectively corresponding to the following operations:

   V_1: $= f_{10}$ $(A \times VM)$; f_{10} indicates an operation to compress or to expand a vector A using VM.

An example is illustrated to clarify the meaning of masking instruction in Figure 10.33 for compressing a long vector into a short index vector.

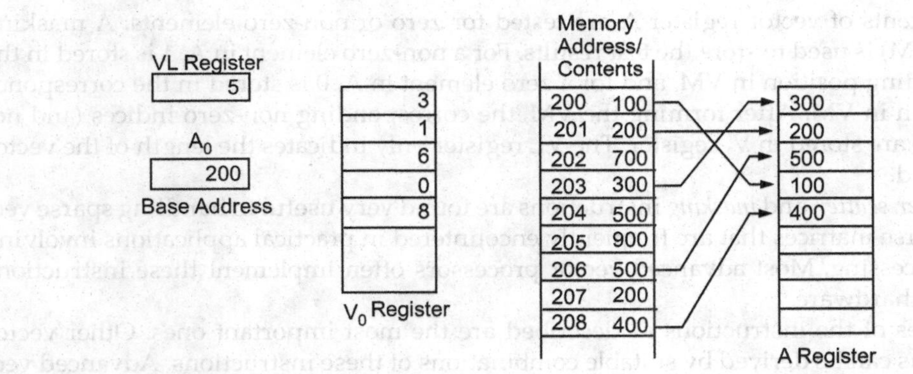

FIGURE 10.31
A scheme of Gather instruction.

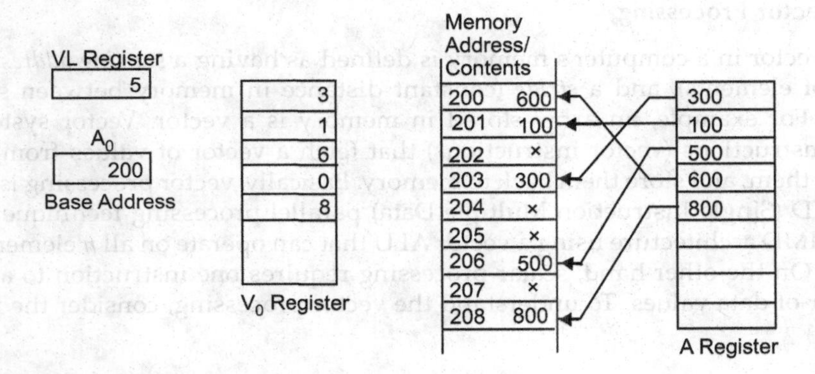

FIGURE 10.32
A scheme of Scatter instruction.

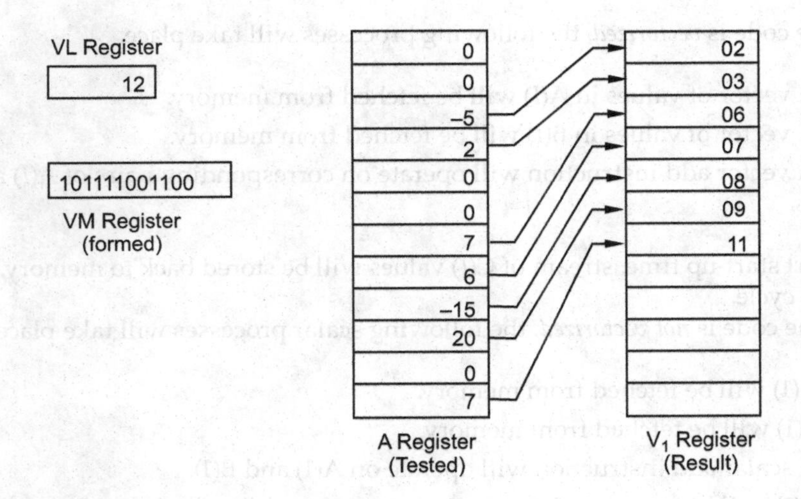

FIGURE 10.33
A scheme of Masking Instruction.

The contents of vector register A are tested for zero or non-zero elements. A masking register (VM) is used to store the test results. For a non-zero element in A, 1 is stored in the corresponding position in VM, and for a zero element in A, 0 is stored in the corresponding position in VM. After forming the VM, the corresponding non-zero indices (and not the values) are stored in V_1 register. The VL register only indicates the length of the vector being tested.

The *gather, scatter,* and *masking* instructions are found very useful in handling sparse vectors or sparse matrices that are frequently encountered in practical applications involving vector processing. Most advanced vector processors often implement these instructions directly in hardware.

The types of the instructions as described are the most important ones. Other vector instructions can be derived by suitable combinations of these instructions. Advanced vector processors often include a subset or even a superset of the above instructions into their instruction set to implement these instructions directly in hardware.

10.10.7 Vector Processing

A simple vector in a computer's memory is defined as having a *starting address*, a *length* (number of elements), and a *stride* (constant distance in memory between successive elements). For example, an array stored in memory is a vector. Vector systems have machine instructions (vector instructions) that fetch a vector of values from memory, operate on them, and store them back to memory. Basically, vector processing is a version of the SIMD (Single Instruction Multiple Data) parallel processing technique and well suited to SIMD architecture using a vector ALU that can operate on all *n* elements simultaneously. On the other hand, scalar processing requires one instruction to act on one or one pair of data values. To understand the vector processing, consider the following example:

```
   DO 50 I = 1, N
     C(I) = A(I) + B(I)
50 CONTINUE
```

If the above code is *vectorized*, the following processes will take place:

Step 1. A vector of values in A(*I*) will be fetched from memory.

Step 2. A vector of values in B(*I*) will be fetched from memory.

Step 3. A vector add instruction will operate on corresponding pairs of *A*(*I*) and *B*(*I*) values.

After a short start-up time, stream of C(*I*) values will be stored back to memory, one value every clock cycle.

And, if the code is *not vectorized*, the following scalar processes will take place:

Step 1: A(1) will be fetched from memory.

Step 2: B(1) will be fetched from memory.

Step 3: A scalar add instruction will operate on A(1) and B(1).

Step 4: C(1) will be stored back to memory

Steps (1) to (4) will be repeated N times.

Vector processing allows a vector of values to be fed continuously to the vector processor. If the value of N is large enough making the start-up time even negligible in comparison, on an average, the vector processor is capable of producing close to one result per clock cycle. If the same code is not vectorized, for every I iteration, e.g., $I = 1$, a clock cycle each is needed to fetch A(1) and B(1), about another 4 clock cycles are needed to complete a floating-point add operation, and one more clock cycle is needed to store the value C(1). Thus a minimum of six clock cycles are needed to produce one result, showing a speed up of about six times for this example, if the code is vectorized.

Vector operations like add or multiply are done in a vector pipeline in the processor, and can be attached to any scalar processor, whether it is superpipelined, superscalar, or both. The example given above contains codes which are inherently vectorizable. But in reality, not all codes are vectorizable, at least not all are readily vectorizable.

10.10.8 Vectorization Inhibitor

Having a DO loop is, by no means, an assurance of vectorization. The existence of certain codes in the DO loop may prevent the compiler from converting the entire or part of the DO loop for vector processing. They are collectively known as the *vectorization inhibitors.* To ensure greater vectorization, a programmer will have to avoid putting vectorization inhibitors in the DO loop, or to reconstruct the DO loop by replacing vectorization inhibitors with an equivalent code which can then be taken up for vectorization.

Recursive data dependencies are one of the most *"destructive"* vectorization inhibitors. Recursion occurs when any address set in one iteration of a DO loop is referenced in another iteration. Single-dimension recursion, as shown below, will prevent any vectorization of arithmetic operations in a DO loop.

$$X(I) = X(I - 1) + Y(I)$$

Here, the correct computation requires the values of X(I) to be updated before they are used in the next iteration. If the above code is attempted to be vectorized forcibly (this can be done by using vector directives), all the old values of X(I) would be fetched from memory by the vector instruction before any new values had been updated. Consequently, the results thus generated would be incorrect. Other commonly found vectorization inhibitors include subroutine CALLs, assigned GOTO statements, certain nested IF blocks, backward transfers within loops, any input/output statements, and references to external functions for which the compiler does not know any of their vector versions as such (some vectorized arithmetic functions are available on the Cray system). These types of inhibitors can be removed by expanding the function, or in-lining subroutine at the point of reference. If the DO loop satisfies the conditions for vectorization after in-line expansion, it could be vectorized. In fact, there exist many other restructuring techniques that can be effectively used to increase the degree of vectorization.

10.10.9 Vectorizing Compilers

In the vector processing system, a vectorizing compiler contributes a lot, and plays a dominant role. The task of a vectorizing compiler is to take sequential (scalar) program and convert it to an equivalent set of vector instructions. In order to accomplish this, the vectorizing compiler needs the support of a number of language constructs of the underlying language. A few of them are:

1. Instruction involving "For loops" can be easily vectorized. As for example, a loop, such as,

```
        for i = 1, N do
        A(i) = B(i) + C(i)
    may be vectorized as a single instruction: A (1: N) = B (1: N) + C
    (1: N)
```

2. Scalar computation involving recurrence can also be converted to an equivalent vector form. For example,

```
        A(0) = m
        for i = 1, N do
        A(i) = A(i - 1) * B(i) + C(i + 1)
    may be vectorized as;
        A(0) = m
        A (1: N) = A (0: N - 1) * B(1: N) + C (2: N + 1)
```

3. Instruction involving "if" can be coded as "where" statements. For example,

```
        for i = 1, N do
        if M(i) = 0 then X(i) = X(i) - 1
    can be vectorized as;
        where [ M(1: N) = 0, X(1: N) = X(1: N) - 1 ]
```

4. The execution sequence can be exchanged without affecting the program semantics, to yield efficient vector code. For example,

```
        for i = 1, N do
        X(i) = Y(i - 1)
        Y(i) = Y(i) + 2
    can be vectorized as;
        Y(1: N) = Y(1: N) + 2
        X(1: N) = Y(0: N - 1)
```

Similarly, many other forms of useful vectorization can also be derived, of course, if the underlying language has the provision of the appropriate language construct facilities in place.

10.10.10 Different Types of Vector Processor Organisations

Vector processor can be organised in many different ways using various approaches that have been, and are being pursued to realize vector processing. Of these, *three* fundamental categories are of primary interest. They are:

- Pipelined ALU
- Parallel ALUs
- Parallel processors

Pipelined ALU approach is simply an extension of traditional pipeline concept in which the relatively time-consuming execution unit (EX stage, the operation of the ALU, in the traditional pipeline) itself is then made of several pipelined stages instead of a single stage to keep the entire pipeline operation with the other existing stages in balance. Consider, a floating-point operation, which is rather complex, and also time-consuming than those of the other stages, is now executed in the ALU. This operation usually needs to be broken down into several substages so that the amount of workload now entrusted on each such substage will be reduced and would be in balance with the other existing stages. These different substages within the ALU having relatively lesser amount of workload can now operate on different sets of data concurrently, and each one can then complete its execution within the stipulated time interval, that in turn, keeps the pipeline operation smooth without any such expected stall. This is illustrated in Figure 10.34. For example, floating-point addition is usually broken up into four pipeline stages: *compare*, *shift*, *add*, and *normalize*. A vector of numbers is presented sequentially to the first stage. As the execution proceeds, four different numbers in different states will now be operated concurrently within the prescribed pipeline.

It is important to note that this organisation is extremely conducive to vector processing with achieving an expected level of speedup, since the ALU is here constantly fed with vector of operands in the form of a stream of data from sequential locations. If a single, isolated floating-point operation is executed on such a pipeline, it certainly cannot yield a noticeable speedup, since all the stages of the ALU are not really filled with their respective work. However, the control unit here constantly cycles the data through the ALU until the entire vector is processed.

In order to further enhance the pipeline operation, Figure 10.34 illustrates the use of the vector registers in which vector operands are made available before hand to feed the ALU for the required execution rather than to fetch them directly from relatively slower main memory at the time of execution. In fact, the elements of each vector operand are loaded as a block into a vector register, which is simply a large bank of identical registers. The result after computation is also placed in a vector register. Thus, most operations involve only the use of the registers with essentially *load* and *store* operations. Moreover, if vector registers are used, intermediate results need not be stored into memory, and can be used immediately for subsequent operation as soon as the first elements of the result vector in vector registers are available. This means that the next operation with the result vector elements can be started even before the vector operation that created them runs to completion. This approach is often referred to as ***chaining***, and is found in use in Cray

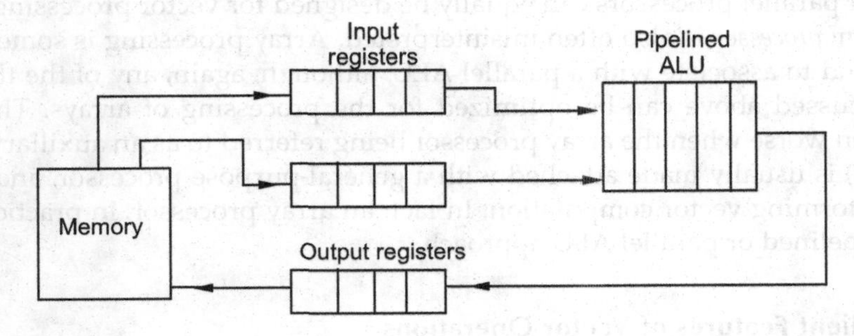

FIGURE 10.34
A schematic approach to vector computation using Pipelined ALU.

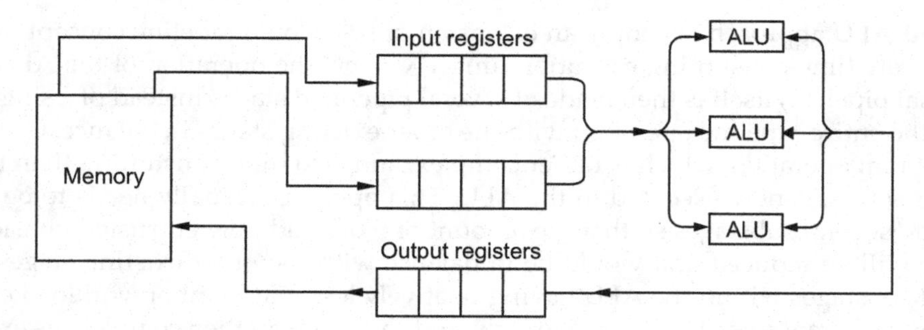

FIGURE 10.35
A schematic approach to vector computation using Parallel ALUs.

supercomputers. Thus, only the beginning (i.e. when the elements of vector operands are fetched from memory to *load* into registers) and end (i.e. when the result of vector operation is finally written back to memory from vector register) of a vector operation require access to memory, and repeated back and forth visits to slower memory can then be avoided by the use of this approach.

Parallel ALUs is simply using of multiple ALUs within a single processor, operated under the control of a single control unit. In this case, the control unit routes data to ALUs so that they can function in parallel. This is depicted in Figure 10.35, in which three ALUs are operated in parallel. It is, however, also possible to use pipelining on each of these parallel ALUs (already discussed in *first approach*). A *parallel ALU organisation*, similar to pipelined organisation, is also equally conducive for vector processing. Here, the control unit routes vector elements to ALUs in a round-robin manner until all elements of vector operands are processed. This type of organisation, however, is naturally more complex than a single-CPU pipelined ALU organisation (Figure 10.34).

Parallel processors' approach in the organisation of vector processors is simply by way of using *multiple parallel processors* (not ALUs). In this case, the task is actually broken into several processes which are to be executed in parallel on multiple processors. The success of this organisational approach and its viability, however, depends mostly on the *effective coordination* of the related software with the underlying hardware comprising of parallel processors being driven.

Lastly, it is to be cautioned in the context of the wrong terminology that is often in use. The term *vector processor* is sometimes equated with a pipelined ALU organisation, although it has been just shown that a parallel ALU organisation and even an organisation consisting of parallel processors can equally be designed for vector processing. Similarly, the term *array processor* is also often misinterpreted. Array processing is sometimes used to refer to, and to associate with a parallel ALU, although, again, any of the three organisations discussed above can be optimized for the processing of arrays. The situation becomes even worse when the array processor being referred to as an auxiliary processor (coprocessor) is usually made attached with a general-purpose processor, and is used to assist in performing vector computation. In fact, an array processor, in practice, may use either the pipelined or parallel ALU approach.

10.10.11 Salient Features of Vector Operations

- The computation of each result (in vector processor) is independent of the computation of previous results.

- A single vector instruction specifies a great deal of work; it is equivalent to executing an entire loop. This substantially reduces the software overheads incurred in controlling the loop.
- Since, an entire loop is replaced by a vector instruction whose behaviour is predetermined, control hazards that would normally arise from the loop branch do not occur here.
- Vector processing nicely matches with the pipeline approach by pipelining the operations that are carried out on the individual elements of a vector.
- Vector instructions that access memory have a known access pattern. If the vector's elements are all adjacent, then fetching the vector from a set of heavily *interleaved memory banks* works nicely.
- Since a single access to memory is initiated for the entire vector rather than to a single word, memory latency is paid for only once for the entire vector, rather than once for each word of the vector.

For these reasons, vector processing is faster, and more efficient than a sequence of equivalent scalar operations on the same number data items. Both pipelined processors and SIMD computers can perform vector operations. Use of ***vectorizing compiler*** (or vectorizer) along with additional hardware support could enhance the performance of a vector machine to a large extent. Most of the vector processors are found to exploit pipeline structures in their implementation. Some of the main reasons are:

- Identical functions (or processes) are repeatedly invoked, each of which can again be subdivided into a sequence of smaller sub-functions, and each such sub-function can then be assigned and executed one after another by the corresponding stage of a pipelined processor. Thus, the executions of functions in vector processing altogether are observed to be nicely matched with the methodology used in pipelined organisation.
- Successive operands that need to be fed in vector processing are provided through the pipeline segments that usually require very few buffers and also lesser local controls.
- While operations are executed by distinct pipelines (or, distinct stages of a pipeline), they are able to share critical costly resources, such as memory and buses, available in the system.

10.10.12 Basic Vector Processor Architecture

Vector computers also need to perform scalar operations and mixed vector–scalar operations. Thus, a vector processor typically consists of an ordinary pipelined scalar unit plus a vector unit. The scalar unit is basically a type of an advanced pipelined CPU as discussed earlier. The major components of a vector unit are detailed below. Several vector processors are found to have different clock rates, separately in the vector and scalar units. However, the basic architecture of vector processors is primarily of two fundamental types:

- Register–Register Vector Processors
- Memory–Memory Vector Processors

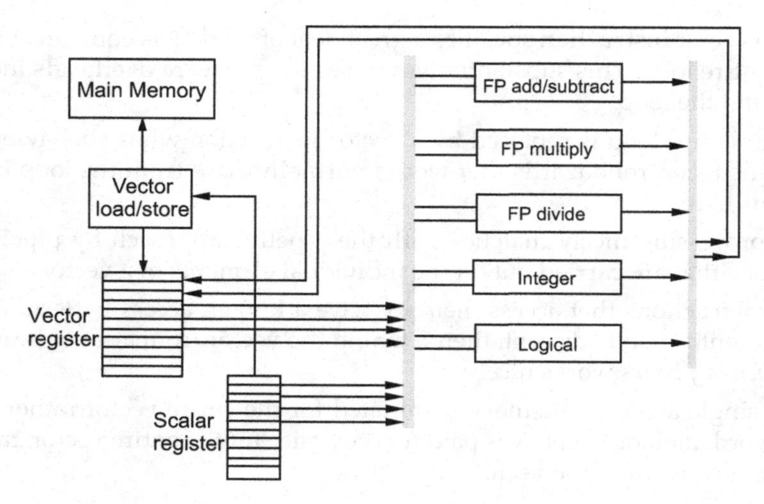

FIGURE 10.36
Basic structure of a vector-register processor.

Most of today's vector machines are *register–register vector machine.* All vector operations in this processor are among vector registers, except load and store. In a *memory–memory vector processor,* all vector operations are memory to memory. Figure 10.36 illustrates the basic structure of a vector–register processor (register–register vector processor) in which the major components are:

- **Vector Registers**: Each vector register is essentially a fixed-length bank holding a single vector with multiple elements (e.g. 64, or multiple of 64), each element is 64-bit in length. Different vector registers can involve in different vector operations in an overlapped manner. To minimize structural hazards, many read ports and write ports are available, and are connected to the inputs or outputs of a functional unit by a pair of crossbars.
- **Vector functional units**: Each unit is fully pipelined, and can start a new operation on every clock cycle. This unit includes all operation units used by the vector instructions. The functional units are usually floating-point, unless stated otherwise.
- **A control unit**: This unit is responsible for decoding, and also carrying out coordination among functional units with the responsibility of detecting both structural hazards (conflicts in functional units) and data hazards (conflicts in register accesses).
- **Vector load / store unit**: Their operations can be pipelined, and the unit is capable of performing an independent, overlapped transfer to or from the vector registers. This unit would also handle scalar loads and stores.
- **A set of scalar registers**: These can be used for various purposes, for example, as a buffer for input data to the vector registers, compute addresses to pass to the vector load / store unit. Normally, 32 general-purpose registers (GPR) and 32 floating-point registers (FPR) are used. The scalar registers are usually connected to the functional units by a pair of crossbars.

Various types of architecture have been developed by different manufacturers in the implementation of different register – register vector processors, aiming towards meeting their own target. However, some common strategy has been followed in the design by all such processors. If this processor is a multiprocessor, all the characteristics being found correspond to one processor. Most processors use the vector FP multiply and divide units for vector integer multiply and divide, and several of the processors use the same units for FP scalar and FP vector operations. Clock rates used in the vector and scalar units have been found different in several processors. Above all, vector processors used in multicomputer / multiprocessor, and in supercomputer, however, differ even in their fundamental characteristics.

10.10.13 Memory-Access Schemes

The execution time required by a sequence of vector operations for the execution of a single vector instruction primarily depends mainly on the following three factors:

- The length of the vectors being operated on,
- Structural hazards (mainly due to simultaneous access to memory for vector operands) among the operations, and
- The data dependencies.

Apart from the many other factors influencing the performance of a vector processor, one of the major issues involved, is the nature of the flow of the vector operands from the main memory to the vector registers for required operations. While the operand *flow mechanism* exploits pipeline approach or parallel access with multiple access paths, the *operand-access mechanism* often uses different schemes to access operands, normally from interleaved memory modules in an overlapped manner. In fact, the memory system is always designed especially for a vector processor, so that it can quickly supply the operands by means of fast access. A few popular schemes of memory organisation befitting to access vector operands by a vector processor are mentioned below.

When the memory is organised in m-way low-order interleaved fashion where m is the number of interleaved memory modules, m memory word can be accessed concurrently in an *overlapped* manner (pipelined approach). If the memory address is of n bits, then the low-order x-bits of n bits select the modules, and the rest high-order y-bits of n bits select the word within each module, where $m = 2^x$ and $n = x + y$. This is called ***concurrent access (C-access)***. It is observed that C-access generally yields a maximum throughput of m words per memory cycle, if the vector stride is made relatively prime to m.

The low-order interleaved memory organisation having the same approach as mentioned above can be further refined so that all the memory modules can be accessed in *parallel* (simultaneously) in a synchronized manner. This is called ***simultaneous access (S-access)***.

Another approach is to have a hybrid memory organisation which can be realized by combining C-access and S-access together, giving rise to ***C/S-access***. Here, n access-buses are used, and m modules of m-way interleaved memory are attached to each access-bus to allow C-access. The n buses, however, again operate in parallel to allow S-access. This organisation thus collectively offers *C/S-access*. Parallel pipelined access (C/S-access) of a vector data set offers a high bandwidth. As a result, at most $m \times n$ words are fetched in each memory cycle, provided n buses are fully utilized.

Each of the schemes as described has its own merits and drawbacks. While C-access is easy to implement and found suitable in single vector processor configurations, C/S-access is well matched in vector multiprocessor systems for smooth data movement between the memory and multiple vector processors, although requiring extra hardware support in its implementation. Numerous other techniques have been also devised to improve the performance of vector processors, such as, to make arrangement so that a sequence of dependent vector operations can run faster (known as ***chaining* method**), arrangement to combat and minimize the harsh effects of conditional execution in a loop, and to effectively handle sparse matrices, etc.

Cray 1 pioneered *multivector system* development in 1975 / 1976. The CDC 7600 was, however, the first vector *dual-processor* system. ***Cray 2*** did not remain far behind. The next version ***Cray / MPP*** is a massively multivector system with distributed shared memory. A vector processor with its numerous varieties is nowadays an inevitable constituent in almost all ranges of computers, from mini to mainframe and supercomputers, such as ***VAX 9000, IBM 390/ VF*** models (Z-Series). Large multiprocessors (vector supercomputers) like ***Convex C3*** series, ***CRAY Y-MP*** family, ***NEC SX*** series, ***Hitachi S-810/20,*** and the ***Fujitsu VP 2000*** are all equipped with vector pipelines. In fact, the number of vector processors with numerous characteristics available in a system is considered as one of the dominant parameters in categorizing the mainframe and full-scale supercomputers from the viewpoint of its performance and allied costs.

10.10.14 Implementation: The CRAY 1 Architecture

CRAY 1 introduced in 1975 was considered the world's first vector supercomputer. Both CDC 6600 and CRAY 1 have a lot of similarities, since both the machines were the brainchild of Seymour Cray, one of the founders of CDC and the chief architect until he left and formed his own company, Cray Research. An enhanced version of this machine, the CRAY 1S was launched in 1979. It was the first ECL-based supercomputer with a 12.5ns clock cycle and a typically *register-oriented* RISC-like machine requiring all operands to be in the registers for all vector operations. High degrees of pipelining (superpipelining) and vector processing in a uniprocessor were among the most notable features of these machines.

A brief description of this machine with appropriate figure is given on the website: http://routledge.com/9780367255732.

10.11 Array Processors and SIMD Parallel Computers

An alternative approach to the SIMD form of parallel processing is the ***array processor;*** the initial form of *parallel processing* that was implemented first by the heavily used SIMD machine, the ILLIAC IV, sometime in the early 1970s. It was also used by NASA for years mainly for computation-intensive problems. Later, this approach has been modified and further extended in the system built by Burroughs Corporation. Due to continuous emergence of numerous specific application areas requiring extensive data parallelism during late 1980s, renewed attention was paid towards this form of processing, first by Thinking Machines Corporation (TMC) in their CM-2 machine, and followed by MasPar in their MP-1 machine.

An *array processor* is essentially a synchronous array of parallel processors, where each processor consists of multiple processing elements (PEs, which are basically ALUs) operated under the supervision of one control unit. Since both vector processors and array processors execute vector type data, one often fails to differentiate between an array processor and a vector processor. A vector processor does not operate on all the elements of a vector at the same time; it only operates on multiple elements concurrently. An array processor performs vector computations over large matrices or arrays of data, and unlike a vector processor, it uses parallel ALUs (multiple PEs), which may or may not be pipelined. Unlike vector processors, the *array processors* do not include scalar processing; they are mostly configured as peripheral devices by both mainframes and minicomputers to execute the vectorized portion of the program running in the general-purpose environment.

The basic design of the array processor as shown in Figure 10.37 illustrates the structure of an ILLIAC IV type array processor in which a limited number of synchronized PEs as square grid (processor and memory) elements are organised in a two-dimensional fashion. Each PE is somehow connected to its nearest neighbours for the purpose of exchanging data. End-around connections, of course, may be provided in both rows and columns. All PEs work under one central control unit (CU). The local memory attached with each PE is used for the storage of its own distributed data, usually vector operands. These data are loaded from an external source (usually from a general-purpose host computer) via the system data bus or via the CU in a broadcast mode using the control bus before the

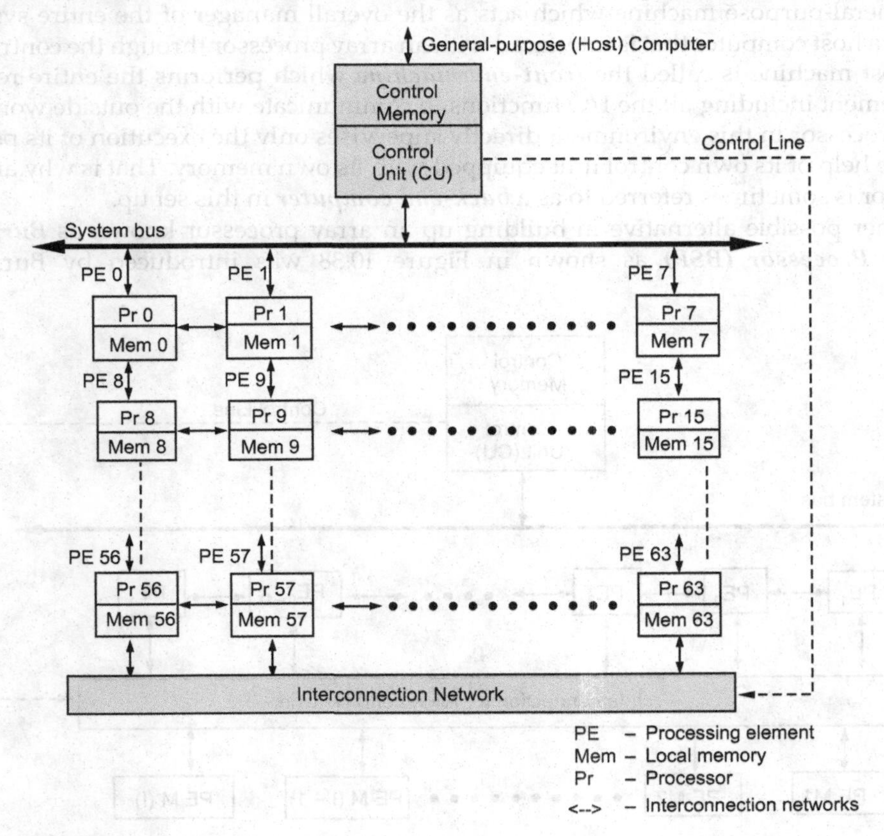

FIGURE 10.37
An Array Processor of ILLIAC IV type.

commencement of parallel execution in the array of PEs. The central control processor has its own memory for the storage of all programs to be executed. The user programs are loaded into the CU memory from the external source; usually from a general-purpose host machine. The function of the CU is as usual to decode all the instructions and determine where the decoded instructions are to be executed. Scalar or branch-type instructions are directly executed inside the CU. Vector instructions are broadcast by the CU to the PEs for distributed execution; all the PEs perform the same function synchronously in a lockstep fashion, each one is using its own data from its own memory (loaded during the initialization phase), thereby achieving spatial parallelism.

Arithmetic / logic operations are performed as usual in an instruction cycle that we have already described. Masking schemes are employed by the control unit using a masking vector for keeping track of the status of each PE while a vector instruction is executed. Masking operations are commonly used to *enable* or *disable* a subset of PEs under the direct control of hardware. In other words, not all the PEs, but only the enabled PEs are involved in the computation process. Data exchanges among the PEs are, however, carried out via an inter-PE communication network which performs all data-routing and data manipulation functions, and is dynamically operative under program control by hardware. These functions include *broadcast, multicast, various rotate and shift operations,* and numerous *permutations.* The interconnection network itself is again operated under the supervision of the control unit.

A general-purpose machine which acts as the overall manager of the entire system is used as a host computer that is interfaced with an array processor through the control unit. This host machine is called the ***front-end machine*** which performs the entire resource management including all the I/O functions to communicate with the outside world. The array processor in this environment directly supervises only the execution of its portions with the help of its own control unit equipped with its own memory. That is why an array processor is sometimes referred to as a ***back-end computer*** in this set up.

Another possible alternative in building up an array processor known as ***Burroughs System Processor (BSP)*** as shown in Figure 10.38 was introduced by Burroughs

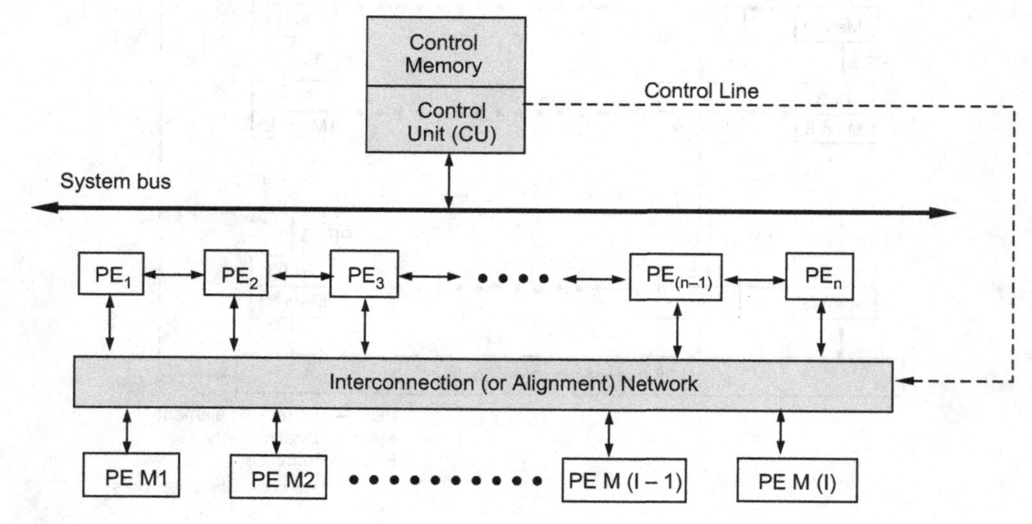

FIGURE 10.38
A block diagram of BSP (Burroughs System Processor) Array Processor.

Corporation. This design configuration although is almost same as that of ILLIAC IV type, but they differ mainly in two aspects. First, the parallel memory modules being shared here by all the PEs through a network called ***alignment network*** replace the local memories attached to the PEs as used in the design configuration of ILLIAC IV type. Secondly, the inter-PE permutation network as used in ILLIAC IV type is replaced here by the inter-PE memory alignment network which is under the direct control of the CU. The alignment network, being a path-switching network between the PEs and parallel memories, offers conflict-free accesses to the shared memories by as many PEs as possible, of course, within a specified limit.

SIMD machines with array processors are highly specialized machines being used for solving special types of applications. They offer excellent performance while solving numerical problems expressed in the form of matrix or vector. These computers execute vector instructions for *arithmetic, logic, masking, data-routing* operations over vector operands of equal length *n*, where *n* is the number of available PEs.

The key design strategy that is to be chalked out while building an array processor depends mainly on the decision as to whether to employ a relatively small number of comparatively powerful processors (processing elements) or a large number of very simple processors (processing elements). The first approach has been chosen at the time of designing ***ILLIAC IV*** where only 64 processors in the form of 8 × 8 grid had a 64-bit internal structure. Due to renewed attention paid by different vendors in the late 1980s, the latter approach, however, has started to gradually dominate. The introduction of ***CM-2*** machine from Thinking Machines Corporation (TMC) eventually could accommodate a large number of up to 65,536 very simple processing elements, each one is only one-bit (binary digits) wide (bit-slice), that tenders the most desirable fine-grained data parallelism which offers output of high resolution. The ***MasPar's MP-1*** series included a maximum of 16,384 processing element, each one is 4-bit wide. Other representative processors in this class, however, have chosen various number of processing elements of numerous width such as byte-sized or bit-sized. Modified versions of **C**, **Fortran 90**, **Lisp**, and other synchronous programming languages have been developed to program SIMD machines. SIMD instructions are, however, similar in both vector-based machines as well as in machines with array processors.

10.11.1 Applications

To demonstrate the working of an SIMD machine, and to understand its strength to handle problems of special types, let us define an application that will operate on data originated from different points over a space of two-dimension. The two-dimensional space may be the *ocean surface*, and the application is to calculate the effect of eddy and boundary currents on the large-scale flow of the fluid in the ocean to forecast about any forthcoming disaster (tsunami). The two-dimensional space may also be a *heat conducting surface* of a boiler where temperature gradients at different points over the surface are to be measured, and the outcome would accordingly determine the appropriate actions that would have to be taken to operate the boiler. These two examples clearly indicate a situation that involves an extensive data parallelism.

The approach to the solution starts with the assumption that the 2D surface is consisting of a set of some arbitrary points. Each arbitrary selective interior point over the 2D surface represents each grid element, and the array of the grid can be used to compute the temperature at different points over the surface of the boiler. If more number of grids (PEs) is available in a processor then more number of points over the surface can be accommodated to involve in this execution. Consequently, finer resolution can be attained in the outcome of the computation. It is also assumed that the outer edges of the surface are initialized at some specified fixed temperatures (boundary condition). All interior points are also initialized to some arbitrary values, not necessarily the same (initial condition). Iterations are then executed *in parallel* by the ALU at each grid element. Each of these iterations may consist of some calculations using necessary formulae (or equations). A sequence of program instructions including arithmetic instructions in this regard is broadcast by the control processor repeatedly to implement the iterative loop. The execution will give the current estimate of temperature at *each point* taking the current value of, say, its four nearest neighbours or by using some other methods. The calculation process (iteration) will continue, and will be stopped only when changes in the estimates during successive iterations fall below some predefined small quantity (δt). Other applications that refer to the identical situation where the similar operations (instruction or program) are to be executed over a large array of data (operands) follow almost the same line of actions as mentioned. It is to be noted that more number of available PEs in a processor will enable accommodation of more number of selective points to involve in the calculation process, which will ultimately offer finer results that tends to be more accurate with high resolution.

10.11.2 Implementation: Connection Machine CM 2 Architecture

The Connection Machine CM 2 is a fine-grained MPP (Massive Parallel Processing) computer system that uses a large number of up to one-bit wide 65,536 (64K) simple processing elements (PEs), and these PEs have been divided into four quadrants, each with 16,384 (16K) PEs. Each quadrant is again divided into two 8192 (8K) PEunits for I/O and other purposes. These PEs are packaged in clusters in one 68-pin custom *VLSI chip*, each cluster contains 16 PEs (ALUs), numbered 0 through 15, and a router, and four standard 32K × 8 (= 256K × 1 bit) RAM chips for a total of 1M × 1 bit (4 × 256K × 1 bit = 1M × 1 bit). This one megabit of memory is equally distributed among 16 ALUs, so that each ALU gets a private memory of size 64K × 1 bits, [(1M × 1 bits) ÷ 16 = 64K × 1 bits] with each bit having a separate address. Thirty two such clusters are packaged to a board (a board then contains 32 × 16 = 512 PEs). Sixteen such boards fit into a backplane called a ***subcube*** for a total of 8K (16 × 512 = 8K) number of processors. Two such subcubes are piled vertically to form a quadrant and four such quadrants are put together to form a ***cube*** with an edge of about 1.5 m. All the quadrants are then controlled by a sequencer (Tucker, L.W., etal.).

The connection machine CM2 is essentially a *back-end machine* for data parallel computation with no storage, and hence, is attached with one or more *front-end* ***hosts*** up to a maximum of four hosts to provide all sorts of main memory supports. All programs are developed, edited, compiled, and then executed on the hosts, but they issues microinstructions to the back-end processing array via a 4 × 4 crossbar switch (or by other methods) for execution, only when the data-parallel operations are wanted. The CM-2 machine offers bit-sliced fine-grained data parallelism giving rise to a peak performance of about 10 Gflops. The use of 16,384 PEs in a single-array configuration requires a lower degree of array segmentation, and thus offers a higher flexibility in programming. Supporting software environment has been created to extract this massive parallelism that includes array processing languages and its compilers for data parallelism (C*, A Pascal-based language *Actus*, etc.), and appropriate Operating system to control the machine including data routing between PEs.

A brief description of CM-2 machine with appropriate figures is given on the website: http://routledge.com/9780367255732.

10.11.3 Vector Processor Versus Array Processor: A Rough Comparison

Both vector processors and array processors are specialized to operate on vectors. However, the ***key differences*** that exist between them are:

- A vector processor includes multiple pipeline modules within the ALU of the processor (pipelined vector processor) to achieve high performance through heavy use of pipelining, giving rise to *temporal parallelism* in pipelines, whereas the array processor includes multiple PEs (processing elements), which may or may not be pipelined, is controlled by a single control unit providing extensive parallelism in the form of *spatial parallelism* in multiple PEs.
- In a vector processor, different pipelines may or may not be synchronized for their operations. But, in array processors, operations of different PEs must be synchronized.
- In a vector processor, one pipeline performs one vector instruction, but in an array processor, multiple PEs may be involved to execute one vector instruction.
- An array processor offers more fine-grained data parallelism than a vector processor.
- Usually, an array processor is relatively faster than its counterpart vector processor.
- Cost-wise, a vector processor is comparatively cheaper than an equivalent array processor.

However, data parallelism is achieved both in pipelined vector processors as well as in SIMD array processors, and similarly in SPMD (Single Program Multiple Data) or MPMD (Multiple Program Multiple Data) multicomputer systems.

10.12 Massively Parallel Processing (MPP) System

The idea of massive parallelism is rather old; the technology is new, and the definition of massive parallelism constantly changes as evolution and revolution in computer technology continuously takes place. Fifth-generation computers were mainly targeted to emphasize more on *massive parallel processing* (MPP) by way of achieving towards a performance of Teraflops (10^{12} floating-point operations per seconds). VLSI silicon, GaAs (Gallium Arsenide), optical technologies, high-density packaging, etc. are the main constituents in building scalable and latency-tolerant architectures for MPP systems. Supercomputers once were considered to execute applications that involve massively parallel processing.

Today's definition of massive parallelism is altogether different, and that any machine which has hundreds or thousands of processors (multiprocessors or multicomputers) is considered to be a massively parallel processing (MPP) system. These machines are targeted providing high-performance computing and communications (1 Teraflops), highspeed and high capacity of hierarchical memory module (1 Terabyte), extensive I/O support with high bandwidth (1 Terabyte / seconds), and synchronization features that are added to efficient MPP systems to meet the continuously growing forthcoming challenges. Higher degree of parallelism is becoming an obvious requirement. Thus, MPP systems while providing parallel processing demand both *hardware parallelism* as well as *software parallelism*.

Hardware parallelism in a system usually refers to the type of parallelism defined by the machine architecture and hardware multiplicity. But the parallelism in an individual processor, however, is generally characterized by the number of instruction issued simultaneously in parallel per machine cycle. If a processor issues k instructions per machine cycle, then it is called a k-issue processor, in contrast to a traditional processor that issues only a single instruction and that too taking one or more machine cycles. This parallelism is, however, accomplished by using multiple functional units in the way of arranging a number of parallel pipelines within the processor. This type of a design approach is called *machine parallelism* that gives rise to what is known as *superscalar architecture* of processor. For example, Intel Pentium 4 is such a *six-issue* (six pipeline) processor, while IBM RS /6000 is a *four-issue* (four pipeline) processor. Hardware parallelism in a system is actually implemented as a joint venture of using special hardware, and an appropriate operating system that drives the MPP system, providing the necessary support during run-time.

Software parallelism is realized by means of suitable system software like *parallelizing compilers, Assemblers, Loaders* along with supporting *system utilities* to extract the potential power of the physical machine resources for program parallelism. Architecture-independent *parallel languages* have been developed aiming towards implementation of parallelism in programming. However, the software domain still remains relatively unexplored.

One type of software parallelism in an individual processor is ***control parallelism.*** Control parallelism appears in the form of pipelining or multiple functional units (*superscalar architecture*) in which several instructions (e.g. k instructions) are simultaneously *issued* or *dispatched* to these different functional units per machine cycle. Such a *k-issue* processor demands an adequate *control mechanism* in the form of software (compiler / OS) which is required to preserve the execution order of dependent instructions to ensure valid results and to avoid pipeline stalling. A multiprocessor system, when built with n numbers of such k-issue processors, could be able to handle a maximum number of nk threads (or processes) of instructions concurrently. However, both pipelining and functional parallelism are handled by the hardware; programmers need not to take any specific actions to invoke them.

The other type of software parallelism is called ***data parallelism,*** in which processors (PEs) work on multiple data concurrently. Data parallelism, found in both SIMD and MIMD modes of MPP systems, is however, much intricate but finer than instruction parallelism. MPP system while built up with multiple processors thus must impose a balance between hardware parallelism and software parallelism in an individual processor or in a total system of processors by the use of optimizing compiler, operating system, and / or through hardware redesign to enable the intelligent compiler to exploit the machine resources more efficiently.

Emergence of ILLIAC IV in 1968 is considered the beginning of MPP systems. Subsequently, many **advanced SIMD systems** like DEC VAX 9000, CM-2, IBM GF/11, MasPar MP1, CRAY 1, and their other models including CRAY Y/MP and C-90, Fujitsu VP 2000, NEC SX, Hitachi S-810 /20, and many other contemporary systems from other reputed manufacturers are all considered massively parallel processors. These machines used their own specialized operating systems and array processing languages to extract the maximum possible data parallelism with explicit data routing between PEs. Application languages include some notable features such as application flexibility, hardware transparency, explicit control structures in both program structuring, data typing operations, etc.

The focus of the definition of MPP system changes with the advancement of computer technology. Instruction-level parallelism became the major issue on demand. Computers operating in MIMD mode, both multiprocessor and multicomputer, now started to redefine the MPP system. A few of the representative systems that belong to this category drawn from SIMD, MIMD, and SIMD–MIMD are as follows:

IBM RP3 was designed to include 512 processors, but only a version of a 64 processors was built. **IBM MPP model** was declared as targeting a configuration of 1024 processor, each processor was an RS / 6000 as building block with a projected peak speed of 50 Gflops.

BBN TC 2000 was configured with a maximum configuration of 512 processors, but was striving for an even larger machine.

Intel Touchstone Delta, a multicomputer system with 672 node processors was launched in 1991. **Intel Paragon**, a host-free multicomputer built with i860 XP microprocessors using 2D Mesh connection along with wormhole routers was introduced in mid-1992 with a peak performance of about 300 Gflops.

CRAY MPP model (T 3D) with a heterogeneous MIMD scalable microarchitecture using 3D Torus network to connect DEC Alpha microprocessor chips used as building block was available in 1993. This architecture with larger configuration eventually attained a speed of *teraflops*. This machine was supposed to work as a back-end accelerator engine compatible with the existing Cray Y-MP Series.

The **CM-5 machine** from Thinking Machine Corporation launched in mid-1992 used a hybrid *SIMD–MIMD* architecture that was equipped with 32 to 16,384 ***processing nodes*** (not elements); each such node contains a 32-MHz SPARC processor, 32-Mbytes of memory, and a 128-Mflops vector processing unit capable of executing 64-bit floating-point operations as well as integer operations. High-performance networks, and high-bandwidth I/O interfaces supported with voluminous mass secondary storage as a data vault promoted this machine architecture to a level fit for a massively parallel processing environment.

Fujitsu VPP (Vector Parallel Processor) **500**, a *MIMD vector system* was launched in last part of 1993. It was a 222-PE (processors with attached memory) system with large crossbar (224 × 224) interconnect. **VP 2000** machine with shared distributed memories was used as host in solving large-scale problems.

KSR-1 model introduced by Kendall Square Research was a multiprocessor with a configuration of 1088 custom-designed processors using *ring interconnect.* It attained a peak performance of about 45 Gflops.

MPP systems use *interconnection networks* (ICN) to communicate between processing elements as in SIMD mode, and between the processors as in MIMD mode of multiprocessors and multicomputers. The selected ICN should support scalable and latency-tolerant architecture of an MPP system. Some of the static topologies are found to be more scalable in specific applications. Crossbar switch, although found to be most expensive due to its hardware complexity, however has the highest bandwidth and routing capability. For MPP systems with lesser number of powerful *processors,* the network size is small, and hence, crossbar switch is the ultimate choice for these systems. For MPP systems built with more number of simple *processors* or *processing elements*, large-scale MINs (Multistage Interconnection Network) or crossbar networks may become more reasonable from a cost/performance point of view for establishing dynamic connections in general-purpose MPP systems.

MPP systems when realized by off-the-shelf microprocessors as building blocks have certain deficiencies for being unable to create a balanced system that could match fast processor speed with fast memory access, fast I/O interface, and capable (befitting) system software. Although most of today's RISC processors satisfy some of these parameters, they mostly suffer from several drawbacks relating to communication, memory access, and synchronization features that are considered as primary requirements of an efficient MPP system. However, making use of thousands of off-the-shelf RISC microprocessors with powerful and costly communication hardware, along with exploiting supercomputer packaging and cooling can somehow circumvent these shortcomings to ultimately produce a supercomputer-class MPP system. In building MPP systems, multiprocessors with centralized shared memory are not suitable due to lack of scalability. On a *message-passing multicomputer,* the parallelism is, however, more scalable than in a centralized shared-memory multiprocessor. Distributed-memory multicomputers, on other hand, while found to be even more scalable, are less programmable due to the presence of added communication protocols.

Many of the MPP systems introduced by different manufacturers thus ultimately put more stress on MIMD-oriented concept. Apart from concentrating on hardware development, more attention is now devoted to the development of software environment including versatile operating systems, and modification of existing language compilers to negotiate increasingly broadened application spectrum of future large MPP applications.

10.12.1 Representative MPP System: Connection Machine CM-5

Connection Machine, model CM-5 machine pioneered an innovative design of an MPP system implementing a universal hybrid SIMD–MIMD architecture by way of combining the prominent features of both SIMD and MIMD machines. Parallel computation of large and complex data-parallel processing has been supported in the form of data parallelism which is implemented in either SIMD mode, or in multiple SIMD mode or synchronized MIMD mode. The basic architecture of CM-5 is, however, shown in Figure 10.39.

This architecture supports a range of 32 to 16,384 *processing nodes* to be available in the machine; each such node comprises a 32-MHz SPARC processor, 32-Mbytes of memory, and a 128-Mflops vector processing unit capable of executing 64-bit floating-point as well as integer operations. The basic structure of a processing node is shown in Figure 10.40.

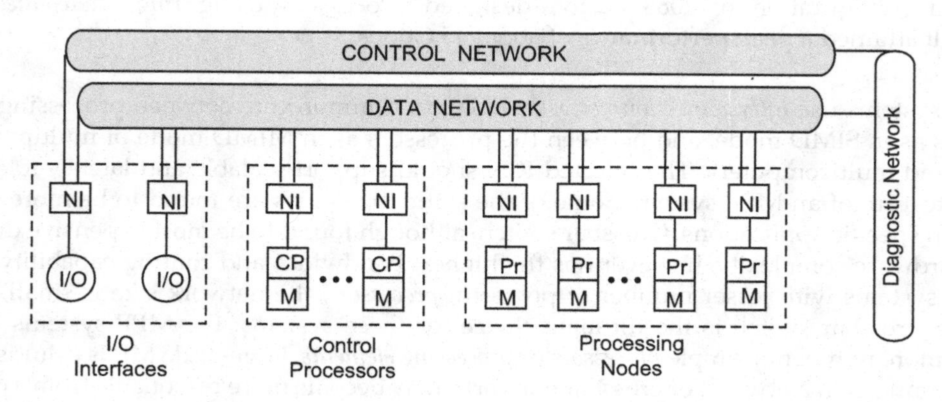

FIGURE 10.39
Block diagram of Network architecture of Connection Machine CM 5.

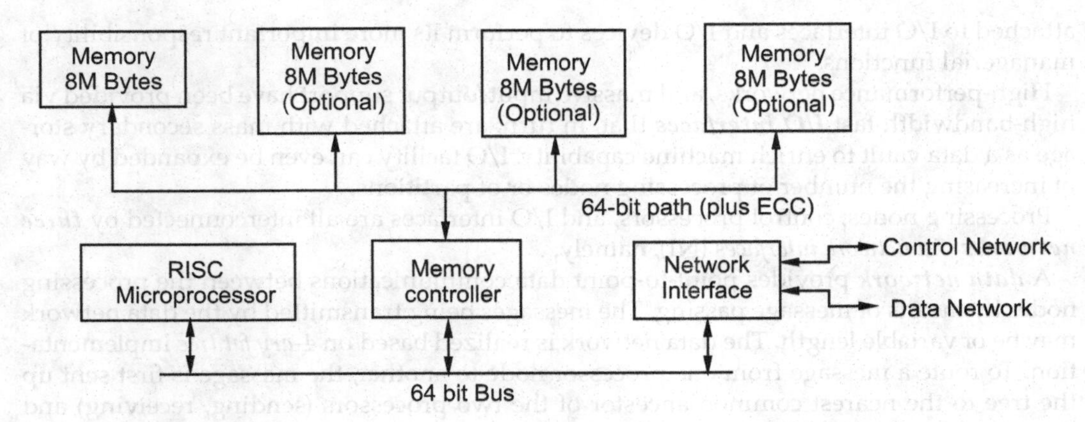

FIGURE 10.40
Block diagram of a processing node in CM 5 Machine.

The processing nodes execute the data parallel part of the code, and are connected to the rest of the system through control networks and data networks. Although SPARC processors have been used here for its multi-window feature to facilitate fast context switching, the architecture is, however, free from any constraint of using a specific type of processors. Moreover, more modern processors, as and when would be available due to the advancement in technology, can be incorporated with ease into the existing architecture.

A number of ***control processors*** have been deployed depending on the configuration of the machine, instead of a single sequencer as used in CM-2 to handle this voluminous workload. The control processor used here is essentially a SUN Microsystems workstation that comprises a RISC microprocessor (CPU), memory subsystems, I/O interfaces to connect local disks and Ethernet connections, and network interfaces to connect the control processor to the rest of the system through the control network and data network. Figure 10.41 shows the basic block diagram of the control processor used in CM-5. The control processors execute the scalar part of the code. Some control processors are dedicated to manage the computational resources in each partition, and some are exclusively

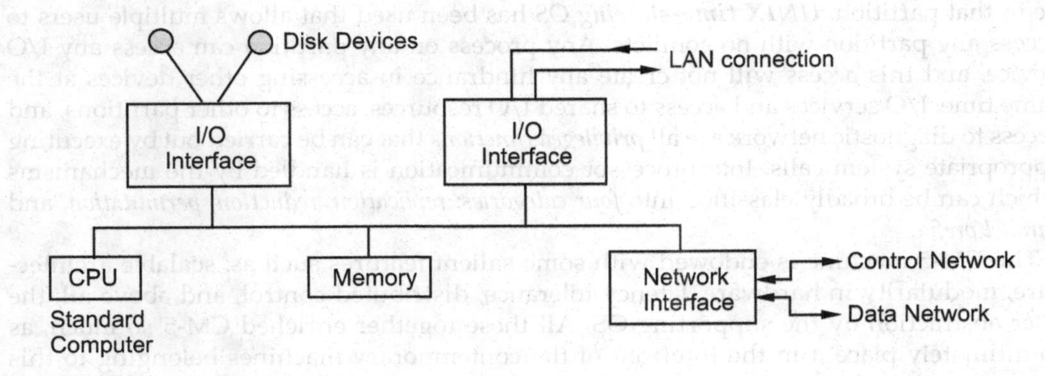

FIGURE 10.41
The control processor in the CM-5 machine.

attached to I/O interfaces and I/O devices to perform its more important responsibility of managerial functions.

High-performance networks, and massive input/output support have been provided via high-bandwidth fast ***I/O interfaces*** that, in turn, are attached with mass secondary storage as a data vault to enrich machine capability. I/O facility can even be expanded by way of increasing the number of processing nodes or of partitions.

Processing nodes, control processors, and I/O interfaces are all interconnected by ***three networks*** via *network interfaces* (NI), namely:

A ***data network*** provides point-to-point data communications between the processing nodes by means of message passing. The messages being transmitted by the data network may be of variable length. The data network is realized based on *4-ary fat tree* implementation. To route a message from one processor node to another, the message is first sent up the tree to the nearest common ancestor of the two processors (sending, receiving) and then down to the destination processor.

The ***control network*** supervises cooperative operations in the system including broadcast, synchronization, and scans, as well as all system management functions. While performing these functions, the control network sent messages in the form of packets having a fixed length of 65 bits. The control network provides the needed mechanism allowing data parallel code to be executed in processing nodes efficiently, and supports MIMD execution for general-purpose applications. Individual network interface is scheduled based on *FIFO algorithm* for each type control operation. The *data network* and *control network* are both scalable, and can be expanded to accommodate higher configuration.

The ***diagnostic network*** tests all system hardware to ensure system integrity, detect any system failure, and identify and isolate system errors. This network itself is fully testable and can be diagnosed. A *binary-tree pattern* is utilized in the organisation of this network for the sake of simplicity in its addressing. The root of the tree contains one or more *diagnostic processors.* The leaves of the tree are called *pods* and each such pod is a physical system. There is only one path from the root to each leaf being tested. Faulty processor nodes and links are identified by this network, and can be mapped out of the system to keep the entire system uninfected. The rest of the system remains functional even when the mapped-out portion is tested and serviced.

The entire system can be operated with one or more ***physical partitions.*** Each partition consists of a control processor, a handful of processing nodes, and dedicated portions of the data and control networks. Each user process is executed on a single partition under the supervision of a control processor which acts as a *partition manager* to manage the traffic in that partition. ***UNIX time-sharing*** *OS* has been used that allows multiple users to access any partition with no conflicts. Any process on any partition can access any I/O device, and this access will not create any hindrance in accessing other devices at the same time. I/O services and access to shared I/O resources, access to other partitions, and access to diagnostic network are all *privileged functions* that can be carried out by executing appropriate system calls. Interprocessor communication is handled by the mechanisms which can be broadly classified into *four categories*: *replication, reduction, permutation,* and *parallel prefix.*

The CM-5 machine is endowed with some salient features such as: scalable architecture, modularity in hardware, latency tolerance, distributed control, and above all, the user abstraction by the supporting OS. All these together enriched CM-5 so much, as to ultimately place it in the forefront of the contemporary machines belonging to this category.

10.13 Scalable Parallel Computer Architecture

During the past two decades, many different computer systems providing high performance computing have constantly evolved. Their taxonomy is essentially based on how their processors, memory, and interconnect have been organised and laid out to realize such systems. The most commonly used systems with different architectural design in this domain are:

- Symmetric Multiprocessors (SMP)
- Cache-Coherent Non-Uniform Memory Access (CC-NUMA) System
- Distributed Computing Systems (Multicomputers)
- Clusters
- Massively Parallel Processing (MPP) System

Figure 10.42 demystifies a modified version comparing the basic architectural and functional characteristics of these machines as originally proposed by Hwang and Xu.

SMP systems of today have 2–64 processors and can be considered to have shared-everything architecture. In these systems, all processors share all the global resources

Characteristics	SMP / CC–NUMA	Cluster	Distributed Comp. Sys.	MPP
Number of Nodes	O(10) – O(100)	O(100) or less	O(10) –O(1000)	O(10) –O(1000)
Nodes Complexity	Medium or Coarse–grained	Medium grain	Wide range	Fine grain or Medium
Internode Communication	Centralized and Distributed shared memory	Message passing	Shared files, RPC, Message passing, and IPC	Message passing / shared variables for distributed shared memory
Job Scheduling	Single run queue mostly	Multiple queue but coordinated	Independent queues	Single run queue on host
SSI Support	Always in SMP and some NUMA	Desired	No	Partially
Node OS Copies and type	One monolithic SMP and many for NUMA	NOS platforms homogeneous or Micro kernel	NOS platforms and mostly homogeneous	N micro kernels Monolithic or Layered OS
Address Space	Single	Multiple or Single	Multiple	Multiple Single for DSM
Internode Security	Not Applicable	Required if Exposed	Required	Not Applicable
Ownership	One organisation	One or more Organisations	Many organisations	One organisation

FIGURE 10.42
Key characteristics of scalable parallel computers.

(bus, memory, I/O systems) available; a single copy of the operating system runs to drive this type of system, always giving a single system image (SSI).

CC-NUMA is a scalable multiprocessor system having cache-coherent non-uniform memory access architecture. Like an SMP, every processor in a CC-NUMA system has a global view of all of the memories. This type of system gets its name (NUMA) from the non-uniform times that are needed to access the local (nearest), global (attached to other processors), and the common memory (most remote parts) shared by all processors.

Distributed computing system architecturally can be roughly considered as often built on conventional *networks of independent computers,* usually called *nodes, hosts, sites,* or *end-systems.* Each node does have its own local memory, and may also have other hardware and software resources. A distributed computing system has *multiple system images,* as each node runs under its own operating system. Such an arrangement is often called, and popularly known as ***computer network***. The nodes often vary in size and function; size-wise, a node may be as small as a personal computer, a workstation, an SMP, a mini-computer, a mainframe, a cluster, a MPP, or even a large supercomputer. Function-wise, a node may be a single-user personal computer, a general-purpose time-sharing system, or a dedicated system (such as, a database server, a file server, or a print server) usually without having any capability to handle interactive users.

A **cluster** at its basic level is a collection of workstations or PCs that are interconnected via some network technology. For parallel computing purposes, a cluster generally consists of high performance workstations or PCs interconnected by a high-speed network. A cluster while working as an integrated collection of resources, can also cast a *single system image* spanning all its nodes. In fact, cluster architecture belongs in the domain of computer architecture somewhere in between distributed computing system (computer networks) and true parallel processing system. Detailed descriptions of clusters, however, have been already provided earlier in this chapter.

A **massively parallel processing (MPP)** system is one that consists of hundreds or thousands of processors (multiprocessors, multicomputers, or some combination of both) and is targeted at providing high-performance computing and communications (1 Teraflops), high-speed high capacity (1 terabyte) of hierarchical memory module along with extensive I/O support with high bandwidth (1 Terabyte / seconds), and synchronization features that are added to efficient MPP systems to meet the continuously growing future challenge. MPP systems, while providing parallel processing, demand both *hardware parallelism* and *software parallelism.* Hardware parallelism is realized by a joint venture of using special hardware, and an appropriate operating system to drive the MPP system, providing the required support during run-time. Software parallelism is realized by appropriate system software like *parallelizing compilers, Assemblers, Loaders,* along with other *system utilities* to extract the potentials of these physical machine resources for program parallelism. Architecture-independent *parallel languages* have been developed aiming at implementation of parallelism in programming.

10.14 Supercomputers

Supercomputers, at the time of its inception, sometime in the 1960s, have been defined in terms of speed of computation, and were simply very fast scalar processors, some ten times the speed of the contemporary fastest machines offered by other companies. In the 1970s, the existing concept gradually changed, and a *vector processor* eventually became the standard. By mid-1980s, many such machines came out, each of which was equipped with a modest number (in the range of 4–16) of vector processors (multivector) working in parallel, mainly to handle special-purpose high-performance applications. In the late 1980s, the focus in the design of supercomputers, however, further changed. Vector processors have been replaced by SIMD machines which consisted of tens of thousands of processing elements (PEs), offering *vector processing* and *data parallelism* that significantly differ from *parallel processing*, and it persisted till mid-1990s.

In the last part of 1990s, attention once again turned from vector processors / SIMD machines to *massive parallel processing* systems with thousands of "ordinary" CPUs; some being off-the-shelf units, and others being custom designs. Today, parallel designs are based on "off-the-shelf" RISC microprocessors, such as the PowerPC, PA-RISC or UltraSPARC, and most modern supercomputers of today are now highly-tuned computer clusters (discussed earlier) using commodity processors combined with fast custom-interconnects. The *speed of a supercomputer* is, however, generally measured now in terms of *"FLOPS"* (***FL****oating–Point* ***O****perations* ***P****er* ***S****econd*) or *TFLOPS* (10^{12} FLOPS) using a particular benchmark.

Basically, there are ***three main classes*** of general-purpose supercomputers:

i. ***Vector processing machines:*** These represent both Vector processor and SIMD vector supercomputers as well as Array processor and SIMD parallel supercomputers comprising a wide range in the number of PEs. These machines allow the same (numerical) operation to be carried out on a large volume of data simultaneously.
ii. ***Tightly connected cluster computers:*** These systems employed specially-developed interconnects to have many processors with shared / distributed shared memory (NUMA) architecture. Processors and networking components are engineered from the ground up for the supercomputer. The fastest range of general-purpose supercomputers in the world today follows this approach.
iii. ***Commodity clusters:*** They use a large number of commodity PCs, interconnected by high-bandwidth, low-latency local area networks.

Many problems being handled by supercomputers are particularly suitable for parallelization, and, especially, fairly coarse-grained parallelization that limits the amount of information required to be transferred between independent processing units. For this reason, traditional supercomputers can be replaced, for many applications, by ***"clusters"*** of computers of standard design which can be programmed to act as one large computer.

Supercomputers are used for highly calculation-intensive tasks such as *weather forecasting, climate research* (including research into *global warming*), *molecular modelling* (computing the structures and properties of chemical compounds, biological macromolecules, polymers, and crystals), *physical simulations* (such as simulation of airplanes in *wind tunnels*, simulation of the detonation of *nuclear weapons*, and research into *nuclear fusion*), *cryptanalysis*, and many other similar problems. Besides, Military and R&D scientific agencies are among the heavy users.

10.14.1 The Contemporary Fastest Supercomputer System

Till the beginning of 2004, the fastest supercomputer was the **NEC Earth Simulator** introduced in 2002 at the *Yokohama Institute for Earth Sciences*, Japan. It is a cluster of 640 custom-designed *8-way vector processor computers* based on the ***NEC SX-6*** architecture (a total of 5,120 processors) to run at 35.86 TFLOPS. It uses a customized version of the UNIX operating system.

On March 25, 2005, IBM's Blue Gene/L prototype, a customized version of IBM's PowerPC architecture, became the fastest supercomputer in a single installation using its 1,31,072 processors to run at 280.6 TFLOPS (10^{12} FLOPS). On October 28, 2005, the machine reached 280.6 TFLOPS, but the system is expected to achieve at least 360 TFLOPS, and a future update is targeted at attaining a peak performance in the region of *petaflop*, i.e., 1 PFLOP = 10^{15} FLOPS. In November, 2005, ***IBM Blue Gene/L*** became the number one on TOP 500's most powerful supercomputer list. **The IBM Blue Gene** series is at present the fastest range of supercomputer systems in the world.

More details about supercomputers and their different types, the environment they require, mini supercomputers, special-purpose supercomputers, and finally the specifications of today's largest supercomputers are all given on the website: http://routledge.com/9780367255732.

10.15 Summary

Computer architectures have been passed through many notable evolutionary changes over the last few decades, yielding many different forms of architectural design, mainly due to rapid progress in superior VLSI technology that contributed to have more modern sophisticated hardware resources with enormous processing capabilities at reduced cost and in size. As a result, various forms of computer architecture evolved that consisted of many CPUs, more memories, and abundant I/Os. To classify these diverse forms of different computer architectures, numerous schemes have been tried, the distinguished one being that of Michael J. Flynn who classified computers into *four broad groups*, namely, SISD, SIMD, MISD, and MIMD. While SISD architecture consists of single CPU along with all its structural variants provides instruction-level parallelism, the MIMD architecture includes multiple CPUs, giving rise to two different forms, namely, multiprocessor and multicomputer, to obtain processor-level parallelism. Multiprocessor is again distinguished by shared-memory multiprocessor (UMA model), and distributed-shared memory (DSM) multiprocessor (NUMA model). Multicomputer is a collection of individual computers containing a single (or multiple) CPU(s) connected with one another by suitable interconnection networks, thereby giving rise to an entity known as network of computers, also called distributed computing systems. Many variations of this basic scheme are also possible; the two notable ones are computer networks and distributed system that project different views to the user based on how they are being organised and also to be driven.

With the proliferation of a diverse spectrum of application areas, certain classes of typically special-purpose problems have been evolved that requires special types of processors known as SIMD processor (according to Flynn's definition), which again is classified into two broad categories, namely; *Vector processors*, and *Array processors* offering more fine-grained data parallelism.

Relentless rapid progress in VLSI technology has further made it feasible to construct *massively parallel distributed-memory machines, large multicomputer configurations, clusters of computers* as well as massively *parallel processing systems* and *supercomputers using* various types of static interconnection networks having different topologies, namely, Ring, Tree, Mesh, Torus, Hypercube (nCube), etc. and also with numerous types of dynamic interconnection structure, such as, multistage interconnection networks. These machines are entwined with improved processor technology providing high-performance computing and communications (1 Teraflops), high-speed high capacity (1 Terabyte) of a hierarchical memory module, extensive I/O support with high bandwidth (1 Terabyte / seconds), and synchronization features, while all working in an orchestra realizing effectively massively parallel processing (MPP) to meet the forthcoming challenges. MPP systems while providing parallel processing require both *hardware parallelism* and *software parallelism* that can, however, be accomplished by various other means. Today's supercomputer is essentially a massively parallel processing systems with thousands of *"ordinary"* CPUs; some being off-the-shelf units, such as, PowerPC, PA-RISC or UltraSPARC (RISC), and others being custom designs. Most modern supercomputers are now highly-tuned computer clusters using commodity processors combined with fast custom-interconnects.

Exercises

10.1 What do you mean by parallel organisation? What type of parallelism you observed in this organisation? What are the parameters to be considered in parallel organisation?

10.2 Discuss the major issues need to be addressed in the design of parallel computers.

10.3 State and explain Flynn's proposal for the classification of parallel structures. What are the limitations of this proposal?

10.4 A multiprocessor is using a shared bus. What happens if two processors try to access the global memory at the same instant? What do you mean by loosely-coupled and tightly-coupled multiprocessors?

10.5 "Various forms of parallel processing at least differ in what is sometimes called *grain size*" – Discuss with example "grain size" with reference to coarse-grained parallelism and fine-grained parallelism.

10.6 Compare between centralized shared memory and distributed shared memory architecture? Which one do you find better than the other and why?

10.7 Distinguish between multiprocessors and multicomputers based on their structures, resource sharing, and interprocessor communications.

10.8 Indicate whether each of the following statements is *right* or *wrong* and justify your answer with reasons giving examples in support of or against it:

a. In an MIMD computer, all processors must execute the same instruction at the same time synchronously.

b. The CPU computations and I/O operations cannot be overlapped in a multiprogrammed computer.

c. Synchronization in all PEs in an SIMD machine is done by hardware rather than by software as is often done in most MIMD machines.

d. As far as scalability is concerned, multicomputers with distributed memory are more scalable than shared-memory multiprocessors.

10.9 What are the parameters that determine the important characteristics of different interconnection networks? Discuss the role of these parameters at the time of selecting a particular interconnection network while a system is designed.

10.10 Discuss the situations where hierarchical common (shared) bus system can be advantageously used as interconnection networks in building up multiprocessor systems. What are the different bus systems used here and what are their role during execution?

10.11 How many switch points are there in a crossbar switch network that connects p processors to m memory modules? Discuss the strength and drawbacks of a crossbar networks. Compare between crossbar switch and multiport memory in multiprocessor hardware organisation.

10.12 How many stages and number of switches in each stage are required to build up a $n \times n$ omega switching network? Construct an 8×8 omega switching network. What are the merits and demerits of omega switching network?

10.13 In an 8×8 omega switching network, show the switch setting required to connect input 4 to output 7. In omega network, suppose the wire between switch 2A and switch 3B breaks (consult the figure given in the text). Who is cut off from whom?

10.14 "One of the major drawback of omega network is its blocking problem" – Discuss this issue and describe a method where it has been modified. (Hint: Benes Network)

10.15 *Hot spots* are clearly a major problem in multistage switching networks. Explain the problem and give one solution approach. Are they also a problem in bus-based systems? Explain.

10.16 Compare dynamic networks (buses, crossbar switches, and multistage networks) for building a multiprocessor system with n processors and m shared memory modules. The comparison is required to be carried out separately in each of the following four categories:

a. Minimum latency in unit data transfer between the processor and memory module.

b. Bandwidth range available to each processor.

c. Hardware complexities such as switching, arbitration, wires, connectors, or cable requirements.

d. Communication capabilities such as permutations, data broadcast, blocking handling, etc.

10.17 What are the distinct advantages of a static connection networks over dynamic networks? What are their limitations?

10.18 "Hypercubes network have several distinct attractive features" – Discuss these features and explain their significance. Give some name of the systems along with the parameters where it is practically implemented.

10.19 "Mesh-based architectures have a lot of variations by way of allowing wrap-around connections". Discuss the various types of connections being observed that give rise to various forms of mesh-based topology.

10.20 State and explain the characteristics that you consider most important while selecting a particular interconnection network at the time of designing a multiple-processor system.

10.21 Multicomputer is having distinct advantages over its counterpart multiprocessor system. Discuss the salient features of a multicomputer system.

10.22 From the view of design issues of multicomputers made in the past, answer the following:

a. Why were large number of low-cost processors chosen over small number of expensive processors as processing nodes?

b. Why was distributed memory chosen over shared memory?

c. Why was message passing chosen over address switching?

10.23 "Multicomputers have gone through a series of generations of developmental stages" – Discuss each such generation mentioning the salient features of each generation.

10.24 Simultaneous execution of tasks that interact with each other can be achieved both by shared-memory multiprocessors and message-passing multicomputers. Which of these two architectures can emulate the action of the other more easily? Justify your answer.

10.25 "Local area networks (LAN) comprising of small computers belong to MIMD category". Justify this statement. What are the several advantages being observed in LAN–based arrangement of computers over the large mainframe (centralized) computers?

10.26 What is meant by clustering of computers? What are the key advantages that can be derived from such clustering?

10.27 State the different classes of clusters. What are the different factors that determine such classification?

10.28 State the different methods that are used to build up clustering.

10.29 State the major components that are involved in the general architecture of a cluster. Explain the role of the middleware in this architecture and state the different responsibilities that are carried out by this middleware.

10.30 Discuss the specific characteristics of the operating system that are used to drive a computer cluster.

10.31 State the difference between failover and failback.

10.32 State and briefly explain the different commonly-used methods that can be used to realize scalable parallel computer architecture.

10.33 Discuss the different models that are used in a generalized SIMD computer organisation.

10.34 Answer the following questions related to vector processing:

a. What are the differences between scalar instructions and vector instructions?

b. Give the classification of vector instructions. State and explain with examples the different types of vector instructions that are commonly used.

10.35 Give the block diagram to indicate the architecture of a typical vector processor or with all its major components.

10.36 Define vector stride and strip mining. What are the responsibilities that are performed by vectorizing compilers?

10.37 State and explain the basic instruction format that are used by generic vector processor.

10.38 Explain with examples the different types of vector instructions that are most commonly used.

10.39 Vector processor can be organised in many different ways. What are the different methods that can be employed to realize a vector processor?
State and explain one of these methods.

10.40 Describe with a diagram the basic vector processor architecture showing the major components that are involved in general.

10.41 Both vector processors and array processors are specialized to operate on vectors. What are the main differences lying between them?

10.42 Distinguish between register-to-register and memory-to-memory architectures used in vector processor for building conventional multivector supercomputers.

10.43 What do you mean by vector stride? What is meant by vectorization inhibitor? Explain the following memory organisations for vector accesses:

a. S-access memory organisation.

b. C-access memory organisation.

c. C/S-access memory organisation.

10.44 The following program segment, consisting of six instructions, needs to be executed 32 times for the evaluation of vector arithmetic expression: $Z(I) = U(I) + V(I) \times W(I)$ for $0 \leq I \leq 31$

```
Load      R1, V(I)    / R1 ← Memory contents of (b + 1) /
Load      R2, W(I)    / R1 ← Memory contents of (c + 1) /
Multiply  R1, R2      / R1 ← (R1) x (R2) /
Load      R3, U(I)    / R3 ← Memory contents of (a + 1) /
Add       R3, R1      / R3 ← (R3) + (R1) /
Store     Z(I), R3    / Memory contents of (d +1) ← (R3) /
```

Where R1, R2, and R3 are the CPU registers, (R1) is the contents of R1, and a, b, c, and d are the starting memory addresses of array U(I), V(I), W(I), and Z(I), respectively. Assume five clock cycles for each Load and Store, two clock cycles for the Add, and nine clock cycles for the Multiply on either a uniprocessor or a single PE in an SIMD machine.

a. Calculate the total number of CPU cycles needed to execute the above program segment repeatedly 32 times on an SISD machine sequentially, ignoring all other time delays.

b. Consider an SIMD machine with 32 PEs to execute the above vector operations in six synchronized vector instructions over 32-component vector data and both driven by the same speed clock. Calculate the total execution time on the SIMD machine, ignoring instruction broadcast and other delays.

c. What is the speedup gain of the SIMD machine over the SISD machine?

10.45 Following the architecture and operations of the Connection Machine CM2. Answer the following questions:

a. The connection machine CM2 is not a complete system. Justify the statement.

b. Describe the processing node architecture, including the processor, memory, floating-point unit, and network interface.

c. Describe the hypercube router and explain its uses
d. Explain the scanning and spread mechanisms and their applications on CM2.
e. Explain the concepts of broadcasting, global combining, and virtual processors in the use of the CM2.

10.46 Massively parallel processing (MPP) systems are fifth-generation computers: Justify the statement in terms of the salient features of an MPP system. Name some MPP systems and give your reasons for why they are being considered in the category of MPP system.

10.47 What are the minimum criteria that must be fulfiled by a system for to be a massively parallel (MPP) one?

10.48 "Connection Machine CM-5 does not belong to any class as proposed by Flynn in his classification of computers". Justify the statement and give your comments. Give a comparative studies and comment on the improvement made in CM-5 over the CM-2 from the viewpoints of computer architect and the user of machine level language programming.

10.49 A certain computation is highly sequential – that is, each step depends on the one preceding it. Would an array processor or a pipeline processor be more appropriate for this computation? Explain.

10.50 State briefly the commonly-used different architectural design concept that are exploited in building up scalable parallel computer architecture.

10.51 What is the modern definition of a supercomputer? "A change in the concept of supercomputer has been observed" What are these changes you detect? What are the reasons behind such changes to happen?

10.52 State and explain the different classes of general-purpose supercomputer. What are the different application areas where supercomputers can be advantageously employed?

10.53 Distinguish among the following vector processing machines in terms of architecture, performance range, and cost-effectiveness:
a. Full-scale vector supercomputers.
b. High-end mainframe or near-supercomputers.
c. Mini supercomputers and supercomputing workstations.

Suggested References and Websites

Buyya, R. *High Performance Cluster Computing: Architecture and Systems.* Upper Saddle River, NJ: Prentice Hall, 1999.

Feng, T. Y. "A survey of interconnection networks. "*IEEE Computer*, vol. 14, pp. 12–27, December 1981.

Jerraya, A. and Wolf, W., eds. *Multiprocessor Systems–on–Chips.* San Francisco, CA: Morgan Kauffmann, 2005.

Kapp, C. "Managing cluster computers. "*Dr. Dobb's Journal*, July 2000.

Lovett, T. and Clapp, R. "Implementation and performance of a CC–NUMA system. "*Proceedings, 23rd Annual International Symposium on Computer Architecture*, May 1996.

Quinn, M. J. *Parallel Computing: Theory and Practice*, 2nd ed. New York: McGraw-Hill, 1994.

Renaud, P. *Introduction to Client–Server Systems.* New York: Wiley, 1993.

Russell, R. M. "The CRAY-1 computer system. " *Communications of the ACM*, vol. 21, pp. 63–78, January 1978 (Reprinted in Ref. 24, pp. 743–752).

Scheurich, C. and Dubois, M. "Correct memory operation of cache-based multiprocessors." *Proceedings 14th Annual International Symposium on Computer Architecture*, ACM, pp. 224–234, 1987.

Short, R., Gamache, R., Vert, J., and Massa, M. "Windows NT clusters for availability and scalability." *Proceedings, COMPCON Spring 97*, February 1997.

IEEE Computer Society Task Force on Cluster Computing: An international forum to promote and extend cluster computing research and education.

Multicore Association: A vendor organisation aiming to promote the development of multicore culture and use of multicore technology.

IEEE Micro, vol. 17, no. 4, *Eight articles on ARM*, July/August, 1997.

Additional Reading

Adams, G. B. III, Aarwal, G. P. and Siegel, H. J. "A survey and comparison of fault-tolerant multistage interconnection networks." *IEEE Computer Magazine*, vol. 20, pp. 14–27, June 1987.

Agarwal, A. *Analysis of Cache Performance for Operating Systems and Multiprogramming*. Boston, MA: Kluwer Academic Publishers, 1989.

Agarwal, A., Simoni, R., Hennessy, J. and Horowitz, M. "An evaluation of directory schemes for cache coherence." *Proceedings 15th Annual International Symposium on Computer Architecture*, ACM, pp. 280–289, 1988.

Archibald, J. and Baer J. L. "An economical solution to the cache coherence problem." *Proceedings 11th Annual International Symposium on Computer Architecture*, ACM, pp. 355–362, 1984

Athas, W. C. and Seitz, C. L. "Multicomputers: Message-passing concurrent computers." *IEEE Computer Magazine*, vol. 21, pp. 9–24, August 1988.

Atkins, M. "PC software performance tuning." *IEEE Computer*, August 1996.

Baer, J. L. and Chen, T. F. "An effective on-chip preloading scheme to reduce data access penalty." *Proceedings of Supercomputing '91*, pp. 176–186, 1991.

Baker, M. and Buyya, R. "Cluster computing at a glance." pdf.semanticscholar.org/9bc0/7a198.pdf

Bal, H. E. and Tanenbaum, A. S. "Distributed programming with shared data." *Proceedings 1988 International Conference on Computer Languages*, IEEE, pp. 82–91, 1988.

Becker, M. C., et al. "The PowerPC 601 microprocessor." *IEEE Micro*, vol. 13, pp. 54–68, October 1993.

Benham, J. "A geometric approach to presenting computer representations of integers." *SIGCSE Bulletin*, December 1992.

Berson, A. *Client–Server Architecture*. New York: McGraw-Hill, 1992.

Bhurchandi, K. M., et al. *Advanced Microprocessors and Peripherals*, 3rd ed. McGraw-Hill Education (India) Pvt. Ltd., 2019.

Borkar, S. "Getting gigascale chips: Challenges and opportunities in continuing Moore's law." *ACM Queue*, October 2003.

Borrill, P. L. "Microprocessor bus structures and standards." *IEEE Micro Magazine*, vol. 1, pp. 84–95, February 1981.

Borrill, P. L. "A comparison of 32-bit buses." *IEEE Micro Magazine*, vol. 5, pp. 71–79, December 1985.

Brewer, E. "Clustering: Multiply and conquer." *Data Communications*, July 1997.

Burks, A. W., Goldstine, H. H. and Von Neumann, J. "Preliminary discussion of the logical design of an electronic computing instrument." Report prepared for U. S. Army Ordnance Department, 1946 (Reprinted in Ref. 26, vol. 5, pp. 34–79.)

Buyya, R. *High Performance Cluster Computing: Programming and Applications*. Upper Saddle River, NJ: Prentice Hall, 1999.

Catanzaro, B. *Multiprocessor System Architecture*. Mountain View, CA: Sun-Press, 1994.

Chakraborty, P. *Operating Systems*. India: Jaico Publishing House, 2011.

Chen, P. M., Lee, E. K., Gibson, G. A., Katz, R. H. and Patterson, D. A. "RAID: High-performance reliable secondary storage." *ACM Computing Surveys*, vol. 26, no. 2, pp. 145–185, June 1994.

Chen, S. and Towsley, D. "A performance evaluation on RAID architectures." *IEEE Transactions on Computers*, October 1996.

Cheriton, D. R. "The V distributed systems." *Communications of the ACM*, vol. 31, no. 3, pp. 314–333, 1988. (© ACM. Inc. 1988.)

Clements, A. *The Principles of Computer Hardware*, 3rd ed. New York: Oxford University Press, 2000.

Cocke, J. and Markstein, V. "The evolution of RISC technology at IBM." *IBM Journal of Research and Development; IBM*, vol. 44, no. 1.2, pp. 48–55, 2000.

Colwell, R. P., et al. "A VLIW architecture for a trace scheduling compiler." *IEEE Transactions on Computers*, vol. 37, no. 8, pp. 967–979, August 1988.

Coulouris, G. F., Dollimore, J. and Kindberg, T. *Distributed Systems, Concepts and Design,* 1st≈Impression. Pearson Education, 2006.

Culler, D., Singh, J. P. and Gupta, A. *Parallel Computer Architecture––A Hardware / Software Approach.* San Fransisco, CA: Morgan Kaufmann, 1988.

Diefendorf, K., Oehler, R. and Hochsprung, R. "Evolution of PowerPC architecture." *IEEE Micro,* vol. 14, pp. 34–49, April 1994.

Dowd, K. and Severance, C. *High Performance Computing.* Sebastopol, CA: O'Reilly, 1998.

Dubey, P. and Flynn, M. "Branch strategies and modeling optimization." *IEEE Transactions on Computers,* October 1991.

Edmondson, J. H., et al. "Superscalar execution in the 21164 alpha microprocessor." *IEEE Micro,* vol. 15, no. 2, pp. 33–43, April 1995.

Evans, J. and Trimper, G. *Itanium Architecture for Programmers.* Upper Saddle River, NJ: Prentice-Hall, 2003.

Farmwald, M. and Mooring, D. "A fast path to one memory." *IEEE Spectrum,* October 1992.

Flynn, M. J. "Very high-speed computing systems." *Proceedings of the IEEE,* vol. 54, pp. 1901–1909, December 1966

Flynn, M. J. and Oberman, S. *Advanced Computer Architecture Design.* New York: Wiley, 2001.

Fujitani, L. "Laser optical disk: The coming revolution in on-line storage." *Communications of the ACM,* vol. 27, pp. 546–566, June 1984.

Gibbs, W. "A split at the core." *Scientific American,* November 2004.

Gifford, D. and Spector, A. "Case study: IBM's System / 360–370 Architecture." *Communications of the ACM,* April 1987.

Gimarc, C. E. and Milutinovic, V. M. "A survey of RISC processors and computers of the Mid-1980s." *IEEE Computer Magazine,* vol. 20, pp. 59–69, September 1987.

Grindlay, R., et al. "The NUMAchine multiprocessor." *Proceedings of the International Conference on Parallel Processing,* Toronto, ON, pp. 487–496, August 2000.

Hayes, J. P. and T. N. Mudge. "Hypercube computers." *Proceedings of the IEEE,* vol. 77, pp. 1829–1841, December 1989.

Hennessy, J. L. and Jouppi, N. "Computer technology and architecture: An evolving interaction." *IEEE Computer,* September 1991.

Hennessy, J. L. and Patterson, D. A. *Computer Organisation and Design – The Hardware / Software Interface,* 2nd ed. San Mateo, CA: Morgan Kaufmann, 1998.

Hitchcock, C. Y. III and Sprunt, H. M. B. "Analyzing multiple register sets." *Proceedings 12th Annual International Symposium on Computer Architecture,* ACM, pp. 55–63, 1985.

Hwang, K. *Advanced Computer Architecture.* New York: McGraw-Hill, 1993.

Hwang, K. and Briggs, F. A. *Computer Architecture and Parallel Processing.* New York: McGraw-Hill, 1984.

Hwang, K. and DeGroot, D. eds. *Parallel Processing for Supercomputers and Artificial Intelligence.* New York: McGraw-Hill, 1989.

Hwang, K., et al. "Designing SSI clusters with hierarchical checkpointing and single I/O space." *IEEE Concurrency,* January–March 1999.

Jermoluk, T. *Multiprocessor UNIX.* Santa Clara, CA: Silicon Graphics Inc., 1990.

Kain, R. Y. *Computer Architecture: Software and Hardware.* Englewood Cliffs, NJ: Prentice-Hall, 1989.

Katevenis, M. *Reduced Instruction Set Architectures for VLSI.* Cambridge, MA: MIT Press, 1985.

Kong, C. "A hardware overview of the nonstop Himalaya K10000 server." *Tandem Systems Review,* vol. 10, January 1994.

Krishnaiyer, R., et al. "An advanced optimizer for the IA-64 architecture." *IEEE Micro,* vol. 20, no. 6, pp. 60–68, November / December 2000.

Lee, R. L., Yew, P. C. and Lawrie, D. H. "Multiprocessor cache design consideration." *Proceedings 14th Annual International Symposium on Computer Architecture,* ACM, pp. 253–262, 1987.

Lewis, T. G. and El-Rewini, H. *Introduction to Parallel Computing.* Englewood Cliffs, NJ: Prentice-Hall, 1992.

Li, K. "A shared virtual memory system for parallel computing." *Proceedings 5th Annual ACM Symposium on Principles of Distributed Computing*, ACM, pp. 229–239, 1986.

Lilja, D. J. "Reducing the branch penalty in pipelined processors." *IEEE Computer Magazine*, vol. 21, July 1988.

Lilja, D. J. "Cache coherence in large-scale shared memory multiprocessors: Issues and comparisons." *ACM Computing Surveys*, September 1993.

Lipovski, G. J. and Malek, M. *Parallel Computing*. New York: John Wiley & Sons, 1987.

Mak, P. et al. "Processor subsystem interconnect for a large symmetric multiprocessing system." *IBM Journal of Research and Development*, May / July 2004.

Mansuripur, M. and Sincerbox, G. "Principles and techniques of optical data storage." *Proceedings of the IEEE*, November 1997.

McDougall, R. and Laudon, J. "Multi-core multiprocessors are here." *Login*, October 2006.

McNairy, C. and Soltis, D. "Itanium 2 Processor microarchitecture." *IEEE Micro*, March–April 2003.

Mee, C. and Daniel, E. eds. *Magnetic Recording Technology*. New York: McGraw-Hill, 1996.

Milenkovic, A. "Achieving high performance in bus-based shared-memory multiprocessors." *IEEE Concurrency*, July–September 2000.

Moudgill, M. and Vassilliadis. "Precise interrupts." *IEEE Micro*, vol. 16, pp. 58–87, February 1996.

Mullender, S. J., Tanenbaum, A. S. and others. "Amoeba: A distributed operating system for the 1990s." *IEEE Computer*, vol. 23, no. 5, pp. 44–53, 1990.

nCUBE Corporation: *nCUBE 2 Supercomputers*. Beaverton, OR, 1990.

Ni, L. "A layered classification of parallel computers." *Proceedings 1991, International Conference for Young Computer Scientists*, pp. 28–33, Beijing, China, May 1991.

Nickolls, J. R. "The design of MasPar MP-1: A cost-effective massively parallel computer." in *IEEE Digest of Papers, -- Comcom*, pp. 25–28, IEEE Computer Society Press, Los Alamitos, CA, 1990.

Peleg, A., Wilkie, S. and Weiser, U. "Intel MMX for multimedia PCs." *Communications of the ACM*, January 1997.

Pfister, G. *In Search of Clusters*. Upper Saddle River, NJ: Prentice-Hall, 1998.

Pollack, F. "New Microarchitecture challenges in the coming generations of CMOS Process Technologies (keynote address)." *Proceedings of the 32nd Annual ACM / IEEE International Symposium on Microarchitecture*, 1999.

"RAID technology white paper." Dell Computer Corporation, 1999.

Schlichting, R. D. and Schneider, F. B. "Fail-stop processors: An approach to designing fault-tolerant computing systems." *ACM Transactions on Computer Systems*, vol. 1, no. 3, pp. 222–238, 1983.

Sequent Computer Systems Inc. *Symmetry 5000 Series*. Beaverton, OR, 1996.

Shanley, T. and Anderson, D. *PCI System Architecture*, 3rd ed. Reading, MA: Addison-Wesley, 1995.

Shen, J. and Lipasti, M. *Modern Processor Design: Fundamentals of Superscalar Processors*. New York: McGraw-Hill, 2005.

Siegel, H. J. *Interconnection Networks for Large-Scale Parallel Processing*, 2nd ed. New York: McGraw-Hill, 1990.

Siegel, T., Pfeffer, E. and Magee, A."The IBM z990 Microprocessor." *IBM Journal of Research and Development*, May / July 2004.

Siewiorek, D. P. and Swarz, R. S. eds. *Reliable Computer Systems*, 2nd ed. Burlington, MA: Digital Press, 1992.

Sima, D. "Superscalar instruction issue." *IEEE Micro*, September / October 1997.

Smith, A. J. "Cache memories." *Computing Surveys*, vol. 14, pp. 473–530, September 1982.

Smith, J. and Sohi, G. "The microarchitecture of superscalar processors." *Proceedings of IEEE*, December 1995.

Solomon, D. A. *Inside Windows NT*, 2nd ed. Redmond, WA: Microsoft Press, 1998.

Stallings, W. *Operating Systems, Internals and Design Principles*, 6th ed. Upper Saddle River, NJ: Prentice-Hall, 2009.

Stenstrom, P. "A survey of cache coherence schemes for multiprocessors." *IEEE Computer*, vol. 23, no. 6, pp. 12–25, 1990.

Stenstrom, P., Joe, T. and Gupta, A. "Comparative performance evaluation of cache-coherent NUMA and COMA architecture." *Proceedings 19th Annual International Symposium on Computer Architecture*, ACM, 1992.
Stone, H. S. "Parallel processing with the perfect shuffle." *IEEE Transactions on Computers*, C20, pp. 153–61, 1971.
Stone, H. S. *High Performance Computer Architecture*, 3rd ed. Reading, MA: Addison-Wesley, 1993.
Tabak, D. *RISC Architecture*. New York: John Wiley & Sons, 1987.
Tabak, D. *Advanced Microprocessors*. New York: McGraw-Hill, 1991.
Tanenbaum, A. S. *Structured Computer Organisation*. New Delhi: Prentice-Hall of India, 1997.
Thacker, S. S., et al. "New directions in scalable shared-memory multiprocessor architectures." *IEEE Computer*, vol. 23, no. 6, pp. 71–83, 1990.
Tremblay, M. and O'Connor, J. M. "UltraSparc 1: A four-issue processor supporting multimedia." *IEEE Micro*, vol. 16, no. 2, pp. 42–50, April 1996.
Tucker, L. W. and Robertson, G. G. "Architecture and applications of the connection machine." *IEEE, Computer*, vol. 21, no. 8, pp. 26–38, 1988.
Ungerer, T., Rubie, B. and Silc, J. "Multithreaded processors." *The Computer Journal*, no. 3, 2002.
Ungerer, T., Rubie, B. and Silc, J. "A survey of processors with explicit multithreading." *ACM Computing Surveys*, March 2003.
Veen, A. H. "Dataflow machine architecture." *Computing Surveys*, vol. 18, pp. 365–396, December 1986.
Vernon, M. K., Lazowska, E. D. and Zahorjan, J. "Snooping cache-consistency protocols." *Computing Surveys*, vol. 18, pp. 365–396, December 1986.
Vranesic, Z. G., Stumm, M., Lewis, D. M. and White, R. "Hector: A hierarchically structured shared-memory multiprocessor." *Computer*, vol. 24, pp. 72–79, January 1991.
Wallace, C. S. "A suggestion for fast multiplier." *IEEE Transactions on Electronic Computers*, vol. EC-13, pp. 14–17, February 1964.
Wang, W. H., Baer, J. L. and Levy, H. M. "Organisation and performance of a two-level virtual-real cache hierarchy." *Proceedings 16th Annual International Symposium on Computer Architecture*, ACM, pp. 140–148, 1989.
Weiss, S. and Smith, J. E. "Instruction issue logic in pipelined supercomputers." *IEEE Transactions on Computers*, pp. 1013–1022, 1984.
Weygant, P. *Clusters for High Availability*. Upper Saddle River, NJ: Prentice-Hall, 2001.
Whitney, S., et al. "The SGI origin software environment and application performance." *Proceedings, COMPCON Spring '97*, February 1997.
Wickelgren, I. "The facts about fire wire." *IEEE Spectrum*, April 1997.
Wu, C. L. and Feng, T. Y. "On a class of multistage interconnection networks." *IEEE Transactions on Computers*, pp. 696–702, August 1980.
Yew, P. C. and Wah, B. W. eds. "Special issue on shared-memory multiprocessors." *Journal of Parallel and Distributed Computing*, June 1991.
Zima, H. and Chapman, B. *Supercompiler for Parallel and Vector Computers*. Reading, MA: Addison-Wesley, 1990.
Zorpetta, G. "The power of parallelism." *IEEE Spectrum*, vol. 29, no. 9, pp. 28–33, 1992.

Suggested Websites

ACM Special Interest Group on Operating Systems: Contains information on SIGOPS publications and conferences.

IBM Journal of Research and Development, vol. 34, pp. 4–11, January 1990.

IBM website: ibm.com

IEEE Technical Committee on Operating Systems and Applications: Contains an on-line newsletter and links to other sites.

InfiniBand Trade Association: Provides technical information and vendor lists on InfiniBand.

Intel Developer's Page: Provides the Intel Technology Journal and includes a starting point for obtaining Pentium information.

Intel website: intel.com

Operating System Resource Center: A good collection of documents and papers on a wide range of OS topics.

PCI Special Interest Group: Describes information about PCI specifications and products.

Standard Performance Evaluation Corporation: SPEC is recognized and a widely accepted organisation in the computer industry for its development of standardized benchmarks used to evaluate and compare performance of different computer systems introduced by different vendors in the market.

Sun Microsystems website: sun.com

TOP 500 Supercomputer Site: Provides brief description of architecture and organisation of current supercomputer products along with their comparisons.

Index

Printed in the United States
By Bookmasters